ADVANCED MATHEMATICAL METHODS IN SCIENCE AND ENGINEERING

Second Edition

ADVANCED MATHEMATICAL METHODS IN SCIENCE AND ENGINEERING

Second Edition

S. I. Hayek
Pennsylvania State University
University Park, U.S.A.

CRC Press
Taylor & Francis Group
Boca Raton London New York

CRC Press is an imprint of the
Taylor & Francis Group an **informa** business
A CHAPMAN & HALL BOOK

Chapman & Hall/CRC
Taylor & Francis Group
6000 Broken Sound Parkway NW, Suite 300
Boca Raton, FL 33487-2742

Printed in the United States of America on acid-free paper
10 9 8 7 6 5 4 3 2 1

International Standard Book Number: 978-1-4200-8197-8 (Hardback)

Visit the Taylor & Francis Web site at
http://www.taylorandfrancis.com

and the CRC Press Web site at
http://www.crcpress.com

TABLE OF CONTENTS

10 NUMERICAL METHODS 585

APPENDIX A
INFINITE SERIES 669

APPENDIX B
SPECIAL FUNCTIONS 683

APPENDIX C
ORTHOGONAL COORDINATE SYSTEMS 709

APPENDIX D
DIRAC DELTA FUNCTIONS 719

APPENDIX E
PLOTS OF SPECIAL FUNCTIONS 735

APPENDIX F
VECTOR ANALYSIS 739

APPENDIX G
MATRIX ALGEBRA 751

PREFACE

This book is intended to cover many topics in mathematics at a level more advanced than a junior level course in differential equations. The book evolved from a set of notes for a three-semester course in the application of mathematical methods to science and engineering problems. The courses attract graduate students majoring in engineering mechanics, engineering science, mechanical, petroleum, electrical, nuclear, civil and aeronautical engineering, as well as physics, meteorology, geology, and geophysics.

The book assumes knowledge of differential and integral calculus and an introductory level of ordinary differential equations. Thus, the book is intended for advanced senior and graduate students. Each chapter of the text contains many solved examples and many problems with answers. Those chapters which cover boundary value problems and partial differential equations also include derivation of the governing differential equations in many fields of applied physics and engineering such as wave mechanics, acoustics, heat flow in solids, diffusion of liquids and gasses, and fluid flow.

Chapter 1 briefly reviews methods of integration of ordinary differential equations. Chapter 2 covers series solutions of ordinary differential equations. This is followed by methods of solution of singular differential equations. Chapter 3 covers Bessel functions and Legendre functions in detail, including recurrence relations, series expansion, integrals, integral representations and generating functions.

Chapter 4 covers the derivation and methods of solution of linear boundary value problems for physical systems in one spatial dimension governed by ordinary differential equations. The concepts of eigenfunctions, orthogonality, and eigenfunction expansions are introduced, followed by an extensive treatment of adjoint and self-adjoint systems. This is followed by coverage of Sturm-Liouville system for second- and fourth-order ordinary differential equations. The chapter concludes with methods of solution of non-homogeneous boundary value problems.

Chapter 5 covers complex variables, calculus, and integrals. The method of residues is fully applied to proper and improper integrals, followed by integration of multi-valued functions. Examples are drawn from Fourier sine, cosine and exponential transforms as well as the Laplace transform.

Chapter 6 covers linear partial differential equations in classical physics and engineering. The chapter covers derivation of the governing partial differential equations for wave equation in acoustics, membranes, plates and beams; strength of materials; heat flow in solids and diffusion of gasses; temperature distribution in solids and flow of incompressible ideal fluids. These equations are then shown to obey partial differential equations of the type; Laplace, Poisson, Helmholtz, wave and diffusion equations. Uniqueness theorems for these equations are then developed. Solutions by eigenfunction expansions are explored fully. These are followed by special methods for non--homogeneous partial differential equations with temporal and spatial source fields.

Chapter 7 covers the derivation of integral transforms such as Fourier complex, sine and cosine, Generalized Fourier, Laplace, and Hankel transforms. The calculus of each of these transforms is then presented together with special methods for inverse transformations. Each transform also includes applications to solutions of partial differential equations for engineering and physical systems.

Chapter 8 covers Green's functions for ordinary and partial differential equations. The Greens functions for adjoint and self-adjoint systems of ordinary differential equations are then presented by use of generalized functions or by construction. These methods are applied to physical examples in the same fields covered in Chapter 6. These are then followed by derivation of fundamental solutions for the Laplace, Helmholtz, Wave and diffusion equations in one, two, and three-dimensional space. Finally, the Green's functions for bounded and semi-infinite media such as half and quarter spaces, in cartesian, cylindrical, and spherical geometry are developed by the method of images with examples in physical systems.

Chapter 9 covers asymptotic methods aimed at the evaluation of integrals as well as the asymptotic solution of ordinary differential equations. This chapter covers asymptotic series and convergence. This is then followed by asymptotic series evaluation of definite and improper integrals. These include the stationary phase method, the steepest descent method, the modified saddle point method, method of the subtraction of poles, and Ott's and Jones' methods. The chapter then covers asymptotic solutions of ordinary differential equations, formal solutions, normal and sub-normal solutions and the WKBJ method.

Chapter 10 covers an extensive treatment of numerical methods that could be used in a three-credit course on the subject. It includes numerical solution of non-linear equations, finite difference differentiation and integration, solution of initial value and boundary value ODE, and solution of PDE in mathematical physics.

Appendix A covers infinite series and convergence tests, infinite series of functions, regions of convergence, with examples and problems.

Appendix B presents a compendium of special functions such as Beta, Gamma, Zeta, Laguerre, Hermite, Hypergeometric, Chebychev and Fresnel. These include differential equations, series solutions, integrals, recurrence formulae and integral representations.

Appendix C presents a compendium of formulae for spherical, cylindrical, ellipsoidal, oblate and prolate spheroidal coordinate system including the divergence, gradient, Laplacian and Scalar and vector wave operators.

Appendix D covers calculus of generalized functions such as the Dirac delta functions in n-dimensional space of zero and higher ranks, with examples and problems.

Appendix E presents plots of special functions.

Appendix F covers vector algebra and calculus, including Divergence and Stokes theorems, and the index notation. Appendix G covers an extensive treatment of matrix algebra.

The aim of this book is to present methods of applied mathematics that are particularly suited for the application of mathematics to physical problems in science and engineering, with numerous examples that illustrate the methods of solution being explored. The problems have answers listed at the end of the book.

The book is currently used in a three-semester course sequence. The author recommends Chapter 1, 2, 3, and 4 and Appendix A in the first course, with emphasis on ordinary differential equations. The second semester would include Chapters 5, 6, and 7 with emphasis on partial differential equations. The third course would include Appendix D, and Chapters 8 and 9. A fourth course could include chapter 10 with Appendix G.

ACKNOWLEDGMENTS

This book evolved from course notes written in the early seventies for a two-semester course at Penn State University. It was completely revamped and retyped in the mid eighties. The course notes were rewritten in the format of a manuscript for a book for the last two years. I would like to acknowledge the many people who had profound influence on me over the last forty years.

I am indebted to my former teachers who instilled in me the love of applied mathematics. In particular, I would like to mention Professors Morton Friedman, Melvin Barron, Raymond Mindlin, Mario Salvadori, and Frank DiMaggio, all of Columbia University's Department of Engineering Mechanics. I am also indebted to the many graduate students who made suggestions on improving the course-notes over the last thirty-five years.

I am also indebted to Prof. Richard McNitt, former head and to Prof. Judy Todd, the current head of the Department of Engineering Science and Mechanics, Penn State University, for their support on this project during the last eight years. I am also grateful to Ms. Kathy Joseph, whose knowledge of the subject matter led to many invaluable technical suggestions while typing of the final manuscript and to Ms. Stacy Smith who typed the additional material for the second edition. I am also grateful for the encouragement and support of Mr. B.J. Clark, Executive Acquisitions Editor at Marcel-Dekker, Inc, and to David Grubbs, Acquisitions Editor, Chapman & Hall/CRC, Taylor and Francis Group.

I am grateful to my wife, Guler, for her moral support during the writing of the first draft of the course-notes, the first and second editions and for freeing me from many responsibilities at home to allow me to work on the first manuscript and first and second editions over the last thirty years. I am also grateful to my children, Emil and Dina, for their moral support when their father could not be there for them during the first edition, and for my son Emil who proofread parts of the first course-notes manuscript.

S. I. Hayek

1

ORDINARY DIFFERENTIAL EQUATIONS

1.1 Definitions

A linear ordinary differential equation is defined as one that relates a dependent variable, an independent variable, and derivatives of the dependent variable with respect to the independent variable. Thus the equation:

$$Ly = a_0(x)\frac{d^n y}{dx^n} + a_1(x)\frac{d^{n-1}y}{dx^{n-1}} + ... + a_{n-1}(x)\frac{dy}{dx} + a_n(x)\,y = f(x) \tag{1.1}$$

relates the dependent variable y and its derivatives up the nth to the independent variable x, where the coefficient $a_0(x)$ does not vanish in $a \le x \le b$, and $a_0(x)$, $a_1(x)$,..., $a_n(x)$ are continuous and bounded in $a \le x \le b$.

The **order** of a differential equation is defined as the order of the highest derivative in the differential equation. Equation (1.1) is an nth order differential equation. A homogeneous linear differential equation is one where a function of the independent variable does not appear explicitly without being multiplied by the dependent variable or any of its derivatives. Equation (1.1) is a **homogeneous equation** if f(x) = 0 and is a **non-homogeneous equation**, if $f(x) \neq 0$ for some $a \le x \le b$. A **homogeneous solution** of a differential equation y_h is the solution that satisfies a homogeneous differential equation:

$$Ly_h = 0 \tag{1.2}$$

with L representing an nth order linear differential operator of the form:

$$L = a_0(x)\frac{d^n}{dx^n} + a_1(x)\frac{d^{n-1}}{dx^{n-1}} + ... + a_{n-1}(x)\frac{d}{dx} + a_n(x)$$

If a set of n functions y_1, y_2,..., y_n, continuous and differentiable n times, satisfies eq. (1.2), then by superposition, the homogeneous solution of eq. (1.2) is:

$$y_h = C_1y_1 + C_2y_2 + ... + C_ny_n$$

with C_1, C_2,..., C_n being arbitrary constants, so that y_h also satisfies eq. (1.2).

A **particular solution** y_p is any solution that satisfies a nonhomogeneous differential equation, such as eq. (1.1), and contains no arbitrary constants, i.e.:

$$Ly_p = f(x) \tag{1.3}$$

The **complete** solution of a differential equation is the sum of the homogeneous and particular solutions, i.e.:

$$y = y_h + y_p$$

Example 1.1

The linear differential equation:

$$\frac{d^2y}{dx^2} + 4y = 2x^2 + 1$$

has a homogeneous solution $y_h = C_1 \sin 2x + C_2 \cos 2x$ and a particular solution $y_p = x^2/2$. Each of the functions $y_1 = \sin 2x$ and $y_2 = \cos 2x$ satisfy the equation $(d^2y)/(dx^2) + 4y = 0$, and the constants C_1 and C_2 are arbitrary.

1.2 Linear Differential Equations of First Order

A linear differential equation of the first order has the form:

$$\frac{dy}{dx} + \phi(x)\, y = \Psi(x) \tag{1.4}$$

where

$$\phi(x) = \frac{a_1(x)}{a_0(x)} \qquad \text{and} \qquad \Psi(x) = \frac{f(x)}{a_0(x)}$$

The homogeneous solution, involving one arbitrary constant, can be obtained by direct integration:

$$\frac{dy}{dx} + \phi(x)\, y = 0$$

or

$$\frac{dy}{y} = -\phi(x)\, dx$$

Integrating the resulting equation gives the homogeneous solution:

$$y_h = C_1 \exp\left(-\int \phi(x)\, dx\right) \tag{1.5}$$

with C_1 an arbitrary constant.

To obtain the particular solution, one uses an integrating factor $\mu(x)$, such that:

$$\mu(x)\left[\frac{dy}{dx} + \phi(x)\, y\right] = \frac{d}{dx}\left(\mu(x)\, y\right) = \frac{d\mu}{dx}\, y + \mu \frac{dy}{dx} \tag{1.6}$$

Thus $\mu(x)$ can be obtained by equating the two sides of eq. (1.6) as follows:

$$\frac{d\mu}{\mu} = \phi(x)\, dx$$

resulting in a closed form for the integrating factor:

$$\mu(x) = \exp\left(\int \phi(x)\, dx\right) \tag{1.7}$$

Using the integrating factor, eq. (1.4) can be rewritten in the form:

$$\frac{d}{dx}\left(\mu(x)\, y_p(x)\right) = \Psi(x)\, \mu(x)$$

or

$$y_p = \frac{1}{\mu(x)} \int \Psi(x)\, \mu(x)\, dx \tag{1.8}$$

Thus, the complete solution of eq. (1.4) can be written as:

$$y = C_1 \exp\left(-\int \phi(x)\, dx\right) + \exp\left(-\int \phi(x)\, dx\right) \int \Psi(x)\, \mu(x)\, dx \tag{1.9}$$

1.3 Linear Independence and the Wronskian

Consider a set of functions $[y_i(x)]$, $i = 1, 2,..., n$. A set of functions are termed linearly independent on (a, b) if there is no nonvanishing set of constants $C_1, C_2,..., C_n$ which satisfies the following equation identically:

$$C_1 y_1(x) + C_2 y_2(x) + ... + C_n y_n(x) = 0 \tag{1.10}$$

If $y_1, y_2,..., y_n$ satisfy eq. (1.1), and if there exists a set of constants such that eq. (1.10) is satisfied, then derivatives of eq. (1.10) are also satisfied, i.e.:

$$\begin{array}{llll} C_1 y_1' & + C_2 y_2' & + ... + C_n y_n' & = 0 \\ C_1 y_1'' & + C_2 y_2'' & + ... + C_n y_n'' & = 0 \\ ... & ... & ... & ... \\ ... & ... & ... & ... \\ ... & ... & ... & ... \\ C_1 y_1^{(n-1)} & + C_2 y_2^{(n-1)} & + ... + C_n y_n^{(n-1)} & = 0 \end{array} \tag{1.11}$$

For a non-zero set of constants $[C_i]$ of the homogeneous algebraic eqs. (1.10) and (1.11), the determinant of the coefficients of $C_1, C_2,..., C_n$ must vanish. The determinant, generally referred to as the Wronskian of $y_1, y_2,..., y_n$, becomes:

$$W(y_1, y_2, ..., y_n) = \begin{vmatrix} y_1 & y_2 & ... & y_n \\ y_1' & y_2' & ... & y_n' \\ y_1^{(n-1)} & y_2^{(n-1)} & ... & y_n^{(n-1)} \end{vmatrix} \tag{1.12}$$

If the Wronskian of a set of functions is not identically zero, the set of functions $[y_i]$ is a linearly independent set. The non-vanishing of the Wronskian is a necessary and sufficient condition for linear independence of $[y_i]$ for all x.

Example 1.2

If $y_1 = \sin 2x$, $y_2 = \cos 2x$:

$$W(y_1, y_2) = \begin{vmatrix} \sin 2x & \cos 2x \\ 2\cos 2x & -2\sin 2x \end{vmatrix} = -2 \neq 0$$

Thus, y_1 and y_2 are linearly independent.

If the set $[y_i]$ is linearly independent, then another set $[z_i]$ which is a linear combination of $[y_i]$ is defined as:

$$z_1 = \alpha_{11} y_1 + \alpha_{12} y_2 + \ldots + \alpha_{1n} y_n$$
$$z_2 = \alpha_{21} y_1 + \alpha_{22} y_2 + \ldots + \alpha_{2n} y_n$$
$$\ldots \quad \ldots \quad \ldots \quad \ldots \quad \ldots$$
$$z_n = \alpha_{n1} y_1 + \alpha_{n2} y_2 + \ldots + \alpha_{nn} y_n$$

with α_{ij} being constants, is also linearly independent provided that:

$$\mathbf{det}\left[\alpha_{ij}\right] \neq 0, \text{ because } W(z_i) = \mathbf{det}\left[\alpha_{ij}\right] \cdot W(y_i)$$

1.4 Linear Homogeneous Differential Equation of Order n with Constant Coefficients

Differential equations of order n with constant coefficients having the form:

$$Ly = a_0 y^{(n)} + a_1 y^{(n-1)} + \ldots + a_{n-1} y' + a_n y = 0 \tag{1.13}$$

where $a_0, a_1, \ldots, a_n$ are constants, with $a_0 \neq 0$, can be readily solved.

Since functions e^{mx} can be differentiated many times without a change of its functional dependence on x, then one may try:

$$y = e^{mx}$$

where m is a constant, as a possible solution of the homogeneous equation. Thus, operating on y with the differential operator L, results:

$$Ly = \left(a_0 m^n + a_1 m^{n-1} + \ldots + a_{n-1} m + a_n\right) e^{mx} = 0 \tag{1.14}$$

which is satisfied by setting the coefficient of e^{mx} to zero. The resulting polynomial equation of degree n:

$$a_0 m^n + a_1 m^{n-1} + \ldots + a_{n-1} m + a_n = 0 \tag{1.15}$$

is called the **characteristic** equation.

If the polynomial in eq. (1.15) has n distinct roots, $m_1, m_2, \ldots, m_n$, then there are n solutions of the form:

$$y_i = e^{m_i x}, \; i = 1, 2, \ldots, n \tag{1.16}$$

each of which satisfies eq. (1.13). The general solution of the homogeneous equation (1.13) can be written in terms of the n independent solutions of eq. (1.16):

$$y_h = C_1 e^{m_1 x} + C_2 e^{m_2 x} + \ldots + C_n e^{m_n x} \tag{1.17}$$

where C_i are arbitrary constants.

The differential operator L of eq. (1.13) can be written in an expanded form in terms of the characteristic roots of eq. (1.15) as follows:

$$Ly = a_0 \left(D - m_1\right)\left(D - m_2\right) \ldots \left(D - m_n\right) y = 0 \tag{1.18}$$

where D = d/dx. It can be shown that any pair of components of the operator L can be interchanged in their order of appearance in the expression for L in eq. (1.18), i.e.:

$$(D - m_i)(D - m_j) = (D - m_j)(D - m_i)$$

such that:

$$Ly = a_0 (D - m_1)(D - m_2) \ldots (D - m_{j-1})(D - m_{j+1}) \ldots (D - m_n)(D - m_j) y = 0$$

Thus, if:

$$(D - m_j) y = 0 \qquad j = 1, 2, 3, \ldots, n$$

then

$$y_j = e^{m_j x} \qquad j = 1, 2, 3, \ldots,$$

satisfies eq. (1.18).

If the roots m_i are distinct, then the solutions in eq. (1.16) are distinct and it can be shown that they constitute an independent set of solutions of the differential equation. If there exist repeated roots, for example the jth root is repeated k times, then there are n - k + 1 independent solutions, and a method must be devised to obtain the remaining k - 1 solutions. In such a case, the operator L in eq. (1.18) can be rewritten as follows:

$$Ly = a_0 (D - m_1)(D - m_2) \ldots (D - m_{j-1})(D - m_{j+k}) \ldots (D - m_n)(D - m_j)^k y = 0 \tag{1.19}$$

To obtain the missing solutions, it would be sufficient to solve the equation:

$$(D - m_j)^k y = 0 \tag{1.20}$$

A trial solution of the form $x^r e^{m_j x}$ can be substituted in eq. (1.20):

$$(D - m_j)^k (x^r e^{m_j x}) = r(r-1)(r-2) \ldots (r-k+2)(r-k+1) x^{r-k} e^{m_j x} = 0$$

which can be satisfied if r takes any of the integer values:

$$r = 0, 1, 2, \ldots, k - 1$$

Thus, solutions of the type:

$$y_{j+r} = x^r e^{m_j x} \qquad r = 0, 1, 2, \ldots, k-1$$

satisfy eq. (1.19) and supply the missing k - 1 solutions, such that the total homogeneous solution becomes:

$$y_h = C_1 e^{m_1 x} + C_2 e^{m_2 x} + \ldots + (C_j + C_{j+1} x + C_{j+2} x^2 + \ldots + C_{j+k-1} x^{k-1}) e^{m_j x}$$
$$+ C_{j+k} e^{m_{j+k} x} + \ldots + C_n e^{m_n x} \tag{1.21}$$

Example 1.3

Obtain the solution to the following differential equation:

$$\frac{d^3 y}{dx^3} - 3\frac{d^2 y}{dx^2} + 4y = 0$$

Let $y = e^{mx}$, then the characteristic equation is given by $m^3 - 3m^2 + 4 = 0$ such that $m_1 = -1$, $m_2 = +2$, $m_3 = +2$, and:

$$y_h = C_1 e^{-x} + (C_2 + C_3 x) e^{2x}$$

1.5 Euler's Equation

Euler's Equation is a special type of a differential equation with non-constant coefficients which can be transformed to an equation with constant coefficients and solved by the techniques developed in Section 1.4. The differential equation:

$$Ly = a_0 x^n \frac{d^n y}{dx^n} + a_1 x^{n-1} \frac{d^{n-1} y}{dx^{n-1}} + \ldots + a_{n-1} x \frac{dy}{dx} + a_n y = f(x) \qquad (1.22)$$

is such an equation, generally known as **Euler's Equation**, where the a_i's are constants.

Transforming the independent variable x to z by the following transformation:

$$z = \log x \qquad x = e^z \qquad (1.23)$$

then the first derivative transforms to:

$$\frac{d}{dx} = \frac{d}{dz}\frac{dz}{dx} = \frac{1}{x}\frac{d}{dz} = e^{-z}\frac{d}{dz}$$

or

$$\overline{D} = \frac{d}{dz} = x\frac{d}{dx}$$

The second derivative transforms to:

$$\frac{d^2}{dx^2} = \frac{d}{dx}\left(\frac{d}{dx}\right) = \left(e^{-z}\frac{d}{dz}\right)\left(e^{-z}\frac{d}{dz}\right) = e^{-2z}\left(\frac{d^2}{dz^2} - \frac{d}{dz}\right)$$

or

$$x^2\frac{d^2}{dx^2} = \frac{d^2}{dz^2} - \frac{d}{dz} = \overline{D}^2 - \overline{D} = \overline{D}\left(\overline{D} - 1\right)$$

Similarly:

$$x^3\frac{d^3}{dx^3} = \overline{D}\left(\overline{D} - 1\right)\left(\overline{D} - 2\right)$$

and by induction:

$$x^n\frac{d^n}{dy^n} = \overline{D}\left(\overline{D} - 1\right)\left(\overline{D} - 2\right)\ldots\left(\overline{D} - n + 1\right)$$

Using the transformation in eq. (1.23), one is thus able to transform eq. (1.22) with variable coefficients on the independent variable x to one with constant coefficients on z. The solution is then obtained in terms of z, after which an inverse transformation is performed to obtain the solution in terms of x.

Example 1.4

$$x^3\frac{d^3 y}{dx^3} - 2x\frac{dy}{dx} + 4y = 0$$

Letting $x = e^z$, then the equation transforms to:

$$\bar{D}(\bar{D}-1)(\bar{D}-2)y - 2\bar{D}y + 4y = 0$$

which can be written as:

$$\frac{d^3y}{dz^3} - 3\frac{d^2y}{dz^2} + 4y = 0$$

The homogeneous solution of the differential equation in terms of z is:

$$y_h(z) = C_1e^{-z} + (C_2 + C_3z)\,e^{+2z}$$

which, after transforming z to x, one obtains the homogeneous solution in terms of x:

$$y_h(x) = C_1x^{-1} + (C_2 + C_3\log x)\,x^2$$

1.6 Particular Solutions by Method of Undetermined Coefficients

The particular solution for non-homogeneous differential equations of the first order was discussed in Section 1.2. Particular solutions to general n^{th} order linear differential equations can be obtained by the method of variation of parameters to be discussed later in this chapter. However, there are simple means for obtaining particular solutions to non-homogeneous differential equations with constant coefficients such as eq. (1.13), if f(x) is an elementary function:

(a) $f(x) = \sin ax$ or $\cos ax$ — try $y_p = A \sin ax + B \cos ax$

(b) $f(x) = e^{\beta x}$ — try $y_p = Ce^{\beta x}$

(c) $f(x) = \sinh ax$ or $\cosh ax$ — try $y_p = D \sinh ax + E \cosh ax$

(d) $f(x) = x^m$ — try $y_p = F_0x^m + F_1x^{m-1} + \ldots + F_{m-1}x + F_m$

If f(x) is a product of the functions given in (a) – (d), then a trial solution can be written in the form of the product of the corresponding trial solutions. Thus if:

$$f(x) = x^2\,e^{-2x}\sin 3x$$

then one uses a trial particular solution:

$$Y_p = (F_0x^2 + F_1x + F_2)(e^{-2x})(A\sin 3x + B\cos 3x)$$

$$= e^{-2x}(H_1x^2\sin 3x + H_2x^2\cos 3x + H_3x\sin 3x + H_4x\cos 3x + H_5\sin 3x + H_6\cos 3x)$$

If a factor or term of f(x) happens to be one of the solutions of the homogeneous differential eq. (1.14), then the portion of the trial solution y_p corresponding to that term or factor of f(x) must be multiplied by x^k, where an integer k is chosen such that the portion of the trial solution is one power of x higher than any of the homogeneous solutions of eq. (1.13).

Example 1.5

$$\frac{d^3y}{dx^3} - 3\frac{d^2y}{dx^2} + 4y = 40\sin 2x + 27x^2\,e^{-x} + 18x\,e^{2x}$$

where

$$y_h = C_1 e^{-x} + \left(C_2 + C_3 x\right) e^{2x}$$

For sin (2x) try A sin (2x) + B cos (2x).

For x^2e^{-x} try:

$$y_p = \left(Cx^2 + Dx + E\right) xe^{-x}$$

since e^{-x} is a solution to the homogeneous equation.

For xe^{2x} try:

$$y_p = (Fx + G)x^2 e^{2x}$$

since e^{2x} and xe^{2x} are both solutions of the homogeneous equation. Thus, the trial particular solution becomes:

$$y_p = A \sin(2x) + B\cos(2x) + Cx^3e^{-x} + Dx^2e^{-x} + Exe^{-x} + Fx^3e^{2x} + Gx^2e^{2x}$$

Substitution of y_p into the differential equation and equating the coefficients of like functions, one obtains:

A = 2 B = 1 C = 1 D = 2 E = 2 F = 1 G = -1

Thus:

$$y = C_1e^{-x} + \left(C_2 + C_3x\right)e^{2x} + 2\sin(2x) + \cos(2x)$$
$$+ (x^3 + 2x^2 + 2x)\, e^{-x} + (x^3 - x^2)\, e^{2x}$$

Example 1.6

Obtain the solution to the following equation:

$$x^3 \frac{d^3y}{dx^3} - 2x\frac{dy}{dx} + 4y = 6x^2 + 16 \log x$$

This equation can be solved readily by transformation of the independent variable as in Section 1.5, such that:

$$\frac{d^3y}{dz^3} - 3\frac{d^2y}{dz^2} + 4y = 6e^{2z} + 16z$$

where $y_h(z) = C_1e^{-z} + (C_2 + C_3z)\, e^{+2z}$.

For e^{2z} try Az^2e^{2z} since e^{2z} and ze^{2z} are solutions of the homogeneous equation, and for z try Bz + C. Substitution in the equation on z, one obtains:

A = 1 B = 4 C = 0

$$y_p(z) = z^2e^{2z} + 4z$$

$$y_p(x) = (\log x)^2\, x^2 + 4 \log x$$

and

$$y = C_1x^{-1} + \left(C_2 + C_3 \log x\right)x^2 + x^2(\log x)^2 + 4\log x$$

1.7 Particular Solutions by the Method of Variations of Parameters

Except for differential equations with constant coefficients, it is very difficult to guess at the form of the particular solution. This section gives a treatment of a general method by which a particular solution can be obtained.

The homogeneous differential equation (1.2) has n independent solutions, i.e.:

$$y_h = C_1 y_1 + C_2 y_2 + \dots + C_n y_n$$

Assume that the particular solution y_p of eq. (1.1) can be obtained from n products of these solutions with n unknown functions $v_1(x), v_2(x), \dots, v_n(x)$, i.e.:

$$y_p = v_1 y_1 + v_2 y_2 + \dots + v_n y_n \tag{1.24}$$

Differentiating eq. (1.24) once results in:

$$y_p' = (v_1' y_1 + v_2' y_2 + \dots + v_n' y_n) + (v_1 y_1' + v_2 y_2' + \dots + v_n y_n')$$

Since y_p in eq. (1.24) must satisfy one equation, i.e., eq. (1.1), one can arbitrarily specify (n-1) more relationships. Thus, let:

$$v_1' y_1 + v_2' y_2 + \dots + v_n' y_n = 0$$

so that:

$$y_p' = v_1 y_1' + v_2 y_2' + \dots + v_n y_n'$$

Differentiating y_p' once again gives:

$$y_p'' = (v_1' y_1' + v_2' y_2' + \dots + v_n' y_n') + (v_1 y_1'' + v_2 y_2'' + \dots + v_n y_n'')$$

Again let:

$$v_1' y_1' + v_2' y_2' + \dots + v_n' y_n' = 0$$

resulting in:

$$y_p'' = v_1 y_1'' + v_2 y_2'' + \dots + v_n y_n''$$

Carrying this procedure to the (n - 1)st derivative one obtains:

$$y_p^{(n-1)} = \left(v_1' y_1^{(n-2)} + v_2' y_2^{(n-2)} + \dots + v_n' y_n^{(n-2)}\right) + \left(v_1 y_1^{(n-1)} + v_2 y_2^{(n-1)} + \dots + v_n y_n^{(n-1)}\right)$$

and letting:

$$v_1' y_1^{(n-2)} + v_2' y_2^{(n-2)} + \dots + v_n' y_n^{(n-2)} = 0$$

then

$$y_p^{(n-1)} = v_1 y_1^{(n-1)} + v_2 y_2^{(n-1)} + \dots + v_n y_n^{(n-1)}$$

Thus far (n - 1) conditions have been specified on the functions $v_1, v_2, \dots, v_n$. The nth derivative is obtained in the form:

$$y_p^{(n)} = v_1' y_1^{(n-1)} + v_2' y_2^{(n-1)} + \dots + v_n' y_n^{(n-1)} + v_1 y_1^{(n)} + v_2 y_2^{(n)} + \dots + v_n y_n^{(n)}$$

Substitution of the solution y and its derivatives into eq. (1.1), and grouping together derivatives of each solution, one obtains:

$$v_1\left[a_0y_1^{(n)}+a_1y_1^{(n-1)}+...+a_ny_1\right]+v_2\left[a_0y_2^{(n)}+a_1y_2^{(n-1)}+...+a_ny_2\right]+...$$
$$+v_n\left[a_0y_n^{(n)}+a_1y_n^{(n-1)}+...+a_ny_n\right]+a_0\left[v_1'y_1^{(n-1)}+v_2'y_2^{(n-1)}+...+v_n'y_n^{(n-1)}\right]=f(x)$$

The terms in the square brackets, which have the form Ly vanish since each y_i is a solution of $Ly_i = 0$, resulting in:

$$v_1'y_1^{(n-1)}+v_2'y_2^{(n-1)}+...+v_n'y_n^{(n-1)}=\frac{f(x)}{a_0(x)}$$

The system of algebraic equations on the unknown functions $v_1', v_2',..., v_n'$ can now be written as follows:

$$\begin{aligned}
&v_1'y_1 && +v_2'y_2 && +...+v_n'y_n && =0\\
&v_1'y_1' && +v_2'y_2' && +...+v_n'y_n' && =0\\
&... && ... && ... && ...\\
&... && ... && ... && ...\\
&v_1'y_1^{(n-2)} && +v_2'y_2^{(n-2)} && +...+v_n'y_n^{(n-2)} && =0\\
&v_1'y_1^{(n-1)} && +v_2'y_2^{(n-1)} && +...+v_n'y_n^{(n-1)} && =\frac{f(x)}{a_0(x)}
\end{aligned} \tag{1.25}$$

The determinant of the coefficients of the unknown functions $[v_i']$ is the Wronskian of the system, which does not vanish for a set of independent solutions $[y_i]$. Equations in (1.25) give a unique set of functions $[v_i']$, which can be integrated to give $[v_i]$, thereby giving a particular solution y_p.

The method of variation of the parameters is now applied to a general 2nd order differential equation. Let:

$$a_0(x)\,y''+a_1(x)\,y'+a_2(x)\,y=f(x)$$

such that the homogeneous solution is given by:

$$y_h=C_1y_1(x)+C_2y_2(x)$$

and a particular solution can be found in the form:

$$y_p=v_1y_1+v_2y_2$$

where the functions v_1 and v_2 are solutions of the two algebraic equations:

$$v_1'y_1+v_2'y_2=0$$

and

$$v_1'y_1'+v_2'y_2'=\frac{f(x)}{a_0(x)}$$

Solving for v_1' and v_2', one obtains:

$$v_1'=\frac{-y_2\,f(x)/a_0(x)}{y_1y_2'-y_1'y_2}=-\frac{y_2\,f(x)}{a_0(x)\,W(x)}$$

and

$$v_2' = \frac{y_1 f(x)/a_0(x)}{y_1 y_2' - y_1' y_2} = +\frac{y_1 f(x)}{a_0(x)\, W(x)}$$

Direct integration of these two expressions gives:

$$v_1 = -\int^{x} \frac{y_2(\eta)\, f(\eta)}{a_0(\eta)\, W(\eta)}\, d\eta$$

and

$$v_2 = +\int^{x} \frac{y_1(\eta)\, f(\eta)}{a_0(\eta)\, W(\eta)}\, d\eta$$

The unknown functions v_1 and v_2 are then substituted into y_p to give:

$$y_p = -y_1(x) \int^{x} \frac{y_2(\eta)\, f(\eta)}{a_0(\eta)\, W(\eta)}\, d\eta + y_2(x) \int^{x} \frac{y_1(\eta)\, f(\eta)}{a_0(\eta)\, W(\eta)}\, d\eta$$

$$= \int^{x} \frac{y_1(\eta)\, y_2(x) - y_1(x)\, y_2(\eta)}{W(\eta)} \frac{f(\eta)}{a_0(\eta)}\, d\eta$$

Example 1.7

Obtain the complete solution to the following equation:

$$y'' - 4y = e^x$$

The homogeneous solution is given by:

$$y_h = C_1 e^{2x} + C_2 e^{-2x}$$

where $y_1 = e^{2x}$, $y_2 = e^{-2x}$, $a_0(x) = 1$, and the Wronskian is given by:

$$W(x) = y_1 y_2' - y_1' y_2 = -4$$

The particular solution is thus given by the following integral:

$$y_p = \int^{x} \frac{e^{2\eta}\, e^{-2x} - e^{2x}\, e^{-2\eta}}{(-4)}\, e^{\eta}\, d\eta = -\frac{1}{3} e^x$$

The complete solution becomes:

$$y = C_1 e^{2x} + C_2 e^{-2x} - \frac{1}{3} e^x$$

1.8 Abel's Formula for the Wronskian

The Wronskian for a set of functions $[y_i]$ can be evaluated by using eq. (1.12). However, one can obtain the Wronskian in a closed form when the set of functions $[y_i]$

are solutions of an ordinary differential equation. Differentiating the determinant in eq. (1.12) is equivalent to summing n-determinants where only one row is differentiated in each determinant, i.e.:

$$\frac{dW}{dx} = \begin{vmatrix} y_1' & y_2' & \cdots & y_n' \\ y_1' & y_2' & \cdots & y_n' \\ y_1'' & y_2'' & \cdots & y_n'' \\ \cdot & \cdot & \cdot & \cdot \\ y_1^{(n-2)} & y_2^{(n-2)} & \cdots & y_n^{(n-2)} \\ y_1^{(n-1)} & y_2^{(n-1)} & \cdots & y_n^{(n-1)} \end{vmatrix} +$$

$$+ \begin{vmatrix} y_1 & y_2 & \cdots & y_n \\ y_1'' & y_2'' & \cdots & y_n'' \\ y_1'' & y_2'' & \cdots & y_n'' \\ y_1^{(n-1)} & y_2^{(n-1)} & \cdots & y_n^{(n-1)} \\ y_1^{(n-1)} & y_2^{(n-1)} & \cdots & y_n^{(n-1)} \end{vmatrix} + \begin{vmatrix} y_1 & y_2 & \cdots & y_n \\ y_1' & y_2' & \cdots & y_n' \\ \cdot & \cdot & \cdot & \cdot \\ \cdot & \cdot & \cdot & \cdot \\ y_1^{(n-2)} & y_2^{(n-2)} & \cdots & y_n^{(n-2)} \\ y_1^{(n)} & y_2^{(n)} & \cdots & y_n^{(n)} \end{vmatrix} \tag{1.26}$$

Since there are two identical rows in the first (n - 1) determinants, each of these determinants vanish, thereby leaving only the non-vanishing last determinant:

$$\frac{dW}{dx} = \begin{vmatrix} y_1 & y_2 & \cdots & y_n \\ y_1' & y_2' & \cdots & y_n' \\ \cdot & \cdot & \cdot & \cdot \\ \cdot & \cdot & \cdot & \cdot \\ y_1^{(n-2)} & y_2^{(n-2)} & \cdots & y_n^{(n-2)} \\ y_1^{(n)} & y_2^{(n)} & \cdots & y_n^{(n)} \end{vmatrix} \tag{1.27}$$

Substitution of eq. (1.2) for $y_i^{(n)}$, i.e.:

$$y_i^{(n)} = -\frac{a_1(x)}{a_0(x)} y_i^{(n-1)} - \frac{a_2(x)}{a_0(x)} y_i^{(n-2)} - \ldots - \frac{a_{n-1}(x)}{a_0(x)} y_i' - \frac{a_n(x)}{a_0(x)} y_i$$

into the determinant of (dW)/(dx), and manipulating the determinant, by successively multiplying the first row by a_n/a_0, the second row by a_{n-1}/a_0, etc., and add them to the last row, one obtains:

$$\frac{dW}{dx} = -\frac{a_1(x)}{a_0(x)} W$$

which can be integrated to give a closed form formula for the Wronskian:

$$W(x) = W_0 \exp\left(\int -\frac{a_1(x)}{a_0(x)}\, dx \right) \tag{1.28}$$

with W_0 = constant. This is known as **Abel's Formula**.

It should be noted that W(x) cannot vanish in a region $a \leq x \leq b$ unless W_0 vanishes identically, $a_1(x) \to \infty$ or $a_0(x) \to 0$ at some point in $a \leq x \leq b$. Since the last two are ruled out, then W(x) cannot vanish.

Example 1.8

Consider the differential equation of Example 1.3. The Wronskian is given by:

$$W(x) = W_0 \exp\left(\int -3\, dx\right) = W_0\, e^{-3x}$$

which is the Wronskian of the solutions of the differential equation. To evaluate the constant W_0, one can determine the dominant term(s) of each solutions' Taylor series, find the leading term of the resulting Wronskian and then take a limit as $x \to 0$ in this special case, resulting in $W_0 = 9$ and $W(x) = 9e^{-3x}$.

1.9 Initial Value Problems

For a unique solution of an ordinary differential equation of order n, whose complete solution contains n arbitrary constants, a set of n-conditions on the dependent variable is required. The set of n-conditions on the dependent variable is a set of the values that the dependent variable and its first (n - 1) derivatives take at a point $x = x_0$, can be given as:

$$\begin{aligned} y(x_0) &= \alpha_0 \\ y'(x_0) &= \alpha_1 \\ &\vdots \\ y^{(n-1)}(x_0) &= \alpha_{n-1} \end{aligned} \qquad a \le x, x_0 \le b \tag{1.29}$$

A unique solution for the set of constants $[C_i]$ in the homogeneous solution y_h can be obtained. Such problems are known as **Initial Value Problems**. To prove uniqueness, let there exist two solutions y_I and y_{II} satisfying the system eq. (1.29) such that:

$$y_I = C_1 y_1 + C_2 y_2 + \ldots + C_n y_n + y_p$$

$$y_{II} = B_1 y_1 + B_2 y_2 + \ldots + B_n y_n + y_p$$

then, the difference of the two solutions also satisfies the same homogeneous equation:

$$L(y_I - y_{II}) = 0$$

and

$$\begin{aligned} y_I(x_0) - y_{II}(x_0) &= 0 \\ y_I'(x_0) - y_{II}'(x_0) &= 0 \\ &\cdot \\ y_I^{(n-1)}(x_0) - y_{II}^{(n-1)}(x_0) &= 0 \end{aligned}$$

which results in the following homogeneous algebraic equations:

$$\begin{aligned} A_1 y_1(x_0) + A_2 y_2(x_0) + \ldots + A_n y_n(x_0) &= 0 \\ A_1 y_1'(x_0) + A_2 y_2'(x_0) + \ldots + A_n y_n'(x_0) &= 0 \\ &\vdots \\ A_1 y_1^{(n-1)}(x_0) + A_2 y_2^{(n-1)}(x_0) + \ldots + A_n y_n^{(n-1)}(x_0) &= 0 \end{aligned} \tag{1.30}$$

where the constants A_i are defined by:

$A_i = C_i - B_i \qquad i = 1, 2, 3, ..., n$

Since the determinant of the coefficients of $[A_i]$ is the Wronskian of the system, which does not vanish for the independent set $[y_i]$, then $A_i = 0$, and the two solutions y_I and y_{II}, satisfying the system eq. (1.29), must be identical.

Example 1.9

Obtain the solution of the following system:

$y'' + 4y = 0$

$y(0) = 1 \qquad x \geq 0$

$y'(0) = 4$

$y = C_1 \sin(2x) + C_2 \cos(2x)$

$y(0) = C_2 = 1$

$y'(0) = 2C_1 = 4 \qquad C_1 = 2$

such that:

$y = 2\sin(2x) + \cos(2x)$

PROBLEMS

Section 1.2

1. Solve the following differential equations: SEP OF VAR | INT FACTOR

   (a) $\dfrac{dy}{dx} + xy = e^{-x^2/2}$ (b) $x\dfrac{dy}{dx} + 2y = x^2$

   (c) $\dfrac{dy}{dx} + 2y \cot anx = \cos x$ (d) $\dfrac{dy}{dx} + y \tanh x = e^x$

   (e) $\sin x \cos x \dfrac{dy}{dx} + y = \sin x$ (f) $\dfrac{dy}{dx} + y = e^{-x}$

Section 1.3

2. Examine the following sets for linear independence: WRONSKIAN

   (a) $u_1(x) = e^{ix}$ $u_2(x) = e^{-ix}$ $u_3 = \sin x$

   (b) $u_1(x) = e^{-x}$ $u_2(x) = e^x$

   (c) $u_1(x) = 1 + x^2$ $u_2(x) = 1 - x^2$

   (d) $v_1(x) = \dfrac{u_1 - u_2}{2}$ $v_2(x) = \dfrac{u_1 + u_2}{2}$

   $u_1(x)$ and $u_2(x)$ are defined in (c).

Section 1.4

NTH ORDER

3. Obtain the homogeneous solution to the following differential equations:

   (a) $\dfrac{d^2y}{dx^2} - \dfrac{dy}{dx} - 2y = 0$ (b) $\dfrac{d^3y}{dx^3} - \dfrac{d^2y}{dx^2} - \dfrac{1}{4}\dfrac{dy}{dx} + \dfrac{1}{4}y = 0$

   (c) $\dfrac{d^3y}{dx^3} - 3\dfrac{dy}{dx} + 2y = 0$ (d) $\dfrac{d^4y}{dx^4} - 8\dfrac{d^2y}{dx^2} + 16y = 0$

   (e) $\dfrac{d^4y}{dx^4} - 16y = 0$ (f) $\dfrac{d^2y}{dx^2} + iy = 0$ $i = \sqrt{-1}$

   (g) $\dfrac{d^4y}{dx^4} + 16y = 0$ (h) $\dfrac{d^5y}{dx^5} - \dfrac{d^4y}{dx^4} - 2\dfrac{d^3y}{dx^3} + 2\dfrac{d^2y}{dx^2} + \dfrac{dy}{dx} - y = 0$

   (i) $\dfrac{d^3y}{dx^3} + 8a^3y = 0$ (j) $\dfrac{d^3y}{dx^3} - a\dfrac{d^2y}{dx^2} + 2a^3y = 0$

   (k) $\dfrac{d^4y}{dx^4} + 2a^2\dfrac{d^2y}{dx^2} + a^4y = 0$ (l) $\dfrac{d^6y}{dx^6} + 64y = 0$

4. If a third order differential equation, with constant coefficients, has three repeated roots $= m$, show that e^{mx}, xe^{mx}, and x^2e^{mx} make up a linearly independent set.

Section 1.5

5. Obtain the solution to the following differential equation:

(a) $x^2 \dfrac{d^2y}{dx^2} + x\dfrac{dy}{dx} - y = 0$ (b) $x^2 \dfrac{d^2y}{dx^2} + 3x\dfrac{dy}{dx} + y = 0$

(c) $x^2 \dfrac{d^2y}{dx^2} + x\dfrac{dy}{dx} + 4y = 0$ (d) $x^3 \dfrac{d^3y}{dx^3} + x^2\dfrac{d^2y}{dx^2} - 2x\dfrac{dy}{dx} + 2y = 0$

(e) $x^3 \dfrac{d^3y}{dx^3} + 3x^2 \dfrac{d^2y}{dx^2} - 2x\dfrac{dy}{dx} + 2y = 0$

(f) $x^4 \dfrac{d^4y}{dx^4} + 6x^3 \dfrac{d^3y}{dx^3} + 7x^2 \dfrac{d^2y}{dx^2} + x\dfrac{dy}{dx} - 16y = 0$

(g) $4x^2 \dfrac{d^2y}{dx^2} + y = 0$ (h) $4x^2 \dfrac{d^2y}{dx^2} + 5y = 0$

Section 1.6

6. Obtain the total solution for the following differential equations: $Ly = f(x)$

(a) L as in Problem 3a and $f(x) = 10\sin x + 4e^x + 9xe^{-x}$

(b) L as in Problem 3c and $f(x) = 2x^2 + 4e^{-x} + 27x^2e^x$

(c) L as in Problem 3d and $f(x) = 16\sin 2x + 8\sinh 2x$

(d) L as in Problem 5a and $f(x) = 3x^2 + 4x$

(e) L as in Problem 5e and $f(x) = 12x + 4x^2$

Section 1.7

7. Obtain the general solution to the following differential equations:

(a) $\dfrac{d^2y}{dx^2} + k^2y = f(x)$ $1 \le x \le 2$

(b) $x^2 \dfrac{d^2y}{dx^2} + x\dfrac{dy}{dx} - y = f(x)$ $x \ge 1$

(c) $x^3 \dfrac{d^3y}{dx^3} - 2x^2 \dfrac{d^2y}{dx^2} + 4x\dfrac{dy}{dx} - 4y = f(x)$ $1 \le x \le 2$

Section 1.8

8. Obtain the total solution to the following systems:

   (a) Problem 6(a)

   $y(0) = 2$ $\qquad y'(0) = -1$ $\qquad x \geq 0$

   (b) Problem 3(b)

   $y(1) = 3$ $\qquad y'(1) = 0$

   $y''(2) = 3$ $\qquad 0 \leq x \leq 2$

2

SERIES SOLUTIONS OF ORDINARY DIFFERENTIAL EQUATIONS

2.1 Introduction

In many instances, it is not possible to obtain the solution of an ordinary differential equation of the type of eq. (1.2) in a closed form. If the differential equation (1.2) has $a_0(x)$ as a non-vanishing bounded functions and $a_1(x), a_2(x), \ldots, a_n(x)$ are bounded in the interval $a \leq x \leq b$, satisfying the system in eq. (1.29), then there exists a set of n solutions $y_i(x)$, $i = 1, 2, \ldots, n$. Such a solution can be expanded into a Taylor series about a point x_0, $a < x_0 < b$, such that:

$$y(x) = \sum_{n=0}^{\infty} c_n (x - x_0)^n \tag{2.1}$$

where

$$c_n = \frac{y^{(n)}(x_0)}{n!} \tag{2.2}$$

This series is referred to as a **Power Series** about the point $x = x_0$, refer to Appendix A.

In general, one does not know y(x) *a priori*, so that the coefficients of the series c_n are not determinable from eq. (2.2). However, one can assume that the solution to eq. (1.2) has a power series of the form in eq. (2.1) and then the unknown constants c_n can be determined by substituting the solution of eq. (2.1) into eq. (1.2).

The power series in eq. (2.1) converges in a certain region. Using the ratio test (Appendix A), then:

$$\lim_{n \to \infty} \left| \frac{c_{n+1} (x - x_0)^{n+1}}{c_n (x - x_0)^n} \right| \quad \begin{matrix} < 1 \text{ series converges} \\ > 1 \text{ series diverges} \end{matrix}$$

or

$$|x - x_0| \quad \begin{matrix} < \rho \text{ series converges} \\ > \rho \text{ series diverges} \end{matrix} \qquad \rho = \lim_{n \to \infty} \left| \frac{a_n}{a_{n+1}} \right|$$

where ρ is known as the **Radius of Convergence**.

Thus the series converges for $x_0 - \rho < x < x_0 + \rho$, and diverges outside this region. The series may or may not converge at the end points, i.e., at $x = x_0 + \rho$ and $x = x_0 - \rho$, where:

$$\lim_{n \to \infty} \left| \frac{c_{n+1}}{c_n} (x - x_0) \right| = 1$$

and the ratio test fails. To test the convergence of the series at the end points, refer to the tests given in Appendix A.

The radius of convergence for series solutions of a differential equation is limited by the existence of singularities, i.e., points where $a_0(x)$ vanishes. If x_1 is the *closest zero* of $a_0(x)$ to x_0, then the radius of convergence $\rho = |x_1 - x_0|$.

2.2 Power Series Solutions

Power series solutions about $x = x_0$ of the form in eq. (2.1) can be transformed to a power series solution about $z = 0$.

Let $z = x - x_0$ then eq. (1.1) transforms to:

$$a_0(z + x_0)\frac{d^n y}{dz^n} + a_1(z + x_0)\frac{d^{n-1} y}{dz^{n-1}} + \ldots + a_{n-1}(z + x_0)\frac{dy}{dz} + a_n(z + x_0)\, y = f(z + x_0)$$

Thus, power series homogeneous solutions about $x = x_0$ become series solutions about $z = 0$; i.e.:

$$y(z) = \sum_{m=0}^{\infty} c_m z^m$$

Henceforth, one only needs to discuss power series solutions about the origin, which will be taken to be $x = 0$ for simplicity, i.e.:

$$y(x) = \sum_{m=0}^{\infty} c_m x^m \tag{2.3}$$

Substitution of the series in eq. (2.3) into the differential equation (1.2) and equating the coefficient of each power of x to zero, results in an infinite number of algebraic equations, each one gives the constant c_m in terms of $c_{m-1}, c_{m-2}, \ldots, c_1$ and c_0, for $m = 1, 2, \ldots$. Since the homogeneous differential equation is of order n, then there will be n arbitrary constants, i.e., the constants $c_0, c_1, \ldots, c_n$ are arbitrary constants. The constants $c_{n+1}, c_{n+2} \ldots$ can then be computed in terms of the arbitrary constants $c_0, \ldots, c_n$.

Example 2.1

Obtain the solution valid in the neighborhood of $x_0 = 0$, of the following equation:

$$\frac{d^2 y}{dx^2} - xy = 0$$

Note that $a_0(x) = 1$, $a_1(x) = 0$, and $a_2(x) = -x$, all bounded and $a_o(0) \neq 0$.

Let the solution to be in the form of a power series about $x_0 = 0$:

$$y = \sum_{n=0}^{\infty} c_n x^n \qquad y' = \sum_{0}^{\infty} n c_n x^{n-1} \qquad y'' = \sum_{0}^{\infty} n(n-1) c_n x^{n-2}$$

which, when substituted into the differential equation gives:

$$Ly = \sum_{0}^{\infty} n(n-1)\, c_n x^{n-2} - \sum_{0}^{\infty} c_n x^{n+1} = 0$$

Writing out the two series in a power series of ascending powers of x results in:

$$0 \cdot c_0 x^{-2} + 0 \cdot c_1 x^{-1} + 2c_2 + (6c_3 - c_0)x + (12c_4 - c_1)x^2$$
$$+ (20c_5 - c_2)x^3 + (30c_6 - c_3)x^4 + (42c_7 - c_4)x^5 + \ldots = 0$$

Since the power series of a null function has zero coefficients, then equating the coefficient of each power of x to zero, one obtains:

$$c_0 = \frac{0}{0} = \textit{indeterminate} \qquad c_1 = \frac{0}{0} = \textit{indeterminate} \qquad c_2 = 0$$

$$c_3 = \frac{c_0}{6} = \frac{c_0}{2 \cdot 3} \qquad c_4 = \frac{c_1}{12} = \frac{c_1}{3 \cdot 4} \qquad c_5 = \frac{c_2}{5 \cdot 4} = 0$$

$$c_6 = \frac{c_3}{6 \cdot 5} = \frac{c_0}{2 \cdot 3 \cdot 5 \cdot 6} \qquad c_7 = \frac{c_4}{6 \cdot 7} = \frac{c_0}{3 \cdot 4 \cdot 6 \cdot 7}$$

Thus, the series solution becomes:

$$y = c_0 + c_1 x + \frac{c_0}{2 \cdot 3} x^3 + \frac{c_1}{3 \cdot 4} x^4 + \frac{c_0}{2 \cdot 3 \cdot 5 \cdot 6} x^6 + \frac{c_1}{3 \cdot 4 \cdot 6 \cdot 7} x^7$$

$$= c_0 \left(1 + \frac{x^3}{6} + \frac{x^6}{6 \cdot 30} + \ldots\right) + c_1 \left(x + \frac{x^4}{12} + \frac{x^7}{12 \cdot 42} + \ldots\right)$$

Since c_0 and c_1 are arbitrary constants, then:

$$y_1 = 1 + \frac{x^3}{6} + \frac{x^6}{6 \cdot 30} + \ldots$$

and

$$y_2 = x + \frac{x^4}{12} + \frac{x^7}{12 \cdot 42} + \ldots$$

are the two independent solutions of the homogeneous differential equation.

It is more advantageous to work out the relationship between c_n and $c_{n-1}, c_{n-2}, \ldots, c_1, c_0$ in a formula known as the **Recurrence Formula**. Rewriting $Ly = 0$ again in expanded form and separating the first few terms of each series, such that the remaining terms of each series start at the same power of x, i.e.:

$$0 \cdot c_0 x^{-2} + 0 \cdot c_1 x^{-1} + 2c_2 + \sum_{n=3}^{\infty} n(n-1) c_n x^{n-2} - \sum_{n=0}^{\infty} c_n x^{n+1} = 0$$

where the first term of each power series starts with x^1.

Letting n = m + 3 in the first series and n = m in the second series, so that the two series start with the same index m = 0 and the power of x is the same for both series, one obtains:

$c_0 = \textit{indeterminate} \qquad c_1 = \textit{indeterminate} \qquad c_2 = 0$

and

$$\sum_{m=0}^{\infty} \left[(m+2)(m+3)\, c_{m+3} - c_m\right] x^{m+1} = 0$$

Equating the coefficient of x^{m+1} to zero gives the recurrence formula:

$$c_{m+3} = \frac{c_m}{(m+2)(m+3)} \qquad m = 0, 1, 2, \ldots$$

which relates c_{m+3} to c_m and results in the same constants evaluated earlier. The recurrence formula reduces the amount of algebraic manipulations needed for evaluating the coefficients c_m.

Note: Henceforth, the coefficient of the power series c_n will be replaced by a_n, which are not to be confused with $a_n(x)$.

Example 2.2

Solve the following ordinary differential equation about $x_0 = 0$:

$$x\frac{d^2y}{dx^2} + 3\frac{dy}{dx} + xy = 0 \qquad y = \sum_{n=0}^{\infty} a_n x^n$$

Note that $a_0(x) = x$, $a_1(x) = 1$, and $a_2(x) = x$ and $a_0(0) = 0$, which means that the equation is singular at x = 0. Attempting a power series solution by substituting into the differential equation and combining the three series gives:

$$Ly = \sum_{n=0}^{\infty} n(n+2)\, a_n x^{n-1} + \sum_{n=0}^{\infty} a_n x^{n+1} = 0$$

$$= 0 \cdot a_0 x^{-1} + 3a_1 + \sum_{n=2}^{\infty} n(n+2)\, a_n x^{n-1} + \sum_{n=0}^{\infty} a_n x^{n+1} = 0$$

Substituting n = m + 2 in the first and n = m in the second series, one obtains:

$$= 0 \cdot a_0 x^{-1} + 3a_1 + \sum_{m=0}^{\infty} \left[(m+2)(m+4)\, a_{m+2} + a_m\right] x^{m+1} = 0$$

Thus, equating the coefficient of each power of x to zero gives:

$a_0 = \textit{indeterminate} \qquad a_1 = 0$

as well as the recurrence formula:

$$a_{m+2} = -\frac{a_m}{(m+2)(m+4)} \qquad m = 0, 1, 2, \ldots$$

The recurrence formula can be used to evaluate the remaining coefficients:

$$a_2 = -\frac{a_0}{2^2\ 2!\,1!} \qquad a_3 = -\frac{a_1}{15} = 0 \qquad a_4 = -\frac{a_2}{24} = \frac{a_0}{2^4\ 3!\,2!}$$

$$a_5 = -\frac{a_3}{35} = 0 = a_7 = a_9 = \ldots \qquad a_6 = -\frac{a_0}{2^6\ 4!\,2!}, \text{ etc.}$$

Thus, the solution obtainable in the form of a power series is:

$$y = a_0\left(1 - \frac{x^2}{2^2\ 2!\,1!} + \frac{x^4}{2^4\ 3!\,2!} - \frac{x^6}{2^6\ 4!\,3!} + \ldots\right)$$

This solution has only one arbitrary constant, thereby giving one solution. The missing second solution cannot be obtained in a power series form due to the fact that $a_0(x) = x$ vanishes at the point about which the series is expanded, i.e., $x = 0$ is a singular point of the differential equation. To obtain the full solution, one needs to deal with differential equations having singular points at the point of expansion x_0.

2.3 Classification of Singularities

Dividing the second-order differential equation by $a_o(x)$, then it becomes:

$$Ly = \frac{d^2y}{dx^2} + \bar{a}_1(x)\frac{dy}{dx} + \bar{a}_2(x)y = 0 \qquad (2.4)$$

where $\bar{a}_1(x) = a_1(x)/a_o(x)$ and $\bar{a}_2(x) = a_2(x)/a_o(x)$.

If either of the two coefficients $\bar{a}_1(x)$ or $\bar{a}_2(x)$ are unbounded at a point x_0, then the equation has a singularity at $x = x_0$.

(i) If $\bar{a}_1(x)$ and $\bar{a}_2(x)$ are both regular (bounded) at x_0, then x_0 is called a **Regular Point (RP)**.

(ii) If $x = x_0$ is a singular point and if:

$$\left.\begin{aligned} &\lim_{x \to x_0} (x - x_0)\bar{a}_1(x) \to \text{finite} \\ &\text{and} \\ &\lim_{x \to x_0} (x - x_0)^2\,\bar{a}_2(x) \to \text{finite} \end{aligned}\right\} \quad x_0 \text{ is a } \textbf{Regular Singular Point (RSP)}$$

(iii) If $x = x_0$ is a singular point and either:

$$\left.\begin{aligned} &\lim_{x \to x_0} (x - x_0)\bar{a}_1(x) \to \text{unbounded} \\ &\text{or} \\ &\lim_{x \to x_0} (x - x_0)^2\,\bar{a}_2(x) \to \text{unbounded} \end{aligned}\right\} \quad x_0 \text{ is an } \textbf{Irregular Singular Point (ISP)}$$

Example 2.3

Classify the behavior of each of the following differential equations at x = 0 and at all the singular points of each equation.

(a) $x\frac{d^2y}{dx^2} + \sin x\frac{dy}{dx} + x^2y = 0$

Here, $\bar{a}_1(x) = \frac{\sin x}{x}$ and $\bar{a}_2(x) = x$

Both coefficients are regular at x = 0, thus x = 0 is a RP.

(b) $x\frac{d^2y}{dx^2} + 3\frac{dy}{dx} + xy = 0$

$\bar{a}_1(x) = \frac{3}{x}$ $\qquad\qquad$ $\bar{a}_2(x) = 1$

Here, x = 0 is the only singular point. Classifying the singularity at x = 0:

$\lim_{x \to 0} x\left(\frac{3}{x}\right) = 3$ $\qquad\qquad$ $\lim_{x \to 0} x^2(1) = 0$

Thus x = 0 is a RSP.

(c) $x^2(x^2 - 1)\frac{d^2y}{dx^2} + (x - 1)^2\frac{dy}{dx} + x^2y = 0$

$\bar{a}_1(x) = \frac{(x-1)}{x^2(x+1)}$ $\qquad\qquad$ $\bar{a}_2(x) = \frac{1}{(x-1)(x+1)}$

Here, there are three singular points; x = -1, 0, and +1. Examining each singularity:

<u>$x_0 = -1$</u>

$$\lim_{x \to -1} (x + 1)\frac{(x - 1)}{x^2(x + 1)} = -2$$

$x_0 = -1$ is a RSP.

$$\lim_{x \to -1} (x + 1)^2 \frac{1}{(x - 1)(x + 1)} = 0$$

<u>$x_0 = 0$</u>

$$\lim_{x \to 0} x\frac{(x - 1)}{x^2(x + 1)} = -\infty$$

$x_0 = 0$ is an ISP

$$\lim_{x \to 0} x^2 \frac{1}{(x - 1)(x + 1)} = 0$$

<u>$x_0 = +1$</u>

$$\lim_{x \to 1} (x-1)\frac{(x-1)}{x^2(x+1)} = 0$$

$x_0 = +1$ is a RSP

$$\lim_{x \to 1} (x-1)^2\frac{1}{(x-1)(x+1)} = 0$$

2.4 Frobenius Solution

If the differential equation (2.4) has a Regular Singular Point at x_0, then one or both solution(s) may not be obtainable by the power series expansion in eq. (2.3). If the equation has a singularity at $x = x_0$, one can perform a linear transformation (discussed in Section 2.2), $z = x - x_0$, and seek a solution about $z = 0$. For simplicity, a solution valid in the neighborhood of $x = 0$ is presented.

For equations that have a RSP at $x = x_0$, a solution of the form:

$$y(x) = \sum_{n=0}^{\infty} a_n (x - x_0)^{n+\sigma} \tag{2.5}$$

can be used, where σ is an unknown constant. If x_0 is a RSP, then the constant σ cannot be a positive integer or zero for at least one solution of the homogeneous equation. This solution is known as the **Frobenius Solution**.

Since $\bar{a}_1(x)$ and $\bar{a}_2(x)$ can, at most, be singular to the order of $(x-x_0)^{-1}$ and $(x-x_0)^{-2}$, then:

$$(x - x_0)\,\bar{a}_1(x) \qquad \text{and} \qquad (x - x_0)^2\,\bar{a}_2(x)$$

are regular functions in the neighborhood of $x = x_0$. Thus, expanding the above functions into a power series about $x = x_0$ results in:

$$(x - x_0)\bar{a}_1(x) = \alpha_0 + \alpha_1(x - x_0) + \alpha_2(x - x_0)^2 + \ldots = \sum_{k=0}^{\infty} \alpha_k (x - x_0)^k \tag{2.6}$$

and

$$(x - x_0)^2\bar{a}_2(x) = \beta_0 + \beta_1(x - x_0) + \beta_2(x - x_0)^2 + \ldots = \sum_{k=0}^{\infty} \beta_k (x - x_0)^k$$

Transforming the equation by $z = x - x_0$ and replacing z by x, one can discuss solutions about $x_0 = 0$. The Frobenius solution in eq. (2.5) and the series expansions of $a_1(x)$ and $a_2(x)$ about $x_0 = 0$ of eq. (2.6) are substituted into the differential equation (2.4), such that:

$$Ly = \sum_{n=0}^{\infty} (n + \sigma - 1)(n + \sigma) a_n x^{n+\sigma-2} + \left[\sum_{k=0}^{\infty} \alpha_k x^{k-1}\right]\left[\sum_{n=0}^{\infty} (n + \sigma) a_n x^{n+\sigma-1}\right]$$

$$+\left[\sum_{k=0}^{\infty}\beta_k x^{k-2}\right]\left[\sum_{n=0}^{\infty}a_n x^{n+\sigma}\right]=0 \tag{2.7}$$

The second term in eq. (2.7) can be written in a Taylor series form as follows:

$$\left[\sum_{k=0}^{\infty}\alpha_k x^{k-1}\right]\left[\sum_{n=0}^{\infty}(n+\sigma)\,a_n x^{n+\sigma-1}\right]=x^{\sigma-2}\Bigg[\sigma\alpha_0 a_0+(\sigma a_0\alpha_1+(\sigma+1)a_1\alpha_0)x$$

$$+(\sigma a_0\alpha_2+(\sigma+1)a_1\alpha_1+(\sigma+2)a_2\alpha_0)x^2+\ldots+\left(\sum_{k=0}^{k=n}(\sigma+k)a_k\alpha_{n-k}\right)x^n+\ldots\Bigg]$$

$$=\sum_{n=0}^{\infty}c_n x^{n+\sigma-2}$$

where

$$c_n=\sum_{k=0}^{k=n}(\sigma+k)a_k\alpha_{n-k}$$

The third term in eq. (2.7) can be expressed in a Taylor series form in a similar manner:

$$\left[\sum_{n=0}^{\infty}\beta_k x^{k-2}\right]\left[\sum_{n=0}^{\infty}a_n x^{n+\sigma}\right]=\sum_{n=0}^{\infty}d_n x^{n+\sigma-2}$$

where

$$d_n=\sum_{k=0}^{k=n}a_k\,\beta_{n-k}$$

Eq. (2.7) then becomes:

$$Ly=x^{\sigma-2}\left[\sum_{n=0}^{\infty}(n+\sigma-1)(n+\sigma)\,a_n x^n+\sum_{n=0}^{\infty}c_n x^n+\sum_{n=0}^{\infty}d_n x^n\right] \tag{2.8}$$

$$=x^{\sigma-2}\Big\{[\sigma(\sigma-1)+\sigma\alpha_0+\beta_0]a_0+[(\sigma(\sigma+1)+(\sigma+1)\alpha_0+\beta_0)a_1+(\sigma\alpha_1+\beta_1)a_0]x$$

$$+[((\sigma+1)(\sigma+2)+(\sigma+2)\alpha_0+\beta_0)a_2+((\sigma+1)\alpha_1+\beta_1)a_1+(\sigma\alpha_2+\beta_2)a_0]x^2$$

$$+\ldots+[((n+\sigma-1)(n+\sigma)+(\sigma+n)a_0+\beta_0)a_n+((\sigma+n-1)a_1+\beta_1)a_{n-1}$$

$$+((\sigma+n-2)\alpha_2+\beta_2)a_{n-2}+\ldots+((\sigma+1)\alpha_{n-1}+\beta_{n-1})a_1+(\sigma\alpha_n+\beta_n)a_0]x^n+\ldots\Big\}$$

Defining the quantities:

$$f(\sigma)=\sigma(\sigma-1)+\sigma\alpha_0+\beta_0$$

and

$$f_n(\sigma)=\sigma\alpha_n+\beta_n$$

then eq. (2.8) can be rewritten in a condensed form:

$$Ly = x^{\sigma-2}\{f(\sigma)a_0 + [f(\sigma+1)a_1 + f_1(\sigma)a_0]x + [f(\sigma+2)a_2 + f_1(\sigma+1)a_1 + f_2(\sigma)a_0]x^2 + \ldots + [f(\sigma+n)a_n + f_1(\sigma+n-1)a_{n-1} + \ldots + f_n(\sigma)]x^n + \ldots\}$$

$$= x^{\sigma-2}\left\{f(\sigma)a_0 + \sum_{n=1}^{\infty}\left[f(\sigma+n)\,a_n + \sum_{k=1}^{n} f_k(\sigma+n-k)\,a_{n-k}\right]x^n\right\} \tag{2.9}$$

Each of the constants $a_1, a_2, \ldots, a_n, \ldots$ can be written in terms of a_0, by equating the coefficients of $x, x^2, \ldots$ to zero as follows:

$$a_1(\sigma) = -\frac{f_1(\sigma)}{f(\sigma+1)}a_0$$

$$a_2(\sigma) = -\frac{f_1(\sigma+1)a_1 + f_2(\sigma)a_0}{f(\sigma+2)}$$

$$= -\frac{-f_1(\sigma)f(\sigma+1) + f_2(\sigma)f(\sigma+1)}{f(\sigma+1)f(\sigma+2)}a_0 = -\frac{g_2(\sigma)}{f(\sigma+2)}a_0$$

$$a_3(\sigma) = -\frac{f_1(\sigma+2)a_2 + f_2(\sigma+1)a_1 + f_3(\sigma)a_0}{f(\sigma+3)} = -\frac{g_3(\sigma)}{f(\sigma+3)}a_0$$

and by induction:

$$a_n(\sigma) = -\frac{g_n(\sigma)}{f(\sigma+n)}a_0 \qquad n \geq 1 \tag{2.10}$$

Substitution of $a_n(\sigma)$ $n = 1, 2, 3,\ldots$ in terms of the coefficient a_0 into eq. (2.9) results in the following expression for the differential equation:

$$Ly = x^{\sigma-2}f(\sigma)a_0 \tag{2.11}$$

and consequently the series solution can be written in terms of $a_n(\sigma)$, which is a function of σ and a_0:

$$y(x,\sigma) = a_0x^{\sigma} + \sum_{n=1}^{\infty} a_n(\sigma)x^{n+\sigma} \tag{2.12}$$

For a non-trivial solution; $a_0 \neq 0$, eq. (2.7) is satisfied if:

$$f(\sigma) = \sigma(\sigma-1) + a_0\sigma + \beta_0 = 0 \tag{2.13}$$

Equation (2.13) is called the **Characteristic Equation**, which has two roots σ_1 and σ_2. Depending on the relationship of the two roots, there are three different cases.

Case (a): *Two roots are distinct and do not differ by an integer.*

If $\sigma_1 \neq \sigma_2$ and $\sigma_1 - \sigma_2 \neq$ integer, then there exists two solutions to eq. (2.7) of the form:

$$y_1(x) = \sum_{n=0}^{\infty} a_n(\sigma_1)\,x^{n+\sigma_1}$$

and (2.14)

$$y_2(x) = \sum_{n=0}^{\infty} a_n(\sigma_2) x^{n+\sigma_2}$$

Example 2.4

Obtain the solutions of the following differential equation about $x_0 = 0$:

$$x^2 \frac{d^2y}{dx^2} + x\frac{dy}{dx} + \left(x^2 - \frac{1}{9}\right) y = 0$$

Since x = 0 is a RSP, use a Frobenius solution about $x_0 = 0$:

$$y = \sum_{n=0}^{\infty} a_n x^{n+\sigma}$$

such that when substituted into the differential equation results in:

$$\sum_{n=0}^{\infty} \left[(n+\sigma)^2 - \frac{1}{9}\right] a_n x^{n+\sigma-2} + \sum_{n=0}^{\infty} a_n x^{n+\sigma} = 0$$

Extracting the first two lowest powered terms of the first series, such that each of the remaining series starts with x^σ one obtains:

$$\left(\sigma^2 - \frac{1}{9}\right) a_0 x^{\sigma-2} + \left[(\sigma+1)^2 - \frac{1}{9}\right] a_1 x^{\sigma-1}$$

$$+ \sum_{n=2}^{\infty} \left[(n+\sigma)^2 - \frac{1}{9}\right] a_n x^{n+\sigma-2} + \sum_{n=0}^{\infty} a_n x^{n+\sigma} = 0$$

Changing the indices n to m + 2 in the first series and to m in the second and combining the two resulting series:

$$\left(\sigma^2 - \frac{1}{9}\right) a_0 x^{\sigma-2} + \left[(\sigma+1)^2 - \frac{1}{9}\right] a_1 x^{\sigma-1}$$

$$+ \sum_{m=0}^{\infty} \left\{\left[(m+\sigma+2)^2 - \frac{1}{9}\right] a_{m+2} + a_m\right\} x^{m+\sigma} = 0$$

Equating the coefficients of $x^{\sigma-1}$ and $x^{m+\sigma}$ to zero and assuming $a_0 \neq 0$, there results the following recurrence formulae:

$$\left[(\sigma+1)^2 - \frac{1}{9}\right] a_1 = 0$$

$$a_{m+2} = -\frac{a_m}{(m+\sigma+2)^2 - 1/9} \qquad m = 0, 1, 2,...$$

and the characteristic equation:

$$\left(\sigma^2 - \frac{1}{9}\right) = 0$$

The two roots are $\sigma_1 = 1/3$ and $\sigma_2 = -1/3$. Note that $\sigma_1 \neq \sigma_2$ and $\sigma_1 - \sigma_2$ is not an integer.

Since $\sigma = \pm 1/3$, then $(\sigma+1)^2 - 1/9 \neq 0$ so that the odd coefficients vanish:

$$a_1 = a_3 = a_5 = ... = 0$$

and

$$a_{m+2} = -\frac{a_m}{\left(m+\sigma+5/3\right)\left(m+\sigma+7/3\right)} \qquad m = 0, 2, 4, \ldots$$

with

$$a_2(\sigma) = -\frac{a_0}{\left(\sigma+5/3\right)\left(\sigma+7/3\right)}$$

$$a_4(\sigma) = -\frac{a_2}{\left(\sigma+11/3\right)\left(\sigma+13/3\right)} = +\frac{a_0}{\left(\sigma+5/3\right)\left(\sigma+7/3\right)\left(\sigma+11/3\right)\left(\sigma+13/3\right)}$$

and by induction:

$$a_{2m}(\sigma) = \frac{(-1)^m a_0}{\left(\sigma+\frac{5}{3}\right)\left(\sigma+\frac{11}{3}\right)\ldots\left(\sigma+\frac{6m-1}{3}\right)\cdot\left(\sigma+\frac{7}{3}\right)\left(\sigma+\frac{13}{3}\right)\ldots\left(\sigma+\frac{6m+1}{3}\right)}$$

These coefficients are substituted in the Frobenius series:

$$y(x,\sigma) = \sum_{m=0}^{\infty} a_{2m}(\sigma)\, x^{2m+\sigma}$$

For the first solution corresponding to the larger root $\sigma_1 = 1/3$:

$$y_1(x) = a_0 x^{1/3} + \sum_{m=1}^{\infty} a_{2m}\left(\frac{1}{3}\right) x^{2m+1/3}$$

where

$$a_{2m}\left(\frac{1}{3}\right) = (-1)^m \frac{a_0}{2^m m! \left(2/3\right)^m \cdot 4\cdot 7\cdot 10\cdot \ldots \cdot (3m+1)} \qquad m \geq 1$$

Letting $\sigma = \sigma_2 = -1/3$ gives the second solution:

$$y_2(x) = a_0 x^{-1/3} + \sum_{m=1}^{\infty} a_{2m}\left(-\frac{1}{3}\right) x^{2m-1/3}$$

$$a_{2m}\left(-\frac{1}{3}\right) = (-1)^m \frac{a_0}{2^m m! \left(2/3\right)^m \cdot 2\cdot 5\cdot 8\cdot \ldots \cdot (3m-1)} \qquad m \geq 1$$

The final solution y(x), setting $a_0 = 1$ in each series gives:

$$y(x) = c_1 y_1(x) + c_2 y_2(x)$$

Case (b): *Two identical roots* $\sigma_1 = \sigma_2 = \sigma_0$

If $\sigma_1 = \sigma_2 = \sigma_0$, then only one possible solution can be obtained by the method of Case (a), eq. (2.14), i.e.:

$$y_1(x) = \sum_{n=0}^{\infty} a_n(\sigma_0)\, x^{n+\sigma_0}$$

where $a_0 = 1$.

To obtain the second solution, one must utilize eqs. (2.11) and (2.12). If $\sigma_1 = \sigma_2 = \sigma_0$, then the characteristic equation has the form:

$$f(\sigma) = (\sigma - \sigma_0)^2$$

and eq. (2.11) becomes:

$$Ly(x,\sigma) = x^{\sigma-2}(\sigma - \sigma_0)^2 a_0 \tag{2.15}$$

where $y(x,\sigma)$ is given in eq. (2.12). First differentiate eq. (2.15) partially with σ:

$$\frac{\partial}{\partial\sigma} Ly = L\frac{\partial y(x,\sigma)}{\partial\sigma} = a_0\left[2(\sigma - \sigma_0) + (\sigma - \sigma_0)^2 \log x\right]x^{\sigma-2}$$

where

$$\frac{d}{d\sigma} x^{\sigma} = x^{\sigma} \log x$$

If $\sigma = \sigma_0$, then:

$$L\left[\frac{\partial y(x,\sigma)}{\partial\sigma}\right]_{\sigma=\sigma_0} = 0$$

Thus, the second solution satisfying the homogeneous differential equation is given by:

$$y_2(x) = \left.\frac{\partial y(x,\sigma)}{\partial\sigma}\right|_{\sigma=\sigma_0}$$

Using the form of the Frobenius solution:

$$y(x,\sigma) = a_0 x^{\sigma} + \sum_{n=1}^{\infty} a_n(\sigma)\, x^{n+\sigma}$$

then differentiating the expression for $y(x,\sigma)$ with σ results in:

$$\frac{\partial y(x,\sigma)}{\partial\sigma} = a_0 x^{\sigma} \log x + \sum_{n=1}^{\infty} a_n'(\sigma)\, x^{n+\sigma} + \sum_{n=1}^{\infty} a_n(\sigma)\, x^{n+\sigma} \log x$$

$$= \log x \sum_{n=0}^{\infty} a_n(\sigma)\, x^{n+\sigma} + \sum_{n=1}^{\infty} a_n'(\sigma)\, x^{n+\sigma}$$

Thus, the second solution for the case of equal roots takes the form with $a_0 = 1$:

$$y_2(x) = \log x \sum_{n=0}^{\infty} a_n(\sigma_0)\, x^{n+\sigma_0} + \sum_{n=1}^{\infty} a_n'(\sigma_0)\, x^{n+\sigma_0}$$

$$= y_1(x)\log x + \sum_{n=1}^{\infty} a'_n(\sigma_0)x^{n+\sigma_0} \tag{2.16}$$

Example 2.5

Solve the following differential equation about $x_0 = 0$:

$$x^2\frac{d^2y}{dx^2} - 3x\frac{dy}{dx} + (4 - x)y = 0$$

Since $x_0 = 0$ is a RSP, then assume a Frobenius series solution which, when substituted into this differential equation results in:

$$\sum_{n=0}^{\infty}(n + \sigma - 2)^2\, a_n x^{n+\sigma-2} - \sum_{n=0}^{\infty} a_n x^{n+\sigma-1} = 0$$

or, upon removing the first term and substituting n = m + 1 in the first series and n = m in the second series results in the following equation:

$$(\sigma - 2)^2\, a_0 x^{\sigma-2} + \sum_{m=0}^{\infty}\left[(m + \sigma - 1)^2\, a_{m+1} - a_m\right]x^{m+\sigma-1} = 0$$

Equating the coefficient of $x^{\sigma-2}$ to zero, one obtains with $a_0 \neq 0$:

$$(\sigma - 2)^2 = 0 \qquad \text{or} \qquad \sigma_1 = \sigma_2 = 2 = \sigma_0$$

Equating the coefficient of $x^{m+\sigma-1}$ to zero, one obtains the recurrence formula in the form:

$$a_{m+1} = \frac{a_m}{(m + \sigma - 1)^2} \qquad m = 0, 1, 2,...$$

where

$$a_1 = \frac{a_0}{(\sigma - 1)^2} \qquad a_2 = \frac{a_1}{\sigma^2} = \frac{a_0}{(\sigma - 1)^2\sigma^2}$$

and by induction:

$$a_n(\sigma) = \frac{a_0}{(\sigma - 1)^2\sigma^2(\sigma + 1)^2...(\sigma + n - 2)^2} \qquad n = 1, 2,...$$

Thus, the first solution corresponding to $\sigma = \sigma_0$ becomes:

$$y_1(x) = y(x,\sigma)_{\sigma = \sigma_0 = 2} = a_0 x^2 + \sum_{n=1}^{\infty}\frac{a_0}{1^2 \cdot 2^2 \cdot ... \cdot n^2}\, x^{n+2}$$

$$= \sum_{n=0}^{\infty}\frac{x^{n+2}}{(n!)^2}$$

where 0! = 1 and a_0 was set = 1.

To obtain the second solution, in the form of eq. (2.16), one needs $a'_n(\sigma)$:

$$\frac{da_n(\sigma)}{d\sigma} = \frac{-2a_0}{(\sigma-1)^2\sigma^2...(\sigma+n-2)^2}\left[\frac{1}{\sigma-1}+\frac{1}{\sigma}+\frac{1}{\sigma+1}+...+\frac{1}{\sigma+n-2}\right]$$

$$a'_n(\sigma)\Big|_{\sigma=\sigma_0=2} = -\frac{2a_0}{1^2\cdot 2^2\cdot...\cdot n^2}\left[\frac{1}{1}+\frac{1}{2}+\frac{1}{3}+...+\frac{1}{n}\right]$$

Defining $g(n) = 1 + 1/2 + ... + 1/n$, with $g(0) = 0$, then:

$$a'_n(\sigma_0) = -\frac{2a_0}{(n!)^2}\, g(n) \qquad\qquad n = 1, 2,...$$

Thus, setting $a_0 = 1$, the second solution of the differential equation takes the form:

$$y_2(x) = y_1(x)\log x - 2\sum_{n=1}^{\infty}\frac{x^{n+2}}{(n!)^2}\, g(n)$$

Case (c): *Distinct roots that differ by an integer.*

If $\sigma_1 - \sigma_2 = k$, a positive integer, then the characteristic equation becomes:

$$f(\sigma) = (\sigma-\sigma_1)(\sigma-\sigma_2) = (\sigma-\sigma_2-k)(\sigma-\sigma_2)$$

First, one can obtain the solution corresponding to the larger root σ_1 in the form given in eq. (2.14). The second solution corresponding to $\sigma = \sigma_2$ may have the constant $a_k(\sigma_2)$ unbounded, because, from eq. (2.10), the expression for $a_k(\sigma_2)$ is:

$$a_k(\sigma_2) = \frac{-g_k(\sigma)}{f(\sigma+k)}\bigg|_{\sigma=\sigma_2}$$

where the denominator vanishes at $\sigma = \sigma_2$:

$$f(\sigma+k)\big|_{\sigma=\sigma_2} = (\sigma+k-\sigma_2-k)(\sigma+k-\sigma_2)\big|_{\sigma=\sigma_2}$$

$$= (\sigma-\sigma_2)(\sigma+k-\sigma_2)\big|_{\sigma=\sigma_2} = 0$$

Thus, unless the numerator $g_k(\sigma_2)$ also vanishes, the coefficient $a_k(\sigma_2)$ becomes unbounded.

If $g_k(\sigma_2)$ vanishes, then $a_k(\sigma_2)$ is indeterminate and one may start a new infinite series with a_k, i.e.:

$$y_2(x) = a_0 \sum_{n=0}^{k-1 \text{ or } \infty}\left(\frac{a_n(\sigma_2)}{a_0}\right)x^{n+\sigma_2} + a_k\sum_{n=k}^{\infty}\left(\frac{a_n(\sigma_2)}{a_k}\right)x^{n+\sigma_2}$$

$$= a_0 \sum_{n=0}^{k-1 \text{ or } \infty}\left(\frac{a_n(\sigma_2)}{a_0}\right)x^{n+\sigma_2} + a_k\sum_{m=0}^{\infty}\left(\frac{a_{m+k}(\sigma_2)}{a_k}\right)x^{m+\sigma_1} \tag{2.17}$$

It can be shown that the solution preceded by the constant a_k is identical to $y_1(x)$, thus one can set $a_k = 0$ and $a_0 = 1$. The first part of the solution with a_0 may be a finite polynomial or an infinite series, depending on the order of the recurrence formula and on the integer k.

If $g_k(\sigma_2)$ does not vanish, then one must find another method to obtain the second solution. A new solution similar to Case (b) is developed next by removing the constant $\sigma - \sigma_2$ from the denominator of $a_n(\sigma)$. Since the characteristic equation in eq. (2.11) is given by:

$$Ly(x,\sigma) = a_0 x^{\sigma-2} f(\sigma) = a_0 x^{\sigma-2}(\sigma-\sigma_1)(\sigma-\sigma_2) \tag{2.18}$$

then multiplying eq. (2.18) by $(\sigma - \sigma_2)$ and differentiating partially with σ, one obtains:

$$\frac{\partial}{\partial\sigma}[(\sigma-\sigma_2)Ly] = \frac{\partial}{\partial\sigma}[L(\sigma-\sigma_2)y(x,\sigma)] = L\left[\frac{\partial}{\partial\sigma}(\sigma-\sigma_2)y(x,\sigma)\right]$$

$$= a_0 \frac{\partial}{\partial\sigma}\left\{x^{\sigma-2}(\sigma-\sigma_1)(\sigma-\sigma_2)^2\right\}$$

$$= a_0\left\{(\sigma-\sigma_1)(\sigma-\sigma_2)^2 x^{\sigma-2}\log x + x^{\sigma-2}(\sigma-\sigma_2)^2 + 2x^{\sigma-2}(\sigma-\sigma_1)(\sigma-\sigma_2)\right\}$$

Thus, the function that satisfies the homogenous differential equation:

$$L\left[\frac{\partial}{\partial\sigma}(\sigma-\sigma_2)y(x,\sigma)\right]_{\sigma=\sigma_2} = 0$$

gives an expression for the second solution, i.e.:

$$y_2(x) = \frac{\partial}{\partial\sigma}(\sigma-\sigma_2)y(x,\sigma)\Big|_{\sigma=\sigma_2} \tag{2.19}$$

The Frobenius solution can be divided into two parts:

$$y(x,\sigma) = \sum_{n=0}^{\infty} a_n(\sigma)\,x^{n+\sigma} = \sum_{n=0}^{n=k-1} a_n(\sigma)\,x^{n+\sigma} + \sum_{n=k}^{\infty} a_n(\sigma)\,x^{n+\sigma}$$

so that the coefficient a_k is the first term of the second series. Differentiating the expression as given in eq. (2.19) one obtains:

$$\frac{\partial}{\partial\sigma}[(\sigma-\sigma_2)y(x,\sigma)] = \frac{\partial}{\partial\sigma}\left[\sum_{n=0}^{n=k-1}(\sigma-\sigma_2)a_n(\sigma)x^{n+\sigma} + \sum_{n=k}^{\infty}(\sigma-\sigma_2)a_n(\sigma)x^{n+\sigma}\right]$$

$$= \log x \sum_{n=0}^{n=k-1}(\sigma-\sigma_2)a_n(\sigma)x^{n+\sigma} + \sum_{n=0}^{n=k-1}(\sigma-\sigma_2)a_n'(\sigma)x^{n+\sigma} + \sum_{n=0}^{n=k-1}a_n(\sigma)x^{n+\sigma}$$

$$+ \sum_{n=k}^{\infty}[(\sigma-\sigma_2)a_n(\sigma)]'\,x^{n+\sigma} + \log x\sum_{n=k}^{\infty}[(\sigma-\sigma_2)a_n(\sigma)]x^{n+\sigma}$$

It should be noted that $a_n(\sigma) = -(g_n(\sigma))/(f(\sigma+n))$ does not contain the term $(\sigma-\sigma_2)$ in its denominator until n = k, thus:

$$\left.\begin{array}{l}(\sigma-\sigma_2)\,a_n(\sigma)\big|_{\sigma=\sigma_2}=0 \\ \text{and} \\ (\sigma-\sigma_2)\,a_n'(\sigma)\big|_{\sigma=\sigma_2}=0\end{array}\right\} \qquad \text{for } n=0,1,2,\ldots,k-1$$

Therefore, the second solution takes the form:

$$y_2(x)=\frac{\partial}{\partial\sigma}\left[(\sigma-\sigma_2)y(x,\sigma)\right]_{\sigma=\sigma_2}=\sum_{n=0}^{n=k-1}a_n(\sigma_2)\,x^{n+\sigma_2}$$

$$+\sum_{n=k}^{\infty}\left[(\sigma-\sigma_2)\,a_n(\sigma)\right]'_{\sigma=\sigma_2}x^{n+\sigma_2}+\log x\sum_{n=k}^{\infty}\left[(\sigma-\sigma_2)\,a_n(\sigma)\right]_{\sigma=\sigma_2}x^{n+\sigma_2} \tag{2.20}$$

It can be shown that the last infinite series is proportional to $y_1(x)$.

Example 2.6

Obtain the solutions of the following differential equation about $x_0 = 0$:

$$x^2\frac{d^2y}{dx^2}+x\frac{dy}{dx}+\left(x^2-\frac{9}{4}\right)y=0$$

Since $x_0 = 0$ is a RSP, then substituting the Frobenius solution into the differential equation results in:

$$\sum_{n=0}^{\infty}\left[(n+\sigma)^2-\frac{9}{4}\right]a_n x^{n+\sigma-2}+\sum_{n=0}^{\infty}a_n x^{n+\sigma}=0$$

which, upon extracting the two terms with the lowest powers of x, gives:

$$\left(\sigma^2-\frac{9}{4}\right)a_0x^{\sigma-2}+\left[(\sigma+1)^2-\frac{9}{4}\right]a_1x^{\sigma-1}+$$

$$+\sum_{m=0}^{\infty}\left\{\left[(m+2+\sigma)^2-\frac{9}{4}\right]a_{m+2}+a_m\right\}x^{m+\sigma}=0$$

Thus, equating the coefficient of each power of x to zero; one obtains:

$$\left(\sigma^2-\frac{9}{4}\right)a_0=0$$

$$\left[(\sigma+1)^2-\frac{9}{4}\right]a_1=0$$

and the recurrence formula:

$$a_{m+2}=-\frac{a_m}{(m+2+\sigma)^2-9/4}=-\frac{a_m}{(m+\sigma+1/2)(m+\sigma+7/2)} \qquad m=0,1,2,\ldots$$

Solving for the roots of the characteristic equation gives:

$$\sigma_1 = \frac{3}{2} \qquad \sigma_2 = -\frac{3}{2} \qquad \sigma_1 - \sigma_2 = 3 = k$$

Using the recurrence formula to evaluate higher-ordered coefficients, one obtains:

$$a_2 = -\frac{a_0}{(\sigma + 1/2)(\sigma + 7/2)}$$

$$a_3 = -\frac{a_1}{(\sigma + 3/2)(\sigma + 9/2)}$$

$$a_4 = -\frac{a_2}{(\sigma + 5/2)(\sigma + 11/2)} = \frac{a_0}{(\sigma + 1/2)(\sigma + 5/2)(\sigma + 7/2)(\sigma + 11/2)}$$

$$a_5 = -\frac{a_3}{(\sigma + 7/2)(\sigma + 13/2)} = \frac{a_1}{(\sigma + 3/2)(\sigma + 7/2)(\sigma + 9/2)(\sigma + 13/2)}$$

Thus, the odd and even coefficients a_n can be written in terms of a_0 and a_1 by induction as follows:

$$a_{2m} = (-1)^m \frac{a_0}{(\sigma + 1/2)(\sigma + 5/2)\ldots(\sigma + 2m - 3/2)\cdot(\sigma + 7/2)(\sigma + 11/2)\ldots(\sigma + 2m + 3/2)}$$

$$a_{2m+1} = (-1)^m \frac{a_1}{\left(\sigma + \frac{3}{2}\right)\left(\sigma + \frac{7}{2}\right)\ldots\left(\sigma + 2m - \frac{1}{2}\right)\cdot\left(\sigma + \frac{9}{2}\right)\left(\sigma + \frac{13}{2}\right)\ldots\left(\sigma + 2m + \frac{5}{2}\right)}$$

for $m = 1, 2, 3, \ldots$

To obtain the first solution corresponding to the larger root $\sigma_1 = 3/2$:

a_0 = indeterminate

$a_1 = a_3 = a_5 = \ldots = 0$

and

$$a_{2m}(3/2) = (-1)^m \frac{3a_0(2m+2)}{(2m+3)!} \qquad m = 1, 2, 3, \ldots$$

and by setting $6a_0 = 1$:

$$y_1(x) = \frac{1}{6}x^{3/2} + \sum_{m=1}^{\infty} (-1)^m \frac{(m+1)\, x^{2m+3/2}}{(2m+3)!} = \sum_{m=0}^{\infty} (-1)^m \frac{(m+1)\, x^{2m+3/2}}{(2m+3)!}$$

To obtain the solution corresponding to the smaller root:

$$\sigma_2 = -\frac{3}{2}, \qquad \text{where} \qquad \sigma_1 - \sigma_2 = 3 = k$$

a_0 = indeterminate

$a_1 = 0$

$$a_{2m}(-3/2) = (-1)^m \frac{(-a_0)(2m-1)}{(2m)!} \qquad m = 1, 2, \ldots$$

The coefficient $a_k = a_3$ must be calculated to decide on whether to use the second form of the solution in eq. (2.20). Using the recurrence formula for $\sigma_2 = -3/2$ gives:

$$a_3 = \frac{0}{0} \qquad \text{(indeterminate)}$$

So that the coefficient a_3 is not unbounded and can be used to start a new series:

$$a_{2m+1} = \left. \frac{(-1)^{m+1}\, a_3}{(\sigma + 7/2) \ldots (\sigma + 2m - 1/2)\cdot(\sigma + 13/2)\ldots(\sigma + 2m + 5/2)} \right|_{\sigma_2 = -3/2}$$

$$= (-1)^{m+1} \frac{6 a_3 m}{(2m+1)!} \qquad m = 2, 3, 4, \ldots$$

Thus, the second solution is obtained in the form:

$$y_2(x) = a_0\, x^{-3/2} - a_0 \sum_{m=1}^{\infty} (-1)^m \frac{(2m-1)\, x^{2m-3/2}}{(2m)!} + a_3 x^{3/2}$$

$$+ 6 a_3 \sum_{m=2}^{\infty} (-1)^{m+1} \frac{m\, x^{2m-1/2}}{(2m+1)!}$$

$$= a_0 \sum_{m=0}^{\infty} (-1)^m \frac{(2m-1)\, x^{2m-3/2}}{(2m)!} - 6 a_3 \sum_{m=0}^{\infty} (-1)^m \frac{(m+1)\, x^{2m+3/2}}{(2m+3)!}$$

Note that the solution starting with $a_k = a_3$ is $y_1(x)$, which is extraneous. Letting $a_0 = 1$ and $a_3 = 0$, the second solution becomes:

$$y_2(x) = \sum_{m=0}^{\infty} (-1)^m \frac{(2m-1)\, x^{2m-3/2}}{(2m)!}$$

Example 2.7

Obtain the solutions of the differential equation about $x_0 = 0$:

$$x^2 \frac{d^2y}{dx^2} - (x+2)\, y = 0$$

Since $x_0 = 0$ is a RSP, then substituting the Frobenius solution in the differential equation gives:

$$(\sigma - 2)(\sigma + 1)\, a_0 x^{\sigma - 2} + \sum_{m=0}^{\infty} \{(m + \sigma - 1)(m + \sigma + 2)\, a_{m+1} - a_m\} x^{m+\sigma-1} = 0$$

Equating the two terms to zero gives the characteristic equation:

$$(\sigma - 2)(\sigma + 1)\, a_0 = 0$$

with roots

$$\sigma_1 = 2 \qquad \sigma_2 = -1 \qquad \sigma_1 - \sigma_2 = 3 = k$$

and the recurrence formula:

$$a_{m+1} = \frac{a_m}{(m+\sigma-1)(m+\sigma+2)} \qquad m = 0, 1, 2, ...$$

Using the recurrence formula, one obtains:

$$a_1 = \frac{a_0}{(\sigma-1)(\sigma+2)}$$

$$a_2 = \frac{a_1}{\sigma(\sigma+3)} = \frac{a_0}{(\sigma-1)\,\sigma\,(\sigma+2)(\sigma+3)}$$

$$a_3 = \frac{a_2}{(\sigma+1)(\sigma+4)} = \frac{a_0}{(\sigma-1)\,\sigma\,(\sigma+1)(\sigma+2)(\sigma+3)(\sigma+4)}$$

and by induction:

$$a_n(\sigma) = \frac{a_0}{(\sigma-1)\,\sigma ... (\sigma+n-2)\cdot(\sigma+2)(\sigma+3)\,...\,(\sigma+n+1)} \qquad n = 1, 2, 3, ...$$

The solution corresponding to the larger root $\sigma_1 = 2$:

$$a_n(2) = \frac{6a_0}{n!\,(n+3)!}$$

so that the first solution corresponding to the larger root is:

$$y_1(x) = \sum_{n=0}^{\infty} \frac{x^{n+2}}{n!\,(n+3)!}$$

where $6a_0$ was set equal to 1.

The solution corresponding to the smaller root $\sigma_2 = -1$ can be obtained after checking a_3 (-1):

$$a_3(-1) \rightarrow \infty$$

Using the expression for the second solution in eq. (2.20) one obtains:

$$y_2(x) = \sum_{n=0}^{n=2} a_n(-1)\,x^{n-1} + \sum_{n=3}^{\infty} \left[(\sigma+1)\,a_n(\sigma)\right]'_{\sigma=-1} x^{n-1}$$

$$+ \log x \sum_{n=3}^{\infty} \left[(\sigma+1)\,a_n(\sigma)\right]_{\sigma=-1} x^{n-1}$$

Substituting for $a_n(\sigma)$ and performing differentiation with σ results in:

$$(\sigma+1)\,a_n(\sigma) = \frac{a_0}{(\sigma-1)\,\sigma(\sigma+2)\,...\,(\sigma+n-2)(\sigma+2)(\sigma+3)\,...\,(\sigma+n+1)}$$

$$(\sigma+1)\,a_n(\sigma)\Big|_{\sigma=-1} = \frac{a_0}{(-2)(-1)\,1\cdot 2\cdot ...\cdot(n-3)\,1\cdot 2\cdot 3\cdot ...\cdot n} = \frac{a_0}{2(n-3)!\,n!}$$

$$\{(\sigma+1)\,a_n(\sigma)\}' = \frac{-a_0}{(\sigma-1)\,\sigma(\sigma+2)\,...\,(\sigma+n-2)(\sigma+2)(\sigma+3)\,...\,(\sigma+n+1)}$$

$$\cdot\left[\frac{1}{\sigma-1}+\frac{1}{\sigma}+\frac{1}{\sigma+2}+\ldots+\frac{1}{\sigma+n-2}+\frac{1}{\sigma+2}+\frac{1}{\sigma+3}+\ldots+\frac{1}{\sigma+n+1}\right]$$

$$\left[(\sigma+1)a_n(\sigma)\right]'\Big|_{\sigma=-1}=\frac{-a_0}{-2\cdot-1\cdot1\cdot2\cdot\ldots\cdot(n-3)\,1\cdot2\cdot\ldots\cdot n}$$

$$\cdot\left[-\frac{1}{2}-1+1+\frac{1}{2}+\ldots+\frac{1}{n-3}+1+\frac{1}{2}+\ldots+\frac{1}{n}\right]$$

$$=-\frac{a_0}{2(n-3)!\,n!}\left[-\frac{3}{2}+g(n-3)+g(n)\right] \qquad n=3,4,5,\ldots$$

where $g(n) = 1 + 1/2 + 1/3 + \ldots + 1/n$ and $g(0) = 0$.

The second solution can thus be written in the form:

$$y_2(x)=x^{-1}-\frac{1}{2}+\frac{x}{4}-\frac{1}{2}\sum_{n=3}^{\infty}\frac{x^{n-1}}{(n-3)!\,n!}\left[-\frac{3}{2}+g(n-3)+g(n)\right]$$

$$+\frac{1}{2}\log x\sum_{n=3}^{\infty}\frac{x^{n-1}}{n!\,(n-3)!}$$

which, upon shifting the indices in the infinite series gives:

$$y_2(x)=x^{-1}-\frac{1}{2}+\frac{x}{4}-\frac{1}{2}\sum_{n=0}^{\infty}\frac{x^{n+2}}{n!\,(n+3)!}\left[-\frac{3}{2}+g(n)+g(n+3)\right]$$

$$+\frac{1}{2}\log x\sum_{n=0}^{\infty}\frac{x^{n+2}}{(n+3)!\,n!}$$

The first series can be shown to be $3y_1(x)/4$ which can be deleted from the second solution, resulting in a final form for $y_2(x)$ as:

$$y_2(x)=x^{-1}-\frac{1}{2}+\frac{x}{4}-\frac{1}{2}\sum_{n=0}^{\infty}\frac{x^{n+2}}{n!\,(n+3)!}\left[g(n)+g(n+3)\right]+\frac{1}{2}\log(x)\,y_1(x)$$

PROBLEMS

Section 2.1

1. Determine the region of convergence for each of the following infinite series, and determine whether they will converge or diverge at the two end points.

(a) $\sum_{n=0}^{\infty} (-1)^n \frac{x^n}{n!}$

(b) $\sum_{n=0}^{\infty} (-1)^n \frac{x^{2n}}{(2n)!}$

(c) $\sum_{n=1}^{\infty} (-1)^n n\, x^n$

(d) $\sum_{n=1}^{\infty} (-1)^n \frac{x^n}{n^2}$

(e) $\sum_{n=0}^{\infty} \frac{x^{2n}}{n^2 + n + 2}$

(f) $\sum_{n=1}^{\infty} (-1)^n \frac{n\, x^n}{2^n}$

(g) $\sum_{n=1}^{\infty} \frac{n+3}{n\, 2^n} x^n$

(h) $\sum_{n=0}^{\infty} (-1)^n \frac{(n!)^2 x^n}{(2n)!}$

(i) $\sum_{n=1}^{\infty} (-1)^n \frac{n(x-1)^n}{2^n(n+1)}$

(j) $\sum_{n=1}^{\infty} (-1)^n \frac{(x+1)^n}{3^n n^2}$

Section 2.2

2. Obtain the solution to the following differential equations, valid near x = 0.

(a) $\frac{d^2y}{dx^2} + x^2 \frac{dy}{dx} + xy = 0$

(b) $(x^2+1)\frac{d^2y}{dx^2} + x\frac{dy}{dx} - y = 0$

(c) $\frac{d^2y}{dx^2} - x\frac{dy}{dx} - y = 0$

(d) $\frac{d^2y}{dx^2} - 4x\frac{dy}{dx} - (x^2+2)y = 0$

(e) $\frac{d^2y}{dx^2} - x\frac{dy}{dx} - (x+2)\,y = 0$

(f) $\frac{d^3y}{dx^3} + x^2\frac{d^2y}{dx^2} + 3x\frac{dy}{dx} + y = 0$

(g) $(x^2+1)\frac{d^2y}{dx^2} + 6x\frac{dy}{dx} + 6y = 0$

(h) $(x^2-1)\frac{d^2y}{dx^2} - 6y = 0$

(i) $(x-1)\frac{d^2y}{dx^2} + y = 0$

(j) $x\frac{d^2y}{dx^2} - \frac{dy}{dx} + 4x^3y = 0$

3. Obtain the general solution to the following differential equations about $x = x_0$ as indicated:

(a) $\frac{d^2y}{dx^2} - (x-1)\frac{dy}{dx} + y = 0$ about $x_0 = -1$

(b) $\frac{d^2y}{dx^2} - (x-1)^2 y = 0$ about $x_0 = 1$

(c) $x(x-2)\frac{d^2y}{dx^2} + 6(x-1)\frac{dy}{dx} + 6y = 0$ about $x_0 = 1$

(d) $x(x+2)\frac{d^2y}{dx^2} + 8(x+1)\frac{dy}{dx} + 12y = 0$ about $x_0 = -1$

Section 2.3

4. Classify all the finite singularities, if any, of the following differential equations:

(a) $x^2\frac{d^2y}{dx^2} + x\frac{dy}{dx} + (x^2 - 4)y = 0$

(b) $x^2\frac{d^2y}{dx^2} + (1+x)\frac{dy}{dx} + y = 0$

(c) $(1-x^2)\frac{d^2y}{dx^2} - 2x\frac{dy}{dx} + 6y = 0$

(d) $x(1-x^2)\frac{d^2y}{dx^2} - 2x\frac{dy}{dx} + (1+x)^2 y = 0$

(e) $\sin x\frac{d^2y}{dx^2} + \cos x\frac{dy}{dx} + y = 0$

(f) $x^2 \tan x\frac{d^2y}{dx^2} + x\frac{dy}{dx} + 3y = 0$

(g) $(x-1)^2\frac{d^2y}{dx^2} + (x^2-1)\frac{dy}{dx} + x^2 y = 0$

(h) $x(1-x)\frac{d^2y}{dx^2} + (2-x)\frac{dy}{dx} + 4y = 0$

Section 2.4

5. Obtain the solution of the following differential equations, valid in the neighborhood of $x = 0$:

(a) $x^2(x+2)y'' + x(x-3)y' + 3y = 0$

(b) $2x^2\frac{d^2y}{dx^2} + [3x + 2x^2]\frac{dy}{dx} - 3y = 0$

(c) $x^2\frac{d^2y}{dx^2} + [x + x^2]\frac{dy}{dx} + \left[-\frac{1}{4} + \frac{x}{2}\right]y = 0$

(d) $x^2\frac{d^2y}{dx^2} + [x - x^2]\frac{dy}{dx} - y = 0$

(e) $x^2\frac{d^2y}{dx^2} + [x^3 - 4x]\frac{dy}{dx} + [6 - x^2]y = 0$

(f) $x^2\frac{d^2y}{dx^2} - x\frac{dy}{dx} - (x-1)y = 0$

(g) $4x^2\frac{d^2y}{dx^2} + 4x^3\frac{dy}{dx} + (6x^2 + 1)y = 0$

(h) $x^2\frac{d^2y}{dx^2} + (x^2 + x)\frac{dy}{dx} + (2x - 1)y = 0$

(i) $x^2\frac{d^2y}{dx^2} + (x^3 + x)\frac{dy}{dx} + (5x^2 - 9)y = 0$

(j) $x^2\frac{d^2y}{dx^2} + 7x\frac{dy}{dx} + (10 - x)y = 0$

(k) $x^2y'' + x(1 + x)y' - y = 0$

(l) $x^2(1 - x)\frac{d^2y}{dx^2} + (x^2 + x)\frac{dy}{dx} - y = 0$

(m) $x(x^2 - 1)\frac{d^2y}{dx^2} + 2(4x^2 - 1)\frac{dy}{dx} + 12xy = 0$

(n) $x(x^2 - 1)\frac{d^2y}{dx^2} + 2(2x^2 - 1)\frac{dy}{dx} + 2xy = 0$

(o) $x(x + 1)\frac{d^2y}{dx^2} + \frac{dy}{dx} - 2y = 0$

(p) $x\frac{d^2y}{dx^2} + \frac{dy}{dx} - 4xy = 0$

(q) $x\frac{d^2y}{dx^2} + \frac{dy}{dx} + y = 0$

(r) $x\frac{d^2y}{dx^2} + 3\frac{dy}{dx} + 4xy = 0$

(s) $x(x - 1)\frac{d^2y}{dx^2} + 3\frac{dy}{dx} - 2y = 0$

(t) $x^2\frac{d^2y}{dx^2} - x\frac{dy}{dx} + (4x^2 + 1)y = 0$

3

SPECIAL FUNCTIONS

3.1 Bessel Functions

Bessel functions are solutions to the second-order differential equation:

$$x^2 \frac{d^2y}{dx^2} + x\frac{dy}{dx} + \left(x^2 - p^2\right)y = 0 \tag{3.1}$$

where x = 0 is a regular singular point and p is a real constant.

Substituting a Frobenius solution into the differential equation results in the series:

$$\left(\sigma^2 - p^2\right)a_0 x^{\sigma-2} + \left[(\sigma+1)^2 - p^2\right]a_1 x^{\sigma-1}$$

$$+ \sum_{m=0}^{\infty} \left\{\left[(m+2+\sigma)^2 - p^2\right]a_{m+2} + a_m\right\}x^{m+\sigma} = 0$$

For $a_0 \neq 0$, $\sigma^2 - p^2 = 0$, $\sigma_1 = p$, $\sigma_2 = -p$ and $\sigma_1 - \sigma_2 = 2p$:

$$\left[(\sigma+1)^2 - p^2\right]a_1 = 0$$

and

$$a_{m+2} = -\frac{a_m}{(m+2+\sigma-p)(m+2+\sigma+p)} \qquad m = 0, 1, 2, \ldots \tag{3.2}$$

The solution corresponding to the larger root $\sigma_1 = p$ can be obtained first. Excluding the case of p = -1/2, then:

$$a_1 = a_3 = a_5 = \ldots 0$$

$$a_{m+2} = -\frac{a_m}{(m+2)(m+2+2p)} \qquad m = 0, 1, 2, \ldots$$

$$a_2 = -\frac{a_0}{2^2 1!\,(p+1)}$$

$$a_4 = -\frac{a_2}{4\,(4+2p)} = \frac{a_0}{2^4 2!\,(p+1)(p+2)}$$

$$a_6 = -\frac{a_0}{2^6 3!\,(p+1)(p+2)(p+3)}$$

and, by induction:

$$a_{2m} = (-1)^m \frac{a_0}{2^{2m}\, m!\,(p+1)(p+2)\ldots(p+m)} \qquad m = 1, 2,\ldots$$

Thus, the solution corresponding to $\sigma_1 = p$ becomes:

$$y_1(x) = a_0 x^p + a_0 \sum_{m=1}^{\infty} (-1)^m \frac{x^{2m+p}}{2^{2m}\, m!\,(p+1)(p+2)\ldots(p+m)}$$

Using the definition of the Gamma function in Appendix B1, then one can rewrite the expression for $y_1(x)$ as:

$$y_1(x) = a_0 x^p + a_0 \sum_{m=1}^{\infty} (-1)^m \frac{\Gamma(p+1)\, x^{2m+p}}{2^{2m}\, m!\,\Gamma(p+m+1)}$$

$$= a_0 \Gamma(p+1)\, 2^p \left\{ \frac{(x/2)^p}{\Gamma(p+1)} + \sum_{m=1}^{\infty} (-1)^m \frac{(x/2)^{2m+p}}{m!\,\Gamma(p+m+1)} \right\}$$

Define the bracketed series as:

$$J_p(x) = \sum_{m=0}^{\infty} (-1)^m \frac{(x/2)^{2m+p}}{m!\,\Gamma(p+m+1)} \tag{3.3}$$

where $a_0\, \Gamma(p+1)\, 2^p$ was set equal to 1 in $y_1(x)$. The solution $J_p(x)$ in eq. (3.3) is known as the **Bessel function of the first kind of order p.**

The solution corresponding to the smaller root $\sigma_2 = -p$ can be obtained by substituting -p for +p in eq. (3.3) resulting in:

$$y_2(x) = J_{-p}(x) = \sum_{m=0}^{\infty} (-1)^m \frac{(x/2)^{2m-p}}{m!\,\Gamma(-p+m+1)} \tag{3.4}$$

$J_{-p}(x)$ is known as the **Bessel function of the second kind of order p.**

If $p \neq$ integer, then:

$$y_h = c_1 J_p(x) + c_2 J_{-p}(x)$$

The expression for the Wronskian can be obtained from the form given in eq. (1.28):

$$W(x) = W_0 \exp\left(\int^x -\frac{d\eta}{\eta} \right) = W_0\, e^{-\log x} = \frac{W_0}{x}$$

$$W(J_p(x), J_{-p}(x)) = J_p(x)\, J'_{-p}(x) - J'_p(x)\, J_{-p}(x) = \frac{W_0}{x}$$

Thus:

$$\lim_{x \to 0} x\, W(x) \to W_0$$

To calculate W_0, it is necessary to account for the leading terms only, since the form of $W \sim 1/x$. Thus:

$$J_p \sim \frac{\left(x/2\right)^p}{\Gamma(p+1)} \qquad J'_p \sim \frac{p/2\left(x/2\right)^{p-1}}{\Gamma(p+1)}$$

$$J_{-p} \sim \frac{\left(x/2\right)^{-p}}{\Gamma(1-p)} \qquad J'_{-p} \sim \frac{p/2\left(x/2\right)^{-p-1}}{\Gamma(1-p)}$$

$$\lim_{x \to 0} x\, W\left(J_p, J_{-p}\right) = W_0 = \frac{-2p}{\Gamma(p+1)\,\Gamma(1-p)} = -\frac{2}{\Gamma(p)\,\Gamma(1-p)}$$

Since:

$$\Gamma(p)\,\Gamma(1-p) = \frac{\pi}{\sin p\pi} \qquad \text{(Appendix B1)}$$

then, the Wronskian is given by:

$$W\left(J_p, J_{-p}\right) = \frac{-2\sin p\pi}{\pi x} \tag{3.5}$$

Another solution that also satisfies eq. (3.1), first introduced by **Weber**, takes the form:

$$Y_p(x) = \frac{\cos p\pi\, J_p(x) - J_{-p}(x)}{\sin p\pi} \qquad p \neq \text{integer} \tag{3.6}$$

such that the general solution can be written in the form known as **Weber function**:

$$y = c_1\, J_p(x) + c_2\, Y_p(x) \qquad p \neq \text{integer}$$

Using the linear transformation formula, the Wronskian becomes:

$$W\left(J_p, Y_p\right) = \det\left[\alpha_{ij}\right] W\left(J_p, J_{-p}\right)$$

as given by eq. (1.13), where:

$$\alpha_{11} = 1 \qquad \alpha_{12} = 0$$

$$\alpha_{21} = \cot p\pi \qquad \alpha_{22} = -1/\sin p\pi \qquad \det\left[\alpha_{ij}\right] = -1/\sin p\pi$$

so that:

$$W\left(J_p, Y_p\right) = J_p Y'_p - J'_p Y_p = \frac{2}{\pi x} \tag{3.7}$$

which is independent of p.

3.2 Bessel Function of Order Zero

If $p = 0$ then $\sigma_1 = \sigma_2 = 0$ (repeated root), which results in a solution of the form:

$$J_0(x) = \sum_{m=0}^{\infty} (-1)^m \frac{\left(x/2\right)^{2m}}{(m!)^2} \tag{3.8}$$

To obtain the second solution, the methods developed in Section (2.4) are applied. From the recurrence formula, eq. (3.2), one obtains the following by setting p = 0:

$$a_{m+2} = -\frac{a_m}{(m+\sigma+2)^2} \qquad m = 0, 1, 2,...$$

Again, by induction, one can show that the even indexed coefficients are:

$$a_{2m} = (-1)^m \frac{a_0}{(\sigma+2)^2(\sigma+4)^2 \ldots (\sigma+2m)^2} \qquad m = 1, 2,...$$

and

$$y(x,\sigma) = a_0 x^\sigma + a_0 \sum_{m=1}^{\infty} (-1)^m \frac{x^{2m+\sigma}}{(\sigma+2)^2(\sigma+4)^2 \ldots (\sigma+2m)^2}$$

Using the form for the second solution given in eq. (2.16), one obtains:

$$y_2(x) = \left.\frac{\partial y(x,\sigma)}{\partial\sigma}\right|_{\sigma_0=0} = a_0 x^\sigma \log x + a_0 \log x \sum_{m=1}^{\infty} \frac{(-1)^m x^{2m+\sigma}}{(\sigma+2)^2(\sigma+4)^2 \ldots (\sigma+2m)^2}$$

$$-2a_0 \sum_{m=1}^{\infty} (-1)^m \frac{x^{2m+\sigma}}{(\sigma+2)^2(\sigma+4)^2 \ldots (\sigma+2m)^2} \left[\frac{1}{\sigma+2} + \frac{1}{\sigma+4} + \ldots + \frac{1}{\sigma+2m}\right]_{\sigma=0}$$

which results in the second solution y_2 as:

$$y_2(x) = \log x \, J_0(x) + \sum_{m=0}^{\infty} (-1)^{m+1} \frac{(x/2)^{2m}}{(m!)^2} g(m)$$

Define:

$$Y_0(x) = \frac{2}{\pi}\left[y_2(x) + (\gamma - \log 2) J_0(x)\right]$$

$$= \frac{2}{\pi}\left\{\left[\log(x/2) + \gamma\right] J_0(x) + \sum_{m=0}^{\infty} (-1)^{m+1} \frac{(x/2)^{2m}}{(m!)^2} g(m)\right\} \tag{3.9}$$

where the Euler Constant $\gamma = \lim_{n\to\infty} (g(n) - \log n) = 0.5772 \ldots\ldots$

Since $Y_0(x)$ is a linear combination of $J_0(x)$ and $y_2(x)$, it is also a solution of the eq. (3.1) as was discussed in Sec. (1.1). $Y_0(x)$ is known as the **Bessel function of the second kind of order zero** or the **Neumann function of order zero**.

Thus, the complete solution of the homogeneous equation is:

$$y_h = c_1 J_0(x) + c_2 Y_0(x) \qquad \text{if } p = 0$$

3.3 Bessel Function of an Integer Order n

If p = n = integer $\neq 0$ then $\sigma_1 - \sigma_2 = 2n$ is an even integer. The solution corresponding to $\sigma_1 = +n$ can be obtained from eq. (3.3) by substituting p = n, resulting in:

$$J_n(x) = \sum_{m=0}^{\infty} (-1)^m \frac{(x/2)^{2m+n}}{m!\,(m+n)!} \tag{3.10}$$

To obtain the second solution for $\sigma_2 = -n$, it is necessary to check $a_{2n}(-n)$ for boundedness. Substituting p = n in the recurrence formula (3.2) gives:

$$a_{m+2} = -\frac{a_m}{(m+2+\sigma-n)(m+2+\sigma+n)} \qquad m = 0, 1, 2, ...$$

and

$$a_1 = a_3 = ... = 0$$

so that the even indexed coefficients are given by:

$$a_{2m} = \frac{(-1)^m a_0}{(\sigma+2-n)...(\sigma+2m-n)\cdot(\sigma+2+n)...(\sigma+2m+n)} \qquad m = 1, 2, 3, ...$$

It is seen that the coefficient $a_{2n}(\sigma = -n)$ becomes unbounded, so that the methods of solution outlined in Section (2.4) must now be followed.

$$y(x,\sigma) = a_0 x^{\sigma} + a_0 \sum_{m=1}^{\infty} (-1)^m \frac{x^{2m+\sigma}}{(\sigma+2-n)...(\sigma+2m-n)\cdot(\sigma+2+n)...(\sigma+2m+n)}$$

Then, the second solution for the case of an integer difference k = 2n is:

$$y_2(x) = \frac{\partial}{\partial\sigma}\{(\sigma-\sigma_2)\,y(x,\sigma)\}_{\sigma=\sigma_2} = \frac{\partial}{\partial\sigma}\{(\sigma+n)\,y(x,\sigma)\}_{\sigma=-n}$$

Using the formula for $y_2(x)$ in eq. (2.20), an expression for y_2 results:

$$y_2(x) = a_0 \sum_{m=0}^{m=n-1} (-1)^m \frac{x^{2m-n}}{(2-2n)(4-2n)...(2m-2n)\cdot 2\cdot 4\cdot ...(2m)}$$

$$+ a_0 \sum_{m=n}^{\infty} \left[\frac{(-1)^m(\sigma+n)}{(\sigma+2-n)...(\sigma+2m-n)\cdot(\sigma+2+n)...(\sigma+2m+n)}\right]'_{\sigma=-n} x^{2m-n}$$

$$+ a_0 \log x \sum_{m=n}^{\infty} \left[\frac{(-1)^m(\sigma+n)}{(\sigma+2-n)...(\sigma+2m-n)\cdot(\sigma+2+n)...(\sigma+2m+n)}\right]_{\sigma=-n} x^{2m-n}$$

Thus, the solution corresponding to the second root $\sigma_2 = -n$ becomes:

$$y_2(x) = -\frac{1}{2}\sum_{m=0}^{n-1} \frac{(x/2)^{2m-n}}{m!}(n-m-1)! + \log x\, J_n(x) + \frac{1}{2} g(n-1)\, J_n(x)$$

$$-\frac{1}{2}\sum_{m=0}^{\infty} (-1)^m \frac{(x/2)^{2m+n}}{m!\,(m+n)!}\left[g(m) + g(m+n)\right]$$

where $-\frac{a_0 2^{-n+1}}{(n-1)!}$ was set equal to one.

The second solution includes the first solution given in eq. (3.10) multiplied by 1/2 $g(n - 1)$, which is a superfluous part of the second solution, thus, removing this component results in an expression for the second solution:

$$y_2(x) = \log x\, J_n(x) - \frac{1}{2}\sum_{m=0}^{m=n-1} \frac{(x/2)^{2m-n}}{m!}(n-m-1)!$$

$$-\frac{1}{2}\sum_{m=0}^{\infty} (-1)^m \frac{(x/2)^{2m+n}}{m!\,(m+n)!}\left[g(m) + g(m+n)\right]$$

Define:

$$Y_n(x) = \frac{2}{\pi}\left[(\gamma - \log 2)\, J_n(x) + y_2(x)\right]$$

$$= \frac{2}{\pi}\left\{\left[\gamma + \log(x/2)\right] J_n(x) - \frac{1}{2}\sum_{m=0}^{m=n-1} \frac{(x/2)^{2m-n}}{m!}(n-m-1)!\right.$$

$$\left.-\frac{1}{2}\sum_{m=0}^{\infty} (-1)^m \frac{(x/2)^{2m+n}}{m!\,(m+n)!}\left[g(m) + g(m+n)\right]\right\} \tag{3.11}$$

where $Y_n(x)$ is known as the **Bessel function of the second kind of order n**, or the **Neumann function of order n**. Thus, the solutions for p = n is:

$$y_h = c_1 J_n(x) + c_2 Y_n(x) \qquad \text{if } p = n = \text{integer}$$

The solutions of eq. (3.1) are also known as **Cylindrical Bessel** functions.

The second solution $Y_n(x)$ as given by Neumann corresponds to that given by Weber for non-integer orders defined in eq. (3.6). Since $\sin p\pi \to 0$ as $p \to n$ = integer, $\cos(p\pi) \to (-1)^n$ as $p \to n$, and:

$$J_{-n}(x) = (-1)^n J_n(x)$$

then the form in eq. (3.6) results in an indeterminate function. Thus:

$$Y_n(x) = \lim_{p \to n} \frac{\cos p\pi\, J_p(x) - J_{-p}(x)}{\sin p\pi}$$

$$= \left.\frac{-\pi \sin p\pi\, J_p(x) + \cos p\pi\, \partial/\partial p\, J_p(x) - \partial/\partial p\, J_{-p}(x)}{\pi \cos p\pi}\right|_{p=n}$$

$$= \frac{1}{\pi}\left\{\frac{\partial}{\partial p} J_p(x) - (-1)^n \frac{\partial}{\partial p} J_{-p}(x)\right\}_{p=n} \tag{3.12}$$

It can be shown that this solution is also a solution to eq. (3.1). It can be shown that the expression in eq. (3.12) gives the same expression given by eq. (3.11). The form given by Weber is most useful in obtaining an expression for the Wronskian, which is identical to the expression given in eq. (3.7).

3.4 Recurrence Relations for Bessel Functions

Recurrence relations between Bessel functions of various orders are of importance because of their use in numerical computations of high-ordered Bessel functions.

Starting with the definition of $J_p(x)$ in eq. (3.3), then differentiating the expression given in eq. (3.3) one obtains:

$$J'_p(x) = \frac{1}{2}\sum_{m=0}^{\infty} (-1)^m \frac{[2(m+p)-p](x/2)^{2m+p-1}}{m!\,\Gamma(m+p+1)}$$

$$= \sum_{m=0}^{\infty} (-1)^m \frac{(x/2)^{2m+p-1}(m+p)}{m!\,\Gamma(m+p+1)} - \frac{p}{2}\sum_{m=0}^{\infty} (-1)^m \frac{(x/2)^{2m+p}(x/2)^{-1}}{m!\,\Gamma(p+m+1)}$$

Using $\Gamma(m+p+1) = (m+p)\,\Gamma(m+p)$, (Appendix B1) then:

$$J'_p(x) = J_{p-1}(x) - \frac{p}{x} J_p(x) \tag{3.13}$$

Another form of eq. (3.13) can be obtained, again starting with $J'_p(x)$:

$$J'_p(x) = \sum_{m=0}^{\infty} (-1)^m \frac{(x/2)^{2m+p-1}}{(m-1)!\,\Gamma(m+p+1)} + \frac{p}{2}\sum_{m=0}^{\infty} (-1)^m \frac{(x/2)^{2m+p-1}}{m!\,\Gamma(p+m+1)}$$

Since $(m-1)! \to \infty$ for $m = 0$. Then:

$$J'_p(x) = \sum_{m=1}^{\infty} (-1)^m \frac{(x/2)^{2m+p-1}}{(m-1)!\,\Gamma(m+p+1)} + \frac{p}{x} J_p(x)$$

$$= \sum_{m=0}^{\infty} (-1)^{m+1} \frac{(x/2)^{2m+p+1}}{m!\,\Gamma(m+p+2)} + \frac{p}{x} J_p(x)$$

$$J'_p(x) = -J'_{p+1}(x) + \frac{p}{x} J_p(x) \tag{3.14}$$

Combining eqs. (3.13) and (3.14), one obtains another expression for the derivative:

$$J'_p(x)=\frac{1}{2}\left[J_{p-1}(x)-J_{p+1}(x)\right] \tag{3.15}$$

Equating (3.13) to (3.14) one obtains a recurrence formula for Bessel functions of order (p + 1) in terms of orders p and p - 1:

$$J_{p+1}(x)=\frac{2p}{x}J_p(x)-J_{p-1}(x) \tag{3.16}$$

Multiplying eq. (3.14) by x^{-p}, and rearranging the resulting expression, one obtains:

$$-\frac{1}{x}\frac{d}{dx}\left[x^{-p}J_p(x)\right]=x^{-(p+1)}J_{p+1}(x) \tag{3.17}$$

If p is substituted by p + 1 in the form given in eq. (3.17) results in:

$$-\frac{1}{x}\frac{d}{dx}\left[x^{-(p+1)}J_{p+1}\right]=x^{-(p+2)}J_{p+2}$$

then upon substitution of eq. (3.17) one obtains:

$$(-1)^2\left(\frac{1}{x}\frac{d}{dx}\right)^2\left[x^{-p}J_p\right]=x^{-(p+2)}J_{p+2}$$

Thus, by induction, one obtains a recurrence formula for Bessel Functions:

$$(-1)^r\left(\frac{1}{x}\frac{d}{dx}\right)^r\left[x^{-p}J_p\right]=x^{-(p+r)}J_{p+r} \qquad r\geq 0 \tag{3.18}$$

Substitution of p by -p in eq. (3.18) results in another recurrence formula:

$$(-1)^r\left(\frac{1}{x}\frac{d}{dx}\right)^r\left[x^{p}J_{-p}\right]=x^{p-r}J_{-(p+r)} \qquad r\geq 0 \tag{3.19}$$

Substitution of p by -p in eq. (3.13) one obtains:

$$J'_{-p}-p\,x^{-1}J_{-p}=J_{-(p+1)} \tag{3.20}$$

Multiplying eq. (3.20) by x^{-p}, one obtains new recurrence formula:

$$\frac{1}{x}\frac{d}{dx}\left[x^{-p}J_{-p}(x)\right]=x^{-(p+1)}J_{-(p+1)}(x) \tag{3.21}$$

Substitution of p + 1 for p in eq. (3.20) results in the following equation:

$$\frac{1}{x}\frac{d}{dx}\left[x^{-(p+1)}J_{-(p+1)}\right]=x^{-(p+2)}J_{-(p+2)}$$

or upon substitution of eq. (3.21) one gets:

$$\left(\frac{1}{x}\frac{d}{dx}\right)^2\left[x^{-p}J_{-p}\right]=x^{-(p+2)}J_{-(p+2)}$$

and, by induction, a recurrence formula for negative-ordered Bessel functions is obtained:

$$\left(\frac{1}{x}\frac{d}{dx}\right)^r\left[x^{-p}J_{-p}\right]=x^{-(p+r)}J_{-(p+r)} \qquad r\geq 0 \tag{3.22}$$

Substitution of p by -p in eq. (3.22) results in the following equation:

$$\left(\frac{1}{x}\frac{d}{dx}\right)^r \left[x^p J_p\right] = x^{p-r} J_{p-r} \qquad r \geq 0 \tag{3.23}$$

To obtain the recurrence relationships for the $Y_p(x)$, it is sufficient to use the form of $Y_p(x)$ given in eq. (3.6) and the recurrence equations given in eqs. (3.18), (3.19), (3.22), and (3.23). Starting with eqs. (3.18) and (3.22) and setting r = 1, one obtains:

$$\frac{1}{x}\frac{d}{dx}\left[x^{-p} J_p\right] = x^{-(p+1)} J_{p+1} \qquad \frac{1}{x}\frac{d}{dx}\left[x^{-p} J_{-p}\right] = x^{-(p+1)} J_{-(p+1)}$$

Then, using the form in eq. (3.6) for $Y_p(x)$:

$$\frac{1}{x}\frac{d}{dx}\left[x^{-p} Y_p\right] = \frac{1}{x}\frac{d}{dx}\left[x^{-p}\left(\frac{\cos(p\pi) J_p - J_{-p}}{\sin(p\pi)}\right)\right]$$

$$= -x^{-(p+1)}\left[\frac{\cos((p+1)\pi) J_{p+1} - J_{-(p+1)}}{\sin((p+1)\pi)}\right] = -x^{-(p+1)} Y_{p+1}$$

such that:

$$x\, Y_p' - p\, Y_p = -x\, Y_{p+1}$$

Similarly, use of eqs. (3.19) and (3.23) results in the following recurrence formula:

$$x\, Y_p' + p\, Y_p = x\, Y_{p-1}$$

Combining the preceding formulæ, the following recurrence formulae can be derived:

$$Y_{p-1} + Y_{p+1} = \frac{2p}{x} Y_p \qquad Y_{p-1} - Y_{p+1} = 2\, Y_p'$$

The recurrence relationships developed for Y_p are also valid for integer values of p, since $Y_n(x)$ can be obtained from $Y_p(x)$ by the expression given in eq. (3.12).

The recurrence formulae developed in this section can be summarized as follows:

$$Z_p' = -Z_{p+1} + \frac{p}{x} Z_p \tag{3.24}$$

$$Z_p' = Z_{p-1} - \frac{p}{x} Z_p \tag{3.25}$$

$$Z_p' = \frac{1}{2}\left(Z_{p-1} - Z_{p+1}\right) \tag{3.26}$$

$$Z_{p+1} = -Z_{p-1} + \frac{2p}{x} Z_p \tag{3.27}$$

where $Z_p(x)$ denotes $J_p(x)$, $J_{-p}(x)$ or $Y_p(x)$ for all values of p.

3.5 Bessel Functions of Half Orders

If the parameter p in eq. (3.1) happens to be an odd multiple of 1/2, then it is possible to obtain a closed form of Bessel functions of half orders.

Starting with the lowest half order, i.e., $p = 1/2$, then using the form in eq. (3.3) one obtains:

$$J_{1/2}(x)=\sum_{m=0}^{\infty}(-1)^m\frac{(x/2)^{2m+1/2}}{m!\,\Gamma(m+3/2)}=\left(\frac{x}{2}\right)^{1/2}\sum_{m=0}^{\infty}(-1)^m\frac{x^{2m}}{2^m\left(2^m\,m!\right)\Gamma(m+3/2)}$$

which can be shown to result in the following closed form:

$$J_{1/2}(x)=\left(\frac{2}{\pi x}\right)^{1/2}\sum_{m=0}^{\infty}(-1)^m\frac{x^{2m+1}}{(2m+1)!}=\left(\frac{2}{\pi x}\right)^{1/2}\sin x \tag{3.28}$$

Similarly, it can be shown that:

$$J_{-1/2}(x)=\left(\frac{2}{\pi x}\right)^{1/2}\cos x \tag{3.29}$$

To obtain the higher-ordered half-order Bessel functions $J_{n+1/2}$ and $J_{-(n+1/2)}$, one can use the recurrence formulæ in eqs. (3.24) – (3.27). One can also obtain these expressions by using eqs. (3.18) and (3.22) by setting p = 1/2, resulting in the following expressions:

$$J_{n+1/2}=(-1)^n\sqrt{2/\pi}\;x^{n+1/2}\left(\frac{1}{x}\frac{d}{dx}\right)^n\left(\frac{\sin x}{x}\right) \tag{3.30}$$

$$J_{-(n+1/2)}=\sqrt{2/\pi}\;x^{n+1/2}\left(\frac{1}{x}\frac{d}{dx}\right)^n\left(\frac{\cos x}{x}\right) \tag{3.31}$$

3.6 Spherical Bessel Functions

Bessel functions of half-order often show up as part of solutions of Laplace, Helmholtz, or the wave equations in the radial spherical coordinate. Define the following functions, known as the **spherical Bessel functions of the first and second kind of order** ν:

$$j_\nu=\sqrt{\frac{\pi}{2x}}\;J_{\nu+1/2}$$

$$y_\nu=\sqrt{\frac{\pi}{2x}}\;(-1)^{\nu+1}\;J_{-(\nu+1/2)} \tag{3.32}$$

These functions satisfy a different differential equation than Bessel's having the form:

$$x^2\frac{d^2y}{dx^2}+2x\frac{dy}{dx}+\left(x^2-\nu^2-\nu\right)y=0 \tag{3.33}$$

For ν = integer = n, the first two functions j_n and y_n have the following form:

$$j_0=\frac{\sin x}{x} \qquad j_1=\frac{1}{x}\left(\frac{\sin x}{x}-\cos x\right)$$

$$y_0=-\frac{\cos x}{x} \qquad y_1=-\frac{1}{x}\left(\frac{\cos x}{x}+\sin x\right)$$

Recurrence relations for the spherical Bessel functions can be easily developed from those developed for the cylindrical Bessel functions in eqs. (3.24) to (3.27) by setting $p = \nu + 1/2$ and $-\nu - 1/2$. Thus, the following recurrence formulæ can be obtained:

$$z'_\nu = -z_{\nu+1} + \frac{\nu}{x} z_\nu \tag{3.34}$$

$$z'_\nu = z_{\nu-1} - \frac{\nu+1}{x} z_\nu \tag{3.35}$$

$$(2\nu+1)\, z'_\nu = \nu z_{\nu-1} - (\nu+1)\, z_{\nu+1} \tag{3.36}$$

$$z_{\nu+1} = -z_{\nu-1} + \frac{2\nu+1}{x} z_\nu \tag{3.37}$$

where z_ν represents j_ν or y_ν.

The Wronskian of the spherical Bessel functions y_ν and j_ν takes the following form:
$W(j_\nu, y_\nu) = x^{-2}$

3.7 Hankel Functions

Hankel functions are complex linear combinations of Bessel functions of the form:

$$H_p^{(1)}(x) = J_p(x) + iY_p(x) \tag{3.38}$$

$$H_p^{(2)}(x) = J_p(x) - iY_p(x) \tag{3.39}$$

where $i^2 = -1$. $H_p^{(1)}(x)$ and $H_p^{(2)}(x)$ are respectively known as the **Hankel functions of first and second kind of order p**. They are also independent solutions of eq. (3.1), since, (See Section 1.3):

$$\alpha_{11} = 1 \qquad \alpha_{12} = i$$

$$\alpha_{21} = 1 \qquad \alpha_{22} = -i$$

and the determinant of the transformation matrix does not vanish:

$$|\alpha_{ij}| = \begin{vmatrix} 1 & i \\ 1 & -i \end{vmatrix} = -2i \neq 0$$

so that the Wronskian of the Hankel functions can be found from the Wronskian of J_p and Y_p in the form:

$$W\left(H_p^{(1)}, H_p^{(2)}\right) = \frac{-4i}{\pi x}$$

The form of $H_p^{(1)}(x)$ and $H_p^{(2)}(x)$, given in eqs. (3.38) and (3.39) respectively, can be written in terms of J_p and J_{-p} by the use of the expression for Y_p given in eq. (3.6), thus:

$$H_p^{(1)} = J_p + i\frac{\cos(p\pi)\, J_p - J_{-p}}{\sin(p\pi)} = \frac{J_{-p} - e^{-ip\pi}\, J_p}{i\sin(p\pi)}$$

$$H_p^{(2)} = \frac{e^{ip\pi}\, J_p - J_{-p}}{i\sin(p\pi)}$$

The general solution of eq. (3.1) then may be written in the form:

$$y = c_1 H_p^{(1)} + c_2 H_p^{(2)}$$

Recurrence formulæ for Hankel functions take the same forms given in eqs. (3.24) to (3.27), since they are linear combinations of J_p and Y_p.

Similar expression for the spherical Hankel functions can be written in the following form:

$$h_v^{(1)} = j_v + iy_v = \sqrt{\pi/2x}\; H_{v+1/2}^{(1)}(x) \tag{3.40}$$

$$h_v^{(2)} = j_v - iy_v = \sqrt{\pi/2x}\; H_{v+1/2}^{(2)}(x) \tag{3.41}$$

These are known as the **spherical Hankel function of first and second kind of order ν**.

3.8 Modified Bessel Functions

Modified Bessel functions are solutions to a differential equation different from that given in eq. (3.1), specifically they are solutions to the following differential equation:

$$x^2 \frac{d^2y}{dx^2} + x\frac{dy}{dx} - \left(x^2 + p^2\right)y = 0 \tag{3.42}$$

Performing the transformation:

$$z = ix$$

then the differential equation (3.42) transforms to:

$$z^2 \frac{d^2y}{dz^2} + z\frac{dy}{dz} + \left(z^2 - p^2\right)y = 0$$

which has two solutions of the form given in eqs. (3.3) and (3.4) if $p \neq 0$ and $p \neq$ integer. Using the form in eq. (3.3) one obtains:

$$J_p(z) = \sum_{m=0}^{\infty} (-1)^m \frac{\left(z/2\right)^{2m+p}}{m!\,\Gamma(m+p+1)} \qquad p \neq 0, 1, 2, \ldots$$

$$J_p(ix) = \sum_{m=0}^{\infty} (-1)^m \frac{\left(ix/2\right)^{2m+p}}{m!\,\Gamma(m+p+1)} = (i)^p \sum_{m=0}^{\infty} \frac{\left(x/2\right)^{2m+p}}{m!\,\Gamma(m+p+1)}$$

and

$$J_{-p}(ix) = (i)^{-p} \sum_{m=0}^{\infty} \frac{\left(x/2\right)^{2m-p}}{m!\,\Gamma(m-p+1)}$$

Define:

$$I_p(x) = \sum_{m=0}^{\infty} \frac{\left(x/2\right)^{2m+p}}{m!\,\Gamma(m+p+1)} = (i)^{-p}\, J_p(ix) \tag{3.43}$$

$$I_{-p}(x) = \sum_{m=0}^{\infty} \frac{(x/2)^{2m-p}}{m!\,\Gamma(m-p+1)} = (i)^p\, J_{-p}(ix) \qquad p \neq 0, 1, 2, \ldots \tag{3.44}$$

$I_p(x)$ and $I_{-p}(x)$ are known, respectively, as the **modified Bessel function of the first and second kind of order p.**

The general solution of eq. (3.42) takes the following form:

$$y = c_1 I_p(x) + c_2 I_{-p}(x)$$

If p takes the value zero or an integer n, then:

$$I_n(x) = \sum_{m=0}^{\infty} \frac{(x/2)^{2m+n}}{m!\,(m+n)!} \qquad n = 0, 1, 2, \ldots \tag{3.45}$$

is the first solution. The second solution must be obtained in a similar manner as described in Sections 3.2 and 3.3 giving:

$$K_n(x) = (-1)^{n+1}\left[\log(x/2) + \gamma\right] I_n(x) + \frac{1}{2}\sum_{m=0}^{m=n-1}(-1)^m \frac{(n-m-1)!}{m!}(x/2)^{2m-n}$$

$$+\frac{(-1)^n}{2}\sum_{m=0}^{\infty}\frac{(x/2)^{2m+n}}{m!\,(m+n)!}\left[g(m)+g(m+n)\right] \qquad n = 0, 1, 2, \ldots \tag{3.46}$$

The second solution can also be obtained from a definition given by Macdonald:

$$K_p = \frac{\pi}{2}\left[\frac{I_{-p} - I_p}{\sin(p\pi)}\right] \tag{3.47}$$

K_p is known as the **Macdonald function**. If p is an integer equal to n, then taking the limit $p \to n$:

$$K_n = \frac{1}{2}(-1)^n \left[\frac{\partial I_{-p}}{\partial p} - \frac{\partial I_p}{\partial p}\right]_{p=n} \tag{3.48}$$

The Wronskian of the various solutions for the modified Bessel's equation can be obtained in a similar manner to the method of obtaining the Wronskians of the modified Bessel functions in eqs. (3.5) and (3.7):

$$W(I_p, I_{-p}) = -\frac{2\sin(p\pi)}{\pi x} \tag{3.49}$$

and

$$W(I_p, K_p) = -\frac{1}{x} \tag{3.50}$$

Following the development of the recurrence formulæ for J_p and Y_p detailed in Section 3.4, one can obtain the following formulae for I_p and K_p:

$$I'_p = I_{p+1} + \frac{p}{x} I_p \tag{3.51}$$

$$I'_p = I_{p-1} - \frac{p}{x} I_p \tag{3.52}$$

$$I'_p = \frac{1}{2}\left(I_{p+1} + I_{p-1}\right) \tag{3.53}$$

$$I_{p+1} = I_{p-1} - \frac{2p}{x} I_p \tag{3.54}$$

$$K'_p = -K_{p+1} + \frac{p}{x} K_p \tag{3.55}$$

$$K'_p = -K_{p-1} - \frac{p}{x} K_p \tag{3.56}$$

$$K'_p = -\frac{1}{2}\left(K_{p+1} + K_{p-1}\right) \tag{3.57}$$

$$K_{p+1} = K_{p-1} + \frac{2p}{x} K_p \tag{3.58}$$

If p is 1/2, then the modified Bessel functions of half-orders can be developed in a similar manner as presented in Section 3.2, resulting in:

$$I_{1/2} = \sqrt{2/\pi x}\,\sinh x \tag{3.59}$$

$$I_{-1/2} = \sqrt{2/\pi x}\,\cosh x \tag{3.60}$$

3.9 Generalized Equations Leading to Solutions in Terms of Bessel Functions

The differential equation given in eq. (3.1) leads to solutions $Z_p(x)$, with $Z_p(x)$ representing J_p, Y_p, J_{-p}, $H_p^{(1)}$, and $H_p^{(2)}$. One can obtain the solutions of different and more complicated equations in terms of Bessel functions.

Starting with an equation of the form:

$$x^2 \frac{d^2y}{dx^2} + (1 - 2a)\, x \frac{dy}{dx} + \left(k^2x^2 - r^2\right) y = 0 \tag{3.61}$$

a solution of the form:

$$y = x^{\nu}\, u(x)$$

can be tried, resulting in the following differential equation:

$$x^2 \frac{d^2u}{dx^2} + x \frac{du}{dx} + \left\{k^2x^2 - \left[r^2 + a^2\right]\right\} u = 0$$

where ν was set equal to a.

Furthermore, if one lets $z = kx$, then $\frac{d}{dx} = k \frac{d}{dz}$ and:

$$z^2\frac{d^2u}{dz^2}+z\frac{du}{dz}+\left(z^2-p^2\right)u=0 \qquad \text{with } p^2=r^2+a^2$$

whose solution becomes:

$$u=c_1\,J_p(z)+c_2\,Y_p(z)$$

Thus, the solution to eq. (3.61) becomes:

$$y(x)=x^a\left\{c_1\,J_p(kx)+c_2\,Y_p(kx)\right\} \tag{3.62}$$

where $p^2 = r^2 + a^2$.

A more complicated equation can be developed from eq. (3.61) by assuming that:

$$z^2\frac{d^2y}{dz^2}+(1-2a)\,z\frac{dy}{dz}+\left(z^2-r^2\right)y=0 \tag{3.63}$$

which has solutions of the form:

$$y=z^a\left\{c_1\,J_p(z)+c_2\,Y_p(z)\right\} \tag{3.64}$$

with $p^2 = r^2 + a^2$.

If one lets $z = f(x)$, then eq. (3.63) transforms to:

$$\frac{d^2y}{dx^2}+\left[(1-2a)\frac{f'}{f}-\frac{f''}{f'}\right]\frac{dy}{dx}+\frac{(f')^2}{f^2}\left(f^2-r^2\right)y=0 \tag{3.65}$$

whose solutions can be written as:

$$y=f^a(x)\left[c_1\,J_p(f(x))+c_2\,Y_p(f(x))\right]$$

with $p^2 = r^2 + a^2$.

Eq. (3.65) may have many solutions depending on the desired form of f(x), e.g.:

(i) If $f(x) = kx^b$, then the differential equation may be written as:

$$x^2\frac{d^2y}{dx^2}+(1-2ab)\,x\frac{dy}{dx}+b^2\left(k^2x^{2b}-r^2\right)y=0 \tag{3.66}$$

whose solutions are given by:

$$y=x^{ab}\left\{c_1\,J_p\left(kx^b\right)+c_2\,Y_p\left(kx^b\right)\right\} \tag{3.67}$$

(ii) If $f(x) = ke^{bx}$, then the differential equation may be written as:

$$\frac{d^2y}{dx^2}-2ab\frac{dy}{dx}+b^2\left(k^2e^{2bx}-r^2\right)y=0 \tag{3.68}$$

whose solutions are given by:

$$y=e^{abx}\left\{c_1\,J_p\left(ke^{bx}\right)+c_2\,Y_p\left(ke^{bx}\right)\right\} \tag{3.69}$$

Another type of a differential equation that leads to Bessel function type solutions can be obtained from the form developed in eq. (3.65).

If one lets y to be transformed as follows:

$$y(x) = \frac{u(x)}{g(x)}$$

then

$$\frac{d^2u}{dx^2} + \left[(1-2a)\frac{f'}{f} - \frac{f''}{f'} - 2\frac{g'}{g}\right]\frac{du}{dx}$$

$$+\left\{\frac{(f')^2}{f^2}\left(f^2 - r^2\right) - \frac{g''}{g} - \frac{g'}{g}\left[(1-2a)\frac{f'}{f} - \frac{f''}{f'} - 2\frac{g'}{g}\right]\right\}u = 0 \tag{3.70}$$

whose solutions are given in the form:

$$u(x) = g(x)\, f^a(x)\, \{c_1\, J_p(f(x)) + c_2\, Y_p(f(x))\} \tag{3.71}$$

with $p^2 = r^2 + a^2$. If one lets:

$$g(x) = e^{cx} \qquad\qquad f(x) = kx^b$$

then the differential equation has the form:

$$x^2\frac{d^2u}{dx^2} + [1 - 2ab - 2cx]\,x\frac{du}{dx} + \left[b^2\left(k^2x^{2b} - r^2\right) + c^2x^2 - cx(1-2ab)\right]u = 0 \tag{3.72}$$

whose solutions are expressed in the form:

$$u = e^{cx}x^{ab}\left[c_1\, J_p\left(kx^b\right) + c_2\, Y_p\left(kx^b\right)\right] \tag{3.73}$$

3.10 Bessel Coefficients

In the preceding sections, Bessel functions were developed as solutions of second-order linear differential equations. Two other methods of development are available, one is the **Generating Function** representation and the other is the **Integral Representation**. In this section the **Generating Function** representation will be discussed.

The generating function of the Bessel coefficients is represented by:

$$f(x,t) = e^{x(t-1/t)/2} \tag{3.74}$$

Expanding the function in eq. (3.74) in a Laurent's series of powers of t, one obtains:

$$f(x,t) = \sum_{n=-\infty}^{\infty} t^n\, J_n(x) \tag{3.75}$$

Expanding the exponential $e^{xt/2}$ about t = 0 results in:

$$e^{xt/2} = \sum_{k=0}^{\infty} \frac{(x/2)^k}{k!}\, t^k$$

Expanding the exponential $e^{-x/2t}$ about $t = \infty$ results in:

$$e^{-x/2t} = \sum_{m=0}^{\infty} \frac{\left(-x/2t\right)^m}{m!} = \sum_{m=0}^{\infty} (-1)^m \frac{(x/2)^m\, t^{-m}}{m!}$$

Thus, the product of the two series gives the desired expansion:

$$f(x,t) = \sum_{m=-\infty}^{\infty} t^n J_n(x) = \left(\sum_{k=0}^{\infty} \frac{\left(\frac{x}{2}\right)^k t^k}{k!} \right) \left(\sum_{m=0}^{\infty} \frac{(-1)^m \left(\frac{x}{2}\right)^m t^{-m}}{m!} \right)$$

The term that is the coefficient of t^n is the one where k - m = n, with k and m range from 0 to ∞. Thus the coefficient of t^n becomes:

$$J_n(x) = \sum_{m=0}^{\infty} \frac{(-1)^m \left(\frac{x}{2}\right)^{2m+n}}{m!\,(m+n)!}$$

having the same form given in eq. (3.10).

The generating function can be used to advantage when one needs to obtain recurrence formulæ. Differentiating eq. (3.74) with respect to t, one obtains:

$$\frac{df(x,t)}{dt} = e^{x(t-1/t)/2}\left[x/2\left(1+t^{-2}\right)\right] = x/2 \sum_{n=-\infty}^{\infty} t^n J_n + x/2 \sum_{n=-\infty}^{\infty} t^{n-2} J_n$$

$$= \sum_{n=-\infty}^{\infty} n t^{n-1} J_n(x)$$

The above expression can be rewritten in the following way:

$$\frac{x}{2} \sum_{n=-\infty}^{\infty} t^n J_n + \frac{x}{2} \sum_{n=-\infty}^{\infty} t^n J_{n+2} = \sum_{n=-\infty}^{\infty} (n+1) t^n J_{n+1}(x)$$

where the coefficient of t^n can be factored out, such that:

$$\frac{x}{2} J_n + \frac{x}{2} J_{n+2} = (n+1) J_{n+1}$$

or, letting n-1 replace n, one obtains:

$$\frac{x}{2} J_{n-1} + \frac{x}{2} J_{n+1} = n J_n$$

which is the recurrence relation given in eq. (3.16).

The other recurrence formulæ given in Section 3.4 can be derived also by manipulating the generating function in a similar manner.

If one substitutes t = - 1/y, then:

$$e^{x(y-1/y)/2} = \sum_{n=-\infty}^{\infty} (-1)^n y^{-n} J_n(x) = \sum_{n=-\infty}^{\infty} (-1)^n y^n J_{-n}(x)$$

also

$$e^{x(y-1/y)/2} = \sum_{n=-\infty}^{\infty} y^n J_n(x)$$

then, equating the two expressions, one gets the relationship:

$$(-1)^n J_{-n}(x) = J_n(x)$$

Rewriting the series for the generating function in eq. (3.75) into two parts:

$$e^{x(t-1/t)/2} = \sum_{n=-\infty}^{\infty} t^n J_n(x) = \sum_{n=-\infty}^{n=-1} t^n J_n + J_0 + \sum_{n=1}^{\infty} t^n J_n(x)$$

$$= \sum_{n=1}^{\infty} t^{-n} J_{-n} + J_0 + \sum_{n=1}^{\infty} t^n J_n = \sum_{n=1}^{\infty} t^{-n} (-1)^n J_n + J_0 + \sum_{n=1}^{\infty} t^n J_n$$

$$= J_0 + \sum_{n=1}^{\infty} \left[t^n + (-1)^n t^{-n} \right] J_n \tag{3.76}$$

If $t = e^{\pm i\theta}$:

$$e^{x\left(e^{\pm i\theta} - e^{\mp i\theta}\right)/2} = e^{\pm ix\sin\theta} = J_0 + \sum_{n=1}^{\infty} \left[e^{\pm in\theta} + (-1)^n e^{\pm in\theta} \right] J_n(x)$$

$$= J_0 + 2 \sum_{n=2,4,6,\ldots}^{\infty} \cos(n\theta)\, J_n(x) \pm 2i \sum_{n=1,3,5,\ldots}^{\infty} \sin(n\theta)\, J_n(x)$$

$$= \sum_{n=0}^{\infty} \varepsilon_{2n} \cos(2n\theta)\, J_{2n}(x) \pm i \sum_{n=0}^{\infty} \varepsilon_{2n+1} \sin((2n+1)\theta)\, J_{2n+1}(x) \tag{3.77}$$

where ε_n, generally known as the **Neumann Factor** is defined as:

$$\varepsilon_n = \begin{cases} 1 & n = 0 \\ 2 & n \geq 1 \end{cases}$$

Replacing θ by $\theta + \pi/2$ in eq. (3.77) then the following expansion results:

$$e^{\pm ix\cos\theta} = \sum_{n=0}^{\infty} (\pm i)^n \varepsilon_n \cos n\theta\, J_n(x) \tag{3.78}$$

Further manipulation of eq. (3.78) results in the following two expressions:

$$\cos(x\sin\theta) = \sum_{n=0}^{\infty} \varepsilon_{2n} \cos(2n\theta)\, J_{2n}(x) \tag{3.79a}$$

$$\sin(x\sin\theta) = \sum_{n=0}^{\infty} \varepsilon_{2n+1} \sin((2n+1)\theta)\, J_{2n+1}(x) \tag{3.79b}$$

One can also obtain a Bessel function series for any power of x. If θ is set to zero in the form given in eq. (3.79a) one obtains the expression for a unity:

$$1 = \sum_{n=0}^{\infty} \varepsilon_{2n} J_{2n}(x) \tag{3.80}$$

Again, differentiating eq. (3.79b) with respect to θ:

$$(x\cos\theta)\cos(x\sin\theta) = 2\sum_{n=0}^{\infty}(2n+1)\cos((2n+1)\theta)J_{2n+1}(x)$$

Setting θ = 0 one obtains an expansion for x results:

$$x = 2\sum_{n=0}^{\infty}(2n+1)\,J_{2n+1}(x) \tag{3.81}$$

Differentiating eq. (3.79a) twice with respect to θ results in the following expression for x^2 by setting θ = 0:

$$x^2 = 4\sum_{n=0}^{\infty}\varepsilon_{2n}n^2J_{2n}(x) = 8\sum_{n=1}^{\infty}n^2J_{2n}(x) \tag{3.82}$$

Thus, a similar procedure can be followed to show that all powers of x can be expanded in a series of Bessel functions. It should be noted that even/odd powers of x are represented by even-/odd-ordered Bessel functions.

Setting θ = π/2 in eqs. (3.79a) and (3.79b), the following Bessel function series representations for sin x and cos x results:

$$\cos x = \sum_{n=0}^{\infty}\varepsilon_{2n}(-1)^nJ_{2n}(x) \tag{3.83}$$

$$\sin x = 2\sum_{n=0}^{\infty}(-1)^nJ_{2n+1}(x) \tag{3.84}$$

Differentiating eqs. (3.79a) and (3.79b) twice with respect to θ and setting θ = π/2 results in the following Bessel series representations for x sin x and x cos x:

$$x\sin x = 8\sum_{n=1}^{\infty}(-1)^n n^2J_{2n}(x) \tag{3.85}$$

$$x\cos x = 2\sum_{n=0}^{\infty}(-1)^n(2n+1)^2J_{2n+1}(x) \tag{3.86}$$

The generating functions can also be utilized to obtain formulae in terms of products or squares of Bessel functions, usually known as the **Addition Theorem**. Starting with the forms given in eq. (3.75):

$$e^{x/(2(t-1/t))}e^{z/(2(t-1/t))} = e^{(x+z)/(2(t-1/t))} = \sum_{n=-\infty}^{\infty}t^nJ_n(x+z)$$

$$= \left(\sum_{k=-\infty}^{\infty}t^kJ_k(x)\right)\left(\sum_{l=-\infty}^{\infty}t^lJ_l(z)\right)$$

$$= \sum_{n=-\infty}^{\infty} t^n \left(\sum_{l=-\infty}^{\infty} J_l(x)\, J_{n-l}(z) \right)$$

$$= \sum_{n=-\infty}^{\infty} t^n \left(\sum_{l=-\infty}^{\infty} J_{n-l}(x)\, J_l(z) \right)$$

Thus, the coefficient of t^n results in the representation for the Bessel function of sum arguments, known as the **Addition Theorem**:

$$J_n(x+z) = \sum_{l=-\infty}^{\infty} J_l(x)\, J_{n-l}(z) = \sum_{l=-\infty}^{\infty} J_l(z) J_{n-l}(x) \tag{3.87}$$

Manipulating the terms in the expression in eq. (3.87) which have Bessel functions of negative orders one obtains:

$$J_n(x+z) = \sum_{l=0}^{n} J_l(x) J_{n-l}(z) + \sum_{l=1}^{\infty} (-1)^l \left[J_l(x)\, J_{n+l}(z) + J_{n+l}(x)\, J_l(z) \right] \tag{3.88}$$

Special cases of the form of the addition theorem given in eq. (3.88) can be utilized to give expansions in terms of products of Bessel functions. If x = z:

$$J_n(2x) = \sum_{l=0}^{n} J_l(x) J_{n-l}(x) + 2 \sum_{l=1}^{\infty} (-1)^l J_l(x)\, J_{n+l}(x) \tag{3.89}$$

If one sets z = -x in eq. (3.88), one obtains new series expansions in terms of squares of Bessel functions:

$$1 = J_0^2(x) + 2 \sum_{l=1}^{\infty} (-1)^l J_l^2(x) \qquad n = 0 \tag{3.90}$$

$$0 = \sum_{l=0}^{2n+1} (-1)^{l-1} J_l(x)\, J_{2n+1-l}(x) \qquad n = 0, 1, 2, \ldots \tag{3.91}$$

$$0 = \sum_{l=0}^{2n} (-1)^l J_l(x)\, J_{2n-l}(x) + 2 \sum_{l=1}^{\infty} J_l(x)\, J_{2n+l}(x) \qquad n = 0, 1, 2, \ldots \tag{3.92}$$

3.11 Integral Representation of Bessel Functions

Another form of representation of Bessel functions is an integral representation. This representation is useful in obtaining asymptotic expansions of Bessel functions and in integral transforms as well as source representations. To obtain an integral representation, it is useful to use the results of Section 3.9.

Integrating eq. (3.79a) on θ over (0,2π), one obtains:

$$\int_0^{2\pi} \cos(x\sin\theta)\, d\theta = \sum_{n=0}^{\infty} \varepsilon_{2n}\, J_{2n}(x) \int_0^{2\pi} \cos(2n\theta)\, d\theta = 2\pi\, J_0(x) \tag{3.93}$$

Multiplication of the expression in eq. (3.79a) by cos 2mθ and then integrating on θ over (0,2π) results:

$$\int_0^{2\pi} \cos(x\sin\theta)\cos(2m\theta)\, d\theta = \sum_{n=0}^{\infty} \varepsilon_{2n}\, J_{2n} \int_0^{2\pi} \cos(2n\theta)\cos(2m\theta)\, d\theta = 2\pi\, J_{2m}(x)$$

$$m = 0, 1, 2,... \tag{3.94}$$

Multiplication of eq. (3.79b) by sin (2m + 1) θ and integrating on θ, one obtains:

$$\int_0^{2\pi} \sin(x\sin\theta)\sin((2m+1)\theta)\, d\theta = 2\pi\, J_{2m+1}(x) \qquad m = 0, 1, 2,... \tag{3.95}$$

The forms given in eqs. (3.93) to (3.95) can thus be transformed to an integral representation of Bessel functions:

$$J_m = \frac{1}{2\pi} \int_{-\pi}^{\pi} \cos(x\sin\theta)\cos(m\theta)\, d\theta = \frac{1}{\pi} \int_0^{\pi} \cos(x\sin\theta)\cos(m\theta)\, d\theta \qquad m = \text{even}$$

and since the following integral vanishes:

$$\int_0^{\pi} \cos(x\sin\theta)\cos(m\theta)\, d\theta = 0 \qquad m = \text{odd}$$

then an integral representation for the Bessel function results as:

$$J_m = \frac{1}{\pi} \int_0^{\pi} \sin(x\sin\theta)\sin(m\theta)\, d\theta$$

and since the following integral vanishes:

$$\int_0^{\pi} \sin(x\sin\theta)\sin(m\theta)\, d\theta = 0 \qquad m = \text{even}$$

then, one can combine the two definitions for odd-/even-ordered Bessel functions J_m as a real integral representation:

$$J_m = \frac{1}{\pi} \int_0^{\pi} \cos(x\sin\theta)\cos(m\theta)\, d\theta + \frac{1}{\pi} \int_0^{\pi} \sin(x\sin\theta)\sin(m\theta)\, d\theta$$

$$= \frac{1}{2\pi} \int_{-\pi}^{\pi} \cos(x\sin\theta - m\theta)\, d\theta \tag{3.96}$$

Since the following integral vanishes identically:

$$\int_{-\pi}^{\pi} \sin(x\sin\theta - m\theta)\, d\theta = 0$$

then one can also find a complex form of the Bessel integral representation:

$$J_m = \frac{1}{2\pi}\int_{-\pi}^{\pi} \cos(x\sin\theta - m\theta)\, d\theta + \frac{i}{2\pi}\int_{-\pi}^{\pi} \sin(x\sin\theta - m\theta)\, d\theta$$

$$= \frac{1}{\pi}\int_{0}^{\pi} \exp\left(i(x\sin\theta - m\theta)\right) d\theta \qquad (3.97)$$

Another integral representation of Bessel functions, similar to those given in eq. (3.96) was developed by Poisson. Noting that the Taylor expansion of the trigonometric function:

$$\cos(x\cos\theta) = \sum_{m=0}^{\infty} (-1)^m \frac{(x\cos\theta)^{2m}}{(2m)!} = \sum_{m=0}^{\infty} (-1)^m \frac{x^{2m}(\cos\theta)^{2m}}{(2m)!}$$

has terms x^{2m}, similar to Bessel functions, one can integrate this trigonometric functions over θ to give another integral representation of Bessel functions. Multiplying this expression by $(\sin\theta)^{2n}$ and integrating on θ:

$$\int_0^{\pi} \cos(x\cos\theta)(\sin\theta)^{2n}\, d\theta = \int_0^{\pi} \sum_{m=0}^{\infty} (-1)^m \frac{x^{2m}}{(2m)!}(\cos\theta)^{2m}(\sin\theta)^{2n}\, d\theta$$

$$= \sum_{m=0}^{\infty} (-1)^m \frac{x^{2m}}{(2m)!}\int_0^{\pi} (\cos\theta)^{2m}(\sin\theta)^{2n}\, d\theta$$

The integration and summation operations can be exchanged, since the Taylor expansion of $\cos(x\cos\theta)$ is uniformly convergent for all values of the argument $x\cos\theta$ (Refer to Appendix A). The integral in the summation can be evaluated as:

$$\int_0^{\pi} (\cos\theta)^{2m}(\sin\theta)^{2n}\, d\theta = \frac{(2m-1)!}{2^{m-1}(m-1)!}\cdot\frac{(2n-1)!}{2^{n-1}(n-1)!}\frac{\pi}{2^{m+n}(m+n)!}$$

and hence

$$\int_0^{\pi} \cos(x\cos\theta)(\sin\theta)^{2n} d\theta = \frac{\pi(2n-1)!}{2^{n-1}(n-1)!}\sum_{m=0}^{\infty}(-1)^m \frac{x^{2m}(2m-1)!}{(2m)!(m-1)!(m+n)!\,2^{2m+n-1}}$$

$$= \frac{\pi(2n-1)!}{2^{n-1}x^n(n-1)!}J_n(x)$$

Thus, from this expression a new integral representation can be developed in the form:

$$J_n(x) = \frac{2(x/2)^n}{\Gamma(n+1/2)\,\Gamma(1/2)} \int_0^{\pi/2} \cos(x\cos\theta)\,(\sin\theta)^{2n}\,d\theta \tag{3.98}$$

Transforming θ by π/2 - θ in the representation of eq. (3.98), one obtains a new representation:

$$J_n(x) = \frac{2(x/2)^n}{\Gamma(n+1/2)\,\Gamma(1/2)} \int_0^{\pi/2} \cos(x\sin\theta)\,(\cos\theta)^{2n}\,d\theta \tag{3.99}$$

Since the following integral vanishes:

$$\int_0^{\pi} \sin(x\cos\theta)\,(\sin\theta)^{2n}\,d\theta = 0 \tag{3.100}$$

due to the fact that sin (x cos θ) is an odd function of θ in the interval $0 \le \theta \le \pi$, then adding eqs. (3.98) and i times eq. (3.100) results in the following integral representation:

$$J_n(x) = \frac{(x/2)^n}{\Gamma(n+1/2)\,\Gamma(1/2)} \int_0^{\pi} e^{ix\cos\theta}\,(\sin\theta)^{2n}\,d\theta \tag{3.101}$$

The integral representations of eqs. (3.98) to (3.101) can also be shown to be true for non-integer values of $p > -1/2$.

Performing the following transformation on eq. (3.101):

$\cos\theta = t$

there results a new integral representation for $J_n(x)$ as follows:

$$J_p(x) = \frac{(x/2)^p}{\Gamma(p+1/2)\,\Gamma(1/2)} \int_{-1}^{+1} e^{ixt}\left(1-t^2\right)^{p-1/2}\,dt \qquad p > -1/2 \tag{3.102}$$

The integral representations given in this section can also be utilized to develop the recurrence relationships already derived in Section 3.4.

3.12 Asymptotic Approximations of Bessel Functions for Small Arguments

Asymptotic approximation of the various Bessel functions for small arguments can be developed from their ascending powers infinite series representations. Thus letting $x \ll 1$, the following approximations are obtained:

$$J_p \sim \frac{\left(x/2\right)^p}{\Gamma(p+1)}, \qquad J_{-p} \sim \frac{\left(x/2\right)^{-p}}{\Gamma(-p+1)}$$

$$Y_0 \sim \frac{2}{\pi}\log x, \qquad Y_p \sim -\frac{1}{\pi}\Gamma(p)\left(x/2\right)^{-p}$$

$$H_0^{(1)(2)} \sim \pm i\frac{2}{\pi}\log x, \qquad H_p^{(2)(1)} \sim \pm\frac{i}{\pi}\Gamma(p)\left(\tfrac{x}{2}\right)^{-p}$$

$$I_p \sim \frac{\left(\tfrac{x}{2}\right)^p}{\Gamma(p+1)}, \qquad K_0 \sim -\log x$$

$$K_p \sim \frac{\Gamma(p)}{2}\left(\tfrac{x}{2}\right)^{-p}, \qquad j_n \sim \frac{x^n}{1\cdot 3\cdot 5\ldots(2n+1)}$$

$$y_n \sim -\frac{1\cdot 3\cdot 5\ldots(2n-1)}{x^{n+1}}, \qquad h_n^{(2)(1)} \sim \pm i\frac{1\cdot 3\cdot 5\ldots(2n-1)}{x^{n+1}}$$

3.13 Asymptotic Approximations of Bessel Functions for Large Arguments

Asymptotic approximations for large arguments can be obtained by asymptotic techniques using their integral representation. These are enumerated below:

$$\underset{x>>1}{J_p(x)} \sim \sqrt{\tfrac{2}{\pi x}}\cos\left(x-\tfrac{\pi}{4}-p\tfrac{\pi}{2}\right)$$

$$\underset{x>>1}{Y_p(x)} \sim \sqrt{\tfrac{2}{\pi x}}\sin\left(x-\tfrac{\pi}{4}-p\tfrac{\pi}{2}\right)$$

$$\underset{x>>1}{H_p^{(1)(2)}(x)} \sim \sqrt{\tfrac{2}{\pi x}}\exp\left(\pm i\left(x-\tfrac{\pi}{4}-p\tfrac{\pi}{2}\right)\right)$$

$$\underset{x>>1}{I_p(x)} \sim \frac{e^x}{\sqrt{2\pi x}}, \qquad \underset{x>>1}{K_p(x)} \sim \sqrt{\tfrac{\pi}{2x}}\,e^{-x}$$

$$\underset{x>>1}{j_n(x)} \sim \tfrac{1}{x}\sin\left(x-n\tfrac{\pi}{2}\right), \qquad \underset{x>>1}{y_n(x)} \sim -\tfrac{1}{x}\cos\left(x-n\tfrac{\pi}{2}\right)$$

$$\underset{x>>1}{h_n^{(1)(2)}(x)} \sim \frac{e^{\pm ix}}{x}$$

3.14 Integrals of Bessel Functions

Integrals of Bessel functions can be developed from the various recurrence formulae in eqs. (3.13) to (3.27). A list of useful indefinite integrals is given below:

$$\int x^{p+1} J_p\,dx = x^{p+1} J_{p+1} \tag{3.103}$$

$$\int x^{-p+1} J_p\,dx = -x^{-p+1} J_{p-1} \tag{3.104}$$

$$\int x^{r+1} J_p \, dx = x^{r+1} J_{p+1} + (r-p) x^r J_p - (r^2 - p^2) \int x^{r-1} J_p \, dx \tag{3.105}$$

$$\int \left[(\alpha^2 - \beta^2) x - \frac{p^2 - r^2}{x} \right] J_p(\alpha x) J_r(\beta x) dx = x \left[J_p(\alpha x) \frac{dJ_r(\beta x)}{dx} - J_r(\beta x) \frac{dJ_p(\alpha x)}{dx} \right] \tag{3.106}$$

If α and β are set = 1 in eq. (3.106) one obtains:

$$\int J_p(x) J_r(x) \frac{dx}{x} = \frac{x}{p^2 - r^2} \left(J_r \frac{dJ_p}{dx} - J_p \frac{dJ_r}{dx} \right) = \frac{J_p + J_r}{p + r} - \frac{x}{p^2 - r^2} \left(J_{p+1} J_r - J_p J_{r+1} \right) \tag{3.107}$$

If one sets p = r in eq. (3.106), one obtains:

$$(\alpha^2 - \beta^2) \int x J_p(\alpha x) J_p(\beta x) dx = x \left[J_p(\alpha x) \frac{dJ_p(\beta x)}{dx} - J_p(\beta x) \frac{dJ_p(\alpha x)}{dx} \right] \tag{3.108}$$

If one lets α → β in eq. (3.108) one obtains the integral of the squared Bessel function:

$$\int x J_p^2(x) \, dx = \frac{1}{2} \left[(x^2 - p^2) J_p^2 + x^2 \left(\frac{dJ_p(x)}{dx} \right)^2 \right] \tag{3.109}$$

Few other integrals of products of Bessel functions and polynomials are presented here:

$$\int x^{-r-p+1} J_r(x) J_p(x) dx = -\frac{x^{-r-p+2}}{2(r+p-1)} \left[J_{r-1}(x) J_{p-1}(x) + J_r(x) J_p(x) \right] \tag{3.110}$$

If one substitutes p and r by -p and -r respectively in eq. (3.110), one obtains a new integral:

$$\int x^{r+p+1} J_r(x) J_p(x) dx = \frac{x^{r+p+2}}{2(r+p+1)} \left[J_{r+1}(x) J_{p+1}(x) + J_r(x) J_p(x) \right] \tag{3.111}$$

If one lets r = -p in eq. (3.110) the following indefinite integral results:

$$\int x J_p^2(x) \, dx = \frac{x^2}{2} \left[J_p^2(x) - J_{p-1}(x) J_{p+1}(x) \right] \tag{3.112}$$

If one sets r = p in eqs. (3.110) and (3.111), one obtains the following indefinite integrals:

$$\int x^{-2p+1} J_p^2(x) \, dx = \frac{x^{-2p+2}}{2(2p-1)} \left[J_{p-1}^2(x) + J_p^2(x) \right] \tag{3.113}$$

and

$$\int x^{2p+1} J_p^2(x) \, dx = \frac{x^{2p+2}}{2(2p+1)} \left[J_{p+1}^2(x) + J_p^2(x) \right] \tag{3.114}$$

3.15 Zeroes of Bessel Functions

Bessel functions $J_p(x)$ and $Y_p(x)$ have infinite number of zeroes. Denoting the sth root of $J_p(x)$, $Y_p(x)$, $J'_p(x)$ and $Y'_p(x)$ by $j_{p,s}, y_{p,s}, j'_{p,s}, y'_{p,s}$, then all the zeroes of these functions have the following properties:

1. That all the zeroes of these Bessel functions are real if p is real and positive.
2. There are no repeated roots, except at the origin.
3. $j_{p,0} = 0$ for $p > 0$
4. The roots of J_p and Y_p interlace, such that:

$$p < j_{p,1} < j_{p+1,1} < j_{p,2} < j_{p+1,2} < j_{p,3} < \ldots$$

$$p < y_{p,1} < y_{p+1,1} < y_{p,2} < y_{p+1,2} < y_{p,3} < \ldots$$

$$p \le j'_{p,1} < y'_{p,1} < y'_{p,1} < j'_{p,1} < j'_{p,2} < y'_{p,2} < y'_{p,2} < j'_{p,2} < \ldots$$

5. The roots $j_{p,1}$ and $j'_{p,1}$ can be bracketed such that:

$$\begin{aligned} \sqrt{p(p+2)} &< j_{p,1} < \sqrt{2(p+1)(p+3)} \\ \sqrt{p(p+2)} &< j'_{p,1} < \sqrt{2p(p+1)} \end{aligned} \qquad (3.115)$$

6. The large roots of Bessel functions for a fixed order p take the following asymptotic form:

$$\begin{aligned} &\underset{s\to\infty}{j_{p,s}} \to \left(s+\frac{p}{2}-\frac{1}{4}\right)\pi \\ &\underset{s\to\infty}{y_{p,s}} \to \left(s+\frac{p}{2}-\frac{3}{4}\right)\pi \\ &\underset{s\to\infty}{j'_{p,s}} \to \left(s+\frac{p}{2}-\frac{3}{4}\right)\pi \\ &\underset{s\to\infty}{y'_{p,s}} \to \left(s+\frac{p}{2}-\frac{1}{4}\right)\pi \end{aligned} \qquad (3.116)$$

The roots as given in these expressions are spaced at an interval = π. The roots of J_p, Y_p, and J'_p and Y'_p are also well tabulated, Ref. [Abramowitz and Stegun]. All roots of $H_p^{(1)}, H_p^{(2)}$, I_p, I_{-p}, and K_p are complex for real and positive orders p.

The roots of products of Bessel functions, usually appearing in boundary value problems of the following form:

$$\begin{aligned} &J_p(x)\,Y_p(ax) - J_p(ax)\,Y_p(x) = 0 \\ &J'_p(x)\,Y'_p(ax) - J'_p(ax)\,Y'_p(x) = 0 \\ &J_p(x)\,Y'_p(ax) - J_p(ax)\,Y'_p(x) = 0 \end{aligned} \qquad (3.117)$$

can be obtained from published tables, Ref. [Abramowitz and Stegun].

The large zeroes of the spherical Bessel functions of order n are the same as the zeroes of J_p, Y_p, J'_p and Y'_p with $p = n + 1/2$. Spherical Hankel functions have no real zeroes.

TABLE OF ZEROES OF BESSEL FUNCTIONS

	s = 1	s = 2	s = 3	s = 4
$j_{0,s}$	2.405	5.520	8.654	11.79
$j_{1,s}$	3.832	7.016	10.17	13.32
$j_{2,s}$	5.136	8.417	11.62	14.80
$y_{0,s}$	0.894	3.958	7.086	10.22
$y_{1,s}$	2.197	5.430	8.596	11.75
$y_{2,s}$	3.384	6.794	10.02	13.21
$j'_{0,s}$	0.000	3.832	7.016	10.17
$j'_{1,s}$	1.841	5.331	8.536	11.71
$j'_{2,s}$	3.054	6.706	9.970	13.17
$y'_{0,s}$	2.197	5.430	8.596	11.75
$y'_{1,s}$	3.683	6.941	10.12	13.29
$y'_{2,s}$	5.003	8.351	11.57	14.76

3.16 Legendre Functions

Legendre functions are solutions to the following ordinary differential equation:

$$\left(1-x^2\right)\frac{d^2y}{dx^2} - 2x\frac{dy}{dx} + r(r+1)y = 0 \qquad (3.118)$$

where r is a real constant.

The differential equation (3.118) has two regular singular points located at x = +1 and x = -1. Since the point x = 0 is classified as a regular point, then an expansion of the solution y(x) into an infinite series of the type (2.3) can be made. Such an expansion results in the following recurrence relationship:

$$a_{m+2} = -\frac{(r-m)(r+m+1)}{(m+1)(m+2)} a_m \qquad m = 0, 1, 2, \ldots$$

with a_0 and a_1 being indeterminate.

The recurrence relation results in the following expression for the coefficients a_m:

$$a_{2m} = (-1)^m \frac{(r-2m+2)(r-2m+4)(r-2m+6)\ldots r\cdot(r+1)(r+3)\ldots(r+2m-1)}{(2m)!} a_0$$

$$m = 1, 2, 3, \ldots$$

and

$$a_{2m+1} = (-1)^m \frac{(r-2m+1)(r-2m+3)\dots(r-1)\cdot(r+2)(r+4)\dots(r+2m)}{(2m+1)!} a_1$$

$$m = 1, 2, 3, \dots$$

Thus, the two solutions of eq. (3.118) become:

$$p_r(x) = 1 - \frac{r(r+1)}{2!}x^2 + \frac{(r-2)r(r+1)(r+3)}{4!}x^4 - \frac{(r-4)(r-2)r(r+1)(r+3)(r+5)}{6!}x^6 + \dots + (-1)^m \frac{[r-(2m-2)][r-(2m-4)]\dots r\cdot(r+1)\dots(r+2m-1)}{(2m)!}x^{2m} + \dots \tag{3.119}$$

$$q_r(x) = x - \frac{(r-1)(r+2)}{3!}x^3 + \frac{(r-3)(r-1)(r+2)(r+4)}{5!}x^5 - \frac{(r-5)(r-3)(r-1)(r+2)(r+4)(r+6)}{7!}x^7 + \dots + (-1)^m \frac{(r-2m+1)(r-2m-1)\dots(r-1)\cdot(r+2)\dots(r+2m)}{(2m+1)!}x^{2m+1} + \dots \tag{3.120}$$

and the final solution is given as:

$$y = c_1 p_r(x) + c_2 q_r(x)$$

The infinite series solutions have a radius of convergence $\rho = 1$, such that $p_r(x)$ and $q_r(x)$ converge in $-1 < x < 1$. At the two end points $x = \pm 1$, both series diverge.

If r is an even integer = 2n, the infinite series in eq. (3.119) becomes a polynomial of degree 2n, having the form:

$$p_{2n}(x) = (-1)^n \frac{2^{2n}(n!)^2}{(2n)!} P_{2n}(x)$$

where

$$P_{2n}(x) = \frac{(4n-1)(4n-3)\dots 5\cdot 3\cdot 1}{(2n)!}\left[x^{2n} - \frac{(2n)(2n-1)}{2(4n-1)}x^{2n-2} + \dots + (-1)^n \frac{((2n)!)^2}{2^{2n}(n!)^2(4n-1)\dots 5\cdot 3}\right] \qquad n = 0, 1, 2, \dots \tag{3.121}$$

The second solution q_{2n} is an infinite series, which diverges at $x = \pm 1$.

If r is an odd integer = 2n + 1, then it can be shown that the infinite series in eq. (3.120) becomes a polynomial of degree 2n+1, having the form:

$$q_{2n+1} = (-1)^n \frac{2^{2n}(n!)^2}{(2n+1)!} P_{2n+1}(x)$$

where

$$P_{2n+1}(x) = \frac{(4n+1)(4n-1)\ldots 3\cdot 1}{(2n+1)!}\left[x^{2n+1} - \frac{(2n+1)(2n)}{2(4n+1)}x^{2n-1} + \right.$$

$$\left. + \frac{(2n+1)(2n)(2n-1)(2n-2)}{2\cdot 4(4n+1)(4n-1)}x^{2n-3} - \ldots + (-1)^n \frac{((2n+1)!)^2 x}{2^{2n}(n!)^2(4n+1)\ldots 5\cdot 3}\right] \qquad n = 0, 1, 2,\ldots$$

(3.122)

The first solution p_{2n+1} it still an infinite series, which is divergent at $x = \pm 1$.

If one defines:

$$Q_{2n}(x) = (-1)^n \frac{2^{2n}(n!)^2}{(2n)!} q_{2n}(x) \qquad n = 0, 1, 2,\ldots \qquad (3.123)$$

and

$$Q_{2n+1}(x) = (-1)^{n+1} \frac{2^{2n}(n!)^2}{(2n+1)!} p_{2n+1}(x) \qquad n = 0, 1, 2,\ldots \qquad (3.124)$$

then the solution to eq. (3.118) for all integer values of r becomes:

$$y = c_1 P_n + c_2 Q_n(x) \qquad n = 0, 1, 2,\ldots$$

where the infinite series expansion for $Q_n(x)$ is convergent in the region $|x| < 1$, and P_n is a polynomial of degree n.

A general form for $P_m(x)$ can be developed for all integer values m by setting $2n = m$ in eq. (3.121) and $2n + 1 = m$ in eq. (3.122), giving the following polynomial expression for $P_m(x)$:

$$P_m(x) = \frac{(2m-1)(2m-3)\ldots 3\cdot 1}{m!}\left[x^m - \frac{m(m-1)}{2\cdot(2m-1)}x^{m-2} + \frac{m(m-1)(m-2)(m-3)}{2\cdot 4\cdot(2m-1)(2m-3)}x^{m-4}\right.$$

$$\left. - \frac{m(m-1)(m-2)(m-3)(m-4)(m-5)}{2\cdot 4\cdot 6\cdot(2m-1)(2m-3)(2m-5)}x^{m-6} + \ldots\right] \qquad m = 0, 1, 2, \ldots \qquad (3.125)$$

The functions $P_n(x)$ and $Q_n(x)$ are known as **Legendre** functions of the first and second kind of degree n.

The Legendre polynomials $P_n(x)$ take the following special values:

$$P_n(1) = 1$$

$$P_n(-1) = (-1)^n$$

$$P_n(0) = (-1)^{n/2} \frac{(n)!}{2^n((n/2)!)^2} \qquad \text{if } n = \text{even integer}$$

$$= 0 \qquad \text{if } n = \text{odd integer}$$

A list of the first few Legendre polynomials is given below:

$$P_0(x) = 1 \qquad P_1(x) = x$$

$$P_2(x) = (3x^2 - 1)/2 \qquad P_3(x) = (5x^3 - 3x)/2$$

$$P_4(x) = (35x^4 - 30x^2 + 3)/8 \qquad P_5(x) = (63x^5 - 70x^3 + 15x)/8$$

Noting that:

$$\frac{d^n}{dx^n}\left(x^{2n}\right) = (2n)(2n-1)(2n-2)\ldots(n+1)\,x^n$$

$$\frac{d^n}{dx^n}\left(x^{2n-2}\right) = (2n-2)(2n-3)\ldots(n+1)\,x^{n-2}$$

then the polynomial form of $P_n(x)$ in eq. (3.125) becomes:

$$P_n = \frac{1}{2^n n!}\frac{d^n}{dx^n}\left[x^{2n} - \frac{n\,x^{2n-2}}{1!} + \frac{n(n-1)}{2!}x^{2n-4} - \ldots\right]$$

Examination of the terms inside the square brackets shows that they represent the binomial expansion of $(x^2 - 1)^n$. Thus, $P_n(x)$ can be defined by the formula:

$$P_n(x) = \frac{1}{2^n n!}\frac{d^n}{dx^n}\left(x^2 - 1\right)^n \tag{3.126}$$

This representation of $P_n(x)$ is known as **Rodrigues' formula.**

The infinite series expansion for $Q_n(x)$ can be written in a closed form in terms of $P_n(x)$. Assuming that the second solution $Q_n(x) = Z(x)\,P_n(x)$, then:

$$\frac{Z''}{Z'} = \frac{2x\,P_n(x) - 2\left(1-x^2\right)P_n'}{\left(1-x^2\right)P_n}$$

resulting in an indefinite integral for Z(x), such that the second solution $Q_n(x)$ becomes:

$$Q_n(x) = P_n(x)\int^{x}\frac{d\eta}{\left(1-\eta^2\right)P_n^2(\eta)} \tag{3.127}$$

Since $P_n(\eta)$ is a polynomial of degree n, then $P_n(\eta)$ can be factored such that:

$$P_n(\eta) = (\eta-\eta_1)(\eta-\eta_2)\ldots(\eta-\eta_n)$$

Thus, the integrand in eq. (3.127) can be factored to give:

$$\frac{a_0}{1-\eta} + \frac{b_0}{1+\eta} + \frac{c_1}{\eta-\eta_1} + \ldots + \frac{c_n}{\eta-\eta_n} + \frac{d_1}{(\eta-\eta_1)^2} + \frac{d_2}{(\eta-\eta_2)^2} + \ldots + \frac{d_n}{(\eta-\eta_n)^2}$$

where

$$a_0 = \frac{1}{2} \qquad\qquad b_0 = \frac{1}{2}$$

$$c_i = \frac{d}{d\eta}\left.\frac{(\eta-\eta_i)^2}{\left(1-\eta^2\right)P_n^2(\eta)}\right|_{\eta=\eta_i} = \frac{d}{d\eta}\frac{1}{\left(1-\eta^2\right)R_i^2} = \left.\frac{2\left(\eta R_i - \left(1-\eta^2\right)R_i'\right)}{\left(1-\eta^2\right)^2 R_i^3}\right|_{\eta=\eta_i}$$

where

$$R_i(\eta) = \frac{P_n(\eta)}{\eta-\eta_i}.$$

Substitution of $P_n(\eta) = (\eta - \eta_i)\, R_i(\eta)$ into eq. (3.118), then:

$$\left(1-\eta^2\right)P_n'' - 2nP_n' + n(n+1)\,P_n\Big|_{\eta=\eta_i}$$

$$= \left(\eta - \eta_i\right)\left[\left(1-\eta^2\right)R_i'' - 2\eta R_i'\right] + 2\left[\left(1-\eta^2\right)R_i' - \eta R_i\right]_{\eta=\eta_i} = 0$$

Thus, R_i satisfies the differential equation:

$$\left(1-\eta^2\right)R_i' - nR_i\Big|_{\eta=\eta_i} = 0$$

hence:

$$c_i \equiv 0$$

and

$$d_i = \frac{\left(\eta-\eta_i\right)^2}{\left(1-\eta^2\right)P_n^2}\Bigg|_{\eta=\eta_i} = \frac{1}{\left(1-\eta_i^2\right)R_i^2(\eta_i)}$$

Thus, the closed form solution for $Q_n(x)$:

$$Q_n(x) = P_n(x)\left[-\frac{1}{2}\log(1-\eta) + \frac{1}{2}\log(1+\eta) - \sum_{i=1}^{n}\frac{d_i}{\eta-\eta_i}\right]_{\eta=x}$$

$$= \frac{1}{2}P_n(x)\log\frac{1+x}{1-x} - P_n(x)\sum_{i=1}^{n}\frac{d_i}{x-x_i} \tag{3.128}$$

Thus, the first few Legendre functions of the second kind have closed form:

$$Q_0 = \frac{1}{2}P_0(x)\log\frac{1+x}{1-x}$$

$$Q_1 = \frac{1}{2}P_1(x)\log\frac{1+x}{1-x} - 1$$

$$Q_2 = \frac{1}{2}P_2(x)\log\frac{1+x}{1-x} - \frac{3}{2}x$$

$$Q_3 = \frac{1}{2}P_3(x)\log\frac{1+x}{1-x} - \frac{5}{2}x^2 + \frac{2}{3}$$

The functions $Q_n(x)$ converge in the region $|x| < 1$.

Another solution of eq. (3.118), for integer values of r, which is valid in the region $|x| > 1$ can be developed. Starting with the recurrence relationship with $r = n$, $n = 0, 1, 2,...$

$$a_{m+2} = -\frac{(n-m)(n+m+1)}{(m+1)(m+2)}\,a_m \qquad m = 0, 1, 2,...$$

$a_{m+2}, a_{m+4}, a_{m+6},...$ can be made to vanish if $m = n$ or $-n - 1$ with the coefficient $a_m \neq 0$ to be taken as the arbitrary constant. For the integer value $r = n$, the recurrence relationship can be rewritten as follows:

$$a_{m-2} = -\frac{m(m-1)}{(n-m+2)(n+m-1)} a_m$$

thus

$$a_{m-4} = \frac{-(m-2)(m-3)a_{m-2}}{(n-m+4)(n+m-3)} = \frac{m(m-1)(m-2)(m-3)a_m}{(n-m+2)(n-m+4)(n+m-1)(n+m-3)}$$

Setting m = n:

$$a_{n-2} = -\frac{n(n-1)}{2\cdot(2n-1)} a_n$$

$$a_{n-4} = +\frac{n(n-1)(n-2)(n-3)}{2\cdot 4\cdot(2n-1)(2n-3)} a_n$$

Thus, the first solution can be written as:

$$y_1(x) = a_n\left[x^n - \frac{n(n-1)}{2\cdot(2n-1)} x^{n-2} + \frac{n(n-1)(n-2)(n-3)}{2\cdot 4\cdot(2n-1)(2n-3)} x^{n-4} - \ldots\right]$$

where a_n can be set to:

$$\frac{(2n-1)(2n-3)\ldots 5\cdot 3\cdot 1}{n!}$$

such that $y_1(x)$ becomes $P_n(x)$. Setting m = -n - 1, then:

$$a_{-n-3} = +\frac{(n+1)(n+2)}{(2n+3)\cdot 2} a_{-n-1}$$

$$a_{-n-5} = \frac{(n+1)(n+2)(n+3)(n+4)}{(2n+3)(2n+5)\cdot 2\cdot 4} a_{-n-1}$$

such that:

$$y_2(x) = a_{-n-1}\left[x^{-n-1} + \frac{(n+1)(n+2)}{2\cdot(2n+3)} x^{-n-3} + \frac{(n+1)(n+2)(n+3)(n+4)}{2\cdot 4\cdot(2n+3)(2n+5)} x^{-n-5} + \ldots\right]$$

Setting the coefficient:

$$a_{-n-1} = \frac{n!}{(2n+1)(2n-1)\ldots 5\cdot 3\cdot 1}$$

then the second solution $Q_n(x)$ can be written in an infinite series form with descending powers of x as follows:

$$Q_n(x) = \frac{n!}{(2n+1)(2n-1)\ldots 5\cdot 3\cdot 1}\left[x^{-n-1} + \frac{(n+1)(n+2)}{2\cdot(2n+3)} x^{-n-3} + \frac{(n+1)(n+2)(n+3)(n+4)}{2\cdot 4\cdot(2n-3)(2n+5)} x^{-n-5} + \ldots\right] \qquad |x| > 1 \qquad (3.129)$$

The Wronskian of $P_n(x)$ and $Q_n(x)$ can be evaluated from the differential equation (3.118).

$$W(P_n, Q_n) = P_n Q_n' - P_n' Q_n = W_0 \exp\left(-\int^{x} \frac{-2\eta\, d\eta}{1-\eta^2} \right) = \frac{W_0}{1-x^2}$$

Using the form for Q_n in eq. (3.128), the following expression approaches unity as $x \to \pm 1$:

$$W_0 = \lim_{x \to \pm 1} (1-x^2)[P_n Q_n' - P_n' Q_n] \to 1$$

and

$$W(P_n, Q_n) = \frac{1}{1-x^2}$$

3.17 Legendre Coefficients

Expanding the following generating function by the binomial series:

$$\frac{1}{(1-2tx+t^2)^{1/2}} = \frac{1}{[1-t(2x-t)]^{1/2}} = 1 + \frac{1}{2}(2x-t)t + \frac{1\cdot 3}{2\cdot 4}(2x-t)^2 t^2 +$$

$$+ \frac{1\cdot 3\cdot 5}{2\cdot 4\cdot 6}(2x-t)^3 t^3 + \ldots + \frac{1\cdot 3\cdot 5 \ldots (2n-1)}{2\cdot 4\cdot 6 \ldots 2n}(2x-t)^n t^n + \ldots \tag{3.130}$$

then one can extract the coefficient of t^n having the form:

$$\frac{1\cdot 3\cdot 5 \ldots (2n-1)}{n!}\left[x^n - \frac{n(n-1)}{2(2n-1)} x^{n-2} + \frac{n(n-1)(n-2)(n-3)}{2\cdot 4\cdot (2n-1)(2n-3)} x^{n-4} + \ldots \right]$$

which is the representation for $P_n(x)$ given in eq. (3.125). Thus, the binomial expansion gives:

$$\frac{1}{(1-2tx+t^2)^{1/2}} = \sum_{n=0}^{\infty} t^n P_n(x) \tag{3.131}$$

The generating function can be used to evaluate the Legendre polynomials at special values. At x = 1:

$$\frac{1}{(1-2t+t^2)^{1/2}} = \frac{1}{1-t} = 1 + t + t^2 + \ldots = \sum_{n=0}^{\infty} t^n P_n(1) = \sum_{n=0}^{\infty} t^n$$

which gives the value:

$$P_n(1) = 1$$

At x = -1:

$$\frac{1}{(1+2t+t^2)^{1/2}} = \frac{1}{1+t} = 1 - t + t^2 - t^3 + \ldots = \sum_{n=0}^{\infty} t^n P_n(-1) = \sum_{n=0}^{\infty} (-1)^n t^n$$

which gives the value:

$$P_n(-1) = (-1)^n$$

At x = 0, the generating function gives:

$$\frac{1}{\left(1+t^2\right)^{1/2}} = \left[1 - \frac{1}{2}t^2 + \frac{1\cdot 3}{2\cdot 4}t^4 + \ldots + (-1)^n \frac{1\cdot 3\cdot 5\ldots(2n-1)}{2\cdot 4\cdot 6\ldots 2n}t^{2n} + \ldots\right]$$

which results in a formula for $P_n(0)$:

$$P_n(0) = (-1)^{n/2}\frac{1\cdot 3\cdot 5\ldots(n-1)}{2\cdot 4\cdot 6\ldots n} = (-1)^{n/2}\frac{n!}{2^n\left[(n/2)!\right]^2} \qquad n = \text{even}$$

$$= 0 \qquad n = \text{odd}$$

Substituting t by -t in eq. (3.131) one obtains:

$$\frac{1}{\left(1+2tx+t^2\right)^{1/2}} = \sum_{n=0}^{\infty}(-1)^n t^n P_n(x) = \sum_{n=0}^{\infty} t^n P_n(-x)$$

results in the following identity:

$$P_n(-x) = (-1)^n P_n(x)$$

Other forms of Legendre polynomials can be obtained by manipulating eq. (3.131).

Letting x = cos θ, then:

$$\frac{1}{\left(1-2t\cos\theta+t^2\right)^{1/2}} = \frac{1}{\left(1-te^{i\theta}\right)^{1/2}\left(1-te^{-i\theta}\right)^{1/2}} =$$

$$= \left\{1 + \frac{t}{2}e^{i\theta} + \frac{1\cdot 3}{2\cdot 4}t^2 e^{2i\theta} + \frac{1\cdot 3\cdot 5}{2\cdot 4\cdot 6}t^3 e^{3i\theta} + \ldots + \frac{1\cdot 3\ldots(2n-1)}{2\cdot 4\ldots 2n}t^n e^{ni\theta} + \ldots\right\}$$

$$\cdot\left\{1 + \frac{t}{2}e^{-i\theta} + \frac{1\cdot 3}{2\cdot 4}t^2 e^{-2i\theta} + \frac{1\cdot 3\cdot 5}{2\cdot 4\cdot 6}t^3 e^{-3i\theta} + \ldots + \frac{1\cdot 3\ldots(2n-1)}{2\cdot 4\ldots 2n}t^n e^{-ni\theta} + \ldots\right\}$$

$$= 1 + t\frac{\left\{e^{i\theta} + e^{-i\theta}\right\}}{2} + t^2\left\{\frac{3}{4}\left(\frac{e^{2i\theta} + e^{-2i\theta}}{2}\right) + \frac{1}{4}\right\} + \ldots$$

$$+ 2t^n\left[\frac{1\cdot 3\ldots(2n-1)}{2\cdot 4\ldots 2n}\right]\left\{\left(\frac{e^{ni\theta} + e^{-ni\theta}}{2}\right) + \frac{1\cdot n}{1\cdot(2n-1)}\left(\frac{e^{(n-1)i\theta} + e^{-(n-1)i\theta}}{2}\right) + \ldots\right\}$$

Thus, the coefficient of t^n must be the Legendre polynomial, the first few of which are listed below:

$$P_0 = 1, \qquad P_1(\cos\theta) = \cos\theta$$

$$P_2(\cos\theta) = \frac{1}{4}[3\cos 2\theta + 1], \qquad P_3(\cos\theta) = \frac{1}{8}[5\cos 3\theta + 3\cos\theta]$$

and the Legendre polynomial with cosine arguments is defined by:

$$P_n(\cos\theta) = 2\frac{1\cdot 3\cdot 5\ldots(2n-1)}{2\cdot 4\cdot 6\ldots 2n}\left\{\cos n\theta + \frac{1\cdot n}{1\cdot(2n-1)}\cos(n-2)\theta\right.$$

$$\left. + \frac{1\cdot 3\cdot n(n-1)}{1\cdot 2\cdot(2n-1)(2n-3)}\cos(n-4)\,\theta + \ldots\right\} \tag{3.132}$$

Expansions of $P_n(x)$ about $x = \pm 1$ can be developed from the generating function. The generating function is rewritten in the following form:

$$\frac{1}{\left(1-2xt+t^2\right)^{1/2}} = \frac{1}{(1-t)\left[1+\frac{4t}{(1-t)^2}\left(\frac{1-x}{2}\right)\right]^{1/2}}$$

Expanding the new form by the binomial theorem there results:

$$= \frac{1}{1-t} + \sum_{m=1}^{\infty}(-1)^m \frac{1\cdot 3\cdot 5\ldots(2m-1)}{2\cdot 4\cdot 6\ldots 2m}\frac{4^m t^m}{(1-t)^{2m+1}}\left(\frac{1-x}{2}\right)^m$$

Expanding each of the terms $(1 - t)^{-2m-1}$ by the binomial theorem and collecting the coefficients of t^n, which must, by definition, be the Legendre polynomials, one obtains the following infinite series expansion about $x = 1$:

$$P_n(x) = 1 - \frac{(n+1)!}{(1!)^2(n-1)!}\left(\frac{1-x}{2}\right) + \frac{(n+2)!}{(2!)^2(n-2)!}\left(\frac{1-x}{2}\right)^2 - \frac{(n+3)!}{(3!)^2(n-3)!}\left(\frac{1-x}{2}\right)^3 + \ldots \tag{3.133}$$

Since $P_n(-x) = (-1)^n P_n(x)$, then an expansion about $x = -1$ can be obtained from eq. (3.133) by substituting x by -x:

$$P_n(x) = (-1)^n\left[1 - \frac{(n+1)!}{(1!)^2(n-1)!}\left(\frac{1+x}{2}\right) + \frac{(n+2)!}{(2!)^2(n-2)!}\left(\frac{1+x}{2}\right)^2 -\right.$$

$$\left. - \frac{(n+3)!}{(3!)^2(n-3)!}\left(\frac{1+x}{2}\right)^3 + \ldots\right] \tag{3.134}$$

3.18 Recurrence Formulae for Legendre Polynomials

Recurrence formulae for Legendre polynomials can be developed from the generating function expansion. Differentiating the generating function with respect to x, one obtains:

$$\frac{t}{\left(1-2xt+t^2\right)^{3/2}} = \sum_{n=0}^{\infty} t^n P_n'(x) \tag{3.135}$$

Differentiating the generating function with respect to t, one obtains:

$$\frac{x-t}{\left(1-2xt+t^2\right)^{3/2}} = \sum_{n=0}^{\infty} n\, t^{n-1} P_n(x) \tag{3.136}$$

Multiplying eq. (3.135) by (x - t) and eq. (3.136) by t, equating the resulting expressions and picking out the coefficient of t^n, a recurrence formula is obtained:

$$\sum_{n=0}^{\infty}(x-t)\,t^n\,P_n'(x) = \sum_{n=0}^{\infty} n\,t^n\,P_n(x)$$

or

$$x\,P_n' - P_{n-1}' = n\,P_n \qquad n \geq 1 \qquad (3.137)$$

with

$$P_0' = 0 \qquad n = 0$$

Multiplying eq. (3.136) by $1 - 2xt + t^2$, another recurrence formula is developed, by picking out the coefficient of t^n, as follows:

$$\frac{x-t}{\left(1-2xt+t^2\right)^{1/2}} = (x-t)\sum_{n=0}^{\infty} t^n\,P_n(x) = \left(1-2xt+t^2\right)\sum_{n=0}^{\infty} n\,t^{n-1}\,P_n(x)$$

$$x\,P_n - P_{n-1} = (n+1)\,P_{n+1} - 2nx\,P_n + (n-1)\,P_{n-1}$$

or, rewriting the last equality gives a recurrence formula for the Legendre polynomials:

$$(n+1)\,P_{n+1}(x) = (2n+1)\,x\,P_n(x) - n\,P_{n-1}(x) \qquad n \geq 1 \qquad (3.138)$$

with

$$P_1 = x\,P_0$$

Differentiating eq. (3.138) with respect to x and subtracting (2n + 1) times eq. (3.137) from the resulting expression, one obtains:

$$P_{n+1}' - P_{n-1}' = (2n+1)\,P_n \qquad n \geq 1 \qquad (3.139)$$

with

$$P_1' = P_0$$

Eliminating P_n from eqs. (3.138) and (3.139) results in the following recurrence formula:

$$x\left(P_{n+1}'(x) - P_{n-1}'(x)\right) = (n+1)\,P_{n+1}(x) + n\,P_{n-1}(x) \qquad (3.140)$$

Elimination of P_{n-1}' from eqs. (3.137) and (3.139), one obtains:

$$P_{n+1}'(x) - x\,P_n'(x) = (n+1)\,P_n(x) \qquad (3.141)$$

Substituting n by n - 1 in eq. (3.141), multiplying eq. (3.137) by x, and eliminating x P_{n-1}' from the resulting expression, the following recurrence formula is developed:

$$\left(1-x^2\right)P_n'(x) = n\,P_{n-1}(x) - n\,x\,P_n(x)$$

$$= -(n+1)\left[P_{n+1}(x) - x\,P_n(x)\right] \qquad (3.142)$$

3.19 Integral Representation for Legendre Polynomials

Noting that the definite integral:

$$\int_0^{\pi} \frac{du}{a + b\cos u} = \frac{\pi}{\sqrt{a^2 - b^2}} \tag{3.143}$$

then, by setting:

$a = 1 - xt$, $\qquad b = \pm t\sqrt{x^2 - 1}$ $\qquad$ then

$$\frac{1}{\left(1 - 2xt + t^2\right)^{1/2}} = \frac{1}{\pi}\int_0^{\pi} \frac{du}{1 - xt \pm t\cos u\sqrt{x^2 - 1}} = \sum_{n=0}^{\infty} t^n\, P_n(x) \tag{3.144}$$

Expanding the integrand of eq. (3.144) by the binomial theorem:

$$\frac{1}{1 - t\left(x \pm \cos u\sqrt{x^2 - 1}\right)} = 1 + t\left[x \pm \cos u\sqrt{x^2 - 1}\right] + t^2\left[x \pm \cos u\sqrt{x^2 - 1}\right]^2 + \ldots$$

$$+ t^n\left[x \pm \cos u\sqrt{x^2 - 1}\right]^n + \ldots$$

thus

$$P_n(x) = \frac{1}{\pi}\int_0^{\pi}\left[x \pm \cos u\sqrt{x^2 - 1}\right]^n du \tag{3.145}$$

The last integral is known as **Laplace's First Integral**.

If one substitutes -n - 1 for n in the differential equation (3.118), the equation does not change, thus giving rise to the following identity:

$$P_n(x) = P_{-n-1}(x) \tag{3.146}$$

Substituting -n - 1 for n in eq. (3.145), another integral representation results, generally known as **Laplace's Second Integral**, which has the form:

$$P_n(x) = \frac{1}{\pi}\int_0^{\pi}\left[x \pm \cos u\sqrt{x^2 - 1}\right]^{-n-1} du \tag{3.147}$$

Substitution of $x = \cos\theta$ in eq. (3.145) results in the following integral representation for $P_n(\cos\theta)$:

$$P_n(\cos\theta) = \frac{1}{\pi}\int_0^{\pi}\left(\cos\theta \pm i\sin\theta\cos u\right)^n du$$

Another integral representation can be obtained from the generating function. Setting $t = e^{iu}$ and $x = \cos\theta$ in the generating function, then:

$$\frac{1}{\left[1-2\cos\theta\, e^{iu}+e^{2iu}\right]^{1/2}}=\sum_{n=0}^{\infty} e^{inu}\, P_n(\cos\theta)=\begin{cases} =\dfrac{\left(\sqrt{2}\right)^{-1}}{e^{iu/2}(\cos u-\cos\theta)^{1/2}} & u<\theta \\ \dfrac{\left(\sqrt{2}\right)^{-1}}{e^{i(u-\pi)/2}(\cos u-\cos\theta)^{1/2}} & u>\theta \end{cases}$$

Equating the real and imaginary parts, one obtains:

$$2\sum_{n=0}^{\infty}\cos(n\,u)\,P_n(\cos\theta)=\sqrt{2}\begin{cases}\dfrac{\cos(u/2)}{(\cos u-\cos\theta)^{1/2}} & u<\theta\\ \dfrac{\sin(u/2)}{(\cos\theta-\cos u)^{1/2}} & u>\theta\end{cases} \tag{3.148}$$

and

$$2\sum_{n=0}^{\infty}\sin(n\,u)\,P_n(\cos\theta)=\sqrt{2}\begin{cases}\dfrac{-\sin(u/2)}{(\cos u-\cos\theta)^{1/2}} & u<\theta\\ \dfrac{\cos(u/2)}{(\cos\theta-\cos u)^{1/2}} & u>\theta\end{cases} \tag{3.149}$$

Multiplying eq. (3.148) by cos (n u) and eq. (3.149) by sin (n u) and integrating over u on $(0, \pi)$, there results two integrals for $P_n(\cos\theta)$:

$$P_n(\cos\theta)=\frac{\sqrt{2}}{\pi}\left\{\int_0^{\theta}\frac{\cos(u/2)\cos(n\,u)}{(\cos u-\cos\theta)^{1/2}}\,du+\int_{\theta}^{\pi}\frac{\sin(u/2)\cos(n\,u)}{(\cos\theta-\cos u)^{1/2}}\,du\right\} \tag{3.150}$$

and

$$P_n(\cos\theta)=\frac{\sqrt{2}}{\pi}\left\{-\int_0^{\theta}\frac{\sin(u/2)\sin(n\,u)}{(\cos u-\cos\theta)^{1/2}}\,du+\int_{\theta}^{\pi}\frac{\cos(u/2)\sin(n\,u)}{(\cos\theta-\cos u)^{1/2}}\,du\right\} \tag{3.151}$$

The integral representations of eqs. (3.150) and (3.151) are due to **Dirichlet**.

Adding and subtracting eqs. (3.150) and (3.151) one obtains:

$$P_n(\cos\theta)=\frac{1}{\pi\sqrt{2}}\left\{\int_0^{\theta}\frac{\cos(n+1/2)\,u}{(\cos u-\cos\theta)^{1/2}}\,du+\int_{\theta}^{\pi}\frac{\sin(n+1/2)\,u}{(\cos\theta-\cos u)^{1/2}}\,du\right\} \tag{3.152}$$

and

$$0=\int_0^{\theta}\frac{\cos(n-1/2)\,u}{(\cos u-\cos\theta)^{1/2}}\,du-\int_{\theta}^{\pi}\frac{\sin(n-1/2)\,u}{(\cos\theta-\cos u)^{1/2}}\,du \tag{3.153}$$

Replacing n by n + 1 in the identity in eq. (3.153), and substituting the resulting identity in eq. (3.152) one obtains:

$$P_n(\cos\theta)=\frac{\sqrt{2}}{\pi}\int_0^{\theta}\frac{\cos(n+1/2)\,u}{(\cos u-\cos\theta)^{1/2}}\,du=\frac{\sqrt{2}}{\pi}\int_{\theta}^{\pi}\frac{\sin(n+1/2)\,u}{(\cos\theta-\cos u)^{1/2}}\,du \tag{3.154}$$

The integral representation in (3.153) and (3.154) are due to **Mehler**.

3.20 Integrals of Legendre Polynomials

One of the most important properties of the Legendre polynomials is the orthogonality property. The first integral to be evaluated is an integral of products of Legendre polynomials.

The integral of products of Legendre polynomials can be evaluated by the use of Rodrigues' formula in eq. (3.126):

$$\int_{-1}^{+1} P_n\, P_m\, dx = \frac{1}{2^{n+m}\, n!\, m!} \int_{-1}^{+1} \frac{d^n}{dx^n}\left(x^2 - 1\right)^n \frac{d^m}{dx^m}\left(x^2 - 1\right)^m dx \qquad n \geq m$$

where n is assumed to be larger than m.

Integrating by parts, one can show that:

$$\int_{-1}^{+1} P_n\, P_m\, dx = 0 \qquad n \neq m \tag{3.155}$$

If n = m, then the last integral becomes:

$$\int_{-1}^{+1} P_n^2\, dx = \frac{(-1)^n (2n)!}{2^{2n} (n!)^2} \int_{-1}^{+1} \left(x^2 - 1\right)^n dx$$

Integrating the last integral by parts, one obtains:

$$\int_{-1}^{+1} P_n^2\, dx = \frac{2}{2n+1} \tag{3.156}$$

The orthogonality property can also be proven by integrating the differential equation. The differential equation that P_n and P_m satisfy for $n \neq m$ can be written in the following form:

$$\frac{d}{dx}\left[\left(1 - x^2\right) P_n'\right] + n(n+1)\, P_n = 0$$

$$\frac{d}{dx}\left[\left(1 - x^2\right) P_m'\right] + m(m+1)\, P_m = 0$$

Multiplying the first equation by P_m, the second by P_n, and subtracting and integrating the resulting equations, one obtains:

$$\int_{x_1}^{x_2} \left\{ P_m \frac{d}{dx}\left[\left(1-x^2\right)P_n'\right] - P_n \frac{d}{dx}\left[\left(1-x^2\right)P_m'\right] \right\} dx$$

$$+ \left[n(n+1) - m(m+1)\right] \int_{x_1}^{x_2} P_n \, P_m \, dx = 0 \tag{3.157}$$

Integrating eq. (3.157) by parts, the following expression results:

$$\left[m(m+1) - n(n+1)\right] \int_{x_1}^{x_2} P_n \, P_m \, dx = \left(1-x^2\right)\left(P_m \, P_n' - P_m' \, P_n\right) \Bigg|_{x_1}^{x_2} \tag{3.158}$$

If one sets $x_1 = -1$, $x_2 = +1$, then one obtains another proof of eq. (3.155). Substituting eq. (3.142) into eq. (3.158), one obtains:

$$\int_{x_1}^{x_2} P_n \, P_m \, dx = \frac{nP_m \, P_{n-1} - mP_n \, P_{m-1} + (m-n)\, xP_n \, P_m}{(m-n)(m+n+1)} \Bigg|_{x_1}^{x_2} \qquad m \neq n \tag{3.159}$$

Setting $x_1 = -1$ and $x_2 = x$ in eq. (3.159) one obtains:

$$\int_{-1}^{x} P_n \, P_m \, dx = \frac{nP_m \, P_{n-1} - mP_n \, P_{m-1} + (m-n)\, xP_n \, P_m}{(m-n)(m+n+1)} \qquad m \neq n \tag{3.160}$$

which can be evaluated at $x = 0$ as follows:

$$\int_{-1}^{0} P_n \, P_m \, dx = 0 \qquad \text{if n is odd and m is odd, } n \neq m$$

$$= 0 \qquad \text{if n is even and m is even, } n \neq m$$

$$= \frac{1}{(m-n)(m+n+1)} \cdot \frac{(-1)^{(n+m+1)/2} \, n! \, m!}{2^{m+n-1}\left[(m/2)! \, ((n-1)/2)!\right]^2}$$

$$\text{if n is odd and m is even, } n \neq m$$

$$= \frac{1}{(m-n)(m+n+1)} \cdot \frac{(-1)^{(n+m+1)/2} \, n! \, m!}{2^{m+n-1}\left[(n/2)! \, ((m-1)/2)!\right]^2}$$

$$\text{if n is even and m is odd, } n \neq m \tag{3.161}$$

Setting $x_1 = x$ and $x_2 = 1$ in eq. (3.159) one obtains:

$$\int_{x}^{1} P_n \, P_m \, dx = \frac{1}{(m-n)(m+n+1)} \left\{ mP_n \, P_{m-1} - nP_m \, P_{n-1} - (m-n)\, xP_n \, P_m \right\} \tag{3.162}$$

which can be evaluated at $x = 0$ by using the results given in eq. (3.161) since:

$$\int_0^1 P_n\, P_m\, dx = -\int_{-1}^0 P_n\, P_m\, dx \qquad n + m = \text{odd},\ n \neq m \tag{3.163}$$

$$= 0 \qquad n + m = \text{even},\ n \neq m$$

The integral of x^m times the Legendre polynomial P_n vanishes if the integer m takes values in the range $0 \le m \le n - 1$. Using Rodrigues' formula in eq. (3.126):

$$\int_{-1}^{+1} x^m\, P_n\, dx = \frac{1}{2^n\, n!}\int_{-1}^{+1} x^m \frac{d^n\left(x^2-1\right)^n}{dx^n}\, dx$$

which, on integration by parts m times, one obtains:

$$\int_{-1}^{+1} x^m\, P_n\, dx = 0 \qquad m = 0, 1, 2, \ldots, n - 1 \tag{3.164}$$

The integral of products of powers of x and P_n can be evaluated by the use of Rodrigues' formula in eq. (3.126):

$$\int_0^1 x^m\, P_n\, dx = \frac{1}{2^n\, n!}\int_0^1 x^m \frac{d^n\left(x^2-1\right)^n}{dx^n}\, dx$$

Integration of the integral by parts n times results in the following expression:

$$\int_0^1 x^m\, P_n\, dx = \frac{m(m-1)(m-2)\ldots(m-n+2)}{(m+n+1)(m+n-1)\ldots(m-n+3)} \qquad m \ge n \tag{3.165}$$

The preceding integrals could be transformed to the θ coordinate since $P_n(\cos\theta)$ shows up in problems with spherical geometries. Thus, the orthogonality property in eq. (3.155) becomes:

$$\int_0^\pi P_n(\cos\theta)\, P_m(\cos\theta)\sin\theta\, d\theta = 0 \qquad n \neq m \tag{3.166}$$

$$= \frac{2}{2n+1} \qquad n = m$$

If $0 \le m \le n - 1$, then the integral in eq. (3.164) becomes:

$$\int_0^\pi P_n(\cos\theta)(\cos\theta)^m\, \sin\theta\, d\theta = 0 \qquad m = 0, 1, 2, \ldots, n-1 \tag{3.167}$$

Eq. (3.165) becomes after transformation:

$$\int_0^{\pi/2} P_n(\cos\theta)\, \cos^m\theta\, \sin\theta\, d\theta = \frac{m(m-1)\ldots(m-n+2)}{(m+n+1)(m+n-1)\ldots(m-n+3)} \qquad m \neq n \tag{3.168}$$

Using the trigonometric identity:

$$\sin(m\theta) \equiv \sin\theta\left[m\cos^{m-1}\theta - \frac{m(m-1)(m-2)}{3!}\sin^2\theta\cos^{m-3}\theta\right.$$

then one can evaluate the following integral:

$$\int_0^\pi P_n(\cos\theta)\sin(m\theta)\,d\theta = \int_0^\pi P_n(\cos\theta)\left[m\cos^{m-1}\theta - \frac{m(m-1)(m-2)}{3!}\sin^2\theta\cos^{m-3}\theta\right.$$

$$\left. + \frac{m(m-1)(m-2)(m-3)}{5!}\sin^4\theta\cos^{m-5}\theta - \ldots\right]\sin\theta\,d\theta$$

If $m \le n$, then the highest power of $\cos\theta$ is $n - 1$, thus using the integral of eq. (3.167), each term vanishes identically, such that:

$$\int_0^\pi P_n(\cos\theta)\sin(m\theta)\,d\theta = 0 \qquad m = 0, 1, 2, \ldots, n \qquad (3.169)$$

If $m > n$ and $m + n$ = even integer, then the integrand is an odd function in $(0, \pi)$, and hence, the following integral vanishes:

$$\int_0^\pi P_n(\cos\theta)\sin(m\theta)\,d\theta = 0 \qquad m + n = \text{even} \qquad (3.170)$$

If $m > n$ and $m + n$ = odd integer, then the integral becomes:

$$\int_0^\pi P_n(\cos\theta)\sin(m\theta)\,d\theta = 2\frac{(m-n+1)(m-n+3)\ldots(m+n-1)}{(m-n)(m-n+2)\ldots(m+n)} \qquad (3.171)$$

Similarly one can show that the integral:

$$\int_0^\pi P_n(\cos\theta)\cos(m\theta)\sin\theta\,d\theta =$$

$$\begin{aligned}
&= 0 && m = 0, 1, 2, \ldots, n-1 \\
&= 0 && m - n = \text{odd integer} \ge 0 \qquad (3.172)\\
&= \frac{-2}{(m-1)(m+1)} && n = 0,\ m = \text{even integer} \ge 0 \\
&= \frac{-2m(m-n+2)(m-n+4)\ldots(m+n-2)}{(m-n-1)(m-n+1)\ldots(m+n+1)} && n \ge 1,\ m - n = \text{even integer} \ge 0
\end{aligned}$$

The following integral can be evaluated by the use of the expression for $P_n(\cos\theta)$ in terms of $\cos m\theta$, given in eq. (3.132) as follows:

$$\int_0^{\pi} P_n(\cos\theta)\cos(m\theta)\,d\theta =$$

$$= 0 \qquad m < n$$

$$= 0 \qquad m + n = \text{odd}$$

$$= \frac{\Gamma(m+k+1/2)\Gamma(k+1/2)}{\Gamma(k+1)\Gamma(m+k+1)} \qquad n = m + 2k,\ k = 0, 1, 2,... \tag{3.173}$$

The following integral can be obtained by using the integral in eq. (3.173):

$$\int_0^{\pi} P_n(\cos\theta)\sin m\theta \sin\theta\,d\theta = \frac{1}{2}\int_0^{\pi} P_n(\cos\theta)[\cos(m-1)\theta - \cos(m+1)\theta]d\theta$$

$$= 0 \qquad m > n + 1$$

$$= 0 \qquad n - m = 0 \text{ or an even integer}$$

$$= -\frac{m}{4}\frac{\Gamma(m+k-1/2)\Gamma(k-1/2)}{\Gamma(k+1)\Gamma(m+k+1)} \qquad n - m = 2k - 1,\ k = 0, 1, 2,... \tag{3.174}$$

Integrals involving products of derivatives of Legendre polynomials can be evaluated. Starting with the integral:

$$\int_{-1}^{+1}(1-x^2)P'_n P'_m\,dx = (1-x^2)P'_n P'_m\Big|_{-1}^{+1} - \int_{-1}^{+1}P_m\{(1-x^2)P'_n\}'\,dx$$

$$= n(n+1)\int_{-1}^{+1}P_m P_n\,dx = 0 \qquad n \neq m$$

$$= \frac{2n(n+1)}{2n+1} \qquad n = m \tag{3.175}$$

The preceding integral is an orthogonality relationship for P'_n.

3.21 Expansions of Functions in Terms of Legendre Polynomials

The first function that can be expanded in finite series of Legendre polynomials is $P_n(x)$. Starting with the recurrence formula eq. (3.138) for n, n-2, n-4,..., one gets:

$$n\,P_n = (2n-1)\,x\,P_n - (n-1)P_{n-2}$$

$$(n-2)P_{n-2} = (2n-5)\,x\,P_{n-3} - (n-3)\,P_{n-4}$$

$$(n-4)\,P_{n-4} = (2n-9)\,x\,P_{n-5} - (n-5)\,P_{n-6}$$

Thus, substituting P_{n-2}, P_{n-4}, into the expression for P_n, one obtains:

$$P_n = x\left[\frac{(2n-1)}{n}P_{n-1} - \frac{(n-1)}{n(n-2)}(2n-5)P_{n-3} + \frac{(n-1)(n-3)}{n(n-2)(n-4)}(2n-9)P_{n-5} - \ldots\right] \tag{3.176}$$

Using the recurrence formula eq. (3.139) for P'_n:

$$P'_n = P'_{n-2} + (2n-1)P_{n-1}$$

$$P'_{n-2} = P'_{n-4} + (2n-5)P_{n-3}$$

$$P'_{n-4} = P'_{n-6} + (2n-9)P_{n-5}$$

and substituting for P'_{n-2}, P'_{n-4},..., one obtains the following finite series for P'_n:

$$P'_n = (2n-1)P_{n-1} + (2n-5)P_{n-3} + (2n-9)P_{n-5} + \ldots \tag{3.177}$$

A different expansion for P'_n can be developed from the recurrence formula eq. (3.140):

$$x\,P'_n = x\,P'_{n-2} + n\,P_n + (n-1)\,P_{n-2}$$

$$x\,P'_{n-2} = x\,P'_{n-4} + (n-2)\,P_{n-2} + (n-3)\,P_{n-4}$$

$$x\,P'_{n-4} = x\,P'_{n-6} + (n-4)\,P_{n-4} + (n-5)\,P_{n-6}$$

Thus, a finite expansion for $x\,P'_n$ results:

$$x\,P'_n = n\,P_n + (2n-3)P_{n-2} + (2n-7)P_{n-4} + \ldots \tag{3.178}$$

Differentiating eq. (3.177) and substituting for P'_{n-1}, P'_{n-2},..., from eq. (3.177) one obtains an expansion for P''_n, having the following form:

$$\begin{aligned} P''_n = {} & (2n-3)(2n-1\cdot 1)P_{n-2} + (2n-7)(4n-2\cdot 3)P_{n-4} \\ & + (2n-11)(6n-3\cdot 5)P_{n-6} + \ldots \end{aligned} \tag{3.179}$$

Using the recurrence formula given in eq. (3.138):

$$(2n+1)\,x\,P_n(x) = (n+1)\,P_{n+1}(x) + n\,P_{n-1}(x)$$

$$(2n+1)\,y\,P_n(y) = (n+1)\,P_{n+1}(y) + n\,P_{n-1}(y)$$

and multiplying the first equation by $P_n(y)$ and the second by $P_n(x)$, and subtracting the resulting equalities, one obtains:

$$\begin{aligned} (2n+1)(x-y)P_n(x)P_n(y) = {} & (n+1)\left[P_n(y)P_{n+1}(x) - P_n(x)P_{n+1}(y)\right] \\ & + n\left[P_n(y)P_{n-1}(x) - P_n(x)P_{n-1}(y)\right] \end{aligned}$$

Thus, summing this equation N times, there results:

$$\begin{aligned} (x-y)\sum_{n=0}^{N}(2n+1)P_n(x)P_n(y) = {} & \\ = \sum_{n=0}^{N}(n+1)\left[P_n(y)P_{n+1}(x) - P_n(x)P_{n+1}(y)\right] & - n\left[P_{n-1}(y)P_n(x) - P_{n-1}(x)P_n(y)\right] \\ = (N+1)\left[P_N(y)P_{N+1}(x) - P_N(x)P_{N+1}(y)\right] & \end{aligned} \tag{3.180}$$

The preceding summation formula is known as **Christoffel's First Summation**.

To obtain an expansion in terms of squares of Legendre polynomials, the form given in (3.180) for x = y gives a trivial identity. Dividing eq. (3.180) by x - y and taking the limit as $y \to x$, one obtains:

$$\sum_{n=0}^{N} (2n+1)\, P_n^2(x) = (N+1) \lim_{y \to x} \frac{P_N(y)\, P_{N+1}(x) - P_N(x)\, P_{N+1}(y)}{x-y}$$

$$= (N+1)\left[P_N(x)\, P'_{N+1}(x) - P'_N(x)\, P_{N+1}(x)\right] \tag{3.181}$$

Since Legendre polynomials $P_n(x)$ are polynomials of degree n, then it stands to reason that one can obtain a finite sum of a finite number of Legendre polynomials to give x^m. Expanding x^m into an infinite series:

$$x^m = \sum_{k=0}^{\infty} a_k\, P_k(x)$$

then multiplying both sides by $P_l(x)$ and integrating both sides, one obtains:

$$a_l = \frac{2l+1}{2} \int_{-1}^{+1} x^m\, P_l(x)\, dx \qquad l = 0, 1, 2, \ldots \tag{3.182}$$

Examination of the preceding integral, shows that the constants a_l for $l \le m$ do not vanish, while $a_l = 0$ for $l > m$ (see 3.164). If m - l is an odd integer in eq. (3.182), then the integrand is an odd function of x, then:

$$a_l = 0 \qquad \text{if } m - l = \text{odd integer}$$

If m - l is an even integer, then using eq. (3.165) one obtains:

$$a_l = \frac{2l+1}{2} \int_{-1}^{+1} x^m\, P_l(x)\, dx = (2l+1) \int_{0}^{1} x^m\, P_l(x)\, dx$$

$$= (2l+1) \cdot \frac{m(m-1)(m-2)\ldots(m-l+2)}{(m+l+1)(m+l-1)\ldots(m-l+3)} \tag{3.183}$$

From the preceding argument, it is obvious that only the Legendre polynomials P_m, P_{m-2}, P_{m-4},..., do enter into the expansion of x^m. Thus:

$$x^m = \frac{m!}{1 \cdot 3 \cdot 5 \ldots (2m+1)} \left\{ (2m+1) P_m + (2m-3) \frac{(2m+1)}{2 \cdot 1!} P_{m-2} \right.$$

$$\left. + (2m-7) \frac{(2m+1)(2m-1)}{2^2 \cdot 2!} P_{m-4} + (2m-11) \frac{(2m+1)(2m-1)(2m-3)}{2^3 \cdot 3!} P_{m-6} + \ldots \right\} \tag{3.184}$$

The first few expansions are listed below:

$$1 = P_0, \qquad x = P_1, \qquad x^2 = \frac{2}{3} P_2 + \frac{1}{3} P_0$$

$$x^3 = \frac{2}{5} P_3 + \frac{3}{5} P_1, \qquad x^4 = \frac{8}{35} P_4 + \frac{4}{7} P_2 + \frac{1}{5} P_0 \tag{3.185}$$

Expansions of functions in terms of θ instead of x can be formulated from the definition of $P_n(\cos\theta)$ and from the integrals developed in Section 3.20. One can first start by getting an expansion of $P_n(\cos\theta)$ in terms of Fourier sine series, in the region of $0 < \theta < \pi$, of the following form:

$$P_n(\cos\theta) = \sum_{k=1}^{\infty} a_k \sin k\theta$$

Multiplying the preceding expansion by sin mθ and integrating the resulting expression on $(0,\pi)$, one obtains:

$$a_m = \frac{2}{\pi} \int_0^{\pi} P_n(\cos\theta) \sin m\theta \, d\theta \qquad m = 1, 2, \ldots$$

Examination of the preceding integral and the integrals in eqs. (3.169) to (3.171) shows that:

$$\begin{aligned} a_m &= 0 && m \le n \\ &= 0 && m - n = \text{even integer} \\ &= \frac{4}{\pi} \frac{(m-n+1)(m-n+3)\ldots(m+n-1)}{(m-n)(m-n+2)\ldots(m+n)} && m \ge n+1 \text{ and } m+n = \text{odd integer} \end{aligned}$$

Thus, one obtains an expansion of Legendre polynomial in terms of sine arguments:

$$P_n(\cos\theta) = \frac{2^{2n+2}}{\pi} \frac{(n!)^2}{(2n+1)!} \left[\sin(n+1)\theta + \frac{1}{1!} \frac{(n+1)}{2n+3} \sin(n+3)\theta + \ldots \right] \qquad 0 < \theta < \pi \tag{3.186}$$

Expansion of cos mθ in an infinite series of $P_n(\cos\theta)$ can be developed from the integrals in eqs. (3.166) and (3.172). Assuming an expansion for cos mθ of the following form:

$$\cos(m\theta) = \sum_{k=0}^{\infty} a_k P_k(\cos\theta)$$

and multiplying both sides of the equality by $P_r(\cos\theta) \sin\theta$, integrating both sides on $(0,\pi)$ and using eq. (3.166), one obtains an expression for the constants of expansion a_r as follows:

$$a_r = \frac{2r+1}{2} \int_0^{\pi} P_r(\cos\theta) \cos m\theta \sin\theta \, d\theta$$

Using the integrals developed in eq. (3.172) one obtains:

$$\begin{aligned} a_r &= 0 && r > m \\ &= 0 && m - r = \text{odd integer} \end{aligned}$$

$$= -\frac{1}{(m-1)(m+1)} \qquad r = 0 \text{ and } m = \text{even integer}$$

$$= -(2r+1)\, m \frac{(m-r+2)(m-r+4)\ldots(m+r-2)}{(m-r-1)(m-r+1)\ldots(m+r+1)} \qquad m - r = \text{even integer}, r \geq 1$$

Thus:

$$\cos(m\theta) = \frac{2^{2m-1}(m!)^2}{(2m+1)!}\left\{(2m+1)P_m + (2m-3)\frac{(-1)}{2}\frac{2m+1}{2m-2}P_{m-2} + \right.$$

$$+(2m-7)\frac{(-1)\cdot 1}{2\cdot 4}\frac{(2m+1)(2m-1)}{(2m-2)(2m-4)}P_{m-4} +$$

$$\left. +(2m-11)\frac{(-1)\cdot 1\cdot 3}{2\cdot 4\cdot 6}\frac{(2m+1)(2m-1)(2m-3)}{(2m-2)(2m-4)(2m-6)}P_{m-6} + \ldots\right\}$$

$$m = 1, 2, 3 \ldots \qquad (3.187)$$

and

$$1 = P_0 \qquad m = 0$$

The first few expansions of cos (mθ) in terms of Legendre polynomials are listed below:

$$1 = P_0 \qquad \cos\theta = P_1 \qquad \cos(2\theta) = \frac{4}{3}\left(P_2 - \frac{1}{4}P_0\right)$$

$$\cos(3\theta) = \frac{8}{5}\left(P_3 - \frac{3}{8}P_1\right) \qquad \cos(4\theta) = \frac{64}{35}\left(P_4 - \frac{5}{12}P_2 - \frac{7}{192}P_0\right)$$

The development of an expansion of sin (mθ) follows a similar procedure to that of cos (mθ). Expanding sin mθ in an infinite series, one can show that:

$$\sin m\theta = -\frac{m}{8}\sum_{k=0}^{\infty}\frac{(2m+4k-1)\,\Gamma(m+k-1/2)\Gamma(k-1/2)}{k!\,(m+k)!}P_{m+2k-1}(\cos\theta) \qquad (3.188)$$

3.22 Legendre Function of the Second Kind $Q_n(x)$

The Legendre functions of the second kind $Q_n(x)$ were developed in Section 3.16 in the two regions $|x| < 1$ and $|x| > 1$. The infinite series expansions for $Q_n(x)$ given in eq. (3.123) and eq. (3.124) are limited to the region $|x| < 1$, while the infinite series expansion given in eq. (3.129) is limited to the region $|x| > 1$. A more convenient closed form for $Q_n(x)$, valid in the region $|x| < 1$, was given in eq. (3.128). Since the expression for $Q_n(x)$ in eq. (3.128) has a logarithmic term in addition to a polynomial of degree (n - 1), one can replace the summation terms by a series of $P_k(x)$, k = 0 to n - 1, as can be seen from eq. (3.184). Starting with the expression in eq. (3.128):

$$Q_n(x) = \frac{1}{2} P_n(x) \log\frac{1+x}{1-x} - W_{n-1}$$

and substituting $Q_n(x)$ into the differential equation (3.118), one obtains after simplification:

$$\frac{d}{dx}\left\{(1-x^2)\frac{d\,W_{n-1}}{dx}\right\} + n(n+1)\,W_{n-1} = 2\frac{d\,P_n}{dx} \tag{3.189}$$

Using the expansion for $(d\,P_n)/(dx)$ from eq. (3.177), the right side of eq. (3.189) becomes:

$$2[(2n-1)\,P_{n-1} + (2n-5)\,P_{n-3} + (2n-9)\,P_{n-5} + \ldots]$$

Assuming that:

$$W_{n-1} = \sum_{k=0}^{k \le (n-1)/2} a_k\,P_{n-1-2k}$$

and substituting W_{n-1} into eq. (3.189) and equating the coefficients of P_k, one obtains an expression for a_k as follows:

$$a_k = \frac{2n-4k-1}{(n-k)(2k+1)} \qquad k = 0, 1, 2, .., k \le (n-1)/2$$

Thus, the function W_{n-1} can be expressed in terms of a finite series of Legendre polynomials as:

$$W_{n-1} = \frac{2n-1}{1\cdot n} P_{n-1} + \frac{2n-5}{3\cdot(n-1)} P_{n-3} + \frac{2n-7}{5\cdot(n-2)} P_{n-5} + \ldots \tag{3.190}$$

A formula, similar to Rodrigues' formula for $P_n(x)$, can be developed for $Q_n(x)$. Starting with the binomial expansion of $(x^2 - 1)$, one obtains:

$$\frac{1}{(x^2-1)^{n+1}} = \frac{1}{x^{2n+2}} + \frac{n+1}{1!}\frac{1}{x^{2n-4}} + \frac{n+2}{2!}\frac{1}{x^{2n-6}} + \ldots$$

Integrating the preceding series n + 1 times, the following expression results:

$$\int_x^\infty\int_\eta^\infty\int_\eta^\infty \ldots \int_\eta^\infty \frac{(d\eta)^{n+1}}{(\eta^2-1)^{n+1}} = \frac{1}{(n+1)(n+2)\ldots(2n-1)(2n)(2n+1)}$$

$$\left[x^{-n-1} + \frac{(n+1)(n+2)}{2(2n+3)} x^{-n-3} + \frac{(n+1)(n+2)(n+3)(n+4)}{2\cdot 4\cdot(2n+3)(2n+5)} x^{-n-5} + \ldots\right]$$

Comparison of the preceding infinite series with the series expansion for $Q_n(x)$ for $|x| > 1$ in eq. (3.129) results in the following form for $Q_n(x)$:

$$Q_n(x) = \frac{n!(n+1)(n+2)..(2n-1)(2n)(2n+1)}{(2n+1)(2n-1)...5\cdot3\cdot1} \int_x^\infty \int_\eta^\infty .. \int_\eta^\infty \frac{(d\eta)^{n+1}}{(\eta^2-1)^{n+1}}$$

$$= 2^n n! \int_x^\infty \int_\eta^\infty .. \int_\eta^\infty \frac{(d\eta)^{n+1}}{(\eta^2-1)^{n+1}} \qquad (3.191)$$

Another expression for $Q_n(x)$ that is similar to the one given in eq. (3.191) can be developed from the solution to the following differential equation:

$$\left(1-x^2\right)\frac{d^2u}{dx^2} + 2(n-1)x\frac{du}{dx} + 2nu = 0 \qquad (3.192)$$

one of its solutions being:

$$u_1 = \left(x^2-1\right)^n$$

The second solution of eq. (3.192) can be obtained from $u_1(x)$ by multiplication of $u_1(x)$ by an unknown function v(x) as follows:

$$u_2 = v(x)\left(x^2-1\right)^n$$

Then, the unknown function v satisfies the following differential equation:

$$\frac{v''}{v'} = -\frac{2(n+1)x}{x^2-1}$$

which can be integrated to give:

$$v = \int_x^\infty \frac{d\eta}{\left(\eta^2-1\right)^{n+1}}$$

so that the second solution is given by:

$$u_2 = \left(x^2-1\right)^n \int_x^\infty \frac{d\eta}{\left(\eta^2-1\right)^{n+1}}$$

Differentiating eq. (3.192) n times, then the resulting differential equation becomes:

$$\left(1-x^2\right)\frac{d^{n+2}u}{dx^{n+2}} - 2x\frac{d^{n+1}u}{dx^{n+1}} + n(n+1)\frac{d^n u}{dx^n} = 0$$

which is the Legendre differential equation on $(d^n u)/(dx^n)$, having the solution $P_n(x)$ and $Q_n(x)$. Thus, the solutions $P_n(x)$ and $Q_n(x)$ can be written in the following form:

$$P_n(x) = \frac{1}{2^n n!}\frac{d^n u_1}{dx^n} = \frac{1}{2^n n!}\frac{d^n}{dx^n}\left(x^2-1\right)^n$$

and

$$Q_n(x) = \frac{(-1)^n 2^n n!}{(2n)!}\frac{d^n u_2}{dx^n} = \frac{(-1)^n 2^n n!}{(2n)!}\frac{d^n}{dx^n}\left\{\left(x^2-1\right)^n \int_x^\infty \frac{d\eta}{\left(\eta^2-1\right)^{n+1}}\right\} \quad |x|>1 \qquad (3.193)$$

The constants were adjusted such that $u_1^{(n)}$ and $u_2^{(n)}$ become P_n and Q_n, respectively.

An integral for $Q_n(x)$ valid in $|x| < 1$ can be obtained from eq. (3.193), resulting in the following integral:

$$Q_n(x) = \frac{(-1)^n 2^n n!}{(2n)!} \frac{d^n}{dx^n} \left\{ (1-x^2)^n \int_0^x \frac{d\eta}{(1-\eta^2)^{n+1}} \right\} \qquad |x| < 1 \qquad (3.194)$$

A generating function representation can be formulated from the following binomial expansion:

$$\frac{1}{x-t} = \frac{1}{x} + \frac{t}{x^2} + \frac{t^2}{x^3} + \ldots + \frac{t^n}{x^{n+1}} + \ldots \qquad \text{valid for } \left[\frac{t}{x}\right] < 1$$

Substituting for t^n by a series of Legendre polynomials having the form: (See eq. 3.184)

$$t^n = \frac{2^n (n!)^2}{(2n+1)!} \left\{ (2n+1)\, P_n(t) + (2n-3) \frac{2n+1}{2} P_{n-2}(t) + \right.$$

$$\left. + (2n-7) \frac{(2n+1)(2n-1)}{2 \cdot 4} P_{n-4}(t) + \ldots \right\}$$

Then:

$$\frac{1}{x-t} = \frac{P_0}{x} + \frac{P_1}{x^2} + \frac{1}{x^3}\left[\frac{2}{3} P_2 + \frac{1}{3} P_0\right] + \frac{1}{x^4}\left[\frac{2}{5} P_3 + \frac{3}{5} P_1\right] + \frac{1}{x^5}\left[\frac{8}{35} P_4 + \frac{4}{7} P_2 + \frac{1}{5} P_0\right]$$

$$+ \ldots + \frac{1}{x^{n+1}} \left\{ \frac{2^n (n!)^2}{(2n+1)!} \left[(2n+1)\, P_n + (2n-3) \frac{(2n+1)}{2} P_{n-2} + \ldots \right] \right\} + \ldots$$

Collecting the terms that multiply $P_0, P_1, P_2, \ldots, P_n$, then the coefficient of P_n becomes:

$$\frac{(2n+1)\, 2^n (n!)^2}{(2n+1)!} \left[x^{-n-1} + \frac{(n+1)(n+2)}{2 \cdot (2n+3)} x^{-n-3} + \ldots \right] = (2n+1)\, Q_n(x)$$

Thus:

$$\frac{1}{x-t} = \sum_{n=0}^{\infty} (2n+1)\, P_n(t)\, Q_n(x) \qquad |x| > 1 \qquad (3.195)$$

The expansion given in eq. (3.195) leads to an integral representation for $Q_n(x)$. Multiplying both sides by $P_m(t)$ and integrating on (-1, 1) one obtains:

$$Q_n(x) = \frac{1}{2} \int_{-1}^{+1} \frac{P_n(t)}{x-t}\, dt \qquad |x| > 1 \qquad (3.196)$$

The last integral is known as the **Neumann Integral**.

3.23 Associated Legendre Functions

Associated Legendre functions are solutions to the following differential equation:

$$\left(1-x^2\right)\frac{d^2y}{dx^2}-2x\frac{dy}{dx}+\left[n(n+1)-\frac{m^2}{1-x^2}\right]y=0 \tag{3.197}$$

where m and n are both integers.

Substituting:

$$y=\left(x^2-1\right)^{m/2}u$$

in eq. (3.197) results in a new differential equation:

$$\left(1-x^2\right)\frac{d^2u}{dx^2}-2(m+1)\,x\frac{du}{dx}+(n-m)(n+m+1)\,u=0 \tag{3.198}$$

Differentiating Legendre's eq. (3.118) m times, one obtains:

$$\left(1-x^2\right)\frac{d^{m+2}y}{dx^{m+2}}-2(m+1)\,x\frac{d^{m+1}y}{dx^{m+1}}+(n-m)(n+m+1)\frac{d^my}{dx^m}=0 \tag{3.199}$$

Equations (3.198) and (3.199) are identical, thus, the solutions of eq. (3.198) are the mth derivative of the solutions of eq. (3.118). Thus, the solution of eq. (3.197) becomes:

$$y=\left(x^2-1\right)^{m/2}\left[C_1\frac{d^mP_n}{dx^m}+C_2\frac{d^mQ_n}{dx^m}\right]$$

Define:

$$P_n^m=\left(x^2-1\right)^{m/2}\frac{d^mP_n}{dx^m} \qquad |x|>1 \tag{3.200}$$

and

$$Q_n^m=\left(x^2-1\right)^{m/2}\frac{d^mQ_n}{dx^m} \qquad |x|>1 \tag{3.201}$$

as the **associated Legendre functions of the first and second kind of degree n and order m,** respectively.

Define:

$$T_n^m=(-1)^m\left(1-x^2\right)^{m/2}\frac{d^mP_n}{dx^m} \qquad |x|<1 \tag{3.202}$$

as **Ferrer's function of the first kind of degree n and order m**. It may be convenient to define P_n^m and Q_n^m in the region $|x|<1$ as follows:

$$P_n^m=T_n^m \qquad |x|<1$$

$$Q_n^m=(-1)^m\left(1-x^2\right)^{m/2}\frac{d^mQ_n}{dx^m} \qquad |x|<1 \tag{3.203}$$

Using the expression for $P_n(x)$ given by Rodrigues' formula, then:

$$P_n^m = \frac{\left(x^2-1\right)^{m/2}}{2^n n!} \frac{d^{m+n}}{dx^{m+n}} \left(x^2-1\right)^n \qquad |x|>1 \tag{3.204}$$

$$= \frac{(2n)!}{2^n n!\,(n-m-1)!} \left(x^2-1\right)^{m/2} \left[x^{n-m} - \frac{(n-m)(n-m-1)}{2(2n-1)} x^{n-m-2} + \frac{(n-m)(n-m-1)(n-m-2)(n-m-3)}{2\cdot 4\cdot (2n-1)(2n-3)} x^{n-m-4} - \ldots \right] \tag{3.205}$$

It can be seen from eq. (3.205) that $P_n^m = 0$ if $m \geq n - 1$. The terms contained in the brackets of eq. (3.214) represent a finite polynomial of degree n - m.

A listing of the first few Associated Legendre functions is given below:

$P_1^1 = \left(x^2-1\right)^{1/2}$ $\qquad P_1^2 = 0$ $\qquad P_1^3 = 0$

$P_2^1 = 3x\left(x^2-1\right)^{1/2}$ $\qquad P_2^2 = 3\left(x^2-1\right)$ $\qquad P_2^3 = 0$

$P_3^1 = \frac{3}{2}\left(5x^2-1\right)\left(x^2-1\right)^{1/2}$ $\qquad P_3^2 = 15x\left(x^2-1\right)$ $\qquad P_3^3 = 15x\left(x^2-1\right)^{3/2}$

$P_4^1 = \frac{5}{2}\left(7x^3-3x\right)\left(x^2-1\right)^{1/2}$ $\qquad P_4^2 = \frac{15}{2}\left(7x^2-1\right)\left(x^2-1\right)$ $\qquad P_4^3 = 105x\left(x^2-1\right)^{3/2}$

If n = m, then:

$$P_n^n = \frac{(2n)!}{2^n n!} \left(x^2-1\right)^{n/2} \qquad |x|>1 \tag{3.206}$$

Another expression for P_n^m, similar to eq. (3.200), can be developed in the form:

$$P_n^m = \frac{1}{2^n (n-m)!} \left(\frac{x-1}{x+1}\right)^{m/2} \frac{d^n}{dx^n} \left[(x-1)^{n-m} (x+1)^{n+m}\right] \qquad |x|>1$$

$$= \frac{(-1)^m}{2^n (n-m)!} \left(\frac{1-x}{1+x}\right)^{m/2} \frac{d^n}{dx^n} \left[(x-1)^{n-m} (x+1)^{n+m}\right] \qquad |x|<1 \tag{3.207}$$

3.24 Generating Function for Associated Legendre Functions

Using the generating function for $P_m(x)$ given in eq. (3.131)

$$\frac{1}{\left(1-2xt+t^2\right)^{1/2}} = \sum_{n=0}^{\infty} t^n P_n(x)$$

and differentiating the equality m times, one obtains:

$$\frac{1\cdot 3\cdot 5\cdot \ldots (2m-1)\, t^m}{\left(1-2xt+t^2\right)^{m+1/2}} = \sum_{n=m}^{\infty} t^n \frac{d^m P_n}{dx^m}$$

or

$$\frac{1}{\left(1-2xt+t^2\right)^{m+1/2}} = (-1)^m \frac{2^m m!}{(2m)!} \left(1-x^2\right)^{-m/2} \sum_{n=m}^{\infty} t^{n-m} P_n^m(x) \qquad |x| < 1 \tag{3.208}$$

3.25 Recurrence Formulae for P_n^m

Recurrence formulae for P_n^m and Q_n^m can be developed from those for P_n and Q_n. Starting with eq. (3.197) and noting the definition for P_n^m in eq. (3.200), then eq. (3.200) becomes:

$$\left(1-x^2\right)\left\{\left(x^2-1\right)^{-(m+2)/2} P_n^{m+2}\right\} - 2(m+1)x\left\{\left(x^2-1\right)^{-(m+1)/2} P_n^{m+1}\right\}$$

$$+ (n-m)(n+m+1)\left\{\left(x^2-1\right)^{-m/2} P_n^m\right\} = 0$$

or

$$P_n^{m+2} + \frac{2(m+1)x}{\sqrt{x^2-1}} P_n^{m+1} - (n-m)(n+m+1) P_n^m = 0 \qquad |x| > 1 \tag{3.209}$$

and

$$P_n^{m+2} - \frac{2(m+1)x}{\sqrt{1-x^2}} P_n^{m+1} + (n-m)(n+m+1) P_n^m = 0 \qquad |x| < 1$$

which relates associated Legendre functions of different orders.

Differentiating the recurrence formula on P_n, given in eq. (3.138) m times and differentiating eq. (3.139) (m - 1) times, results in a recurrence formula relating the Associated Legendre functions of different degrees, which has the form:

$$(n-m+1) P_{n+1}^m - (2n+1)\, x\, P_n^m + (n+m) P_{n-1}^m = 0 \qquad \text{for all } x \tag{3.210}$$

Differentiating and then multiplying eq. (3.139) by $(x^2 - 1)^{m/2}$, one obtains:

$$P_{n+1}^m - P_{n-1}^m = (2n+1)\left(x^2-1\right)^{1/2} P_n^{m-1} \qquad |x| > 1 \tag{3.211}$$

$$= -(2n+1)\left(1-x^2\right)^{1/2} P_n^{m-1} \qquad |x| < 1$$

Other recurrence formulae are listed below for completeness:

$$(2n+1)\sqrt{1-x^2}\, P_n^m = (n+m)(n+m-1) P_{n-1}^{m-1} - (n-m+1)(n-m+2) P_{n+1}^{m+1} \qquad |x| < 1 \tag{3.212}$$

$$\left(x^2-1\right)\frac{dP_n^m}{dx} = nx\, P_n^m - (n+m) P_{n-1}^m \tag{3.213}$$

$$\left(x^2-1\right)\frac{dP_n^m}{dx} = -(n+1)\, x\, P_n^m + (n-m+1) P_{n+1}^m \tag{3.214}$$

$$\sqrt{x^2-1}\; P_{n+1}^{m+1} = x(n-m+1)P_{n+1}^{m} - (n+m+1)\, P_n^m \qquad |x|>1 \qquad (3.215)$$

$$(n+m)\sqrt{x^2-1}\; P_n^{m-1} = P_{n+1}^{m} - x\, P_n^m \qquad |x|>1 \qquad (3.216)$$

$$(n-m+1)\sqrt{x^2-1}\; P_n^{m-1} = x\, P_n^{m} - P_{n-1}^m \qquad |x|>1 \qquad (3.217)$$

More recurrence formulae can be found in Prasad, Volume II.

3.26 Integrals of Associated Legendre Functions

Integrals of products of associated Legendre functions are presented in this section. Starting with the differential equation that associated Legendre functions of different degrees and the same order P_n^m and P_r^m satisfy, and multiplying the first equation by P_r^m, the second equation by P_n^m, subtracting the resulting equations and integrating the resulting equation on (-1, +1), one obtains:

$$[r(r+1)-n(n+1)]\int_{-1}^{+1} P_r^m P_n^m\, dx =$$

$$= \int_{-1}^{+1}\left\{P_r^m \frac{d}{dx}\left[(1-x^2)\frac{dP_n^m}{dx}\right] - P_n^m \frac{d}{dx}\left[(1-x^2)\frac{dP_r^m}{dx}\right]\right\} dx$$

$$= (1-x^2)\left[P_r^m \frac{dP_n^m}{dx} - P_n^m \frac{dP_r^m}{dx}\right]_{-1}^{+1} = 0 \qquad n \neq r$$

Starting with the differential equations that associated Legendre of the same degree and different orders P_n^m and P_n^k satisfy, and multiplying the first equation by P_n^k, the second equation by P_n^m, subtracting the resulting equations and integrating the resultant equality, one obtains:

$$(m^2-k^2)\int_{-1}^{+1} \frac{P_n^m P_n^k}{1-x^2}\, dx = \int_{-1}^{+1}\left\{P_n^k \frac{d}{dx}\left[(1-x^2)\frac{dP_n^m}{dx}\right] - P_n^m \frac{d}{dx}\left[(1-x^2)\frac{dP_n^k}{dx}\right]\right\} dx = 0$$

$$m \neq k$$

The integral of squares of associated Legendre functions can be obtained by using the definition of P_n^m.

$$\int_{-1}^{+1} (P_n^m)^2\, dx = \frac{(n+m)!}{(n-m)!}\int_{-1}^{+1} P_n^m P_n^{-m}\, dx$$

$$= \frac{(n+m)!}{(n-m)!}\frac{1}{2^{2n}(n!)^2}\int_{-1}^{+1} \frac{d^{n+m}}{dx^{n+m}}(x^2-1)^n \frac{d^{n-m}}{dx^{n-m}}(x^2-1)^n\, dx$$

Integrating the last integral by parts m times gives:

$$= (1)^m \frac{(n+m)!}{(n-m)!} \frac{2}{2n+1}$$

Summarizing the results of these integrals:

$$\int_{-1}^{+1} P_r^m P_n^m \, dx = 0 \qquad r \neq n \qquad (3.218)$$

$$\int_{-1}^{+1} \frac{P_n^m P_n^k}{1-x^2} \, dx = 0 \qquad k \neq m \qquad (3.219)$$

$$\int_{-1}^{+1} \left(P_n^m\right)^2 dx = (-1)^m \frac{(n+m)!}{(n-m)!} \frac{2}{2n+1} \qquad (3.220)$$

It can be shown that Ferrer's functions give the following integral:

$$\int_{-1}^{+1} \left(T_n^m\right)^2 dx = \frac{(n+m)!}{(n-m)!} \frac{2}{2n+1} \qquad (3.221)$$

3.27 Associated Legendre Function of the Second Kind Q_n^m

The Associated Legendre functions of the second kind Q_n^m can be derived from the definition given in eq. (3.120) as follows:

$$Q_n^m = \left(x^2-1\right)^{m/2} \frac{d^m Q_n}{dx^m} = (-1)^m \frac{2^n n! (n+m)!}{(2n+1)} \left(x^2-1\right)^{m/2}$$

$$\cdot \left\{ x^{-n-m-1} + \frac{(n+m+1)(n+m+2)}{2 \cdot (2n+3)} x^{-n-m-3} \right.$$

$$\left. + \frac{(n+m+1)(n+m+2)(n+m+3)(n+m+4)}{2 \cdot 4 \cdot (2n+3)(2n+5)} x^{-n-m-5} + \ldots \right\} \qquad |x| > 1 \qquad (3.222)$$

Since $Q_n(x)$ was defined by an integral on $P_n(t)$ given in eq. (3.196), then differentiating eq. (3.196) m times results in an integral definition for Q_n^m as follows:

$$Q_n^m = (-1)^m \frac{m!}{2} \left(x^2-1\right)^{m/2} \int_{-1}^{+1} \frac{P_n(t)}{(x-t)^{m+1}} \, dt \qquad |x| > 1 \qquad (3.223)$$

The definition of Q_n^m in eq. (3.223) can be utilized to advantage when recurrence formulae for Q_n^m are to be developed. The recurrence formulae developed for P_n^m in Section 3.25 turn out to be valid for Q_n^m also.

Using the definition of P_n^m in eq. (3.223)

PROBLEMS

Section 3.3

1. Show that the definition for $Y_n(x)$, as defined by eq. (3.12), results in the same expression given in eq. (3.11).

Section 3.4

2. Using the expressions for the Wronskian and the recurrence formulae, prove that:

(a) $J_p Y_{p+1} - J_{p+1} Y_p = -\dfrac{2}{\pi x}$

(b) $J_p J''_{-p} - J_{-p} J''_p = +\dfrac{2\sin p\pi}{\pi x^2}$

(c) $\displaystyle\int \frac{dx}{x J_p^2} = -\frac{\pi}{2\sin p\pi}\frac{J_{-p}(x)}{J_p(x)}$

(d) $\displaystyle\int \frac{dx}{x J_p J_{-p}} = -\frac{\pi}{2\sin p\pi}\log\frac{J_{-p}(x)}{J_p(x)}$

(e) $\displaystyle\int \frac{dx}{x J_p^2} = \frac{\pi}{2}\frac{Y_p(x)}{J_p(x)}$

(f) $\displaystyle\int \frac{dx}{x J_p Y_p} = \frac{\pi}{2}\log\frac{Y_p(x)}{J_p(x)}$

(g) $\displaystyle\int \frac{dx}{x Y_p^2} = -\frac{\pi}{2}\frac{J_p(x)}{Y_p(x)}$

(h) Equations (3.103) and (3.104)

Section 3.6

3. Show that:

$$j_n(x) = \frac{1}{x}\left[\sin\left(x - \frac{n\pi}{2}\right)\sum_{m=0}^{m \le n/2} (-1)^m \frac{(n+2m)!}{(2m)!(n-2m)!(2x)^{2m}} - \cos\left(x - \frac{n\pi}{2}\right)\sum_{m=0}^{m \le \frac{1}{2}(n-1)} (-1)^m \frac{(n+2m+1)!}{(2m+1)!(n-2m-1)!(2x)^{2m+1}}\right]$$

(Hint: Use the form given in (3.30).)

4. Show that:

$$y_n(x) = \frac{(-1)^{n+1}}{x}\left[\cos\left(x + \frac{n\pi}{2}\right)\sum_{m=0}^{m \le n/2} (-1)^m \frac{(n+2m)!}{(2m)!(n-2m)!(2x)^{2m}}\right.$$

$$\left. -\sin\left(x+\frac{n\pi}{2}\right) \sum_{m=0}^{m\leq \frac{1}{2}(n-1)} (-1)^m \frac{(n+2m+1)!}{(2m+1)!\,(n-2m-1)!\,(2x)^{2m+1}} \right]$$

(Hint: Use the form given in (3.31).)

5. Obtain the forms given in Problems 3 and 4 by using $(e^{\pm ix})/x$, instead of the sinusoidal functions that appear in eqs. (3.30) and (3.31).

6. Obtain the expression for the Wronskian $W(j_n, y_n)$ given in Section (3.5). (Hint: Use the definition of j_n and y_n in terms of Bessel functions of half orders).

Section 3.8

7. Obtain the expression for the Wronskians given in (3.49) and (3.50).

8. Obtain the recurrence relationships (3.51) and (3.52) for the Modified Bessel functions.

Section 3.9

9. Obtain the solution to the following differential equations in the form of Bessel functions:

   (a) $x^2y'' + (k^2x^2 - n^2 - n)y = 0$

   (b) $x^2y'' - xy' + \left(k^2x^2 + \frac{8}{9}\right)y = 0$

   (c) $x^2y'' + xy' + 4x^4y = 0$

   (d) $xy'' - y' + 4x^3y = 0$

   (e) $x^2y'' + (5+2x)\,xy' + (9k^2x^6 + x^2 + 5x - 5)y = 0$

   (f) $x^2y'' + 7xy' + (36k^2x^6 - 27)y = 0$

   (g) $x^2y'' + \frac{1}{2}xy' + \left(k^2x^4 - \frac{7}{144}\right)y = 0$

   (h) $x^2y'' + 5xy' + (k^2x^4 - 12)y = 0$

   (i) $y'' - 2y' + (e^{2x} - 3)y = 0$

   (j) $x^2y'' - 2x^2y' + 2(x^2 - 1)y = 0$

(k) $x^2y'' + (4 - x)\,xy' + \left(4k^2x^4 + \dfrac{x^2}{4} - 2x + \dfrac{5}{4}\right)y = 0$

(l) $x^2y'' - (2x^2 + x)y' + xy = 0$

(m) $x^2y'' + (2x^2 - x)y' + x^2y = 0$

(n) $x^2y'' + (2x^2 + x)y' + (5x^2 + x - 4)y = 0$

10. Show that the substitution for g(x) in eq. (3.70) by the following expression:

$$g(x) = (f(x))^{b-a}$$

results in the following differential equation.

$$\frac{d^2u}{dx^2} + \left[(1-2b)\frac{f'}{f} - \frac{f''}{f'}\right]\frac{du}{dx} - \left[f^2 - r^2 + b^2 - a^2\right]\left(\frac{f'}{f}\right)^2 u = 0$$

whose solution becomes:

$$u = (f(x))^b Z_p(f(x))$$

where

$$p^2 = r^2 + a^2$$

11. Show that the substitution for g(x) in eq. (3.70) by the following expression:

$$g(x) = \sqrt{f/f'}$$

results in the following differential equation:

$$\frac{d^2u}{dx^2} - 2a\frac{f'}{f}\frac{du}{dx} + \left[\frac{(f')^2}{f^2}\left(f^2 - r^2 + \frac{1}{4} + a\right) - a\frac{f''}{f} - \frac{3}{4}\frac{(f'')^2}{(f')^2} + \frac{1}{2}\frac{f'''}{f'}\right]u = 0$$

whose solution becomes:

$$u = \sqrt{f/f'}\; f^a Z_p(f)$$

with $p^2 = r^2 + a^2$.

Section 3.10

12. Show that the Bessel Coefficients $J_n(x)$ given in eq. (3.75) satisfy Bessel's differential equation (3.1).

13. Obtain the recurrence formulæ given in eqs. (3.13) to (3.16) by utilizing the generating function.

14. Show, by induction, that:

$$x^{2m} = \sum_{n=m}^{\infty} 2^{2m-1} \frac{(n+m-1)!}{(n-m)!} J_{2n}(x) \qquad m = 1, 2, 3, \ldots$$

$$x^{2m+1} = \sum_{n=m}^{\infty} 2^{2m+1} (2n+1) \frac{(n+m)!}{(n-m)!} J_{2n+1}(x) \qquad m = 0, 1, 2, \ldots$$

Hint: Follow the procedures used in obtaining the forms in eqs. (3.80) to (3.82).

Section 3.11

15. Show that the integral representation for $J_n(x)$ given in eq. (3.97) satisfy Bessel's differential equation.

16. Obtain the recurrence formulae given in eqs. (3.13) to (3.16) by using the integral representation of $J_n(x)$ given in (3.97).

17. Show that the integral representation for $J_n(x)$ given in eq. (3.102) satisfy Bessel's differential equation.

18. Obtain the recurrence formulæ given in eqs. (3.13) to (3.16) by using the integral representation of $J_n(x)$ given in (3.102).

Section 3.12

19. Use the asymptotic behavior of the Bessel functions for small arguments to obtain the limit of the following expressions as $x \rightarrow 0$:

(a) $x^p Y_p(x)$ (b) $x^{-p} J_p(x)$

(c) $x Y_0(x)$ (d) $x^n H_n^{(2)}(x)$

(e) $x^3 h_2^{(1)}(x)$ (f) $x^{-1/2} J_{1/2}(x)$

Section 3.14

20. Prove the equality given in eq. (3.105).

21. Prove the equality given in (3.106).
(Hint: Use the differential equations of $J_p(x)$ and $J_r(x)$.)

22. Prove the equality given in (3.110).
(Hint: Use the integrals given in (3.103) and (3.104).)

Section 3.16

23. Assuming a trial solution for Legendre's equation, having the following form:

$$y = \sum_{n=0}^{\infty} a_n x^{-n+\sigma}$$

obtain the two solutions of Legendre's equation valid in the region $|x| > 1$ (see eq. 3.129).

24. Show that:

$$P_n'(1) = \frac{n(n+1)}{2} \qquad \text{and} \qquad P_n'(-1) = (-1)^{n-1}\frac{n(n+1)}{2}$$

25. Obtain the first three $Q_n(x)$ by utilizing the form given in (3.128).

Section 3.17

26. Show that:

$$P_{2n}(0) = (-1)^n \frac{(2n)!}{2^{2n}(n!)^2} \qquad n = 0, 1, 2,...$$

and

$$P_{2n+1}(0) = 0 \qquad n = 0, 1, 2,...$$

by the use of the generating function.

27. Prove that the Legendre coefficients of the expansion of the generating function satisfy Legendre's equation.

Section 3.18

28. Show that:

$$(2n+1)(1-x^2)P_n' = n(n+1)(P_{n-1} - P_{n+1})$$

29. Show that:

$$(1-x^2)(P_n')^2 = \frac{d}{dx}\left[(1-x^2)P_nP_n'\right] + n(n+1)\,P_n^2$$

Section 3.19

30. Prove the first equality in (3.142) by using the integral representation for $P_n(x)$ in (3.145).

31. Prove the second equality (3.142) by using the integral representation for $P_n(x)$ in (3.147).

Section 3.20

32. Show that:

$$\int_{-1}^{+1} (1-x^2)(P_n')^2\, dx = 2n(n+1)/(2n+1)$$

33. Show that:

$$\int_{-1}^{+1} x\, P_n\, P_{n-1}\, dx = \frac{2n}{4n^2-1}$$

34. Show that:

$$\int_{-1}^{+1} x^2\, P_{n+1} P_{n-1}\, dx = \frac{2n(n+1)}{(4n^2-1)(2n+3)}$$

35. Show that:

$$\int_{-1}^{+1} (1-x^2) P_n' P_{n+1}\, dx = \frac{-2n(n+1)}{(2n+1)(2n+3)}$$

Section 3.21

36. Prove that:

$$\int_{-1}^{+1} (1-x^2)\, P_n'\, P_{n+1}\, dx = \frac{-2n(n+1)}{(2n+1)(2n+3)}$$

37. Prove that:

$$P_{2n+1}' = (2n+1)P_{2n} + 2n\, x\, P_{2n-1} + (2n-1)\, x^2\, P_{2n-2} + (2n-2)\, x^3\, P_{2n-3} + \ldots$$

38. Prove that:

$$P_{n+1}' + P_n' = (2n+1)\, P_n + (2n-1)\, P_{n-1} + (2n-3)\, P_{n-2} + \ldots$$

Section 3.22

39. Show that:

$$(n+1)[Q_n P_{n+1} - Q_{n+1} P_n] = n[Q_{n-1} P_n - Q_n P_{n-1}]$$

40. Show that:

$$P_{n+1} Q_{n-1} - P_{n-1} Q_{n+1} = \frac{2n+1}{n(n+1)} x$$

41. Show that:

$$(1-x^2)[P'_{n+1} Q'_{n+1} - P'_n Q'_n] = (n+1)^2 [P_n Q_n - P_{n+1} Q_{n+1}]$$

42. Show that:

$$x^{-n-1} = \frac{1 \cdot 3 \cdot 5 \cdot \ldots (2n-1)}{n!}\left[(2n+1)Q_n - (2n+5)\frac{2n+1}{2} Q_{n+2} + (2n+9)\frac{(2n+1)(2n+3)}{2 \cdot 4} Q_{n+4} + \ldots\right]$$

(Hint: Differentiate (3.195) n times with respect to t and set t = 0.)

4

BOUNDARY VALUE PROBLEMS AND EIGENVALUE PROBLEMS

4.1 Introduction

Solutions of linear differential equations of order n together with n conditions specified on the dependent variable and its first (n - 1) derivatives at an **initial point** were discussed in Section (1.8) and were referred to as **Initial Value Problems**. It was shown that the solutions to such problems are unique and valid over the range of all values of the independent variable. If the differential equation as well as the **Initial Condition** are homogeneous, then it can be shown that the solutions to such problems vanish identically. In this chapter, solutions to linear differential equations of order n with n conditions specified on **two end points** of a bounded region valid in the closed region between the two end points, will be explored. These points are called **Boundary Points**, and the conditions on the dependent variable and its derivatives up to the $(n - 1)^{st}$ are called **Boundary Conditions (BC)**. Such problems are referred to as **Boundary Value Problems (BVP)**.

To illustrate the primary difference between the two types of problems, the solutions of two simple problems are shown:

Example 4.1

Obtain the solution to the following initial value problem:

Differential Equation (DE): $y'' + 4y = f(x) = 4x$

Initial Conditions (IC): $y(\pi/4) = 2$ $\qquad$ $y'(\pi/4) = 3$

The complete solution to the differential equation becomes:

$$y = C_1 \sin 2x + C_2 \cos 2x + x$$

The two arbitrary constants can be evaluated from the specified two initial conditions at the point $x_0 = \pi/4$, resulting in:

$$C_1 = 2 - \pi/4 \qquad \text{and} \qquad C_2 = -1$$

and the complete solution to the problem becomes:

$$y = (2 - \pi/4)\sin 2x - \cos 2x + x \qquad \text{for all } x$$

If the differential equation is homogeneous, i.e., if $f(x) = 0$, and the initial conditions are non-homogeneous, then the solution becomes:

$$y = 2\sin 2x - \frac{3}{2}\cos 2x \qquad \text{for all } x$$

If the differential equation and the initial conditions are homogeneous, then the solution vanishes identically, i.e.:

$$y \equiv 0$$

Example 4.2

Obtain the solution to the following boundary value problem:

Differential Equation (DE): $y'' + 4y = f(x) = 4x \qquad 0 \le x \le \pi/4$

Boundary Conditions (BC): $y(0) = 2$

$y(\pi/4) = 3$

The complete solution to the differential equation is again:

$$y = C_1 \sin 2x + C_2 \cos 2x + x$$

The two arbitrary constants can be evaluated from the two boundary conditions, one at each of the end points at x = 0 and x = π/4:

$$y(0) = C_2 = 2$$

$$y(\pi/4) = C_1 \sin\frac{\pi}{2} + C_2 \cos\frac{\pi}{2} + \frac{\pi}{4} = 3 \qquad C_1 = 3 - \frac{\pi}{4}$$

Thus, the final solution becomes:

$$y = (3 - \pi/4)\sin 2x + 2\cos 2x + x \qquad 0 \le x \le \pi/4$$

If the differential equation is homogeneous, i.e., if f(x) = 0, but the boundary conditions are not, then the complete solution satisfying these boundary conditions becomes:

$$y = 3\sin 2x + 2\cos 2x \qquad 0 \le x \le \pi/4$$

If the differential equation and the boundary conditions are both homogeneous, the solution vanishes identically:

$$y \equiv 0 \qquad 0 \le x \le \pi/4$$

A special type of a homogeneous boundary value problem that has a non-trivial solution is one whose differential equation has an undetermined parameter. A non-trivial solution exists for such problems if the parameter takes on certain values. Such problems are known as **Eigenvalue Value Problems**, whose non-trivial solutions are referred to as **Eigenfunctions** whenever the undetermined parameter takes on certain values, known as **Eigenvalues**.

Example 4.3

Obtain the solution to the following homogeneous boundary value problem:

DE: $y'' + \lambda y = 0$

BC: $y(0) = 0 \qquad y(\pi/4) = 0$

The complete solution of the differential equation becomes:

$$y = C_1 \sin\sqrt{\lambda}\, x + C_2 \cos\sqrt{\lambda}\, x \qquad \lambda \neq 0$$

$$y_0 = C_3 x + C_4 \qquad \lambda = 0$$

Satisfying the boundary conditions at the two end points yields:

$$y(0) = C_2 = 0 \qquad \text{and} \qquad C_4 = 0$$

and

$$y(\pi/4) = C_1 \sin\sqrt{\lambda}\,\frac{\pi}{4} = 0 \ \text{ and } \qquad C_3 = 0 \qquad \text{or} \qquad y_0 \equiv 0$$

The last equation on C_1 leads to two possible solutions:

(i) For a non-trivial solution, i.e., $C_1 \neq 0$, then $\sin\sqrt{\lambda}\,\pi/4 = 0$, which can be satisfied if the undetermined parameter λ takes any one of the following infinite discrete number of possible values, i.e.:

$$\lambda_1 = 16 \cdot 1^2, \quad \lambda_2 = 16 \cdot 2^2, \quad \lambda_3 = 16 \cdot 3^2, \quad \ldots$$

In other words, $\lambda_n = 16n^2$ n = 1, 2, 3,... are the **Eigenvalues** which satisfy the following **Characteristic Equation**:

$$\sin\sqrt{\lambda}\,\frac{\pi}{4} = 0$$

Thus, the solution, which is nontrivial if λ takes any one of these special values, has the following form:

$$y = C_1 \sin 4nx \qquad n = 1, 2, 3, \ldots$$

which is non-unique, since the constant C_1 is undeterminable.

The functions $\phi_n = \sin 4nx$ are known as **Eigenfunctions**. The value $\lambda = 0$ gives a trivial solution, thus it is not an **Eigenvalue**.

(ii) If λ does not take any one of those values, i.e., if:

$$\lambda \neq 16n^2 \qquad n = 1, 2, \ldots$$

then

$$C_1 \equiv 0$$

and the solution vanishes identically.

4.2 Vibration, Wave Propagation or Whirling of Stretched Strings

Consider a stretched loaded thin string of length L and mass density per unit length ρ in its undeformed state. The string is stretched at its end by a force T_0, loaded by a distributed force f(x) and is being rotated about its axis by an angular speed = ω, as shown in Fig. 4.1.

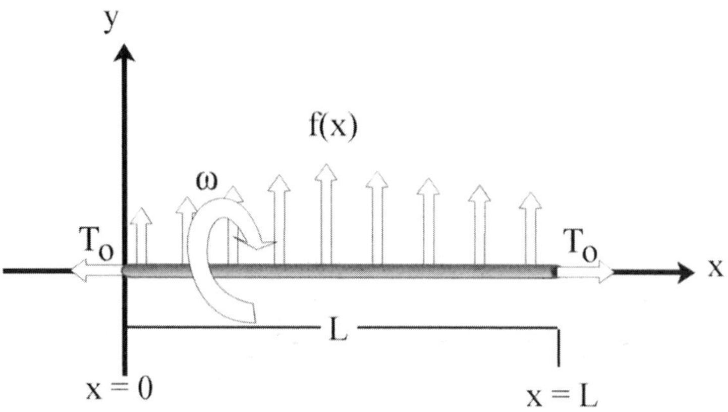

Figure 4.1: *Stretched String in Undeformed State*

Consider an element of length dx of the string in the deformed state, such that its center of gravity is deformed laterally a distance y as shown in Fig. 4.2. The forces at each end of the element are also shown in Fig. 4.2. The equations of equilibrium on the tension T in the x-direction state that:

$$T_{x+dx}\cos\theta_{x+dx} - T_x\cos\theta_x = 0$$

If one assumes that the motion is small, such that $\theta << 1$, then both $\cos\theta_{x+dx} \approx \cos\theta_x \approx 1$, resulting in:

$$T_{x+dx} = T_x = \text{constant} = T_0$$

The equation of equilibrium in the y-direction can then be written as follows:

$$T_0\sin\theta_{x+dx} - T_0\sin\theta_x + \int_x^{x+dx} f(\eta)\,d\eta + \rho\omega^2 y\,dx = 0$$

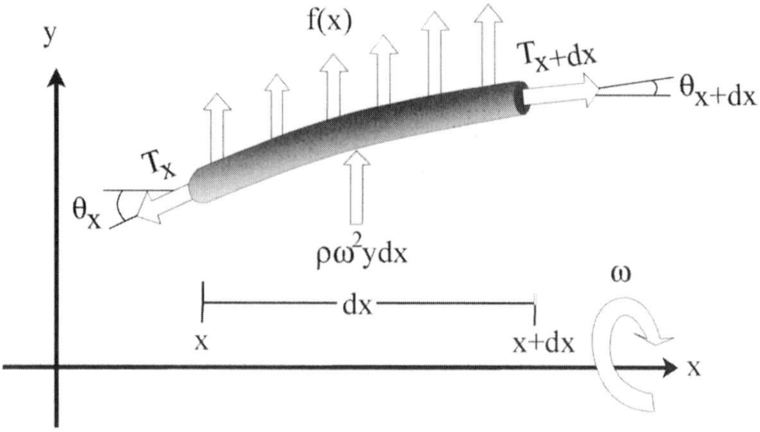

Fig. 4.2: *Element of Vibrating, Stretched String in Deformed State*

Since:

$$\sin\theta_x = \frac{dy}{ds}$$

$$\sin\theta_{x+dx} = \left(\frac{dy}{ds}\right)_{x+dx} = \frac{dy}{ds} + \frac{d}{dx}\left(\frac{dy}{ds}\right)dx + \frac{d^2}{dx^2}\left(\frac{dy}{ds}\right)\frac{(dx)^2}{2} + \ldots$$

Substituting these into the equilibrium equation, and replacing the integral by its average value at x, and neglecting higher-order terms of (dx), the linearized equation becomes:

$$\frac{d}{dx}\left(T_0\frac{dy}{ds}\right) + f(x) + \rho\omega^2\, y = 0$$

Since $dy/dx \ll 1$ was assumed in the derivation of the equation of motion, then:

$$\frac{dy}{ds} = \frac{dy/dx}{\sqrt{1+(dy/dx)^2}} \approx \frac{dy}{dx}$$

and the differential equation of motion becomes:

$$\frac{d^2y}{dx^2} + \frac{\rho\omega^2}{T_0}\, y = -\frac{f(x)}{T_0} \tag{4.1a}$$

or, if ρ is constant $= \rho_0$, then:

$$\frac{d^2y}{dx^2} + \frac{\omega^2}{c^2}\, y = -\frac{f(x)}{T_0} \qquad \text{where} \quad c^2 = T_0/\rho_0 \tag{4.1b}$$

where c is known as the sound speed of waves in the stretched string.

In the case of a vibrating stretched string, then $y = y^*(x, t)$, $f = f^*(x, t)$, and one substitutes $-\rho(\partial^2 y^*/\partial t^2)\,dx$ for the centrifugal force such that the wave equation for the string becomes:

$$\frac{\partial^2 y^*}{\partial x^2} = \frac{1}{c^2}\frac{\partial^2 y^*}{\partial t^2} - \frac{f^*(x,t)}{T_0} \tag{4.2}$$

If one assumes that the applied force field and the displacement are periodic in time, such that:

$$y^*(x, t) = y(x) \sin \omega t$$

$$f^*(x, t) = f(x) \sin \omega t$$

where ω is the circular frequency, then eq. (4.2) becomes the same as eq. (4.1b), which can be rewritten as:

$$\frac{d^2y}{dx^2} + k^2\, y = -\frac{f(x)}{T_0}$$

where $k = \omega/c$ is the wave number.

The natural (physical) boundary conditions are of two types:

(i) fixed end: $y(0) = 0$ or $y(L) = 0$

(ii) free end: $y'(0) = 0$ or $y'(L) = 0$

(iii) elastically supported end (spring)

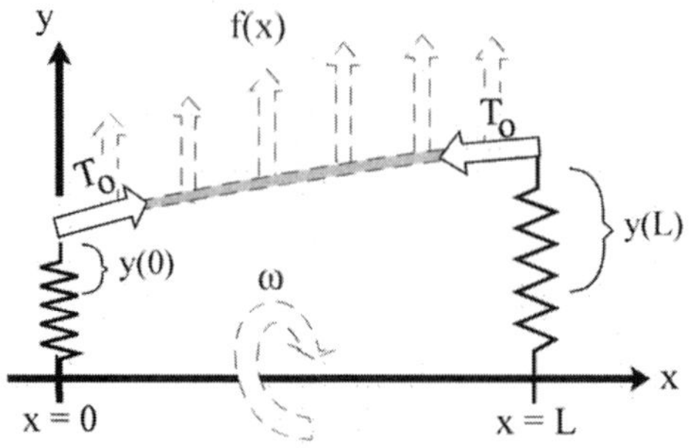

left end: $T_0 \dfrac{dy(0)}{dx} - \gamma\, y(0) = 0$ right end: $T_0 \dfrac{dy(L)}{dx} + \gamma\, y(L) = 0$

where γ = spring constant = force/ unit displacement.

Example 4.4 Vibration of Fixed Stretched String

Obtain the natural frequencies (or the critical angular speeds) of a fixed-fixed stretched string whose length is L:

DE: $\dfrac{d^2y}{dx^2} + k^2 y = 0 \qquad 0 \le x \le L$

BC: $y(0) = 0$ and $y(L) = 0$

The solution of the homogeneous differential equation is given by:

$$y = C_1 \sin kx + C_2 \cos kx$$

The above solution must satisfy the boundary conditions:

$$y(0) = C_2 = 0$$

$$y(L) = C_1 \sin kL = 0$$

For a non-trivial solution:

$\sin kL = 0$ (Characteristic equation)

which is satisfied if k_n takes the following values:

$$k_n = \frac{n\pi}{L} \qquad n = 1, 2, 3, \ldots$$

or

$$\lambda_n = k_n^2 = \frac{n^2\pi^2}{L^2} \qquad n = 1, 2, 3, \ldots \text{ (Eigenvalues)}$$

and the corresponding solution:

$$\phi_n(x) = \sin k_n x = \sin\frac{n\pi}{L}x \qquad n = 1, 2, 3, \ldots \text{ (Eigenfunctions)}$$

Also for $k = 0$, it can be shown that $y \equiv 0$.

The natural frequencies (or the critical angular speeds) are given by:

$$\omega_n = ck_n = \frac{cn\pi}{L} \qquad n = 1, 2, 3,...$$

As the angular speed (or forcing frequency) ω is increased from zero, the deflection stays small until the angular speed (or frequency) reaches ω_n, thus:

$$y = 0 \qquad 0 < \omega < \omega_1$$

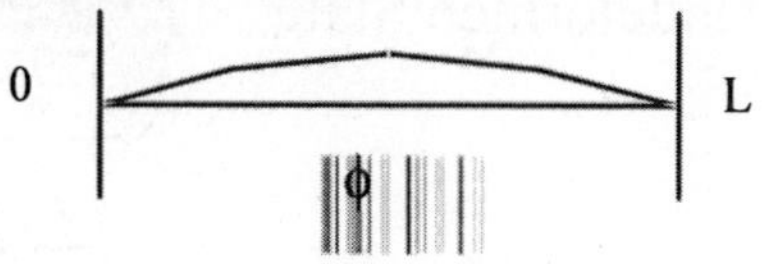

$$y = A_1\phi_1 = A_1 \sin\frac{\pi}{L}x \qquad \omega_1 = \frac{\pi c}{L}$$

$$y = 0 \qquad \omega_1 < \omega < \omega_2$$

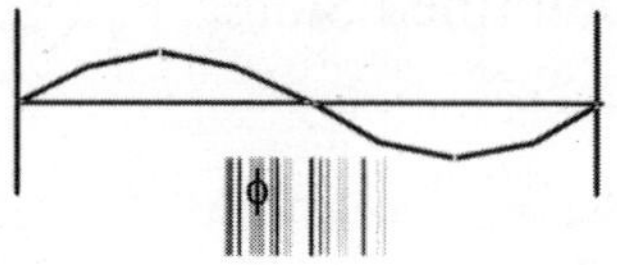

$$y = A_2\phi_2 = A_2 \sin\frac{2\pi}{L}x \qquad \omega_2 = \frac{2\pi c}{L}$$

It should be noted that each eigenfunction satisfies all the boundary conditions and the eigenfunction of order n has one more null than the preceding one, i.e., $(n-1)^{st}$ eigenfunction.

4.3 Longitudinal Vibration and Wave Propagation in Elastic Bars

Consider a bar of cross section A, Young's modulus E and mass density ρ, as shown in Fig. 4.3. Consider an element of the bar of length dx shown in Fig. 4.4.

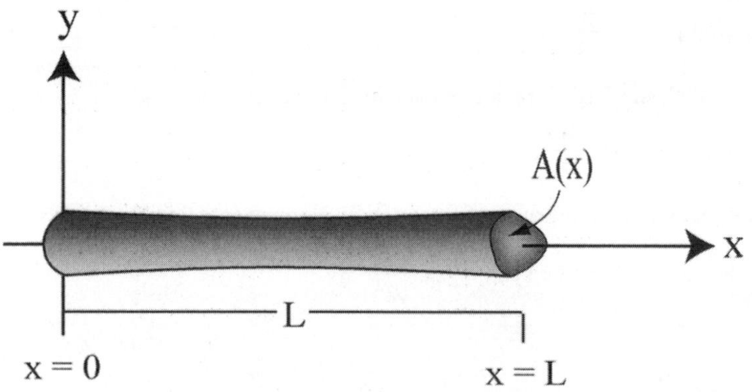

Fig. 4.3: *Elastic Bar*

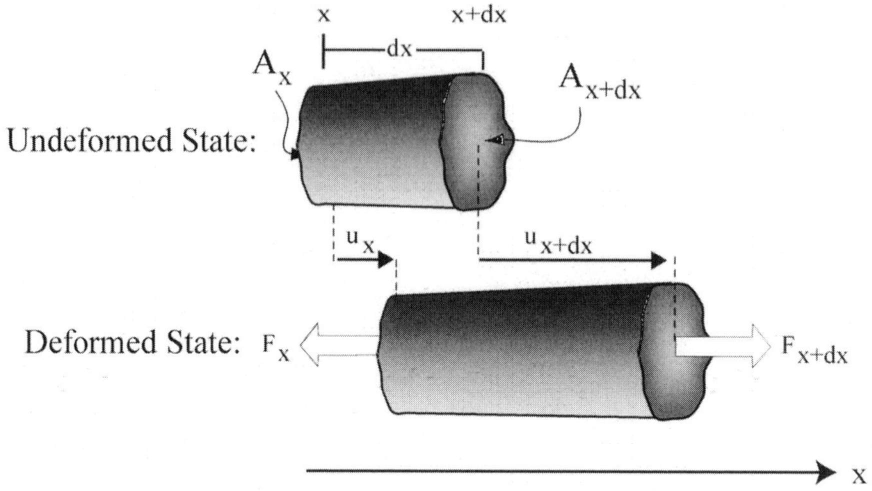

Fig 4.4: *Element of a Vibrating Elastic Bar in Longitudinal Motion*

Each cross section is assumed to deform by u*(x,t) along the axis of the rod as shown in Fig. 4.4. Let u*(x,t) be the deformation at location x and at time t, then the deformation at location x+dx and t is:

$$u^*_{x+dx} \cong u^*_x + \frac{\partial u^*}{\partial x}dx$$

then the elastic strain as defined by:

$$\text{strain } \varepsilon = \frac{\text{deformation}}{\text{original length}} \cong \frac{u^*_x + (\partial u^*/\partial x)\,dx - u^*_x}{dx} \cong \frac{\partial u^*}{\partial x}$$

and the corresponding elastic stress using Hooke's law becomes:

$$\text{stress } \sigma = E\frac{\partial u^*}{\partial x}$$

The total elastic force F on a cross section can be computed as:

$$F = A\sigma = AE\frac{\partial u^*}{\partial x}$$

The equation of equilibrium of forces on an element satisfies Newton's second law:

$$F_{x+dx} - F_x + f^*(x,t)\,dx = \rho A\frac{\partial^2 u^*}{\partial t^2}dx$$

or

$$\left(AE\frac{\partial u^*}{\partial x}\right)_{x+dx} - \left(AE\frac{\partial u^*}{\partial x}\right)_x$$

$$= \frac{\partial}{\partial x}\left(AE\frac{\partial u^*}{\partial x}\right)dx + \frac{\partial^2}{\partial x^2}\left(AE\frac{\partial u^*}{\partial x}\right)\frac{(dx)^2}{2} + \ldots = \rho A\frac{\partial^2 u^*}{\partial t^2}dx - f^*(x,t)\,dx$$

where $f^*(x,t)$ is the distributed load per unit length. Linearizing the equation, one obtains the wave equation for an elastic bar:

$$\frac{\partial}{\partial x}\left(EA\frac{\partial u^*}{\partial x}\right) = \rho A\frac{\partial^2 u^*}{\partial t^2} - f^* \tag{4.3a}$$

If the material of the bar is homogeneous, then the Young's modulus E is constant, and eq. (4.3) becomes:

$$\frac{\partial}{\partial x}\left(A\frac{\partial u^*}{\partial x}\right) = \frac{1}{c^2}A\frac{\partial^2 u^*}{\partial t^2} - \frac{f^*}{E} \tag{4.3b}$$

where the sound speed of longitudinal waves in the bar c is:

$$c^2 = E/\rho$$

If the cross-sectional area is constant (independent of the shape of the area along the length of the bar), then the wave equation (4.4) simplifies to:

$$\frac{\partial^2 u^*}{\partial x^2} = \frac{1}{c^2}\frac{\partial^2 u^*}{\partial t^2} - \frac{f^*}{AE} \tag{4.4}$$

For a bar that is vibrating with a circular frequency ω, under the influence of a time-harmonic load f*, i.e., $f^*(x,t) = f(x)\sin\omega t$, $u^*(x,t) = u(x)\sin\omega t$ eq. (4.5) becomes:

$$\frac{d^2u}{dx^2} + k^2 u = -\frac{f}{AE} \qquad k = \frac{\omega}{c} \tag{4.5}$$

The natural (physical) boundary conditions can be any of the following types:

(i) Fixed end $u^* = 0$

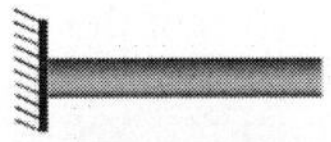

(ii) Free end $AE\,\partial u^*/\partial x = 0$

(iii) Elastically supported by a linear spring:

Left end: $AE\,\partial u^*/\partial x - \gamma u^* = 0$

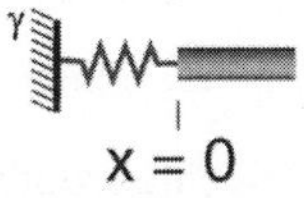

Right end: $AE\,\partial u^*/\partial x + \gamma u^* = 0$

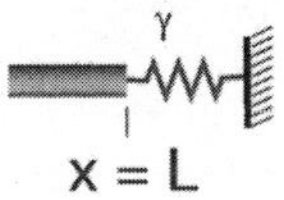

where γ is the elastic constant of the spring.

Example 4.5 Longitudinal Vibration of a Bar

Obtain the natural frequencies and the mode shapes of a longitudinally vibrating uniform homogeneous rod of constant cross section. The rod is fixed at x = 0 and elastically supported at x = L.

DE: $$\frac{d^2u}{dx^2} + k^2 u = 0 \qquad 0 \le x \le L$$

BC: $u(0) = 0$ and $AE\frac{du}{dx} + \gamma u\Big|_{x=L} = 0$

The solution to the homogeneous equation is:

$$u = C_1 \sin kx + C_2 \cos kx$$

which is substituted in the two homogeneous boundary conditions:

$$u(0) = 0 = C_2$$

and

$$C_1\left[k\cos kL + \frac{\gamma}{AE}\sin kL\right] = 0$$

For non-trivial solution, the bracketed expression must vanish resulting in the following characteristic equation:

$$\tan\alpha = -\frac{AE}{\gamma L}\alpha \qquad \text{where} \quad \alpha = kL$$

The roots of the transcendental equation on α_n can only be obtained numerically. An estimate of the location of the roots can be obtained by plotting the two parts of the equation as shown in Fig. 4.5. There is an infinite number of roots $\alpha_1, \alpha_2, \ldots, \alpha_n, \ldots$. Note that the roots for large values of n approach:

$$\alpha_n \underset{n >> 1}{\rightarrow} \frac{2n+1}{2}\pi$$

Thus, the resonant frequencies of the finite rod are given by:

$$\omega_n = ck_n = c\frac{\alpha_n}{L} \qquad n = 1, 2, 3, \ldots$$

the eigenvalues are given in terms of the roots α_n:

$$\lambda_n = k_n^2 = \frac{\alpha_n^2}{L^2} \qquad n = 1, 2, 3, \ldots$$

and the corresponding eigenfunctions (mode shapes) are given by:

$$\phi_n = \sin k_n x = \sin \alpha_n \frac{x}{L} \qquad n = 1, 2, 3, \ldots$$

The root $\alpha_0 = 0$ corresponds to a trivial solution, thus, it is not an eigenvalue.

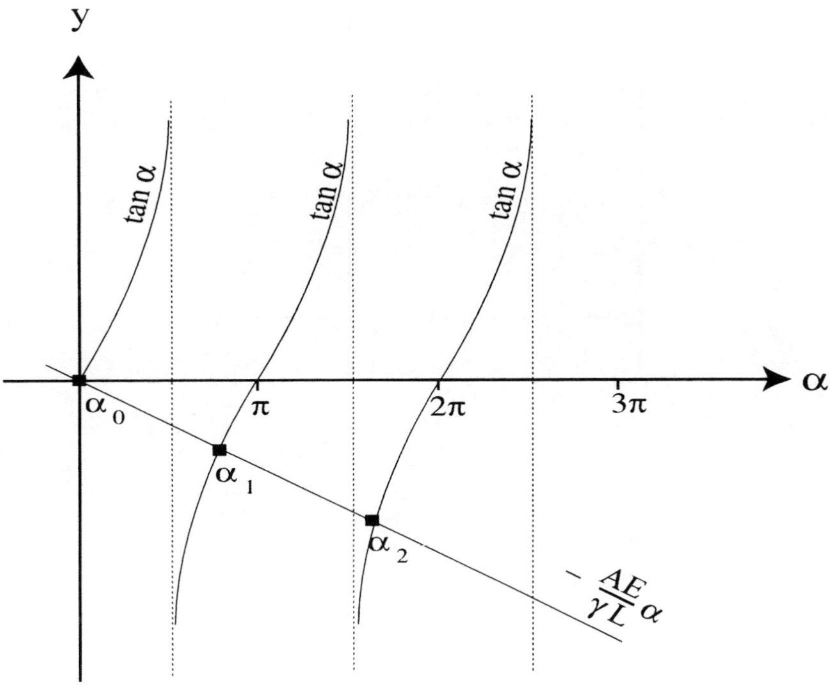

Fig. 4.5 Curves for Characteristic Equation

It should be noted that the eigenfunctions $\phi_n(x)$ have n nulls, which makes sketching them easier.

4.4 Vibration, Wave Propagation, and Whirling of Beams

The vibration of beams or the whirling of shafts can be considered as a similar dynamic system to the vibration or whirling of strings. Consider a beam of mass density ρ, cross-sectional area A and cross-sectional area moment of inertia I, which is acted upon by distributed forces f(x), and is rotated about its axis by an angular speed ω, as shown in Fig. 4.6. If the beam deforms from its straight line configuration, then one considers an element of the deformed beam, where the shear V and the moment M exerted by the other parts of the beam on the element are shown in Fig. 4.7.

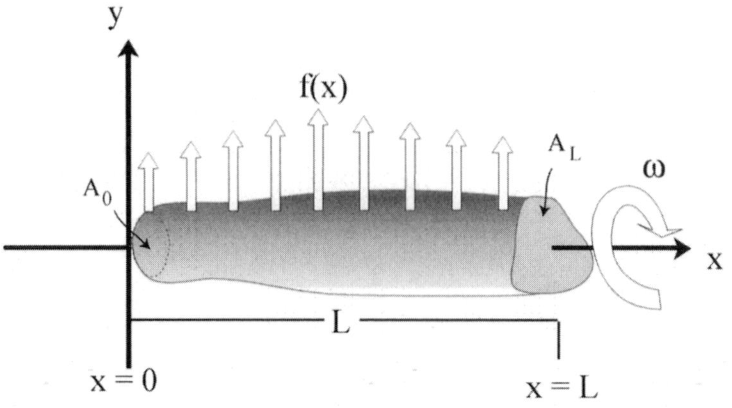

Fig 4.6: *Undeformed Beam*

The equation of equilibrium of forces in the y-direction becomes:

$$V_x + \rho\omega^2 yA\, dx + \int_x^{x+dx} f(\eta)\, d\eta - V_{x+dx} = 0$$

Expanding the shear at x+dx by a Taylor series about x:

$$V_{x+dx} = V_x + \frac{dV_x}{dx} dx + \ldots$$

then an equilibrium equation results of the form:

$$\frac{dV_x}{dx} = \rho\omega^2 Ay + f(x)$$

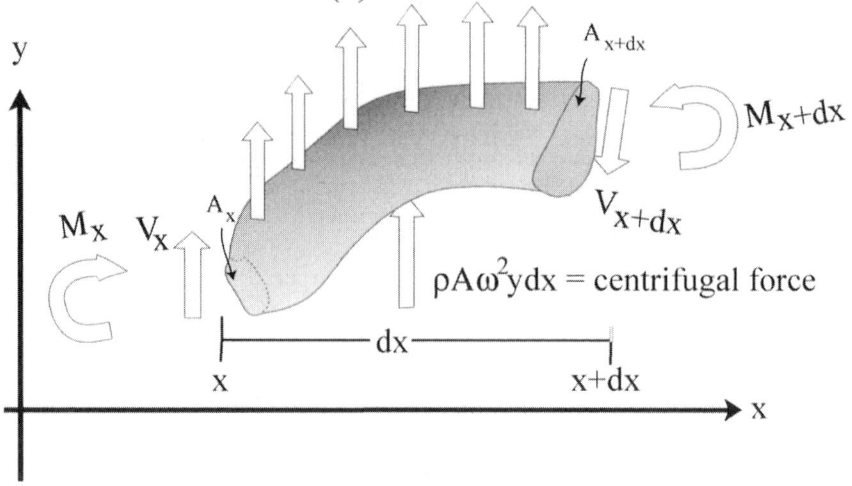

Fig. 4.7: *Element of a Deformed Beam in Flexure*

Taking the equilibrium of the moment about the left end of the element, one obtains:

$$V_{x+dx}dx + M_x - M_{x+dx} - \int_x^{x+dx} f(\eta)(\eta - x)\,d\eta - \rho\omega^2 Ay\frac{(dx)^2}{2} = 0$$

Again, expanding V_{x+dx} and M_{x+dx} by a Taylor's series about x and using the mean value for the integral as $dx \to 0$, results in the following relationship between the moment and the shear:

$$V_x = \frac{dM_x}{dx}$$

Thus, the equation of motion becomes:

$$\frac{d^2M_x}{dx^2} = \rho\omega^2 Ay + f(x)$$

The constitutive relations for the beam under the action of moments M_x and M_{x+dx} can be developed by considering the element in Fig. 4.8 of length s. The element's two crosssections at it's ends undergoes a rotation about the neutral axis, so that the element subtends an angle ($d\theta$) and has a radius of curvature R. The element undergoes rotation ($d\theta$) and elongation Δ at a location z:

$$\frac{ds}{R} = d\theta = \frac{\Delta}{z}$$

Thus, the local strain, defined as the longitudenal deformation at z per unit length is given by:

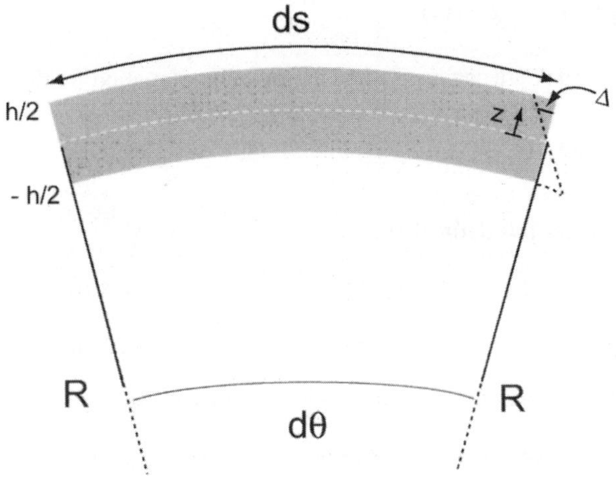

Fig. 4.8: *Element of a Beam Deformed in Flexure*

$$\text{strain } \varepsilon = \frac{\Delta}{ds} = \frac{z}{R}$$

and the local stress is given by Hook's Law:

$$\text{strain } \sigma = E\varepsilon = \frac{Ez}{R}$$

Integrating the moment of the stress, due to the stress field at z over the cross-sectional area of the beam gives:

$$\text{Moment } M_x = \int_A \sigma z \, dA = \frac{E}{R}\int_A z^2 dA = \frac{EI}{R}$$

where $I = \int_A z^2 dA$ is the moment of inertia of the cross-sectional area A.

Since the radius of curvature is defined by:

$$\frac{1}{R} = \frac{d\theta}{ds} \approx \frac{d^2y}{dx^2}$$

for small slopes, then the moment is obtained in terms of the second derivative of the displacement y, i.e.:

$$M_x = EI\frac{d^2y}{dx^2}$$

and the equation of motion for the beam becomes:

$$\frac{d^2}{dx^2}\left(EI\frac{d^2y}{dx^2}\right) = \rho\omega^2 Ay + f(x) \tag{4.6}$$

If the functions EI and A are constants, then the equation of motion for the beam eq. (4.6) simplifies to:

$$\frac{d^4y}{dx^4} - \beta^4 y = \frac{f(x)}{EI} \tag{4.7}$$

where the wave number β is defined by:

$$\beta^4 = \frac{\rho A}{EI}\omega^2$$

The wave equation for a time-dependent displacement of a vibrating beam $y^*(x,t)$ can be obtained by replacing the centrifugal force by the inertial force $\left(-\rho A\frac{\partial^2 y^*}{\partial t^2}dx\right)$.

Replacing d/dx by $\partial/\partial x$ such that eq. (4.6) becomes the wave equation for a beam:

$$\frac{\partial^2}{\partial x^2}\left(EI\frac{\partial^2 y^*}{\partial x^2}\right) + \rho A\frac{\partial^2 y^*}{\partial t^2} = f^*(x,t) \tag{4.8}$$

where $y^* = y^*(x,t)$ and $f^* = f^*(x,t)$.

If the motion as well as the applied force are time-harmonic, i.e.:

$$y^* = y(x)\sin\omega t$$

$$f^* = f(x)\sin\omega t$$

then the ordinary differential equation governing harmonic vibration of the beam reduces to the same equation for whirling of beams eq. (4.6):

The natural boundary conditions for the beams takes any one of the following nine pairs:

(i) fixed end:

$$y = 0 \qquad \frac{dy}{dx} = 0$$

(ii) simply supported:

$$y = 0 \qquad EI\frac{d^2y}{dx^2} = 0$$

(iii) free end:

$$EI\frac{d^2y}{dx^2} = 0 \qquad \frac{d}{dx}\left(EI\frac{d^2y}{dx^2}\right) = 0$$

(iv) free-fixed end:

$$\frac{dy}{dx} = 0 \qquad \frac{d}{dx}\left(EI\frac{d^2y}{dx^2}\right) = 0$$

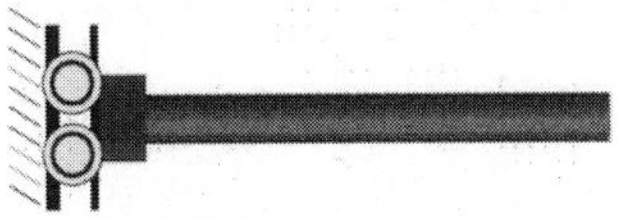

(v) elastically supported end by transverse elastic spring of stiffness γ:

$$\frac{d}{dx}\left(EI\frac{d^2y}{dx^2}\right) \pm \gamma y = 0 \qquad \frac{d^2y}{dx^2} = 0$$

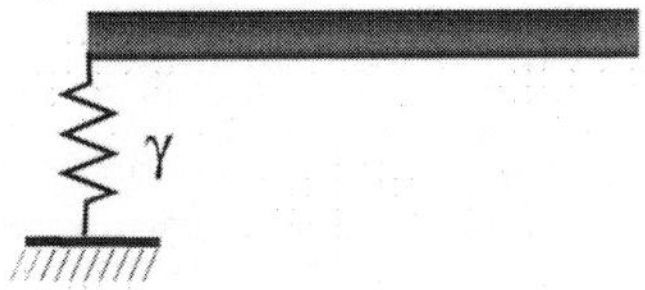

The + and - signs refer to the left and right ends, respectively.

(vi) free-fixed end with a transverse elastic spring of stiffness γ:

$$\frac{d}{dx}\left(EI\frac{d^2y}{dx^2}\right) \pm \gamma y = 0 \qquad \frac{dy}{dx} = 0$$

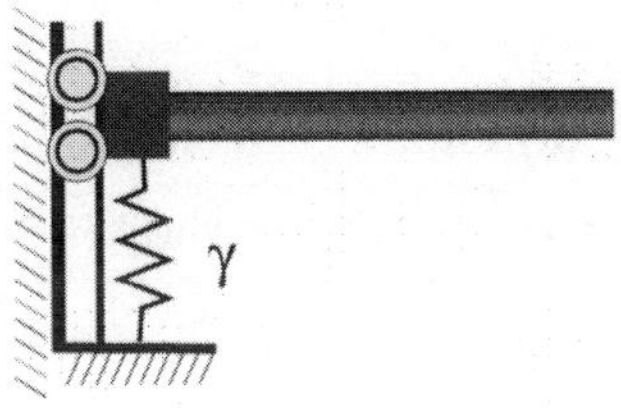

The sign convention as in (v) above.

(vii) free end elastically supported by a helical elastic spring of stiffness α:

$$EI\frac{d^2y}{dx^2} \mp \alpha\frac{dy}{dx} = 0 \qquad \frac{d}{dx}\left(EI\frac{d^2y}{dx^2}\right) = 0$$

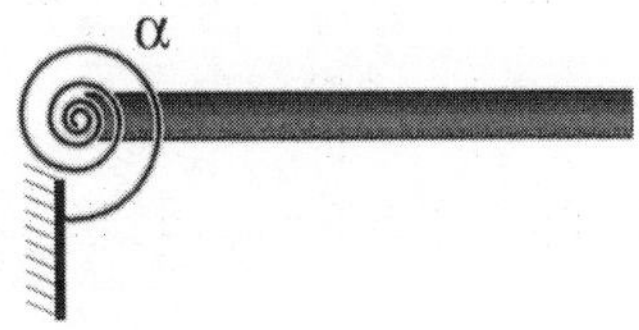

The - and + signs refer to the left and right ends respectively.

(viii) hinged and elastically supported by a helical elastic spring of stiffness α:

$$EI\frac{d^2y}{dx^2} \mp \alpha\frac{dy}{dx} = 0 \qquad y = 0$$

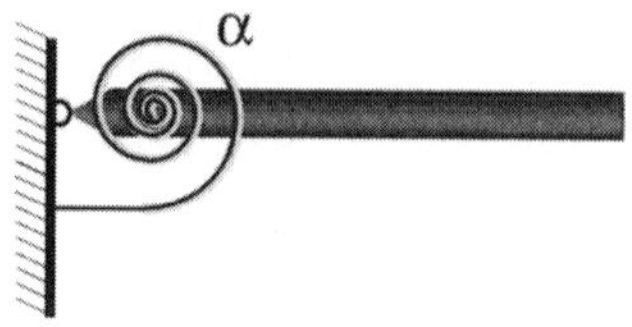

The sign convention as in (vii).

(ix) elastically supported end by transverse and helical springs of stiffnesses γ and α:

$$\frac{d}{dx}\left(EI\frac{d^2y}{dx^2}\right) \pm \gamma y = 0 \qquad EI\frac{d^2y}{dx^2} \mp \alpha\frac{dy}{dx} = 0$$

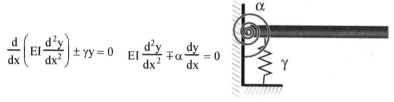

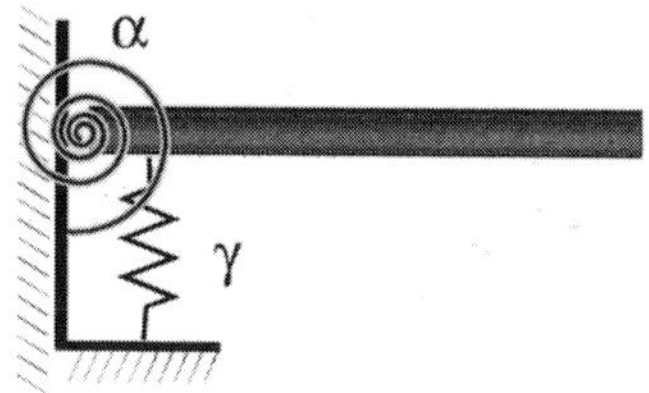

The sign convention is the same as in (vi) & (vii).

Example 4.6 Whirling of a Fixed Shaft

Obtain the critical speeds of a rotating shaft whose length is L and ends are fixed:

DE: $$\frac{d^4y}{dx^4} - \beta^4 y = 0$$

BC: $y(0) = 0 \qquad y'(0) = 0$

$y(L) = 0 \qquad y'(L) = 0$

The solution of the ordinary differential equation with constant coefficients takes the form:

$$y = A\sin\beta x + B\cos\beta x + C\sinh\beta x + D\cosh\beta x$$

Satisfying the four boundary conditions:

$y(0) = 0 \qquad B + D = 0$

$y'(0) = 0 \qquad A + C = 0$

$y(L) = 0 \qquad A\sin\beta L + B\cos\beta L + C\sinh\beta L + D\cosh\beta L = 0$

$y'(L) = 0 \qquad A\cos\beta L - B\sin\beta L + C\cosh\beta L + D\sinh\beta L = 0$

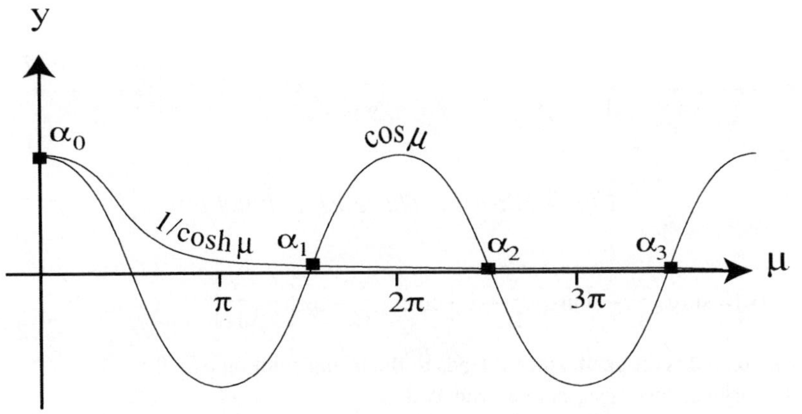

***Fig. 4.9** Curves for Characteristic Equation*

For a non-trivial solution, the determinant of the arbitrary constants A, B, C and D must vanish, i.e.:

$$\begin{vmatrix} 0 & 1 & 0 & 1 \\ 1 & 0 & 1 & 0 \\ \sin\beta L & \cos\beta L & \sinh\beta L & \cosh\beta L \\ \cos\beta L & -\sin\beta L & \cosh\beta L & \sinh\beta L \end{vmatrix} = 0$$

The determinant reduces to the following transcendental equation:

$$\cosh\mu\cos\mu = 1 \quad \text{(Characteristic Equation)} \qquad \text{where} \quad \mu = \beta L$$

The roots can be obtained numerically by rewriting the equation:

$$\cos\mu = 1/\cosh\mu$$

where the two sides of the equality can be sketched as shown in Fig. 4.9.

The roots can be estimated from the sketch above and obtained numerically through the use of numerical methods such as the Newton-Raphson Method. The first four roots of the transcendental equation are listed below:

$$\mu_0 = 0 \qquad \mu_1 = \beta_1 L = 4.730$$

$$\mu_2 = \beta_2 L = 7.853 \qquad \mu_3 = \beta_3 L = 10.966$$

Denoting the roots by μ_n, then $\beta_n = \mu_n/L$, n = 0, 1, 2,... and the eigenvalues become:

$$\lambda_n = \beta_n^4 = \mu_n^4/L^4 \qquad n = 1, 2, 3, \ldots$$

One can obtain the constants in terms of ratios by using any three of the four equations representing the boundary conditions. Thus, the constants B, C, and D can be found in terms of A as follows:

$$\frac{B}{A} = -\frac{\sinh\mu_n - \sin\mu_n}{\cosh\mu_n - \cos\mu_n} = -\frac{D}{A} = \xi_n$$

$$\frac{C}{A} = -1$$

which, when substituted in the solution, results in the eigenfunctions:

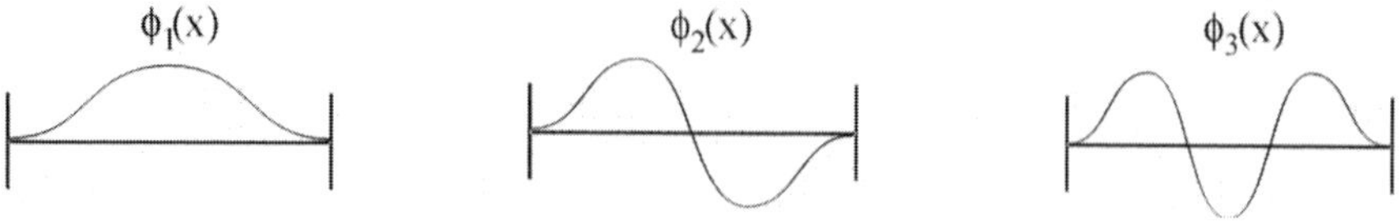

Fig. 4.10: *First Three Eigenfunctions*

$$\phi_n(x) = \sin \mu_n \frac{x}{L} - \sinh \mu_n \frac{x}{L} + \xi_n \left[\cos \mu_n \frac{x}{L} - \cosh \mu_n \frac{x}{L} \right] \qquad n = 1, 2, \ldots$$

The root $\mu_0 = 0$ is dropped, since it leads to the trivial solution $\phi_0 = 0$.

The critical speeds ω_n can be evaluated as:

$$\omega_n = \sqrt{\frac{EI}{\rho A}} \frac{\mu_n^2}{L^2} \qquad n = 1, 2, 3, \ldots$$

A plot of the first three eigenfunctions are shown in Fig. 4.10.

4.5 Waves in Acoustic Horns

Consider a tube (horn) of cross-sectional area A, filled with a compressible fluid, having a density $\rho^*(x,t)$. Let $v^*(x,t)$ and $p^*(x,t)$ represent the particle velocity and the pressure at a cross section x, respectively. Consider an element of the fluid of length dx and a unit cross section, shown in Fig. 4.11.

Then, the equation of motion for the element becomes:

$$p_x^* - p_{x+dx}^* = \frac{d}{dt} \int_x^{x+dx} \rho^*(\eta,t)\, v^*(\eta,t)\, d\eta$$

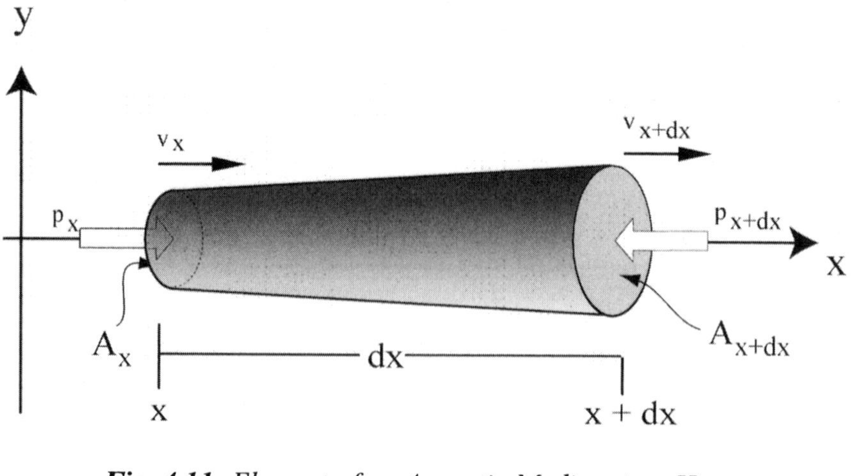

Fig. 4.11*: Element of an Acoustic Medium in a Horn*

Expanding the pressure p^*_{x+dx} by Taylor's series about x and obtaining the mean value of the integral as $dx \to 0$, one obtains:

$$-\frac{\partial p^*}{\partial x} = \rho^* \frac{dv^*}{dt} = \rho^* \left(\frac{\partial v^*}{\partial t} + v^* \frac{\partial v^*}{\partial x} \right) \approx \rho_0 \frac{\partial v^*}{\partial t}$$

where

$$\frac{\partial v^*}{\partial t} >> v^* \frac{\partial v^*}{\partial x}$$

and $\rho_0(x)$ is the quasi-static density. This is known as **Euler's Equation**.

The mass of an element dx inside the tube, as in Fig. 4.11, is conserved, such that:

$$\frac{d}{dt}(A\rho^* \, dx) = \rho_0 A_x v^*_x - \rho_0 A_{x+dx} v^*_{x+dx}$$

or

$$A\frac{d\rho^*}{dt} \approx A\frac{\partial \rho^*}{\partial t} = -\rho_0 \frac{\partial}{\partial x}(Av^*)$$

The constitutive equation relating the pressure in the fluid to its density is given by:

$$p^* = p^*(\rho^*)$$

so that the time rate of change of the pressure is given by:

$$\frac{dp^*}{dt} = \frac{dp^*}{d\rho^*}\frac{d\rho^*}{dt} \approx c^2 \frac{d\rho^*}{dt}$$

where c is the speed of sound in the acoustic medium

$$\frac{dp^*}{d\rho^*} \approx c^2$$

and the pressure is given by:

$$p^* \approx \rho_0 c^2 + p_0$$

with p_0 being the ambient pressure. Thus, the continuity equation becomes:

$$A\frac{\partial \rho^*}{\partial t} = \frac{A}{c^2}\frac{\partial p^*}{\partial t} = -\rho_0 \frac{\partial}{\partial x}(Av^*)$$

Differentiating the last equation with respect to t, it becomes:

$$\frac{A}{c^2}\frac{\partial^2 p^*}{\partial t^2} = -\rho_0 \frac{\partial}{\partial x}\left(A\frac{\partial v^*}{\partial t}\right)$$

Multiplying Euler's equation by A and differentiating it with respect to x, one obtains:

$$-\frac{\partial}{\partial x}\left(A\frac{\partial p^*}{\partial x}\right) = \rho_0 \frac{\partial}{\partial x}\left(A\frac{\partial v^*}{\partial t}\right) = -\frac{A}{c^2}\frac{\partial^2 p^*}{\partial t^2}$$

which, upon rearranging, gives the wave equation for an acoustic horn:

$$\frac{1}{A}\frac{\partial}{\partial x}\left(A\frac{\partial p^*}{\partial x}\right) = \frac{1}{c^2}\frac{\partial^2 p^*}{\partial t^2} \tag{4.9}$$

It can be shown that if $v^* = -\partial\phi^*/\partial x$, where ϕ^* is a velocity potential, then:

$$p^* = \rho_0 \frac{\partial \phi^*}{\partial t}$$

such that velocity potential ϕ^* satisfies the following differential equation:

$$\frac{1}{A}\frac{\partial}{\partial x}\left(A\frac{\partial \phi^*}{\partial x}\right) = \frac{1}{c^2}\frac{\partial^2 \phi^*}{\partial t^2} \tag{4.10}$$

If the motion is harmonic in time, such that:

$$p^*(x,t) = p(x)\,e^{i\omega t} \qquad \text{and} \qquad v^*(x,t) = v(x)\,e^{i\omega t}$$

then the wave equation for an acoustic horn becomes:

$$\frac{1}{A}\frac{d}{dx}\left(A\frac{dp}{dx}\right) + k^2 p = 0 \qquad k = \omega/c \tag{4.11}$$

and

$$v = -\frac{1}{i\omega\rho}\frac{dp}{dx} \tag{4.12}$$

The natural boundary conditions takes one of the two following forms:

(i) open end $\quad p = 0$

(ii) rigid end $\quad v = 0 \quad$ or $\quad dp/dx = 0$

Example 4.7 Resonances of an Acoustic Horn of Variable Cross Section

Obtain the natural frequencies of an acoustic horn, having a length L and a cross-sectional area varying according to the following law:

$$A(x) = A_0 x/L$$

A_0 being a reference area and the end $x = L$ is rigidly closed.

DE: $$\frac{1}{x}\frac{d}{dx}\left(x\frac{dp}{dx}\right) + k^2 p = 0 \qquad k = \omega/c$$

or

$$x^2 p'' + xp' + k^2 x^2 p = 0$$

The end $x = L$ has a zero particle velocity:

BC: $$\frac{dp(L)}{dx} = 0$$

The acoustic pressure is bounded in the horn, so that $p(0)$ must be bounded. The solution to the differential equation is given by:

$$p(x) = C_1 J_0(kx) + C_2 Y_0(kx)$$

Since $Y_0(kx)$ becomes unbounded at $x = 0$, then one must set $C_2 \equiv 0$. The boundary condition at $x = L$ is then satisfied:

$$v(L) = 0 \equiv \frac{dp(L)}{dx} = C_1 k \frac{dJ_0(kL)}{dkL} = -C_1 k J_1(kL) = 0 \qquad \text{(Characteristic equation)}$$

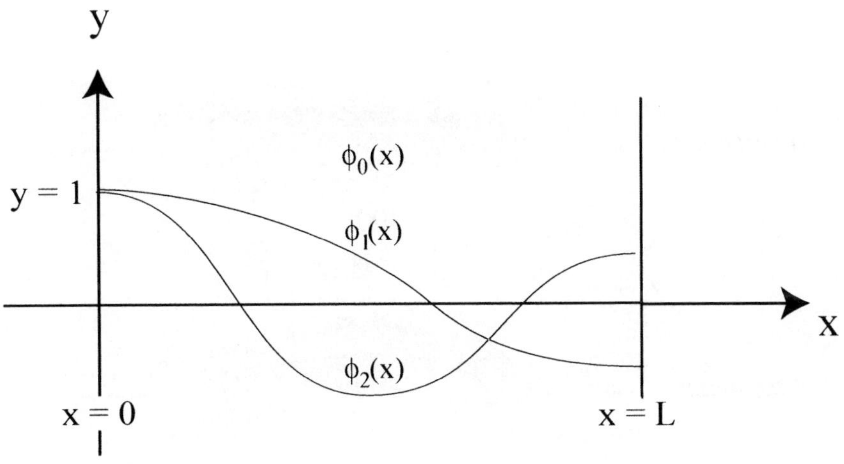

***Fig. 4.12**: First Three Eigenfunctions*

The roots of the characteristic equation (Section 3.15) and the corresponding eigenfunctions become:

$k_0 L = 0$	$\phi_0 = 1$
$k_1 L = 3.832$	$\phi_1 = J_0\left(3.832\, x/L\right)$
$k_2 L = 7.016$	$\phi_2 = J_0\left(7.016\, x/L\right)$
$k_3 L = 10.17$	$\phi_3 = J_0\left(10.17\, x/L\right)$

A plot of the first three modes is shown in Figure 4.12.

4.6 Stability of Compressed Columns

Consider a column of length L, having a cross-sectional area A, and moment of inertia I, being compressed by a force P as shown in Fig. 4.13.

If the beam is displaced laterally from out of its straight shape, then the moment at any cross section becomes:

$$M_x = -Py$$

which, when substituting M_x in Section 4.4 gives the following equation governing the stability of a compressed column:

$$\frac{d^2}{dx^2}\left(EI\frac{d^2y}{dx^2}\right) + P\frac{d^2y}{dx^2} = f(x) \tag{4.13}$$

Equation (4.13) can be integrated twice to give the following differential equation:

$$EI\frac{d^2y}{dx^2} + Py = \iint f(\eta)\, d\eta\, d\eta + C_1 + C_2 x \tag{4.14}$$

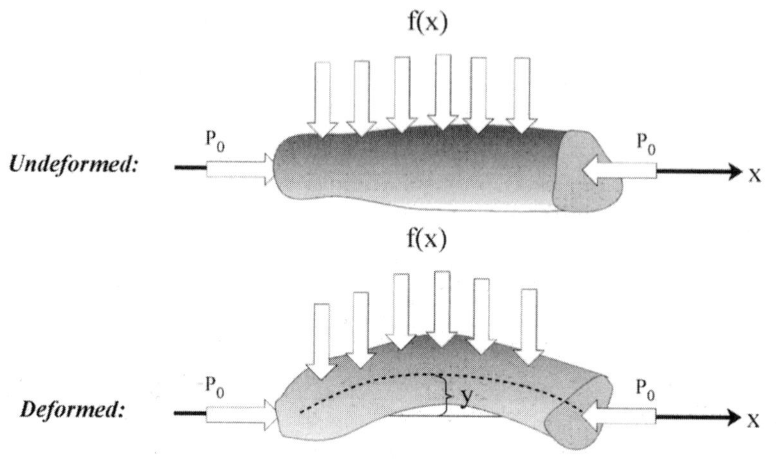

Fig 4.13: Column under P_o Load

Example 4.8 Stability of an Elastic Column

Obtain the critical loads and the corresponding buckling shapes of a compressed column fixed at x = 0 and elastically supported free-fixed end at x = L. The column has a constant cross section. The equation of the compressed column is:

DE: $$\frac{d^2y}{dx^2}+\frac{P}{EI}y = C_1 + C_2x$$

with boundary conditions specified as:

BC: $$y(0) = 0$$

$$y'(0) = 0$$

$$\frac{d^3y}{dx^3}(L)-\frac{\gamma}{EI}y(L)= 0$$

$$y'(L)= 0$$

The solution becomes:

$$y = C_1 + C_2x + C_3 \sin rx + C_4 \cos rx \qquad \text{where} \qquad r^2 = P/EI$$

$$y(0) = 0 \qquad C_1 + C_4 = 0$$

$$y'(0) = 0 \qquad C_2 + rC_3 = 0$$

$$y'(L)= 0 \qquad C_2 + rC_3 \cos rL - rC_4 \sin rL = 0$$

$$y'''(L)-\frac{\xi}{L^3}y(L) = 0 \qquad -\frac{\xi}{L^3}C_1 - \frac{\xi}{L^3}C_2L + C_3\left(-r^3 \cos rL - \frac{\xi}{L^3}\sin rL\right)$$

$$+ C_4\left(r^3 \sin rL - \frac{\xi}{L^3}\cos rL\right) = 0$$

where

$$\xi = \frac{\gamma L^3}{EI}$$

For a non-trivial solution, the determinant of the coefficients of C_1, C_2, C_3, and C_4 must vanish, resulting in the following characteristic equation:

$$\cos\alpha + \left(\frac{\alpha}{2} - \frac{\alpha^3}{2\xi}\right)\sin\alpha = 1 \qquad \text{where } \alpha = rL$$

The characteristic equation can be simplified further as follows:

$$\left(\sin\frac{\alpha}{2}\right)\left[\left(\frac{\alpha}{2} - \frac{\alpha^3}{2\xi}\right)\cos\frac{\alpha}{2} - \sin\frac{\alpha}{2}\right] = 0$$

All possible roots are the roots of either one of the following two characteristic equations:

(i) $\sin\frac{\alpha}{2} = 0 \qquad \text{where} \quad \alpha_n = 2n\pi \qquad n = 0, 1, 2, ...$

and

(ii) $\tan\frac{\alpha}{2} = \frac{\alpha}{2} - \frac{\alpha^3}{2\xi} = \frac{\alpha}{2} - \frac{4}{\xi}\left(\frac{\alpha}{2}\right)^3$

The roots of the second equation are sketched in Fig. 4.14.

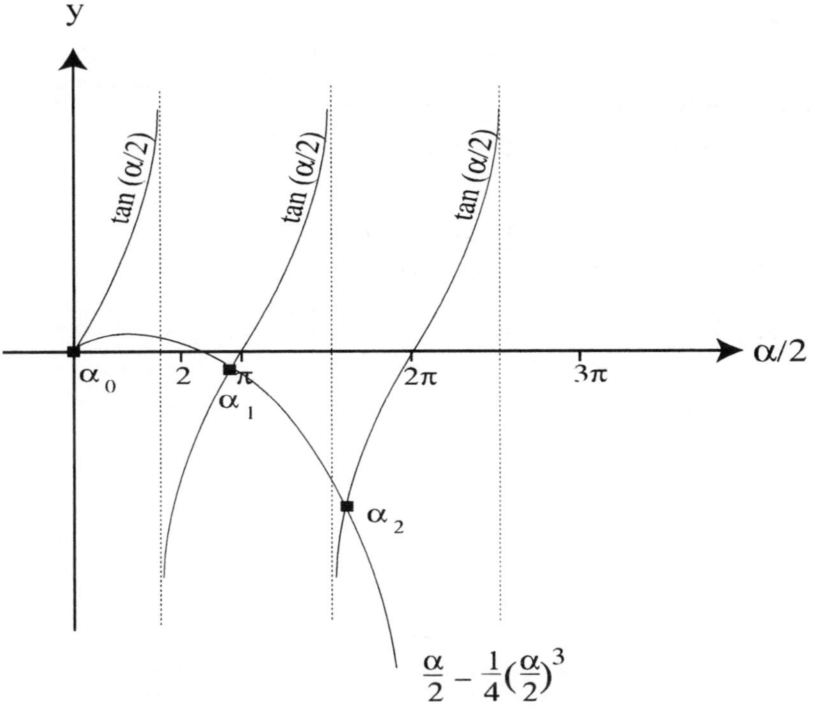

***Fig. 4.14** Curves for Characteristic Equation*

$\alpha_0 = 0$, α_1 falls between $\sqrt{\xi}$ and an integer number of 2π, etc., and

$$\text{Lim}\ \alpha_n \rightarrow (2n - 1)\ \pi \qquad n >> 1$$

For example, if $\xi = 16$, then the roots are:

$$\alpha_0 = 0,\ \alpha_1 = 4.74,\ \alpha_2 = 9.52 \approx 3\pi$$

Thus, the roots resulting from the two equations can be arranged in ascending values as follows:

0, 4.74, 6.28, 9.52, 12.50 ...

The eigenfunctions corresponding to these eigenvalues are:

(i) $\phi_n = 1 - \cos 2n\pi\ x/L \qquad n = 1, 2, 3,...$

for α_n being the roots of (i)

(ii) $\phi_n = 1 - \cos(\alpha_n\ x/L) - \cot(\alpha_n/2)(\alpha_n\ x/L - \sin\alpha_n\ x/L)$

where α_n are the roots of (ii).

Note that if $\alpha_n \rightarrow (2n - 1)\ \pi$ for $n >> 1$, then:

$$\phi_n \rightarrow 1 - \cos\alpha_n\ x/L \qquad n >> 1$$

Also note that $\alpha_0 = 0$ gives a trivial solution in either case.

4.7 Ideal Transmission Lines (Telegraph Equation)

Consider a lossless transmission line carrying an electric current, having an inductance per unit length L and a capacitance per unit length C. Consider an element of the wire of length dx shown in Fig. 4.15, with I and V representing the current and the voltage, respectively. Thus:

$$V_x - V_{x+dx} = \text{voltage drop} = (L dx)\ \partial I/\partial t$$

also

$$I_x - I_{x+dx} = \text{decrease in current} = (C dx)\ \partial V/\partial t$$

Thus, the two equations can be linearized as follows:

$$-\frac{\partial V}{\partial x} = L\frac{\partial I}{\partial t}$$

$$-\frac{\partial I}{\partial x} = C\frac{\partial V}{\partial t}$$

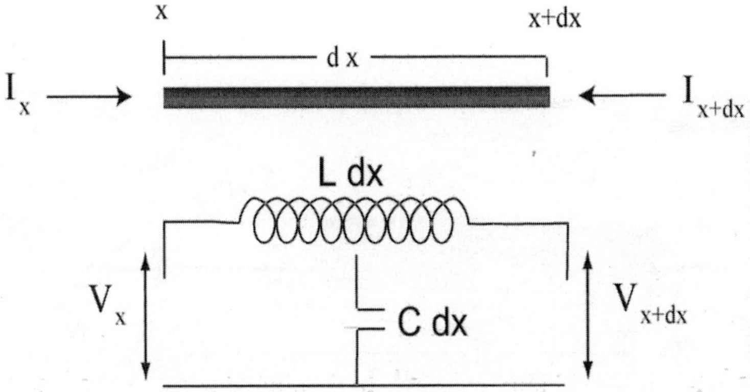

***Fig. 4.15:** Element of an Electrical Transmission Line*

Both equations combine to give differential equations on V and I as follows:

$$\frac{\partial^2 V}{\partial x^2} = LC\frac{\partial^2 V}{\partial t^2} \tag{4.15}$$

and

$$\frac{\partial^2 I}{\partial x^2} = LC\frac{\partial^2 I}{\partial t^2} \tag{4.16}$$

If the time dependence of the voltage and current is harmonic as follows:

$$V(x,t) = \overline{V}(x)e^{i\omega t}$$

$$I(x,t) = \bar{I}(x)\, e^{i\omega t}$$

then eqs. (4.15) and (4.16) become:

$$\frac{d^2\overline{V}}{dx^2} + \frac{\omega^2}{c^2}\overline{V} = 0 \tag{4.17}$$

and

$$\frac{d^2\bar{I}}{dx^2} + \frac{\omega^2}{c^2}\bar{I} = 0 \tag{4.18}$$

where $LC = 1/c^2$.

The natural boundary conditions for transmission lines can be one of the two following types:

(i) shorted end $\overline{V} = 0$ or $\dfrac{d\bar{I}}{dx} = 0$

(ii) open end $\bar{I} = 0$ or $\dfrac{d\overline{V}}{dx} = 0$

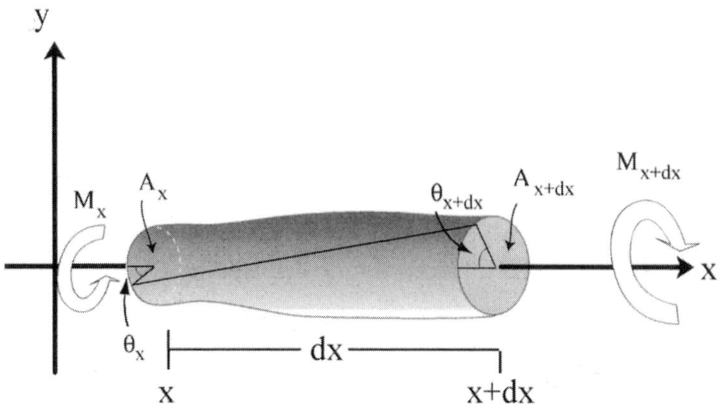

***Fig. 4.16:** Element of a Circular Bar Twisted in Torsion*

4.8 Torsional Vibration of Circular Bars

Consider a bar of cross-sectional area A, polar area moment of inertia J, mass density ρ and shear modulus G. The bar is twisted about its axis by torque M twisting the bar cross section by an angle θ*(x,t) at a station x as shown in Fig. 4.16.

$$\text{Shear strain at } r = \frac{r\,\theta^*(x+dx,t) - r\,\theta^*(x,t)}{dx} = r\frac{\partial \theta^*}{\partial x}$$

$$\text{Shear stress at } r = Gr\frac{\partial \theta^*}{\partial x}$$

$$\text{Torque } M = \int_A \left(G\frac{\partial \theta^*}{\partial x} \right) r^2\, dA = GJ\frac{\partial \theta^*}{\partial x}$$

where the polar moment of inertia J is given by:

$$J = \int_A r^2 dA$$

The equilibrium equation of the twisting element becomes:

$$M_{x+dx} - M_x + f^*(x,t)\, dx = (\rho dx) J \frac{\partial^2 \theta^*}{\partial t^2}$$

Thus:

$$\frac{\partial M}{\partial x} = \frac{\partial}{\partial x}\left(GJ\frac{\partial \theta^*}{\partial x} \right) = \rho J \frac{\partial^2 \theta^*}{\partial t^2} - f^* \tag{4.19}$$

where f*(x,t) is the distributed external torque. If G is constant, then the torsional wave equation becomes:

$$\frac{1}{J}\frac{\partial}{\partial x}\left(J\frac{\partial \theta^*}{\partial x} \right) = \frac{1}{c^2}\frac{\partial^2 \theta^*}{\partial t^2} - \frac{f^*}{GJ} \tag{4.20}$$

where c is the shear sound speed in the bar defined by:

$$c^2 = G/\rho$$

If $f^*(x,t) = f(x) \sin \omega t$ and $\theta^*(x, t) = \theta(x) \sin \omega t$, then eq. (4.20) becomes:

$$\frac{1}{J}\frac{d}{dx}\left(J\frac{d\theta}{dx}\right) + k^2\theta = -\frac{f}{GJ} \tag{4.21}$$

If the polar moment of inertia J is constant, then:

$$\frac{d^2\theta}{dx^2} + k^2\theta = -\frac{f}{GJ} \tag{4.22}$$

The natural boundary conditions take one of the following forms:

(i) fixed end $\qquad \theta = 0$

(ii) free end $\qquad M = GJ\frac{\partial\theta}{\partial x} = 0$

(iii) elastically supported end by helical spring

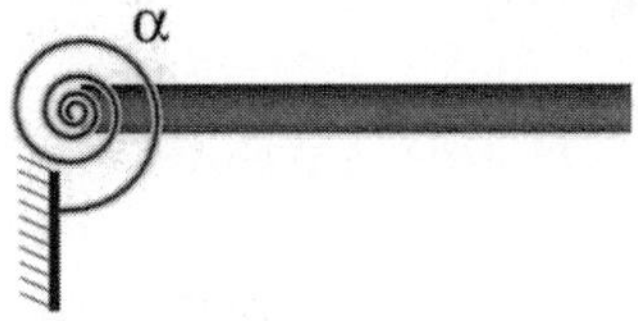

$$GJ\frac{\partial\theta}{\partial x} \mp \alpha\theta = 0$$

The + and - signs refer to the BC's at the right and left sides.

4.9 Orthogonality and Orthogonal Sets of Functions

The concept of orthogonality of a pair of functions $f_1(x)$ and $f_2(x)$ can be defined through an integral over a range [a,b]:

$$\left(f_1(x), f_2(x)\right) = \int_a^b f_1(x)\, f_2(x)\, dx$$

If the functions $f_1(x)$ and $f_2(x)$ are orthogonal, then:

Define the norm of f(x) as:

$$N\left(f(x)\right) = \int_a^b [f(x)]^2 dx$$

A set of orthogonal functions $\{f_i(x)\} i = 1, 2,...$ is one where every pair of functions of the set is orthogonal, i.e., a set $\{f_m(x)\}$ is an orthogonal set if:

$$\int_a^b f_m(x)\, f_n(x)\, dx = 0 \qquad m \neq n$$

$$= N(f_m(x)) \qquad m = n$$

If one defines:

$$g_n(x) = \frac{f_n(x)}{\sqrt{N(f_n(x))}}$$

then the orthogonal set $\{g_n(x)\}$ is called an **Orthonormal set**, since:

$$\int_a^b g_n(x)\, g_m(x)\, dx = \delta_{mn}$$

where the Kronecker delta $\delta_{mn} = 1 \qquad n = m$

$$= 0 \qquad n \neq m$$

In some cases, a set of functions $\{f_n(x)\}$ is orthogonal with respect to a **"Weighting Function"** $w(x)$ if:

$$(f_n, f_m) \equiv \int_a^b w(x)\, f_n(x)\, f_m(x)\, dx = 0 \qquad m \neq n$$

where the norm of $f_n(x)$ is defined as:

$$N(f_n(x)) = \int_a^b w(x)\, f_n^2(x)\, dx$$

A more formal definition of orthogonality, one that can be applied to real as well as complex functions, takes the following form:

$$\int_a^b f_n(z)\, \bar{f}_m(z)\, dz = 0 \qquad n \neq m$$

where $\bar{f}$ is the complex conjugate function of f. The norm is then defined as:

$$N(f_n(z)) = \int_a^b f_n(z)\, \bar{f}_n(z)\, dz = \int_a^b |f_n(z)|^2\, dz$$

Example 4.9

(i) The set $g_n(x) = \dfrac{\sin(\frac{n\pi}{L} x)}{\sqrt{L/2}} \qquad n = 1, 2, 3, \ldots \text{in } 0 \le x \le L$

constitutes an orthogonormal set, where:

$$\frac{2}{L}\int_0^L \sin\left(\frac{n\pi}{L}x\right)\sin\left(\frac{m\pi}{L}x\right)dx = \delta_{mn}$$

(ii) The set $g_n(x) = \dfrac{J_0(\alpha_n x/L)}{LJ_1(\alpha_n)/\sqrt{2}}$ $\qquad n = 1, 2, 3, \ldots$ in $0 \leq x \leq L$

constitutes an orthonormal set, where $\{g_n(x)\}$ is orthogonal with $w(x) = x$:

$$\frac{2}{L^2 J_1^2(\alpha_n)}\int_0^L x\, J_0(\alpha_n x/L)\, J_0(\alpha_m x/L)\, dx = \delta_{mn}$$

where α_n are the roots of $J_0(\alpha_n) = 0$.

(iii) The set $g_n(x) = \dfrac{e^{inx}}{\sqrt{2\pi}}$ $\qquad n = 0, 1, 2, \ldots$ in $-\pi \leq x \leq \pi$

constitutes an orthonormal set where:

$$\frac{1}{2\pi}\int_{-\pi}^{\pi} e^{inx}\, \overline{e^{imx}}\, dx = \frac{1}{2\pi}\int_{-\pi}^{\pi} e^{inx}\, e^{-imx}\, dx = \delta_{mn}$$

4.10 Generalized Fourier Series

Consider an infinite orthonormal set $\{g_n(x)\}$ orthogonal over $[a,b]$. Then one can approximate any arbitrary function $F(x)$ defined on $[a,b]$ in terms of a finite series of the functions $g_m(x)$. Let:

$$F(x) = c_1 g_1 + c_2 g_2 + \ldots + c_N g_N = \sum_{m=1}^{N} c_m g_m(x) \tag{4.23}$$

then multiplying the equality by $g_n(x)$, n being any integer number $1 \leq n \leq N$, and integrating on $[a,b]$, one obtains:

$$\int_a^b F(x)\, g_n(x)\, dx = \sum_{m=1}^{N} c_m \int_a^b g_m(x)\, g_n(x)\, dx = c_n$$

since every term vanishes because of the orthogonality of the set $\{g_n(x)\}$, except for the term $m = n$. Thus the coefficient of the expansion, called the **Fourier Coefficient**, becomes:

$$c_m = \int_a^b F(x)\, g_m(x)\, dx$$

and $F(x)$ can be represented by a series of orthonormal functions as follows:

$$F(x) \sim \sum_{m=1}^{N} g_m(x) \left(\int_a^b F(\eta) g_m(\eta) \, d\eta \right) \qquad a \le x \le b \tag{4.24}$$

The series representation of eq. (4.24) is called the **Generalized Fourier Series** corresponding to F(x). The symbol ~, used for the representation instead of an equality, refers to the possibility that the series may not converge to F(x) at some point or points in [a,b]. If an orthonormal set $\{g_n\}$ extends to an infinite dimensional space, then N extends to infinity.

The Generalized Fourier Series is the best **approximation in the mean** to a function F(x). Consider a finite number of an orthonormal set as follows:

$$\sum_{m=1}^{n} k_m g_m(x)$$

then one can show that the best least square approximation to F(x) is that where $c_m = k_m$.

The square of the error J between the function F(x) and its representation, defined as:

$$J = \int_a^b \left[F(x) - \sum_{m=1}^{n} k_m g_m(x) \right]^2 dx \ge 0$$

must be minimized. The square of the error is expanded as:

$$J = \int_a^b F^2 \, dx - 2 \sum_{m=1}^{n} k_m \int_a^b F(x) g_m(x) \, dx + \int_a^b \left[\sum_{m=1}^{n} k_m g_m(x) \right] \left[\sum_{r=1}^{n} k_r g_r(x) \right] dx$$

Since the set $\{g_m\}$ is an orthonormal set, then J becomes:

$$J = \int_a^b F^2 \, dx - 2 \sum_{m=1}^{n} c_m k_m + \sum_{m=1}^{n} k_m^2 \ge 0$$

$$= \int_a^b F^2 \, dx + \sum_{m=1}^{n} \left[(k_m - c_m)^2 - c_m^2 \right] \ge 0$$

$$= \int_a^b F^2 \, dx - \sum_{m=1}^{n} c_m^2 + \sum_{m=1}^{n} (k_m - c_m)^2 \ge 0$$

To minimize J, which is positive, then one must choose $k_m = c_m$. Thus, the series:

$$\sum_{m=1}^{n} c_m g_m(x)$$

is the **best approximation in the mean** to the function F(x). Since $J \ge 0$, then:

$$\int_a^b F^2 \, dx \ge \sum_{m=1}^{n} c_m^2$$

The above inequality is not restricted to a specific number n, thus:

$$\int_a^b F^2\, dx \geq \sum_{m=1}^{\infty} c_m^2 \tag{4.25}$$

Since $\int_a^b F^2\, dx$ is finite, then the **Fourier Coefficients** c_m must constitute a convergent series, i.e.:

$$\lim_{m \to \infty} c_m = \left[\int_a^b F(x)\, g_m\, dx \right] \to 0$$

A <u>necessary</u> and <u>sufficient</u> condition for an orthonormal set $\{g_n(x)\}$ to be <u>complete</u> is that:

$$\int_a^b F^2(x)\, dx = \sum_{m=1}^{\infty} c_m^2$$

The generalized Fourier series representing a function F(x) is *unique*. Thus two functions represented by the same generalized Fourier series must be equal, if the set $\{g_n\}$ is complete.

If an orthonormal set is complete and continuous and if the generalized Fourier series corresponding to F(x) is uniformly convergent in [a,b], then the series converges uniformly to F(x) on [a,b], if F(x) is continuous.

Similar expansions to eq. (4.24) can be developed, if the orthonormal set $\{g_n(x)\}$ is orthonormal with respect to a weighting function w(x) as follows:

$$F(x) = \sum_{m=1}^{\infty} c_m g_m(x)$$

where

$$c_m = \int_a^b w(x)\, F(x)\, g_m(x)\, dx$$

and

$$\int_a^b w(x)\, g_m(x)\, g_n(x)\, dx = \delta_{mn} \tag{4.26}$$

4.11 Adjoint Systems

Consider the linear n^{th} order differential operator L:

$$Ly = \left[a_0(x)\frac{d^n}{dx^n} + a_1(x)\frac{d^{n-1}}{dx^{n-1}} + \ldots + a_{n-1}(x)\frac{d}{dx} + a_n(x)\right] y = 0 \quad a \le x \le b \tag{4.27}$$

where $a_0(x)$ does not vanish in [a,b] and the coefficients a_i, i = 0, 1, 2,... n are continuous and differentiable (n-i) times, then define the linear n^{th} order differential operator K:

$$Ky = (-1)^n \frac{d^n}{dx^n}[a_0(x)\,y] + (-1)^{n-1}\frac{d^{n-1}}{dx^{n-1}}[a_1(x)\,y] + (-1)^{n-2}\frac{d^{n-2}}{dx^{n-2}}[a_2(x)y]$$

$$-\frac{d}{dx}[a_{n-1}(x)\,y] + a_n(x)y \tag{4.28}$$

as the **Adjoint operator** to the operator L. The differential equation:

$$Ky = 0$$

is the adjoint differential equation of (4.27).

The operator L and its adjoint operator K satisfy the following identify:

$$v\,Lu - u\,Kv = \frac{d}{dx}P(u,v) \tag{4.29}$$

where

$$P(u,v) = \sum_{m=0}^{n-1}\frac{d^m u}{dx^m}\left\{\sum_{k=0}^{n-m-1}(-1)^k\frac{d^k}{dx^k}\left(a_{n-m-k-1}(x)\,v\right)\right\} \tag{4.30}$$

Equation (4.29) is known as **Lagrange's Identity**.

The determinant Δ(x) of the coefficients of the bilinear form of $u^{(i)}\,v^{(j)}$ becomes:

$$\Delta(x) = \pm[a_0(x)]^n \tag{4.31}$$

which does not vanish in $a \le x \le b$.

Integrating eq. (4.29), one obtains **Green's formula** having the form:

$$\int_a^b (v\,Lu - u\,Kv)\,dx = P(u,v)\Big|_a^b \tag{4.32}$$

The determinant of the bilinear form of the right side of eq. (4.32) becomes:

$$\begin{vmatrix}\Delta(a) & 0\\ 0 & \Delta(b)\end{vmatrix} = \Delta(a)\,\Delta(b) = [a_0(a)\,a_0(b)]^n \neq 0$$

If the operators K = L, then the operator L and K are called **Self-Adjoint**.

As an example, take the general second-order differential equation:

$$Ly = a_0(x)\,y'' + a_1(x)\,y' + a_2(x)\,y = 0$$

then the adjoint operator K becomes:

$$Ky = (a_0 y)'' - (a_1 y)' + a_2 y$$
$$= a_0 y'' + (2a_0' - a_1) y' + (a_0'' - a_1' + a_2) y$$

which is not equal to Ly in general and hence the operator L is not self-adjoint. If the operator L is self-adjoint, then the following equalities must hold:

$$a_1 = 2a_0' - a_1 \qquad \text{and} \qquad a_2 = a_0'' - a_1' + a_2$$

which can be satisfied by one relationship, namely:

$$a_0' = a_1$$

which is not true in general. However, one can change the second-order operator L by a suitable function multiplier so that it becomes self-adjoint, an operation that is valid only for the second-order operator. Hence, if one multiplies the operator L_1 by an undetermined function z(x), then:

$$L_1 y = z\, Ly$$

so that L_1 is self-adjoint, then each coefficient is multiplied by z(x). Since the condition for self-adjoincy requires that the differential of the first coefficient of L equals the second, then:

$$(z a_0)' = z a_1$$

which is rewritten as:

$$\frac{z'}{z} = \frac{a_1 - a_0'}{a_0}$$

The function z can be obtained readily by integrating the above differentials:

$$z = \frac{1}{a_0(x)} \exp\left[\int^x \frac{a_1(\eta)}{a_0(\eta)} d\eta\right] = \frac{p(x)}{a_0(x)}$$

Using the multiplier function z(x), the self-adjoint operator L_1 can be rewritten as:

$$L_1 y = p(x)\, y'' + \frac{a_1(x)}{a_0(x)} p(x) y' + \frac{a_2(x)}{a_0(x)} p(x) y$$
$$= \left[\frac{d}{dx}\left(p\frac{d}{dx}\right) + q\right] y$$

where

$$p(x) = \exp\left[\int^x \frac{a_1(\eta)}{a_0(\eta)} d\eta\right]$$

and

$$q(x) = \frac{a_2(x)}{a_0(x)} p(x) \tag{4.33}$$

Thus, any second-order, linear differential equation can be transformed to a form that is self-adjoint. The method used to change a second-order differential operator L to become self-adjoint cannot be duplicated for higher-order equations. In general if the order n is an odd integer, then that operator cannot be self-adjoint, since that requires that $a_0(x) = - a_0(x)$. It should be noted that if the order n of the differential operator L is an odd integer, then the differential equation is not invariant to coordinate inversion, i.e., the operator is not the same if x is changed to (-x). Therefore, if the independent variable x is a spatial coordinate, then the operator L, representing the system's governing equation, would have a change of sign of the coefficient of its highest derivative if x is changed to (-x). This would lead to a solution that is drastically different from that due to an uninverted operator L. Thus, a differential operator L which represent a physical system's governing equation on a spatial coordinate x cannot have an odd order n.

In general, a physical system governed by a differential operator L on a spatial coordinate is self-adjoint if the system satisfies the law of conservation of energy. Thus, if the governing equations are derived from a Lagrangian function representing the total energy of a system, then, the differential operator L is self-adjoint. A general form of a linear, non-homogeneous $(2n)^{th}$ order differential operator L which is self-adjoint can be written as follows:

$$Ly = \sum_{k=0}^{k=n} (-1)^k \frac{d^k}{dx^k}\left[p_{n-k}(x)\frac{d^k y}{dx^k}\right]$$

$$= (-1)^n\left[p_0 y^{(n)}\right]^{(n)} + (-1)^{n-1}\left[p_1 y^{(n-1)}\right]^{(n-1)} + \ldots - \left[p_{n-1}y'\right]' + p_n y = f(x)$$

$$a \le x \le b \qquad (4.34)$$

4.12 Boundary Value Problems

As mentioned earlier, the solution of a system is unique iff n conditions on the function y and its derivatives up to (n - 1) are specified at the end points a and b. Thus, a general form of non-homogeneous boundary conditions can be written as follows:

$$U_i(y) = \sum_{k=0}^{n-1}\left[\alpha_{ik}y^{(k)}(a) + \beta_{ik}y^{(k)}(b)\right] = \gamma_i \qquad i = 1, 2, 3, \ldots, n \qquad (4.35)$$

where α_{ik}, β_{ik} and γ_i are real constants. The boundary conditions in eq. (4.35) must be independent. This means that the determinant:

$$\det\left[\alpha_{ij}, \beta_{ij}\right] \neq 0$$

The non-homogeneous differential equation (4.27) and the non-homogeneous boundary conditions eq. (4.35) constitute a general form of boundary value problems. A *necessary* and *sufficient* condition for the solution of such problems to be unique is that the equivalent homogeneous system:

$$Ly = 0$$

$$U_i(y) = 0 \qquad i = 1, 2, \ldots, n$$

has only the trivial solution $y \equiv 0$. Thus, an $(n)^{th}$ order self-adjoint operator given in eq. (4.27) has n independent solutions $\{y_i(x)\}$. Thus, since the set of n homogeneous conditions given in eq (4.35) are independent, then the solution of the differential eq. (4.34) can be written as:

$$y = y_p(x) + \sum_{i=1}^{n} C_i y_i(x)$$

where C_i are arbitrary constants. Since the set of n non-homogeneous boundary conditions in eq. (4.35) are independent, then there exists a non-vanishing unique set of constants $[C_i]$ which satisfies these boundary conditions.

A homogeneous boundary value problem consists of an nth differential operator and a set of n linear boundary conditions, i.e.:

$$Lu = 0 \tag{4.27}$$

$$U_i(u) = 0 \qquad i = 1, 2, \ldots n \tag{4.35}$$

An adjoint system to that defined above is defined by:

$$Kv = 0 \tag{4.28}$$

$$V_i(v) = 0 \qquad i = 1, 2, \ldots n \tag{4.36}$$

where the homogeneous boundary conditions in eq. (4.36) are obtained by substituting the boundary conditions $U_i(u) = 0$ in eq. (4.35) into:

$$P(u, v)\Big|_a^b = 0 \tag{4.37}$$

with the bilinear form P(u, v) being given in eq. (4.30). If the operator L is a self-adjoint operator, i.e., if K = L, then the boundary conditions can be shown to be identical, i.e.:

$$U_i(u) = V_i(u) \tag{4.38}$$

Example 4.10

For the operator:

$$Ly = a_0(x)\, y'' + a_1(x) y' + a_2(x) y = 0 \qquad a \le x \le b$$

the adjoint operator is given by:

$$Ky = (a_0 y)'' - (a_1 y)' + a_2 y = 0$$

The bilinear form P(u, v) is given by:

$$P(u, v)\Big|_a^b = u\left[a_1 v - (a_0 v)'\right] + u'\left[a_0 v\right]\Big|_a^b = 0$$

(i) Consider the boundary condition pair on u given by:

$u(a) = 0$ and $u(b) = 0$

and substitute into eq. (4.37) results in the following:

$$u'(a_0 v)\Big|_a^b = u'(b)[a_0(b)\, v(b)] - u'(a)[a_0(a)\, v(a)] = 0$$

Since u(b) = 0 then $u'(b)$ is an arbitrary constant. Similarly since u(b) = 0, then $u'(a)$ is also arbitrary. For arbitrary constants $u'(a)$ and $u'(b)$, the relation can be satisfied iff:

$$v(a) = 0 \qquad \text{and} \qquad v(b) = 0$$

(ii) If $u'(a) = 0$ and $u(b) = 0$
then one obtains the following when substituted into P(u,v) = 0:

$$u'(b)[a_0(b)\, v(b)] - u(a)\left[a_1(a)\, v(a) - (a_0(a)\, v(a))'\right] = 0$$

Since $u'(a) = 0$ then u(a) is arbitrary. Similarly, since u(b) = 0, then $u'(b)$ is arbitrary. Thus, the boundary conditions $V_i(v) = 0$ are:

$$a_1(a)\, v(a) - [a_0(a)\, v(a)]' = 0$$

and

$$v(b) = 0$$

4.13 Eigenvalue Problems

An eigenvalue problem is a system that satisfies a differential equation with an unspecified arbitrary constant λ and satisfying a homogenous or non-homogeneous set of boundary conditions.

Consider a general form of a homogeneous eigenvalue problem:

$$Ly + \lambda My = 0 \tag{4.39}$$

$$U_i(y) = 0 \qquad i = 1, 2, ..., n$$

where L is given by eq. (4.27) and the boundary conditions by eq. (4.35). The operator M is an mth order differential operator where m < n and λ is an arbitrary constant.

A general form of a self-adjoint homogeneous eigenvalue problem takes the following form:

$$Ly + \lambda My = 0 \qquad a \le x \le b$$

$$U_i(y) = 0 \qquad i = 1, 2, ..., 2n \tag{4.40}$$

where L and M are linear self-adjoint operators of order 2n and 2m respectively, where:

$$Ly = \sum_{k=0}^{n} (-1)^k \frac{d^k}{dx^k}\left[p_{n-k} \frac{d^k y}{dx^k}\right] \tag{4.41}$$

$$My = \sum_{k=0}^{m} (-1)^k \frac{d^k}{dx^k}\left[q_{m-k} \frac{d^k y}{dx^k}\right] \qquad n > m$$

λ is an undetermined parameter, and $U_i(y) = 0$ are 2n homogeneous boundary conditions having the form given in eq. (4.35).

Define a **Comparison Function** u(x) as an arbitrary function that has 2n continuous derivatives and *satisfies the boundary conditions* $U_i(u) = 0$, i = 1, 2,..., 2n. For **self-adjoint** eigenvalue problems the following integrals vanish:

$$\int_a^b (u\,Lv - v\,Lu)\,dx = 0$$

and

$$\int_a^b (u\,Mv - v\,Mu)\,dx = 0 \tag{4.42}$$

where u and v are arbitrary comparison functions.

The expression for P(u,v) in eq. (4.30) that corresponds to a differential operator L or M given in eq. (4.40) becomes:

$$\int_a^b (u\,Lv - v\,Lu)\,dx = P(v,u)\Big|_a^b = \sum_{k=1}^{n} \sum_{r=0}^{k-1} (-1)^{k+r} \Big\{ u^{(r)} \left[p_{n-k} v^{(k)}\right]^{(k-r-1)} - v^{(r)} \left[p_{n-k} u^{(k)}\right]^{(k-r-1)} \Big\}\Big|_a^b = 0 \tag{4.43}$$

Similar expression for P(v,u) for the differential operator M can be developed by substituting m and q_i in eq. (4.43) for n and p_i, respectively. It is obvious that the right side of eq. (4.43) must vanish for the system to be self-adjoint.

An eigenvalue problem is called **Positive Definite** if, for every non-vanishing comparison function u, the following inequalities hold:

$$\int_a^b u\,Lu\,dx < 0 \qquad \text{and} \qquad \int_a^b u\,Mu\,dx > 0 \tag{4.44}$$

Example 4.11

Examine the following eigenvalue problem for self-adjointness and positive-definiteness:

$$y'' + \lambda r(x)\,y = 0 \qquad r(x) > 0 \qquad a \le x \le b$$

$$y(a) = 0 \qquad y(b) = 0$$

For this problem the operators L and M, defined as:

$$L = \frac{d^2}{dx^2} \qquad M = r(x)$$

are self-adjoint. Let u and v be comparison functions, such that:

$$u(a) = v(a) = 0 \qquad u(b) = v(b) = 0$$

Thus, to establish if the system is self-adjoint, one substitutes into eq. (4.42):

$$\int_a^b (uv'' - vu'')\,dx = uv' - vu'\Big|_a^b - \int_a^b (u'v' - v'u')\,dx = 0$$

and

$$\int_a^b (u\,rv - v\,ru)\,dx = 0$$

which proves that the eigenvalue problem is *self-adjoint*. To establish that the problem is also positive definite, substitute L and M into eq. (4.44):

$$\int_a^b u\,u''\,dx = u\,u'\Big|_a^b - \int_a^b (u')^2\,dx = -\int_a^b (u')^2\,dx < 0$$

$$\int_a^b u\,ru = \int_a^b ru^2\,dx > 0 \qquad \text{since } r(x) > 0$$

which indicates that the eigenvalue problem is also positive definite.

4.14 Properties of Eigenfunctions of Self-Adjoint Systems

Self-adjoint eigenvalue problems have few properties unique to this system.

(i) **Orthogonal Eigenfunctions**

If the eigenvalue problem is self-adjoint, then the *eigenfunctions are orthogonal.* Let ϕ_n and ϕ_m be any two eigenfunctions corresponding to different eigenvalues λ_n and λ_m, then each satisfies its respective differential equation, i.e.:

$$L\phi_n + \lambda_n M\phi_n = 0 \qquad \text{and} \qquad L\phi_m + \lambda_m M\phi_m = 0$$

where $n \neq m$ and $\lambda_n \neq \lambda_m$.

Multiplying the first equation by ϕ_m, the second by ϕ_n, subtracting the resulting equations and integrating the final expression on [a,b], one obtains:

$$\int_a^b [\phi_m L\phi_n - \phi_n L\phi_m]\,dx + \lambda_n \int_a^b \phi_m M\phi_n\,dx - \lambda_m \int_a^b \phi_n M\phi_m\,dx = 0$$

Since the system of differential operators and boundary conditions is self-adjoint, and since $\lambda_n \neq \lambda_m$, then the integral:

$$\int_a^b \phi_m M\phi_n\,dx = 0 \qquad n \neq m \tag{4.45}$$

$$= N_n \qquad n = m$$

is a generalized form of an orthogonality integral, with N_n being the normalization constant.

(ii) **Real Eigenfunctions and Eigenvalues**

If the system is self-adjoint and positive definite, then the *eigenfunctions are real* and the *eigenvalues are real and positive*. Assuming that a pair of eigenfunctions and eigenvalues are complex conjugates, i.e., let:

$$\phi_n = u_n(x) + i v_n(x) \qquad \lambda_n = \alpha_n + i\beta_n$$

$$\phi_n^* = u_n(x) - i v_n(x) \qquad \lambda_n^* = \alpha_n - i\beta_n$$

then the orthogonality integral in eq. (4.45) results in the following integral:

$$(\lambda - \lambda^*) \int_a^b \phi_n M \phi_n^* \, dx = 0$$

Since the eigenvalues are complex, i.e., $\beta_n \neq 0$, then:

$$\int_a^b \phi_n M \phi_n^* \, dx = 0$$

which results in the following real integral:

$$\int_a^b (u_n M u_n + v_n M v_n) \, dx = 0$$

Invoking the definition of a positive definite system, both of these integrals are positive, which indicates that the only complex eigenfunction possible is the null function, i.e., $u_n = v_n = 0$. One can also show that the eigenvalues λ_n are real and positive. Starting out with the differential equation satisfied by either ϕ_n or ϕ_n^*, i.e.:

$$L\phi_n + \lambda_n M \phi_n = 0$$

and multiplying this equation by ϕ_n^* and integrating over [a,b], one obtains an expression for λ_n:

$$\lambda_n = \alpha_n + i\beta_n = -\frac{\int_a^b \phi_n^* L \phi_n \, dx}{\int_a^b \phi_n^* M \phi_n \, dx} = -\frac{\int_a^b (u_n L u_n + v_n L v_n) \, dx}{\int_a^b (u_n M u_n + v_n M v_n) \, dx}$$

Since the system is positive definite and the integrands are real, then these integrals are real, which indicates that $\beta_n \equiv 0$ and λ_n is real. Since the system is positive definite, then the eigenvalues λ_n are also positive. Having established that the eigenvalues of a self-adjoint positive definite system are real and positive one can obtain a formula for λ_n. Starting with the equation satisfied by ϕ_n:

$$L\phi_n + \lambda_n M \phi_n = 0$$

and multiplying the equation by ϕ_n and integrating the resulting equation on [a,b], one obtains:

$$\lambda_n = -\frac{\int_a^b \phi_n L\phi_n \, dx}{\int_a^b \phi_n M\phi_n \, dx} > 0 \tag{4.46}$$

(iii) **Rayleigh Quotient**

The eigenvalues λ_n obtained from eq. (4.46) require the knowledge of the exact form of the eigenfunction $\phi_n(x)$, which of course could have been obtained only if λ_n is already known. However, one can obtain an approximate upper bound to these eigenvalues if one can estimate the form of the eigenfunction. Define the Rayleigh quotient R(u) as:

$$R(u) = -\frac{\int_a^b uLu \, dx}{\int_a^b uMu \, dx} \tag{4.47}$$

where u is a non-vanishing comparison function. It can be shown that for a self-adjoint and positive definite system:

$$\lambda_1 = \min R(u)$$

where u runs through all possible non-vanishing comparison functions. It can also be shown that if u runs through all possible comparison functions that are orthogonal to the first r eigenfunctions, i.e.:

$$\int_a^b uM\phi_i \, dx = 0 \qquad i = 1, 2, 3, \ldots r$$

then

$$\lambda_{r+1} = \min R(u)$$

Example 4.12

Obtain approximate values of the first two eigenvalues of the following system:

$$y'' + \lambda y = 0 \qquad 0 \le x \le \pi$$

$$y(0) = 0 \qquad y(\pi) = 0$$

For this system, $L = d^2/dx^2$ and $M = 1$ and hence the system is self-adjoint and also positive definite. Solving the problem exactly, one can show that it has the following eigenfunctions and eigenvalues:

$$\phi_n(x) = \sin(nx)$$

and

$$\lambda_n = n^2 \qquad\qquad n = 1, 2, 3, \ldots$$

Using the definition of L and M, one can show that Rayleigh's quotient becomes:

$$R(u) = -\frac{\int_a^b u\, u''\, dx}{\int_a^b u^2\, dx} = \frac{\int_a^b (u')^2\, dx}{\int_a^b u^2\, dx}$$

where min $[R(u)] = \lambda_1 = 1.00$

One can choose the following comparison functions which satisfies $u(0) = u(\pi) = 0$ and has no other null between 0 and π, approximating $\phi_1(x)$:

$$u_1(x) = \begin{cases} x/\pi & 0 \le x \le \pi/2 \\ 1 - x/\pi & \pi/2 \le x \le \pi \end{cases}$$

which is not a proper comparison function, because u' is discontinuous. The Rayleigh quotient gives:

$$R_1(u) = +\frac{\int_0^{\pi/2} (1/\pi)^2\, dx + \int_{\pi/2}^{\pi} (-1/\pi)^2\, dx}{\int_0^{\pi/2} (x/\pi)^2\, dx + \int_{\pi/2}^{\pi} (1 - x/\pi)^2\, dx} = \frac{12}{\pi^2} = 1.23 > 1.00$$

If one was to use a comparison function that is at least once differentiable, again approximating $\phi_1(x)$ such as:

$$u_1(x) = x(\pi - x)$$

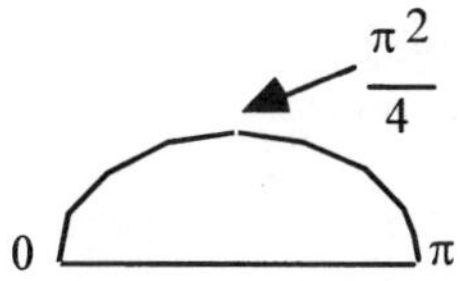

$$R_1(u) = \frac{\int_0^{\pi} (\pi - 2x)^2\, dx}{\int_0^{\pi} x^2(\pi - x)^2\, dx} = \frac{10}{\pi^2} = 1.03 > 1.00$$

which represents an error of 3 percent.

It can be seen that $R(u) > \lambda_1 = 1$, i.e., it is an upper bound to λ_1 and that the closer u_1 comes to sin x, the closer the Rayleigh quotient approaches λ_1.

To obtain an approximate value for $\lambda_2 = 4.00$, one can use a comparison function $u_2(x)$ that has one more null than $u_1(x)$, e.g.:

$$u_2(x) = 4x/\pi \qquad 0 \le x \le \pi/4$$
$$= 2 - 4x/\pi \qquad \pi/4 \le x \le 3\pi/4$$
$$= -4 + 4x/\pi \qquad 3\pi/4 \le x \le \pi$$

whose u' is not continuous. Substituting $u_2(x)$ into $R(u)$, one obtains:

$$R_2(u) = 4.86 > 4.00$$

which has a 21 percent error. Using a comparison function which is at least once differentiable, e.g.:

$$u_2(x) = x(\pi/2 - x) \qquad 0 \le x \le \pi/2$$
$$= (x - \pi)(x - \pi/2) \qquad \pi/2 \le x \le \pi$$

then the quotient gives:

$$R_2(u) = 4.053 > 4.00$$

One should note that the error is down to 1.3 percent for a comparison function which is at least once differentiable.

4.15 Sturm-Liouville System

The Sturm-Liouville (S-L) system is a special case of eq. (4.40) limited to a second-order eigenvalue problem. Starting with a general, second-order operator with an arbitrary parameter:

$$a_0(x)\, y'' + a_1(x)\, y' + a_2(x)\, y + \lambda a_3(x)\, y = 0 \qquad a \le x \le b \tag{4.48}$$

then one rewrites eq. (4.48) in a self-adjoint form by using a multiplier function to the differential equation in the form:

$$\mu(x) = \frac{p(x)}{a_0(x)}$$

where

$$p(x) = \exp\left(\int a_1(x)/a_0(x)\, dx \right)$$

then the differential equation can be rewritten in the form:

$$\left[p(x)\, y'\right]' + q(x)\, y + \lambda r(x)\, y = 0 \qquad a \le x \le b \tag{4.49}$$

where

$$q(x) = a_2(x)\, p(x)/a_0(x)$$

and

$$r(x) = a_3(x)\, p(x)/a_0(x)$$

The two general boundary conditions that can be imposed on $y(x)$ may take the form:

$$\alpha_1 y(a) + \alpha_2 y(b) + \alpha_3 y'(a) + \alpha_4 y'(b) = 0$$
$$\beta_1 y(a) + \beta_2 y(b) + \beta_3 y'(a) + \beta_4 y'(b) = 0$$

The differential equation (4.49) is self-adjoint, i.e., the operators:

$$L = \frac{d}{dx}\left[p\frac{d}{dx}\right] + q \qquad \text{and} \qquad M = r(x) \text{ are self-adjoint.}$$

In order that the system has orthogonal eigenfunctions and positive eigenvalues, the problem must be self-adjoint and positive definite (see 4.42 to 4.44). The problem is self-adjoint, if:

$$\int_a^b \left\{u\left[(pv')' + qv\right] - v\left[(pu')' + qu\right]\right\} dx = P(v,u)\Big|_a^b = p(x)\left[uv' - vu'\right]\Big|_a^b = 0$$

$$= p(b)\left[u(b)\, v'(b) - u'(b)\, v(b)\right] - p(a)\left[u(a)\, v'(a) - u'(a)\, v(a)\right] = 0 \qquad (4.50)$$

Eliminating in turn y(a) and $y'(a)$ from the boundary conditions, one obtains:

$$\gamma_{13}y(a) + \gamma_{23}y(b) - \gamma_{34}y'(b) = 0$$

$$\gamma_{13}y'(a) + \gamma_{12}y(b) + \gamma_{14}y'(b) = 0 \qquad (4.51a)$$

Eliminating in turn y(b) and $y'(b)$ one obtains:

$$\gamma_{24}y(b) + \gamma_{14}y(a) + \gamma_{34}y'(a) = 0$$

$$\gamma_{24}y'(b) + \gamma_{12}y(a) - \gamma_{23}y'(a) = 0 \qquad (4.51b)$$

where

$$\gamma_{ij} = \alpha_i\beta_j - \alpha_j\beta_i = -\gamma_{ji} \qquad i, j = 1, 2, 3, 4$$

If one substitutes for y(a) and $y'(a)$ from eq. (4.51) into the self-adjoint condition eq. (4.50), one obtains:

$$\left[p(b) - p(a)\frac{\gamma_{24}}{\gamma_{13}}\right]\left[u(b)\, v'(b) - u'(b)\, v(b)\right] = 0$$

which can be satisfied if:

$$\gamma_{24}\,p(a) = \gamma_{13}\,p(b) \qquad (4.52)$$

where the identity:

$\gamma_{14}\gamma_{23} + \gamma_{34}\gamma_{12} = \gamma_{13}\gamma_{24}$ was used.

(i) If $\gamma_{13} = 0$, then $\gamma_{24} = 0$, and eq. (4.51) becomes:

$$y(b) - \frac{\gamma_{34}}{\gamma_{23}}\, y'(b) = 0 \qquad y(a) + \frac{\gamma_{34}}{\gamma_{14}}\, y'(a) = 0$$

$$y(b) + \frac{\gamma_{14}}{\gamma_{12}}\, y'(b) = 0 \qquad y(a) - \frac{\gamma_{23}}{\gamma_{12}}\, y'(a) = 0$$

which indicates that:

$$\frac{\gamma_{34}}{\gamma_{23}} = -\frac{\gamma_{14}}{\gamma_{12}} \qquad \text{and} \qquad \frac{\gamma_{34}}{\gamma_{14}} = -\frac{\gamma_{23}}{\gamma_{12}}$$

Denoting the ratio $\frac{\gamma_{23}}{\gamma_{12}} = \theta_1 > 0$, and $\frac{\gamma_{14}}{\gamma_{12}} = \frac{\gamma_{14}}{\gamma_{23}}\theta_1 = \theta_2 > 0$, then the boundary conditions become:

$$y(a) - \theta_1 y'(a) = 0$$

$$y(b) + \theta_2 y'(b) = 0 \tag{4.53}$$

In particular:

if θ_1 and $\theta_2 = 0$	then	$y(a) = 0$ and $y(b) = 0$	
if θ_1 and $\theta_2 \to \infty$	then	$y'(a) = 0$ and $y'(b) = 0$	
if $\theta_1 = 0$ and $\theta_2 \to \infty$	then	$y(a) = 0$ and $y'(b) = 0$	(4.54)
if $\theta_1 \to \infty$ and $\theta_2 = 0$	then	$y'(a) = 0$ and $y(b) = 0$	

(ii) If $\gamma_{13} \neq 0$, then the boundary condition in eq. (4.51) can be written as follows:

$$y(a) = \tau_1 y(b) + \tau_2 y'(b) \qquad \tau_1 = -\frac{\gamma_{23}}{\gamma_{13}} \text{ and } \tau_2 = \frac{\gamma_{34}}{\gamma_{13}}$$

$$y'(a) = \tau_3 y(b) + \tau_4 y'(b) \qquad \tau_3 = -\frac{\gamma_{12}}{\gamma_{13}} \text{ and } \tau_4 = -\frac{\gamma_{14}}{\gamma_{13}} \tag{4.55}$$

such that the condition of self-adjoincy eq. (4.52) becomes:

$$(\tau_1 \tau_4 - \tau_2 \tau_3)\, p(a) = p(b)$$

In particular, if $\tau_2 = \tau_3 = 0$ and $\tau_1 = \tau_4 = 1$, then:

$$y(a) = y(b)$$

$$y'(a) = y'(b) \tag{4.56}$$

and

$$p(a) = p(b)$$

The boundary conditions in eq. (4.56) are known as **Periodic Boundary Conditions.**

(iii) If p(x) vanishes at an end-point, then there is no need for a boundary condition at that end point, provided that the product:

$$\lim_{x \to a \text{ or } x \to b} p y y' \to 0$$

which can be restricted to y being bounded and $py' \to 0$ at the specific end point(s). Thus the S-L system composed of the differential equation (4.49) and any one of the possible sets of boundary conditions eqs. (4.53) to (4.56), is a self-adjoint system.

The eigenfunctions ϕ_n of the system are thus orthogonal, satisfying the following orthogonality integral, eq. (4.45), i.e.:

$$\int_a^b r(x)\,\phi_n(x)\,\phi_m(x)\,dx = 0 \qquad n \neq m \tag{4.57}$$

$$= N_n \qquad n = m$$

In order to insure that the eigenvalues are real and positive, the system must be positive definite (see 4.44). Thus:

$$\int_a^b u\,Lu\,dx = \int_a^b u\left[(pu')' + qu\right]dx = \int_a^b\left[-p(x)(u')^2 + q(x)u^2\right]dx < 0$$

and

$$\int_a^b u\,Mu\,dx = \int_a^b ru^2\,dx > 0$$

Thus, it is *sufficient (but not necessary)* that the functions p, q and r satisfy the following conditions for positive-definiteness:

$$p(x) > 0$$

$$a < x < b$$

$$q(x) \leq 0 \tag{4.58}$$

and

$$r(x) > 0$$

to guarantee real and positive eigenvalues.

It can be shown that the set of orthogonal eigenfunctions of the proper S-L system with the conditions imposed on p, q and r constitute a complete orthogonal set and hence may be used in a Generalized Fourier series.

Example 4.13 Longitudinal Vibration of a Free Bar

Obtain the eigenfunction and the eigenvalues for the longitudinal vibration of a free bar, giving explicitly the orthogonality conditions and the normalization constants.

$$y'' + \lambda y = 0 \qquad 0 \leq x \leq L$$

$$y'(0) = 0 \qquad y'(L) = 0$$

The system is S-L form already, since it can be readily rewritten as:

$$\frac{d}{dx}\left(\frac{dy}{dx}\right) + \lambda y = 0$$

where

$$p = 1 \qquad q = 0 \qquad \text{and} \qquad r = 1$$

The system is a proper S-L system since the differential equation, boundary conditions, as well as the conditions on p, q and r are those of a proper S-L system:

$$y = C_1 \sin(\sqrt{\lambda}x) + C_2 \cos(\sqrt{\lambda}x)$$

$$y'(0) = C_1 = 0$$

$$y'(L) = -C_2\sqrt{\lambda}\sin(\sqrt{\lambda}L) = 0$$

Thus, the characteristic equation becomes:

$$\alpha \sin\alpha = 0 \qquad \text{where} \qquad \alpha = \sqrt{\lambda}\, L$$

having roots $\alpha_n = n\pi$, n = 0, 1, 2,...:

$$\lambda_n = \frac{\alpha_n^2}{L^2} \qquad n = 0, 1, 2,...$$

The eigenfunction becomes:

$$\phi_n(x) = \cos(\alpha_n \frac{x}{L}) \qquad n = 0, 1, 2,...$$

Note that $\alpha_0 = 0$ is an eigenvalue corresponding to $\phi_0 = 1$. The orthogonality condition becomes (see 4.57):

$$\int_0^L 1 \bullet \cos(\alpha_n \frac{x}{L}) \cos(\alpha_m \frac{x}{L})\, dx = 0 \qquad n \neq m$$

and the normalization factor becomes:

$$N_n = N\left(\cos(\alpha_n \frac{x}{L})\right) = \int_0^L 1 \bullet \cos^2(\alpha_n \frac{x}{L})\, dx = \int_0^L \cos^2(\frac{n\pi}{L} x)\, dx = \frac{L}{2} \qquad n \geq 1$$

$$= L \qquad n = 0$$

which can be written as $N = L/\varepsilon_n$,where the Neumann constant is $\varepsilon_n = 1$ for n = 0 and 2 for $n \geq 1$.

Example 4.14 Vibration of a Stretched String with Variable Density

A vibrating stretched string is fixed at x = 0 and x = L, whose density ρ varies as:

$$\rho = \rho_0 x^2/L^2$$

The differential equation governing the motion of the string can be written as:

$$\frac{d^2y}{dx^2} + \frac{\rho_0 x^2}{T_0 L^2}\omega^2 y = 0 \qquad 0 \leq x \leq L$$

with the boundary conditions:

$$y(0) = 0 \qquad y(L) = 0$$

Let $\frac{\rho_0}{T_0}\omega^2 = \lambda$, then the differential equation becomes:

$$y'' + \lambda x^2 y/L^2 = 0$$

The system is in S-L form, with:

$p(x) = 1 > 0 \qquad q(x) = 0 \text{ and } \qquad r(x) = x^2/L^2 > 0$

which indicates that it is a proper S-L system.

The solution to the differential equation (see 3.66) can be written in terms of Bessel functions of fractional order:

$$y = \sqrt{x}\{C_1 J_{-1/4}(\sqrt{\lambda}x^2/(2L^2)) + C_2 J_{-1/4}(\sqrt{\lambda}x^2/(2L^2))\}$$

Since:

$$\lim_{x \to 0} \sqrt{x}\, J_{1/4}\left(\frac{\sqrt{\lambda}x^2}{2L^2}\right) \approx \lim_{x \to 0} \sqrt{x}\left(\frac{\sqrt{\lambda}x^2}{2L^2}\right)^{1/4} = \lim_{x \to 0}\left(\frac{\sqrt{\lambda}}{2L^2}\right)^{1/4} x \to 0$$

$$\lim_{x\to 0} \sqrt{x}\, J_{-1/4}\left[\frac{\sqrt{\lambda}x^2}{2L^2}\right] = \lim_{x\to 0} \sqrt{x}\left[\frac{\sqrt{\lambda}x^2}{2L^2}\right]^{-1/4} = \left[\frac{\sqrt{\lambda}}{2L^2}\right]^{-1/4}$$

then both homogeneous solutions are *finite* at $x = 0$. Satisfying the first boundary conditions yields $C_2 = 0$ and satisfying the second boundary condition yields:

$$y(L) = 0 = C_1\sqrt{L}\, J_{1/4}\left(\sqrt{\lambda}/2\right) = 0$$

which results in the following characteristic equation:

$$J_{1/4}(\alpha) = 0 \qquad \text{where} \qquad \alpha = \frac{\sqrt{\lambda}}{2}$$

The number of the roots α_n of the preceding transcendental equation are infinite with $\alpha_0 = 0$ being the first root. Thus, the eigenfunctions and the eigenvalues become:

$$\phi_n(x) = \sqrt{x}\, J_{1/4}\left(\alpha_n\, x^2/L^2\right) \qquad n = 1, 2, 3,...$$

$$\lambda_n = 4\alpha_n^2/L^4 \qquad n = 1, 2, 3,...,$$

where the $\alpha_0 = 0$ root is not an eigenvalue. The orthogonality integral is defined as:

$$\int_0^L x^2 \phi_n(x)\phi_m(x)\,dx/L^2 = \int_0^L x^3 J_{1/4}\left(\alpha_m \frac{x^2}{L^2}\right) J_{1/4}\left(\alpha_n \frac{x^2}{L^2}\right) dx/L^2 = 0 \qquad n \neq m$$

and the norm is:

$$N(\phi_n(x)) = \int_0^L x^3 J_{1/4}^2\left(\alpha_n \frac{x^2}{L^2}\right) dx/L^2 = -\frac{L^2}{4} J_{-3/4}(\alpha_n) J_{5/4}(\alpha_n)$$

Example 4.15 Tortional Vibration of a Bar of Variable Cross Section

A circular rod whose polar moment of inertia J varies as:

$J(x) = I_0 x$, where I_0 = constant

with the end L fixed and the torsional displacement at $x = 0$ is bounded is undergoing torsional vibration. The system satisfied by the deflection angle θ becomes:

$$\frac{d}{dx}\left(I_0 x \frac{d\theta}{dx}\right) + \lambda\, I_0 x\, \theta = 0$$

where $\lambda = \omega^2/c^2$ with the conditions that $\theta(0)$ is bounded and $\theta(L) = 0$.

The system is in S-L form where:

$$p(x) = x > 0 \qquad q(x) = 0, \qquad r(x) = x > 0$$

Since $p(0) = 0$, only one boundary condition at $x = L$ is required, provided that $\theta(0)$ is bounded, and:

$$\lim_{x \to 0} p\theta' \to 0$$

The solution of the differential equation can be written in terms of Bessel functions:

$$\theta = C_1 J_0\left(\sqrt{\lambda}\, x\right) + C_2 Y_0\left(\sqrt{\lambda}\, x\right)$$

Since $\theta(0)$ must be bounded, set $C_2 = 0$, and:

$$\theta(L) = C_1 J_0\left(\sqrt{\lambda}\, L\right) = 0$$

which results in the characteristic equation:

$$J_0(\alpha) = 0 \qquad \text{where} \qquad \alpha = \sqrt{\lambda}\, L$$

where the roots α_n are (see Section 3.13):

$$\alpha_1 = 2.405, \qquad \alpha_2 = 5.520, \qquad \alpha_3 = 8.654, \ldots$$

The eigenvalues are defined in terms of the roots α_n as:

$$\lambda_n = \alpha_n^2/L^2 \qquad n = 1, 2, \ldots$$

and the corresponding eigenfunctions are expressed as:

$$\phi_n = J_0\left(\alpha_n \frac{x}{L}\right) \qquad n = 1, 2, 3, \ldots$$

For the S-L system, the orthogonality integral can be written as:

$$\int_0^L x\, J_0\left(\alpha_n \frac{x}{L}\right) J_0\left(\alpha_m \frac{x}{L}\right) dx = 0 \qquad n \neq m$$

with the normalization constant defined as:

$$N\left(J_0\left(\alpha_n \frac{x}{L}\right)\right) = \int_0^L x\, J_0^2\left(\alpha_n \frac{x}{L}\right) dx = \frac{L^2}{\alpha_n^2} \int_0^{\alpha_n} z\, J_0^2(z)\, dz$$

$$= \frac{L^2}{2}\left[J_0'(\alpha_n)\right]^2 = \frac{L^2}{2} J_1^2(\alpha_n)$$

since $J_0(\alpha_n) = 0$ and the integral in eq. (3.109) was used.

4.16 Sturm-Liouville System for Fourth-Order Equations

Consider a general fourth-order linear differential equation of the type that governs vibration of beams:

$$a_0(x)\,y^{(iv)} + a_1(x)\,y''' + a_2(x)\,y'' + a_3(x)y' + a_4(x)y + \lambda a_5(x)\,y = 0$$

It can be shown that for this equation to be self-adjoint, the following equalities must hold:

$$a_1 = 2a_0'$$

$$a_2' - a_3 = a_0'''$$

It can also be shown that there is no single integrating function that can render this equation self-adjoint, as was the case of a second-order differential operator. Assuming that these relationships hold and denoting:

$$s(x) = \exp\frac{1}{2}\int^{x}\frac{a_1(\eta)}{a_0(\eta)}\,d\eta$$

and

$$p(x) = \int^{x}\frac{a_3(\eta)}{a_0(\eta)}\,s(\eta)\,d\eta$$

then the fourth-order equation can be written in self-adjoint form as:

$$Ly + \lambda My = [sy'']'' + [py']' + [q + \lambda r]y = 0 \tag{4.59}$$

where

$$q = \frac{a_4(x)}{a_0(x)}\,s(x)$$

$$r(x) = \frac{a_5(x)}{a_0(x)}\,s(x)$$

For the fourth-order S-L system to have orthogonal and real eigenfunctions and positive eigenvalues, the system must be self-adjoint and positive definite (see eqs. 4.42 to 4.44).

In the notation of eq. (4.40), the operators L and M are:

$$L = \frac{d^2}{dx^2}\left[s(x)\frac{d^2}{dx^2}\right] + \frac{d}{dx}\left[p(x)\frac{d}{dx}\right] + q(x)$$

and

$$M = r(x)$$

The system is self adjoint, so that P(u,v) given by eq. (4.43), is given by:

$$P(u,v)\Big|_a^b = u(sv'')' - v(su'')' - s(u'v'' - u''v') + p(uv' - vu')\Big|_a^b = 0 \tag{4.60}$$

Boundary conditions on y, and consequently on the comparison functions u and v, can be prescribed such that eq. (4.60) is satisfied identically. The five pairs of boundary conditions are listed below:

$$\begin{array}{lll}\text{(i)}\ \ y=0 & y'=0 & \\ \text{(ii)}\ \ y=0 & sy''=0 & \\ \text{(iii)}\ \ y'=0 & (sy'')'=0 & (4.61)\\ \text{(iv)}\ \ (sy'')'\mp\gamma y=0 & y'=0 & \\ \text{(v)}\ \ sy''\mp\alpha y'=0 & y=0 & \end{array}$$

where +ve sign for x = a and -ve sign for x = b.

If p(a) or p(b) vanishes (singular boundary conditions), then at the end point where p(x) vanishes, the boundedness condition is invoked i.e.:

$$\lim_{x\to a\text{ or }b} pyy' \to 0$$

(which can be restricted to y being finite and $py' \to 0$), as well as the following pairs of boundary conditions *in addition* to those given in eq. (4.61), can be specified at the end where p = 0:

$$\begin{array}{lll}\text{(i)}\ \ (sy'')'=0 & sy''=0 & \\ \text{(ii)}\ \ (sy'')'\mp\gamma y=0 & sy''=0 & (4.62)\\ \text{(iii)}\ \ sy''\mp\alpha y'=0 & (sy'')'=0 & \\ \text{(iv)}\ \ (sy'')'\mp\gamma y=0 & sy''\mp\alpha y'=0 & \end{array}$$

where +/- refer to the boundaries x = a or b, respectively.

If s(x) vanishes at one end (singular boundary conditions), then, together with the requirement that:

$$\lim_{x\to a\text{ or }b} sy'y'' \to 0$$

the following boundary conditions can be prescribed at the end where s(x) vanishes:

$$\begin{array}{lll}\text{(i)}\ \ y=0 & y'=0 & \\ \text{(ii)}\ \ (sy'')'=0 & y'=0 & (4.63)\\ \text{(iii)}\ \ (sy'')'\mp\gamma y=0 & y'=0 & \\ \text{(iv)}\ \ (sy'')'=0 & y'\mp\alpha y=0 & \end{array}$$

The +/- signs refer to the boundaries x = a or b, respectively.

If both p(x) and s(x) vanish at one end (singular boundary conditions) then, together with the requirement that:

$$\lim_{x\to a\text{ or }b} pyy' \to 0 \qquad \lim_{x\to a\text{ or }b} sy'y'' \to 0 \qquad \text{and} \qquad \lim_{x\to a\text{ or }b} syy''' \to 0$$

one can prescribe the following condition at the end where both p(x) and s(x) vanish having the form:

$$\begin{aligned}&\text{(i)}\quad y = 0\\ &\text{(ii)}\quad (sy'')' = 0\\ &\text{(iii)}\quad (sy'')' \mp \gamma y = 0\end{aligned} \tag{4.64}$$

The +/- signs refer to the boundaries x = a or b, respectively.

If p(x), s(x) and $s'(x)$ vanish at one end point, then there are no boundary conditions at those ends provided that:

$$\lim_{x \to a \text{ or } b} syy''' \to 0 \qquad \lim_{x \to a \text{ or } b} s'yy'' \to 0$$

and

$$\lim_{x \to a \text{ or } b} sy'y'' \to 0 \qquad \lim_{x \to a \text{ or } b} pyy' \to 0$$

If $p(x) \equiv 0$ in $a \le x \le b$, (see Section 4.4), then the nine boundary conditions specified in eqs.(4.61) and (4.62) satisfy eq. (4.60), as was shown for beam vibrations.

More complicated boundary conditions of the type:

$$\begin{aligned}&\alpha_{i1}y'''(a)+\alpha_{i2}y''(a)+\alpha_{i3}y'(a)+\alpha_{i4}y(a)\\ &\qquad +\beta_{i1}y'''(b)+\beta_{i2}y''(b)+\beta_{i3}y'(b)+\beta_{i4}y(b)=0\end{aligned} \qquad i = 1, 2, 3, 4$$

can be postulated, but it would be left to the reader to develop the conditions on α_{ij} and β_{ij} under which such boundary conditions satisfy eq. (4.60).

To guarantee positive eigenvalues, the system must be positive definite. Then the following inequalities must hold (see eq. 4.44).

$$\int_a^b u\left[(su'')'' + (pu')' + qu\right]dx = \int_a^b \left[qu^2 - p(u')^2 + s(u'')^2\right]dx < 0$$

and

$$\int_a^b uru\,dx = \int_a^b ru^2\,dx > 0$$

where the boundary conditions specified in eqs. (4.61) to (4.64) were used. Thus, *sufficient (but not necessary)* conditions on the functions can be imposed to satisfy positive definiteness:

$$\begin{aligned}&p \ge 0\\ &r > 0\\ &q \le 0 \qquad\qquad a < x < b\\ &s < 0\end{aligned} \tag{4.65}$$

4.17 Solution of Non-Homogeneous Eigenvalue Problems

Consider the following non-homogeneous system:

$$Ly + \lambda My = F(x) \qquad a \le x \le b$$

$$U_i(y) = \gamma_i \qquad i = 1, 2, ..., 2n \tag{4.66}$$

where L and M are self-adjoint operators and U_i were given in eqs. (4.40) and (4.41) and λ is given constant.

Due to the linearity of the system in eq. (4.66), one can split the solution into two parts. The first solution satisfies the homogeneous differential equation with non-homogeneous boundary conditions and the second system satisfies the non-homogeneous equation with homogeneous boundary condition. The sum of the two solutions satisfies the original system of eq. (4.66).

Let $y = y_I(x) + y_{II}(x)$ such that:

$$Ly_I + \lambda My_I = 0 \qquad Ly_{II} + \lambda My_{II} = F(x) \tag{4.67}$$

$$U_i(y_I) = \gamma_i \qquad U_i(y_{II}) = 0 \qquad i = 1, 2, ..., 2n$$

The solution to $y_I(x)$ in eq. (4.67) can be obtained by solving the homogeneous differential equation on y_I and substituting the (2n) independent solutions into the non-homogeneous boundary conditions for y_I. It should be noted that if $\gamma_i \equiv 0$, then $y_I \equiv 0$.

The solution y_{II} in eq. (4.67) can be developed by utilizing the eigenfunctions of the system. The eigenfunctions $\phi_n(x)$ of the system must be obtained first, satisfying the following homogeneous systems:

$$L\phi_m + \lambda_m M\phi_m = 0 \tag{4.68}$$

where each eigenfunction satisfies the homogeneous boundary conditions:

$$U_i(\phi_m) = 0 \qquad m = 1, 2, \ldots \qquad i = 1, 2, ..., 2n$$

The set of eigenfunctions $\{\phi_m(x)\}$ satisfy the orthogonality integral in eq. (4.45). The solution $y_{II}(x)$ can be expanded in a generalized Fourier series in the eigenfunctions of eq. (4.68) as follows:

$$y_{II} = \sum_{n=1}^{\infty} a_n \phi_n(x) \tag{4.69}$$

Substituting the series in eq. (4.69) into the differential equation on y_{II}, one obtains:

$$\sum_{n=1}^{\infty} a_n L\phi_n + \lambda \sum_{n=1}^{\infty} a_n M\phi_n = F(x) \tag{4.70}$$

Substituting for $L\,\phi_n$ from eq. (4.68) into eq. (4.70), one obtains:

$$\sum_{n=1}^{\infty} \left[(\lambda - \lambda_n) a_n M\phi_n\right] = F(x) \tag{4.71}$$

Multiplying both sides of eq. (4.71) by $\phi_m(x)$, integrating over [a,b] and invoking the orthogonality relationship in eq. (4.45) one obtains:

$$a_n = \frac{b_n}{(\lambda - \lambda_n)N_n}$$

where N_n is the Norm of the eigenfunctions and:

$$b_n = \int_a^b F(x)\phi_n(x)\,dx \tag{4.72}$$

Thus, the solution to y_{II} becomes:

$$y_{II}(x) = \sum_{n=1}^{\infty} \frac{b_n}{(\lambda - \lambda_n)N_n}\,\phi_n(x) \tag{4.73}$$

The solution due to the source term F(x) can be seen to become unbounded whenever λ becomes equal to any of the eigenvalues λ_n. It should be noted that if the system has inherent absorption, then the constant λ is complex valued, so that $\lambda \neq \lambda_n$, since λ_n are real and positive. So if the real part of λ is equal to λ_n, the solution y_{II} becomes large but still bounded.

Example 4.16 Forced Vibration of a Simply Supported Beam

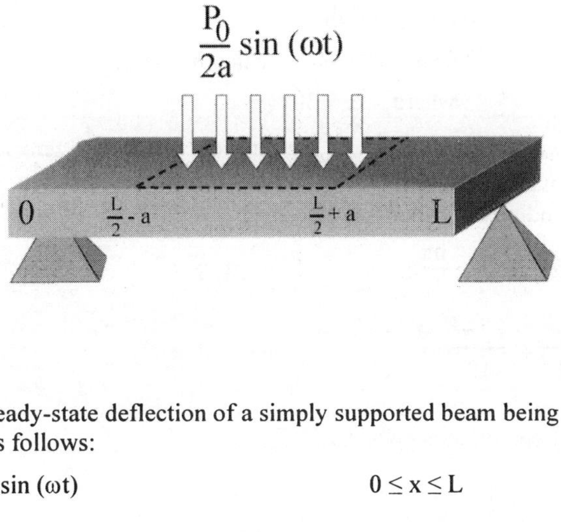

Obtain the steady-state deflection of a simply supported beam being vibrated by a distributed load as follows:

$$f^*(x,t) = f(x)\sin(\omega t) \qquad 0 \le x \le L$$

where

$$f(x) = \begin{cases} P_0/2a & L/2 - a < x < L/2 + a \\ 0 & \text{everywhere else} \end{cases}$$

The beam has a length L and has a constant cross section. It is simply supported at both ends such that:

$$y^*(0,t) = 0 \qquad y^{*\prime\prime}(0,t) = 0$$

$$y^*(L,t) = 0 \qquad y^{*\prime\prime}(L,t) = 0$$

Letting y*(x,t) = y(x) sin (ωt), then:

$$-y^{(iv)} + \beta^4 y = -\frac{f(x)}{EI} \qquad \beta^4 = \frac{\rho A \omega^2}{EI}$$

$$y(0) = 0 \qquad y''(0) = 0$$

$$y(L) = 0 \qquad y''(L) = 0$$

One must find the eigenfunctions of the system first:

$$-u^{(iv)} + \lambda u = 0 \qquad \text{where} \quad \lambda = \beta^4, \qquad L = -d^4/dx^4, \qquad M = 1$$

$$u(0) = 0 \qquad u''(0) = 0$$

$$u(L) = 0 \qquad u''(L) = 0$$

The solution of the fourth-order differential equation with constant coefficients is:

$$u = C_1 \sin \beta x + C_2 \cos \beta x + C_3 \sinh \beta x + C_4 \cosh \beta x$$

Satisfying the boundary conditions:

$$u(0) = 0 = C_2 + C_4 = 0$$

$$u''(0) = 0 = -C_2 + C_4 = 0$$

which means that $C_2 = C_4 = 0$

$$u(L) = 0 = C_1 \sin \beta L + C_3 \sinh \beta L$$

$$u''(L) = 0 = -C_1 \sin \beta L + C_3 \sinh \beta L$$

which results in $C_3 = 0$. The characteristic equation becomes:

$$\sin \alpha = 0 \qquad \text{where} \quad \alpha = \beta L$$

which has roots $\alpha_n = n\pi$, n = 0, 1, 2,... . The zero root results in a zero solution, so that $\alpha_o = 0$ is not an eigenvalue.

The corresponding eigenfunctions become:

$$\phi_n(x) = \sin \alpha_n \frac{x}{L} = \sin \frac{n\pi}{L} x \qquad n = 1, 2, 3$$

$$\lambda_n = \beta_n^4 = \frac{\alpha_n^4}{L^4} = \frac{n^4 \pi^4}{L^4} \qquad n = 1, 2, 3$$

The orthogonality condition is given by:

$$\int_0^L \sin(\frac{n\pi}{L} x) \sin(\frac{m\pi}{L} x)\, dx = \begin{cases} 0 & n \neq m \\ L/2 & n = m \end{cases}$$

Since the boundary conditions are homogeneous, then $y_I = 0$ and $y = y_{II}$. Expanding the function y(x) into an infinite series of the eigenfunctions, then the constant b_n is given by:

$$b_n = \int_{L/2-a}^{L/2+a} \left(-\frac{P_0}{2a}\right) \sin(\frac{n\pi}{L}x)\, dx = -P_0 \frac{\sin(\frac{n\pi}{2})\sin(\frac{n\pi}{L}a)}{EI\,(n\pi a/L)}$$

Thus, the solution becomes:

$$y(x) = -\frac{2P_0}{EI\,L} \sum_{n=1}^{\infty} \frac{\sin(\frac{n\pi}{2})\sin(\frac{n\pi}{L}a)}{\left(\beta^4 - n^4\pi^4/L^4\right)\left(n\pi a/L\right)} \sin(\frac{n\pi}{L}x)$$

If $a \to 0$, the distributed forcing function becomes a concentrated force, P_0, then the limit of the solution approaches:

$$\underset{a \to 0}{y(x)} \to -\frac{2P_0}{EI\,L} \sum_{n=1}^{\infty} \frac{\sin(\frac{n\pi}{2})}{(\beta^4 - \frac{n^4\pi^4}{L^4})} \sin(\frac{n\pi}{L}x)$$

For concentrated point sources and forces, one can represent them by Dirac delta functions (appendix D). Thus, one can represent f(x) by:

$$f(x) = P_0\, \delta(x - L/2)$$

The constant b_n can now be found using the sifting property of Dirac delta functions (D.4):

$$b_n = -P_0 \int_0^L \delta(x - L/2) \sin(\frac{n\pi}{L}x)\, dx = -P_0 \sin(\frac{n\pi}{2})$$

4.18 Fourier Sine Series

Consider the following S-L system:

$$y'' + \lambda y = 0 \qquad 0 \le x \le L$$

$$y(0) = 0 \qquad y(L) = 0$$

In this case $p = r = 1$ and $q = 0$. The eigenfunctions and eigenvalues of the system are:

$$\phi_n(x) = \sin(\frac{n\pi}{L}x) \qquad n = 1, 2, 3$$

$$\lambda_n = \frac{n^2\pi^2}{L^2}$$

the orthogonality integral becomes:

$$\int_0^L \sin(\frac{n\pi}{L}x)\sin(\frac{m\pi}{L}x)\, dx = 0 \qquad n \ne m$$

and the Norm becomes:

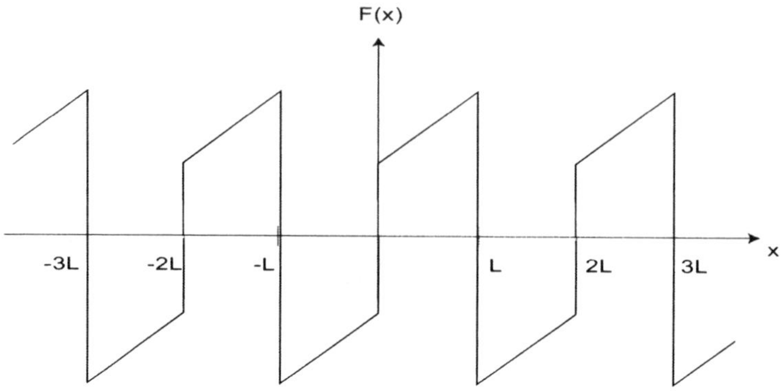

Fig. 4.17 Periodic Function

$$N\left(\sin(\frac{n\pi}{L}x)\right) = \int_0^L \left(\sin(\frac{n\pi}{L}x)\right)^2 dx = \frac{L}{2} \qquad n = 1, 2, 3, \ldots$$

A function F(x) can be expanded into an infinite Fourier sine series as follows:

$$\frac{1}{2}\left[F(x^+) + F(x^-)\right] = \sum_{n=1}^{\infty} a_n \sin(\frac{n\pi}{L}x) \qquad 0 \le x \le L$$

where the Fourier coefficients a_n are given by:

$$a_n = \frac{2}{L}\int_0^L F(x)\sin\frac{n\pi}{L}x\, dx \tag{4.74}$$

The function F(x) is represented by the series at all points in the region $0 < x < L$. The series represents an odd function in the region $-L < x < L$, since:

$$\sin(-\frac{n\pi}{L}x) = -\sin(\frac{n\pi}{L}x)$$

Thus, the series also represents -F(-x) in the region $-L < x < 0$. The series also represents a periodic function in the open region $-\infty < x < \infty$ with periodicity = 2L, since:

$$\sin\left(\frac{n\pi}{L}(x \mp 2mL)\right) = \sin\left(\frac{n\pi}{L}x\right)\cos(2nm\pi) \mp \cos\left(\frac{n\pi}{L}x\right)\sin(2nm\pi)$$

$$= \sin(\frac{n\pi}{L}x) \qquad \text{for all integers m}$$

Thus, the Fourier sine series represents a periodic function every 2L, with the function being odd within each region of periodicity = 2L as shown in Fig 4.17.

At the two ends point x = 0 and x = L, each term of the series vanishes, even though the function it represents may not vanish at either point. This is due to the fact that since the series represents an odd function in the periodic regions = 2L, there will be an ordinary discontinuity at x = 0, $\pm$ L, $\pm$ 2L,..., such that the function averages to zero at the end points (see 4.74), i.e.:

$$\frac{1}{2}\left[F(0^+)+F(0^-)\right]=0$$

$$\frac{1}{2}\left[F(L^+)+F(L^-)\right]=0$$

Example 4.17

Expand the following function in a Fourier sine series.

$$f(x) = L - \frac{x}{2} \qquad 0 \le x \le L$$

$$a_n = \frac{2}{L}\int_0^L \left(L - \frac{x}{2}\right) \sin\frac{n\pi}{L} x\, dx = \frac{2L}{n\pi}\left[1 - \frac{(-1)^n}{2}\right] \qquad n = 1, 2, 3, \ldots$$

and

$$f(x) \sim \frac{2L}{\pi} \sum_{n=1}^{\infty} \frac{\left[1 - \frac{(-1)^n}{2}\right]}{n} \sin\left(\frac{n\pi}{L} x\right)$$

If one sets L = 1 and x = 1/2:

$$\frac{3}{4} = \frac{2}{\pi} \sum_{n=1}^{\infty} \frac{\left[1 - \frac{(-1)^n}{2}\right]}{n} \sin\left(\frac{n\pi}{2}\right) = \frac{3}{\pi} \sum_{n=1,3,5}^{\infty} \frac{1}{n} \sin\left(\frac{n\pi}{2}\right)$$

or

$$\frac{\pi}{4} = 1 - \frac{1}{3} + \frac{1}{5} - \frac{1}{7} + \ldots$$

The last series can be used to calculate the series for π.

4.19 Fourier Cosine Series

The Fourier cosine series can be developed in a similar manner to the Fourier sine series.

Consider the following S-L system:

$$y'' + \lambda y = 0 \qquad 0 \le x \le L$$

$$y'(0) = 0 \qquad y'(L) = 0$$

In this case, p = r = 1 and q = 0.

The eigenfunctions and eigenvalues of the system become:

$$\phi_n(x) = \cos\left(\frac{n\pi}{L} x\right) \qquad n = 0, 1, 2, \ldots$$

$$\lambda_n = \frac{n^2\pi^2}{L^2}$$

with the orthogonality integral defined by:

$$\int_0^L \cos(\frac{n\pi}{L}x)\cos(\frac{m\pi}{L}x)\,dx = 0 \qquad n \neq m$$

and the norm given by:

$$N\left(\cos(\frac{n\pi}{L}x)\right) = \int_0^L \left(\cos(\frac{n\pi}{L}x)\right)^2 dx = \frac{L}{\varepsilon_n}$$

where ε_n is Neumann's Factor, $\varepsilon_0 = 1$ and $\varepsilon_n = 2$, $n \geq 1$.

A function F(x) can be expanded into an infinite Fourier cosine series as follows:

$$\frac{1}{2}\left[F(x^+) + F(x^-)\right] = \sum_{n=0}^{\infty} b_n \cos(\frac{n\pi}{L}x)$$

where the Fourier coefficients b_n are given by:

$$b_n = \frac{\varepsilon_n}{L}\int_0^L F(x)\cos(\frac{n\pi}{L}x)\,dx \tag{4.75}$$

The function F(x) is represented by the series at all points in the region $0 < x < L$. The series represents an even function in the region $-L < x < L$, since:

$$\cos\left(-\frac{n\pi}{L}x\right) = \cos\left(\frac{n\pi}{L}x\right)$$

Thus, the series represents F(-x) in the region $-L < x < 0$. The series also represents a periodic function in the open region $-\infty < x < \infty$, with a periodicity = 2L, since:

$$\cos\left(\frac{n\pi}{L}(x + 2mL)\right) = \cos(\frac{n\pi}{L}x)\cos(2nm\pi) \mp \sin(\frac{n\pi}{L}x)\sin(2nm\pi)$$

$$= \cos(\frac{n\pi}{L}x) \qquad \text{for all integer values of m}$$

Thus the Fourier cosine series represents a periodic function every 2L, with the function being even in the periodic regions = 2L as shown in Fig. 4.18:

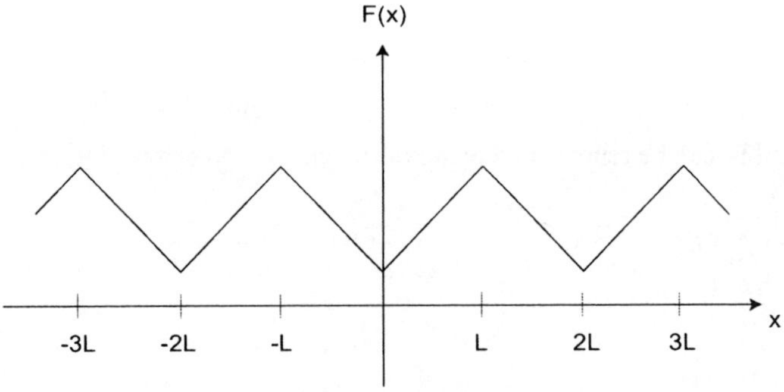

Fig 4.18 *Saw Tooth Function*

Since the series represents an even function, then the series does represent the function F(x) at the end points x = 0 and x = L.

Example 4.18

Expand the following function in a Fourier cosine series:

$$f(x) = L - \frac{x}{2} \qquad 0 \le x \le L$$

$$b_n = \frac{\varepsilon_n}{L}\int_0^L \left(L - \frac{x}{2}\right)\cos\left(\frac{n\pi}{L}x\right)dx = \begin{cases} \frac{3}{4}L & n = 0 \\ \frac{L}{(n\pi)^2}\left[1 - (-1)^n\right] & n \ge 1 \end{cases}$$

Thus

$$f(x) = \frac{3}{4}L + \frac{2L}{\pi^2}\sum_{n=1,3,5}^{\infty}\frac{1}{n^2}\cos\left(\frac{n\pi}{L}x\right)$$

4.20 Complete Fourier Series

Since the Fourier sine and cosine series represent an odd and an even function respectively in the region -L < x < L, then it can be shown that an asymmetric function F(x) can be expanded in both series in the region -L < x < L. Let F(x) be a function defined in -L ≤ x ≤ L, then:

$$F(x) = \frac{1}{2}\left[F(x) + F(-x)\right] + \frac{1}{2}\left[F(x) - F(-x)\right] \qquad -L \le x \le L$$

Denoting:

$$F_1(x) = \frac{1}{2}\left[F(x) + F(-x)\right]$$

and

$$F_2(x) = \frac{1}{2}\left[F(x) - F(-x)\right]$$

then $F_1(x)$ and $F_2(x)$ represent even and odd functions, respectively, since:

$$F_1(-x) = F_1(x) \qquad \text{and} \qquad -F_2(x) = -F_2(-x)$$

Hence, F_1 and F_2 can be represented by a Fourier cosine and sine series, respectively:

$$F_1(x) = \sum_{n=0}^{\infty} b_n \cos(\frac{n\pi}{L}x) \qquad -L < x < L$$

where

$$b_n = \frac{\varepsilon_n}{L}\int_0^L F_1(x)\cos(\frac{n\pi}{L}x)\,dx = \frac{\varepsilon_n}{2L}\int_{-L}^{L} F_1(x)\cos(\frac{n\pi}{L}x)\,dx$$

and

$$F_2(x) = \sum_{n=1}^{\infty} a_n \sin(\frac{n\pi}{L}x) \qquad -L < x < L$$

where

$$a_n = \frac{2}{L}\int_0^L F_2(x)\sin(\frac{n\pi}{L}x)\,dx = \frac{1}{L}\int_{-L}^{L} F_2(x)\sin(\frac{n\pi}{L}x)\,dx$$

Thus, one can rewrite the integrals for the Fourier coefficients as:

$$b_n = \frac{\varepsilon_n}{2L}\int_{-L}^{+L} [F_1(x) + F_2(x)]\cos(\frac{n\pi}{L}x)\,dx = \frac{\varepsilon_n}{2L}\int_{-L}^{L} F(x)\cos(\frac{n\pi}{L}x)\,dx$$

and

$$a_n = \frac{1}{L}\int_{-L}^{L} [F_1(x) + F_2(x)]\sin(\frac{n\pi}{L}x)\,dx = \frac{1}{L}\int_{-L}^{L} F(x)\sin(\frac{n\pi}{L}x)\,dx$$

In these integrals, use was made of the fact that:

$$\int_{-L}^{L} F_1(x)\sin(\frac{n\pi}{L}x)\,dx = 0 \qquad \text{and} \qquad \int_{-L}^{L} F_2(x)\cos(\frac{n\pi}{L}x)\,dx = 0$$

due to the fact that the integrands are odd functions.

Finally, the function F(x) can be represented by the complete Fourier series as follows:

$$\frac{1}{2}\left[F(x^+) + F(x^-)\right] = \sum_{n=1}^{\infty} a_n \sin(\frac{n\pi}{L}x) + \sum_{n=0}^{\infty} b_n \cos(\frac{n\pi}{L}x)$$

where

$$a_n = \frac{1}{L}\int_{-L}^{L} F(x)\sin(\frac{n\pi}{L}x)\,dx$$

and

$$b_n = \frac{\varepsilon_n}{2L}\int_{-L}^{L} F(x)\cos(\frac{n\pi}{L}x)\,dx \qquad (4.76)$$

Note that the eigenfunctions are completely orthogonal in (-L, L) since:

$$\int_{-L}^{L} \sin(\frac{n\pi}{L}x)\sin(\frac{n\pi}{L}x)\,dx = \begin{cases} 0 & n \neq m \\ L & n = m \end{cases}$$

$$\int_{-L}^{L} \cos(\frac{n\pi}{L}x)\cos(\frac{n\pi}{L}x)\,dx = \begin{cases} 0 & n \neq m \\ 2L/\varepsilon_n & n = m \end{cases}$$

and

$$\int_{-L}^{+L} \sin\frac{n\pi}{L}x \cos\frac{m\pi}{L}x\,dx = 0 \qquad \text{for all n, m}$$

One can develop the complete Fourier series representation from the S-L system. Let a S-L system given by:

$$y'' + \lambda y = 0 \qquad -L \leq x \leq +L$$

$$y(-L) = y(L)$$

$$y'(-L) = y'(L)$$

The system is a proper S-L system, since the operator is self-adjoint and the boundary conditions are those of the periodic type. The system yields the following set of eigenfunctions and eigenvalues:

$$\phi_n = \begin{cases} \sin(\frac{n\pi}{L}x) & n = 1, 2, 3... \\ \cos(\frac{n\pi}{L}x) & n = 0, 1, 2, ... \end{cases}$$

$$\lambda_n = \frac{n^2\pi^2}{L^2}$$

The entire set of eigenfunctions is orthogonal over [-L, L], as given above.

In a more general form, the complete Fourier series, orthogonal over a range [a, b] can be stated as follows:

$$F(x) \sim \sum_{n=1}^{\infty} a_n \sin\left(\frac{2n\pi(x-a)}{T}\right) + \sum_{n=0}^{\infty} b_n \cos\left(\frac{2n\pi(x-a)}{T}\right) \qquad a \leq x \leq b$$

where T = b – a, and the Fourier coefficients are given by:

$$a_n = \frac{2}{T} \int_a^b F(x) \sin\left(\frac{2n\pi}{T}(x-a)\right) dx$$

and

$$b_n = \frac{\varepsilon_n}{T} \int_a^b F(x) \cos\left(\frac{2n\pi}{T}(x-a)\right) dx \qquad (4.77)$$

Example 4.19

Obtain the expansion of the following function in a complete Fourier series:

$$F(x) = \begin{cases} 0 & -L \le x < 0 \\ L - \dfrac{x}{2} & 0 < x \le L \end{cases}$$

$$b_0 = \frac{1}{2L} \int_0^L \left(L - \frac{x}{2}\right) dx = \frac{3L}{8}$$

$$b_n = \frac{1}{L} \int_0^L F(x) \cos\left(\frac{n\pi}{L} x\right) dx = \frac{L}{2n^2\pi^2}\left[1 - (-1)^n\right] \qquad n \ge 1$$

$$a_n = \frac{1}{L} \int_0^L F(x) \sin\left(\frac{n\pi}{L} x\right) dx = \frac{L}{n\pi}\left[1 - \frac{(-1)^n}{2}\right] \qquad n \ge 1$$

and

$$F(x) \sim \frac{3L}{8} + \frac{L}{\pi^2} \sum_{n=1,3,5}^{\infty} \frac{1}{n^2} \cos\left(\frac{n\pi}{L} x\right) + \frac{L}{\pi} \sum_{n=1}^{\infty} \frac{\left[1 - \frac{(-1)^n}{2}\right]}{n} \sin\left(\frac{n\pi}{L} x\right)$$

In general, the fact that the integrals of the type given in eqs. (4.74) to (4.77) must converge, requires that F(x) must satisfy the following conditions over the range [L, -L]:

(a) piecewise continuous

(b) have a first derivative that is piecewise continuous

(c) have a finite number of maxima and minima

(d) single valued

(e) bounded

The conditions imposed on F(x) listed above are quite relaxed when compared with those imposed on functions to be expanded by Taylor's series.

The following general remarks can be made in regard to expansions of F(x) in a Fourier sine, cosine, or complete series:

(a) The series converges to F(x) at every point where F(x) is continuous

(b) The series converges to $[F(x^+) + F(x^-)]/2$ at a point of ordinary discontinuity, i.e., wherever F(x) is discontinuous but has finite right and left derivatives.

(c) The series represents a periodic function in the open region $-\infty < x < \infty$

(d) The series converges uniformly and absolutely in $-L < x < L$ if F(x) is continuous, $F'(x)$ is piecewise continuous and F(L) = F(-L).

(e) The series can be differentiated term by term if F(x) satisfies the conditions in (d), i.e.:

$$F'(x) \sim \frac{\pi}{L}\sum_{n=1}^{\infty} n\left(a_n \cos\frac{n\pi}{L}x - b_n \sin\frac{n\pi}{L}x\right)$$

(f) If F(x) is piecewise continuous then one may integrate the series term by term any number of times, i.e.:

$$\int F(x)\,dx = -\frac{L}{\pi}\sum_{n=1}^{\infty}\frac{a_n}{n}\cos\frac{n\pi}{L}x + \frac{L}{\pi}\sum_{n=1}^{\infty}\frac{b_n}{n}\sin\frac{n\pi}{L}x + b_0 x$$

This series converges faster than the series for F(x)

4.21 Fourier-Bessel Series

Consider the following system:

$$x^2 y'' + xy' + \left(\alpha^2 x^2 - a^2\right)y = 0 \qquad 0 \le x \le L$$

with y satisfying the following conditions:

$$y(0) \text{ is bounded} \qquad \gamma_1 y(L) + \gamma_2 y'(L) = 0$$

where γ_1, γ_2 are known constants.

The system is first transformed to S-L system, having the form:

$$\frac{d}{dx}\left(x\frac{dy}{dx}\right) + \left(\alpha^2 x - \frac{a^2}{x}\right)y = 0$$

where

$$p(x) = x \qquad q(x) = -a^2/x \qquad r(x) = x \qquad \lambda = \alpha^2$$

The solution to the differential equation becomes:

$$y = C_1 J_a(\alpha x) + C_2 Y_a(\alpha x)$$

Since p(0) = 0, then y(0) must be finite and $py' \to 0$. This requires that C_2 must be set to zero to insure that y(0) is bounded. Thus, the remaining solution:

$$y = C_1 J_a(\alpha x)$$

satisfies the condition that:

$$\lim_{x \to 0} xJ'_a(\alpha x) \to 0$$

The boundary condition at x = L takes the following form:

$$\gamma_1 J_a(\alpha x) + \gamma_2 \frac{dJ_a(\alpha x)}{dx}\bigg|_{x=L} = 0$$

resulting in the following characteristic equation:

$$\gamma_1 J_a(\mu) + \frac{\mu\gamma_2}{L} J'_a(\mu) = 0 \tag{4.78}$$

where $\mu = \alpha L$ and $J'_a(\mu) = dJ_a(\mu)/d\mu$

The characteristic equation can be transformed (see 3.14) to the following form:

$$\left(\gamma_1 + \frac{\gamma_2 a}{L}\right) J_a(\mu) - \frac{\gamma_2}{L}\mu\, J_{a+1}(\mu) = 0 \tag{4.79}$$

a transcendental equation with an infinite number of roots μ_n.

If $a \neq 0$, then the first root is $\mu_0 = 0$ but it is not an eigenvalue, since $\phi_a(0) = 0$. If $a = 0$, then there is a root $\mu_0 = 0$ only if $\gamma_1 = 0$, otherwise $\mu_0 = 0$ is not a root in general if $a = 0$. The eigenfunctions and eigenvalues become:

$$\lambda_n = \frac{\mu_n^2}{L^2}$$

(a) For $a \neq 0$

$$\phi_n(x) = J_a\left(\mu_n \frac{x}{L}\right) \qquad n = 1, 2, ... \tag{4.80}$$

(b) For $a = 0$ and $\gamma_1 = 0$

$$\phi_n(x) = J_0\left(\mu_n \frac{x}{L}\right) \qquad n = 0, 1, 2, ... \tag{4.81}$$

(c) For $a = 0$ and $\gamma_1 \neq 0$

$$\phi_n(x) = J_0\left(\mu_n \frac{x}{L}\right) \qquad n = 1, 2, 3, ... \tag{4.82}$$

The norm of the eigenfunctions can be computed form (3.109) as follows:

$$N_n = N(\phi_n) = \int_0^L xJ_a^2\left(\mu_n \frac{x}{L}\right) dx = \frac{L^2}{2\mu_n^2}\left\{\left(\mu_n^2 - a^2\right)J_a^2(\mu_n) + \mu_n^2\left[J'_a(\mu_n)\right]^2\right\} \tag{4.83}$$

Substituting in turn for $J'_a(\mu_n)$ and $J_a(\mu_n)$ in eq. (4.81) one obtains:

$$N_n = \frac{L^2}{2}\left\{\left(\mu_n^2 - a^2\right)\frac{\gamma_2^2}{\gamma_1^2 L^2} + 1\right\}\left[J'_a(\mu_n)\right]^2 \tag{4.84}$$

or

$$N_n = \frac{L^2}{2\mu_n^2}\left\{\mu_n^2 - a^2 + \frac{\gamma_1^2 L^2}{\gamma_2^2}\right\} J_a^2(\mu_n)$$

Thus, if $\gamma_1 = 0$, hence $J'_a(\mu_n) = 0$, then the norm becomes:

$$N_n = \frac{L^2}{2\mu_n^2}\left(\mu_n^2 - a^2\right) J_a^2(\mu_n) \qquad n \geq 1 \tag{4.85}$$

and if $\gamma_2 = 0$, hence $J_a(\mu_n) = 0$, then the norm becomes:

$$N_n = \frac{L^2}{2}\left[J'_a(\mu_n)\right]^2 \qquad n \geq 1 \tag{4.86}$$

Expansion of a function F(x) defined over the range $0 < x < L$ into an infinite series of the Fourier-Bessel orthogonal functions $J_a(\mu_n x/L)$ can be made as follows:

$$F(x) = \sum_{n=0 \text{ or } 1}^{\infty} b_n J_a\left(\mu_n \frac{x}{L}\right)$$

where

$$b_n = \frac{1}{N_n}\int_0^L x\, F(x)\, J_a\left(\mu_n \frac{x}{L}\right) dx \tag{4.87}$$

Example 4.20

Obtain an expansion of the following function:

$F(x) = 1 \qquad\qquad 0 \leq x \leq L$

in a Fourier-Bessel series:

$$\phi_n(x) = J_0\left(\mu_n \frac{x}{L}\right) \qquad \text{where} \quad J_0(\mu_n) = 0$$

The Fourier coefficients are given by:

$$b_n = \frac{1}{N_n}\int_0^L x\, J_0\left(\mu_n \frac{x}{L}\right) dx = \frac{2}{\mu_n}\frac{1}{J_1(\mu_n)} \qquad n = 1, 2, 3, \ldots$$

and $b_0 = 0$, where eqs. (3.14), (3.103) and (4.86) were used.

Thus, the Fourier-Bessel series representation of $F(x) = 1$ is:

$$1 = 2\sum_{n=1}^{\infty} \frac{J_0\left(\mu_n x/L\right)}{\mu_n J_1(\mu_n)}$$

4.22 Fourier–Legendre Series

Consider the following differential equation:

$$\left(1 - x^2\right) y'' - 2xy' + \nu(\nu + 1)\, y = 0 \qquad -1 \leq x \leq +1$$

where y(1) and y(-1) are bounded.

The equation can be transformed to an S-L system as follows:

$$\frac{d}{dx}\left[\left(1-x^2\right)\frac{dy}{dx}\right]+\nu(\nu+1)\,y=0 \qquad \nu = \text{constant}$$

where

$$p(x) = 1\text{-}x^2 \qquad q(x) = 0 \qquad r(x) = 1 \qquad \lambda = \nu(\nu+1)$$

The solution of the equation becomes:

$$y = C_1 P_\nu(x) + C_2 Q_\nu(x)$$

Since $p(\pm 1) = 0$, then $y(\pm 1)$ must be bounded and hence one must set $C_2 = 0$ since $Q_\nu(\pm 1)$ is unbounded for all ν. In addition, $P_\nu(\pm 1)$ is bounded only if ν is an integer = n. Thus, the eigenfunctions and eigenvalues of the system are:

$$\phi_n = P_n(x) \qquad \lambda_n = n(n+1) \qquad n = 0, 1, 2,...$$

It should be noted that:

$$\lim_{x\to\mp 1} p(x)\,y' = \lim_{x\to\mp 1}\left(1-x^2\right)P_n'(x) \to 0$$

It should be noted that the eigenfunctions and eigenvalues were obtained without the satisfaction of boundary conditions. For these eigenfunctions, the orthogonality integral becomes:

$$\int_{-1}^{+1} P_n(x)\,P_m(x)\,dx = 0 \qquad n \neq m$$

which was established earlier (see 3.155), and the norm was obtained in (3.156) as follows:

$$N_n = N(P_n(x)) = \int_{-1}^{+1} P_n^2\,dx = \frac{2}{2n+1}$$

A function F(x) can be expanded in a Fourier-Legendre series as follows:

$$F(x) = \sum_{n=0}^{\infty} a_n P_n(x) \qquad -1 \leq x \leq 1$$

where

$$a_n = \frac{2n+1}{2}\int_{-1}^{+1} F(x)\,P_n(x)\,dx \tag{4.88}$$

Example 4.21

Expand the following function by Fourier-Legendre series:

$$F(x) = 0 \qquad -1 \leq x \leq 0$$

$$= 1 \qquad 0 \leq x \leq 1$$

$$a_n = \frac{2n+1}{2}\int_0^1 P_n(x)\,dx = \frac{2n+1}{2(n+1)}\,P_{n-1}(0) \qquad n \geq 1$$

$$= \frac{1}{2} \qquad n = 0$$

where the integral in (3.162) was used,

$$a_n = \frac{1}{2} \qquad n = 0$$

$$= 0 \qquad n = \text{even}, \geq 2$$

$$= (-1)^{n-1/2}\,\frac{2n+1}{2^n(n+1)}\,\frac{(n-1)!}{\left[\left(n-1/2\right)!\right]^2} \qquad n = \text{odd}$$

Thus:

$$a_0 = \frac{1}{2} \qquad a_1 = \frac{3}{4} \qquad a_2 = 0 \qquad a_3 = -\frac{7}{16}, \ldots$$

and

$$f(x) = \frac{1}{2} + \frac{3}{4}\,P_1(x) - \frac{7}{16}\,P_3(x) + \ldots$$

PROBLEMS

Section 4.2

1. Obtain the natural frequencies and mode shapes of a vibrating string, elastically supported at both ends, x = 0 and x = L by springs of stiffness $\gamma = T_0/L$.

2. Obtain the natural frequencies and mode shapes of a composite string made of two strings of densities ρ_1 and ρ_2 and having lengths = L/2 joined at one end (x = L/2) and the terminal end of each string being fixed, i.e., at x = 0 and x = L.

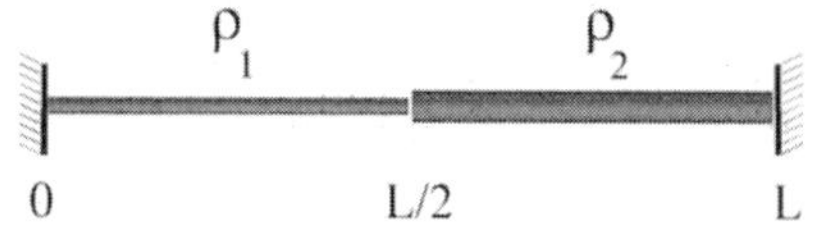

3. Obtain the natural frequencies and mode shapes of string whose density varies as:

$$\rho = \rho_0 \left(1 + \frac{x}{L}\right)^2 \qquad 0 \le x \le L$$

and whose ends are fixed.
Hint: Let z = 1 + x/L, such that the equation of motion becomes:

$$\frac{d^2y}{dz^2} + \lambda z^2 y = 0 \qquad 1 \le z \le 2$$

where

$$\lambda = \frac{\rho_0 \omega^2 L^2}{T_0}$$

4. A uniform stretched string of mass density ρ and length L has a point mass equal to the total mass of the string attached at x = L/2 such that:

$$2T_0 \left.\frac{\partial y}{\partial x}\right|_{L/2} = -m \left.\frac{\partial^2 y}{\partial t^2}\right|_{L/2} = +m\omega^2 y\Big|_{L/2}$$

Obtain the natural frequencies and mode shapes.

Section 4.3

5. Obtain the natural frequencies and mode shapes of a uniform rod vibrating in a longitudinal mode, such that:

   (a) the rod is free at both ends x = 0 and x = L.

   (b) the rod is fixed at both ends x = 0 and x = L.

(c) the rod is fixed at x = 0 and free at x = L.

(d) the rod is free at x = 0 and supported by linear spring of stiffness γ at x = L, such that:

$$\frac{du(0)}{dx} = 0 \qquad \frac{du}{dx}(L) + au(L) = 0 \qquad a = \gamma/(AE) > 0$$

(e) the rod is fixed at x = 0 and have a concentrated mass M at x = L, such that:

$$u(0) = 0 \qquad \frac{du}{dx}(L) - ak^2u(L) = 0 \qquad a = M/(\rho A) > 0$$

(f) the rod is elastically supported at x = o by a spring of constant γ and has a concentrated mass M at x = L such that:

$$\frac{du}{dx}(0) - au(0) = 0 \qquad \frac{du}{dx}(L) - bk^2u(L) = 0$$

$$a = \gamma/(AE) > 0 \qquad b = M/(A\rho) > 0$$

6. A uniform rod has a mass M attached to each of its ends. Obtain the natural frequencies and mode shapes of such bar vibrating in a longitudinal mode.

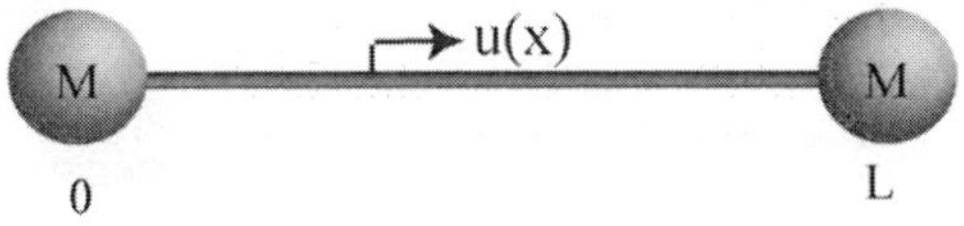

Hint: The boundary condition at x = 0 and L becomes:

$$AE\left.\frac{\partial u}{\partial x}\right|_{x=0} = +M\left.\frac{\partial^2 u}{\partial t^2}\right|_{x=0} = -M\omega^2 u\big|_{x=0}$$

$$AE\left.\frac{\partial u}{\partial x}\right|_{x=L} = -M\left.\frac{\partial^2 u}{\partial t^2}\right|_{x=L} = +M\omega^2 u\big|_{x=L}$$

7. Obtain the natural frequencies and mode shapes of a longitudinally vibrating bar whose cross-sectional area varies as:

$$A(x) = A_0\left(1 + \frac{x}{L}\right) \qquad 0 \le x \le L$$

and whose ends are fixed.
Hint: Let z = 1 + x/L and transform the equation of motion.

Section 4.4

8. Obtain the natural frequencies (or critical speeds) and the corresponding mode shape of a vibrating (rotating) beam having the following boundary conditions:

(a) simply supported at x = 0 and x = L

(b) fixed at x = 0 and free at x = L

(c) free at x = 0 and x = L

(d) free-fixed at x = 0 and x = L

(e) simply supported at x = 0 and fixed at x = L

(f) simply supported at x = 0 and free at x = L

(g) simply supported at x = 0 and elastically supported at free end x = L by a linear spring of stiffness γ

(h) simply supported at x = 0 and elastically supported at free end x = L by a helical spring of stiffness η

(i) fixed at x = 0 and elastically supported at free end x = L by a linear spring of stiffness γ

(j) fixed at x = 0 and elastically supported at free end x = L by a helical spring of stiffness η

9. Obtain the natural frequencies and mode shapes of a vibrating beam of length L with a mass M attached to its end. The beam is fixed at x = 0 and free at x = L.

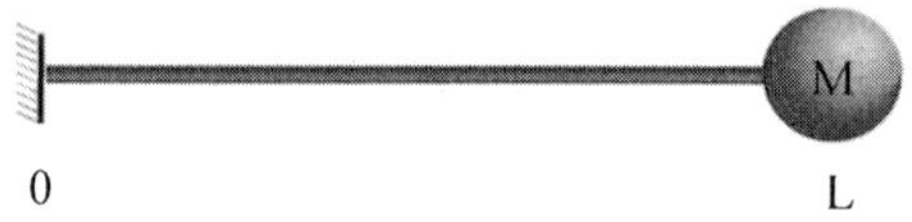

Hint: The boundary condition at x = L becomes:

$$y''' + k\beta^4 L y\Big|_{x=L} = 0$$

$$y''(L) = 0 \qquad k = \frac{M}{\rho A L}$$

10. Obtain the natural frequencies and mode shapes of a vibrating beam of length L with a mass M attached at its center. The beam is simply supported at x = 0 and x = L.

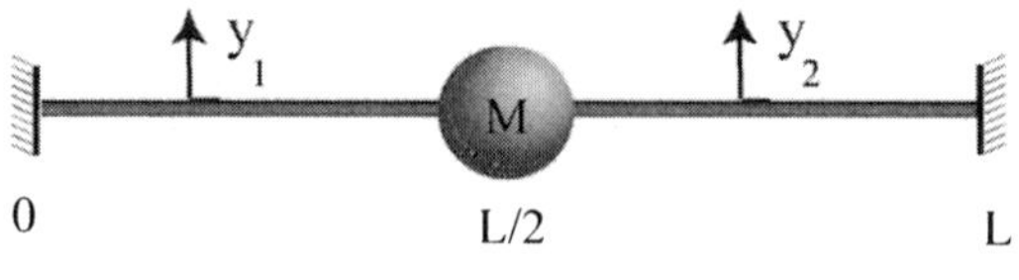

Hint: The conditions at x = L/2 become:

$$y_1(L/2) = y_2(L/2) \qquad y_1'(L/2) = y_2'(L/2)$$

$$y_1''(L/2) - y_2''(L/2) = 0 \qquad EI\left(y_1'''(L/2) - y_2'''(L/2)\right) + M\omega^2 y_2(L/2) = 0$$

11. Obtain the natural frequencies and mode shapes of a non-uniform beam of length L having the following properties:

$$A(x) = A_0\left(\frac{x}{L}\right)^n$$

$$I(x) = I_0\left(\frac{x}{L}\right)^{n+2} \qquad n = \text{positive integer}$$

The beam's motion is bounded at x = 0 and fixed at x = L.
Hint: The equation of motion can be factored as follows:

$$\left\{\frac{1}{x^n}\frac{d}{dx}\left(x^{n+1}\frac{d}{dx}\right)+\beta^2 L\right\}\left\{\frac{1}{x^n}\frac{d}{dx}\left(x^{n+1}\frac{d}{dx}\right)-\beta^2 L\right\}y = 0$$

$$\beta^4 = \frac{\rho A_0}{EI_0}\omega^2 \qquad 0 \le x \le L$$

Section 4.5

12. Obtain the natural frequencies and mode shapes of standing waves in a tapered acoustic horn of length L whose cross-sectional area varies parabolically as follows:

$$A(x) = A_0 x^2$$

such that the pressure is finite at x = 0 and at the end x = L is:

(a) open end, or
(b) rigid

13. Obtain the natural frequencies and mode shapes of standing waves in a parabolic acoustic horn, whose cross-sectional area varies as follows:

$$A(x) = A_0 x^4$$

where the pressure at x = 0 is finite and the end x = L is open end.

14. Obtain the natural frequencies and mode shapes of standing waves in an exponential horn of length L whose cross-sectional area varies as:

$$A(x) = A_0 e^{2ax}$$

such that the end x = 0 is rigid and the end x = L is open end.

Section 4.6

15. Obtain the critical buckling loads and the corresponding buckling shape of compressed columns, each having length L and a constant cross section and the following boundary conditions:

(a) fixed at $x = 0$ and $x = L$

(b) simply supported at $x = 0$ and $x = L$

(c) fixed at $x = 0$ and simply supported at $x = L$

(d) simply supported at $x = 0$ and elastically supported free end by a linear spring, having a stiffness γ, at $x = L$

(e) simply supported at $x = 0$ and fixed-free at $x = L$

(f) simply supported at $x = 0$ and elastically supported simply supported end by a helical spring of stiffness η at $x = L$

(g) fixed at $x = 0$ and free-fixed at $x = L$

(h) fixed at $x = 0$ and elastically supported simply supported end by a helical spring of stiffness η at $x = L$

16. Obtain the critical buckling loads and the corresponding buckling shape of a compressed tapered column whose moment of inertia varies as follows:

$$I(x) = I_0\left(\frac{x}{b}\right)^2$$

such that

$$\frac{d^2y}{dx^2} + \frac{Pb^4}{EI_0}x^{-2}y = 0 \qquad a \le x \le b$$

and the boundary conditions become:

$$y(a) = 0 \qquad y'(b) = 0$$

17. Obtain the critical buckling loads and the corresponding buckling shape of a column buckling under its own weight, such that the deflection satisfies the following differential equation:

$$EI\frac{d^3y}{dx^3} + qx\frac{dy}{dx} = 0 \qquad 0 \le x \le L$$

where q represents the weight of the column per unit length. The column is fixed at $x = 0$ and free at $x = L$ such that:

$$y''(0) = 0 \qquad y'''(0) = 0 \qquad y'(L) = 0$$

Hint: let $y'(x) = u(x)$

$u'(0) = 0 \qquad u(L) = 0 \qquad u''(0) = 0$ is satisfied identically.

18. Obtain the critical buckling loads and the corresponding buckling shapes of a compressed column which is elastically supported along its entire length by linear spring of stiffness k per unit length. The equation of stability becomes:

$$EI\frac{d^4y}{dx^4} + P\frac{d^2y}{dx^2} + ky = 0 \qquad 0 \le x \le L$$

where $p^2 > 4k\,EI$. The column is simply supported at both ends.

Section 4.11

19. Show that the differential operators given in eq. (4.34) are self-adjoint.

20. Obtain the conditions that the coefficients of a linear fourth-order differential operator must satisfy so that the operator can be transformed to a self-adjoint operator.

Section 4.15

21. Transform the following differential operators to the self-adjoint Sturm-Liouville form given in eq. (4.49):

(a) $(1 - x^2)y'' - 2xy' + \lambda y = 0$

(b) $(1 - x^2)y'' - xy' + \lambda y = 0$

(c) $(1 - x^2)^2 y'' + [\lambda(1 - x^2) + 1]y = 0$

(d) $xy'' + (a + 1 - x)\,y' + \lambda y = 0$

(e) $y'' - 2xy' + \lambda y = 0$

(f) $(1 - x^2)y'' - (2a + 1)\,xy' + \lambda y = 0$

(g) $(1 - x^2)y'' - [b - a - (a + b + 2)\,x]\,y' + \lambda y = 0$

(h) $x(1 - x)\,y'' + [c - (a + b + 1)\,x]\,y' + \lambda y = 0$

(i) $x^2y'' + x^2y' + (\lambda x^2 - 2)y = 0$

(j) $x^2y'' + xy' + (\lambda x^2 - n^2)y = 0$

(k) $(ax + b)\,y'' + 2ay' + \lambda(ax + b)\,y = 0$

(l) $y'' + 2a \cot an\; ax + \lambda y = 0$

(m) $xy'' + \frac{3}{2}y' + \lambda y = 0$

(n) $y'' + ay' + \lambda y = 0$

(o) $y'' - 2a \tan ax y' + \lambda y = 0$

(p) $y'' + 2a \tanh ax\, y' + \lambda y = 0$

(q) $y'' - a \tan ax\, y' + \lambda \cos^2 ax\, y = 0$

(r) $y'' + 2axy' + a^2x^2 y + \lambda y = 0$

(s) $y'' - a^2 y + \lambda e^{-4ax} y = 0$

(t) $x^{4a} y'' - a(a-1) x^{4a-2} y + \lambda y = 0$ $\qquad a < 0$

(u) $x^4 y'' + \lambda y = 0$

(v) $xy'' + \lambda y = 0$

(w) $xy'' + 4y' + \lambda xy = 0$

(x) $y'' + 4y' + (\lambda + 4) y = 0$

(y) $x^2 y'' - 2xy' + \frac{9}{4} \lambda x^3 y = 0$

(z) $x^2 y'' - xy' + (\lambda + 1) y = 0$

(aa) $xy'' + 2y' + \lambda xy = 0$

(bb) $x^2 y'' + 3xy' + [\lambda x^8 - 3] y = 0$

(cc) $xy'' + 3y' + \lambda x^{-1/3} y = 0$

(dd) $xy'' + 6y' + \lambda xy = 0$

(ee) $xy'' + 4y' + \lambda x^3 y = 0$

(ff) $xy'' + 2y' + \lambda x^3 y = 0$

(gg) $x^2 y'' + \frac{11}{2} xy' + \frac{9}{4}\left(\lambda x^3 - \frac{7}{4}\right) y = 0$

(hh) $xy'' + \frac{9}{7} y' + \lambda x^3 y = 0$

22. Obtain the eigenfunctions $\phi_n(x)$, eigenvalues λ_n and write down the orthogonality integral for the following differential systems:

(a) Problem 21a	$0 \le x \le 1$	$y(0) = 0$	$y(1)$ finite
(b) Problem 21a	$0 \le x \le 1$	$y'(0) = 0$	$y(1)$ finite
(c) Problem 21b	$-1 \le x \le 1$		$y(\pm 1)$ finite

(d) Problem 21c	$-1 \le x \le 1$		$y(\pm 1) = 0$
(e) Problem 21k	$0 \le x \le L$	$y(0) = y(L) = 0$	
(f) Problem 21l	$0 \le x \le L$	$y(0) = y(L) = 0$	
(g) Problem 21m	$0 \le x \le L$	$y(L) = 0$	$y(0)$ finite
(h) Problem 21n	$0 \le x \le L$	$y(0) = y(L) = 0$	
(i) Problem 21o	$0 \le x \le L$	$y(0) = y(L) = 0$	
(j) Problem 21p	$0 \le x \le L$	$y(0) = y(L) = 0$	
(k) Problem 21r	$0 \le x \le L$	$y(0) = y(L) = 0$	
(l) Problem 21s	$0 \le x \le L$	$y(0) = y(L) = 0$	
(m) Problem 21t	$0 \le x \le L$	$y(L) = 0$	$y(0)$ finite
(n) Problem 21u	$1 \le x \le 2$	$y(1) = y(2) = 0$	
(o) Problem 21v	$0 \le x \le L$	$y(0) = y(L) = 0$	
(p) Problem 21w	$0 \le x \le L$	$y'(L) = 0$	$y(0)$ finite
(q) Problem 21x	$0 \le x \le 1$	$y(0) = 0$	$y(1) = 0$
(r) Problem 21y	$0 \le x \le L$	$y(0) = 0$	$y(L) = 0$
(s) Problem 21z	$1 \le x \le e$	$y(1) = 0$	$y(e) = 0$
(t) Problem 21aa	$0 \le x \le L$	$y(L) = 0$	$y(0)$ finite
(u) Problem 21bb	$0 \le x \le L$	$y(L) = 0$	$y(0)$ finite
(v) Problem 21cc	$0 \le x \le L$	$y(L) = 0$	$y(0)$ finite
(w) Problem 21dd	$0 \le x \le L$	$y(L) = 0$	$y(0)$ finite
(x) Problem 21ee	$0 \le x \le L$	$y(L) = 0$	$y(0)$ finite
(y) Problem 21ff	$0 \le x \le L$	$y(L) = 0$	$y(0)$ finite
(z) Problem 21gg	$0 \le x \le L$	$y(L) = 0$	$y(0)$ finite
(aa) Problem 21hh	$0 \le x \le L$	$y(L) = 0$	$y(0)$ finite

Section 4.17

23. Obtain the solution to the following systems:

(a) $y'' + \lambda y = f(x)$ $\quad y(L) = 0 \quad y(0) = 0$

(b) $y'' + \frac{1}{x} y' + \lambda y = 1$ $\quad y(L) = 0 \quad y(0)$ finite

(c) $(1 - x^2)y'' - 2xy' + \lambda y = f(x)$ $\quad y(\pm 1)$ finite

(d) $y'' - 2y' + (1+\beta)\, y = e^x$ $\quad y(0) = 3 \quad y(1) = 0$

$0 \le x \le 1$ $\quad \beta$ is a fixed constant

(e) $xy'' + \left(\frac{3}{2} - x\right) y' + \left(\frac{x}{4} - \frac{3}{4} + \lambda\right) y = x^{-1/2} e^{x/2}$ $\quad y(0)$ finite (bounded)

$0 \le x \le 1$ $\quad y(1) = 0$

(f) $xy'' + (3 - 2x)\, y' + (\alpha^2 x^3 + x - 3)y = xe^x$ $\quad y(0)$ finite

$0 \le x \le 1$ $\quad y(1) = 0$

(g) $xy'' + 2y' + k^2 xy = 1$ $\quad y(1) = 0 \quad y(0)$ finite

(h) $xy'' + (3 - 2x)\, y' + (\lambda x + (x^2 - 3x))y = \frac{e^x}{x}$ $\quad y(0)$ finite (bounded)

$0 \le x \le 1$ $\quad y(1) = 0$

(i) $x^2 y'' + [3 - 6x]\, xy' + [9x^2 - 9x - 15 + \lambda x^4]y = x^3 e^{3x}$

$0 \le x \le 1$ $\quad y(1) = 0 \quad y(0)$ finite (bounded)

(j) $x^2 y'' + 2(1 - 2x)\, xy' + \left(\lambda x^4 + 4x^2 - 4x - \frac{3}{4}\right) y = x^{5/2} e^{2x}$

$0 \le x \le 1$ $\quad y(1) = 0 \quad y(0)$ finite (bounded)

(k) $x^2 y'' + \left(\frac{5}{2} - 2x\right) xy' + \left(\lambda x^4 + x^2 - \frac{5}{2} x - \frac{7}{16}\right) y = e^x x^{9/4}$

$0 \le x \le 1$ $\quad y(1) = 0 \quad y(0)$ finite (bounded)

(l) $x^2 y'' + 2(1 + x)\, xy' + (\lambda x^4 + x^2 + 2x)y = e^{-x} x^3$

$0 \le x \le L$ $\quad y(L) = 0 \quad y(0)$ finite (bounded)

(m) $xy'' + (4 - 2x)\, y' + (\lambda x^5 + x - 4)y = x^2 e^x$

$0 < x < 1 \qquad y(1) = 0 \qquad y(0)$ finite

(n) $xy'' + (5 - 2x)\,y' + (\lambda x^7 + x - 5)\,y = x^3 e^x$

$0 \le x \le 1 \qquad y(1) = 0 \qquad y(0)$ finite

Section 4.18 and 4.19

24. Expand the following function in a Fourier sine series over the specified range:

(a) $f(x) = x^2 \qquad 0 < x < \pi$

(b) $f(x) = 1 \qquad 0 < x < \pi/2$

$\quad = 0 \qquad \pi/2 < x < \pi$

(c) $f(x) = x \qquad 0 < x < \pi$

(d) $f(x) = x - x^2 \qquad 0 < x < 1$

(e) $f(x) = e^x \qquad 0 < x < \pi$

(f) $f(x) = \sin x \qquad 0 < x < \pi$

25. Expand the functions of Problem 24 in Fourier cosine series.

Section 4.20

26. Expand the following by a complete Fourier series in the specified range:

(a) $f(x) = \sin x \qquad 0 \le x \le \pi$

$\quad = 0 \qquad \pi \le x \le 2\pi$

(b) $f(x) = \cos ax \qquad -\pi < x < \pi$

a = non-integer

(c) $f(x) = x - x^2 \qquad -1 < x < 1$

(d) $f(x) = \sin ax \qquad -\pi < x < \pi$

a = non-integer

(e) $f(x) = 1 \qquad -L < x < L/2$

$\quad = 0 \qquad L/2 < x < L$

Section 4.21

27. Expand the function:

$f(x) = 1 \qquad 0 < x < 1$

$\quad = 0 \qquad 1 < x < 2$

in a series of $J_0(\mu_n x)$ where μ_n are roots of $J_0(2\mu_n) = 0$

28. Expand the function:

$f(x) = x^2 \qquad 0 < x < 1$

in a series of $J_2(\mu_n x)$ where μ_n are the root of:

$J_2(\mu_n) = 0$

29. Expand the function:

$f(x) = 1 \qquad 0 < x < L$

in a series of $J_2(\mu_n x)$, where μ_n are the roots of:

$\mu_n L J_1(\mu_n L) - a J_0(\mu_n L) = 0$

Section 4.22

30. Expand the function:

$f(x) = 0 \qquad -1 < x < 0$

$\quad = 1 \qquad 0 < x < 1$

in the series of Legendre Polynomials.

31. Expand the function:

$f(x) = 0 \qquad -1 < x < 0$

$\quad = x \qquad 0 < x < 1$

in a series of Legendre Polynomials.

5

FUNCTIONS OF A COMPLEX VARIABLE

5.1 Complex Numbers

A complex number z can be defined as an ordered pair of real numbers x and y:

$$z = (x,y)$$

The complex number (1,0) is a real number = 1. The complex number (0,1) = i, is an imaginary number. The components of z are: the **real** part *Re* (z) = x and the **imaginary** part *Im* (z) = y. Thus, the number z can be expressed conveniently as follows:

$$z = x + iy$$

The number z = 0 iff x = 0 and y = 0. New operational rules and laws must be specified for the new number system. Let the complex numbers a, b, c be defined by their components (a_1, a_2), (b_1, b_2) and (c_1, c_2), respectively.

Equality: $a = b$ iff $a_1 = a_2$ and $b_1 = b_2$

Thus it can be written in complex notation as follows:

$$a = a_1 + ia_2 = b = b_1 + ib_2 \quad \text{iff} \quad a_1 = b_1 \text{ and } a_2 = b_2$$

Addition: $c = a + b = (a_1 + b_1, a_2 + b_2)$

$$c = c_1 + ic_2 = (a_1 + ia_2) + (b_1 + ib_2) = (a_1 + b_1) + i(a_2 + b_2)$$

Subtraction: $c = a - b = (a_1 - b_1, a_2 - b_2)$

$$c = c_1 + ic_2 = (a_1 + ia_2) - (b_1 + ib_2) = (a_1 - b_1) + i(a_2 - b_2)$$

Multiplication: $c = ab = (a_1 b_1 - a_2 b_2, a_1 b_2 + a_2 b_1)$

$$c = c_1 + ic_2 = (a_1 + ia_2)(b_1 + ib_2)$$

If one defines $i^2 = -1$, then:

$$c = (a_1 b_1 - a_2 b_2) + i(a_1 b_2 + a_2 b_1)$$

Division: if $a \neq 0$

$$\frac{1}{a} = \frac{1}{a_1 + ia_2}$$

Multiplying the numerator and denominator by $(a_1 - ia_2)$:

$$\frac{1}{a} = \frac{a_1 - ia_2}{a_1^2 + a_2^2}$$

Furthermore, a division of two complex numbers gives:

$$c = \frac{b}{a} = \frac{a_1 b_1 + a_2 b_2}{a_1^2 + a_2^2} + i\frac{a_1 b_2 - a_2 b_1}{a_1^2 + a_2^2}$$

The preceding operations satisfy the following laws:

1. **Associative Law:** $(a + b) + c = a + (b + c)$

   $(ab)c = a(bc)$

2. **Commutative Law:** $a + b = b + a$

   $ab = ba$

3. **Distributive Law:** $(a + b)c = ac + bc$
4. For every a, $a + 0 = a$
5. For every a, there exists -a, such that $a + (-a) = 0$
6. For every a, $a \cdot 1 = a$
7. For every a, there exists a^{-1} such that $a \cdot a^{-1} = 1$, $a \neq 0$

5.1.1 Complex Conjugate

Define **Complex Conjugate** "$\bar{a}$ " of "a" as follows:

$$a = a_1 + ia_2 \qquad \bar{a} = a_1 - ia_2$$

Thus:

$$\overline{a + b} = \bar{a} + \bar{b} \qquad \overline{ab} = \bar{a}\bar{b} \qquad \overline{\left(\frac{a}{b}\right)} = \frac{\bar{a}}{\bar{b}}$$

If $a = \bar{a}$, then a is a real number.

5.1.2 Polar Representation

Define the **Absolute Value (Modulus)** $|a|$ of "a" as follows:

$$|a| = \sqrt{a_1^2 + a_2^2} \geq 0 \qquad \text{a real positive number}$$

Since complex numbers are ordered pairs of real numbers, a geometric (vector) representation of such numbers (**Argand Diagram**) can be constructed as shown in Fig. 5.1, where:

$$x_0 = r\cos\theta$$

$$y_0 = r\sin\theta$$

$$z_0 = r(\cos\theta + i\sin\theta) = re^{i\theta}$$

In this system, the radius r is:

$$r = \sqrt{x_0^2 + y_0^2} = |z_0|$$

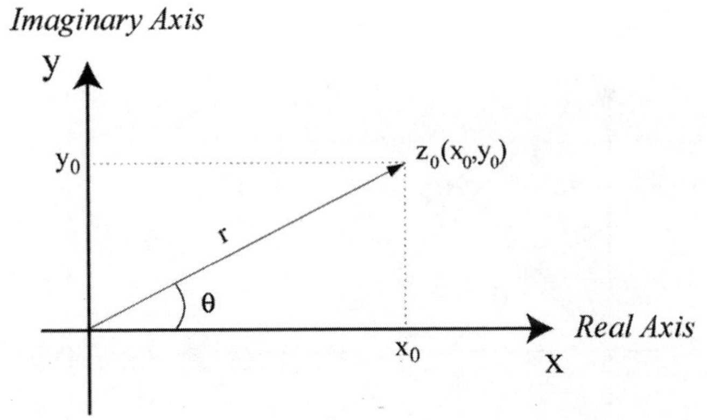

***Fig 5.1**: Vector Representation of the Complex Plane*

$$\tan\theta = \frac{y_0}{x_0}, \qquad 0 \le \theta \le 2\pi \qquad \text{or} \qquad -\pi \le \theta \le \pi$$

and the angle θ is called the **Argument** of z_0.

The complex number z_0 does not change value if θ is increased or decreased by an integer number of 2π, i.e.:

$$z_0 = re^{i\theta} = re^{i(\theta \pm 2n\pi)} \qquad n = 1, 2, 3$$

Thus, in the polar form, let $a = r_1 e^{i\theta_1}$, $b = r_2 e^{i\theta_2}$ then their product is:

$$ab = r_1 r_2 e^{i(\theta_1+\theta_2)} = r_1 r_2 \left[\cos(\theta_1+\theta_2) + i\sin(\theta_1+\theta_2)\right]$$

and their quotient is given by:

$$\frac{a}{b} = \frac{r_1}{r_2} e^{i(\theta_1-\theta_2)} = \frac{r_1}{r_2}\left[\cos(\theta_1-\theta_2) + i\sin(\theta_1-\theta_2)\right]$$

In polar representation the expression $|z - z_0| = c$ represents a circle centered at z_0 and whose radius is "c".

5.1.3 Absolute Value

The absolute value of z_0 represents the distance of point z_0 from the origin. The absolute value of the difference between two complex numbers, is:

$$|a - b| = \sqrt{(a_1 - b_1)^2 + (a_2 - b_2)^2}$$

and represents the distance between a and b.

The absolute value of the products and quotients become:

$$|abc| = |a||b||c|$$

$$\left|\frac{a}{b}\right| = \frac{|a|}{|b|}$$

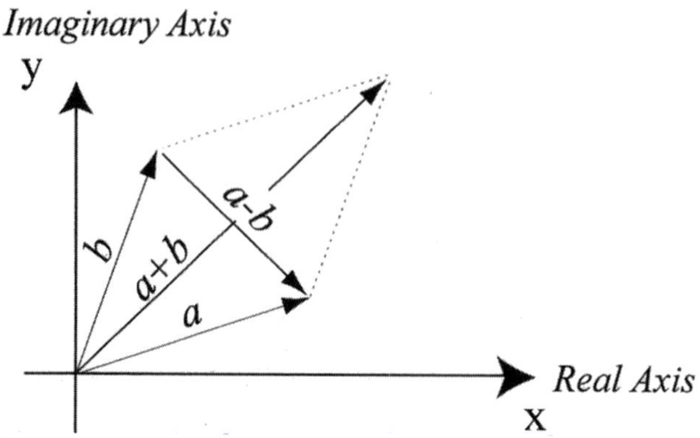

Fig. 5.2: Geometric Argument for Inequalities

The following inequalities can be obtained from geometric arguments as shown in Fig. 5.2.

$$|a + b| \le |a| + |b| \qquad |a - b| \le |a| + |b|$$

$$|a - b| \ge |a| - |b| \qquad |a + b| \ge |a| - |b|$$

5.1.4 Powers and Roots of a Complex Number

The nth power of a complex number with n being integer becomes:

$$a^n = \left(re^{i\theta}\right)^n = r^n e^{in\theta}$$

The nth root of a complex numbers:

$$a^{1/n} = \left[r \exp i(\theta + 2m\pi)\right]^{1/n} = r^{1/n} \exp\left(i\frac{\theta + 2m\pi}{n}\right) \qquad m = 0, 1, 2,...$$

There are n different roots of (a) as follows:

$$\left(a^{1/n}\right)_1 = r^{1/n} \exp\left(i\frac{\theta}{n}\right) \qquad m = 0$$

$$\left(a^{1/n}\right)_2 = r^{1/n} \exp\left(i\frac{\theta + 2\pi}{n}\right) \qquad m = 1$$

$$\left(a^{1/n}\right)_{n-1} = r^{1/n} \exp\left(i\frac{\theta + 2n\pi - 4\pi}{n}\right) \qquad m = n - 2$$

$$\left(a^{1/n}\right)_n = r^{1/n} \exp\left(i\frac{\theta + 2n\pi - 2\pi}{n}\right) \qquad m = n - 1$$

$$\left(a^{1/n}\right)_{n+1} = r^{1/n} \exp\left(i\frac{\theta + 2n\pi}{n}\right) = r^{1/n} \exp\left(i\frac{\theta}{n}\right) = \left(a^{1/n}\right)_1 \qquad m = n$$

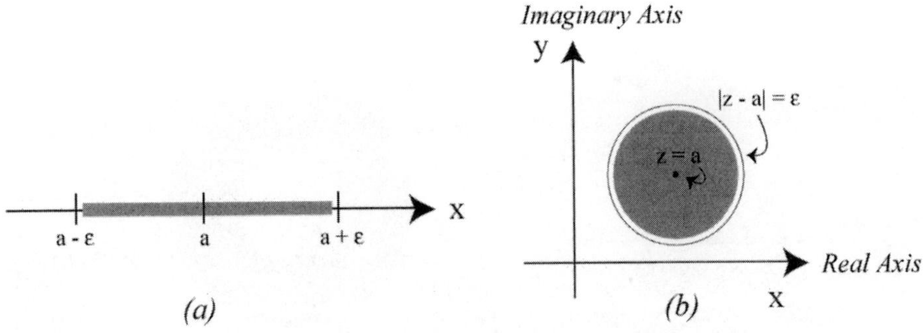

Fig. 5.3 Illustration for the Neighborhood of a Point

Succeeding roots repeat the first n roots. Hence, $a^{1/n}$ has n distinct roots. In polar form, the n roots fall on a circle whose radius is $r^{1/n}$ and whose arguments are equally spaced by $2\pi/n$.

5.2 Analytic Functions

One must develop the calculus of complex variables in a treatment that parallels the calculus of real variables. Thus, one must define a neighborhood of a point, regions, functions, limits, continuity, derivatives, and integrals. In each case, the corresponding treatment of real variables will be presented to give a clearer picture of the ideas being presented.

5.2.1 Neighborhood of a Point

In real variables, the neighborhood of a point $x = a$ represents all the points inside the segment of the real axis $a - \varepsilon < x < a + \varepsilon$, with $\varepsilon > 0$, as shown in the shaded section in Fig. 5.3a. This can be written in more compact form as $|x - a| < \varepsilon$.

In complex variable, all the points inside a circle of radius ε centered at $z = a$, but not including points on the circle, make up the neighborhood of $z = a$, i.e.:

$$|z - a| < \varepsilon$$

This is shown as the shaded area in Figure 5.3b.

5.2.2 Region

A **closed region** in real variables contains all interior as well as boundary points, e.g., the closed region:

$$|x - 1| \le 1$$

contains all points $0 \le x \le 2$, see Fig. 5.4a. The closed region in the complex plane contains all interior points as well as the boundary points, e.g., the closed region:

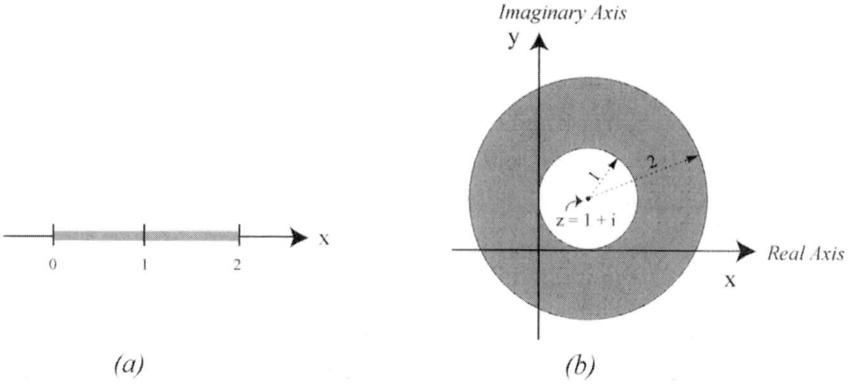

Fig. 5.4 *Illustration for Closed Regions*

$$1 \le |z - 1 - i| \le 2$$

represents all the interior points contained inside the annular circular ring defined by an inner and outer radii of 1 and 2, respectively, as well as all the points on the outer and inner circles, as shown in Fig. 5.4(b).

An **Open Region** is one that includes all the interior points, but does not include the boundary points, e.g., the following regions are open:

$$|x - 1| < 1 \qquad \text{or} \qquad 0 < x < 2$$

as well as:

$$1 < |z - 1 - i| < 2$$

A region is called a **semi-closed Region** if it includes all the interior points as well as points on part of the boundary, e.g.:

$$1 < |z - 1 - i| \le 2$$

A **simply connected** region R is one where every closed contour within it encloses only points belonging to R. A region that is not simply connected is called **multiply-connected**. Thus, the region inside a circle is simply connected, the region outside a circle is multiply-connected. The order of the multiply-connectiveness of a region can be defined by the number of independent closed contours that cannot be collapsed to zero plus one. Thus, the region inside an annular region (e.g. the region between two concentric circles) is doubly connected.

5.2.3 Functions of a Complex Variable

A function of a real variable $y = f(x)$ maps each point x in the region of definition of x on the real x-axis onto one or more corresponding point(s) in another region of definition of y on the real y-axis. A single-valued function is one where each point x maps into one point y. For example, the function:

$$y = f(x) = \frac{1}{x^3} \qquad 0 < |x| \le 1$$

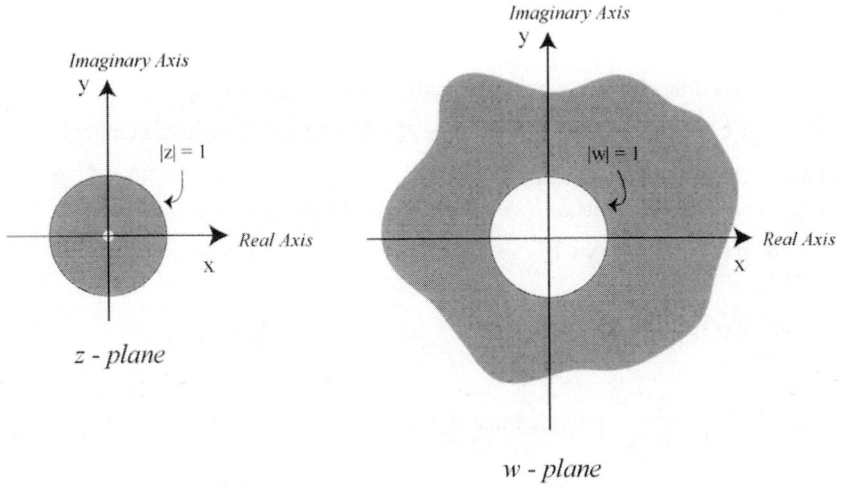

***Fig. 5.5**: Mapping of the Function $w = z^2$*

maps every x in the region $0 < |x| \leq 1$ onto a point y in the region $|y| \geq 1$.

The region of definition of x is called the **Domain** of the function f(x), the set of values of y = f(x), $x \varepsilon D$, is called the **Range** of f(x), e.g., in the example above:

The Domain is $0 < |x| \leq 1$

The Range is $|y| \geq 1$

A function of a complex variable w = f(z) maps each point z in the domain of f(z) onto one or more points w in the range of w. A single-valued function maps one point z onto one point w. For example, the function:

$$w = \frac{1}{z^3} \qquad\qquad 0 < |z| \leq 1$$

maps all the points inside and on a circle of radius = 1, but not the point z = 0, onto the region outside and on the circle of radius = 1, see Fig. 5.5.

The function w = f(z) of a complex variable is also a complex variable, which can be written as follows:

$$w = f(z) = u(x, y) + iv(x, y) \tag{5.1}$$

where u and v are real functions of x and y. For example:

$$w = z^2 = (x + iy)^2 = x^2 - y^2 + i(2xy)$$

where

$$u(x, y) = x^2 - y^2 \qquad \text{and} \qquad v(x, y) = 2xy$$

5.2.4 Limits

If the function f(z) is defined in the neighborhood of a point z_0, except possibly at the point itself, then the limit of the function as z approaches z_0 is a number A, i.e.:

$$\lim_{z \to z_0} f(z) = A \tag{5.2}$$

This means that if there exists a positive small number ε such that:

$|z - z_0| < \varepsilon$ then $|w - A| < \delta$ for a small positive number δ

The limit of the function is **unique**.

Let A and B be the limits of f(z) and g(z) respectively as $z \to z_0$, then:

$$\lim_{z \to z_0} [f(z) + g(z)] = A + B$$

$$\lim_{z \to z_0} [f(z)g(z)] = AB$$

$$\lim_{z \to z_0} \frac{f(z)}{g(z)} = \frac{A}{B} \qquad \text{provided that } B \neq 0$$

Since the limit is unique, then the limit of a function as z approaches z_0 by any path C must be unique. If a function posses more than one limit as $z \to z_0$, when the limiting process is performed along different paths, then the function has no limit as $z \to z_0$.

Example 5.1

Find the limit of the following function as $z \to 0$:

$$f_1(z) = \frac{xy^2}{x^2 + y^4}$$

Let y = mx be the path C along which a limit of the function $f_1(z)$ as $z \to 0$ is to be obtained:

$$\lim_{z \to 0} f_1(z) = \lim_{x \to 0} \frac{m^2 x^3}{x^2 + m^2 x^4} \to 0$$

independent of the value of m. This is not conclusive because, on the curve $x = my^2$, the limit of f(z) as $z \to 0$ on C is $m/(m^2 + 1)$, which depends on m for its limit. Thus, if $f_1(z)$ has many limits, then $f_1(z)$ has no limit as $z \to 0$.

5.2.5 Continuity

A function is continuous at a point $z = z_0$, if $f(z_0)$ exists, and $\lim_{z \to z_0} f(z)$ exists, and if $\lim_{z \to z_0} f(z) = f(z_0)$. A complex function f(z) is continuous at $z = z_0$ iff, both u(x,y) and v(x,y) are continuous functions at $z = z_0$.

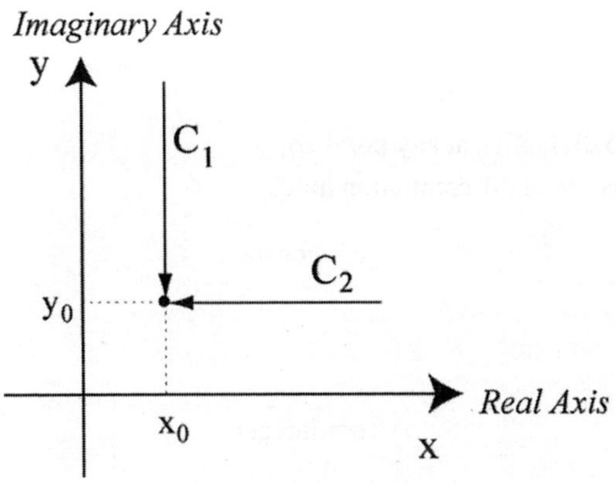

Fig. 5.6: *Two Paths for Differentiation of f(z)*

5.2.6 Derivatives

Let z be a point in the neighborhood of a point z_0, then one defines Δz as:

$\Delta z = z - z_0$ a complex number

The derivative of a function f(z) is defined as follows:

$$f'(z_0) = \left.\frac{df(z)}{dz}\right|_{z=z_0} = \lim_{\Delta z \to 0} \frac{f(z_0 + \Delta z) - f(z_0)}{\Delta z} = \lim_{z \to z_0} \frac{f(z) - f(z_0)}{z - z_0} \tag{5.3}$$

Thus, the derivative is defined only if the limit exists, which indicates that the derivative must be unique. If a function possesses more than one derivative at a point $z = z_0$ depending on the path along which a limit was taken, then it has no derivative at the point $z = z_0$.

Example 5.2

(i) $f(z) = z^2$

$$f'(a) = \lim_{\Delta z \to 0} \frac{(a + \Delta z)^2 - a^2}{\Delta z} = \lim_{\Delta z \to 0} (2a + \Delta z) = 2a$$

(ii) The function $f(z) = R\, z = x$ has no derivative at $z = z_0$, since one can show that it possesses more than one derivative. If one takes the limit along path parallel to the y-axis at (x_0, y_0) (see Fig. 5.6, path C_1):

$$f'(z_0) = \lim_{\Delta z \to 0} \frac{\Delta f}{\Delta z} = \lim_{z \to z_0} \frac{f(z) - f(z_0)}{z - z_0} = \lim_{y \to y_0} \frac{x_0 - x_0}{i(y - y_0)} = 0$$

If one takes the path parallel to the x-axis (see Fig. 5.6, path C_2):

$$f'(z_0) = \lim_{z \to z_0} \frac{\Delta f}{\Delta z} = \lim_{x \to x_0} \frac{x - x_0}{x - x_0} = 1$$

Thus f(z) has no derivative at any point z_0.

The following properties of differentiation holds:

$$\frac{d}{dz} c = 0 \qquad c = \text{constant}$$

$$\frac{d}{dz} z = 1$$

$$\frac{d}{dz} z^n = nz^{n-1} \qquad n = \text{integer}$$

$$\frac{d}{dz}(cf) = c\frac{df}{dz} \qquad c = \text{constant}$$

$$\frac{d}{dz}(f_1 + f_2) = f_1' + f_2'$$

$$\frac{d}{dz}(f_1 f_2) = f_1 f_2' + f_2 f_1'$$

$$\frac{d}{dz}\left(\frac{f_1}{f_2}\right) = \frac{f_1' f_2 - f_1 f_2'}{f_2^2} \qquad f_2 \neq 0$$

$$\frac{d}{dz} f(g(z)) = \frac{df}{dg}\frac{dg}{dz} \tag{5.4}$$

5.2.7 Cauchy-Reimann Conditions

If f(z) has a derivative at z_0 and if $\frac{\partial u}{\partial x}, \frac{\partial u}{\partial y}, \frac{\partial v}{\partial x}$, and $\frac{\partial v}{\partial y}$ are continuous at z_0, then it can be shown that:

$$\frac{\partial u}{\partial x} = \frac{\partial v}{\partial y} \qquad \text{and} \qquad \frac{\partial u}{\partial y} = -\frac{\partial v}{\partial x}$$

or in polar coordinates: (5.5)

$$\frac{\partial u}{\partial r} = \frac{1}{r}\frac{\partial v}{\partial \theta} \qquad \text{and} \qquad \frac{1}{r}\frac{\partial u}{\partial \theta} = -\frac{\partial v}{\partial r}$$

These are known as the **Cauchy-Reimann conditions**.

The derivative computed along path C_1 (see Fig. 5.6):

$$f'(z_0)=\lim_{\substack{z\to z_0\\ \text{on } C_1}}\frac{f(z)-f(z_0)}{z-z_0}$$

$$=\lim_{\substack{y\to y_0\\ \text{on } C_1}}\frac{[u(x_0,y)+iv(x_0,y)]-[u(x_0,y_0)+iv(x_0,y_0)]}{(x_0+iy)-(x_0+iy_0)}$$

$$=\lim_{\substack{y\to y_0\\ \text{on } C_1}}\frac{u(x_0,y)-u(x_0,y_0)}{i(y-y_0)}+\lim_{\substack{y\to y_0\\ \text{on } C_1}}\frac{v(x_0,y)-v(x_0,y_0)}{y-y_0}$$

$$=-i\frac{\partial u}{\partial y}(x_0,y_0)+\frac{\partial v}{\partial y}(x_0,y_0)$$

The derivative computed along path C_2 becomes:

$$f'(z_0)=\lim_{\substack{z\to z_0\\ \text{on } C_2}}\frac{f(z)-f(z_0)}{z-z_0}$$

$$=\lim_{\substack{x\to x_0\\ \text{on } C_2}}\frac{[u(x,y_0)+iv(x,y_0)]-[u(x_0,y_0)+iv(x_0,y_0)]}{(x+iy_0)-(x_0+iy_0)}$$

$$=\lim_{\substack{x\to x_0\\ \text{on } C_2}}\frac{u(x,y_0)-u(x_0,y_0)}{x-x_0}+i\lim_{\substack{x\to x_0\\ \text{on } C_2}}\frac{v(x,y_0)-v(x_0,y_0)}{x-x_0}$$

$$=\frac{\partial u}{\partial x}(x_0,y_0)+i\frac{\partial v}{\partial x}(x_0,y_0)$$

Thus, equating the two expressions for $f'(z)$, one obtains the Cauchy-Riemann conditions given in eq. (5.5). The Cauchy-Riemann conditions can also be written in the polar form given in eq. (5.5). The derivative can thus be evaluated by:

$$f'(z)=\frac{\partial u}{\partial x}+i\frac{\partial v}{\partial x}=\frac{\partial v}{\partial y}-i\frac{\partial u}{\partial y} \tag{5.6}$$

Example 5.3

(i) The function:

$$f(z)=z^2=(x^2-y^2)+i(2xy)$$

$$u(x,y)=x^2-y^2 \qquad v(x,y)=2xy$$

has a derivative:

$$f'(z) = 2x + i2y = 2(x + iy) = 2z.$$

The partial derivation of u and v are continuous:

$$\frac{\partial u}{\partial x} = 2x \qquad \frac{\partial u}{\partial y} = -2y \qquad \frac{\partial v}{\partial x} = 2y \qquad \frac{\partial v}{\partial y} = 2x$$

Note that the partial derivatives satisfy the Cauchy-Riemann conditions.

(ii) The function:

$$f(z) = Re\,(z) = x$$

has no derivative:

$$u = x \qquad \frac{\partial u}{\partial x} = 1 \qquad \frac{\partial u}{\partial y} = 0$$

$$v = 0 \qquad \frac{\partial v}{\partial x} = 0 \qquad \frac{\partial v}{\partial y} = 0$$

The partial derivatives do not satisfy the Cauchy-Riemann conditions in eq. (5.5).

If u and v are single valued functions, whose partial derivatives of the first order are continuous and if the partial derivatives satisfy the Cauchy-Riemann conditions in eq. (5.5), then $f'(z)$ exists. This is a **necessary and sufficient** condition for existence of continuous derivative $f'(z)$.

If one differentiates eq. (5.5) partially once with respect to x and once with respect to y, one cans show that:

$$\nabla^2 u = \frac{\partial^2 u}{\partial x^2} + \frac{\partial^2 u}{\partial y^2} = 0$$

$$\nabla^2 v = \frac{\partial^2 v}{\partial x^2} + \frac{\partial^2 v}{\partial y^2} = 0 \tag{5.7}$$

These equations are known as **Laplace's equations**. Functions that satisfy eq. (5.7) are called **Harmonic Functions**.

The Cauchy-Riemann conditions can be used to obtain one of the two components of a complex function w = f(z) to within an additive complex constant if the other component is known. Thus, if v is known, then the total derivative becomes:

$$du = \frac{\partial u}{\partial x} dx + \frac{\partial u}{\partial y} dy = \frac{\partial v}{\partial y} dx - \frac{\partial v}{\partial x} dy$$

Example 5.4

If v = xy, then one can obtain u(x, y) as follows:

$$\frac{\partial v}{\partial y} = x = \frac{\partial u}{\partial x} \qquad \text{then} \qquad u = \frac{x^2}{2} + g(y)$$

$$\frac{\partial u}{\partial y} = -\frac{\partial v}{\partial x} = -y = g'(y) \qquad \text{then} \qquad g(y) = -\frac{y^2}{2} + c$$

Thus, the real part u(x, y) is given by:

$$u = \frac{1}{2}\left(x^2 - y^2\right) + c$$

so that the function f(z) is:

$$f(z) = \frac{1}{2}\left(x^2 - y^2\right) + c + ixy = \frac{1}{2} z^2 + c$$

5.2.8 Analytic Functions

A function f(z) is **analytic** at a point z_0 if its derivative $f'(z)$ exists and is continuous at z_0 and at every point in the neighborhood of z_0. An **entire** function is one that is analytic at every point in the entire complex z plane. If the function is analytic everywhere in the neighborhood of a point z_0 but not a $z = z_0$, then $z = z_0$ is called an **isolated singularity** of f(z).

Example 5.5

(i) The function:

$$f(z) = z^n \qquad n = 0, 1, 2, ...$$

is an analytic function since $f'(z) = nz^{n-1}$ exists and is continuous everywhere. It is also an entire function.

(ii) The function:

$$f(z) = \frac{1}{(z-1)^2}$$

is analytic everywhere, except at $z_0 = 1$, since $f'(z) = -2/(z-1)^3$ does not exist at $z_0 = 1$. The point $z_0 = 1$ is an isolated singularity of the function f(z).

5.2.9 Multi-Valued Functions, Branch Cuts and Branch Points

Some complex functions can be multivalued in the complex z-plane and hence, are not analytic over some region. In order to make these functions single-valued, one can define the range of points z in the z-plane in a way that the function is single-valued for those points. For example, the function $z^{1/2}$ is multivalued since:

$$w = z^{1/2} = [re^{i(\theta \pm 2n\pi)}]^{1/2} = r^{1/2} e^{i(\theta \pm 2n\pi)/2} \qquad r \geq 0, 0 < \theta < 2\pi$$

Therefore, for n = 0:

$$z^{1/2} = r^{1/2} e^{i\theta/2}$$

and for n = 1:

$$z^{1/2} = r^{1/2} e^{i(\theta + 2\pi)/2}$$

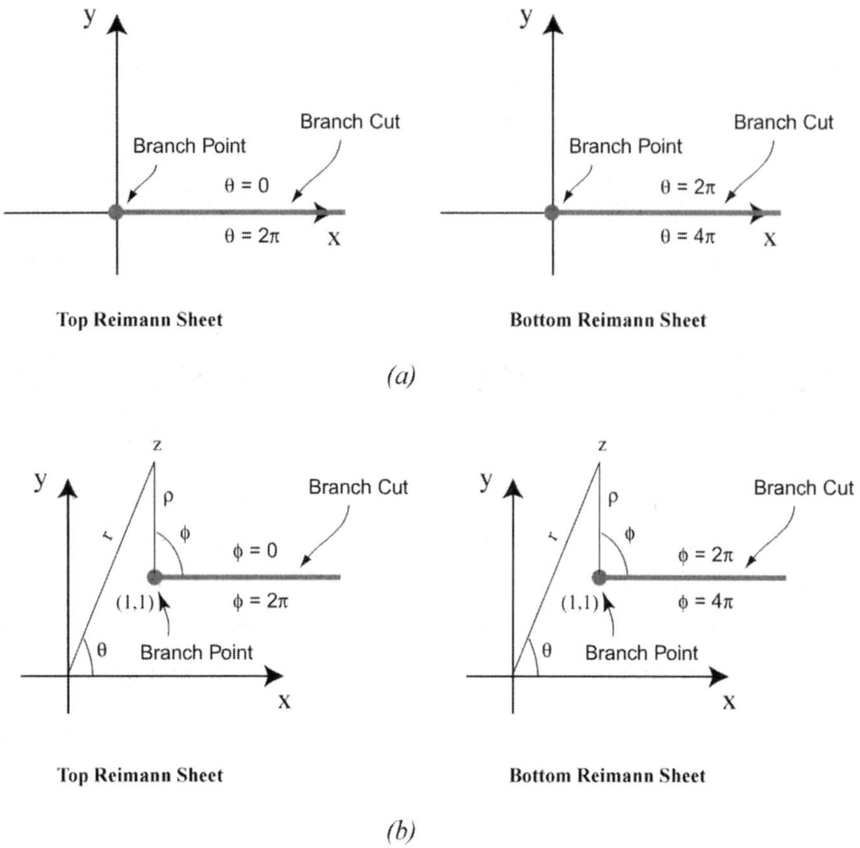

***Fig. 5.7**: Branch Cuts and Reimann Sheets*

For $n = 2, 3, \ldots$, the value of w is equal to those for $n = 0$ and $n = 1$. Thus, there are two distinct values of the function $w = z^{1/2}$ for every point z in the z-plane. Instead of letting w have two values on the z-plane, one can create two planes where w is single-valued in each. This can be done by defining in one plane:

$$z^{1/2} = r^{1/2} e^{i\theta/2} \qquad r \geq 0,\ 0 < \theta < 2\pi$$

and in a second plane:

$$z^{1/2} = r^{1/2} e^{i\theta/2} \qquad r \geq 0,\ 2\pi < \theta < 4\pi$$

Thus, the function w is single valued in each plane. It should be noted that θ is limited to one range in each plane. This can be achieved by making a cut of the 0/2π ray from the origin $r = 0$ to ∞ in such a way that θ cannot exceed 2π or be less than zero in the first plane. The same cut from the origin $r = 0$ to ∞ is made in the other plane at 2π/4π ray so that θ cannot exceed 4π or be less than 2π, see Fig. 5.7(a). Each of these planes is called a **Reimann Sheet**. The cut is called a **Branch cut**. The origin point where the

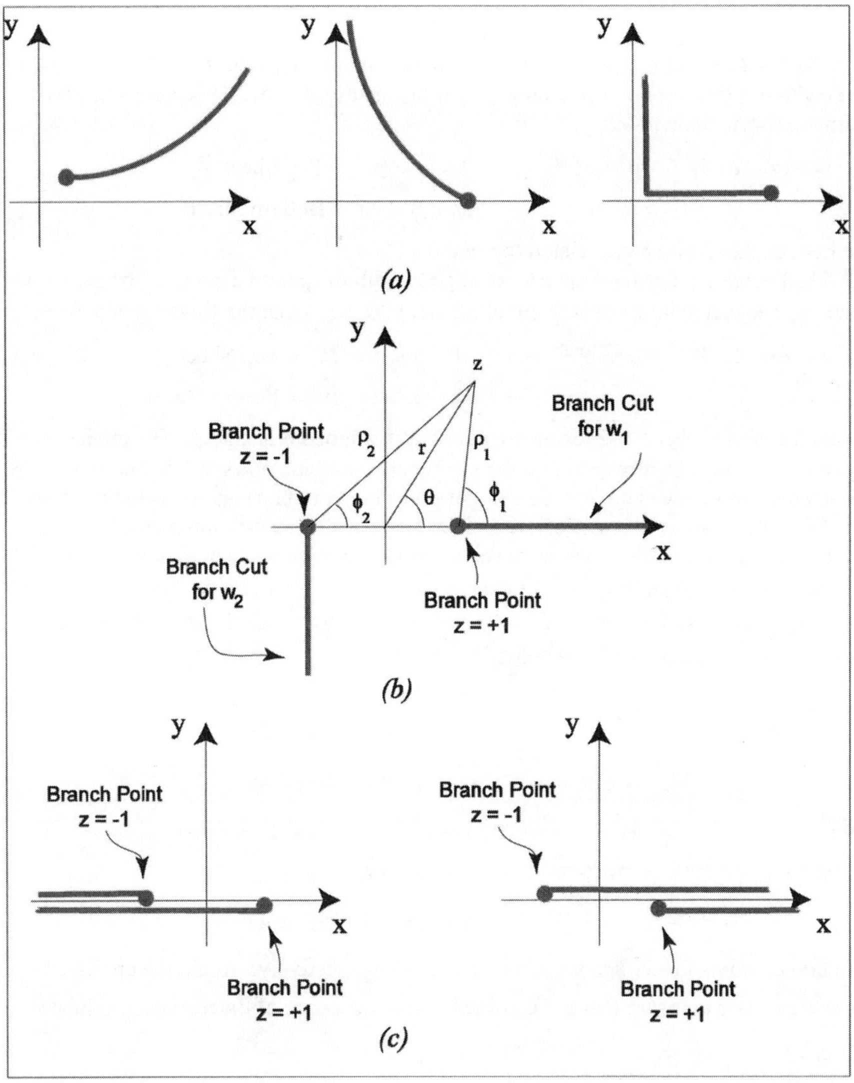

Fig 5.8 Examples of Branch Cuts (a) Non-linear, (b) Multiple, and (c) Co-linear Branch Cuts

cut starts at r = 0 is called the **Branch Point**. Since the function w is continuous at $\theta = 2\pi$ in both sheets and is continuous at $\theta = 0$ and 4π, one can envision joining these two Reimann sheets at the 2π and at $0/4\pi$ rays.

For the function $w = (z - 1 - i)^{1/2}$, one must first express it in cylindrical coordinates in order to calculate the function. Let the origin of the z-plane be at (0,0), such that:

$$z = r\, e^{i\theta}$$

Let the origin of the coordinate system for the function w be (1,1), such that:

$z - 1 - i = \rho\, e^{i\phi} \qquad \rho > 0$

To make the function w single-valued, one needs to cut the plane from the branch point at (1+i) with $\rho = 0$ to ∞ at $\phi = 0/2\pi$ and $\phi = 2\pi/4\pi$, see Fig. 5.7(b). This results in two Reimann sheets defined by:

$$w = (z - 1 - i)^{1/2} = \rho^{1/2}\, e^{i\phi/2} \qquad 0 < \phi < 2\pi \quad : \text{Top Sheet}$$
$$2\pi < \phi < 4\pi \quad : \text{Bottom Sheet}$$

one can see that ρ and ϕ are related to r and θ.

The branch cut does not have to be aligned with the positive x-axis. For the above function, one can define a branch cut along a ray, α, such that the function is defined by:

$$w = (z - 1 - i)^{1/2} = \rho^{1/2}\, e^{i\phi/2} \qquad \alpha < \phi < \alpha \pm 2\pi \quad : \text{Top Sheet}$$
$$\alpha \pm 2\pi < \phi < \alpha \pm 4\pi \quad : \text{Bottom Sheet}$$

so that the choice of $\alpha = \pi/2$ results in a vertically aligned branch cut. The choice of $\pm 2\pi$ depends on α, in such a way so that the top sheet should include $\phi = 0$ in its range. Branch cuts do not even have to be straight lines, but could be curved, as long as they start from the branch point and end at $z \rightarrow \infty$, see Fig. 5.8(a) for examples.

Sometimes, a function may have two or more components that are multi-valued. For example, the function $w = (z^2 - 1)^{1/2}$ can be written as $w = (z-1)^{1/2}(z+1)^{1/2}$ which contains two multivalued functions $w_1 = (z-1)^{1/2}$ and $w_2 = (z+1)^{1/2}$. Both functions require branch cuts to make them single valued.

Let:

$$w_1 = (z-1)^{1/2} = \rho_1^{1/2} e^{i\phi_1/2} \qquad \alpha_1 < \phi_1 < \alpha_1 \pm 2\pi$$
$$\alpha_1 \pm 2\pi < \phi_1 < \alpha_1 \pm 4\pi$$

and

$$w_2 = (z+1)^{1/2} = \rho_2^{1/2} e^{i\phi_2/2} \qquad \alpha_2 < \phi_2 < \alpha_2 \pm 2\pi$$
$$\alpha_2 \pm 2\pi < \phi_2 < \alpha_2 \pm 4\pi$$

with branch poins for w_1 and w_2 at z = -1 and +1, respectively. Again the choice of $\pm 2\pi$ is made in order to insure that $\phi = 0$ is included in the range of the top sheet. Thus:

$$w = w_1 w_2 = (\rho_1 \rho_2)^{1/2} e^{i(\phi_1 + \phi_2)/2}$$

where ϕ_1 and ϕ_2 could take any of the angles given above, i.e., four possible Reimann sheets. Thus, one can choose, see Fig. 5.8(b):

$0 < \phi_1 < 2\pi,\ -\pi/2 < \phi_2 < 3\pi/2$: Sheet 1

$0 < \phi_1 < 2\pi,\ 3\pi/2 < \phi_2 < 7\pi/2$: Sheet 2

$2\pi < \phi_1 < 4\pi,\ -\pi/2 < \phi_2 < 3\pi/2$: Sheet 3

$2\pi < \phi_1 < 4\pi,\ 3\pi/2 < \phi_2 < 7\pi/2$: Sheet 4

It should be noted that ρ_1, ϕ_1, ρ_2 and ϕ_2 are related to r, θ.

In many instances, it may be advantageous to make the branch cuts colinear. In those cases, the function may become single valued along the position that is common to those branch cuts. For example, the choice of $\alpha_1 = \alpha_2 = 0$ or $\alpha_1 = \alpha_2 = -\pi$ for both branch cuts may work better than example 5.86 (see Fig. 5.8(c)). For a point slightly above the two branch cuts, $\phi_1 = \phi_2 \cong 0$ so that:

$$w = w_1 w_2 = (\rho_1 \rho_2)^{1/2}$$

For a point slightly below the two branch cuts, $\phi_1 = \phi_2 \cong 2\pi$ so that:

$$w = w_1 w_2 = (\rho_1 \rho_2)^{1/2} e^{i(2\pi + 2\pi)/2} = (\rho_1 \rho_2)^{1/2}$$

Thus the function w is continuous across both branch cuts over the segment from z = 1 to ∞. Similarly, one can show the same for the other pair of branch cuts in Fig. 5.8(c).

5.3 Elementary Functions

5.3.1 Polynomials

An nth degree polynomial can be defined as follows:

$$f(z) = \sum_{k=0}^{k=n} a_k z^k \qquad a_k \text{ complex number} \tag{5.8}$$

A polynomial function is an entire function. The derivative can be obtained as follows:

$$f'(z) = \sum_{k=1}^{k=n} k\, a_k z^{k-1}$$

The polynomial function has n complex zeroes.

5.3.2 Exponential Function

Define the exponential function:

$$e^z = e^x (\cos y + i \sin y) \tag{5.9}$$

The exponential function is an entire function, since u and v:

$$u = e^x \cos y \qquad v = e^x \sin y$$

together with their first partial derivatives are continuous everywhere and:

$$\frac{d}{dz}\left(e^z\right) = e^z$$

exists everywhere. One can write e^z in a polar form as follows:

$$e^z = \rho(\cos\phi + i \sin\phi)$$

where:

$$\rho = e^x \qquad \text{and} \qquad \phi = y$$

The exponential function has no zeros, since $\left|e^z\right| > 0$. The complex exponential function follows the same rules of calculus as the real exponential functions. Thus:

$$
\begin{aligned}
&e^{z_1} e^{z_2} = e^{z_1 + z_2} \\
&\frac{1}{e^z} = e^{-z} \\
&\left(e^z\right)^n = e^{nz} \qquad\qquad (5.10) \\
&e^{\bar{z}} = \overline{\left(e^z\right)} \\
&e^z = e^{z+2\pi i} \qquad\qquad \text{periodicity} = 2\pi i
\end{aligned}
$$

The periodicity of the complex exponential function in $2i\pi$ is a property of the complex function only.

5.3.3 Circular Functions

From the definition of an exponential function, one can define the complex circular functions as:

$$\sin z = \frac{e^{iz} - e^{-iz}}{2i} = -i \sinh(iz) \qquad (5.11)$$

$$\cos z = \frac{e^{iz} + e^{-iz}}{2} = \cosh(iz)$$

The functions sin z and cos z are entire functions. From the definition in eq. (5.11), one can obtain the real and imaginary components:

$$\cos z = \cos x \cosh y - i \sin x \sinh y$$

$$\sin z = \sin x \cosh y + i \cos x \sinh y$$

$$|\sin z|^2 = \sin^2 x + \sinh^2 y$$

$$|\cos z|^2 = \cos^2 x + \sinh^2 y$$

It should be noted that the magnitude of the complex functions cos z and sin z can be unbounded in contrast to their real counterparts, cos x and sin x.

Define:

$$\tan z = \frac{\sin z}{\cos z} \qquad\qquad \sec z = \frac{1}{\cos z}$$

$$\cot z = \frac{\cos z}{\sin z} = \frac{1}{\tan z} \qquad\qquad \operatorname{cosec} z = \frac{1}{\sin z} \qquad (5.12)$$

The function tan z and sec z are analytic everywhere except at points where $\cos z = 0$. The functions cot z and cosec z are analytic everywhere except at points where $\sin z = 0$. The circular functions in eq. (5.11) and (5.12) are periodic in 2π, i.e., $f(z+2\pi) = f(z)$. Furthermore, it can also be shown that:

$$\cos(z+\pi) = -\cos z$$
$$\sin(z+\pi) = -\sin z$$
$$\tan(z+\pi) = \tan z$$

The derivative formulae for the circular function are listed below:

$$\frac{d}{dz}(\sin z) = \cos z \qquad \frac{d}{dz}(\cot z) = -\mathrm{cosec}^2 z$$
$$\frac{d}{dz}(\cos z) = -\sin z \qquad \frac{d}{dz}(\sec z) = \sec z \tan z$$
$$\frac{d}{dz}(\tan z) = \sec^2 z \qquad \frac{d}{dz}(\mathrm{cosec}\, z) = -\mathrm{cosec}\, z\, \mathrm{cotan}\, z \tag{5.13}$$

The trigonometric identities have the same form for complex variables as in real variables, a few of which are listed below:

$$\sin^2 z + \cos^2 z = 1$$
$$\sin(z_1 \pm z_2) = \sin z_1 \cos z_2 \pm \cos z_1 \sin z_2$$
$$\cos(z_1 \pm z_2) = \cos z_1 \cos z_2 \mp \sin z_1 \sin z_2$$
$$\cos 2z = 2\cos^2 z - 1 = \cos^2 z - \sin^2 z$$
$$\sin 2z = 2 \sin z \cos z \tag{5.14}$$

The only zeros of cos z and sin z are the real zeros, i.e.:

$$\cos z_0 = 0 \qquad z_0 = \left(\pm\frac{2n+1}{2}\pi, 0\right) \qquad n = 0, 1, 2, \ldots$$
$$\sin z_0 = 0 \qquad z_0 = (\pm n\pi, 0) \qquad n = 0, 1, 2, \ldots$$

The function tan z (cot z) has zeros corresponding to the zeroes of sin z (cos z).

5.3.4 Hyperbolic Functions

Define the complex hyperbolic functions in the same way as real hyperbolic functions, i.e.:

$$\sinh z = \frac{e^z - e^{-z}}{2} \qquad \coth z = \frac{1}{\tanh z}$$
$$\cosh z = \frac{e^z + e^{-z}}{2} \qquad \mathrm{sech}\, z = \frac{1}{\cosh z} \tag{5.15}$$
$$\tanh z = \frac{\sinh z}{\cosh z} \qquad \mathrm{cosech}\, z = \frac{1}{\sinh z}$$

The functions sinh z and cosh z are entire functions. The function tanh z (coth z) is analytic everywhere except at the zeroes of cosh z (sinh z). The components of the hyperbolic functions u and v can be obtained from the definitions in eq. (5.15).

$$\sinh z = \sinh x \cos y + i \cosh x \sin y$$

$$\cosh z = \cosh x \cos y + i \sinh x \sin y$$

$$|\sinh z|^2 = \sinh^2 x + \sin^2 y = \cosh^2 x - \cos^2 y$$

$$|\cosh z|^2 = \sinh^2 x + \cos^2 y = \cosh^2 x - \sin^2 y$$

Unlike real hyperbolic functions, complex hyperbolic functions are periodic in $2i\pi$ and have infinite number of zeroes. The zeros of cosh z and sinh z are:

$$\sinh z_0 = 0 \qquad z_0 = (0, \pm n\pi) \qquad n = 0, 1, 2,...$$

$$\cosh z_0 = 0 \qquad z_0 = \left(0, \pm \frac{2n+1}{2}\pi\right) \qquad n = 0, 1, 2,...$$

The derivative formulae for the hyperbolic functions are listed below:

$$\frac{d}{dz}(\sinh z) = \cosh z \qquad \frac{d}{dz}(\coth z) = -\text{cosech}^2 z$$

$$\frac{d}{dz}(\cosh z) = \sinh z \qquad \frac{d}{dz}(\text{sech } z) = -\text{sech } z \tanh z \tag{5.16}$$

$$\frac{d}{dz}(\tanh z) = \text{sech}^2 z \qquad \frac{d}{dz}(\text{cosech } z) = -\text{cosech } z \coth z$$

A few identities for complex hyperbolic functions are listed below:

$$\cosh^2 z - \sinh^2 z = 1$$

$$\sinh(z_1 \pm z_2) = \sinh z_1 \cosh z_2 \pm \cosh z_1 \sinh z_2$$

$$\cosh(z_1 \pm z_2) = \cosh z_1 \cosh z_2 \pm \sinh z_1 \sinh z_2 \tag{5.17}$$

$$\sinh (2z) = 2 \sinh z \cosh z$$

$$\cosh (2x) = \cosh^2 z + \sinh^2 z = 2 \cosh^2 z - 1$$

5.3.5 Logarithmic Function

Define the logarithmic function log z as follows:

$$\log z = \log r + i\theta \qquad \text{for } r > 0$$

where $z = re^{i\theta}$. Since z is a periodic in 2π, i.e.:

$$z(r,\theta) = z(r,\theta \pm 2n\pi) \qquad n = 1, 2,...$$

then the function log z is a multivalued function. To make the function single-valued, make a branch cut along a ray $\theta = \alpha$, starting from the branch point at $z_0 = 0$. Thus, define:

$$\log z = \log r + i\theta \qquad \alpha + 2n\pi < \theta < \alpha + (2n+2)\pi \qquad n = 0, \pm 1, \pm 2, \ldots \tag{5.18}$$

where $-\pi \le \alpha \le 0$ then there is an infinite number of Reimann sheets. The Reimann sheet with n = 0 is called the **Principal Reimann sheet** of log z, i.e.:

$$\log z = \log r + i\theta \qquad r > 0 \qquad \alpha < \theta < \alpha \pm 2\pi \tag{5.19}$$

where the choice of $\pm 2\pi$ is made in order to include the angle $\theta = 0$ in the Principal Reimann sheet. The function log z as defined by eq. (5.19) is thus single-valued. The function log z is not continuous along the rays defined by $\theta = \alpha$ and $\theta = \alpha \pm 2\pi$, because the function jumps by a value equal to $2\pi i$ when θ crosses these rays. Since the function is not single-valued on $\theta = \alpha$ and $\theta = \alpha \pm 2\pi$, the logarithmic function has no derivative on the branch cut defined by the ray $\theta = \alpha$, as well as at the branch point $z_0 = 0$. Hence, all the points on the ray $\theta = \alpha$ are non-isolated singular points.

A few other formulae for the complex function are listed below:

$$\frac{d}{dz}(\log z) = \frac{1}{z} \qquad z \neq 0 \qquad r > 0 \qquad \alpha < \theta < \alpha \pm 2\pi$$

$$e^{\log z} = e^{\log r + i\theta} = e^{\log r} e^{i\theta} = re^{i\theta} = z$$

$$\log e^z = \log\left(e^x\right)\left[e^{i(y \pm 2n\pi)}\right] = x + iy \pm 2in\pi = z \pm 2in\pi$$

$$\log z_1 z_2 = \log z_1 + \log z_2$$

$$\log \frac{z_1}{z_2} = \log z_1 - \log z_2$$

$$\log z^m = m \log z$$

5.3.6 Complex Exponents

Define the function z^a, where a is a real or complex constant as:

$$z^a = e^{a \log z} = e^{a[\log r + i(\theta \pm 2n\pi)]} \qquad \alpha < \theta \le \alpha \pm 2\pi \tag{5.20a}$$

The inverse function can also be defined as follows:

$$z^{-a} = \frac{1}{z^a}$$

The function z^a is a multi-valued function when the constant "a" is not an integer, unless one specifies a particular branch.

To achieve this, one can follow the same method of making the function single-valued on each of many Reimann sheets.

Defining:

$$z^a = e^{a \log z} = e^{a[\log r + i\theta]} \qquad \alpha \pm 2n\pi < \theta \le \alpha \pm 2(n+1)\pi \tag{5.20b}$$

where , n = 0, 1, 2, …, then the function z^a is single-valued in each Reimann sheet, numbered n = 0 (principal), n = 1, 2, … .

For example, let a = 1/3 and let $\alpha = 0$, then:

$$z^{1/3} = r^{1/3}\, e^{i\theta/3} \qquad \text{where } 2n\pi < \theta \le 2(n+1)\pi$$

Therefore, for $n = 0$, $0 < \theta < 2\pi$, for $n = 1$, $2\pi < \theta < 4\pi$, and for $n = 2$, $4\pi < \theta < 6\pi$. For $n = 3$, the value of $z^{1/3}$ is the same as defined by $n = 0$. Thus, there are *only* three Reimann sheets $n = 0$, 1 and 2.

The derivative of z^a can be evaluated as follows:

$$\frac{d}{dz} z^a = \frac{d}{dz} e^{a \log z} = \frac{a}{z} e^{a \log z} = az^{a-1}$$

The exponential function with a base "a", where "a" is a complex constant, can be defined as follows:

$$a^z = e^{z \log a}$$

$$\frac{d}{dz} a^z = \frac{d}{dz} e^{z \log a} = (\log a) e^{z \log a} = a^z \log a \tag{5.21}$$

5.3.7 Inverse Circular and Hyperbolic Functions

Define the inverse function arcsin z:

$$w = \arcsin z \qquad \text{or} \qquad z = \sin w = \frac{e^{iw} - e^{-iw}}{2i}$$

$$\left(e^{iw}\right)^2 - 2iz\left(e^{iw}\right) - 1 = 0$$

or

$$e^{iw} = iz + \sqrt{1 - z^2}$$

where $\sqrt{1 - z^2}$ is a multi-valued function. Thus:

$$w = \arcsin z = -i \log\left[iz + \sqrt{1 - z^2}\right] = -i \operatorname{arcsinh}(iz)$$

Similarly:

$$w = \arccos z = -i \log\left[z + \sqrt{z^2 - 1}\right] = -i \operatorname{arccosh}(iz)$$

$$\arctan z = \frac{i}{2} \log \frac{1 - iz}{1 + iz} = \frac{i}{2} \log \frac{i + z}{i - z} = -i \operatorname{arctanh}(iz) \tag{5.22}$$

Since the definitions involve multivalued functions, all the inverse functions are also multivalued functions.

The inverse hyperbolic functions can be defined as follows:

$$\operatorname{arcsinh} z = \log\left[z + \sqrt{z^2 + 1}\right] = -i \arcsin(iz)$$

$$\operatorname{arccosh} z = \log\left[z + \sqrt{z^2 - 1}\right] = i \arccos(iz) \tag{5.23}$$

$$\operatorname{arctanh} z = \frac{1}{2} \log \frac{1 + z}{1 - z} = -i \arctan(iz)$$

5.4 Integration in the Complex Plane

Integration of real functions is a process of a limiting summation. Thus, integration in the Reimann sense can be defined as:

$$\int_a^b f(x)dx = \lim_{N\to\infty} \sum_{j=1}^{N} f(x_j)(\Delta x_j)$$

where $\Delta x_j = x_j - x_{j-1}$ and $x_0 = a$ and $x_N = b$, N being the number of segments.

Integration of a real function f(x,y) along a path C defined by the following equation:

$$y = g(x) \qquad \text{on C}$$

can be performed as follows:

$$\int_C f(x,y)\, ds = \int_{x_a}^{x_b} f[x, g(x)]\sqrt{(g')^2 + 1}\, dx$$

One can perform the preceding integration by a parametric substitution, i.e., if one lets $x = \xi(t)$ and hence $y = g(x) = g(\xi(t)) = \eta(t)$, where $t_a \le t \le t_b$ correspond to the limits a and b, then the integral is transformed to:

$$\int_C f(x,y)\, ds = \int_{t_a}^{t_b} f[\xi(t), \eta(t)]\sqrt{(\xi')^2 + (\eta')^2}\, dt$$

Integration of a real function f(x,y) on two variables (area integrals) can be performed as follows:

$$\int_A f(x,y)\, dx\, dy = \lim_{N,M\to\infty} \sum_{i=1}^{N}\sum_{j=1}^{M} f(x_i, y_j)(\Delta x_i)(\Delta y_j)$$

where $\Delta x_i = x_i - x_{i-1}$ and $\Delta y_j = y_j - y_{j-1}$.

5.4.1 Green's Theorem

A theorem that transforms an area integral to a line integral can be stated as follows: If two functions, f(x,y) and g(x,y), together with their first partial derivatives are continuous in a region R, and on the curve C that encloses R, then (see Fig. 5.9).

$$\int_R \left(\frac{\partial f}{\partial x} - \frac{\partial g}{\partial y}\right) dx\, dy = \oint_C [g(x,y)\, dx + f(x,y)\, dy] \tag{5.24}$$

where the closed contour integration on C is taken in the *Positive* (counter-clockwise) sense.

Similarly, one can define an integration in the complex plane on a path C (Fig. 5.10) by:

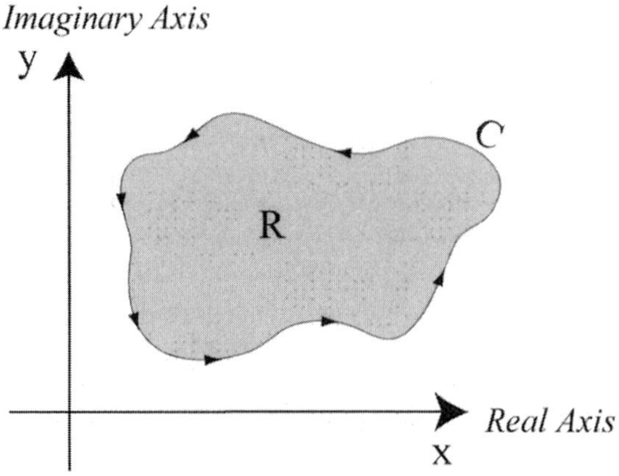

Fig. 5.9: *Green's Theorem*

$$\int_{z_1 \text{ on } C}^{z_2} f(z)\,dz = \lim_{N\to\infty} \sum_{j=1}^{N} f(z_j)(\Delta z_j)$$

where the increments $\Delta z_j = z_j - z_{j-1}$ are taken on C.

Since the complex function is written in terms of u and v, i.e.:

$$f(z) = u(x, y) + iv(x, y)$$

and defining z_j as:

$$z_j = x_j + iy_j$$

then the function $f(z_j)$ is given by:

$$f(z_j) = u(x_j, y_j) + iv(x_j, y_j)$$

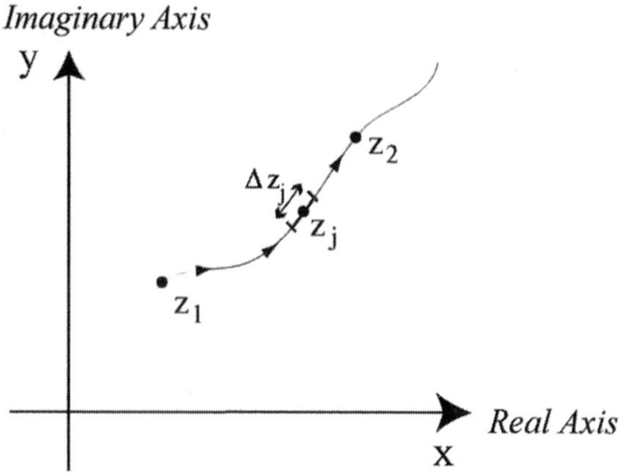

Fig. 5.10: *Complex Integration of a Path C*

with

$$\Delta z_j = \Delta x_j + i\Delta y_j$$

The integral can now be defined as a limit of a sum:

$$\int_{\substack{z_1 \\ \text{on } C}}^{z_2} f(z)\,dz = \lim_{N\to\infty} \sum_{j=1}^{N} \left[u(x_j,y_j) + iv(x_j,y_j)\right]\left[\Delta x_j + i\Delta y_j\right]\Bigg|_{x_j,y_j \text{ on } C}$$

$$= \lim_{N\to\infty} \sum_{j=1}^{N} u(x_j,y_j)\Delta x_j - v(x_j,y_j)\Delta y_j + i \lim_{N\to\infty} \sum_{j=1}^{N} u(x_j,y_j)\Delta y_j + v(x_j,y_j)\Delta x_j$$

$$= \int_{\substack{x_1,y_1 \\ \text{on } C}}^{x_2,y_2} \left[u(x,y)\,dx - v(x,y)\,dy\right] + i \int_{\substack{x_1,y_1 \\ \text{on } C}}^{x_2,y_2} \left[u(x,y)\,dy + v(x,y)\,dx\right] \tag{5.25}$$

The integration of a complex function on path C as defined in eq.(5.25) has the following properties:

(i) $$\int_{\substack{a \\ \text{on } C}}^{b} f(z)\,dz = -\int_{\substack{b \\ \text{on } C}}^{a} f(z)\,dz$$

(ii) $$\int_{\substack{a \\ \text{on } C}}^{b} c\,f(z)\,dz = c\int_{\substack{a \\ \text{on } C}}^{b} f(z)\,dz \qquad c = \text{constant}$$

(iii) $$\int_{\substack{a \\ \text{on } C}}^{b} \left[f(z) + g(z)\right] dz = \int_{\substack{a \\ \text{on } C}}^{b} f(z)\,dz + \int_{\substack{a \\ \text{on } C}}^{b} g(z)\,dz$$

(iv) $$\int_{\substack{a \\ \text{on } C}}^{b} f(z)\,dz = \int_{\substack{a \\ \text{on } C}}^{c} f(z)\,dz + \int_{\substack{c \\ \text{on } C}}^{b} f(z)\,dz \qquad c \text{ on } C$$

(v) $$\left|\int_{\substack{a \\ \text{on } C}}^{b} f(z)\,dz\right| \le ML \tag{5.26}$$

where M is the maximum value of $|f(z)|$ on C in the range [a,b] and:

$$L = \int_{\substack{a \\ \text{on}\,C}}^{b} |dz| = \int_{\substack{a \\ \text{on}\,C}}^{b} ds = \text{length of the path on C}$$

Example 5.6

Obtain the integral in the clockwise direction of f(z) = 1/(z-a) on a path that is a semi-circle centered at z = a and having a radius = 2 units.

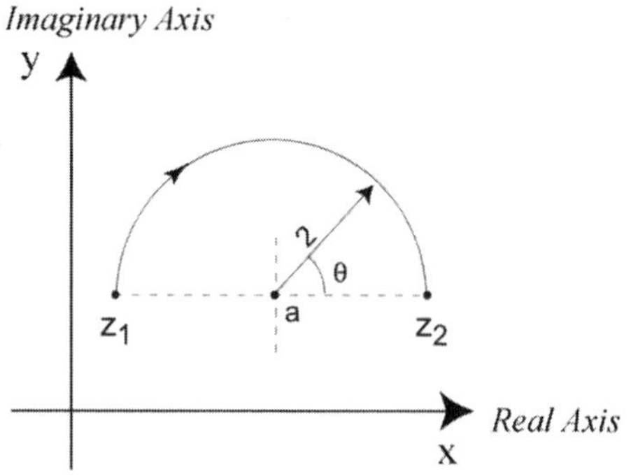

To perform the integration, one can use parametric representation:

$z - a = 2e^{i\theta}$ $\qquad dz = 2ie^{i\theta}\, d\theta$

$z_1 = a + 2e^{+i\pi} = a - 2$ $\qquad z_2 = a + 2$

$$\int_{z_1}^{z_2} f(z)\, dz = \int_{\pi}^{0} \frac{1}{2e^{i\theta}}\, 2ie^{i\theta}\, d\theta = -\pi i$$

If one integrates over a complete circle of radius = 2 in counter-clockwise direction, then:

$$\oint \frac{dz}{z-a} = \int_{-\pi}^{\pi} (i\, d\theta) = 2\pi i$$

where the integral symbol $\oint$ indicates a closed path in the positive (counter-clockwise) sense. Note that the integral over a closed path, where the upper and lower limit are the same, is *not* zero.

5.5 Cauchy's Integral Theorem

If a function is analytic inside a simply connected region R and on the closed contour C containing R, then:

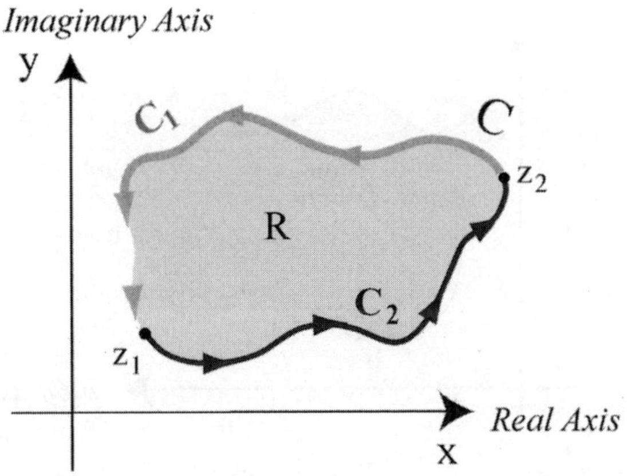

Fig. 5.11 *Paths for Cauchy Integral Theorem*

$$\oint_C f(z)dz = 0 \tag{5.27}$$

Using the form given in Green's Theorem in eqs. (5.25) and (5.24), one can transform the closed path integral to an area integral:

$$\int_C f(z)dz = \int_C (u\,dx - v\,dy) + i\int_C (u\,dy + v\,dx)$$

$$= \int_R \left(-\frac{\partial v}{\partial x} - \frac{\partial u}{\partial y}\right) dx\,dy + i\int_R \left(\frac{\partial u}{\partial x} - \frac{\partial v}{\partial y}\right) dx\,dy = 0$$

The integrands vanish by the use of the Cauchy-Riemann condition in eq. (5.5). As a consequence of Cauchy's Integral Theorem, eq.(5.27), one can show that the integral of an analytic function in a simply connected region is independent of the path taken (see Fig. 5.11). The integral over a closed path C can be divided over two segments C_1 and C_2:

$$\oint_{C_1+C_2} f(z)dz = \int_{z_1 \text{ on } C_1}^{z_2} f(z)\,dz + \int_{z_2 \text{ on } C_2}^{z_1} f(z)dz = 0$$

Using the integral relationship in eq. (5.26):

$$\int_{z_1 \text{ on } C_1}^{z_2} f(z)\,dz = -\int_{z_2 \text{ on } C_2}^{z_1} f(z)\,dz = \int_{z_1 \text{ on } C_1}^{z_2} f(z)dz \tag{5.28}$$

Thus, the integral of an analytic function is independent of the path taken within a simply connected region. As a consequence of eq. (5.28), the indefinite integral of an analytic function f(z):

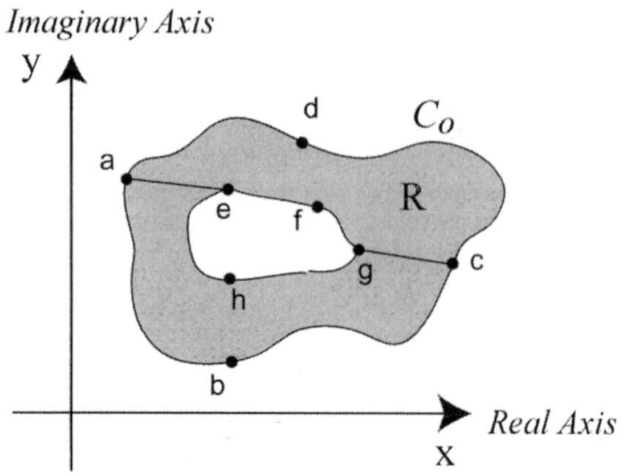

***Fig 5.12:** Integration on Closed Path C_0 of a Doubly Connected Region*

$$F(z) = \int_{z_0}^{z} f(\xi)\,d\xi$$

independent of C in R, is also an analytic function. Furthermore, it can be shown that:

$$\frac{dF(z)}{dz} = f(z) \tag{5.29}$$

The Cauchy Integral theorem can be extended to multiply-connected regions. Consider a complex function f(z) which is analytic in a doubly connected region between the closed paths C_0 and C_1 as in Fig. 5.12. One can connect the inner and outer paths by line segments, (af) and (dc), such that two simply-connected regions are created. Invoking Cauchy's Integral, eq. (5.27), on the two closed paths, one finds that:

$$\oint_{C = abcghea} f(z)\,dz = 0 \qquad \text{and} \qquad \oint_{C = aefgcd\,a} f(z)\,dz = 0$$

Adding the two contour integrals and canceling out the line integrals on (ae) and (gc), one obtains:

$$\oint_{C_0} f(z)\,dz = \oint_{C_1} f(z)\,dz \tag{5.30}$$

where C_0 and C_1 represent contours outside and inside the region R.

If the region is N-tuply connected, see Fig. 5.13, then one can show that:

$$\oint_{C_0} f(z)\,dz = \sum_{j=1}^{N-1} \oint_{C_j} f(z)\,dz \tag{5.31}$$

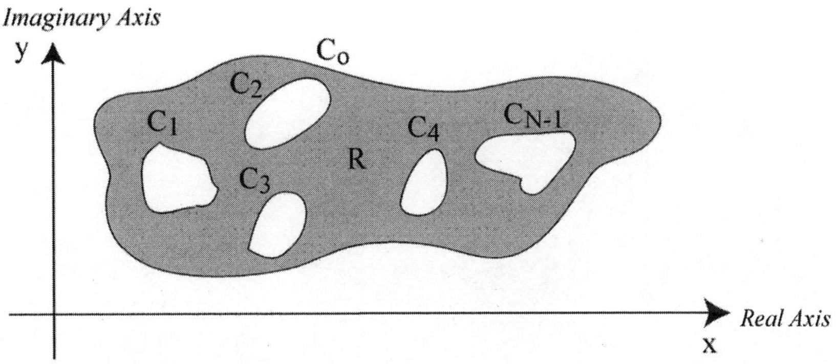

***Fig 5.13:** Integration on Closed Path C_0 in a Multiply Connected Region*

Example 5.7

Obtain the integral of $f(z) = 1/(z-a)$ on a circle centered at $z = a$ and having a radius of 4 units. Since the integral on a circle of radius = 2 was obtained in Example 5.6, then:

$$\oint\limits_{\substack{C_0 \\ \text{on}\,\rho=4}} \frac{dz}{z-a} = \oint\limits_{\substack{C_1 \\ \text{on}\,\rho=2}} \frac{dz}{z-a} = 2\pi i$$

5.6 Cauchy's Integral Formula

Let the function f(z) be analytic within a region R and on the closed contour C containing R. If z_0 is any point in R, then:

$$f(z_0) = \frac{1}{2\pi i} \oint\limits_C \frac{f(z)}{z - z_0}\, dz \tag{5.32}$$

Proof:

Since the function f(z) is analytic everywhere in R, then $f(z)/(z-z_0)$ is analytic everywhere in R except at the point $z = z_0$. Thus, one can surround the point z_0 by a closed contour C_1, such as a circle of radius ε, so that the function $f(z)/(z-z_0)$ is analytic everywhere in the region between C and C_1 (see Fig. 5.14). Invoking Cauchy's integral theorem in eq. (5.30):

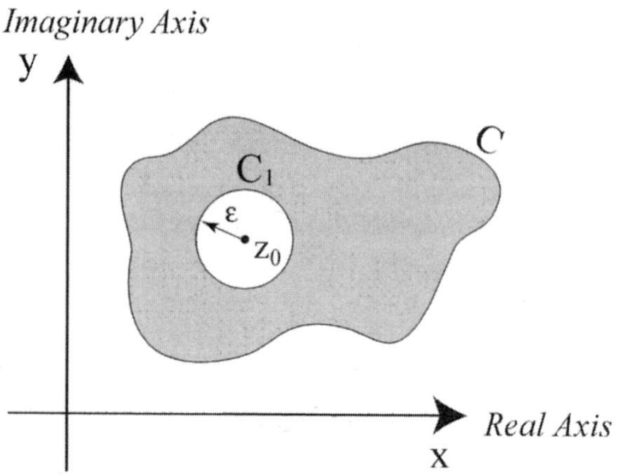

***Fig. 5.14**: Complex Integration over a Closed Circular Path*

$$\oint_C \frac{f(z)}{z-z_0}\,dz = \oint_{C_1} \frac{f(z)}{z-z_0}\,dz = \oint_{C_1} \frac{f(z)-f(z_0)}{z-z_0}\,dz + f(z_0)\oint_{C_1}\frac{dz}{z-z_0}$$

$$= \oint_{C_1} \frac{f(z)-f(z_0)}{z-z_0}\,dz + 2\pi i\, f(z_0)$$

by the use of results in Example 5.6.

The remainder integral must be evaluated as $\varepsilon \to 0$, using the results of eq. (5.26):

$$\lim_{\varepsilon\to 0}\left|\oint_{C_1} \frac{f(z)-f(z_0)}{z-z_0}\,dz\right| \le \lim_{\substack{\varepsilon\to 0\\ \text{on } C_1}}\left(\frac{|f(z)-f(z_0)|}{\varepsilon}\right)2\pi\varepsilon = \lim_{\substack{\varepsilon\to 0\\ \text{on } C_1}} 2\pi|f(z)-f(z_0)| \to 0$$

since f(z) is continuous and analytic everywhere inside R. Cauchy's integral formula can be used to obtain integral representation of a derivative of an analytic function. Using the definition of $f'(z_0)$ in eq. (5.3), and the representation of $f(z_0)$ in eq. (5.32):

$$f'(z_0) = \lim_{\Delta z\to 0}\left(\frac{f(z_0+\Delta z)-f(z_0)}{\Delta z}\right)$$

$$= \frac{1}{2\pi i}\lim_{\Delta z\to 0}\frac{1}{\Delta z}\oint_C\left[\frac{f(z)}{z-(z_0+\Delta z)} - \frac{f(z)}{z-z_0}\right]dz$$

$$= \frac{1}{2\pi i}\lim_{\Delta z\to 0}\oint_C \frac{f(z)}{(z-z_0)(z-z_0-\Delta z)}\,dz \to \frac{1}{2\pi i}\oint_C \frac{f(z)}{(z-z_0)^2}\,dz$$

Similarly, it can be shown that the nth derivative of f(z) can be represented by the integral:

$$f^{(n)}(z_0) = \frac{n!}{2\pi i}\oint_C \frac{f(z)}{(z-z_0)^{n+1}}\,dz \qquad (5.33)$$

Example 5.8

(i) Integrate the following function:

$$f(z) = \frac{1}{z^2 + 1}$$

on a closed contour defined by $|z-i| = 1$ in the counter-clockwise (positive) sense.

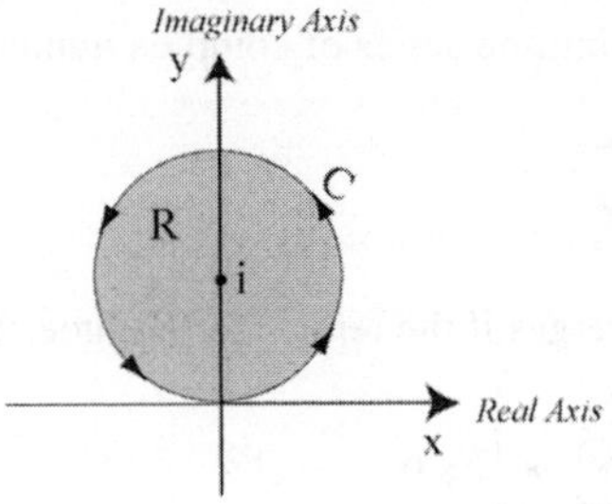

Since:

$$\frac{1}{z^2+1} = \frac{1}{(z+i)(z-i)}$$

let:

$$g(z) = \frac{1}{z+i}$$

then by Cauchy's Integral Formula:

$$g(z_0) = \frac{1}{2\pi i} \oint_C \frac{g(z)}{z - z_0}\, dz$$

where g(z) is analytic everywhere within R and on C.

Thus:

$$\oint_C \frac{1}{z^2+1}\, dz = \oint_C \frac{g(z)}{z-i}\, dz = 2\pi i\, g(i) = 2\pi i \left. \frac{1}{z+i} \right|_{z=i} = \pi$$

(ii) Integrate the following function:

$$f(z) = \frac{1}{(z^2+1)^2}$$

on the closed contour described in (i).

Let $g(z) = (z+i)^{-2}$ which is analytic in R and on C, then using eq. (5.33):

$$\oint_C \frac{1}{(z^2+1)^2}\, dz = \oint_C \frac{g(z)}{(z-i)^2}\, dz = \frac{2\pi i}{1!} g'(i) = 2\pi i \left. \frac{-2}{(z+i)^3} \right|_{z=i} = \frac{\pi}{2}$$

Morera's Theorem

If a function f(z) is continuous in a simply-connected region R and if:

$$\oint_C f(z)dz = 0$$

for all possible closed contours C inside R, then f(z) is an analytic function in R.

5.7 Infinite Series

Define the sum Z of an infinite series of complex numbers as:

$$Z = \sum_{n=1}^{\infty} z_n = \lim_{N\to\infty} \sum_{n=1}^{N} z_n \tag{5.34}$$

The series in eq. (5.34) converges if the remainder R_N goes to zero as $N\to\infty$ i.e.:

$$\lim_{N\to\infty} R_N = \lim_{N\to\infty} \left| Z - \sum_{n=1}^{N} z_n \right| \to 0$$

If the series in eq. (5.34) converges, then the two series $\sum_{n=1}^{\infty} x_n$ and $\sum_{n=1}^{\infty} y_n$ also converge. An infinite series is **Absolutely Convergent** if the series, $\sum_{n=1}^{\infty} |z_n|$ converges. If a series is absolutely convergent, then the series also converges.

A series of functions of a complex variable is defined as:

$$F(z) = \sum_{j=1}^{\infty} f_j(z)$$

where each function $f_j(z)$ is defined throughout a region R. The series is said to converge to $F(z_0)$ if:

$$F(z_0) = \sum_{j=1}^{\infty} f_j(z_0)$$

The region where the series converges is called the **Region of Convergence**. Finally, define a **Power Series** about $z = z_0$ as follows:

$$f(z) = \sum_{n=0}^{\infty} a_n (z - z_0)^n$$

The radius of convergence ρ is defined as:

$$\rho = \lim_{n\to\infty} \left| \frac{a_n}{a_{n+1}} \right|$$

such that the power series converges if $|z - z_0| < \rho$, and diverges if $|z - z_0| > \rho$.

If a power series about z_0 converges for $z = z_1$, then it converges absolutely for

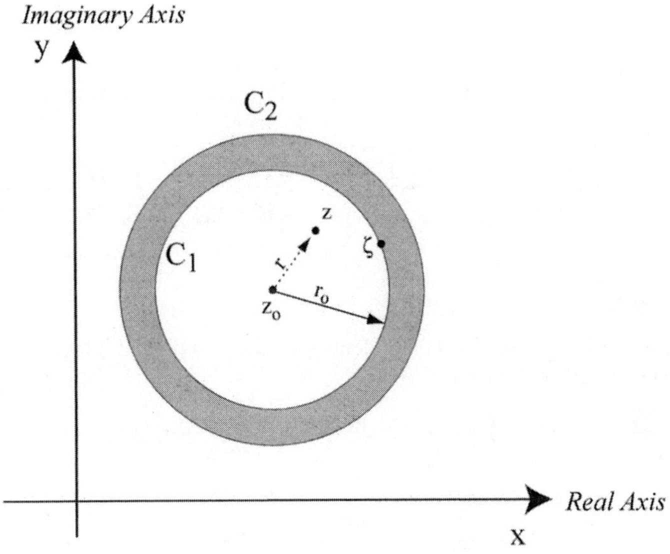

Fig. 5.15: *Closed Path for Taylor's Series*

$z = z_2$ where:

$$|z_1 - z_0| < |z_2 - z_0|$$

5.8 Taylor's Expansion Theorem

If f(z) is an analytic function at z_0, then there is a power series that converges inside a circle C_2 centered at z_0 and represents the function f(z) inside C_1, i.e.:

$$f(z) = \sum_{n=0}^{\infty} a_n (z - z_0)^n$$

where:

$$a_n = \frac{f^{(n)}(z_0)}{n!} \tag{5.35}$$

Proof:

Consider a point z_0 where the function f(z) is analytic, (see Fig. 5.15). Let points z and ζ be interior to a circle C_2 , ζ being a point on a circle C_1 centered at z_0 whose radius = r_0 . Consider the term:

$$\frac{1}{\zeta - z} = \frac{1}{(\zeta - z_0) - (z - z_0)} = \frac{1}{(\zeta - z_0)[1 - \frac{z - z_0}{\zeta - z_0}]}$$

Using the following identity, which can be obtained by direct division:

$$\frac{1}{1-u} = 1 + u + u^2 + \ldots + \frac{u^n}{1-n} \qquad \text{for} \qquad |u| < 1$$

then:

$$\frac{1}{\zeta - z} = \frac{1}{\zeta - z_0} + \frac{z - z_0}{(\zeta - z_0)^2} + \frac{(z - z_0)^2}{(\zeta - z_0)^3} + \ldots + \frac{(z - z_0)^{n-1}}{(\zeta - z_0)^n} + \frac{(z - z_0)^n}{(\zeta - z)(\zeta - z_0)^n}$$

since:

$$|(z - z_0)/(\zeta - z_0)| < 1$$

Multiplying both sides of the preceding identity by $f(\zeta)/2\pi i \, d\zeta$ and integrating on the closed contour C_1, one obtains:

$$\frac{1}{2\pi i}\oint_{C_1} \frac{f(\zeta)}{\zeta - z} d\zeta = \frac{1}{2\pi i}\oint_{C_1} \frac{f(\zeta)}{\zeta - z_0} d\zeta + \frac{(z - z_0)}{2\pi i}\oint_{C_1} \frac{f(\zeta)}{(\zeta - z_0)^2} d\zeta + \ldots$$

$$+ \frac{(z - z_0)^{n-1}}{2\pi i}\oint_{C_1} \frac{f(\zeta)}{(\zeta - z_0)^n} d\zeta + \frac{(z - z_0)^n}{2\pi i}\oint_{C_1} \frac{f(\zeta)}{(\zeta - z)(\zeta - z_0)^n} d\zeta$$

Using Cauchy's integral formula eq. (5.33), one can show that:

$$f(z) = f(z_0) + \frac{(z - z_0)}{1!} f'(z_0) + \frac{(z - z_0)^2}{2!} f''(z_0) + \ldots + \frac{(z - z_0)^{n-1}}{(n-1)!} f^{(n-1)}(z_0) + R_n$$

where:

$$R_n = \frac{(z - z_0)^n}{2\pi i}\oint_{C_1} \frac{f(\zeta)}{(\zeta - z)(\zeta - z_0)^n} d\zeta$$

Taking the absolute value of R_n, then:

$$|R_n| \le \frac{r^n}{2\pi} \frac{2\pi M r_0}{(r_0 - r) r_0^n} = \frac{r_0}{r_0 - r} M \left(\frac{r}{r_0}\right)^n$$

The remainder R_n vanishes as n increases:

$$\lim_{n \to \infty} |R_n| \to 0 \qquad \text{since} \qquad r / r_0 < 1$$

Finally, the Taylor series representation is given by:

$$f(z) = \sum_{n=0}^{\infty} \frac{f^{(n)}(z_0)}{n!} (z - z_0)^n$$

The Taylor's series representation has the following properties:

1. The series represents an analytic function inside its circle of convergence.

2. The series is uniformly convergent inside its circle of convergence.
3. The series may be differentiated or integrated term by term.
4. There is only one Taylor series that represents an analytic function f(z) about a point z_0.
5. Since the function is analytic at z_0, then the circle of convergence has a radius ρ that extends from the center at z_0 to the nearest singularity.

Example 5.9

(i) Expand the function e^z in a Taylor's series about $z_0 = 0$. Since:

$$f^{(n)}(z_0) = e^z\Big|_{z=0} = 1$$

then the Taylor series about $z_0 = 0$ is given by:

$$e^z = \sum_{n=0}^{\infty} \frac{z^n}{n!}$$

The radius of convergence, ρ, is ∞, since:

$$\rho = \lim_{n\to\infty} \left|\frac{a_n}{a_{n+1}}\right| = \lim_{n\to\infty} \left|\frac{(n+1)!}{n!}\right| \to \infty$$

(ii) Obtain the Taylor's series expansion of the following function about $z_0 = 0$:

$$f(z) = \frac{1}{z^2 - 4}$$

Using the series expansion for $(1 - u)^{-1}$, with $u = z^2/4$:

$$f(z) = -\frac{1}{4}\frac{1}{1-(\frac{z}{2})^2} = -\frac{1}{4}\sum_{n=0}^{\infty} (\frac{z}{2})^{2n}$$

which is convergent in the region $|z| < 2$. It should be noted that the radius of convergence is the distance from $z_o = 0$ to the closest singularities at $z = \pm 2$, i.e., $\rho = 2$.

(iii) Obtain the Taylor's series expansion of the function in (ii) about $z_0 = 1$ and about $z_0 = -1$.

To find the series about $z_0 = 1$, let $\zeta = z - 1$, then the function f(z) transforms to $\hat{f}(\zeta)$:

$$\hat{f}(\zeta) = \frac{1}{(\zeta+3)(\zeta-1)} = \frac{1}{4}\left[\frac{1}{\zeta-1} - \frac{1}{\zeta+3}\right]$$

Expanding f(z) about $z_0 = 1$ is equivalent to expanding $\hat{f}(\zeta)$ about $\zeta = 0$. The Taylor series for the functions are as follows:

$$\frac{1}{\zeta-1}=-\sum_{n=0}^{\infty}\zeta^n \qquad \text{convergent in } |\zeta|<1$$

$$\frac{1}{\zeta+3}=\frac{1}{3}\sum_{n=0}^{\infty}(-1)^n\frac{\zeta^n}{3^n} \qquad \text{convergent in } |\zeta|<3$$

Thus, the two series have a common region of convergence $|\zeta|<1$:

$$\hat{f}(\zeta)=-\frac{1}{4}\left[\sum_{n=0}^{\infty}\left(\zeta^n+\frac{(-1)^n}{3^{n+1}}\zeta^n\right)\right] \qquad \text{convergent in } |\zeta|<1$$

and the Taylor series representation of f(z) about $z_0 = 1$ becomes:

$$f(z)=-\frac{1}{4}\sum_{n=0}^{\infty}\left[1+\frac{(-1)^n}{3^{n+1}}\right](z-1)^n \qquad \text{convergent in } |z-1|<1$$

It should be noted that the radius of convergence represents the distance between $z_0 = 1$ and the closest singularity at $z = 2$.

To find the Taylor series representation of f(z) about $z_0 = -1$, let $\zeta = z + 1$, then the function transforms to $\hat{f}(\zeta)$:

$$\hat{f}(\zeta)=\frac{1}{(\zeta+1)(\zeta-3)}=\frac{1}{4}\left[\frac{1}{\zeta-3}-\frac{1}{\zeta+1}\right]$$

Expanding f(z) about $z_0 = -1$ is equivalent to expanding $\hat{f}(\zeta)$ about $\zeta = 0$. The Taylor series for the functions are as follows:

$$\frac{1}{\zeta-3}=\frac{1}{3}\sum_{n=0}^{\infty}\frac{\zeta^n}{3^n} \qquad \text{convergent in } |\zeta|<3$$

$$\frac{1}{\zeta+1}=\sum_{n=0}^{\infty}(-1)^n\zeta^n \qquad \text{convergent in } |\zeta|<1$$

Thus the two series, when added, converge in the common region $|\zeta|<1$:

$$\hat{f}(\zeta)=-\frac{1}{4}\sum_{n=0}^{\infty}\left[\frac{1}{3^{n+1}}+(-1)^n\right]\zeta^n \qquad \text{convergent in } |\zeta|<1$$

and the Taylor series representation of f(z) about $z_0 = -1$ is:

$$f(z)=-\frac{1}{4}\sum_{n=0}^{\infty}\left[\frac{1}{3^{n+1}}+(-1)^n\right](z+1)^n \quad \text{convergent in } |z+1|<1$$

Again, note that the radius of convergence represents the distance between $z_0 = -1$, and the closest singularity at $z = -2$.

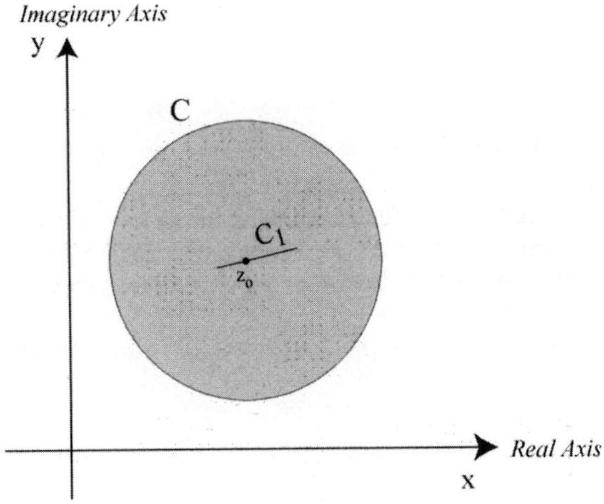

***Fig. 5.16:** Path for Identity Theorem*

(iv) Expand the function $1/z$ by a Taylor's series about $z_0 = -1$.

$$f^{(n)}(z_0) = \left.\frac{(-1)^n}{z^{n+1}}\, n!\right|_{z_0=-1} = -(n!)$$

Thus:

$$\frac{1}{z} = \sum_{n=0}^{\infty} \frac{-(n!)}{n!}(z+1)^n = -\sum_{n=0}^{\infty} (z+1)^n$$

The region of convergence becomes $|z+1| < 1$ since the closest singularity to $z_0 = -1$ is $z = 0$, which is one unit away from $z_0 = -1$.

Identity Theorem

As a consequence of Taylor's expansion theorem, one can show that if $f(z)$ and $g(z)$ are two analytic functions inside a circle C, centered at z_0 and if $f(z) = g(z)$ along a segment passing through z_0, then $f(z) = g(z)$ everywhere inside C. This can be shown by expanding both functions in a Taylor's series about z_0 as follows: (see Fig. 5.16):

$$f(z) = \sum_{n=0}^{\infty} a_n (z - z_0)^n \qquad \text{where} \qquad a_n = \frac{f^{(n)}(z_0)}{n!}$$

and

$$g(z) = \sum_{n=0}^{\infty} b_n (z - z_0)^n \qquad \text{where} \qquad b_n = \frac{g^{(n)}(z_0)}{n!}$$

At $z = z_0$, $f(z_0) = g(z_0)$, thus $a_0 = b_0$. The derivatives of f(z) and g(z) at z_0 can be taken as a limiting process along C_1. Thus:

$$f'(z_0) = g'(z_0) \quad \text{on } C_1$$

which means that $a_1 = b_1$, etc. Thus, one can show that $a_n = b_n$, n = 0, 1, 2, ... and f(z) = g(z) everywhere in C.

The identity theorem can be used to extend Taylor series representations in real variables to those in complex variables. If a real function is analytic in a segment on the real axis, then one can show that the extension to the complex plane of the equivalent complex function is analytic inside a certain region. Thus, all the Taylor series expansions of functions on the real axis can be extended to the complex plane.

Example 5.10

(i) The function:

$$e^x = \sum_{n=0}^{\infty} \frac{x^n}{n!}$$

is analytic everywhere on the real axis. One can extend the function into the complex plane where e^z is equal to e^x on the entire x-axis. Since the function e^z and e^x are equal on the entire x-axis, then they must be equal in the entire z-plane. Hence, the Taylor series representation of the complex function e^z:

$$e^z = \sum_{n=0}^{\infty} \frac{z^n}{n!}$$

is analytic in the entire complex plane.

(ii) The function:

$$\frac{1}{1-x} = \sum_{n=0}^{\infty} x^n$$

is analytic on the segment of the real axis, $|x| < 1$, then the extended function $(1 - z)^{-1}$ has an expansion $\sum_{n=0}^{\infty} z^n$ which is analytic in the region $|z| < 1$.

5.9 Laurent's Series

If a function is analytic on two concentric circles C_1 and C_2 centered at z_0 and in the interior region between them, then there is an infinite series expansion with positive and negative powers of $z - z_0$ about $z = z_0$ (see Fig. 5.17), representing this function in this region called the **Laurent series**. Thus, the Laurent's series can be written as:

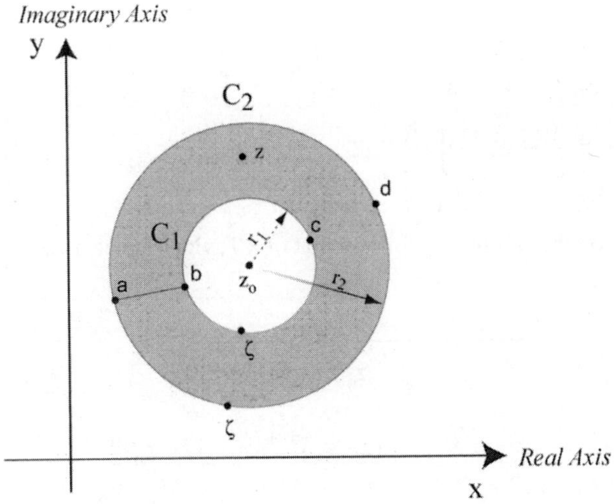

Fig. 5.17: *Closed Paths for Laurent's Series*

$$f(z)=\sum_{n=0}^{\infty} a_n (z-z_o)^n+\sum_{n=1}^{\infty} \frac{b_n}{(z-z_o)^n} \tag{5.36}$$

where the coefficients a_n and b_n are given by:

$$a_n=\frac{1}{2\pi i}\oint_{C_2} \frac{f(\zeta)}{(\zeta-z_o)^{n+1}}\,d\zeta \qquad n=0, 1, 2, \ldots$$

and

$$b_n=\frac{1}{2\pi i}\oint_{C_1} f(\zeta)(\zeta-z_o)^{n-1}\,d\zeta \qquad n=1, 2, 3, \ldots$$

The Laurent Series can also be written in more compact form as:

$$f(z)=\sum_{n=-\infty}^{\infty} c_n(z-z_o)^n$$

where:

$$c_n=\frac{1}{2\pi i}\oint_{C_2} \frac{f(\zeta)}{(\zeta-z_o)^{n+1}}\,d\zeta \qquad n=0, \pm 1, \pm 2, \ldots$$

where C is a circular contour inside the region between C_1 and C_2 and is centered at z_0.

Proof:

Consider a cut (ab) between the two circles C_1 and C_2 as shown on Fig. 5.17. Then let the closed contour for use in the Cauchy integral formula be (ba da bc b). Thus, writing out the integral over the closed contour becomes:

$$2\pi i\, f(z) = \oint_{C_2} \frac{f(\zeta)}{\zeta - z} d\zeta + \int_a^b \frac{f(\zeta)}{\zeta - z} d\zeta - \oint_{C_1} \frac{f(\zeta)}{\zeta - z} d\zeta + \int_b^a \frac{f(\zeta)}{\zeta - z} d\zeta$$

$$= \oint_{C_2} \frac{f(\zeta)}{\zeta - z} d\zeta - \oint_{C_1} \frac{f(\zeta)}{\zeta - z} d\zeta$$

The expansion on the contour C_2 follows that of a Taylor's series, i.e., for ζ on C_2:

$$\frac{1}{\zeta - z} = \frac{1}{(\zeta - z_0) - (z - z_0)} = \frac{1}{(\zeta - z_0)[1 - \frac{z - z_0}{\zeta - z_0}]}$$

$$= \frac{1}{\zeta - z_0} + \frac{z - z_0}{(\zeta - z_0)^2} + \ldots + \frac{(z - z_0)^{n-1}}{(\zeta - z_0)^n} + \frac{(z - z_0)^n}{(\zeta - z)(\zeta - z_0)^n}$$

where the division was performed on 1/(1-u) with:

$$|u| = \left|\frac{z - z_0}{\zeta - z_0}\right| < 1 \quad \text{for } \zeta \text{ on } C_2$$

The expansion on the contour C_1 can be made as follows:

$$\frac{1}{\zeta - z} = \frac{1}{(\zeta - z_0) - (z - z_0)} = -\frac{1}{z - z_0} \frac{1}{[1 - \frac{\zeta - z_0}{z - z_0}]}$$

$$= -\frac{1}{z - z_0} - \frac{\zeta - z_0}{(z - z_0)^2} - \frac{(\zeta - z_0)^2}{(z - z_0)^3} - \ldots - \frac{(\zeta - z_0)^{n-1}}{(z - z_0)^n} - \frac{(\zeta - z_0)^n}{(z - z_0)^n (z - \zeta)}$$

where the division was performed on 1/(1-u) with

$$|u| = \left|\frac{\zeta - z_0}{z - z_0}\right| < 1 \quad \text{for } \zeta \text{ on } C_1$$

Thus, substituting these terms in the expansion for f(z):

$$2\pi i\, f(z) = \oint_{C_2} \frac{f(\zeta)}{\zeta - z_0} d\zeta + (z - z_0) \oint_{C_2} \frac{f(\zeta)}{(\zeta - z_0)^2} d\zeta + \ldots + (z - z_0)^{n-1} \oint_{C_2} \frac{f(\zeta)}{(\zeta - z_0)^n} d\zeta$$

$$+ {}_1R_n + \frac{1}{(z - z_0)} \oint_{C_1} f(\zeta)\, d\zeta + \frac{1}{(z - z_0)^2} \oint_{C_1} f(\zeta)(\zeta - z_0)\, d\zeta$$

$$+ \frac{1}{(z - z_0)^3} \oint_{C_1} f(\zeta)(\zeta - z_0)^2\, d\zeta +$$

$$\ldots + \frac{1}{(z - z_0)^n} \oint_{C_1} f(\zeta)(\zeta - z_0)^{n-1}\, d\zeta + {}_2R_n$$

Where the remainder R_n can be shown to vanish as $n \rightarrow \infty$:

$$\lim_{n\to\infty} \left| {}_1R_n \right| = \lim_{n\to\infty} \left| (z-z_0)^n \oint_{C_2} \frac{f(\zeta)}{(\zeta - z_0)^{n+1}(\zeta - z)} d\zeta \right| \to 0$$

and

$$\lim_{n\to\infty} \left| {}_2R_n \right| = \lim_{n\to\infty} \left| \frac{1}{(z-z_0)^n} \oint_{C_1} \frac{f(\zeta)(\zeta - z_0)^n}{(z-\zeta)} d\zeta \right| \to 0$$

Example 5.11

(i) Obtain the Laurent's series of the following function about $z_0 = 0$:

$$f(z) = \frac{1+z}{z^3}$$

The function f(z) is analytic everywhere except at $z = 0$.

The function f(z) can be rewritten as:

$$f(z) = \frac{1}{z^2} + \frac{1}{z^3}$$

In this case, it is already in Laurent's series form where $a_n = 0$, $b_1 = 0$, $b_2 = 1$, $b_3 = 1$, and $b_n = 0$ for $n \geq 4$.

(ii) Obtain Laurent's series for the following function about $z_0 = 0$ valid in the region $|z| > 1$:

$$f(z) = \frac{1}{1-z}$$

Since the region is defined by $|z| > 1$, then $1/|z| < 1$, thus, letting $\zeta = 1/z$, then:

$$f(z) = f(\zeta^{-1}) = \frac{1}{1-\dfrac{1}{\zeta}} = \frac{\zeta}{\zeta - 1} = -\zeta \sum_{n=0}^{\infty} \zeta^n = -\sum_{n=0}^{\infty} \zeta^{n+1}$$

which is convergent for $|\zeta| < 1$. Thus:

$$f(z) = -\sum_{n=0}^{\infty} z^{-n-1}$$

which is convergent in the region $|z| > 1$.

(iii) Obtain the Laurent's series for the following function about $z_0 = 0$, valid in the region $|z| > 2$:

$$f(z) = \frac{1}{z^2 - 4}$$

The function has two singularities at $z = \pm 2$. Since the function is analytic inside the circle $|z| = 2$, a Taylor's series can be obtained (see Example 5.9). For the region outside $|z| = 2$, one needs a Laurent's series representation. Factoring out z^2 from f(z):

$$f(z) = \frac{1}{z^2 (1 - 4/z^2)}$$

then one can use the division of $1/(1-u)$ where $u = 4/z^2$:

$$f(z) = \frac{1}{z^2} \sum_{n=0}^{\infty} (4/z^2)^n$$

convergent over the region, $|2/z| < 1$. This can be rewritten as:

$$f(z) = \frac{1}{4} \sum_{n=0}^{\infty} (z/2)^{-2(n+1)}$$

convergent over the region, $|z| > 2$.

(iv) Obtain the Laurent's series for the function in (iii) about $z_0 = 2$ valid in the regions:

(a) $0 < |z-2| < 4$ (b) $|z-2| > 4$

To obtain the series expansion, transfer the origin of the expansion to $z_0 = 2$, i.e., let $\eta = z - 2$ such that the function f(z) transforms to $\hat{f}(\eta)$:

$$\hat{f}(\eta) = \frac{1}{\eta(\eta + 4)}$$

which has two singularities at $\eta = 0$ and $\eta = -4$. Thus, two Laurent's series corresponding to $\hat{f}(\eta)$ are required, one for $0 < |\eta| < 4$ and one for $|\eta| > 4$ as shown in Fig. 5.18.

(a) In the region R_1, where $0 < |\eta| < 4$, one can expand $1/(\eta + 4)$ as follows:

$$\frac{1}{(\eta + 4)} = \frac{1}{4(1 + \eta/4)} = \frac{1}{4} \sum_{n=0}^{\infty} \frac{(-\eta)^n}{4^n}$$

convergent in $|\eta/4| < 1$. Thus, the Laurent series representation for $\hat{f}(\eta)$ becomes:

$$\hat{f}(\eta) = \frac{1}{16} \sum_{n=0}^{\infty} (-1)^n \frac{(\eta)^{n-1}}{4^{n-1}}$$

convergent in $0 < |\eta| < 4$, since η^{-1} is not analytic at $\eta = 0$.

Thus, the Laurent's series about $z_0 = 2$ becomes:

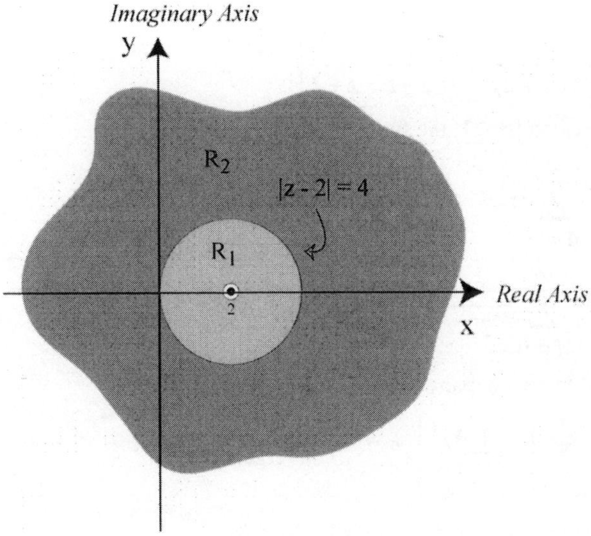

***Fig. 5.18:** Laurent's Series Expansions in Two Regions*

$$f(z)=\frac{1}{16}\sum_{n=0}^{\infty}(-1)^n\frac{(z-2)^{n-1}}{4^{n-1}}$$

convergent in $0<|z-2|<4$.

(b) In the region R_2, where $|\eta|>4$, or $4/|\eta|<1$, one may factor out η from the function, such that:

$$\hat{f}(\eta)=\frac{1}{\eta^2(1+4/\eta)}=\frac{1}{\eta^2}\sum_{n=0}^{\infty}\frac{(-4)^n}{\eta^n}$$

convergent in $|4/\eta|<1$. Thus, the Laurent's series representation about $\eta=0$ is:

$$\hat{f}(\eta)=\frac{1}{16}\sum_{n=0}^{\infty}(-1)^n\frac{4^{n+2}}{\eta^{n+2}}$$

convergent in $|\eta|>4$, or, about the point $z_0=2$ is represented by:

$$f(z)=\frac{1}{16}\sum_{n=0}^{\infty}(-1)^n\frac{4^{n+2}}{(z-2)^{n+2}}$$

convergent in $|z-2|>4$.

(v) Obtain the Laurent's series of the following function about $z_0=0$:

$$f(z)=\frac{1}{(z-1)(z+2)}$$

valid in the entire complex plane: i.e., $|z|<1$, $1<|z|<2$, and $|z|>2$.

The function f(z) can be factored out in terms of its two components:

$$f(z)=\frac{1}{(z-1)(z+2)}=\frac{1}{3}\left[\frac{1}{z-1}-\frac{1}{z+2}\right]$$

(a) In the region |z| < 1, the function is analytic thus:

$$\frac{1}{(z-1)}=-\sum_{n=0}^{\infty} z^n \qquad \text{convergent in } |z|<1$$

$$\frac{1}{(z+2)}=\frac{1}{2}\sum_{n=0}^{\infty}\frac{(-z)^n}{2^n} \qquad \text{convergent in } |z|<2$$

and, the sum of the two expansions becomes:

$$f(z)=-\frac{1}{3}\sum_{n=0}^{\infty}\left[1+\frac{(-1)^n}{2^{n+1}}\right]z^n \qquad \text{convergent in } |z|<1$$

(b) Expansion of f(z) in the region 1 < |z| < 2:

Since 1/(z+2) is analytic inside |z| = 2, then a Taylor's series is needed, while 1/(z-1) is not analytic inside |z| = 2, which requires a Laurent's series:

$$\frac{1}{(z-1)}=\sum_{n=0}^{\infty} z^{-n-1} \qquad \text{convergent in } |z|>1$$

$$\frac{1}{(z+2)}=\frac{1}{2}\sum_{n=0}^{\infty}\frac{(-z)^n}{2^n} \qquad \text{convergent in } |z|<2$$

Thus, the addition of the two series converge in the common region of convergence:

$$f(z)=\frac{1}{3}\sum_{n=0}^{\infty}\left[z^{-n-1}+\frac{(-z)^n}{2^{n+1}}\right] \qquad \text{convergent in } 1<|z|<2$$

(c) Expansion of f(z) in the region |z|> 2:

The function 1/(z+2) and 1/(z-1) are not analytic inside and on |z| = 2, thus a Laurent's series is necessary for both:

$$\frac{1}{(z-1)}=\sum_{n=0}^{\infty} z^{-n-1} \qquad \text{convergent in } |z|>1$$

$$\frac{1}{(z+2)}=\sum_{n=0}^{\infty}(-1)^n 2^n z^{-n-1} \qquad \text{convergent in } |z|>2$$

Thus, the series resulting from the addition of the two series converges in the common region |z|> 2 becomes:

$$f(z) = \frac{1}{3} \sum_{n=0}^{\infty} \left[1 - (-2)^n\right] z^{-n-1} \qquad \text{convergent in } |z| > 2$$

5.10 Classification of Singularities

An **Isolated singularity** of a function f(z) was previously defined as a point z_0 where $f(z_0)$ is not analytic and where f(z) is analytic at all the neighborhood points of z_0. If f(z) has an isolated singularity at z_0, then $f(z_0)$ can be represented by a Laurent's series about z_0, convergent in the ring $0 < |z - z_0| < a$ where the real constant (a) signifies the distance from z_0 to the nearest isolated singularity.

The part of Laurent's series that has negative powers of $(z-z_0)$ is called the **Principal Part** of the series:

$$\sum_{n=1}^{\infty} \frac{b_n}{(z-z_0)^n}$$

If the principal part has a finite number of terms, then $z = z_0$ is called a **Pole** of f(z). If the lowest power of $(z-z_0)$ in the principal part is m, then $z = z_0$ is called a **Pole of Order m**, i.e., the principal part looks like:

$$\sum_{n=1}^{m} \frac{b_n}{(z-z_0)^n}$$

If m = 1, z_0 is known as a **Simple Pole**. If the principal part contains all negative powers of $(z-z_0)$, then $z = z_0$ is called an **Essential Singularity**. If the function f(z) is not defined at $z = z_0$, but its Laurent's series representation about z_0 has no principal part, then $z = z_0$ is called a **Removable Singularity**.

Example 5.12

(i) The function:

$$f(z) = \frac{1}{z^2 - 4}$$

has two isolated singularities $z_0 = \pm 2$. Both singularities are simple poles (see Example 5.11-iv-a).

(ii) The function:

$$f(z) = \frac{1+z}{z^3} = \frac{1}{z^2} + \frac{1}{z^3}$$

has an isolated singularity at z = 0. The singularity is a pole of order 3.

(iii) The function:

$$f(z) = \sin(1/z) = \frac{1}{z} - \frac{1}{3!}z^{-3} + \frac{1}{5!}z^{-5} - ....$$

has an essential singularity at z = 0.

(iv) The function:

$$f(z) = \frac{\sin(z)}{z}$$

has a removable singularity at z = 0, since its Laurent's series representation about z = 0 has the form:

$$f(z) = \sum_{n=0}^{\infty} (-1)^n \frac{z^{2n}}{(2n+1)!}$$

with no principal part.

The points $z_0 = \infty$ in the complex plane would represent points on a circle whose radius is unbounded. One can classify the behavior of a function at infinity by first performing the following mapping:

$$\zeta = \frac{1}{z}$$

such that the points at infinity map into the origin at $\zeta = 0$.

Example 5.13

(i) The function:

$$f(z) = \frac{1}{z^2 - 4}$$

is transformed by $z = 1/\zeta$ such that:

$$f(z) = f(\zeta^{-1}) = \frac{1}{\frac{1}{\zeta^2} - 4} = \frac{\zeta^2}{1 - 4\zeta^2} = \frac{1}{4} \sum_{n=0}^{\infty} (4\zeta^2)^{n+1}$$

which is analytic at $\zeta = 0$. Thus f(z) is analytic at infinity.

(ii) The function:

$$f(z) = z + z^2$$

transforms to $\hat{f}(\zeta) = \zeta^{-2} + \zeta^{-1}$ where $\zeta = 0$ is a pole of order two. Thus, f(z) has a pole of order two at infinity.

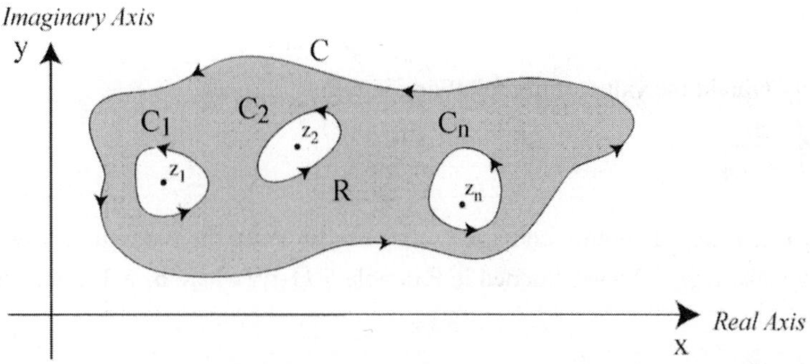

Fig 5.19: *Residue Theorem for a Multiply Connected Region*

(iii) The function:

$$f(z) = e^z = \sum_{n=0}^{\infty} \frac{z^n}{n!}$$

transforms to:

$$f(\zeta^{-1}) = \sum_{n=0}^{\infty} \frac{1}{n!\,\zeta^n}$$

where $\zeta = 0$ is an essential singularity. Thus e^z has an essential singularity at infinity.

5.11 Residues and Residue Theorem

Define the **Residue** of a function f(z) at one of its isolated singularities z_0 as the coefficient b_1 of the term $(z-z_0)^{-1}$ in the Laurent's series representation of f(z) about z_0, where the coefficient is defined by a closed contour integral:

$$b_1 = \frac{1}{2\pi i} \oint_C f(\zeta)\, d\zeta \tag{5.37}$$

and C is closed contour containing only the singularity z_0. The representation in eq. (5.37) can be used to obtain the integral of functions on a closed contour.

Example 5.14

(i) Obtain the value of the following integral:

$$\oint_C \frac{3}{z}\, dz$$

where C is a closed contour containing $z = 0$. Since the function $f(z) = 3/z$ is already in Laurent's series form, where $b_1 = 3$, then:

$$\oint_C \frac{3}{\zeta} d\zeta = 3(2\pi i) = 6\pi i$$

(ii) Obtain the value of the following integral:

$$\oint_C \frac{dz}{z^2 - 4}$$

where C is a closed contour containing $z_0 = 2$ only. Since the Laurent's series of the function about $z_0 = 2$ was obtained in Example 5.11-iv, where $b_1 = 1/4$, then the integral can be solved:

$$\oint_C \frac{dz}{z^2 - 4} = 2\pi i \frac{1}{4} = \frac{\pi}{2} i$$

5.11.1 Residue Theorem

If f(z) is analytic within and on a closed contour C except for a finite number of isolated singularities entirely inside C, then:

$$\oint_C f(z)dz = 2\pi i (r_1 + r_2 + ... + r_n) \tag{5.38}$$

where r_j = Residue of f(z) at the jth singularity.

Proof:

Enclose each singularity z_j with a closed contour C_j, such that f(z) is analytic inside C and outside the regions enclosed by all the other paths as shown by the shaded area in Fig. 5.19. Then, using Cauchy's integral theorem, one obtains:

$$\oint_C f(z)dz = \sum_{j=1}^{n} \oint_{C_j} f(z)dz$$

Since each contour C_j encloses only one pole z_j, then each closed contour integral can be evaluated by the residue at the pole z_j located within C_j:

$$\oint_{C_j} f(z)dz = 2\pi i r_j$$

then the integral over a closed path containing n poles is given by:

$$\oint_C f(z)dz = 2\pi i \sum_{j=1}^{n} r_j$$

Example 5.15

Obtain the value of the following integral:

$$\oint_C \frac{dz}{(z-2)(z-4)}$$

where C is a circle of radius = 3 centered at z_0, where z_0 is: (i) –2, (ii) 0, (iii) 3 and (iv) 6.

The function f(z) has simple poles at z = 2 and 4. The residue of f(z) at z = 2 is - 1/2 and at z = 4 is 1/2.

(i) Since there are no singularities inside this closed contour, then:

$$\oint_C f(z)dz = 0$$

(ii) The contour contains the simple pole at z = 2, thus:

$$\oint_C f(z)dz = 2\pi i(-\frac{1}{2}) = -\pi i$$

(iii) The contour contains both poles, hence the integrals gives:

$$\oint_C f(z)dz = 2\pi i(-\frac{1}{2}+\frac{1}{2}) = 0$$

(iv) The contour contains only the z = 4 simple pole, hence its value is:

$$\oint_C f(z)dz = 2\pi i(\frac{1}{2}) = \pi i$$

To facilitate the computation of the residue of a function, various methods can be developed so that one need not obtain a Laurent's series expansion about each pole in order to extract the value of coefficient b_1.

If f(z) has a pole of order m at z_0, then one can find a function g(z) such that:

$$g(z) = (z-z_0)^m f(z)$$

where $g(z_0) \neq 0$ and is analytic at z_0. Thus, the function g(z) can be expanded in a Taylor's series at z_0 as follows:

$$g(z) = \sum_{n=0}^{\infty} g^{(n)}(z_0)\frac{(z-z_0)^n}{n!}$$

Then, the Laurent's series for f(z) becomes:

$$f(z) = \frac{g(z)}{(z-z_0)^m} = \sum_{n=0}^{\infty} g^{(n)}(z_0)\frac{(z-z_0)^{n-m}}{n!}$$

From this expansion, the coefficient b_1 can be evaluated in terms of the $(m-1)^{st}$ derivative of g:

$$b_1 = \frac{g^{(m-1)}(z_0)}{(m-1)!} \tag{5.39}$$

If f(z) is a quotient of two functions p(z) and q(z):

$$f(z) = \frac{p(z)}{q(z)}$$

where the functions p(z) and q(z) are analytic at z_0 and $p(z_0) \neq 0$, then one can find the residue of f(z) at z_0 if $q(z_0) = 0$. Since the functions p(z) and q(z) are analytic at z_0, then one can find their Taylor series representations about z_0 as follows:

$$p(z) = \sum_{n=0}^{\infty} p^{(n)}(z_0)\frac{(z-z_0)^n}{n!}$$

and

$$q(z) = \sum_{n=0}^{\infty} q^{(n)}(z_0)\frac{(z-z_0)^n}{n!}$$

Various cases can be treated, depending on the form the Taylor series for p(z) and q(z) take where $p(z_0) \neq 0$:

(i) If $q(z_0) = 0$ and $q'(z_0) \neq 0$, then f(z) has a simple pole at z_0, and:

$$g(z) = (z-z_0)f(z) = \frac{p(z_0) + p'(z_0)(z-z_0) + \ldots}{q'(z_0) + q''(z_0)(z-z_0)/2 + \ldots}$$

Thus, the residue for a simple pole can be obtained by direct division of the two series, resulting in:

$$b_1 = g(z_0) = \frac{p(z_0)}{q'(z_0)} \tag{5.40}$$

(ii) If $q(z_0) = 0$ and $q'(z_0) = 0$ and $q''(z_0) \neq 0$, then f(z) has a pole of order 2, where:

$$g(z) = (z-z_0)^2 f(z) = \frac{p(z_0) + p'(z_0)(z-z_0) + \ldots}{q''(z_0)/2 + q'''(z_0)(z-z_0)/6 + \ldots}$$

Dividing the two infinite series to include terms up to $(z-z_0)$ and differentiating the resulting series results in:

$$b_1 = \frac{g'(z_0)}{1!} = \frac{2p'(z_0)}{q''(z_0)} - \frac{2}{3}\frac{p(z_0)q'''(z_0)}{[q''(z_0)]^2} \tag{5.41}$$

(iii) If $q(z_0) = 0$, $q'(z_0) = 0$, ... , $q^{(m-1)}(z_0) = 0$, and $q^{(m)}(z_0) \neq 0$, then f(z) has a pole of order m such that:

$$g(z) = (z - z_0)^m f(z) = \frac{p(z_0) + p'(z_0)(z - z_0) + \dots}{q^{(m)}(z_0)/m! + q^{(m+1)}(z_0)(z - z_0)/(m+1)! + \dots}$$

then, in order to evaluate b_1, one must divide the two infinite series and retain terms up to $(z-z_0)^{m-1}$ in the resulting series. Differentiating the series (m-1) times and setting $z = z_0$ one obtains the value of b_1:

$$b_1 = \frac{g^{(m-1)}(z_0)}{(m-1)!} \tag{5.42}$$

Example 5.16

Obtain the residues of each of the following functions at all its isolated singularities:

(i) $f(z) = \dfrac{z^2}{(z+1)(z-2)}$

At $z_0 = -1$, there is a simple pole, where the residue is as follows:

$$r(-1) = g(-1) = (z+1)f(z)\Big|_{z=-1} = \frac{(-1)^2}{(-1-2)} = -\frac{1}{3}$$

At $z_0 = 2$, there is a simple pole, where the residue is as follows:

$$r(2) = g(2) = (z-2)f(z)\Big|_{z=2} = \frac{(2)^2}{(2+1)} = \frac{4}{3}$$

(ii) $f(z) = \dfrac{e^z}{z^3}$

$z_0 = 0$ is a pole of order 3. Therefore, g(z) is defined as:

$$g(z) = z^3 f(z) = e^z$$

and the residue is as follows:

$$r(0) = \frac{g''(0)}{2!} = \frac{1}{2}e^0 = \frac{1}{2}$$

(iii) $f(z) = \dfrac{z+1}{\sin z}$

This function has an infinite number of simple poles, $z_n = n\pi$, $n = 0, \pm 1, \pm 2, \dots$. Let $p(z) = z+1$ and $q(z) = \sin z$.

Thus, using the formula in eq. (5.40):

$$r(n\pi) = \left.\frac{z+1}{\sin' z}\right|_{n\pi} = \frac{n\pi + 1}{\cos(n\pi)} = (-1)^n (n\pi + 1)$$

5.12 Integrals of Periodic Functions

The residue theorem can be used to evaluate integrals of the following type:

$$I = \int_0^{2\pi} F(\sin\theta, \cos\theta)\, d\theta \tag{5.43}$$

where $F(\sin\theta, \cos\theta)$ is a rational function of $\sin\theta$ and $\cos\theta$, and is bounded on the path of integration.

Using the parametric transformation:

$$z = e^{i\theta}$$

which transforms the integral to one on a unit circle centered at the origin, and using the definition of $\sin\theta$ and $\cos\theta$, one gets:

$$\sin\theta = \frac{1}{2i}\left[z - \frac{1}{z}\right] \qquad \cos\theta = \frac{1}{2}\left[z + \frac{1}{z}\right]$$

and the differential can be written in terms of z:

$$d\theta = -i\frac{dz}{z} \tag{5.44}$$

The integral in eq. (5.43) can be transformed to the following integral:

$$I = \oint_C f(z)\, dz$$

where f(z) is a rational function of z, which is finite on the path C, and C is the unit circle centered at the origin. Let f(z) have N poles inside the unit circle. The integral on the unit circle can be evaluated by the residue theorem, i.e.:

$$I = \oint_C f(z)\, dz = 2\pi i \sum_{j=1}^{N} r_j \tag{5.45}$$

where r_j's are the residues at all the isolated singularities of f(z) inside the unit circle $|z|=1$.

Example 5.17

Evaluate the following integral:

$$\int_0^{2\pi} \frac{2\,d\theta}{2 + \cos\theta}$$

Using the transformation in eq. (5.44), the integral becomes:

$$\oint_C \frac{4}{i(z^2+4z+1)} dz = \oint_C \frac{4}{i(z-z_1)(z-z_2)} dz$$

where $z_1 = -2 + \sqrt{3}$, and $z_2 = -2 - \sqrt{3}$. Therefore, the function f(z) has simple poles at z_1 and z_2. Since $|z_1| < 1$ and $|z_2| > 1$, only the simple pole at z_1 will be considered for computing the residue of poles inside the unit circle $|z| = 1$:

$$r(z_1) = g(-2+\sqrt{3}) = (z - z_1)f(z_1)\Big|_{z=z_1} = \frac{4}{i(z-z_2)} = \frac{2}{i\sqrt{3}}$$

Therefore:

$$\int_0^{2\pi} \frac{2d\theta}{2+\cos\theta} = 2\pi i\left(\frac{2}{i\sqrt{3}}\right) = \frac{4\pi}{\sqrt{3}}$$

5.13 Improper Real Integrals

The residue theorem can be used to evaluate improper real integrals of the type

$$\int_{-\infty}^{\infty} f(x)dx \tag{5.46}$$

where f(x) has no singularities on the real axis. The improper integral can be defined as:

$$\int_{-\infty}^{\infty} f(x)dx = \lim_{A\to\infty} \int_{-A}^{a} f(x)\,dx + \lim_{B\to\infty} \int_{a}^{B} f(x)dx$$

where the limits $A \to \infty$ and $B \to \infty$ of the two integrals are to be taken independently. If either or both limits do not exist, but the limit of the sum exists if $A = B \to \infty$, then the value of such an integral is called **Cauchy's Principal Value**, defined as:

$$\text{P.V.} \int_{-\infty}^{\infty} f(x)dx = \lim_{A\to\infty} \int_{-A}^{A} f(x)\,dx$$

If f(z) has a finite number of poles, n, and if, for $|z| >> 1$ there exists an M and $p > 1$ such that:

$$|f(z)| < M|z|^{-p} \qquad p > 1 \qquad |z| >> 1$$

then:

$$\text{P.V.} \int_{-\infty}^{\infty} f(x)dx = 2\pi i \sum_{j=1}^{n} r_j \tag{5.47}$$

where the r_j are the residues of f(z) at all the poles of f(z) in the upper half-plane. Let C_R be a semi-circle in the upper half plane with its radius R sufficiently large to enclose all the poles of f(z) in the upper half plane (see Fig. 5.20).

Thus, using the Residue Theorem, the integral over the closed path is:

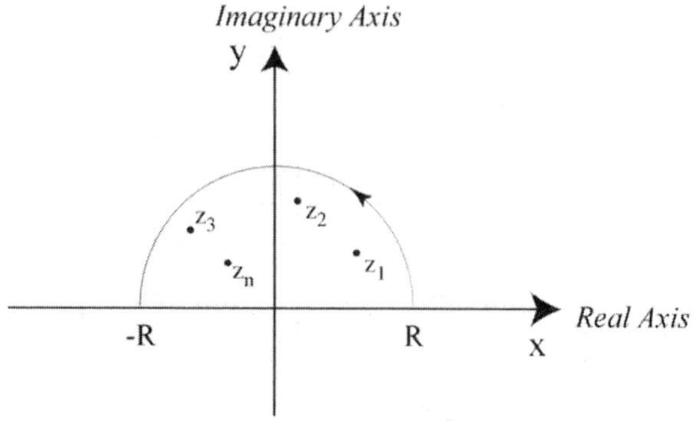

Fig. 5.20: *Closed Path for Improper Integrals*

$$\oint_C f(z)dz = \int_{-R}^{R} f(x)\,dx + \int_{C_R} f(z)\,dz = 2\pi i \sum_{j=1}^{n} r_j$$

The integral on the semi-circular path C_R can be shown to vanish as $R \rightarrow \infty$. On this path, let:

$$z = Re^{i\theta}$$

then the integral over the large circle can be evaluated as:

$$\left|\int_{C_R} f(z)\,dz\right| = \left|\int_0^{\pi} f(Re^{i\theta})i\,Re^{i\theta}\,d\theta\right| \le \pi R\left|f(Re^{i\theta})\right|_{max} \le \frac{\pi M}{R^{p-1}}$$

Thus, since $p > 1$, the integral over C_R vanishes:

$$\lim_{R\to\infty}\left|\int_{C_R} f(z)dz\right| \to 0$$

Example 5.18

Evaluate the following integral:

$$\int_0^{\infty} \frac{x^2}{x^4 + x^2 + 1}\,dx$$

If the function f(z) is defined as:

$$f(z) = \frac{z^2}{z^4 + z^2 + 1}$$

then:

$$|f(z)| \to \frac{1}{|z^2|} \qquad \text{when} \qquad |z| >> 1$$

hence p = 2 and the integral over C_R vanishes. The function f(z) has four simple poles:

$$z_1 = \frac{1+i\sqrt{3}}{2} \qquad z_2 = \frac{-1+i\sqrt{3}}{2} \qquad z_3 = \frac{1-i\sqrt{3}}{2} \qquad z_4 = \frac{-1-i\sqrt{3}}{2}$$

where the first two lie in the upper half plane. The residue of f(z) at the poles z_1 and z_2 becomes:

$$r(z_1 = \frac{1+i\sqrt{3}}{2}) = \lim_{z \to z_1} (z - z_1) f(z) = \frac{1+i\sqrt{3}}{4i\sqrt{3}}$$

$$r(z_2 = \frac{-1+i\sqrt{3}}{2}) = \lim_{z \to z_2} (z - z_2) f(z) = \frac{1-i\sqrt{3}}{4i\sqrt{3}}$$

Thus, using the results of eq. (5.47), the integral can be evaluated:

$$\int_0^\infty \frac{x^2}{x^4 + x^2 + 1} dx = \frac{1}{2} \int_{-\infty}^{\infty} \frac{x^2}{x^4 + x^2 + 1} dx = \frac{1}{2} 2\pi i \left[\frac{1+i\sqrt{3}}{4i\sqrt{3}} + \frac{1-i\sqrt{3}}{4i\sqrt{3}} \right] = \frac{\pi}{2\sqrt{3}}$$

5.14 Improper Real Integral Involving Circular Functions

The residue theorem can also be used to evaluate integrals having the following form:

$$\int_{-\infty}^{\infty} f(x) \cos(ax) dx \qquad \text{for} \qquad a > 0$$

$$\int_{-\infty}^{\infty} f(x) \sin(ax) dx \qquad \text{for} \qquad a > 0$$

and

$$\int_{-\infty}^{\infty} f(x) e^{iax} dx \qquad \text{for} \qquad a > 0 \tag{5.48}$$

where f(x) has no singularities on the real axis and a is positive. Let f(z) be an analytic function in the upper half plane except for isolated singularities, such that:

$$|f(z)| < M|z|^{-p} \qquad \text{where} \qquad p > 0 \qquad \text{for} \qquad |z| >> 1$$

Since the first two integrals of eq. (5.48) are the real and imaginary parts of the integral of eq. (5.48), one needs to treat only the third integral.

Performing the integration on f(z) e^{iaz} on the closed contours shown in Fig. 5.20, then:

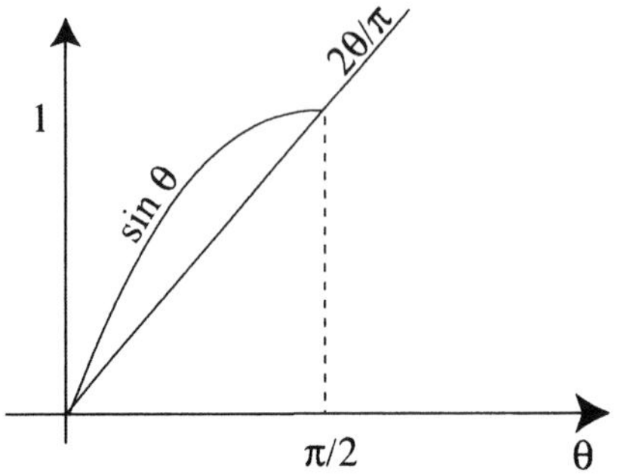

***Fig. 5.21:** Approximation for Jordan's Lemma*

$$\oint_C f(z)e^{iaz}dz = \int_{-R}^{R} f(x)\,e^{iax}dx + \int_{C_R} f(z)e^{iaz}dz$$

One must now show that the integral vanishes as $R \rightarrow \infty$. This proof is known as **Jordan's Lemma**:

$$\lim_{R\to\infty}\left|\int_{C_R} f(z)\,e^{iaz}dz\right| \to 0$$

Let $z = R\,e^{i\theta}$ on C_R, then the integral becomes:

$$\int_0^{\pi} f(Re^{i\theta})\,e^{iaRe^{i\theta}}\,e^{i\theta}\,iR\,d\theta = iR\int_0^{\pi} f(Re^{i\theta})\,e^{iaR[\cos\theta + i\sin\theta]}e^{i\theta}d\theta$$

Thus, the absolute value of the integral on C_R becomes:

$$\left|\int_{C_R} f(z)\,e^{iaz}dz\right| = R\left|\int_0^{\pi} f(Re^{i\theta})\,e^{i[aR\cos\theta+\theta]}e^{-aR\sin\theta}d\theta\right| \le \frac{RM}{R^{p}}\left|\int_0^{\pi} e^{-aR\sin\theta}d\theta\right|$$

The last integral can be evaluated as follows:

$$\int_0^{\pi} e^{-aR\sin\theta}d\theta = 2\int_0^{\pi/2} e^{-aR\sin\theta}d\theta \le 2\int_0^{\pi/2} e^{-2aR\theta/\pi}d\theta = \frac{\pi}{aR}(1-e^{-aR})$$

where the following inequality was used, (see Fig. 5.21):

$$\sin\theta \ge \frac{2\theta}{\pi} \qquad \text{for} \qquad 0 \le \theta \le \pi/2$$

Thus:

$$\left| \int_{C_R} f(z)\, e^{iaz} dz \right| \le \frac{M}{R^p} \frac{\pi}{a} (1 - e^{-aR})$$

which vanishes as R → ∞, since $a > 0$ and $p > 0$. It should be noted that the integral becomes unbounded if $a < 0$, or equivalently, if $a > 0$ and the circular path is taken in the lower half plane. Thus:

$$\text{P.V.} \int_{-\infty}^{\infty} f(x) e^{iax}\, dx = 2\pi i \sum_{j=1}^{N} r_j$$

$$\text{P.V.} \int_{-\infty}^{\infty} f(x) \cos(ax) dx = Re\left[2\pi i \sum_{j=1}^{N} r_j \right] = -2\pi\, Im\left[\sum_{j=1}^{N} r_j \right]$$

$$\text{P.V.} \int_{-\infty}^{\infty} f(x) \sin(ax) dx = Im\left[2\pi i \sum_{j=1}^{N} r_j \right] = 2\pi\, Re\left[\sum_{j=1}^{N} r_j \right] \qquad (5.49)$$

where the r_j's represent the residues of the N poles of $\{f(z)\, e^{iax}\}$ in the upper half plane.

Example 5.19

Evaluate the following integral:

$$I = \int_0^{\infty} \frac{\cos x}{x^4 + 1} dx$$

Since $f(z) = (z^4 + 1)^{-1}$, then $|f(z)| \le R^{-p}$ on C_R for $R >> 1$, where $p = 4$ and $a = 1$.

The function f(z) has four simple poles:

$$z_1 = \frac{1+i}{\sqrt{2}} \qquad z_2 = \frac{-1+i}{\sqrt{2}} \qquad z_3 = \frac{-1-i}{\sqrt{2}} \qquad z_4 = \frac{1-i}{\sqrt{2}}$$

where the first two lie in the upper half plane, note that $z_1^4 = z_2^4 = -1$. Thus, the integral can be obtained by eq. (5.49):

$$I = \frac{1}{2} \int_{-\infty}^{\infty} \frac{\cos x}{x^4 + 1} dx = -\pi Im[r_1 + r_2]$$

where the residue, z_1 is calculated from p/q' for simple poles:

$$r(z_1) = \frac{e^{iz_1}}{4z_1^3} = \frac{z_1 e^{iz_1}}{4z_1^4} = -\frac{z_1 e^{iz_1}}{4} = -\frac{1+i}{4\sqrt{2}} e^{(i-1)/\sqrt{2}}$$

and similarly for the second pole z_2:

$$r(z_2) = -\frac{z_2 e^{iz_2}}{4} = \frac{1-i}{4\sqrt{2}} e^{-(i+1)/\sqrt{2}}$$

Thus, the integral I can be solved:

$$I = \frac{\pi m}{2} e^{-m}[\cos m + \sin m] \qquad m = 1/\sqrt{2}.$$

5.15 Improper Real Integrals of Functions Having Singularities on the Real Axis

Functions that have singularities on the real axis can be integrated by deforming the contour of integration. The following real integral:

$$\int_a^b f(x)dx$$

where f(x) has a singularity on the real axis at $x = c$, $a < c < b$ is defined as follows:

$$\int_a^b f(x)dx = \lim_{\varepsilon \to 0}\left[\int_a^{c-\varepsilon} f(x)dx\right] + \lim_{\delta \to 0}\left[\int_{c+\delta}^{b} f(x)dx\right]$$

The integral on [a,b] exists iff the two partial integrals exist independently. If either or both limits as $\varepsilon \to 0$ and $\delta \to 0$ do not exist but the limit of the sum of the two integrals exists if $\varepsilon = \delta$, that is if the following integral:

$$\lim_{\varepsilon \to 0}\left[\int_a^{c-\varepsilon} f(x)dx + \int_{c+\varepsilon}^{b} f(x)dx\right]$$

exists, then the value of the integral thus obtained is called the **Cauchy Principal Value** of the integral, denoted as:

$$\text{P.V.}\int_a^b f(x)dx$$

Example 5.20

(i) Evaluate the following integral:

$$\int_{-1}^{2} x^{-1/3}dx$$

Note the function $f(x) = x^{-1/3}$ is singular at $x = 0$. Therefore:

$$\int_{-1}^{2} x^{-1/3}\,dx = \underset{\varepsilon\to 0}{\text{Lim}}\left[\int_{-1}^{0-\varepsilon} x^{-1/3}dx\right] + \underset{\delta\to 0}{\text{Lim}}\left[\int_{0+\delta}^{2} x^{-1/3}dx\right]$$

$$= \frac{3}{2}\underset{\varepsilon\to 0}{\text{Lim}}\left[(-\varepsilon)^{2/3} - (-1)^{2/3}\right] + \frac{3}{2}\underset{\delta\to 0}{\text{Lim}}\left[(2)^{2/3} - (\delta)^{2/3}\right] = \frac{3}{2}\left[4^{1/3} - 1\right]$$

(ii) Evaluate the following integral:

$$\int_{-1}^{2} x^{-3}dx$$

The function $f(x) = x^{-3}$ is singular at $x = 0$.

$$\int_{-1}^{2} x^{-3}\,dx = \underset{\varepsilon\to 0}{\text{Lim}}\left[\int_{-1}^{0-\varepsilon} x^{-3}dx\right] + \underset{\delta\to 0}{\text{Lim}}\left[\int_{0+\delta}^{2} x^{-3}dx\right]$$

$$= \frac{1}{2}\underset{\varepsilon\to 0}{\text{Lim}}\left[1 - \frac{1}{\varepsilon^2}\right] + \frac{1}{2}\underset{\delta\to 0}{\text{Lim}}\left[\frac{1}{\delta^2} - \frac{1}{4}\right]$$

Neither integral exists for ε and δ to vanish independently. If one takes the P.V. of the integral:

$$\text{P.V.}\int_{-1}^{2} x^{-3}\,dx = \underset{\varepsilon\to 0}{\text{Lim}}\left[\int_{-1}^{0-\varepsilon} x^{-3}dx + \int_{0+\varepsilon}^{2} x^{-3}dx\right] = \frac{3}{8}$$

Improper integrals of function on the real axis

$$\int_{-\infty}^{\infty} f(x)dx$$

where f(x) has simple poles on the real axis can be evaluated by the use of the residue theorem in the Cauchy Principal Value sense.

Let $x_1, x_2, \ldots x_n$ be the simple poles of f(z) on the real axis, and let $z_1, z_2, \ldots, z_m$ be the poles of f(z) in the upper-half plane. Let C_R be the semi-circular path with a radius R, sufficiently large to include all the poles of f(z) on the real axis and in the upper-half plane. The contour on the real axis is indented such that the contour includes a semi-circle of small radius = ε around each simple pole x_j as shown in Fig. 5.22.

Thus, one can obtain the principal value of the integral as follows:

$$\left\{\int_{-R}^{x_1-\varepsilon} + \int_{C_1} + \int_{x_1+\varepsilon}^{x_2-\varepsilon} + \int_{C_2} + \ldots + \int_{x_n+\varepsilon}^{R} + \int_{C_R}\right\} f(z)\,dz = \oint f(z)dz = 2\pi i\sum_{j=1}^{m} r_j$$

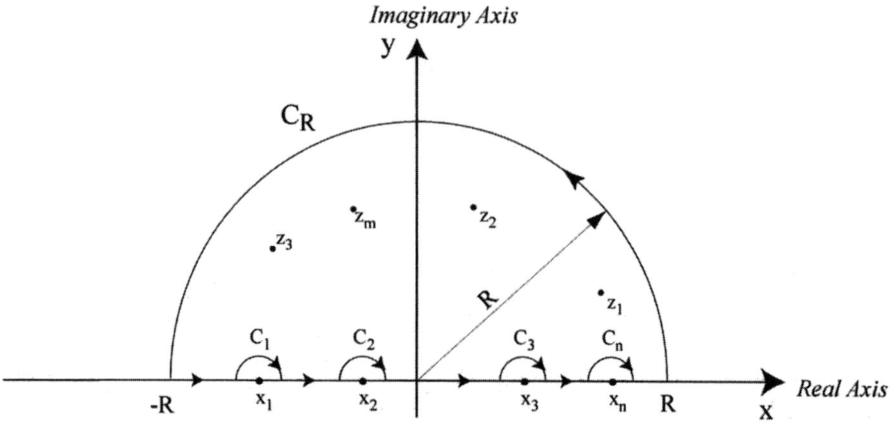

Fig. 5.22: *Closed Path for Improper Integrals with Real and Complex Poles*

where the contours C_j are half-circle paths in the clockwise direction and r_j's are the residues of f(z) at the poles of f(z) in the upper half plane at z_j. The limit as $R \to \infty$ and $\varepsilon \to 0$ must be taken to evaluate the integrals in Cauchy Principal Value form on C_R and on C_j for j = 1, 2, ..., n.

If the function f(z) decays for $|z| \to \infty$ as follows:

$$|f(z)| < M|z|^{-p} \qquad \text{where} \quad p > 1 \qquad \text{for} \qquad |z| >> 1$$

then, it was shown earlier that:

$$\lim_{R\to\infty}\left[\int_{C_R} f(z)dz\right] \to 0$$

Since the function f(z) has simple poles on the real axis, then in the neighborhood of each real simple pole $x_{j,}$ it has one term with a negative power as follows:

$$f(z) = \frac{r_j^*}{z - x_j} + g(z)$$

where g(z) = the part of f(z) that is analytic at x_j, and r_j^* are the residues of f(z) at x_j.

Thus, the integral over a small semi-circular path about x_j becomes in the limit as the radius $\varepsilon \to 0$:

$$\lim_{\varepsilon\to 0}\int_{C_j} f(z)\,dz = \lim_{\varepsilon\to 0}\left[r_j^*\int_{C_j}\frac{dz}{z - x_j} + \int_{C_j} g(z)dz\right] = -\pi i r_j^*$$

where the results of Example 5.6 were used.

Thus, the principal value of the integral is given by:

$$\text{P.V.} \int_{-\infty}^{\infty} f(x)dx = 2\pi i \sum_{j=1}^{m} r_j + \pi i \sum_{j=1}^{n} r_j^* \tag{5.50}$$

Example 5.21

Evaluate the following integral:

$$\int_{-\infty}^{\infty} \frac{\sin x}{x(x^2+a^2)} dx = Im\left[\int_{-\infty}^{\infty} \frac{e^{ix}}{x(x^2+a^2)} dx\right] \qquad \text{for } a > 0$$

The function $\frac{e^{iz}}{z(z^2+a^2)}$ has three simple poles:

$x_1 = 0, \qquad z_1 = ia, \qquad z_2 = -ia$

To evaluate the integral, one needs to evaluate the residues of the appropriate poles:

$$r_1(z_1) = \left.\frac{(z-z_1)e^{iz}}{z(z^2+a^2)}\right|_{z=z_1=ia} = -\frac{1}{2a^2e^a}$$

$$r_1^*(x_1) = \left.\frac{z\,e^{iz}}{z(z^2+a^2)}\right|_{z=x_1=0} = \frac{1}{a^2}$$

Thus:

$$\text{P.V.} \int_{-\infty}^{\infty} \frac{\sin x}{x(x^2+a^2)} dx = Im\left[2\pi i\left\{\frac{-1}{2a^2e^a}\right\} + \pi i\left\{\frac{1}{a^2}\right\}\right] = \frac{\pi}{a^2}\left[1-e^{-a}\right]$$

5.16 Theorems on Limiting Contours

In Sections 5.13 to 5.15 integrals on semi-circular contours in the upper half-plane with unbounded radii were shown to vanish if the integral behaved in a prescribed manner on the contour. In this section, theorems dealing with contours that are not exclusively in the upper half-plane are explored.

5.16.1 Generalized Jordan's Lemma

Consider the following contour integral (see Fig. 5.23).

$$\int_{C_R} e^{az} f(z)\, dz$$

where $a = b\,e^{ic}$, $b > 0$, c real and C_R is an arc of circle described by $z = Re^{i\theta}$, whose radius is R, and whose angle θ falls in the range:

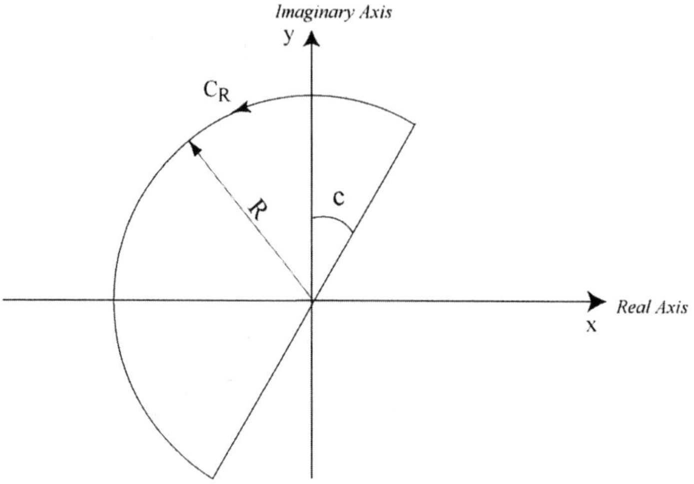

Fig 5.23: *Path for Generalized Jordan's Lemma*

$$\frac{\pi}{2} - c \le \theta \le \frac{3\pi}{2} - c$$

Let $|f(z)| \le M\,|z|^{-p}$, as $|z| >> 1$, where $p > 0$. Then one can show that:

$$\lim_{R\to\infty} \int_{C_R} e^{az} f(z)\,dz \to 0 \tag{5.51}$$

The absolute value of the integral on C_R becomes:

$$\left| \int_{\frac{\pi}{2}-c}^{\frac{3\pi}{2}-c} e^{a\,Re^{i\theta}} f(Re^{i\theta})\, i\,Re^{i\theta}\, d\theta \right|$$

$$= R \left| \int_{\frac{\pi}{2}-c}^{\frac{3\pi}{2}-c} e^{b\,Re^{ic}} e^{i\theta} f(Re^{i\theta})\, d\theta \right| \le R \frac{M}{R^p} \left| \int_{\frac{\pi}{2}-c}^{\frac{3\pi}{2}-c} e^{bR\cos(c+\theta)}\, d\theta \right|$$

$$= \frac{M}{R^{p-1}} \left| \int_0^{\pi} e^{bR\cos(\theta+\pi/2)}\, d\theta \right| = \frac{M}{R^{p-1}} \left| \int_0^{\pi} e^{bR\sin(\theta)}\, d\theta \right|$$

$$\le \frac{M}{R^p} (1 - e^{-bR}) \frac{\pi}{b}$$

as has been shown in Section 5.14. Thus, the integral over any segment of the half circle C_R vanishes as R → ∞. Four special cases of eq. (5.51) can be discussed, due to their importance to integral transforms, see Fig. 5.24:

(i) If c = π/2, eq. (5.51) takes the form:

$$\lim_{R\to\infty} \int_{C_R} e^{ibz} f(z)\, dz \to 0$$

where C_R is an arc in the first and/or the second quadrants.

(ii) If c = - π/2, eq. (5.51) takes the form:

$$\lim_{R\to\infty} \int_{C_R} e^{-ibz} f(z)\, dz \to 0$$

where C_R is an arc in the third and/or the fourth quadrants.

(iii) If c = 0, eq. (5.51) takes the form:

$$\lim_{R\to\infty} \int_{C_R} e^{bz} f(z)\, dz \to 0$$

where C_R is an arc in the second and/or the third quadrants.

(iv) If c = π, eq. (5.51) takes the form:

$$\lim_{R\to\infty} \int_{C_R} e^{-bz} f(z)\, dz \to 0$$

where C_R is an arc in the fourth and/or the first quadrants.

The form of given in (iii) is known as Jordan's Lemma. The form given in eq. (5.51) is the **Generalized Jordan's Lemma**.

5.16.2 Small Circle Theorem

Consider the following contour:

$$\int_{C_\varepsilon} f(z) dz$$

where C_ε is a circular arc of radius = ε, centered at z = a (Fig. 5.24). If the function f(z) behaves as:

$$\lim_{\varepsilon\to 0} |f(z)| \le \frac{M}{\varepsilon^p} \qquad \text{for} \qquad p < 1$$

or if:

$$\lim_{\varepsilon\to 0} \varepsilon\, f(a + \varepsilon\, e^{i\theta}) \to 0$$

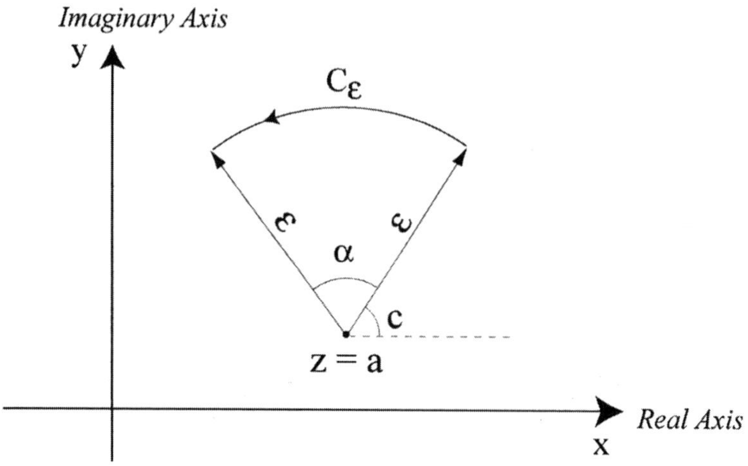

Fig 5.24: Closed Path for Small Circles

then:

$$\lim_{\varepsilon \to 0} \int_{C_\varepsilon} f(z)\,dz \to 0$$

Let $z = a + \varepsilon\, e^{i\theta}$, then:

$$\left| \int_{C_\varepsilon} f(z)\,dz \right| = \left| \int_{c}^{c+\alpha} f(a+\varepsilon\, e^{i\theta})\, i\varepsilon\, e^{i\theta}\, d\theta \right|$$

$$\le \frac{M}{\varepsilon^{p-1}} \left| \int_{c}^{c+\alpha} d\theta \right| = M \alpha\, \varepsilon^{1-p}$$

Thus:

$$\lim_{\varepsilon \to 0} \left| \int_{C_\varepsilon} f(z)\,dz \right| \to 0 \qquad \text{if} \qquad p < 1$$

5.16.3 Small Circle Integral

If f(z) has a simple pole at z = a, then:

$$\lim_{\varepsilon \to 0} \int_{C_\varepsilon} f(z)\,dz = \alpha\, i\, r(a)$$

where C_ε is a circular arc of length $\alpha\varepsilon$, centered at z = a, radius = ε, (Fig. 5.24), r(a) is the residue of f(z) at z = a, and the integration is performed in the counter-clockwise sense.

Since f(z) has a simple pole at z = a, then it can be expressed as a Laurent's series about z = a:

$$f(z) = \frac{r(a)}{z-a} + g(z)$$

where g(z) is analytic (hence bounded) at z = a. Thus:

$$\int_{C_\varepsilon} f(z)dz = r(a)\int_{C_\varepsilon} \frac{dz}{z-a} + \int_{C_\varepsilon} g(z)dz = \alpha\, i\, r(a) + \int_{C_\varepsilon} g(z)\, dz$$

where the results of Example 5.6 were used. Also:

$$\lim_{\varepsilon\to 0}\left|\int_{C_\varepsilon} g(z)\, dz\right| = \lim_{\varepsilon\to 0}\left|\int_c^{c+\alpha} g(a+\varepsilon e^{i\theta})\, i\varepsilon e^{i\theta}\, d\theta\right| \le \lim_{\varepsilon\to 0}(M\,\alpha\,\varepsilon) \to 0$$

then the integral over the small circle becomes:

$$\lim_{\varepsilon\to 0}\left|\int_{C_\varepsilon} f(z)dz\right| = \alpha\, i\, r(a) \tag{5.52}$$

5.17 Evaluation of Real Improper Integrals by Non-Circular Contours

The residue theorem was used in Sections 5.13 to 5.15 to evaluate improper integrals by closing the straight integration path with semi-circular paths. In this section, more convenient and efficient non-circular contours are used to evaluate improper integrals.

If a periodic function has an infinite number of poles in the complex plane, then to use the Residue Theorem and a circular contour, one must resort to summing an infinite number of residues at the poles in the entire half-plane.

However, a more prudent choice of a non-circular contour may yield the desired evaluation of the improper integral by enclosing few poles.

Example 5.22

Evaluate the following integral:

$$I = \int_{-\infty}^{\infty} \frac{e^{ax}}{1+e^x}\, dx \qquad 0 < a < 1$$

The function:

$$f(z) = \frac{e^{az}}{1+e^z}$$

has an infinite number of simple poles at $z = (2n + 1)\,\pi i$, n = 0, 1, 2, ... in the upper half-plane. Choose the contour shown in Fig. 5.25 described by the points -R, R,

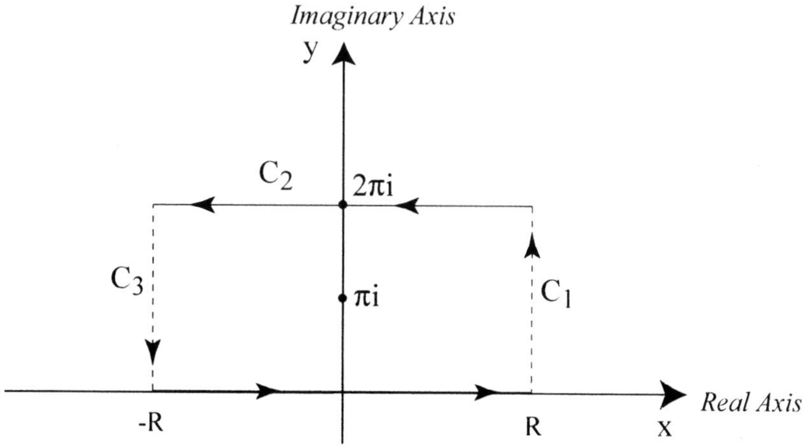

Fig. 5.25: *Closed Path for Periodic Integrals*

R+2πi, -R+2πi, which encloses only one pole. The choice of the contour C_2 was made because of the periodicity of $e^z = e^{z+2\pi i}$.

Thus, the contour of integration results in:

$$\int_{-R}^{R} f(z)\,dx + \int_{C_1} f(z)dz + \int_{C_2} f(z)\,dz + \int_{C_3} f(z)\,dz = 2\pi i\, r(\pi i)$$

The integral on C_1 is given by $z = R + iy$

$$\left|\int_{C_1} \frac{e^{az}}{1+e^z}\,dz\right| = \left|\int_0^{2\pi} \frac{e^{a(R+iy)}}{1+e^{R+iy}}\,i\,dy\right| \le 2\pi\, e^{R(a-1)}$$

and consequently:

$$\lim_{R\to\infty} \int_{C_1} f(z)\,dz \to 0 \qquad \text{since} \qquad 0 < a < 1$$

Similarly, the integral over C_3 also vanishes.

The integral on C_2 can be evaluated by letting $z = x + 2\pi i$:

$$\int_{C_2} \frac{e^{az}}{1+e^z}\,dz = \int_{-R}^{R} \frac{e^{a(x+2\pi i)}}{1+e^{x+2\pi i}}\,dx = -e^{2\pi ia} \int_{-R}^{R} \frac{e^{ax}}{1+e^x}\,dx = -e^{2\pi ia}\, I$$

The residue of f(z) at πi becomes:

$$r(\pi i) = \left.\frac{e^{az}}{e^z}\right|_{z=\pi i} = -e^{a\pi i}$$

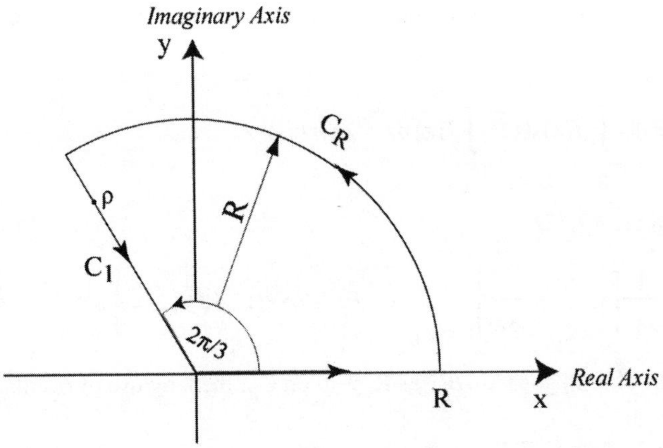

Fig. 5.26: *Closed Path for Periodic Integrals*

Thus, as R → ∞:

$$I - e^{2\pi ia} I = 2\pi i(-e^{\pi ia})$$

$$I = \frac{2\pi i(e^{\pi ia})}{e^{2\pi ia} - 1} = \frac{\pi}{\sin(a\pi)}$$

The evaluation of improper integrals of the form:

$$\int_0^\infty f(x)\, dx$$

where f(x) is not an even function, cannot be evaluated by extending the straight path to [-∞,∞]. Thus, one must choose another contour that would duplicate the original integral, but with a multiplicative constant.

Example 5.23

Evaluate the following integral:

$$I = \int_0^\infty \frac{dx}{x^3 + 1}$$

Since the integral path cannot be extended to (-∞), then it is expedient to choose the contour as shown in Fig. 5.26. The choice of the path C_1 where $\theta = 2\pi/3$ was made since along that path, $z^3 = (\rho\, e^{2\pi i/3})^3 = \rho^3$ and is real, so that the denominator does not change.

The function $f(z) = 1/(z^3+1)$ has three simple poles:

$$z_1 = \frac{1 + i\sqrt{3}}{2}, \qquad z_2 = -1, \qquad \text{and} \qquad z_3 = \frac{1 - i\sqrt{3}}{2}$$

where only the z_1 pole falls within the closed path. The integral over the closed path becomes:

$$\int_0^R f(x)dx + \int_{C_R} f(z)\,dz + \int_{C_1} f(z)\,dz = 2\pi i\, r(z_1)$$

where the residue $r(z_1)$ is:

$$r(z_1) = \left.\frac{1}{3z^2}\right|_{z=z_1} = \left.\frac{z}{3z^3}\right|_{z=z_1} = -\frac{z_1}{3} = -\frac{1}{6}(1+i\sqrt{3}) = -\frac{1}{3}e^{i\pi/3}$$

Since the limit of $|f(z)|$ goes to $1/R^3$ as $R \to \infty$ on C_R, then, by use of results of Section 5.16, with p = 3:

$$\lim_{R\to\infty} \int_{C_R} f(z)dz \to 0$$

The path C_1 is described by $z = \rho\, e^{2\pi i/3}$, the integral across the path becomes:

$$\int_{C_1} f(z)dz = \int_R^0 \frac{e^{2\pi i/3}}{(\rho e^{2\pi i/3})^3 + 1} d\rho = -e^{2\pi i/3}\int_0^R \frac{d\rho}{\rho^3+1} = -e^{2\pi i/3}\, I$$

Thus, as $R \to \infty$:

$$I - e^{2\pi i/3} I = -\frac{2\pi i}{3} e^{\pi i/3}$$

$$I = \frac{2\pi i}{3}\frac{e^{\pi i/3}}{e^{2\pi i/3} - 1} = \frac{\pi}{3}\frac{1}{\sin(\pi/3)} = \frac{2\pi}{3\sqrt{3}}$$

5.18 Integrals of Even Functions Involving log x

Improper integrals involving log x can be evaluated by indenting the contour along the real axis. The following integral can be evaluated:

$$\int_0^\infty f(x) \log x\, dx$$

where f(x) is an even function and has no singularities on the real axis and, as $|z| \to \infty$, $|f(z)| \le M\,|z|^{-p}$, where $p > 1$.

Since the function log z is not single-valued, a branch cut is made, starting from the branch point at z = 0 along the negative y-axis. Define the branch cut:

$$z = \rho\, e^{i\theta}, \qquad \rho > 0, \qquad -\frac{\pi}{2} < \theta < \frac{3\pi}{2}$$

where the choice was made to include the $\theta = 0$ in the range. Because of the branch point, the contour on the real axis must be indented around x = 0 as shown in Fig. 5.27. Let

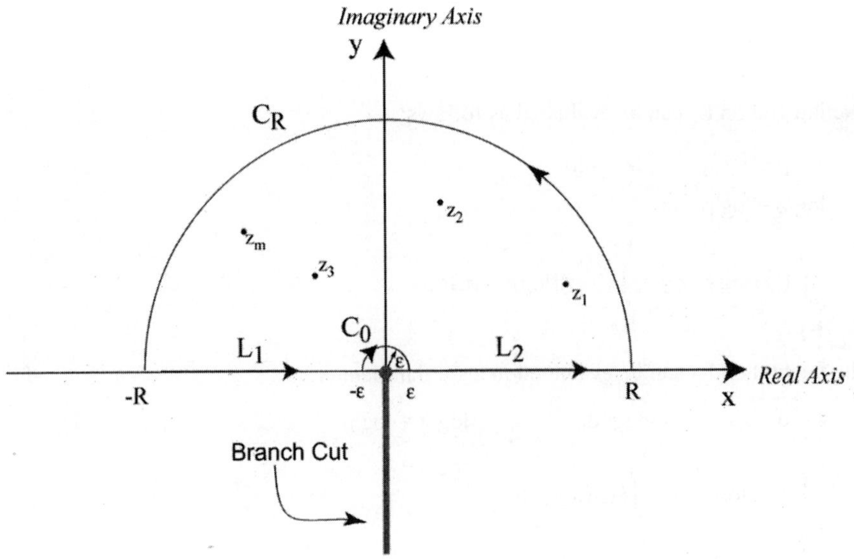

***Fig. 5.27:** Closed Path for Logarithmic Integrals*

the poles of f(z) in the upper-half plane be $z_1, z_2, \ldots, z_m$. Let C_R be a semi-circular contour, radius = R and C_0 be semi-circular contour radius = ε in the counter-clockwise sense. Thus, the integral over the closed path becomes:

$$\oint f(z)\log z\,dz = \left\{ \int_{L_1} - \int_{C_0} + \int_{L_2} + \int_{C_R} \right\} f(z)\log z\,dz = 2\pi i \sum_{j=1}^{m} r_j(z_j)$$

where r_j's are the residues of [f(z) log z] at the poles of f(z) in the upper-half plane. The integral on C_0 in the clockwise direction can be evaluated where $z = \varepsilon\, e^{i\theta}$:

$$\left| \int_{C_0} f(z)\log z\,dz \right| = \left| \int_0^{\pi} f(\varepsilon\, e^{i\theta})(\log\varepsilon + i\theta)\, i\varepsilon\, e^{i\theta}\, d\theta \right| \le \left[\pi\varepsilon|\log\varepsilon| + \frac{\varepsilon\pi^2}{2} \right] |f(\varepsilon\, e^{i\theta})|_{C_0}$$

Thus, since the limit of f(z) is finite as z goes to zero, then:

$$\lim_{\varepsilon\to 0} \int_{C_0} f(z)\log z\,dz \to 0$$

The integral on C_R can be evaluated, where $z = R\, e^{i\theta}$:

$$\left| \int_{C_R} f(z)\log z\,dz \right| = \left| \int_0^{\pi} f(R\,e^{i\theta})(\log R + i\theta)\, iR\, e^{i\theta}\, d\theta \right| \le R\frac{M}{R^p}\left[\pi\log R + \frac{\pi^2}{2} \right]$$

which vanishes when R → ∞, so that:

$$\lim_{R\to\infty} \int_{C_R} f(z) \log z \, dz \to 0 \qquad \text{since} \qquad p > 1$$

The integral on L_1 can be evaluated as follows:

$$z = \rho\, e^{i\pi} = -\rho \qquad dz = -d\rho$$

$$\log z = \log \rho + i\pi$$

$$\int_{L_1} f(z) \log z \, dz = -\int_R^{\varepsilon} f(-\rho)[\log\rho + i\pi]\, d\rho$$

The integral on L_2 can be evaluated in a similar manner:

$$z = \rho \qquad dz = d\rho \qquad \log z = \log \rho$$

$$\int_{L_2} f(z) \log z \, dz = \int_{\varepsilon}^{R} f(\rho) \log\rho \, d\rho$$

The total integral, after substituting $f(-\rho) = f(\rho)$, becomes:

$$2\int_{\varepsilon}^{R} f(\rho)\log\rho \, d\rho + i\pi \int_{\varepsilon}^{R} f(\rho)\, d\rho = 2\pi i \sum_{j=1}^{m} r_j$$

Taking the limits $\varepsilon \to 0$ and $R \to \infty$ one obtains upon substituting x for ρ:

$$\int_0^{\infty} f(x) \log x \, dx = -\frac{i\pi}{2} \int_0^{\infty} f(x)\, dx + \pi i \sum_{j=1}^{m} r_j \tag{5.53}$$

If the function f(x) is real then the integral of f(x) log x must result in a real value. The integral on the right side is also real, hence this term constitutes a purely imaginary number. For a real answer, the imaginary number resulting from the integral term must cancel out the imaginary part of the residue contribution. Thus, one can then simplify finding the final answer by choosing the real part of the residue contributions on the right side of eq. (5.53), i.e.:

$$\int_0^{\infty} f(x) \log x \, dx = Re\left[\pi i \sum_{j=1}^{m} r_j\right] = -\pi\, Im\left[\sum_{j=1}^{m} r_j\right]$$

Example 5.24

Evaluate the following integral:

$$\int_0^{\infty} \frac{\log x}{x^2 + 4}\, dx$$

The function $f(z) = 1/(z^2 + 4)$ has two simple poles at:

$$z_1 = 2i = 2\,e^{i\pi/2}, \qquad z_2 = -2i$$

The residue at z_1 is:

$$r(2i) = \left.\frac{\log z}{2z}\right|_{z=2i} = \frac{\log(2i)}{4i} = \frac{1}{4i}(\log 2 + i\frac{\pi}{2}) = \frac{\pi}{8} - i\frac{\log 2}{4}$$

also f(x) is real and:

$$|f(R)| \le \frac{1}{R^p} \qquad \text{with } p = 2, \qquad \text{for } R >> 1$$

Thus:

$$\int_0^\infty \frac{\log x}{x^2+4}\,dx = -\frac{i\pi}{2}\int_0^\infty \frac{dx}{x^2+4} + i\pi\left[\frac{\pi}{8} - i\frac{\log 2}{4}\right]$$

since:

$$\int_0^\infty \frac{dx}{x^2+4} = \frac{\pi}{4}$$

then:

$$\int_0^\infty \frac{\log x}{x^2+4}\,dx = -\frac{i\pi}{2}[\frac{\pi}{4}] + i\pi\left[\frac{\pi}{8} - i\frac{\log 2}{4}\right] = \frac{\pi}{4}\log 2$$

It is thus shown that the imaginary parts of the answer cancel out since f(x) is a real function. Or, one could use the shortcut, to give:

$$\int_0^\infty \frac{\log x}{x^2+4}\,dx = -\pi\, Im\left[\frac{\pi}{8} - i\frac{\log 2}{4}\right] = \frac{\pi}{4}\log 2$$

If the function f(x) has n simple poles on the real axis, then one indents the contour over the real axis by a small semi-circles of radius ε, so that the path integral becomes:

$$\left\{\int_{-R}^{-\varepsilon} + \sum_{j=1}^{n}\int_{C_j} + \int_{C_0} + \int_{\varepsilon}^{R} + \int_{C_R}\right\} f(z)\log z\,dz = 2\pi i\sum_{j=1}^{m} r_j(z_j)$$

where the function has m poles at z_j in the upper half plane as well as n simple poles at x_j on the real axis, and C_j are the semi-circular paths around x_j in the clockwise direction. Each semi-circular path contributes - $i\pi\, r_j^*(x_j)$ so that the integral becomes:

$$\int_0^\infty f(x)\log x\,dx = -\frac{i\pi}{2}\int_0^\infty f(x)\,dx + \pi i\sum_{j=1}^{m} r_j(z_j) + \frac{\pi i}{2}\sum_{j=1}^{n} r_j^*(x_j) \qquad (5.54)$$

Once again if f(x) is real then the integral of f(x) log x must be real and eq. (5.54) can be rewritten:

$$\int_0^\infty f(x)\log x\,dx = -\pi\,\mathrm{Im}\left[\sum_{j=1}^{m} r_j(z_j) + \frac{1}{2}\sum_{j=1}^{n} r_j^*(x_j)\right]$$

Example 5.25

Evaluate the following integral:

$$\int_0^\infty \frac{\log x}{x^2-4}\,dx$$

The function has two simple poles at $x = \pm 2$ and no other poles in the complex plane. Since the branch cut is defined for $-\pi/2 < \theta < 3\pi/2$, then these are described by:

$$x_1 = -2 = 2\,e^{i\pi}, \qquad x_2 = 2\,e^{i0}$$

The residues at x_1 and x_2 are:

$$r^*(2e^{i\pi}) = \left.\frac{\log z}{z-2}\right|_{z=-2} = -\frac{1}{4}(\log 2 + i\pi) = -i\frac{\pi}{4} - \frac{\log 2}{4}$$

$$r^*(2e^{0}) = \left.\frac{\log z}{z+2}\right|_{z=2} = \frac{1}{4}(\log 2)$$

Since f(x) is real the integral has the following solution:

$$\int_0^\infty f(x)\log x\,dx = -\frac{\pi}{2}\,\mathrm{Im}\left[\sum_{j=1}^{m} r_j^*(x_j)\right] = -\frac{\pi}{2}\,\mathrm{Im}\left[-i\frac{\pi}{4} - \frac{\log 2}{4} + \frac{\log 2}{4}\right] = \frac{\pi^2}{8}$$

Integrals involving $(\log x)^n$ can be obtained from integrals involving $(\log x)^k$, $k = 0, 1, 2, ..., n-1$. The following integral can be evaluated:

$$\int_0^\infty f(x)(\log x)^n\,dx$$

where n = positive integer, f(x) is an even function and has no singularities on the real axis and:

$$|f(z)| \le \frac{M}{|z|^p} \qquad p > 1, \qquad \text{for} \qquad |z| >> 1$$

Using the same contour shown in Fig. 5.27, then one can show that:

$$\left\{\int_{L_1} - \int_{C_o} + \int_{L_2} + \int_{C_R}\right\} f(z)(\log z)^n\,dz = 2\pi i \sum_{j=1}^{m} r_j(z_j)$$

where the r_j's are the residues of $\{f(z)(\log z)^n\}$ at the poles of f(z) in the upper half plane.

The integral on C_0 can be evaluated, where $z = \varepsilon\, e^{i\theta}$:

$$\left| \int_{C_0} f(z)(\log z)^n\, dz \right| = \left| \int_0^{\pi} f(\varepsilon\, e^{i\theta})(\log \varepsilon + i\theta)^n\, i\varepsilon\, e^{i\theta}\, d\theta \right|$$

$$\leq \left| f(\varepsilon\, e^{i\theta}) \right|_{C_0} \varepsilon \sum_{k=0}^{n} \frac{\pi^{n-k+1}\, n!}{(n-k+1)!\, k!} \left| \log \varepsilon \right|^k$$

Since the limit as $z \to 0$ of f(z) is finite and the limit as $\varepsilon \to 0$ of $\varepsilon\,(\log\varepsilon)^k \to 0$, then the integral on the small circle vanishes, i.e.:

$$\lim_{\varepsilon \to 0} \int_{C_0} f(z)(\log z)^n\, dz \to 0$$

The integral on C_R can also be shown to vanish when $R \to \infty$. Let $z = R\, e^{i\theta}$ on C_R:

$$\left| \int_{C_R} f(z)(\log z)^n\, dz \right| = \left| \int_0^{\pi} f(R\, e^{i\theta})(\log R + i\theta)^n\, iR\, e^{i\theta}\, d\theta \right|$$

$$\leq \frac{M}{R^{p-1}} \sum_{k=0}^{n} \frac{\pi^{n-k+1}\, n!}{(n-k+1)!\, k!} (\log R)^k$$

Since:

$$\lim_{R \to \infty} \frac{\left| \log R \right|^k}{R^{p-1}} \to 0 \qquad \text{for} \qquad p > 1$$

then the integral on C_R vanishes:

$$\lim_{R \to \infty} \int_{C_R} f(z)(\log z)^n\, dz \to 0$$

Following same integration evaluation on L_1 and L_2 one obtains:

$$\int_{L_1} f(z)(\log z)^n\, dz = \int_{\varepsilon}^{R} f(-x)(\log x + i\pi)^n\, dx = \int_{\varepsilon}^{R} f(x) \sum_{k=0}^{n} \frac{(i\pi)^{n-k}\, n!}{(n-k)!\, k!} (\log x)^k\, dx$$

and

$$\int_{L_2} f(z)(\log z)^n\, dz = \int_{\varepsilon}^{R} f(x)(\log x)^n\, dx$$

Thus, the total contour integral results in the following relationship:

$$\int_0^\infty f(x)(\log x)^n\, dx = \pi i \sum_{j=1}^{m} r_j - \frac{1}{2}\sum_{k=0}^{n-1} \frac{(i\pi)^{n-k}\, n!}{(n-k)!k!} \int_0^\infty f(x)(\log x)^k\, dx \tag{5.55}$$

Thus the integral in eq. (5.54) can be obtained as a linear combination of integrals involving $(\log x)^k$, with k = 0, 1, 2, ..., n - 1.

Example 5.26

Evaluate the following integral:

$$\int_0^\infty \frac{(\log x)^2}{x^2+4}\, dx$$

f(x) has two simple poles at +2i and –2i. Using the results of Example (5.24), and eq. (5.55):

$$\int_0^\infty \frac{(\log x)^2}{x^2+4}\, dx = i\pi\, r(2i) - \frac{1}{2}\left\{ (i\pi)^2 \int_0^\infty f(x)dx + 2i\pi \int_0^\infty f(x)\log x\, dx \right\}$$

The residue can be derived as:

$$r(2i) = \left.\frac{(\log z)^2}{2z}\right|_{2i} = \frac{(\log 2 + i\pi/2)^2}{4i} = -\frac{i}{4}\left((\log 2)^2 - \frac{\pi^2}{4}\right) + \frac{\pi}{4}\log 2$$

Using the results of Example 5.21 and eq. (5.55):

$$\int_0^\infty \frac{\log x}{x^2+4}\, dx = \frac{\pi}{4}\log 2 \qquad \text{and} \qquad \int_0^\infty \frac{1}{x^2+4}\, dx = \frac{\pi}{4}$$

so that:

$$\int_0^\infty \frac{(\log x)^2}{x^2+4}\, dx = i\pi\left\{-\frac{i}{4}\left((\log 2)^2 - \frac{\pi^2}{4}\right) + \frac{\pi}{4}\log 2\right\} - \frac{1}{2}\left\{(i\pi)^2 [\frac{\pi}{4}] + 2i\pi[\frac{\pi}{4}\log 2]\right\}$$

$$= \frac{\pi}{4}[(\log 2)^2 + \frac{\pi^2}{4}]$$

Or, since f(x) is real, the integral of f(x) must be real, therefore only the real part of the right side needs to be computed, i.e.:

$$\int_0^\infty \frac{(\log x)^2}{x^2+4}\, dx = -\pi Im[r(2i)] + \frac{\pi^2}{2}\int_0^\infty \frac{1}{x^2+4}\, dx = \frac{\pi}{4}\left((\log 2)^2 - \frac{\pi^2}{4}\right) + \frac{\pi^2}{2}\left(\frac{\pi}{4}\right)$$

$$= \frac{\pi}{4}[(\log 2)^2 + \frac{\pi^2}{4}]$$

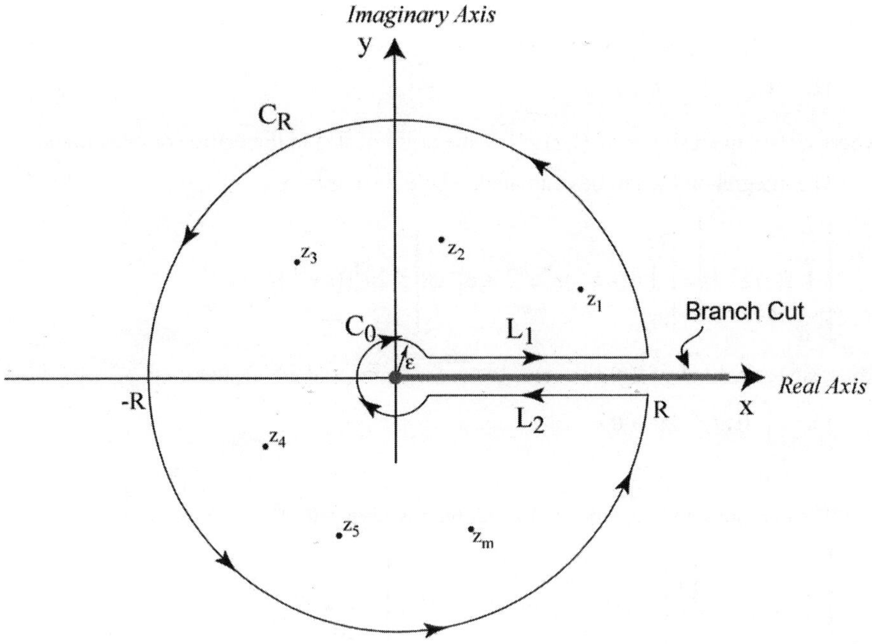

Fig 5.28: Closed Path for Integrals with x^a

5.19 Integrals of Functions Involving x^a

Integrals involving x^a, which is a multi-valued function, can be evaluated by using the residue theorem. Consider the following integral:

$$\int_0^\infty f(x)\,x^a\,dx \qquad a > -1 \qquad (5.56)$$

where f(x) has no singularities on the positive real axis, and a is a non-integer real constant. To evaluate the integral in eq. (5.56), the integrand is made single-valued by extending a branch cut along the positive real axis, as is shown in Fig. 5.28, such that the principal branch is defined in the range $0 < \theta < 2\pi$. Let the poles of f(z) be $z_1, z_2, \ldots z_m$ in the complex plane and:

$$|f(R)| \le \frac{M}{R^p} \qquad \text{with } p > a+1, \text{ for } R >> 1$$

The contour on C_1 is closed by adding a circle of radius = R, a line contour on C_2 and a circle of radius = ε, as shown in Fig. 5.28. The contour is closed as shown in such a way that it does not cross the branch cut and hence, the path integration stays in the principal Reimann sheet. Thus:

$$\left\{\int_{L_1} + \int_{C_R} + \int_{L_2} + \int_{C_o}\right\} f(z)z^a\,dz = 2\pi i \sum_{j=1}^{m} r_j(z_j)$$

where r_j's are the residues of $\{f(z)\,z^a\}$ at the poles of f(z) in the entire complex plane.

The integral on C_0 can be evaluated, where $z = \varepsilon\, e^{i\theta}$:

$$\left|\int_{C_o} f(z)\,z^a\,dz\right| = \left|\int_0^{2\pi} f(\varepsilon\, e^{i\theta})\varepsilon^a e^{ia\theta}\, i\varepsilon\, e^{i\theta}\, d\theta\right| \le 2\pi \left|f(\varepsilon\, e^{i\theta})\right|_{C_o} \varepsilon^{a+1}$$

Thus, since the limit of f(z) as z goes to zero is finite, and a > -1, then:

$$\lim_{\varepsilon\to 0} \int_{C_o} f(z)\,z^a\,dz \to 0$$

The contour on C_R can be evaluated, where $z = z = R\, e^{i\theta}$:

$$\left|\int_{C_R} f(z)\,z^a\,dz\right| = \left|\int_0^{2\pi} f(R\, e^{i\theta})\,R^a e^{ai\theta}\; iR\, e^{i\theta}\, d\theta\right| \le \frac{2\pi M}{R^{p-a-1}}$$

Thus:

$$\lim_{R\to\infty} \int_{C_R} f(z)z^a\,dz \to 0 \qquad \text{since} \qquad p > a+1$$

Since the function $z^a = \rho^a$ on L_1 and $z^a = (\rho\, e^{2i\pi})^a = \rho^a\, e^{2i\pi a}$ on L_2, then the line integrals on L_1 and L_2 become:

$$\int_{L_1} f(z)\,z^a\,dz = \int_{\varepsilon}^{R} f(x)x^a\,dx$$

and

$$\int_{L_2} f(z)\,z^a\,dz = \int_{R}^{\varepsilon} f(\rho e^{2\pi i})\rho^a\, e^{2\pi ia}\,d\rho = -\,e^{2\pi ia}\int_{\varepsilon}^{R} f(x)x^a\,dx$$

Thus, summing the two integrals results in:

$$\int_0^{\infty} f(x)\,x^a\,dx = \frac{2\pi i}{1-e^{2\pi ai}} \sum_{j=1}^{m} r_j = -\frac{\pi}{\sin(a\pi)}\, e^{-a\pi i} \sum_{j=1}^{m} r_j \tag{5.57}$$

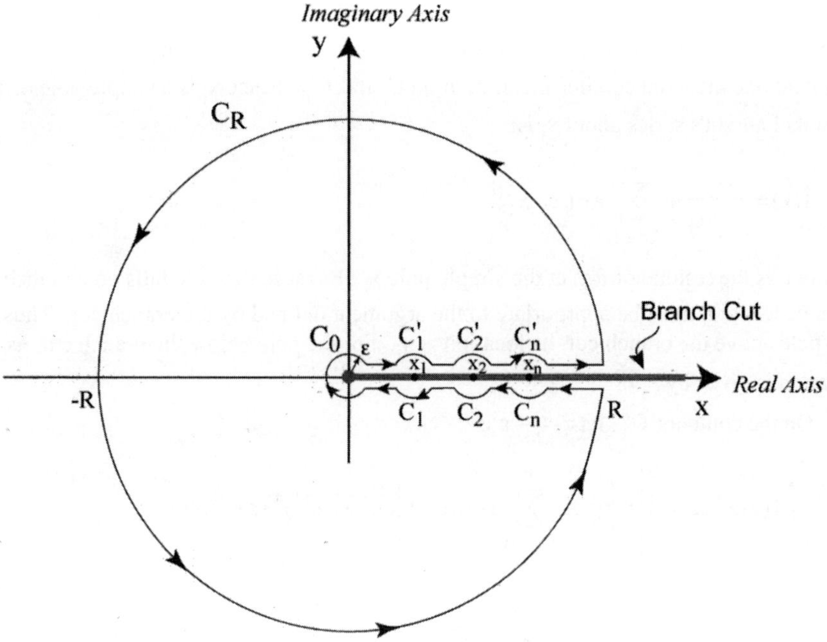

***Fig. 5.29:** Closed Path for Integrals with x^a and Real Poles*

Example 5.27

Evaluate the following integral:

$$\int_0^\infty \frac{x^{1/2}}{x^2+1}\,dx$$

Let $f(z) = 1/(z^2 + 1)$, which has two simple poles:

$$z_1 = i = e^{i\pi/2} \qquad z_2 = -i = e^{3i\pi/2}$$

where the argument was chosen appropriate to the branch cut. The residues become:

$$r_1(i) = \left.\frac{z^{1/2}}{2z}\right|_i = \frac{1}{2i}e^{i\pi/4} \qquad r_2(i) = \left.\frac{z^{1/2}}{2z}\right|_{-i} = -\frac{1}{2i}e^{3i\pi/4}$$

Since $a > -1$ and:

$$|f(z)| \sim \frac{1}{R^p} \qquad \text{as } |z| >> 1 \qquad \text{where } p = 2 > 1/2 + 1$$

then the integrals on C_R and C_0 vanish as $R \to \infty$ and $\varepsilon \to 0$ respectively. Thus:

$$\int_0^\infty \frac{x^{1/2}}{x^2+1}\,dx = \frac{2\pi i}{1-e^{\pi i}}\left[\frac{1}{2i}\left(e^{i\pi/4} - e^{3i\pi/4}\right)\right] = \pi\cos(\pi/4)$$

If f(x) has simple poles at x_j on the positive real axis then one can indent the contour on the positive real axis at each simple pole x_j, j = 1, 2, ... n, as shown on Fig. 5.29. One can treat one indented contour integration on C_j and C'_j. Since x_j is a simple pole of f(x), then its Laurent's series about x_j is:

$$f(z) = \frac{r_j}{z - x_j} + \sum_{k=0}^{\infty} a_k (z - x_j)^k$$

where r_j is the residue of f(z) at the simple pole x_j. Because the pole falls on a branch cut, then its location must be appropriate to the argument defined by the branch cut. Thus, for the pole above the branch cut, its location is x_j. For the pole below the branch cut, its location is given by $x_j\, e^{2i\pi}$.

On the contours C'_j, let $z - x_j = \varepsilon\, e^{i\theta}$:

$$\int_{C'_j} f(z)\, z^a\, dz = \int_{\pi}^{0} \left[\frac{r_j}{\varepsilon\, e^{i\theta}} + \sum_{k=0}^{\infty} a_k (\varepsilon\, e^{i\theta}) \right] (x_j + \varepsilon\, e^{i\theta})^a\, i\varepsilon\, e^{i\theta}\, d\theta$$

and

$$\lim_{\varepsilon \to 0} \int_{C'_j} f(z) z^a\, dz = -i\pi\, (x_j)^a\, r_j$$

The contour on C_j can be treated in the same manner. Let $z - x_j\, e^{2\pi i} = \varepsilon\, e^{i\theta}$ then:

$$\lim_{\varepsilon \to 0} \int_{C_j} f(z) z^a\, dz = -i\pi\, (x_j)^a\, e^{2\pi a i} r_j$$

Thus, the sum of the integrals on C_j and C'_j becomes:

$$\left\{ \int_{C'_j} + \int_{C_j} \right\} f(z) z^a\, dz = -i\pi\, r_j^* (x_j)(1 + e^{2\pi a i})$$

where the r_j^* 's are the residues of $\{f(z)\, z^a\}$ at the poles x_j on L_1. Thus:

$$\int_0^{\infty} f(x)\, x^a\, dx = \frac{2\pi i}{1 - e^{2\pi a i}} \sum_{j=1}^{m} r_j + \frac{\pi i (1 + e^{2\pi a i})}{1 - e^{2\pi a i}} \sum_{j=1}^{n} r_j^*$$

$$= \frac{-\pi}{\sin(a\pi)}\, e^{-a\pi i} \sum_{j=1}^{m} r_j + \pi \cot(\pi a) \sum_{j=1}^{n} r_j^* \qquad (5.58)$$

5.20 Integrals of Odd or Asymmetric Functions

In order to perform integrations of real functions, either the integrand is even or the integral is defined initially over the entire x-axis. Otherwise, one cannot extend the semi-infinite integral to the entire x-axis. To use the residue theorem for odd or asymmetric functions, one can use the logarithmic function to allow for the evaluation of such integrals. Consider a function f(x), a real function without poles on the positive real axis and with n poles in the complex plane behaving as:

$$|f(z)| \sim \frac{1}{|z|^p} \qquad \text{as } |z| >> 1 \quad \text{where } p > 1$$

then one can evaluate the following integral:

$$\int_0^\infty f(x)\,dx$$

by considering first the following integral:

$$\int_0^\infty f(x)\log x\,dx$$

Using the contour in Fig. 5.28, one can write the closed path integral:

$$\oint f(z)\log z\,dz = \left\{\int_{L_1} + \int_{C_R} + \int_{L_2} + \int_{C_\varepsilon}\right\} f(z)\log z\,dz = 2\pi i \sum_{j=1}^{n} r_j(z_j)$$

where r_j are the residue of [f(z) log z] at the poles z_j of f(z).

The integrals on L_1 and L_2 become:

Path L_1:

$$z = \rho \qquad dz = d\rho \qquad \log z = \log \rho$$

Path L_2

$$z = \rho\, e^{2i\pi} = \rho \qquad dz = d\rho \qquad \log z = \log \rho + 2i\pi$$

The integrals over C_R and C_ε vanish as $R \to \infty$ and $\varepsilon \to 0$, respectively. Thus, the integrals are combined to give:

$$\int_0^\infty f(\rho)\log\rho\,d\rho + \int_\infty^0 f(\rho)[\log\rho + 2i\pi]d\rho = 2\pi i \sum_{j=1}^{n} r_j$$

so that the integrals with f(x) log x cancel, leaving an integral on f(x) only:

$$\int_0^\infty f(x)dx = -\sum_{j=1}^{n} r_j \tag{5.59}$$

Example 5.28

Obtain the value of the following integral:

$$\int_0^\infty \frac{dx}{x^5+1}$$

Following the method of evaluating asymmetric integrals, and using the prescribed branch cut, one only needs to find the residues of all the poles of the integrand. The integrand has five poles:

$$z_1 = e^{i\pi/5}, \qquad z_2 = e^{i3\pi/5}, \qquad z_3 = e^{i\pi}, \qquad z_4 = e^{7i\pi/5}, \qquad z_5 = e^{9i\pi/5}$$

The choice of the argument for the simple poles is made to fall between zero and 2π, as defined by the branch cut. The residues are defined as follows:

$$r(z_j) = \left.\frac{\log z}{5z^4}\right|_{z_j} = \left.\frac{z\log z}{5z^5}\right|_{z_j} = -\frac{1}{5} z_j \log z_j$$

Therefore, the integral equals:

$$\int_0^\infty \frac{dx}{x^5+1} = \frac{i\pi}{25}\left\{e^{i\pi/5} + 3e^{3i\pi/5} + 5e^{i\pi} + 7e^{7i\pi/5} + 9e^{9i\pi/5}\right\}$$

$$= \frac{8\pi}{25}\sin(\pi/5)(1+\cos(\pi/5))$$

5.21 Integrals of Odd or Asymmetric Functions Involving log x

In Section 5.18, integrals of even functions involving log x was discussed. Let f(x) be an odd or asymmetric function with no poles on the positive real axis and n poles in the entire complex plane and:

$$|f(R)| \le \frac{1}{R^p} \qquad \text{where } p > 1 \qquad R >> 1$$

To evaluate the integral

$$\int_0^\infty f(x)\log x\, dx$$

one again must start with the following integral:

$$\int_0^\infty f(x)(\log x)^2\, dx$$

evaluated over the contour in Fig. 5.28. Thus, the closed contour integral gives:

$$\oint f(z)(\log z)^2 dz = \left\{ \int_{L_1} + \int_{C_R} + \int_{L_2} + \int_{C_\varepsilon} \right\} f(z)(\log z)^2 dz = 2\pi i \sum_{j=1}^{n} r_j(z_j)$$

where r_j are the residues of $[f(z)(\log z)^2]$ at the poles of f(z) in the entire complex plane.

On the path L_1:

$$z = \rho \qquad dz = d\rho \qquad \log z = \log \rho$$

and on the path L_2:

$$z = \rho\, e^{2i\pi} = \rho \qquad dz = d\rho \qquad \log z = \log \rho + 2i\pi$$

The integrals over C_R and C_ε vanish as $R \to \infty$ and $\varepsilon \to 0$, respectively. Thus:

$$\int_0^\infty f(\rho)(\log \rho)^2 d\rho - \int_0^\infty f(\rho)[\log \rho + 2i\pi]^2 d\rho$$

$$= -4\pi i \int_0^\infty f(\rho) \log \rho \, d\rho + 4\pi^2 \int_0^\infty f(\rho)\, d\rho = 2\pi i \sum_{j=1}^{n} r_j(z_j)$$

Rearranging these terms yields:

$$\int_0^\infty f(x) \log x \, dx = -i\pi \int_0^\infty f(x) dx - \frac{1}{2} \sum_{j=1}^{n} r_j(z_j) \tag{5.60}$$

If f(x) is not real then the integral of f(x) must be evaluated to find the value of the integral of f(x) log x. However, if f(x) is real, then the integral of f(x) must be real as well and hence the imaginary parts of the right hand side must cancel out, then eq. (5.60) can be simplified by taking the real part:

$$\int_0^\infty f(x) \log x \, dx = -\frac{1}{2} \, Re\left[\sum_{j=1}^{n} r_j(z_j) \right]$$

Example 5.29

Evaluate the following integral

$$\int_0^\infty \frac{\log x}{x^3 + 1} dx$$

Following the method of evaluating asymmetric function involving log x above, one needs to find all the poles in the entire complex plane. The function has three simple poles:

$$z_1 = e^{i\pi/3}, \qquad z_2 = e^{i\pi}, \qquad z_3 = e^{5i\pi/3}$$

The choice of the argument for the simple poles is made to fall between zero and 2π, as defined by the branch cut. The residues of $f(z)(\log z)^2$ are defined as follows:

$$r(z_j) = \left.\frac{(\log z)^2}{3z^2}\right|_{z_j} = \left.\frac{z(\log z)^2}{3z^3}\right|_{z_j} = -\frac{1}{3} z_j (\log z_j)^2$$

The sum of the residues is:

$$\sum_{j=1}^{3} r_j(z_j) = \frac{1}{3}\left(e^{i\pi/3}(\frac{\pi^2}{9}) + e^{i\pi}(\pi^2) + e^{5i\pi/3}(25\frac{\pi^2}{9})\right) = \frac{4\pi^2}{27}(1 - 3\sqrt{3}i)$$

and the integral of f(x) is:

$$\int_0^\infty \frac{dx}{x^3+1} = \frac{2\pi}{3\sqrt{3}}$$

Thus, the integral becomes:

$$\int_0^\infty \frac{\log x}{x^3+1} dx = -i\pi\left(\frac{2\pi}{3\sqrt{3}}\right) - \frac{1}{2}\left(\frac{4\pi^2}{27} - i\frac{4\pi^2}{3\sqrt{3}}\right) = -\frac{2\pi^2}{27}$$

However, since the integrand is real, then the integral can also be evaluated as:

$$\int_0^\infty \frac{\log x}{x^3+1} dx = -\frac{1}{2} Re\left\{\frac{4\pi^2}{27}(1 - 3\sqrt{3}i)\right\} = -\frac{2\pi^2}{27}$$

5.22 Inverse Laplace Transforms

More complicated contour integrations around branch points are discussed in the following examples of inverse Laplace transforms:

Example 5.30

Obtain the inverse Laplace transforms of the following function:

$$f(z) = \frac{\sqrt{z}}{z - a^2} \qquad a > 0$$

The inverse Laplace transform is defined as:

$$f(t) = \frac{1}{2\pi i}\int_{\gamma - i\infty}^{\gamma + i\infty} f(z)\, e^{zt}\, dz \qquad t > 0$$

where γ is chosen to the right of all the poles and singularities of f(z), as is shown in Fig. 5.30. Since $\sqrt{z}$ is a multi-valued function, then a branch cut is made along the negative real axis starting with the branch point $z = 0$. Note that the choice of the branch cut must be made so that it falls entirely to the left of the line $x = \gamma$. Hence it could be taken along the negative x-axis (the choice for this example) or along the positive or negative y-axis.

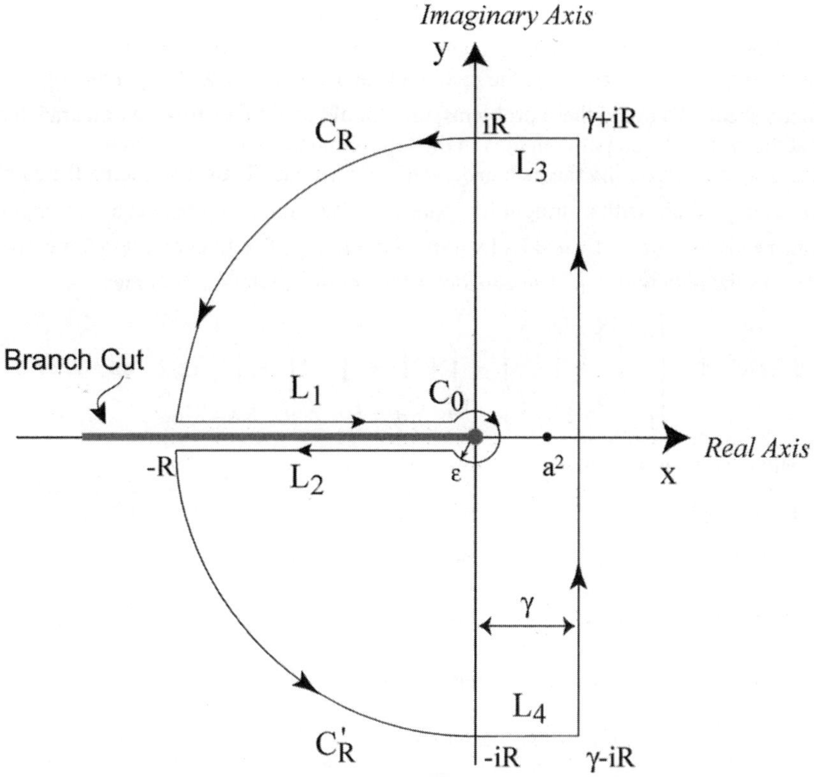

Fig. 5.30: *Closed Path for Inverse Laplace Transform*

The branch cut is thus defined:

$$z = \rho\, e^{i\phi} \qquad \rho > 0 \qquad -2n\pi - \pi < \phi < -2n\pi + \pi \qquad n = 0, 1$$

The angular range is chosen so that the $\phi = 0$ is included in the top Reimann sheet. The top Reimann sheet is defined by n = 0 so that:

$$\sqrt{z} = \rho^{1/2}\, e^{i\phi/2} \qquad \rho > 0 \qquad -\pi < \phi < \pi \qquad n = 0$$

and the bottom Reimann sheet is defined by:

$$\sqrt{z} = \rho^{1/2}\, e^{i\phi/2} \qquad \rho > 0 \qquad -3\pi < \phi < -\pi \qquad n = 1$$

The two sheets are joined at $\phi = -\pi$ as well as the π and -3π rays. This means that as ϕ increases without limit, the $\sqrt{z}$ is located in either the top or bottom Reimann sheet.

The original line path along γ - iR to γ + iR must be closed in the top Reimann sheet to allow the evaluation of the inverse transform by the use of the residue theorem. To close the contour in the top Reimann sheet, one needs to connect $\gamma \pm$ iR with straight line segments L_3 and L_4. These are then to be connected to a semi-circle of radius R to satisfy the Jordan's Lemma (Section 5.16). However, a continuous semi-circle on the third and fourth quadrants would cross the branch cut. Crossing the branch cut would result in the circular path in the second quadrant being continued in the third in the *bottom* Reimann sheet where the function $\sqrt{z}$ would have a different value. Furthermore, one has to continue the path to close it eventually with L_4 in the top Reimann sheet. To avoid these problems, one should avoid the crossing of a branch cut, so that the *entire* closed path remains in the top Reimann sheet. This can be accomplished by rerouting the path *around* the branch cut. Thus, continuing the path C_R in the third quadrant with a straight line path L_1. To continue to connect by a straight line L_2, one needs to connect L_1 and L_2 by a small circle C_0. The final quarter circular path C'_R closes the path with L_4. The equation of the closed path then becomes:

$$\oint f(z)e^{zt}\,dz = \left\{ \int_{\gamma - iR}^{\gamma + iR} + \int_{L_3} + \int_{C_R} + \int_{L_1} + \int_{C_0} + \int_{L_2} + \int_{C'_R} + \int_{L_4} \right\} f(z)\,e^{zt}\,dz = 2\pi i\, r(a^2)$$

The residue at $z = a^2$ becomes:

$$r(a^2) = a\,e^{a^2 t}$$

The integrals on C_R and C'_R vanish, since using Section 5.16.1:

$$|f(R)| \sim \frac{1}{R^p} \qquad \text{as } |R| >> 1 \quad \text{where } p = 1/2 > 0$$

The integral on C_0 vanishes, since using Section 5.16.2:

$$|f(\varepsilon)| \sim \frac{1}{\varepsilon^p} \qquad \text{as } |\varepsilon| \to 0 \quad \text{where } p = -1/2 < 1$$

The two line integrals L_3 and L_4 can be evaluated as follows:

Let $z = x \pm iR$, then:

$$\left| \int_{\gamma \pm iR}^{\pm iR} \frac{\sqrt{z}}{z - a^2} e^{zt}\,dz \right| = \left| \int_{\gamma}^{0} \frac{\sqrt{x \pm iR}}{x \pm iR - a^2} e^{xt} e^{\pm iRt}\,dz \right| \leq \frac{1}{\sqrt{R}}\gamma e^{\gamma t}$$

Thus:

$$\lim_{R\to\infty} \int_{\gamma \pm iR}^{\pm iR} \frac{\sqrt{z}}{z - a^2} e^{zt}\,dz \to 0$$

The line integral on L_1 can be evaluated, where $z = \rho\, e^{i\pi}$, as follows:

$$\int_{R}^{\varepsilon} \frac{\sqrt{\rho}\, e^{i\pi/2}}{\rho e^{i\pi} - a^2} e^{\rho t e^{i\pi}} e^{i\pi}\,d\rho = -i \int_{\varepsilon}^{R} \frac{\sqrt{\rho}}{\rho + a^2} e^{-\rho t}\,d\rho$$

The line integral on L_2 can be evaluated, where $z = \rho\, e^{-i\pi}$, as follows:

$$\int_{\varepsilon}^{R} \frac{\sqrt{\rho}\, e^{-i\pi/2}}{\rho e^{-i\pi} - a^2} e^{\rho t e^{-i\pi}} e^{i\pi}\,d\rho = -i \int_{\varepsilon}^{R} \frac{\sqrt{\rho}}{\rho + a^2} e^{-\rho t}\,d\rho$$

Thus:

$$\int_{\gamma-iR}^{\gamma+iR} \frac{\sqrt{z}}{z-a^2} e^{zt}\, dz - 2i \int_{\varepsilon}^{R} \frac{\sqrt{\rho}}{\rho+a^2} e^{-\rho t}\, d\rho = 2\pi i\, a\, e^{a^2 t}$$

Therefore f(t) becomes:

$$f(t) = a\,e^{a^2 t} + \frac{1}{\pi} \int_0^{\infty} \frac{\sqrt{\rho}\, e^{-\rho t}}{\rho + a^2}\, d\rho$$

Letting $u^2 = \rho t$, the integral transforms to:

$$f(t) = a\,e^{a^2 t} + \frac{2}{\pi\sqrt{t}} \int_0^{\infty} \frac{u^2 e^{-u^2}}{u^2 + a^2 t}\, du = a\,e^{a^2 t} + \frac{2}{\pi\sqrt{t}} \left\{ \int_0^{\infty} e^{-u^2}\, du - a^2 t \int_0^{\infty} \frac{e^{-u^2}}{u^2 + a^2 t}\, du \right\}$$

$$= a\,e^{a^2 t} + \frac{1}{\sqrt{\pi t}} - \frac{2a^2\sqrt{t}}{\pi} \int_0^{\infty} \frac{e^{-u^2}}{u^2 + a^2 t}\, du$$

which can be written in the form of an error function (see eq. B 5.22):

$$f(t) = \frac{1}{\sqrt{\pi t}} + a e^{a^2 t} \operatorname{erf}(a\sqrt{t})$$

Example 5.31

Obtain the inverse Laplace Transform for the following function:

$$f(z) = \frac{1}{\sqrt{z^2 - a^2}}$$

The function f(z) has two singularities which happen to be branch points. The integral:

$$\int_{\gamma-i\infty}^{\gamma+i\infty} \frac{e^{zt}}{\sqrt{z^2 - a^2}}\, dz$$

can be evaluated by closing the contour of integration and using the residue theorem. Two branch cuts must be made at the branch points z = a and -a to make the function $\sqrt{z^2 - a^2}$ single-valued.

One has the freedom to make each of the functions $\sqrt{z-a}$ and $\sqrt{z+a}$ single-valued by a branch cut from z = a and z = -a, respectively, in a straight line in any direction in such a way that both must fall entirely to the left of the line $x = \gamma$. As was mentioned earlier in Section (5.2(i)), it is sometimes advantageous to run branch cuts for a function linearly so that they overlap, as this *may* result in the function becoming single-valued over the overlapping segment. Thus, the cuts are chosen to extend from z = a and z = -a to $-\infty$ on the real axis, as shown in Fig. 5.31(a). The two branch cuts for the top Reimann sheet of each function can be described as follows:

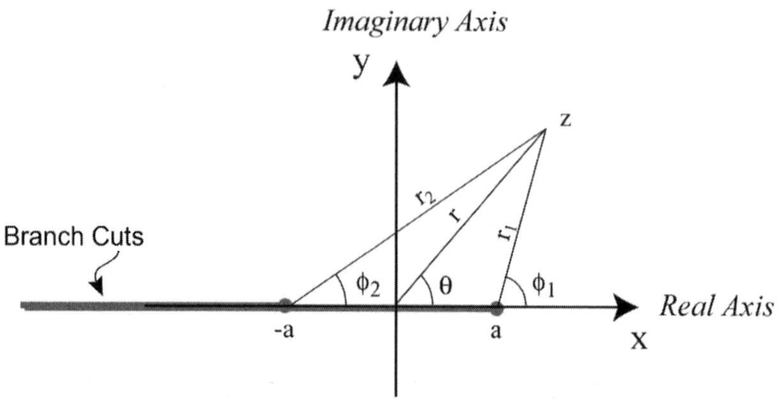

(a)

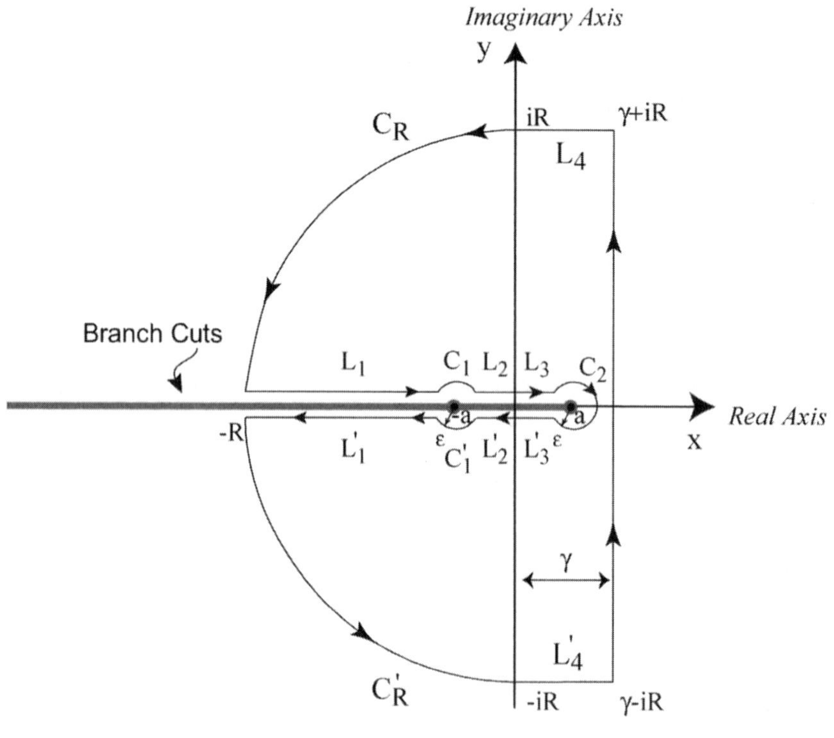

(b)

Fig. 5.31: *(a) Branch Cuts and (b) Integration Contour for Example 5.31*

$$z - a = r_1 e^{i\phi_1}, \qquad -\pi < \phi_1 < \pi, \qquad r_1 > 0$$

$$z + a = r_2 e^{i\phi_2}, \qquad -\pi < \phi_2 < \pi, \qquad r_2 > 0$$

The single-valued function z is described by:

$$z = r e^{i\theta}, \qquad r > 0$$

The closure of the original path from γ - iR to γ + iR in the top Reimann sheet would require first the joining of two straight line segments $\gamma \pm iR$ to $\pm$ iR, see Figure 5.31 (b). Two quarter circular paths, C_R and C'_R are required to avoid crossing both branch cuts. To continue the path closure in the top Reimann sheet, one has to encircle both branch cuts. This takes the form of two straight-line paths above and below the branch cuts from C_R and C'_R to the branch point at z = a. Since the straight line paths cross a singular (branch) point at z = -a, then one must avoid that point by encircling it by two small semi-circular paths C_1 and C'_1. Similarly, the joining of the straight line paths at the branch point z = a requires the joining of the two by a small circle C_2. The line segments between z = -a and z = a is split into two parts, namely, L_2 and L_3 and L'_2 and L'_3. This is done purely to simplify the integrations along these two parts of each line segment, as z on L_3 becomes –z along L_2. The equation of the closed path becomes:

$$\oint f(z)e^{zt}\,dz = \left\{ \int_{\gamma - iR}^{\gamma + iR} + \int_{\gamma + iR}^{iR} + \int_{C_R} + \int_{\substack{-R \\ \text{on } L_1}}^{-a-\varepsilon} + \int_{C_1} + \int_{\substack{-a-\varepsilon \\ \text{on } L_2}}^{0} \right\} f(z)\,e^{zt}\,dz$$

$$+ \left\{ \int_{\substack{0 \\ \text{on} L_3}}^{a-\varepsilon} + \int_{C_2} + \int_{\substack{a-\varepsilon \\ \text{on } L'_3}}^{0} + \int_{\substack{0 \\ \text{on } L'_2}}^{-a+\varepsilon} + \int_{C'_1} + \int_{\substack{-a-\varepsilon \\ \text{on} L'_1}}^{-R} + \int_{C'_R} \right\} f(z)\,e^{zt}\,dz = 0$$

The integrals on C_R and C'_R vanish as R → ∞, since (Section 5.16.1):

$$|f(R)| \sim \frac{1}{R^p} \qquad \text{as } R >> 1 \qquad \text{where } p = 1 > 0$$

The integrals on [$\gamma \pm$ iR to $\pm$ iR] vanish since:

$$\left| \int_{\gamma} \frac{e^{(x \pm iR)t}}{\sqrt{(x \pm iR)^2 - a^2}}\,dx \right| \le \frac{\gamma}{R} e^{\gamma t} \to 0 \text{ as } R \to \infty$$

The integrals on C_1, C_2 and C'_1 vanish, since (Section 5.16.2):

$$\lim_{z \to \pm a} \frac{z \mp a}{\sqrt{z^2 - a^2}} \to 0$$

To facilitate accounting of the integrand of these multi-valued functions, one can evaluate the integrand term by term in tabular form. Thus, the remaining integrals can be evaluated in tabular form (see accompanying table):

Table for Example 5.31

Line	z		z	dz	ezt	z+a		$\sqrt{z+a}$	z-a		$\sqrt{z-a}$	$\sqrt{z^2-a^2}$	Limits
	r	θ				r2	ϕ2		r1	ϕ1			
L1	r	π	-r	-dr	e-rt	r-a	π	$\sqrt{r-a}\,e^{i\pi/2}$	r+a	π	$\sqrt{r+a}\,e^{i\pi/2}$	$-\sqrt{r^2-a^2}$	(∞, a)
L_1'	r	-π	-r	-dr	e-rt	r-a	-π	$\sqrt{r-a}\,e^{-i\pi/2}$	r+a	-π	$\sqrt{r+a}\,e^{-i\pi/2}$	$-\sqrt{r^2-a^2}$	(a, ∞)
L2	r	π	-r	-dr	e-rt	a-r	0	$\sqrt{a-r}$	r+a	π	$\sqrt{r+a}\,e^{i\pi/2}$	$i\sqrt{a^2-r^2}$	(a, 0)
L_2'	r	-π	-r	-dr	e-rt	a-r	0	$\sqrt{a-r}$	r+a	-π	$\sqrt{r+a}\,e^{-i\pi/2}$	$-i\sqrt{a^2-r^2}$	(0, a)
L3	r	0	r	dr	ert	a+r	0	$\sqrt{a+r}$	a-r	π	$\sqrt{a-r}\,e^{i\pi/2}$	$i\sqrt{a^2-r^2}$	(0, a)
L_3'	r	0	r	dr	ert	a+r	0	$\sqrt{a+r}$	a-r	-π	$\sqrt{a-r}\,e^{-i\pi/2}$	$-i\sqrt{a^2-r^2}$	(a, 0)

The sum of branch cut integrals $L_1 + L_1'$ vanishes. This reinforces the stipulation that running overlapping branch cuts may make the function single-valued over the overlapping section. The sum of the integrals over the branch cut integrals $L_2 + L_2'$, and $L_3 + L_3'$ become:

$$\int_{L_2+L_2'} = -2i\int_0^a \frac{e^{rt}\,dr}{\sqrt{a^2 - r^2}} \qquad \int_{L_3+L_3'} = -2i\int_0^a \frac{e^{-rt}\,dr}{\sqrt{a^2 - r^2}}$$

The final result for the inverse Laplace transform:

$$f(t) = \frac{1}{2i\pi}\int_{\gamma - i\infty}^{\gamma + i\infty} \frac{e^{zt}\,dz}{\sqrt{z^2 - a^2}} = \frac{-1}{2i\pi}\int_{L_2+L_2'+L_3+L_3'}$$

$$= \frac{1}{\pi}\left\{\int_0^a \frac{e^{rt}\,dr}{\sqrt{a^2 - r^2}} + \int_0^a \frac{e^{-rt}\,dr}{\sqrt{a^2 - r^2}}\right\} = \frac{1}{\pi}\int_{-a}^{a} \frac{e^{-rt}\,dr}{\sqrt{a^2 - r^2}} = I_0(at)$$

where I_0 (at) is the Modified Bessel Function of the first kind and order zero.

Example 5.32

Obtain the inverse Laplace transform of the following function:

$$f(z) = \log\left(\frac{z^2 - a^2}{z^2}\right)$$

The inverse Laplace transform is defined as:

$$f(t) = \frac{1}{2i\pi}\int_{\gamma - i\infty}^{\gamma + i\infty} \log\left(\frac{z^2 - a^2}{z^2}\right) e^{zt}\,dz$$

$$= \frac{1}{2i\pi}\int_{\gamma - i\infty}^{\gamma + i\infty} [\log(z-a) + \log(z+a) - 2\log z] e^{zt}\,dz$$

The integrand is multi-valued, thus colinear branch cuts starting from the branch points at +a, 0, -a to -∞ must be made to make the logarithmic functions single-valued, as shown in Fig. 5.32(a). The three branch cuts are defined for the top Reimann sheet of each of the three logarithmic functions are:

$$z - a = r_1\, e^{i\phi_1}, \qquad -\pi < \phi_1 < \pi$$

$$z = r_2\, e^{i\phi_2}, \qquad -\pi < \phi_2 < \pi$$

$$z + a = r_3\, e^{i\phi_3}, \qquad -\pi < \phi_3 < \pi$$

The single-valued function z is defined as:

$$z = r\, e^{i\theta}, \qquad r > 0$$

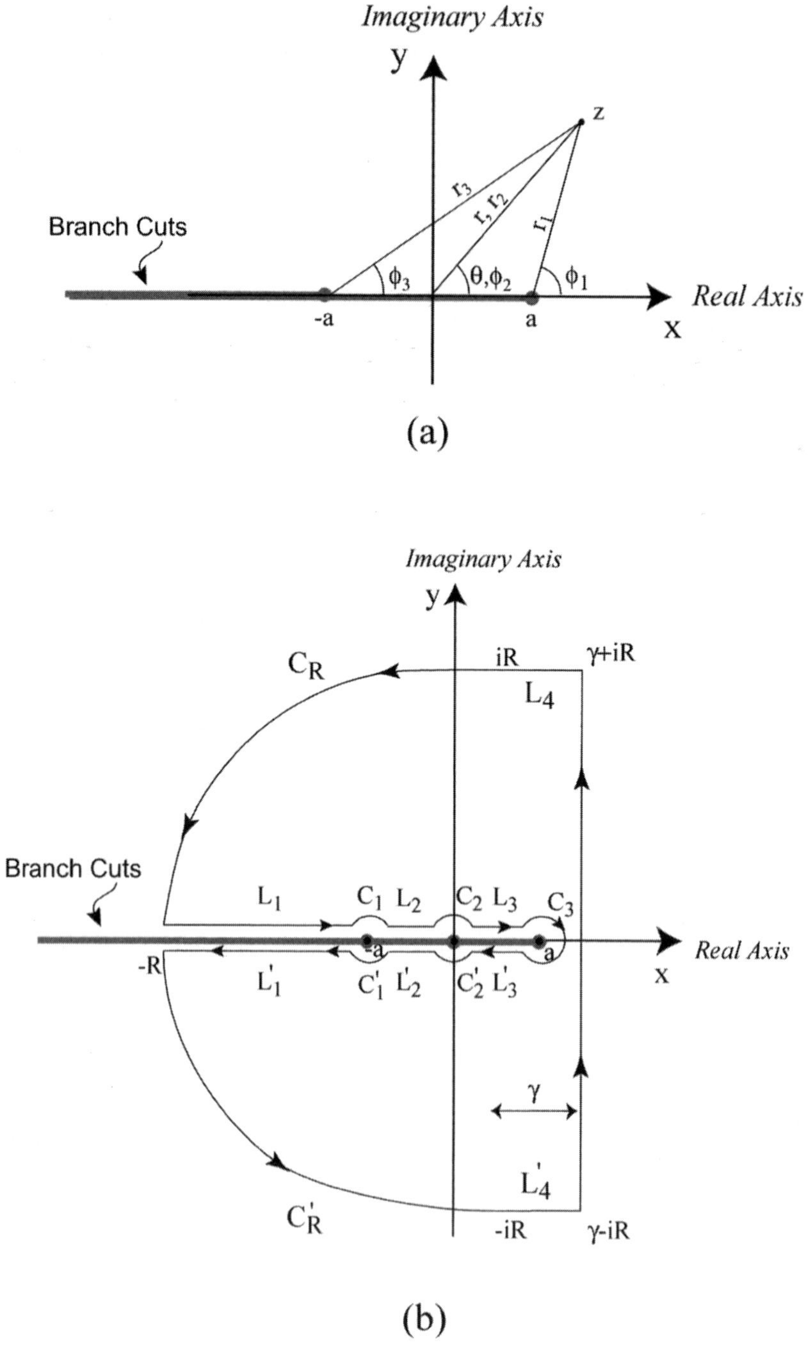

Fig. 5.32: *(a) Branch Cuts and (b) Integration Contour for Example 5.32*

Again, the branch cuts are chosen to be colinear and overlapping extending from x = a, 0 and –a to -∞.

The contour is closed on the complex plane as shown in Fig. 5.32(b). The contour is wrapped around the three branch cuts, with small circular paths near each branch point in such a way to leave the entire path in the top Reimann sheet of all three logarithmic functions. Thus, since there are no poles in the complex plane, the closed path integral is:

$$\oint f(z)e^{zt}\,dz = \left\{ \int_{\gamma - iR}^{\gamma + iR} + \int_{\gamma + iR}^{iR} + \int_{C_R} + \int_{\substack{-R \\ \text{on } L_1}}^{-a-\varepsilon} + \int_{C_1} + \int_{\substack{-a+\varepsilon \\ \text{on } L_2}}^{-\varepsilon} \right\} f(z)\,e^{zt}\,dz$$

$$+ \left\{ \int_{C_2} + \int_{\substack{\varepsilon \\ \text{on } L_3}}^{a-\varepsilon} + \int_{C_3} + \int_{\substack{a-\varepsilon \\ \text{on } L_3'}}^{0} + \int_{C_2'} + \int_{\substack{-\varepsilon \\ \text{on } L_2'}}^{-a+\varepsilon} \right\} f(z)\,e^{zt}\,dz$$

$$+ \left\{ \int_{C_1'} + \int_{\substack{-a-\varepsilon \\ \text{on } L_1'}}^{-R} + \int_{C_R'} + \int_{-iR}^{\gamma - iR} \right\} f(z)\,e^{zt}\,dz = 0$$

The integrals on C_R and C_R' vanish since:

$$\lim_{R\to\infty} \left| \log\left(\frac{R^2 - a^2}{R^2}\right) \right| = \lim_{R\to\infty} \left| \log\left(1 - \frac{a^2}{R^2}\right) \right| \cong \frac{a^2}{R^p} \qquad \text{where } p = 2 > 0$$

The integrals on C_1, C_2, C_3, C_2' and C_3' vanish, since:

$$\lim_{z\to\pm a} \left[(z \mp a)\log\left(\frac{z^2 - a^2}{z^2}\right) \right] \to 0$$

and

$$\lim_{z\to 0} \left[z\log\left(\frac{z^2 - a^2}{z^2}\right) \right] \to 0$$

The integrals on [$\gamma \pm iR$ to $\pm iR$] vanish since, on the line paths:

$$\lim_{R\to\infty} |f(z)| \to \frac{a^2}{R^2} \to 0 \text{ as } R \to \infty$$

The line integrals can be evaluated in tabular form where:

$$f(z) = \log\left(\frac{z^2 - a^2}{z^2}\right) = \log(z - a) + \log(z + a) - 2\log(z)$$

Table for Example 5.32

Line	z		z	dz	ezt	z-a		log(z-a)	z-0		log(z-0)	z+a		log(z+a)	$\log\left(\frac{z^2-a^2}{z^2}\right)$	Limits
	r	θ				r1	$\phi1$		r2	$\phi2$		r3	$\phi3$			
L1	r	π	-r	-dr	e-rt	r+a	π	log(r+a) + $i\pi$	r	π	log r + $i\pi$	r-a	π	log(r-a) + $i\pi$	$\log\left(\frac{r^2-a^2}{r^2}\right)$	(∞, a)
L_1'	r	$-\pi$	-r	-dr	e-rt	r+a	$-\pi$	log(r+a) - $i\pi$	r	$-\pi$	log r - $i\pi$	r-a	$-\pi$	log(r-a) - $i\pi$	$\log\left(\frac{r^2-a^2}{r^2}\right)$	(a, ∞)
L2	r	π	-r	-dr	e-rt	r+a	π	log(r+a) + $i\pi$	r	π	log r + $i\pi$	a-r	0	log(a-r)	$\log\left(\frac{a^2-r^2}{r^2}\right) - i\pi$	(a, 0)
L_2'	r	$-\pi$	-r	-dr	e-rt	r+a	$-\pi$	log(r+a) - $i\pi$	r	$-\pi$	log r - $i\pi$	a-r	0	log(a-r)	$\log\left(\frac{a^2-r^2}{r^2}\right) + i\pi$	(0, a)
L3	r	0	r	dr	ert	a-r	π	log(a-r) + $i\pi$	r	0	log r	a+r	0	log(a+r)	$\log\left(\frac{a^2-r^2}{r^2}\right) + i\pi$	(0, a)
L_3'	r	0	r	dr	ert	a-r	$-\pi$	log(a-r) - $i\pi$	r	0	log r	a+r	0	log(a+r)	$\log\left(\frac{a^2-r^2}{r^2}\right) - i\pi$	(a, 0)

The branch cut integrals L_1 and L'_1 add up to zero. This means that the function becomes single-valued on the overlapped portion of the branch cut integrals. The remaining branch cut integrals give:

$$\int_{L_2+L'_2} = -2i\pi \int_0^a e^{-rt}\,dr \qquad \int_{L_3+L'_3} = 2i\pi \int_0^a e^{rt}\,dr$$

where all the logarithmic parts of the integrands cancel out. Finally, summing the six branch cut integrals with the original integral gives:

$$\int_{\gamma-i\infty}^{\gamma+i\infty} f(z)\,dz - 2i\pi \int_0^a e^{-rt}\,dr + 2i\pi \int_0^a e^{rt}\,dr = 0$$

and

$$f(t) = \frac{1}{2\pi i} \int_{\gamma-i\infty}^{\gamma+i\infty} \log\left(\frac{z^2-a^2}{z^2}\right) e^{zt}dz = -2\int_0^a \sinh(r\,t)\,dr = \frac{2}{t}[1-\cosh(at)]$$

PROBLEMS

Section 5.1

1. Verify that:

(a) $\frac{1+i}{1-i} - \frac{1-i}{1+i} = 2i$ (b) $(i-1)^4 = -4$

(c) $\frac{5(1+i)^3}{(2+i)(1+2i)} = 2(1+i)$ (d) $(1+i)^2 + (1-i)^2 = 0$

2. Verify that the two complex numbers $1 \pm i$ satisfy the equation:

$z^2 - 2z + 2 = 0$

3. Prove that a complex number is equal to the conjugate of its conjugate.

4. Show that:

(a) $\overline{z+5i} = \bar{z} - 5i$ (b) $\overline{iz} = -i\bar{z}$

(c) $\overline{\left(\frac{z}{i}\right)} = i\bar{z}$ (d) $\overline{(1+i)(1+2i)} = -1-3i$

5. Use the polar form to show that:

(a) $i(1+2i)(2+i) = -5$ (b) $\frac{1+i}{1-i} = i$

(c) $(1+i)^4 = -4$ (d) $\frac{i}{-1-i} = \frac{-1-i}{2}$

6. Show that all the roots of:

(a) $(-1)^{1/4}$ are $(2)^{-1/2}(\pm 1 \pm i)$ (b) $(8i)^{1/3}$ are $-2i, \pm\sqrt{3} + i$

(c) $(i)^{1/2}$ are $\pm\frac{1+i}{\sqrt{2}}$ (d) $\left(\frac{i\sqrt{3}-1}{2}\right)^{3/2}$ are ± 1

7. Describe geometrically the region specified below:

(a) $Re(z) > 0$ (b) $|Im(z)| < 3$

(c) $|z-1| \le 1$ (d) $1 < |z-2| < 2$

(e) $|z| > 2 \quad -\pi \le \arg z \le 0$ (f) $|z-2| \le Re(z)$

(g) $|z-1| > |z|$ (h) $|z+1-i| > 1 \quad 0 \le \arg(z+1-i) \le \pi/2$

Section 5.2

8. Apply the definition of the derivative to find the derivative of:

(a) $\frac{1}{z}$ (b) $\frac{z+1}{z+2}$

(c) $z^2 (1 + z)$ (d) $(z^2 + 1)^4$

9. Show that the following functions are nowhere differentiable:

(a) *Im* (z) (b) $\bar{z}$

(c) $|z + 1|^2$ (d) $z\bar{z}$

(e) $\frac{\overline{z+1}}{\overline{z-1}} = i\bar{z}$ (f) $\frac{z}{z+\bar{z}}$

10. Test for analyticity the following functions by use of Cauchy-Reimann conditions:

(a) $\frac{z+1}{z^2+1}$ (b) *Re* (z)

(c) $\bar{z}$ (d) $Re\left(\frac{z}{z+1}\right)$

(e) $z - \bar{z}$ (f) $z^2 + 2$

11. Show that u is harmonic and find the conjugate v, where:

(a) $u = e^x \cos y$ (b) $u = x^3 - 3xy^2$

(c) $u = \cosh x \cos y$ (d) $u = \log (x^2 + y^2),\ x^2 + y^2 \neq 0$

(e) $u = \cos x \cosh y$ (f) $u = x + \frac{x}{x^2+y^2},\ x^2 + y^2 \neq 0$

Section 5.3

12. Prove the identities given in eq. (5.10).

13. Show that if *Im* (z) > 1, then $|e^{iz}| < 1$.

14. Prove the identities given in eq. (5.15).

15. Show that $\overline{f(z)} = f(\bar{z})$ where f(z) is:

(a) exp z (b) sin z

(c) cos z (d) cosh z

16. Find all the roots of:

(a) $\cos z = 1$ (b) $\sin z = 2$

(c) $\sin z = \cosh \alpha$, α = real constant (d) $\sinh z = -i$

(e) $e^z = -2$ (f) $\log z = \pi i$

Section 5.4

17. Evaluate the following integrals:

(a) $\int_1^i (z-1)dz$ on a straight line from 1 to i.

(b) $\int_0^{1+i} (z-1)dz$ on a parabola $y = x^2$.

(c) $\int_0^{1+i} 3(x^2 + iy)\,dz$ on the paths $y = x$ and $y = x^3$.

(d) $\int_C \frac{z+2}{2z}\,dz$ where C is a circle, $|z| = 2$ in the positive direction.

(e) $\int_C \sin z\,dz$ where C is a rectangle, with corners: $(\pi/2, -\pi/2, \pi/2+i, -\pi/2+i)$

Section 5.5

18. Determine the region of analyticity of the following functions and show that:

$$\int_C f(z)dz = 0$$

where the closed contour C is the circle $|z| = 2$.

(a) $f(z) = \frac{z^2}{z-4}$ (b) $f(z) = z\,e^z$

(c) $f(z) = \frac{1}{z^2 - 8i}$ (d) $f(z) = \tan(z/2)$

(e) $f(z) = \frac{\sin z}{z}$ (f) $f(z) = \frac{\cos z}{z+3}$

19. Evaluate the following integrals:

(a) $\int_{0}^{i/2} \sin(2z)\, dz$ (b) $\int_{1-i}^{1+i} (z^2 + 1)\, dz$

(c) $\int_{0}^{3+i} z^2\, dz$ (d) $\int_{0}^{1+i} z^2\, dz$

(e) $\int_{0}^{\pi i} \cosh z\, dz$ (f) $\int_{-i}^{i} e^z\, dz$

20. Use Cauchy's Integral formula to evaluate the following integrals on the closed contour C in the positive sense:

(a) $\int_C \frac{z^3 + 3z + 2}{z}\, dz,$ C is a unit circle $|z| = 1$.

(b) $\int_C \frac{\cos z}{z}\, dz,$ C is a unit circle $|z| = 1$.

(c) $\int_C \frac{\cos z}{(z-\pi)^2}\, dz,$ C is a circle $|z| = 4$.

(d) $\int_C \frac{\sin z}{(z-\pi)^2}\, dz,$ C is a circle $|z| = 4$.

(e) $\int_C \left[\frac{1}{z-1} + \frac{3}{z+2}\right] dz$ C is a circle $|z| = 3$.

(f) $\int_C \frac{dz}{z^4 - 1},$ C is a circle $|z| = 3$.

(g) $\int_C \frac{e^z + 1}{z - i\pi/2}\, dz,$ C is a circle $|z| = 2$.

(h) $\int_C \frac{\tan z}{z^2}\, dz,$ C is a unit circle $|z| = 1$.

Section 5.8

21. Obtain Taylor's series expansion of the following functions about the specified point z_0 and give the region of convergence:

(a) $\cos z, \quad z_0 = 0$ (b) $\frac{\sin z}{z}, \quad z_0 = 0$

(c) $\frac{1}{(z+1)^2}, z_0 = 0$ (d) $\frac{e^z - 1}{z}, \quad z_0 = 0$

(e) $\frac{1}{z}, \quad z_0 = 2$ (f) $\frac{z}{z-2}, \quad z_0 = 1$

(g) $\frac{1}{z^2}, z_0 = -1$ (h) $\frac{z-1}{z+1}, \quad z_0 = 1$

(i) $e^z, \quad z_0 = 2$ (h) $e^z, \quad z_0 = i\pi$

22. Prove L'Hospital's rule:

(a) If $p(z_0) = q(z_0) = 0$, $p'(z_0) \neq 0$, and $q'(z_0) \neq 0$, then:

$$\lim_{z \to z_0} \frac{p(z)}{q(z)} = \frac{p'(z_0)}{q'(z_0)}$$

(a) If $p(z_0) = q(z_0) = 0$, $p'(z_0) = q'(z_0) = 0$, $p''(z_0) \neq 0$, and $q''(z_0) \neq 0$, then:

$$\lim_{z \to z_0} \frac{p(z)}{q(z)} = \frac{p''(z_0)}{q''(z_0)}$$

Section 5.9

23. Obtain the Laurent's series expansion of the following functions about the specified point z_0, convergent in the specified region:

(a) $\frac{e^z}{z^3}, z_0 = 0 \quad |z| > 0$ (b) $e^{1/z}, z_0 = 0 \quad |z| > 0$

(c) $\frac{1}{(z-1)(z-2)}, z_0 = 0, 1 < |z| < 2$ (d) $\frac{1}{(z-1)(z-2)}, z_0 = 1, |z - 1| > 1$

(e) $\frac{1}{(z-1)(z-2)}, z_0 = 1, 0 < |z-1| < 1$ (f) $\frac{1}{(z^2+1)(z+2)}, z_0 = 0, 1 < |z| < 2$

(g) $\frac{1}{z(z-1)}, z_0 = 1, |z - 1| > 1$ (h) $\frac{1}{z(z-1)}, z_0 = -1, 1 < |z + 1| < 2$

Section 5.10

24. Locate and classify all of the singularities of the following functions:

(a) $\tan z$ (b) $\frac{\sin z}{z^2}$ (c) $\frac{e^z}{z^2 + \pi^2}$

(d) $\dfrac{z}{\sin z}$ (e) $\dfrac{z^3-4}{(z^2+1)^2}$ (f) $\dfrac{z^2-4}{z^5-z^3}$

(g) $\dfrac{z+2}{z^2(z-2)}$ (h) $\dfrac{1}{\sin z - z}$ (i) $\dfrac{1}{e^z-1}$ (j) $\dfrac{z^3}{(z+1)^3}$

Section 5.11

25. Find the residue of the function in Problem 5.24 at all the singularities of each function.

Section 5.12

26. Evaluate the following integrals, where n is an integer, and $|a| < 1$.

(a) $\displaystyle\int_0^{2\pi} \frac{\sin(n\theta)}{1+2a\cos\theta+a^2}\,d\theta$ (b) $\displaystyle\int_0^{2\pi} \frac{d\theta}{1-2a\sin\theta+a^2}$

(c) $\displaystyle\int_0^{2\pi} \frac{\cos^3\theta}{1-2a\cos\theta+a^2}\,d\theta$ (d) $\displaystyle\int_0^{\pi} \frac{d\theta}{(1+a\cos\theta)^2}$

(e) $\displaystyle\int_0^{\pi} (\sin\theta)^{2n}\,d\theta$ (f) $\displaystyle\int_0^{2\pi} \frac{\cos(n\theta)}{1+2a\cos\theta+a^2}\,d\theta$

(g) $\displaystyle\int_0^{\pi} \frac{(\sin\theta)^2}{1+a\cos\theta}\,d\theta$ (h) $\displaystyle\int_0^{\pi} (\cos\theta)^{2n}\,d\theta$

(i) $\displaystyle\int_0^{\pi} \frac{1+\cos\theta}{1+\cos^2\theta}\,d\theta$ (j) $\displaystyle\int_0^{2\pi} \frac{\cos(n\theta)}{\cosh a+\cos\theta}\,d\theta$

(k) $\displaystyle\int_0^{\pi} \frac{d\theta}{1+a\cos\theta}$ (l) $\displaystyle\int_0^{2\pi} \frac{d\theta}{1+a\sin\theta}$

(m) $\displaystyle\int_0^{2\pi} \frac{d\theta}{(1+a\sin\theta)^2}$ (n) $\displaystyle\int_0^{\pi} \frac{\cos(2\theta)}{1-2a\cos\theta+a^2}\,d\theta$

Section 5.13

27. Evaluate the following integrals, with $a > 0$, and $b > 0$, unless otherwise stated:

(a) $\displaystyle\int_{-\infty}^{\infty} \frac{dx}{x^2 + ax + b}$ $a^2 < 4b$, a and b real (b) $\displaystyle\int_0^{\infty} \frac{x^2}{(x^2 + a^2)^2}\, dx$

(c) $\displaystyle\int_0^{\infty} \frac{dx}{(x^2 + a^2)^2}$ (d) $\displaystyle\int_{-\infty}^{\infty} \frac{dx}{(x^2 + a^2)^2 (x^2 + b^2)}$

(e) $\displaystyle\int_0^{\infty} \frac{dx}{x^4 + a^4}$ (f) $\displaystyle\int_0^{\infty} \frac{dx}{(x^2 + a^2)^3}$

(g) $\displaystyle\int_0^{\infty} \frac{x^4}{(x^4 + a^4)^2}\, dx$ (h) $\displaystyle\int_0^{\infty} \frac{x^2}{x^6 + a^6}\, dx$

(i) $\displaystyle\int_0^{\infty} \frac{x^2}{x^4 + a^4}\, dx$ (j) $\displaystyle\int_0^{\infty} \frac{x^4}{x^6 + a^6}\, dx$

(k) $\displaystyle\int_0^{\infty} \frac{x^2\, dx}{(x^2 + a^2)^3}$ (l) $\displaystyle\int_0^{\infty} \frac{dx}{(x^2 + a^2)(x^2 + b^2)}$

(m) $\displaystyle\int_0^{\infty} \frac{x^6\, dx}{(x^4 + a^4)^2}$ (n) $\displaystyle\int_0^{\infty} \frac{x^2\, dx}{(x^2 + a^2)(x^2 + b^2)}$

Section 5.14

28. Evaluate the following integrals, where $a > 0$, $b > 0$, $c > 0$, and $b \neq c$:

(a) $\displaystyle\int_0^{\infty} \frac{\cos(ax)}{x^2 + b^2}\, dx$ (b) $\displaystyle\int_0^{\infty} \frac{x \sin(ax)}{x^4 + 4b^4}\, dx$

(c) $\displaystyle\int_0^{\infty} \frac{x^2 \cos(ax)\, dx}{(x^2 + c^2)(x^2 + b^2)}$ (d) $\displaystyle\int_0^{\infty} \frac{\cos x}{(x + b)^2 + a^2}\, dx$

(e) $\displaystyle\int_0^{\infty} \frac{\cos(ax)\, dx}{(x^2 + b^2)^2}$ (f) $\displaystyle\int_0^{\infty} \frac{\cos(ax)\, dx}{(x^2 + c^2)(x^2 + b^2)}$

(g) $\displaystyle\int_0^{\infty} \frac{x \sin(ax)\, dx}{(x^2 + c^2)(x^2 + b^2)}$ (h) $\displaystyle\int_0^{\infty} \frac{x \sin(ax)\, dx}{(x^2 + b^2)^2}$

(i) $\displaystyle\int_0^\infty \frac{x\sin(ax)\,dx}{x^2+b^2}$ (j) $\displaystyle\int_0^\infty \frac{x^3\sin(ax)dx}{(x^2+b^2)^2}$

(k) $\displaystyle\int_0^\infty \frac{x^2\cos(ax)}{x^4+4b^4}dx$ (h) $\displaystyle\int_0^\infty \frac{x\sin(ax)\,dx}{(x^2+b^2)^3}$

Section 5.15

29. Evaluate the following integrals, where $a > 0$, $b > 0$, $c > 0$, and $b \neq c$:

(a) $\displaystyle\int_{-\infty}^\infty \frac{\sin(ax)}{x+b}dx$ (b) $\displaystyle\int_{-\infty}^\infty \frac{\sin x}{x}dx$

(c) $\displaystyle\int_0^\infty \frac{\cos x}{4x^2-\pi^2}dx$ (d) $\displaystyle\int_0^\infty \frac{\sin(ax)}{x(x^2+b^2)^2}dx$

(e) $\displaystyle\int_0^\infty \frac{x^4}{x^6-1}dx$ (f) $\displaystyle\int_{-\infty}^\infty \frac{dx}{x(x^2-4x+5)}$

(g) $\displaystyle\int_0^\infty \frac{\sin x}{x(\pi^2-x^2)}dx$ (h) $\displaystyle\int_{-\infty}^\infty \frac{dx}{(x-1)(x^2+1)}$

(i) $\displaystyle\int_{-\infty}^\infty \frac{dx}{x^3-1}$ (j) $\displaystyle\int_0^\infty \frac{\cos(ax)}{x^2-b^2}dx$

(k) $\displaystyle\int_0^\infty \frac{\cos(ax)}{x^4-b^4}dx$ (l) $\displaystyle\int_0^\infty \frac{x\sin(ax)}{x^2-b^2}dx$

(m) $\displaystyle\int_0^\infty \frac{x^2\cos(ax)}{x^4-b^4}dx$ (n) $\displaystyle\int_0^\infty \frac{\cos(ax)}{(x^2-b^2)(x^2-c^2)}dx$

(o) $\displaystyle\int_0^\infty \frac{\sin(ax)}{x(x^4+4b^4)}dx$ (p) $\displaystyle\int_0^\infty \frac{\sin(ax)}{x(x^4-b^4)}dx$

(q) $\displaystyle\int_0^\infty \frac{x^2\cos(ax)}{(x^2-b^2)(x^2-c^2)}dx$ (r) $\displaystyle\int_0^\infty \frac{\cos(ax)}{(x^2+b^2)(x^4-b^4)}dx$

(s) $\displaystyle\int_0^\infty \frac{x^3 \sin(ax)}{(x^2 - b^2)(x^2 - c^2)}\, dx$ (t) $\displaystyle\int_0^\infty \frac{x^3 \sin(ax)}{x^4 - b^4}\, dx$

Section 5.16

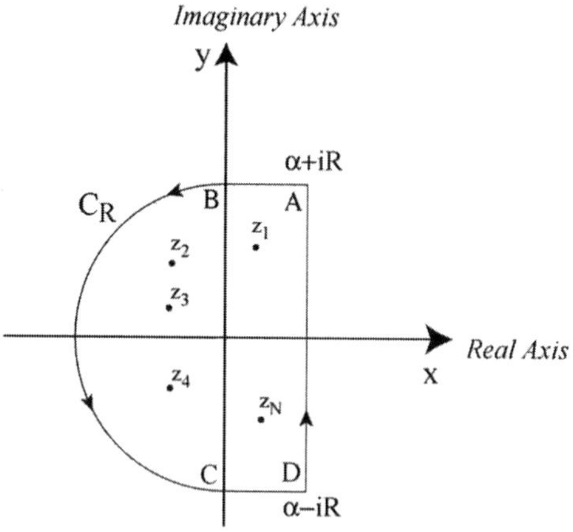

30. The inverse of the Laplace transform is defined as:

$$F(t) = \frac{1}{2\pi i} \int_{\alpha - i\infty}^{\alpha + i\infty} f(p) e^{pt}\, dp$$

where p is a complex variable, α is chosen such that all the poles of f(p) fall to the left of the line $p = \alpha$ as shown in the accompanying figure and:

$$|f(p)| \le \frac{M}{|p|^q} \quad \text{when } |p| >> 1 \text{ and } q > 0$$

Show that one can evaluate the integral by closing the contour shown in the accompanying figure with $R \to \infty$, such that:

$$F(t) = \sum_{j=1}^{N} r_j$$

where r_j's are the residues of the function $\{f(p)\, e^{pt}\}$ at the poles of f(p). Show that the integrals on AB and CD vanish as $R \to \infty$.

31. Obtain the inverse Laplace transforms F(t), defined in Problem 5.30, for the following functions f(p):

(a) $\dfrac{1}{p}$ (b) $\dfrac{1}{(p+a)(p+b)}$ $a \neq b$ (c) $\dfrac{1}{p^2 + a^2}$

(d) $\dfrac{p}{p^2 + a^2}$ (e) $\dfrac{p^2 - a^2}{(p^2 + a^2)^2}$ (f) $\dfrac{1}{(p+b)^2 + a^2}$

(g) $\dfrac{1}{p^{n+1}}$ n = integer > 0

(h) $\dfrac{p}{p^2 - a^2}$

(i) $\dfrac{a^2}{p(p^2 + a^2)}$

(j) $\dfrac{2a^3}{(p^2 + a^2)^2}$

(k) $\dfrac{2ap^2}{(p^2 + a^2)^2}$

(l) $\dfrac{2a^3}{p^4 - a^4}$

(m) $\dfrac{2a^2 p}{p^4 - a^4}$

32. The inverse Fourier Cosine transform is defined as:

$$F(t) = \sqrt{\frac{2}{\pi}} \int_0^\infty f(\omega) \cos(\omega x) d\omega$$

Find the inverse Fourier Transform of the following functions f(ω):

(a) $\sqrt{\dfrac{2}{\pi}} \dfrac{\sin(\omega x)}{\omega}$

(b) $\sqrt{\dfrac{2}{\pi}} \dfrac{1}{\omega^2 + a^2}$

(c) $\dfrac{1}{\omega^4 + 1}$

(d) $\sqrt{\dfrac{2}{\pi}} \dfrac{1}{(\omega^2 + a^2)^2}$

Section 5.17

33. Show that:

$$\int_{-\infty}^{\infty} \frac{e^{ax}}{\cosh x} dx = \frac{\pi}{\cos(\frac{a\pi}{2})} \qquad |a| < 1$$

Use a contour connecting the points –R, R, R + πi, -R + πi and –R, where R → ∞.

34. Show that:

$$\int_{-\infty}^{\infty} \frac{x}{\sinh x - i} dx = \pi$$

Use a contour connecting the points –R, R, R + πi, -R + πi and –R, where R → ∞.

35. Show that:

$$\int_{-\infty}^{\infty} e^{-x^2} \cos(ax) dx = \sqrt{\pi}\, e^{-a^2/4}$$

Use a contour connecting the points –R, R, R + ai/2, -R + ai/2 and –R, where R → ∞. The following integral is needed in the solution:

$$\int_{-\infty}^{\infty} e^{-x^2} dx = \sqrt{\pi}$$

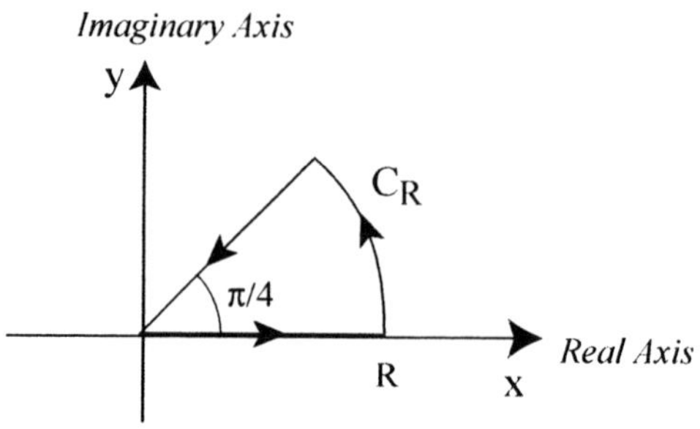

36. Show that::

$$\int_0^\infty \cos(ax^2)\,dx = \frac{\sqrt{\pi}}{2\sqrt{2a}}$$

close the contour by a ray, $z = \rho\, e^{i\pi/4}$, and a circular sector, $z = R\, e^{i\theta}$, $R \to \infty$, and 0 $\le \theta \le \pi/4$, see accompanying figure.

37. Show that::

$$\int_{-\infty}^{\infty} \frac{\sinh(ax)}{\sinh x}\,dx = \frac{1}{\pi}\tan\left(\frac{a\pi}{2}\right) \qquad |a| < 1$$

Use a contour connecting the points $-R$, R, $R + \pi i$, $-R + \pi i$ and $-R$, where $R \to \infty$.

Section 5.18

38. Evaluate the following integrals, where n is an integer, with $a > 0$, $b > 0$ and $a \neq b$:

(a) $\displaystyle\int_0^\infty \frac{(\log x)^2}{(x^2+1)^2}\,dx$

(b) $\displaystyle\int_0^\infty \frac{(\log x)^3}{x^2+1}\,dx$

(c) $\displaystyle\int_0^\infty \frac{(\log x)^4}{x^2+1}\,dx$

(d) $\displaystyle\int_0^\infty \frac{(\log x)^2}{x^4+1}\,dx$

(e) $\displaystyle\int_0^\infty \frac{\log x}{x^4+1}\,dx$

(f) $\displaystyle\int_0^\infty \frac{\log x}{(x^2+1)^2}\,dx$

(g) $\displaystyle\int_0^\infty \frac{(1-x^2)}{(x^2+1)^2}\log x\,dx$

(h) $\displaystyle\int_0^\infty \frac{\log x}{(x^2+a^2)(x^2+b^2)}\,dx$

(i) $\displaystyle\int_0^\infty \frac{\log x}{(x^2+1)^4}\,dx$

(j) $\displaystyle\int_0^\infty \frac{(\log x)^2}{x^{2n}+1}\,dx$

(k) $\displaystyle\int_0^\infty \frac{\log x}{x^{2n}+1}\,dx$ (l) $\displaystyle\int_0^\infty \frac{x^2 \log x}{(x^2+a^2)(x^2+1)}\,dx$

(m) $\displaystyle\int_0^\infty \frac{\log x}{(x^2-a^2)}\,dx$ (n) $\displaystyle\int_0^\infty \frac{(1+x^2)}{(x^2-1)^2}\log x\,dx$

Section 5.19

39. Evaluate the following integrals, where "a" is a real constant, $b>0$, $c>0$ and $b \neq c$:

(a) $\displaystyle\int_0^\infty \frac{x^a}{(x^2+1)^2}\,dx$ $-1<a<3$ (b) $\displaystyle\int_0^\infty \frac{x^a}{x+1}\,dx$ $-1<a<0$

(c) $\displaystyle\int_0^\infty \frac{x^a}{x^2+x+1}\,dx$ $|a|<1$ (d) $\displaystyle\int_0^\infty \frac{x^a}{(x+b)(x+c)}\,dx$ $|a|<1$

(e) $\displaystyle\int_0^\infty \frac{x^a}{x^2+2x\cos b+1}\,dx$ $|a|<1$, $|b|<\pi$ (f) $\displaystyle\int_0^\infty \frac{x^a}{x-1}\,dx$ $-1<a<0$

(g) $\displaystyle\int_0^\infty \frac{x^a}{(x+b)^2}\,dx$ $|a|<1$ (h) $\displaystyle\int_0^\infty \frac{x^a}{(x+b)^n}\,dx$ $-1<a<n-1$

(i) $\displaystyle\int_0^\infty \frac{x^{3/2}}{(x+b^2)^2(x+c^2)^2}\,dx$ $|a|<1$ (j) $\displaystyle\int_0^\infty \frac{x^a}{(x+b)(x-c)}\,dx$ $|a|<1$

(k) $\displaystyle\int_0^\infty \frac{x^a}{(x-b)(x-c)}\,dx$ $|a|<1$ (l) $\displaystyle\int_0^\infty \frac{x^a}{(x^3+1)^2}\,dx$ $-1<a<5$

Section 5.20

40. Obtain the value of the following integrals with $a>0$:

(a) $\displaystyle\int_0^\infty \frac{x}{x^3+a^3}\,dx$ (b) $\displaystyle\int_0^\infty \frac{x^3}{x^5+a^5}\,dx$

(c) $\displaystyle\int_0^\infty \frac{x}{x^4+a^4}\,dx$ (d) $\displaystyle\int_0^\infty \frac{x^2}{x^5+a^5}\,dx$

(e) $\displaystyle\int_0^\infty \frac{x}{x^3 - a^3}\,dx$

(f) $\displaystyle\int_0^\infty \frac{x^3}{x^5 - a^5}\,dx$

(g) $\displaystyle\int_0^\infty \frac{x}{x^5 - a^5}\,dx$

(h) $\displaystyle\int_0^\infty \frac{x}{(x^2 + a^2)^2}\,dx$

(i) $\displaystyle\int_0^\infty \frac{x^2}{(x^3 + a^3)^2}\,dx$

Section 5.21

41. Evaluate the following integrals, with $a > 0$, $b > 0$ and $a \neq b$:

(a) $\displaystyle\int_0^\infty \frac{\log x}{(x+a)(x+b)}\,dx$

(b) $\displaystyle\int_0^\infty \frac{\log x}{(x+a)^2}\,dx$

(c) $\displaystyle\int_0^\infty \frac{\log x}{x^3 + 1}\,dx$

(d) $\displaystyle\int_0^\infty \frac{x \log x}{x^3 + 1}\,dx$

(e) $\displaystyle\int_0^\infty \frac{\log x}{(x+a)(x-1)}\,dx$

(f) $\displaystyle\int_0^\infty \frac{(\log x)^2}{(x+a)(x-1)}\,dx$

(g) $\displaystyle\int_0^\infty \frac{(\log x)^2}{(x-1)^2}\,dx$

(h) $\displaystyle\int_0^\infty \frac{\log x}{x^3 - 1}\,dx$

(i) $\displaystyle\int_0^\infty \frac{x \log x}{x^3 - 1}\,dx$

(j) $\displaystyle\int_0^\infty \frac{x^3 \log x}{x^6 - 1}\,dx$

Section 5.22

42. Obtain the inverse Laplace transform f(t) from the following function F(z) (see definition of f(t) in Problem 30), $a > 0$, $b > 0$, $c > 0$ and $a \neq b$:

(a) $\displaystyle\frac{\sqrt{z}}{z(z - a^2)}$

(b) $\displaystyle\frac{1}{\sqrt{z} + a}$

(c) $\sqrt{z-a} - \sqrt{z-b}$

(d) $\displaystyle\frac{1}{\sqrt{z+a}}$

(e) $\dfrac{1}{z(\sqrt{z}+a)}$ (f) $\dfrac{1}{(z+a)^b}$

(g) $\dfrac{e^{-az}}{z+b}$ (h) $\dfrac{e^{a^2/z}}{\sqrt{z}}$

(i) $e^{\sqrt{az}}$ (j) $\log\left(\dfrac{z+b}{z+a}\right)$

(k) $\log\left(\dfrac{z^2+b^2}{z^2+a^2}\right)$ (l) $\log\left(\dfrac{(z+a)^2+c^2}{(z+b)^2+c^2}\right)$

(m) $\dfrac{z}{(z+b)^{3/2}}$ (n) $\dfrac{1}{\sqrt{z}}$

(o) $\dfrac{1}{\sqrt{z^2+a^2}}$ (p) $\dfrac{\sqrt{z+\sqrt{z^2+a^2}}}{\sqrt{z^2+a^2}}$

(q) $\dfrac{\sqrt{z+\sqrt{z^2-a^2}}}{\sqrt{z^2-a^2}}$ (r) $\dfrac{e^{-\sqrt{az}}}{z}$

(s) $\dfrac{(\log z)^2}{z}$ (t) $z\log\left(\dfrac{\sqrt{z^2+a^2}}{z}\right)$

(u) $\dfrac{e^{-a\sqrt{z^2+b^2}}}{\sqrt{z^2+b^2}}$ (v) $\dfrac{1}{(z-a^2)(\sqrt{z}+b)}$

(w) $\dfrac{1}{\sqrt{(z+2a)(z+2b)}}$ (x) $\dfrac{1}{\sqrt{(z+a)(z+b)^3}}$

(y) $(z^2+a^2)^{-\nu-1/2}$ $\nu > -1/2$ (z) $\dfrac{\left(\sqrt{z^2+a^2}-z\right)^\nu}{\sqrt{z^2+a^2}}$ $\nu > -1$

(aa) $\dfrac{\sqrt{z}}{z(\sqrt{z}+a)}$ (bb) $\dfrac{1}{(z+a)\sqrt{z-b}}$ $b > a$

(cc) $\dfrac{a}{\sqrt{z}(\sqrt{z}+b)(z-a^2)}$ (dd) $\dfrac{\sqrt{z+2a}-\sqrt{z}}{\sqrt{z}}$

(ee) $\log\left(\dfrac{z^2+a^2}{z^2}\right)$ (ff) $e^{-\sqrt{az}}$

(gg) $\dfrac{e^{-\sqrt{az}}}{\sqrt{z}}$ (hh) arctan (a/z)

(e) $\dfrac{1}{z(\sqrt{z}+a)}$

(f) $\dfrac{1}{(z+a)^b}$

(g) $\dfrac{e^{-az}}{z+b}$

(h) $\dfrac{e^{a^2/z}}{\sqrt{z}}$

(i) $e^{\sqrt{az}}$

(j) $\log\left(\dfrac{z+b}{z+a}\right)$

(k) $\log\left(\dfrac{z^2+b^2}{z^2+a^2}\right)$

(l) $\log\left(\dfrac{(z+a)^2+c^2}{(z+b)^2+c^2}\right)$

(m) $\dfrac{z}{(z+b)^{3/2}}$

(n) $\dfrac{1}{\sqrt{z}}$

(o) $\dfrac{1}{\sqrt{z^2+a^2}}$

(p) $\dfrac{\sqrt{z+\sqrt{z^2+a^2}}}{\sqrt{z^2+a^2}}$

(q) $\dfrac{\sqrt{z+\sqrt{z^2-a^2}}}{\sqrt{z^2-a^2}}$

(r) $\dfrac{e^{-\sqrt{az}}}{z}$

(s) $\dfrac{(\log z)^2}{z}$

(t) $z\log\left(\dfrac{\sqrt{z^2+a^2}}{z}\right)$

(u) $\dfrac{e^{-a\sqrt{z^2+b^2}}}{\sqrt{z^2+b^2}}$

(v) $\dfrac{1}{(z-a^2)(\sqrt{z}+b)}$

(w) $\dfrac{1}{\sqrt{(z+2a)(z+2b)}}$

(x) $\dfrac{1}{\sqrt{(z+a)(z+b)^3}}$

(y) $(z^2+a^2)^{-\nu-1/2}$ $\nu > -1/2$

(z) $\dfrac{\left(\sqrt{z^2+a^2}-z\right)^{\nu}}{\sqrt{z^2+a^2}}$ $\nu > -1$

(aa) $\dfrac{\sqrt{z}}{z(\sqrt{z}+a)}$

(bb) $\dfrac{1}{(z+a)\sqrt{z-b}}$ $b > a$

(cc) $\dfrac{a}{\sqrt{z}(\sqrt{z}+b)(z-a^2)}$

(dd) $\dfrac{\sqrt{z+2a}-\sqrt{z}}{\sqrt{z}}$

(ee) $\log\left(\dfrac{z^2+a^2}{z^2}\right)$

(ff) $e^{-\sqrt{az}}$

(gg) $\dfrac{e^{-\sqrt{az}}}{\sqrt{z}}$

(hh) $\arctan(a/z)$

6

Partial Differential Equations of Mathematical Physics

6.1 Introduction

This chapter deals with the derivation, presentation, and methods of solution of partial differential equations of the various fields in mathematical Physics and Engineering. The types of equations treated in this chapter includes: Laplace, Poisson, diffusion, wave, vibration, and Helmholtz. The method of separation of variables will be used throughout this chapter to obtain solutions to boundary value problems, steady-state solutions, as well as transient solutions.

6.2 The Diffusion Equation

6.2.1 Heat Conduction in Solids

Heat flow in solids is governed by the following laws:

(a) Heat is a form of energy, and

(b) Heat flows from bodies with higher temperatures to bodies with lower temperatures.

Consider a volume, V, with surface, S, and surface normal, $\vec{n}$, as in Fig. 6.1. For such a volume, the heat content can be defined as follows:

$$h = cmT^*$$

where c is the specific heat coefficient, m is the mass of V, and T* is the average temperature of V defined by $T^* = \frac{1}{m}\int_V T\rho\, dV$, where ρ is the mass density.

Define q such that:

$$q = \text{negative rate of change of heat flow} = -\frac{\partial h}{\partial t} = -cm\frac{\partial T^*}{\partial t}$$

Since the flow of heat across a boundary is proportional to the temperature differential across that boundary, we know that:

$$dq = -k\frac{\partial T}{\partial n}dS$$

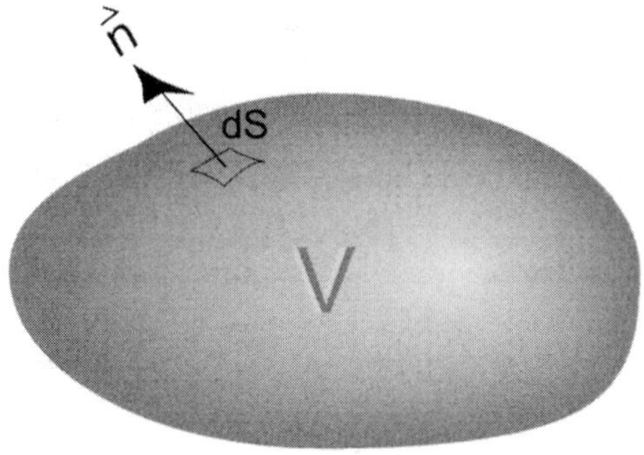

Fig. 6.1 *Normal and Volume for a Body*

where n is the spatial distance along $\vec{n}$, ($\vec{n}$ being the outward normal vector to the surface S, positive in the direction away from V), dS is a surface element and k is the thermal conductivity.

The partial differential equation that governs the conduction of heat in solids can be obtained by applying the above mentioned laws to an element dV, as shown in Fig. 6.2.

Let the rectangular parallelopiped (Fig. 6.2) have one of its vortices at point (x,y,z), whose sides are aligned with x, y, and z axes and whose sides have lengths dx, dy, and dz, respectively. Consider heat flow across the two sides perpendicular to the x-axis, whose surface area is (dy dz):

side at x: $\vec{n} = -\vec{e}_x$ and

$$q_x = -k(dy\,dz)\left.\frac{\partial T}{\partial(-x)}\right|_x$$

side at x + dx: $\vec{n} = \vec{e}_x$ and

$$q_{x+dx} = -k(dy\,dz)\left.\frac{\partial T}{\partial x}\right|_{x+dx}$$

Expanding $\left.\frac{\partial T}{\partial x}\right|_{x+dx}$ in a Taylor's series about x, results:

$$q_{x+dx} = -k(dy\,dz)\left[\left.\frac{\partial T}{\partial x}\right|_x + \left.\frac{\partial^2 T}{\partial x^2}\right|_x (dx) + \frac{1}{2}\left.\frac{\partial^3 T}{\partial x^3}\right|_x (dx)^2 + \ldots\right]$$

Thus, the total heat flux across the two opposite sides of the element at x and x + dx becomes:

$$(dq_x)_{tot} = -k(dx\,dy\,dz)\left[\frac{\partial^2 T}{\partial x^2} + \frac{1}{2}\frac{\partial^3 T}{\partial x^3}(dx) + \ldots\right]$$

Similarly, the total heat flux across the remaining two pairs of sides of the element becomes:

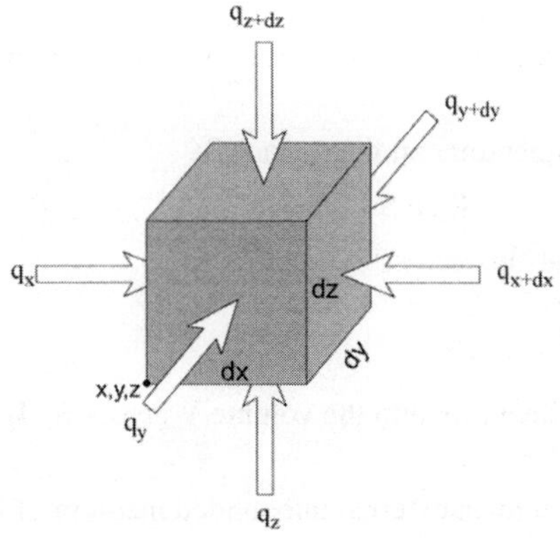

Fig. 6.2 *Heat Flow for an Element*

$$(dq_y)_{tot} = -k(dx\,dy\,dz)\left[\frac{\partial^2 T}{\partial y^2} + \frac{1}{2}\frac{\partial^3 T}{\partial y^3}(dy) + \ldots\right]$$

and

$$(dq_z)_{tot} = -k(dx\,dy\,dz)\left[\frac{\partial^2 T}{\partial z^2} + \frac{1}{2}\frac{\partial^3 T}{\partial z^3}(dz) + \ldots\right]$$

Thus, the total heat flux into the element, to a first-order approximation, becomes:

$$dq = -k(dx\,dy\,dz)\left[\frac{\partial^2 T}{\partial x^2} + \frac{\partial^2 T}{\partial y^2} + \frac{\partial^2 T}{\partial z^2}\right]$$

The time rate of change of heat content of the element becomes:

$$dq = -c\,(\rho\,dx\,dy\,dz)\frac{\partial T}{\partial t}$$

If heat is being generated inside the element at the rate of Q(x,y,z,t) per unit volume then the equation that governs heat flow in solids becomes:

$$\rho c\frac{\partial T}{\partial t} = k\left(\frac{\partial^2 T}{\partial x^2} + \frac{\partial^2 T}{\partial y^2} + \frac{\partial^2 T}{\partial z^2}\right) + Q(x,y,z,t)$$

and the temperature at any point P = P(x,y,z) obeys the equation:

$$\nabla^2 T = \frac{1}{K}\frac{\partial T}{\partial t} - q(P,t) \qquad P \text{ in } V \qquad t > 0 \qquad (6.1)$$

where the material conductivity, K, is defined as $K = k/\rho c$, q is the rate of heat generated divided by thermal conductivity k per unit volume, $q = Q/k$, and the Laplacian operator ∇^2 is defined as:

$$\nabla^2 = \frac{\partial^2}{\partial x^2} + \frac{\partial^2}{\partial y^2} + \frac{\partial^2}{\partial z^2}$$

The sign of q indicates a heat source if positive, a heat sink if negative.

The boundary condition that is required for a unique solution can be one of the following types:

(a) Prescribe the temperature on the surface S:

$$T(P,t) = g(P,t) \qquad \text{P on S}$$

(b) Prescribe the heat flux across the surface S:

$$-k\frac{\partial T}{\partial n}(P,t) = l(P,t) \qquad \text{P on S}$$

where l is the prescribed heat flux into the volume V across S. If $l = 0$, the surface is thermally insulated.

(c) Heat convection into an external unbounded medium of known temperature:

If the temperature in the exterior unbounded region of the body is known and equal to T_O, one may make use of Newton's law of cooling:

$$-k\frac{\partial T}{\partial n}(P,t) = r\,[T(P,t) - T_o(P,t)] \qquad \text{P on S}$$

where r is a constant, which relates the rate of heat convection across S to the temperature differential.

Thus, the boundary condition becomes:

$$\frac{\partial T}{\partial n}(P,t) + b\,T(P,t) = b\,T_o(P,t) \qquad \text{P on S} \qquad \text{where} \qquad b = \frac{r}{k}$$

The type of initial condition that is required for uniqueness takes the following form:

$$T(P, 0^+) = f(P) \qquad \text{P in V}$$

6.2.2 Diffusion of Gases

The process of diffusion of one gas into another is described by the following equation:

$$\nabla^2 C = \frac{1}{D}\frac{\partial C}{\partial t} - q \tag{6.2}$$

where C represents the concentration of the diffusing gas in the ambient gas, D represents the diffusion constant and q represents the additional source of the gas being diffused. If the diffusion process involves the diffusion of an unstable gas, whose decomposition is proportional to the concentration of the gas (equivalent to having sinks of the diffusing gas) then the process is defined by the following differential equation:

$$\nabla^2 C - \alpha C = \frac{1}{D}\frac{\partial C}{\partial t} - q \qquad \alpha > 0 \tag{6.3}$$

where α represents the rate of decomposition of the diffusing gas.

6.2.3 Diffusion and Absorption of Particles

The process of diffusion of electrons in a gas or neutrons in matter can be described as a diffusion process with absorption of the particles by matter proportional to their concentration in matter, a process equivalent to having sinks of the diffusing material in matter. This process is described by the following differential equation:

$$\nabla^2 \rho - \alpha \rho = \frac{1}{D}\frac{\partial \rho}{\partial t} - q \tag{6.4}$$

where:

$\rho = \rho(x,y,z,t)$ = Density of the diffusing particles

α = Mean rate of absorption of particles, $\alpha > 0$

q = Source of particles created (by fission or radioactivity) per unit volume per unit time

D = Diffusion coefficient = $(v_a \lambda_a)/3$, where v_a is the average velocity and λ_a is the mean free path of the particles.

If the process of diffusion is associated with a process of creation of more particles in proportion to the concentration of the particles in matter, the process of chain reaction, eq. (6.4) becomes:

$$\nabla^2 \rho + \alpha \rho = \frac{1}{D}\frac{\partial \rho}{\partial t} - q \qquad \alpha > 0 \tag{6.5}$$

6.3 The Vibration Equation

6.3.1 The Vibration of One-Dimensional Continua

The vibration of homogeneous, non-uniform cross-sectional one-dimensional continua, such as stretched strings, bars, torsional rods, transmission lines, and acoustic horns were adequately covered in Chapter 4. All of these equations have the following form:

$$\frac{\partial}{\partial x}\left(A(x)\frac{\partial y}{\partial x}\right) = \frac{1}{c^2}A(x)\frac{\partial^2 y}{\partial t^2} - \frac{q(x,t)}{ER} \qquad a \le x \le b, \qquad t > 0 \tag{6.6}$$

where y(x,t) is the deformation, c is the characteristic wave speed in the medium, q(x,t) is the external loading per unit length, ER is the elastic restoring modulus, and A(x) is the cross-sectional area of the medium.

The boundary conditions, required for uniqueness, take one of the following forms:

(a) $y = 0$ at a or b

(b) $\frac{\partial y}{\partial x} = 0$ at a or b

(c) $\frac{\partial y}{\partial x} \mp \alpha y = 0$ (-) for a, (+) for b $\alpha > 0$

The initial conditions, required for uniqueness, take the following form:

$$y(x,0^+) = f(x) \qquad \text{and} \qquad \frac{\partial y}{\partial t}(x,0^+) = g(x)$$

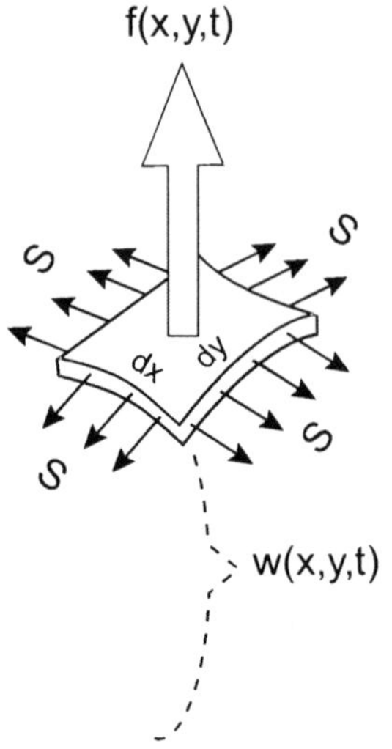

Fig 6.3 *Element of a Stretched Membrane*

The transverse vibration of uniform beams, covered in Section 4.4, is described by differential equation of fourth order in the space coordinate x, as follows:

$$\frac{\partial^2}{\partial x^2}\left(E\,I(x)\frac{\partial^2 y}{\partial x^2}\right) + \rho\,A(x)\frac{\partial^2 y}{\partial t^2} = q(x,t) \qquad a \le x \le b, \qquad t > 0$$

The boundary conditions for a beam were covered in Section 4.4.

6.3.2 The Vibration of Stretched Membranes

Consider a stretched planar membrane whose area A is surrounded by a boundary contour C. The membrane is stretched by in-plane forces S per unit length, acted on by normal forces f(x,y,t) per unit area, and has a density ρ per unit area. Consider an element of the membrane, shown in Fig. 6.3, deformed to a position w(x,y,t) from the equilibrium position. Assuming small slopes, then one can obtain the sum of forces acting on the element, in a manner similar to stretched strings, which equals the inertial forces, as follows:

$$dF \cong (S\,dy)\frac{\partial^2 w}{\partial x^2}dx + (S\,dx)\frac{\partial^2 w}{\partial y^2}dy + f(dx\,dy)$$

$$= S(dx\,dy)\left(\frac{\partial^2 w}{\partial x^2} + \frac{\partial^2 w}{\partial y^2}\right) + f(x,y,t)\,dx\,dy = (\rho\,dx\,dy)\frac{\partial^2 w}{\partial t^2}$$

Thus, the forced vibration of a membrane is described by the following equation:

$$\frac{\partial^2 w}{\partial x^2} + \frac{\partial^2 w}{\partial y^2} = \frac{1}{c^2}\frac{\partial^2 w}{\partial t^2} - \frac{f(P,t)}{S} \qquad P \text{ in } A, \qquad t > 0 \tag{6.7}$$

where $c = \sqrt{S/\rho}$ is the wave speed in the membrane and P = P(x,y).

For uniqueness, the boundary conditions along the contour C, can be one of the following types:

(a) Fixed Boundary: $w(P,t) = 0 \qquad P \text{ on } C, \qquad t > 0$

(b) Free Boundary: $\frac{\partial w}{\partial n}(P,t) = 0 \qquad P \text{ on } C, \qquad t > 0$

(c) Elastically Supported Boundary: $\left.\frac{\partial w}{\partial n} + \frac{\gamma}{S} w\right|_{P,t} = 0 \qquad P \text{ on } C, \qquad t > 0$

where γ is the elastic constant per unit length of the boundary.

For uniqueness, the initial conditions must be prescribed in the following form:

$w(P, 0^+) = f(P) \qquad P \text{ in } A$

and

$\frac{\partial w}{\partial t}(P, 0^+) = g(P) \qquad P \text{ in } A$

6.3.3 The Vibration of Plates

The vibration of uniform plates, occupying an area A, surrounded by a contour boundary C can be analyzed in a similar manner to the vibration of beams, (Fig. 6.4). Let h be the thickness of the plate, ρ be the mass density of the plate material, E be the Young's modulus and ν be the Poisson's ratio. The moments per unit length M_x, M_y, the twisting moment per unit length M_{xy}, and the shear forces per unit length, V_x, and V_y, acting on an element of the plate are shown in Fig. 6.4.

Summing moments and forces on the element (dx dy), the equilibrium equations of the plate are:

$$\frac{\partial M_x}{\partial x} + \frac{\partial M_{xy}}{\partial y} - V_x = 0$$

$$\frac{\partial M_{xy}}{\partial x} + \frac{\partial M_y}{\partial y} - V_y = 0 \qquad (6.8)$$

$$\frac{\partial V_x}{\partial x} + \frac{\partial V_y}{\partial y} + q = \rho h \frac{\partial^2 w}{\partial t^2}$$

where q(x,y,t) is the normal distributed external force per unit area acting on the plate.

The moments M_x, M_y, and M_{xy} can be related to the change of curvatures of the plate as follows:

$$M_x = -D\left(\frac{\partial^2 w}{\partial x^2} + \nu \frac{\partial^2 w}{\partial y^2}\right)$$

$$M_y = -D\left(\frac{\partial^2 w}{\partial y^2} + \nu \frac{\partial^2 w}{\partial x^2}\right) \qquad (6.9)$$

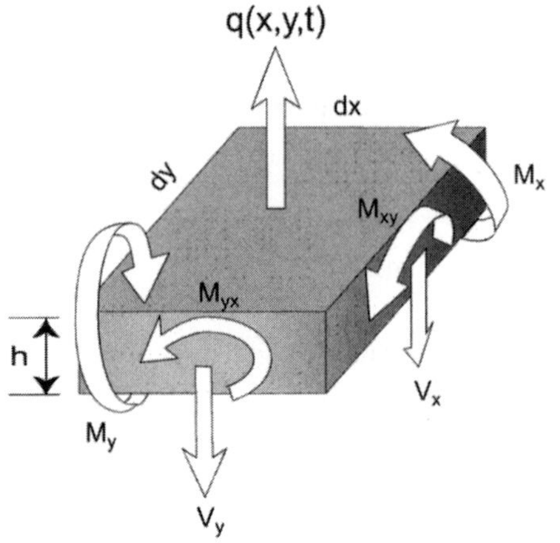

Fig. 6.4 *Element of an Elastic Plate*

$$M_{xy} = -D(1-\nu)\frac{\partial^2 w}{\partial x \partial y}$$

where D = plate stiffness = $E h^3 / [12(1-\nu^2)]$.

The shear forces can be related to the derivatives of the moments, such that:

$$V_x = -D\frac{\partial}{\partial x}\nabla^2 w \quad \text{and} \quad V_y = -D\frac{\partial}{\partial y}\nabla^2 w \tag{6.10}$$

The first two equilibrium eqs. of (6.8) are identically satisfied by expressions for the moments and shear forces given in eqs. (6.9) and (6.10). Substitution of the shear forces of eqs. (6.9) and (6.10) into the third of eq. (6.8) results in the equation of motion of plates on w(P,t):

$$D\nabla^4 w + \rho h \frac{\partial^2 w}{\partial t^2} = q(P,t) \qquad P \text{ in } A, \qquad t > 0 \tag{6.11}$$

where $\nabla^4 = \nabla^2 \nabla^2$ is called the bi-Laplacian.

The boundary conditions on the contour boundary C of the plate can be one of the following pairs:

(a) Fixed Boundary: Displacement w(P,t) = 0

$$\text{Slope}\frac{\partial w}{\partial n}(P,t) = 0 \qquad P \text{ on } C, \qquad t > 0$$

(b) Simply Supported Boundary: Displacement w(P,t) = 0

$$\text{Moment } M_n = -D\frac{\partial^2 w}{\partial n^2}(P,t) = 0 \quad P \text{ on } C, t > 0$$

(c) Free boundary: Moment $M_n = 0$ [See item (b)]

$$V_n = -D\frac{\partial M_{ns}}{\partial n}(P,t) = 0 \qquad P \text{ on } C, \qquad t > 0$$

where s is the distance measured along C.

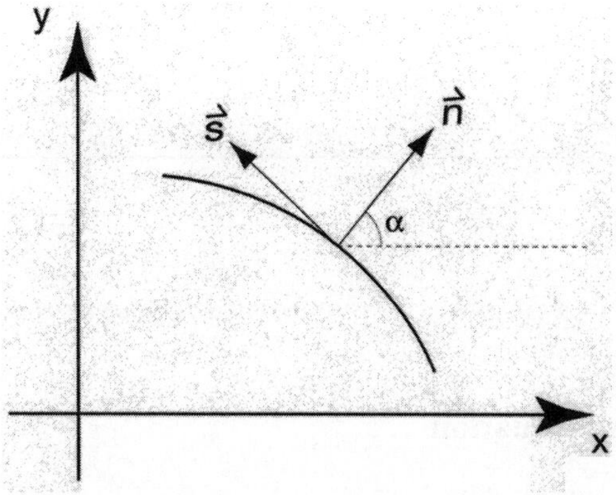

Fig. 6.5 *Element of a Boundary Curve*

More boundary conditions can be specified in a similar manner to those for beams (see Section 4.4).

In the boundary conditions (a) to (c), the partial derivatives $\partial/\partial n$ and $\partial/\partial s$ refer to differentiation with respect to coordinates normal (n) and tangential (s) to the contour C, as shown in Fig. 6.5. Thus:

$$\frac{\partial}{\partial n} = \frac{\partial}{\partial x}\cos\alpha + \frac{\partial}{\partial y}\sin\alpha \quad \text{and} \quad \frac{\partial}{\partial s} = \frac{\partial}{\partial x}\sin\alpha - \frac{\partial}{\partial y}\cos\alpha$$

or

$$\frac{\partial}{\partial x} = \frac{\partial}{\partial n}\cos\alpha + \frac{\partial}{\partial s}\sin\alpha \quad \text{and} \quad \frac{\partial}{\partial y} = \frac{\partial}{\partial n}\sin\alpha - \frac{\partial}{\partial s}\cos\alpha$$

Thus:

$$\nabla^2 w = \frac{\partial^2 w}{\partial n^2} + \frac{\partial^2 w}{\partial s^2}$$

and

$$M_n = M_x \cos^2\alpha + M_y \sin^2\alpha - M_{xy}\sin 2\alpha$$

$$M_{ns} = M_{xy}\cos^2 2\alpha + \frac{M_x - M_y}{2}\sin 2\alpha$$

$$V_n = V_x \cos\alpha + V_y \sin\alpha$$

The initial conditions to be prescribed, for a unique solution, must have the following forms:

$$w(P, 0^+) = f(P) \qquad P \text{ in } A$$

and

$$\frac{\partial w}{\partial t}(P, 0^+) = g(P) \qquad P \text{ in } A$$

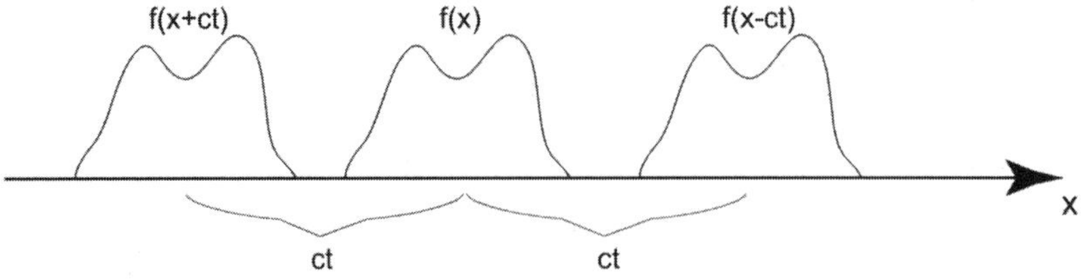

Fig. 6.6 Wave Form Progression

6.4 The Wave Equation

The propagation of a disturbance in a medium is known as wave propagation. The phenomena of wave propagation is best illustrated by propagation of a disturbance in a infinite string.

The equation of motion of a stretched string has the following form:

$$\frac{\partial^2 y}{\partial x^2} = \frac{1}{c^2}\frac{\partial^2 y}{\partial t^2}$$

The solution of such an equation can be obtained in general by transforming the independent variables x and t to u and v, where u = x - ct and v = x + ct. Thus, the equation of motion transforms to:

$$\frac{\partial^2 y}{\partial u\, \partial v} = 0$$

which can be integrated directly, to give the following solution:

$$y = f(u) + g(v) = f(x - ct) + g(x + ct)$$

Functions having the form f(x - ct) and f(x + ct) can be shown to indicate that a function f(x) is displaced to a position (ct) to the right and left, respectively, as shown in Fig. 6.6. Thus, a disturbance having the shape f(x) at t = 0, propagates to the left and to the right without a change in shape, at a speed of c.

A special form of wave functions $f(x \pm ct)$ that occur in physical applications is known as **Harmonic Plane Waves** having the form:

$$f(x \pm ct) = C \exp[ik(x \pm ct)] = C \exp[i(kx \pm \omega t)]$$

where

$$k = \frac{\omega}{c} = \text{wavenumber} = \frac{2\pi}{\lambda}$$

λ = wavelength

ω = circular frequency (rad/sec) = $2\pi f$

f = frequency in cycles per second or Hertz (cps or Hz)

τ = period in time for motion to repeat = $\dfrac{2\pi}{\omega} = \dfrac{1}{f}$

C = amplitude of motion

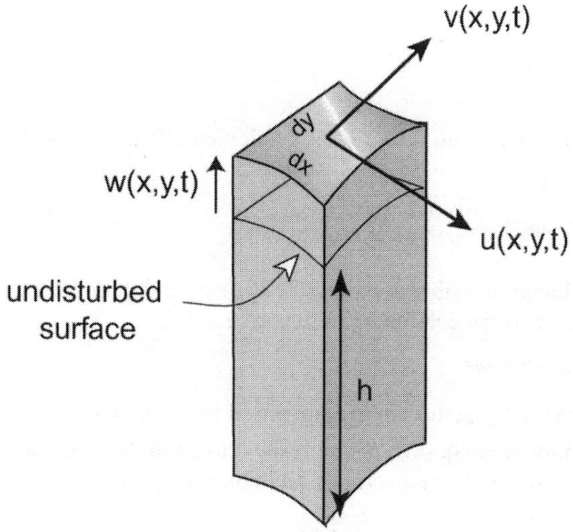

Fig. 6.7 *Surface Element of a Water Basin*

6.4.1 Wave Propagation in One-Dimensional Media

The equation of motion for vibrating stretched strings, bars, torsional rods, acoustic horns, etc., together with the boundary conditions at the end points (if any) and the initial conditions make up the wave propagation system for those media.

6.4.2 Wave Propagation in Two-Dimensional Media

Wave propagation in stretched membranes and in the water surface of basins make up few of the phenomena of wave motion in two-dimensional continua.

The propagation of waves in a stretched membrane obey the same differential equation as the vibration of membranes, with the same type of boundary and initial conditions. The system of differential equations, boundary and initial conditions are the same as those for the vibration problem.

6.4.3 Wave Propagation in Surface of Water Basin

The propagation of waves on the surface of a water basin can be developed by the use of the hydrodynamic equations of equilibrium of an incompressible fluid. Let a free surface basin of a liquid (A) (Fig. 6.7) be surrounded by a rigid wall described by a contour boundary C, whose undisturbed height is h and whose density is ρ.

Let u(x,y,t) and v(x,y,t) represent the components of the vector particle velocity of fluid on the surface in the x and y directions, respectively and w(x,y,t) be the vertical displacement from the level h of the particle in the z-direction.

The law of conservation of mass for an incompressible fluid requires that the rate of change of mass of a column having a volume (h dx dy) must be zero, thus:

$$dx \frac{\partial}{\partial x}(u\,h\,dy) + dy \frac{\partial}{\partial y}(v\,h\,dx) + \frac{\partial}{\partial t}\left[(w + h)\,dx\,dy\right]$$

or

$$\frac{\partial w}{\partial t} + h\left(\frac{\partial u}{\partial x} + \frac{\partial v}{\partial y}\right) = 0 \tag{6.12}$$

Let p be the pressure acting on the sides of an element, then the equation of equilibrium becomes:

$$\rho\frac{\partial u}{\partial t} = -\frac{\partial p}{\partial x} \quad \text{and} \quad \rho\frac{\partial v}{\partial t} = -\frac{\partial p}{\partial y} \tag{6.13}$$

Since the fluid is incompressible, the pressure at any depth z in the basin can be described by the static pressure of the column of fluid above z, i.e.:

$$p = p_0 + \rho\, g(h - z + w) \tag{6.14}$$

where p_0 is the external pressure on the surface of the basin and g is the acceleration due to gravity. Differentiating eq. (6.12) with respect to t and the first and second of eq. (6.13) with respect to x and y respectively and combining the resulting equalities, one obtains:

$$\frac{\partial^2 w}{\partial t^2} = \frac{h}{\rho}\nabla^2 p \tag{6.15}$$

Substitution of p from eq. (6.14) into eq. (6.15) results in the equation of motion of a particle on the surface of a liquid basin as follows:

$$\nabla^2 w = \frac{1}{c^2}\frac{\partial^2 w}{\partial t^2}$$

where $c^2 = gh$. Substituting eq. (6.14) into eq. (6.13) one obtains:

$$\frac{\partial u}{\partial t} = -g\frac{\partial w}{\partial x} \quad \text{and} \quad \frac{\partial v}{\partial t} = -g\frac{\partial w}{\partial y} \tag{6.16}$$

Thus, since the wall surrounding the basin is rigid, then the component of the velocity normal to the boundary C must vanish. Hence, using eq. (6.16) (see Fig. 6.5), the normal component of the velocity v_n becomes:

$$v_n = u\cos\alpha + v\sin\alpha = -g\int\left[\frac{\partial w}{\partial x}\cos\alpha + \frac{\partial w}{\partial y}\sin\alpha\right]dt$$

$$= -g\int\frac{\partial w}{\partial n}dt = 0$$

where n is the normal to the curve C, so that the boundary condition on w becomes:

$$\frac{\partial w}{\partial n}(P,t) = 0 \qquad \text{P on C,} \qquad t > 0$$

6.4.4 Wave Propagation in an Acoustic Medium

Wave propagation in three-dimensional media is a phenomena that covers a variety of fields in Physics and Engineering. Wave propagation in acoustic media is the simplest three-dimensional wave phenomena in physical systems. Let a compressible fluid medium occupy V and be surrounded by a surface S and consider an element of such a field as shown in Fig. 6.8. The law of conservation of mass for the element can be stated

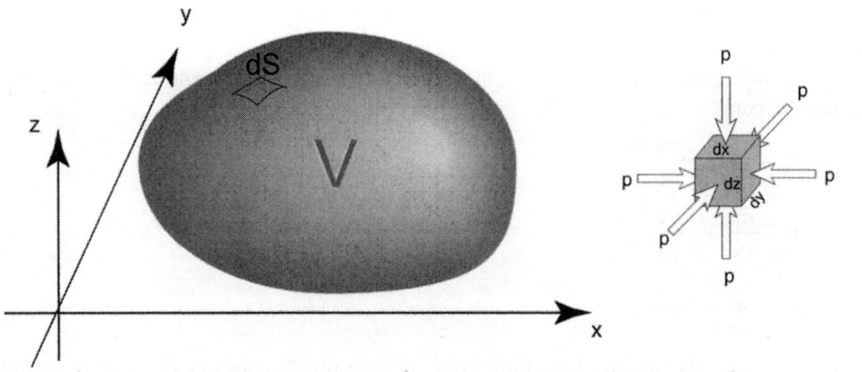

Fig. 6.8 *Element of an Acoustic Medium*

as the rate of change of mass of an element is zero. Thus, the increase in the mass of the element must be equal to the influx of mass through the six sides of the element. Let u, v, and w represent the particle velocity of the fluid in the x, y, and z directions, respectively. Thus, the influx of mass from the element through the two sides perpendicular to the x-axis becomes:

$$\rho\, u(x,y,z)dy\, dx - \rho\, u(x+dx,y,z)dy\, dx \approx -\rho\frac{\partial u}{\partial x}dx\, dy\, dz$$

Similarly, the mass influx from the remaining two pairs in of sides becomes:

$$-\rho\frac{\partial v}{\partial y}dx\, dy\, dz \quad \text{and} \quad -\rho\frac{\partial w}{\partial x}dx\, dy\, dz$$

Thus, the law of conservation of mass of a compressible fluid element requires that:

$$\frac{\partial(\rho\, dx\, dy\, dz)}{\partial t} = -\rho\left(\frac{\partial u}{\partial x}+\frac{\partial v}{\partial y}+\frac{\partial w}{\partial z}\right)dx\, dy\, dz$$

or

$$\frac{\partial \rho}{\partial t}+\rho\left(\frac{\partial u}{\partial x}+\frac{\partial v}{\partial y}+\frac{\partial w}{\partial z}\right)=0 \qquad (6.17)$$

Let p(x,y,z,t) be the fluid pressure acting normal to the six faces of the fluid element. The equations of motion of the element can be written as three equations governing the motion in the x, y, and z directions in terms of the fluid pressure p, by satisfying Newton's second law:

$$x: [p(x,y,z,t) - p(x+dx,y,z,t)]\, dy\, dz + f_x\, \rho\, dx\, dy\, dz = (\rho\, dx\, dy\, dz)\frac{\partial u}{\partial t}$$

$$y: [p(x,y,z,t) - p(x,y+dy,z,t)]\, dx\, dz + f_y\, \rho\, dx\, dy\, dz = (\rho\, dx\, dy\, dz)\frac{\partial v}{\partial t}$$

$$z: [p(x,y,z,t) - p(x,y,z+dz,t)]\, dx\, dy + f_z\, \rho\, dx\, dy\, dz = (\rho\, dx\, dy\, dz)\frac{\partial w}{\partial t}$$

where f_x, f_y, and f_z are distributed forces per unit mass in the x, y, and z directions, respectively.

Expanding the fluid pressure, p, in a Taylor series about x, y, and z, the equations of motion become:

$$\rho f_x - \frac{\partial p}{\partial x} = \rho \frac{\partial u}{\partial t}$$

$$\rho f_y - \frac{\partial p}{\partial y} = \rho \frac{\partial v}{\partial t} \tag{6.18}$$

$$\rho f_z - \frac{\partial p}{\partial z} = \rho \frac{\partial w}{\partial t}$$

For small adiabatic motion of the fluid, let the fluid density to vary linearly with the change in volume of a unit volume element:

$$\rho = \rho_0 (1 + s)$$

and

$$p = p_0 \left(\frac{\rho}{\rho_0} \right) \gamma \approx p_0 (1 + \gamma s) \quad \text{for } |s| << 1 \tag{6.19}$$

where ρ_0 and p_0 are the initial (undisturbed) fluid density and pressure, respectively, s is the condensation (change of volume of a unit volume element) and γ is the ratio of the specific heat constant for the fluid at constant pressure C_p to that at constant volume C_v. Substituting eq. (6.19) into eqs. (6.17) and (6.18) results in:

$$\frac{\partial s}{\partial t} \approx -\left(\frac{\partial u}{\partial x} + \frac{\partial v}{\partial y} + \frac{\partial w}{\partial z} \right)$$

$$\rho_0 f_x - \rho_0 \gamma \frac{\partial s}{\partial x} \approx \rho_0 \frac{\partial u}{\partial t}$$

$$\rho_0 f_y - \rho_0 \gamma \frac{\partial s}{\partial y} \approx \rho_0 \frac{\partial v}{\partial t}$$

$$\rho_0 f_z - \rho_0 \gamma \frac{\partial s}{\partial z} \approx \rho_0 \frac{\partial w}{\partial t} \tag{6.20}$$

Differentiating the four equations of eq. (6.20) with respect to t, x, y, and z, respectively, one obtains the acoustic Wave Equation as follows:

$$\frac{\partial^2 s}{\partial t^2} = c^2 \nabla^2 s - \left(\frac{\partial f_x}{\partial x} + \frac{\partial f_y}{\partial y} + \frac{\partial f_z}{\partial z} \right) \tag{6.21}$$

where $c = \sqrt{\frac{p_0 \gamma}{\rho_0}}$ is the sound speed in the acoustic medium.

If one uses a velocity potential $\phi(x,y,z,t)$ and a source potential $F(x,y,z,t)$, such that:

$$u = -\frac{\partial\phi}{\partial x} \qquad f_x = \frac{\partial F}{\partial x}$$
$$v = -\frac{\partial\phi}{\partial y} \qquad f_y = \frac{\partial F}{\partial y} \qquad (6.22)$$
$$w = -\frac{\partial\phi}{\partial z} \qquad f_z = \frac{\partial F}{\partial z}$$

and

$$p = \rho_0 \frac{\partial\phi}{\partial t} + \rho_0 F + p_0 \qquad s = \frac{1}{c^2}\left[\frac{\partial\phi}{\partial t} + F\right]$$

then the equations (6.18) and the last three equations of eq. (6.20) are satisfied identically. Substitution of eq. (6.22) into the first of eq. (6.20) result in the Wave Equation on the velocity potential ϕ as follows:

$$\nabla^2\phi = \frac{1}{c^2}\frac{\partial^2\phi}{\partial t^2} + \frac{1}{c^2}\frac{\partial F}{\partial t}(P,t) \qquad \text{P in V, } t > 0 \qquad (6.23)$$

The boundary conditions can be one of the following types:

(a) $p(P,t) = g(P, t)$ P on S, $t > 0$

(b) $v_n = -\frac{\partial\phi}{\partial n}$ = normal component of the velocity = $g(P,t)$ P on S, $t > 0$

(c) Elastic boundary: $\frac{\partial p}{\partial t}(P,t) + \gamma\, v_n(P,t) = g(P,t)$ P on S, $t > 0$

where γ represents the elastic constant and v_n is the normal particle velocity.

Wave propagation in elastic media and electromagnetic waves in dielectric materials are governed by vector potentials instead of the one scalar potential for an acoustic medium. Neither of these media will be further explored in this book.

6.5 Helmholtz Equation

Helmholtz equation results from the assumption that the vibration or wave propagation in certain media are time harmonic, i.e. if one lets $e^{-i\omega t}$ be the time dependance, then Helmholtz equation results, having the following form:

$$\nabla^2\phi + k^2\phi = F(P) \qquad \text{P in V} \qquad (6.24)$$

This equation describes a variety of diverse physical phenomena.

6.5.1 Vibration in Bounded Media

One method of obtaining the solution to forced vibration problems is the method of separation of variables. This method assumes that the deformation $\phi(P,t)$ can be written as a product as follows:

$$\phi(P,t) = \psi(P)\, T(t)$$

where the functions $\psi(P)$ and $T(t)$ satisfy the following equations:

$$\nabla^2\psi_n + \lambda_n \psi_n = 0$$
$$T_n'' + c^2\lambda_n T_n = 0 \qquad (6.25)$$

This equation leads to eigenfunctions $\psi_n(P)$ where λ_n are the corresponding eigenvalues. The functions $\psi_n(P)$ are known as **Standing Waves**. The lines (or surfaces) where $\psi_n(P) = 0$ are known as the **Nodal Lines** (or Surfaces).

The general solution can thus be represented by superposition of infinite such standing waves. The boundary conditions required for a unique solution of the Helmholtz equation are the same type specified in Section 6.3.

6.5.2 Harmonic Waves

The solution of wave propagation problems in media where the medium is induced to motion ϕ (P,t) by forces which are periodic in time, i.e., when the forcing function f(P,t) has the form:

$$f(P,t) = g(P)\, e^{i\omega t}$$

can be developed in the form of harmonic waves, i.e.:

$$\phi(P,t) = \psi(P)\, e^{i\omega t} \qquad (6.26)$$

where $\psi(P)$ satisfies eq. (6.24). The function $\phi(P,t)$ would not initially have the form given in (6.26), but if the wave process is given enough time (say, if initiated at $t_o = -\infty$) then the initial transient state decays and the steady state described in eq. (6.26) results, where the solution is periodic in time, i.e., the solution would have the same frequency ω as that of the forcing function. Since the motion is assumed to have been started at an initial instance $t_o = -\infty$, then no initial conditions need be specified.

6.6 Poisson and Laplace Equations

Poisson equation has the following form:

$$\nabla^2\phi = f(P) \qquad \text{P in V} \qquad (6.27)$$

while the Laplace equation has the following form:

$$\nabla^2\phi = 0 \qquad (6.28)$$

Various steady state phenomena in Physics and Engineering are governed by equations of the type (6.27) and (6.28). Non-trivial solutions of (6.27) are due to either the source function f(P) or to non-homogeneous boundary conditions. Non-trivial solutions of (6.28) are due to non-homogeneous boundary conditions.

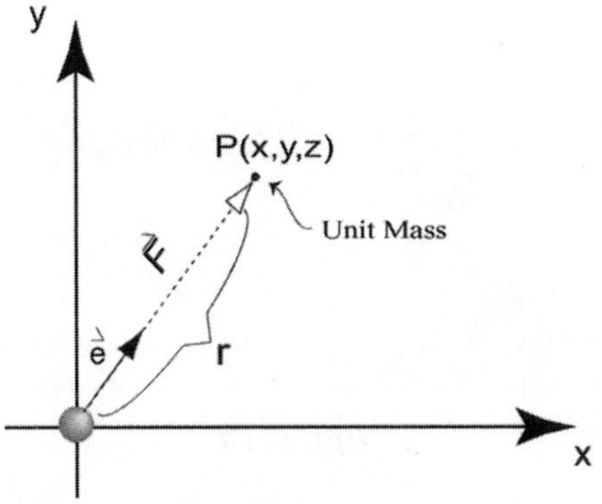

Fig. 6.9

6.6.1 Steady State Temperature Distribution

If the thermal state of a solid is independent of time (steady state), then eq. (6.1) becomes:

$$\nabla^2 T = -q\,(P)$$

The boundary conditions are those specified in Section 6.2.1.

6.6.2 Flow of Ideal Incompressible Fluids

Fluid flow of incompressible fluids can be developed from the formalism of flow of compressible fluids. Since the density of an incompressible fluid is constant, then eq. (6.17) becomes:

$$\frac{\partial u}{\partial x}+\frac{\partial v}{\partial y}+\frac{\partial w}{\partial z}=0 \tag{6.29}$$

If one uses a velocity potential $\phi(P)$ as described in (6.22), then the velocity potential satisfies Laplace's equation. If there are sources or sinks in the fluid medium, then the velocity potential satisfies the Poisson equation.

6.6.3 Gravitational (Newtonian) Potentials

Consider two point masses m_1 and m_2, located at positions x_1 and x_2, respectively, and are separated by a distance r, then the force of attraction (F) between m_1 and m_2 can be stated as follows:

$$\vec{F} = \gamma \frac{m_1 m_2}{r^2}\,\vec{e}_r$$

where $\vec{e}_r$ is a unit base vector pointing from m_2 to m_1 along r. If one sets $m_1 = 1$, and $m_2 = m$, then the force $\vec{F}$ becomes the field-strength at a point P due to a mass m at x defined as (see Fig. 6.9):

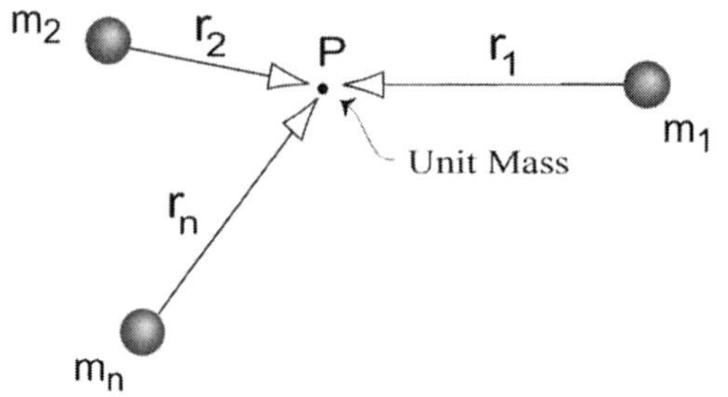

Fig. 6.10

$$\vec{F} = \gamma \frac{m}{r^2} \vec{e}_r$$

or written in terms of its components:

$$F_x = \gamma \frac{m}{r^2} \cos(r, x) = \gamma \frac{m}{r^2} \frac{\partial r}{\partial x}$$

$$F_y = \gamma \frac{m}{r^2} \cos(r, y) = \gamma \frac{m}{r^2} \frac{\partial r}{\partial y}$$

and

$$F_z = \gamma \frac{m}{r^2} \cos(r, z) = \gamma \frac{m}{r^2} \frac{\partial r}{\partial z} \tag{6.30}$$

If one defines a gravitational potential such that the force is represented by:

$$\vec{F} = -\nabla \psi_o$$

such that:

$$F_x = -\frac{\partial \psi_o}{\partial x} \qquad F_y = -\frac{\partial \psi_o}{\partial y} \qquad F_z = -\frac{\partial \psi_o}{\partial z} \tag{6.31}$$

then ψ_o is obtained by comparing eqs. (6.30) and (6.31), giving:

$$\psi_o = \frac{\gamma m}{r} \tag{6.32}$$

It can be shown that ψ_0 in eq. (6.32) satisfies the Laplace Equation. If there is a finite number of masses m_1, m_2, ...m_n, situated at r_1, r_2, ...r_n respectively, away from a unit mass at point P (see Fig. 6.10), then the potential for each mass can be described as follows:

$$\psi_i = \frac{\gamma m_i}{r_i}$$

and the total gravitational potential per unit mass at P becomes:

$$\psi_i = \sum_{i=1}^{n} \psi_i = \gamma \sum_{i=1}^{n} \frac{m_i}{r_i}$$

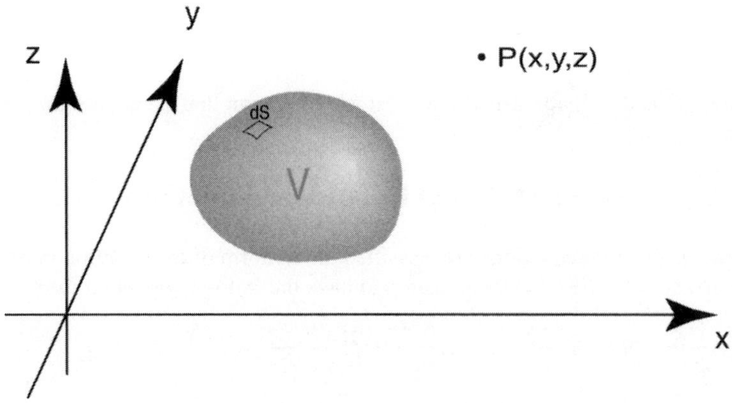

Fig. 6.11 *Gravitational Potential for a Mass*

If the masses are distributed in a volume V, then the total potential due to the mass occupying V becomes (see Fig. 6.11):

$$\psi = \gamma \int_V \frac{\rho(x',y',z')}{r}\, dV' \tag{6.33}$$

where $\rho(x,y,z)$ is the mass density of the material occupying V and ψ satisfies the Poisson equation.

6.6.4 Electrostatic Potential

The electrostatic potential can be defined in a similar manner to gravitational potential. Define the repulsive (attractive) force F between two similar (dissimilar) charges of magnitudes q_1 and q_2, located at positions x_1 and x_2, respectively, as:

$$\vec{F} = \frac{q_1 q_2}{4\pi\varepsilon r^2}\,\vec{e}_r$$

where r is the distance between q_1 and q_2 and ε is the material's dielectric constant. Define the electric field as the force on a unit charge (where $q_2 = 1$) located at a point P due to a charge $q_2 = q$ as:

$$\vec{E} = \frac{q}{4\pi\varepsilon r^2}\,\vec{e}_r = -\frac{q}{4\pi\varepsilon}\nabla\left(\frac{1}{r}\right)$$

If we define an electrostatic potential, ψ, such that $\vec{E} = -\nabla\psi$, then a solution for the potential is:

$$\psi = \frac{q}{4\pi\varepsilon}\, r$$

If there exists distributed charges in a volume V, then the potential can be defined as:

$$\psi = \int_V \frac{\rho(x',y',z')}{4\pi\varepsilon r} dV' \tag{6.34}$$

where $\rho(x,y,z)$ is the charge density in V. It can be shown that ψ satisfies the Poisson equation.

6.7 Classification of Partial Differential Equations

Partial differential equations are classified on the form of the equation in two dimensional coordinates. Let the equation to have the following general form:

$$a(x,t)\frac{\partial^2\phi}{\partial x^2} + 2b(x,t)\frac{\partial^2\phi}{\partial x\partial t} + c(x,t)\frac{\partial^2\phi}{\partial t^2} = f\left(\frac{\partial\phi}{\partial x}, \frac{\partial\phi}{\partial t}, \phi, x, t\right) \tag{6.35}$$

then the equation can be classified into three categories:

(a) Hyperbolic: If $b^2 > ac$ everywhere in [x, t].

Examples: The Wave and Vibration equations.

(b) Elliptic: If $b^2 < ac$ everywhere in [x, t].

Examples: Laplace and Helmholtz equations.

(c) Parabolic: If $b^2 = ac$ everywhere [x, t].

Examples: The Diffusion equation.

The boundary conditions are classified as follows:

(a) Dirichlet: Specify $\phi(P,t) = g(P)$ P on S, $t > 0$

(b) Neumann: Specify $\frac{\partial\phi(P,t)}{\partial n} = g(P)$ P on S, $t > 0$

(c) Robin: Specifiy $\frac{\partial\phi(P,t)}{\partial n} + k\,\phi(P,t) = g(P)$ P on S, $k > 0$, $t > 0$

6.8 Uniqueness of Solutions

6.8.1 Laplace and Poisson Equations

Uniqueness of solutions of the Laplace and Poisson Equations, requires the specification of boundary conditions. To prove uniqueness, assume that there are two different solutions of the differential equation. Let ϕ_1 and ϕ_2 be two different solutions to Poisson's equation (6.27), for a bounded region V with identical boundary conditions. Thus, each solution satisfies the same Poisson equation:

$$\nabla^2\phi_1 = f(P) \quad \text{and} \quad \nabla^2\phi_2 = f(P) \qquad \text{P in V}$$

such that the difference solution satisfies:

$$\nabla^2\phi = 0 \qquad \text{where } \phi = \phi_1 - \phi_2$$

Multiplying the Laplace equation above on ϕ by ϕ and integrating over V, one obtains:

$$\int_V \phi\nabla^2\phi \, dV = -\int_V (\nabla\phi)\bullet(\nabla\phi)\, dV + \int_S \phi\frac{\partial\phi}{\partial n}\, dS = 0$$

Thus,

$$\int_V [(\frac{\partial\phi}{\partial x})^2 + (\frac{\partial\phi}{\partial y})^2 + (\frac{\partial\phi}{\partial z})^2]dV = \int_S \phi\frac{\partial\phi}{\partial n} dS \tag{6.36}$$

To solve for the difference potential given the three boundary conditions described above:

(a) Dirichlet: If the two solutions satisfy the same Dirichlet boundary conditions then:

$$\phi_1(P) = \phi_2(P) = g(P) \qquad \text{P on S}$$

then:

$$\phi(P) = 0 \qquad \text{P on S}$$

and eq. (6.36) becomes:

$$\int_V [(\frac{\partial\phi}{\partial x})^2 + (\frac{\partial\phi}{\partial y})^2 + (\frac{\partial\phi}{\partial z})^2]dV = 0$$

which can be satisfied if and only if $\frac{\partial\phi}{\partial x} = \frac{\partial\phi}{\partial y} = \frac{\partial\phi}{\partial z} = 0$ or $\phi = C =$ constant. However, since ϕ is continuous in V and on S, and since ϕ is zero on the surface, then the constant C must be zero. Thus, ϕ must be zero throughout the volume, and, hence, the solution is unique.

(b) Neumann: If normal derivatives of the potentials satisfy the same boundary condition on the surface, then:

$$\frac{\partial\phi_1(P)}{\partial n} = \frac{\partial\phi_2(P)}{\partial n} = g(P) \qquad \text{P on S}$$

Thus:

$$\frac{\partial\phi(P)}{\partial n} = 0 \qquad \text{P on S}$$

Therefore, the difference solution $\phi = C =$ constant, and the two solutions are unique to within a constant.

(c) Robin: If the two solutions satisfy the same Robin boundary conditions then:

$$\frac{\partial\phi_1(P)}{\partial n} + k\,\phi_1(P) = \frac{\partial\phi_2(P)}{\partial n} + k\,\phi_2(P) = g(P) \qquad \text{P on S, } k > 0$$

Therefore:

$$\frac{\partial\,\phi(P)}{\partial n} = -k\phi(P) \qquad \text{P on S, } k > 0$$

and eq. (6.36) can be rewritten:

$$\int_V [(\frac{\partial \phi}{\partial x})^2 + (\frac{\partial \phi}{\partial y})^2 + (\frac{\partial \phi}{\partial z})^2]dV = -\int_S k\,\phi^2 dS \qquad (6.37)$$

However, since k is positive and ϕ^2 is positive, both integrals of eq. (6.37) must vanish, resulting in ϕ = constant in V and $\phi = 0$ on S. Due to the continuancy of ϕ in V and on S, then ϕ is zero throughout the volume and the solution is unique.

6.8.2 Helmholtz Equation

Helmholtz equation can be solved by eigenfunction expansions. Thus, the eigenfunctions $\phi_M(P)$ satisfy:

$$\nabla^2 \phi_M + \lambda_M \phi_M = 0 \qquad (6.38)$$

and homogeneous boundary conditions of the type Dirichlet, Neumann or Robin. The capitalized index M represents one, two or three dimensional integers and λ_M is the corresponding eigenvalue.

The solution to the non-homogeneous Helmholtz equation (6.24):

$$\nabla^2 \phi + \lambda \phi = F(P) \qquad \text{P in V}$$

can be written as a superposition of the eigenfunctions $\phi_M(P)$.

Let ϕ_1 and ϕ_2 be two solutions of Helmholtz equation (6.24), i.e.:

$$\nabla^2 \phi_1 + \lambda \phi_1 = F(P) \qquad \text{and} \qquad \nabla^2 \phi_2 + \lambda \phi_2 = F(P)$$

If we once again define ϕ as $\phi_1 - \phi_2$, then ϕ satisfies the homogeneous Helmholtz equation:

$$\nabla^2 \phi + \lambda \phi = 0 \qquad (6.39)$$

Expanding the solutions for ϕ_1 and ϕ_2 in a series of the eigenfunctions:

$$\phi_1 = \sum_M a_M \, \phi_M(P)$$

and

$$\phi_2 = \sum_M b_M \, \phi_M(P)$$

then the difference solution ϕ is expressed by:

$$\phi = \sum_M (a_M - b_M)\phi_M(P)$$

Substituting ϕ into eq. (6.39), and using eq. (6.38) one obtains:

$$\sum_M (a_M - b_M)(\lambda - \lambda_M)\, \phi_M(P) = 0$$

which, for $\lambda \neq \lambda_M$ and after using the orthogonality condition (Section 6.12) results in $a_M = b_M$. Therefore, Helmholtz equation has unique solutions for any of the three types of boundary conditions.

6.8.3 Diffusion Equation

Let ϕ_1 and ϕ_2 be two solutions to the Diffusion Equation (6.1) that satisfy the same boundary conditions and initial conditions as follows:

$$\nabla^2\phi_1 = \frac{1}{K}\frac{\partial\phi_1}{\partial t} + f(P,t) \qquad P \text{ in } V, \quad t>0$$

$$\nabla^2\phi_2 = \frac{1}{K}\frac{\partial\phi_2}{\partial t} + f(P,t) \qquad P \text{ in } V, \quad t>0$$

$$\phi_1(P,0^+) = \phi_2(P,0^+) = g(P) \qquad P \text{ in } V$$

Letting $\phi = \phi_1 - \phi_2$ then the difference solution $\phi(P, t)$ satisfies:

$$\nabla^2\phi = \frac{1}{K}\frac{\partial\phi}{\partial t} \qquad P \text{ in } V, \quad t>0$$

and

$$\phi(P,0^+) = 0 \qquad P \text{ in } V$$

and one of the following conditions for points P on S and for $t > 0$:

(a) Dirichlet: $\phi(P,t) = 0 \qquad P \text{ on } S, \quad t>0$

(b) Neuman: $\dfrac{\partial\phi(P,t)}{\partial n} = 0 \qquad P \text{ on } S, \quad t>0$

or

(c) Robin: $\dfrac{\partial\phi(P,t)}{\partial n} + h\,\phi(P,t) = 0 \qquad \text{where } h>0, \; P \text{ on } S, \quad t>0$

Multiplying the homogeneous Diffusion equation on the difference solution ϕ by ϕ and integrating over V, one obtains:

$$\frac{1}{K}\int_V \phi\frac{\partial\phi}{\partial t}\,dV = \frac{1}{2K}\frac{\partial}{\partial t}\int_V \phi^2\,dV = \int_V \phi\nabla^2\phi\,dV$$

$$= -\int_V (\nabla\phi)\cdot(\nabla\phi)\,dV + \int_S \phi\frac{\partial\phi}{\partial n}\,dS \tag{6.40}$$

For Dirichlet and Neumann boundary conditions, the surface integral vanishes and eq. (6.40) becomes:

$$\frac{1}{2K}\frac{\partial}{\partial t}\int_V \phi^2\,dV + \int_V \left[\left(\frac{\partial\phi}{\partial x}\right)^2 + \left(\frac{\partial\phi}{\partial y}\right)^2 + \left(\frac{\partial\phi}{\partial z}\right)^2\right]dV = 0 \tag{6.41}$$

For Robin boundary condition, eq. (6.40) becomes:

$$\frac{1}{2K}\frac{\partial}{\partial t}\int_V \phi^2\,dV + \int_V \left[\left(\frac{\partial\phi}{\partial x}\right)^2 + \left(\frac{\partial\phi}{\partial y}\right)^2 + \left(\frac{\partial\phi}{\partial z}\right)^2\right]dV + h\int_S \phi^2\,dS = 0 \tag{6.42}$$

For either eq. (6.41) or (6.42) to be true:

$$\frac{\partial}{\partial t}\int_V \phi^2 dV \le 0 \tag{6.43}$$

Let:

$$\int_V \phi^2 dV = F(t)$$

Since the time derivative is always negative, due to the inequality in eq. (6.43), we can define a new variable, f(t), such that:

$$\frac{\partial F(t)}{\partial t} = -[f(t)]^2$$

and

$$F(t) = -\int_0^t [f(\eta)]^2\, d\eta + C$$

Since $\phi(P,0^+) = 0$, then $F(0) = 0$ and hence $C = 0$. Thus:

$$F(t) = -\int_0^t [f(\eta)]^2\, d\eta = \int_V \phi^2 dV$$

which is only possible if integrand $\phi = 0$. Therefore, the solution must be unique.

6.8.4 Wave Equation

Let ϕ_1 and ϕ_2 be solutions to the Wave Equation (6.23) which satisfy the same boundary conditions and initial conditions, such that:

$$\nabla^2\phi_1 = \frac{1}{c^2}\frac{\partial^2\phi_1}{\partial t^2} + q(P,t) \qquad \text{P in V,} \quad t > 0$$

$$\nabla^2\phi_2 = \frac{1}{c^2}\frac{\partial^2\phi_2}{\partial t^2} + q(P,t) \qquad \text{P in V,} \quad t > 0$$

$$\phi_1(P,0^+) = \phi_2(P,0^+) = f(P) \qquad \text{P in V}$$

$$\frac{\partial\phi_1}{\partial t}(P,0^+) = \frac{\partial\phi_2}{\partial t}(P,0^+) = g(P) \qquad \text{P in V}$$

for P in V and $t > 0$, such that the difference solution ϕ satisfies:

$$\nabla^2\phi = \frac{1}{c^2}\frac{\partial^2\phi}{\partial t^2} \qquad \text{P in V,} \quad t > 0$$

$$\phi(P,0^+) = 0 \qquad \text{P in V}$$

$$\frac{\partial\phi}{\partial t}(P,0^+) = 0 \qquad\qquad P \text{ in } V$$

and one of the following conditions for P on S and for $t > 0$:

(a) Dirichlet: $\phi(P,t) = 0$

(b) Neuman: $\dfrac{\partial\,\phi(P,t)}{\partial n} = 0$

or

(c) Robin: $\dfrac{\partial\,\phi(P,t)}{\partial n} + h\,\phi(P,t) = 0$ where $h > 0$

Multiplying the homogeneous Diffusion Equation on the difference solution ϕ by $\partial\phi/\partial t$ and integrating over V, one obtains:

$$\frac{1}{c^2}\int_V \frac{\partial\phi}{\partial t}\frac{\partial^2\phi}{\partial t^2}\,dV = \frac{1}{2c^2}\frac{\partial}{\partial t}\int_V \left[\frac{\partial\phi}{\partial t}\right]^2 dV = \int_V \frac{\partial\phi}{\partial t}\nabla^2\phi\,dV \tag{6.44}$$

The last integral can be rearranged so that:

$$\frac{\partial\phi}{\partial t}\left[\left(\frac{\partial^2\phi}{\partial x^2}\right)+\left(\frac{\partial^2\phi}{\partial y^2}\right)+\left(\frac{\partial^2\phi}{\partial z^2}\right)\right]$$

$$= \frac{\partial}{\partial x}\left[\frac{\partial\phi}{\partial t}\frac{\partial\phi}{\partial x}\right]+\frac{\partial}{\partial y}\left[\frac{\partial\phi}{\partial t}\frac{\partial\phi}{\partial y}\right]+\frac{\partial}{\partial z}\left[\frac{\partial\phi}{\partial t}\frac{\partial\phi}{\partial z}\right]-\frac{1}{2}\frac{\partial}{\partial t}[(\frac{\partial\phi}{\partial x})^2+(\frac{\partial\phi}{\partial y})^2+(\frac{\partial\phi}{\partial z})^2]$$

$$= \nabla\bullet\left(\frac{\partial\phi}{\partial t}\nabla\phi\right)-\frac{1}{2}\frac{\partial}{\partial t}|\nabla\phi|^2$$

Thus, the equality in eq. (6.44) becomes:

$$\frac{1}{2c^2}\frac{\partial}{\partial t}\int_V\left[\frac{\partial\phi}{\partial t}\right]^2 dV = \int_V \nabla\bullet\left(\frac{\partial\phi}{\partial t}\nabla\phi\right)dV - \frac{1}{2}\frac{\partial}{\partial t}\int_V |\nabla\phi|^2 dV \tag{6.45}$$

Using the divergence theorem:

$$\int_V \nabla\bullet\vec{F}\,dV = \int_S \vec{n}\bullet\vec{F}\,dS$$

and $\vec{n}\bullet\nabla\phi = \dfrac{\partial\phi}{\partial n}$, one obtains:

$$\int_V \nabla\bullet\left(\frac{\partial\phi}{\partial t}\nabla\phi\right)dV = \int_S \vec{n}\bullet\left(\frac{\partial\phi}{\partial t}\nabla\phi\right)dS = \int_S\left(\frac{\partial\phi}{\partial t}\frac{\partial\phi}{\partial n}\right)dS$$

Thus, eq. (6.45) becomes:

$$\frac{1}{2}\frac{\partial}{\partial t}\int_V\left[\frac{1}{c^2}\left(\frac{\partial\phi}{\partial t}\right)^2+|\nabla\phi|^2\right]dV = \int_S\left(\frac{\partial\phi}{\partial t}\frac{\partial\phi}{\partial n}\right)dS \tag{6.46}$$

Remember, that for the Dirichlet boundary condition, $\phi(P,t) = 0$ for P on S and for $t > 0$. Therefore, the time derivative of the boundary condition vanishes, i.e.:

$$\frac{\partial \phi}{\partial t}(P,t) = 0 \text{ for P on S}$$

For Neumann boundary conditions:

$$\frac{\partial \phi}{\partial n}(P,t) = 0 \text{ for P on S}$$

Thus, for either Dirichlet or Neumann boundary conditions, the surface integral vanishes and eq. (6.46) becomes:

$$\frac{\partial}{\partial t}\int\limits_V \left[\frac{1}{c^2}\left(\frac{\partial \phi}{\partial t}\right)^2 + \left(\frac{\partial \phi}{\partial x}\right)^2 + \left(\frac{\partial \phi}{\partial y}\right)^2 + \left(\frac{\partial \phi}{\partial z}\right)^2\right] dV = 0 \tag{6.47}$$

Integrating eq. (6.47) with respect to t, one obtains:

$$\int\limits_V \left[\frac{1}{c^2}\left(\frac{\partial \phi}{\partial t}\right)^2 + \left(\frac{\partial \phi}{\partial x}\right)^2 + \left(\frac{\partial \phi}{\partial y}\right)^2 + \left(\frac{\partial \phi}{\partial z}\right)^2\right] dV = C \quad \text{(constant for all t)}$$

Substituting $t = 0$ in the integrand, then since $\phi(P,0^+) = 0$ for P in V, then $\nabla\phi(P,0^+) = 0$. Also, since $\frac{\partial \phi}{\partial t}(P,0^+) = 0$ for P in V, then $C = 0$ in V and the integrand must vanish for all t. Thus:

$$\frac{\partial \phi}{\partial t} = 0, \quad \frac{\partial \phi}{\partial x} = 0, \quad \frac{\partial \phi}{\partial y} = 0, \quad \text{and} \quad \frac{\partial \phi}{\partial z} = 0, \quad \text{P in V}, \ t > 0$$

which, when integrated results in:

$$\phi(P,t) = C_1 = \text{constant} \qquad \text{P in V}, \quad t > 0$$

Since $\phi(P,0^+) = 0$, then $C_1 = 0$ and

$$\phi(P,t) \equiv 0 \qquad \text{P in V}, \quad t > 0$$

For Robin boundary condition, eq. (6.46) becomes:

$$\frac{\partial}{\partial t}\int\limits_V \left[\frac{1}{c^2}\left(\frac{\partial \phi}{\partial t}\right)^2 + \left(\frac{\partial \phi}{\partial x}\right)^2 + \left(\frac{\partial \phi}{\partial y}\right)^2 + \left(\frac{\partial \phi}{\partial z}\right)^2\right] dV$$
$$= -h\int\limits_S \phi\frac{\partial \phi}{\partial t}\, dS = -\frac{h}{2}\frac{\partial}{\partial t}\int\limits_S \phi^2 dS \tag{6.48}$$

Integrating equation (6.48) with respect to t results in a constant:

$$\int\limits_V \left[\frac{1}{c^2}\left(\frac{\partial \phi}{\partial t}\right)^2 + \left(\frac{\partial \phi}{\partial x}\right)^2 + \left(\frac{\partial \phi}{\partial y}\right)^2 + \left(\frac{\partial \phi}{\partial z}\right)^2\right] dV + \frac{h}{2}\int\limits_S \phi^2 dS = C$$

Invoking the same arguments as above, $C = 0$, and, one can again show that:

$$\phi(P,t) \equiv 0 \qquad \text{P in V}, \quad t > 0$$

Therefore, all three boundary conditions are sufficient to produce a unique solution in wave functions.

6.9 The Laplace Equation

The method of separation of variables will be employed to obtain solutions to the Laplace equation. The method consists of assuming the solution to be a product of functions, each depending one coordinate variable only. The use of the method can be best illustrated by working out examples in various fields in Physics and Engineering and in various coordinate systems. The method requires the separability of the Laplacian operator into two or three ordinary differential equations. A few of the orthogonal and separable coordinate systems are presented in Appendix C.

Example 6.1 *Steady State Temperature Distribution in a Rectangular Sheet*

Obtain the steady state temperature distribution in a rectangular slab, occupying the space $0 \le x \le L$ and $0 \le y \le H$, where the boundary conditions are specified as follows:

$$T = T(x, y)$$

$$T(0, y) = f(y) \qquad T(x, 0) = 0$$

$$T(L, y) = 0 \qquad T(x, H) = 0$$

Since the sheet is thin, we can assume that the temperature differential is only a function of x and y, i.e. T(x,y). The differential equation on the temperature satisfies the Laplace equation:

$$\nabla^2 T = \frac{\partial^2 T}{\partial x^2} + \frac{\partial^2 T}{\partial y^2} = 0$$

Assume that the solution can be written in the form of a product of two single variable functions as follows:

$$T(x,y) = X(x)\, Y(y) \qquad X \ne 0,\ Y \ne 0,\ 0 < x < L \text{ and } 0 < y < H$$

Substituting T(x,y) into the differential equation, one obtains:

$$Y\frac{d^2X}{dx^2} + X\frac{d^2Y}{dy^2} = 0$$

Dividing out by XY, the equation becomes:

$$\frac{X''}{X} = -\frac{Y''}{Y}$$

Since both sides of the equality in the above equation are functions of one variable only, then the equality must be set equal to a real constant, $\pm a^2$.

Choosing $a^2 \ge 0$, then the Laplace Equation is transformed into two ordinary differential equations:

$$X'' - a^2 X = 0$$

and

$$Y'' + a^2 Y = 0$$

which has the following solutions:

for $a \neq 0$: $X = A \sinh(ax) + B \cosh(ax)$

$Y = C \sin(ay) + D \cos(ay)$

for $a = 0$: $X = A\,x + B$

$Y = C\,y + D$

Applying the boundary conditions to the solution, and assuring non-trivial solutions, one obtains:

$T(x,0) = D\,X(x) = 0 \;\rightarrow\; D = 0$

$T(x,H) = C \sin(aH) \bullet X(x) = 0 \quad \rightarrow \quad \sin(aH) = 0$

To satisfy the characteristic equation, sin (aH) = 0, "a" must take one of the following characteristic values:

$$a_n = \frac{n\pi}{H} \qquad n = 1, 2, 3, \ldots$$

The non-trivial solution of Y(y) consists of an eigenfunction set:

$$\phi_n(y) = \sin(n\pi y/H)$$

where the eigenfunctions $\phi_n(y)$ are orthogonal over [0,H], i.e.:

$$\int_0^H \sin(n\pi y/H)\,\sin(m\pi y/H)\,dy = \begin{cases} 0 & n \neq m \\ H/2 & n = m \end{cases}$$

The a = 0 case results in a trivial solution. Substituting the solution into the second boundary condition:

$$T(L,y) = [A \sinh(a_n L) + B \cosh(a_n L)]\,Y(y) = 0$$

which can be satisfied if:

$$B/A = -\tanh(a_n L)$$

Finally, the solution can be written as:

$$T_n(x,y) = \sin(\frac{n\pi}{H}y)\,[\sinh(\frac{n\pi}{H}x) - \tanh(\frac{n\pi}{H}L)\cosh(\frac{n\pi}{H}x)] \qquad n = 1, 2, \ldots$$

$T_n(x,y)$ satisfies the Laplace Equation and three homogeneous boundary conditions. Due to the linearity of the system, one can use the principle of superposition, such that the temperature in the slab can be written in terms of an infinite Generalized Fourier series in terms of the solutions $T_n(x,y)$, i.e.:

$$T(x,y) = \sum_{n=1}^{\infty} E_n\,T_n(x,y)$$

The remaining non-homogeneous boundary condition can be satisfied by the total solution T(x,y) as follows:

$$T(0,y) = f(y) = \sum_{n=1}^{\infty} -E_n \tanh(\frac{n\pi}{H}L)\sin(\frac{n\pi}{H}y)$$

Using the orthogonality of the eigenfunctions, one obtains an expression for the Fourier constants E_n as:

$$E_n = -\frac{2}{H\tanh(\frac{n\pi}{H}L)}\int_0^H f(y)\sin(\frac{n\pi}{H}y)\,dy$$

Note that the choice of sign for the separation constant is not arbitrary. If $-a^2$ was chosen with $a^2 > 0$, then the above analysis must be repeated:

$$X'' + a^2 X = 0$$

$$Y'' - a^2 Y = 0$$

whose solutions become for $a \neq 0$:

$$X = A \sin(ax) + B \cos(ax)$$

$$Y = C \sinh(ay) + D \cosh(ay)$$

The solution must satisfy the boundary conditions:

$$T(x,0) = D\,X(x) = 0, \text{ or } D = 0$$

$$T(x,H) = C \sinh(aH) \bullet X(x) = 0$$

However, since sinh (aH) cannot vanish unless $a = 0$, then:

$$C = 0 \quad \text{for } a \neq 0.$$

Thus, for $-a^2$, there is no non-trivial solution that can satisfy the differential equation and the boundary conditions. This indicates that the choice of the sign of a^2 leads to the either the existence of non-trivial solutions, or to the trivial solution.

In order to eliminate the guesswork and minimize unnecessary work, the choice of the correct sign of a^2 can be made by examining the boundary conditions. Since the solution involves an expansion in a Generalized Fourier series, then one would need an eigenfunction set. These eigenfunctions must satisfy homogeneous boundary conditions. Furthermore, these eigenfunctions must be non-monotonic functions, specifically, they are oscillating functions with one or more zeroes. Thus, for this example, since the boundary conditions were homogeneous in the y-coordinate, then choose the sign of a^2 to give an oscillating function in y and not in x. This leads to a choice of $a^2 \geq 0$.

If the temperature is prescribed on all four boundaries, one can use the principle of superposition by separating the problem into four problems as follows. Let:

$$T = T_1 + T_2 + T_3 + T_4$$

where $\nabla^2 T_i = 0$, $i = 1, 2, 3, 4, \ldots$. Each solution T_i satisfies one non-homogeneous boundary condition on one side and three homogeneous boundary conditions on the remaining three sides, resulting in four new problems. Each of these problems would resemble the problem above, yielding four different solutions. The solution then would be the sum of the four solutions $T_i(x,y)$.

Example 6.2 *Steady State Temperature Distribution in an Annular Sheet*

Obtain the temperature distribution in an annular sheet with outer and inner radii b and a, respectively. The sheet is insulated at its inner boundary, and the temperature $T = T(r,\theta)$ is prescribed at the outer boundary as follows:

$$T(b,\theta) = f(\theta)$$

Laplace's equation in cylindrical coordinates, where $T = T(r,\theta)$, becomes (Appendix C):

$$\frac{\partial^2 T}{\partial r^2} + \frac{1}{r}\frac{\partial T}{\partial r} + \frac{1}{r^2}\frac{\partial^2 T}{\partial \theta^2} = 0$$

The boundary conditions can be stated as follows:

$$\left.\frac{\partial T}{\partial n}\right|_C = -\frac{\partial T(a,\theta)}{\partial r} = 0$$

and

$$T(b,\theta) = f(\theta)$$

Assuming that the solution is separable and can be written in the form:

$$T = R(r)\, U(\theta)$$

and substituting the solution into Laplace's Equation, one can show that:

$$\frac{r^2 R'' + r R'}{R} = -\frac{U''}{U} = k^2 = \text{constant}$$

The choice of the sign for k^2 is based on the coordinate with the homogeneous boundary conditions. Since the boundary condition on r is non-homogeneous, then choosing $k^2 > 0$ leads to an oscillating function in θ. Thus, two ordinary differential equations result:

$$r^2 R'' + rR' - k^2 R = 0$$

$$U'' + k^2 U = 0$$

If $k = 0$, then the solution becomes:

$$U = A_o + B_o\,\theta$$

$$R = C_o + D_o \log r$$

If $k \neq 0$, then the solution becomes:

$$U = A \sin(k\theta) + B \cos(k\theta)$$

$$R = C\, r^k + D\, r^{-k}$$

The solution must be tested for single-valuedness and for boundedness. Single-valuedness of the solution requires that:

$$T(r,\theta) = T(r,\theta + 2\pi)$$

Thus:

$B_o = 0$ for $k = 0$

$\sin(k\theta) = \sin k(\theta + 2\pi)$ for $k \neq 0$

$\cos(k\theta) = \cos k(\theta + 2\pi)$ for $k \neq 0$

which can be satisfied if k is an integer n = 1, 2, 3, ... Therefore, the solution takes the form:

$k = 0 \qquad T_o = E_o + F_o \log r$

$k = n \qquad T_n = (A_n \sin(n\theta) + B_n \cos(n\theta))(C_n r^n + D_n r^{-n}) \qquad n = 1, 2, 3, ...$

Remember that these solutions must also satisfy the boundary condition $\frac{\partial T}{\partial r}(a,\theta) = 0$.

Therefore:

$$k = 0 \qquad \frac{\partial T_o}{\partial r}(a,\theta) = \frac{F_o}{a} = 0 \;\rightarrow\; F_o = 0$$

$$k = n \qquad \frac{\partial T_n}{\partial r}(a,\theta) = nC_n a^{n-1} - nD_n a^{-n-1} = 0 \;\rightarrow\; D_n = a^{2n} C_n$$

Thus, one can write the general solution in a Generalized Fourier series, i.e.:

$$T(r,\theta) = E_o + \sum_{n=1}^{\infty} (r^n + a^{2n} r^{-n})(E_n \cos(n\theta) + F_n \sin(n\theta))$$

where E_n and F_n are the unknown Fourier coefficients. The last non-homogeneous boundary condition can be satisfied by $T(r, \theta)$ as follows:

$$T(b,\theta) = f(\theta) = E_o + \sum_{n=1}^{\infty} (b^n + a^{2n} b^{-n})(E_n \cos(n\theta) + F_n \sin(n\theta))$$

Then, using the orthogonality of the eigenfunctions, one can obtain expressions for the Fourier coefficients:

$$E_o = \frac{1}{2\pi} \int_0^{2\pi} f(\theta)\, d\theta$$

$$E_n = \frac{1}{\pi(b^n + a^{2n} b^{-n})} \int_0^{2\pi} f(\theta) \cos(n\theta)\, d\theta \qquad n = 1, 2, 3, \ldots$$

and

$$F_n = \frac{1}{\pi(b^n + a^{2n} b^{-n})} \int_0^{2\pi} f(\theta) \sin(n\theta)\, d\theta \qquad n = 1, 2, 3, \ldots$$

If $f(\theta)$ is constant $= T_o$, then:

$$E_n = F_n = 0 \qquad n = 1, 2, 3, \ldots$$

and

$$E_o = T_o$$

Thus, the temperature in the annular sheet is constant and equals T_o.

Example 6.3 *Steady State Temperature Distribution in a Solid Sphere*

Obtain the steady state temperature distribution in a solid sphere of radius = a, where $T = T(r,\theta)$, and has the temperature specified on its surface r = a as follows:

$$T(a,\theta) = f(\theta)$$

Examination of the boundary condition indicates that the temperature distribution in the sphere is axi-symmetric, i.e, $\partial/\partial\phi = 0$. Thus, from Appendix C:

$$\nabla^2 T = \frac{1}{r^2}\left[\frac{\partial}{\partial r}\left(r^2\frac{\partial T}{\partial r}\right) + \frac{1}{\sin\theta}\frac{\partial}{\partial\theta}\left(\sin\theta\frac{\partial T}{\partial\theta}\right)\right] = 0$$

Let $T(r,\theta) = R(r)\, U(\theta)$, then:

$$\frac{1}{R}\frac{d}{dr}\left(r^2\frac{dR}{dr}\right) = -\frac{1}{U\sin\theta}\frac{d}{d\theta}\left(\sin\theta\frac{dU}{d\theta}\right) = k^2$$

Since the non-homogeneous boundary condition is in the r-coordinate, then $k^2 > 0$ results in an eigenfunction in the θ - coordinate. Thus, the two components satisfy the following equations:

$$r^2R'' + 2r\,R' - k^2R = 0$$

and

$$U'' + (\cot\theta)\,U' + k^2U = 0$$

Transforming the independent variable from θ to η, such that:

$$\eta = \cos\theta \qquad -1 \le \eta \le 1$$

then U satisfies the following equation:

$$\frac{d}{d\eta}\{(1-\eta^2)\frac{dU}{d\eta}\} + k^2U = 0$$

Letting $k^2 = \nu(\nu + 1)$, where $\nu \ge 0$, then the solution to the differential equation becomes:

$$U(\eta) = A_\nu\, P_\nu(\eta) + B_\nu\, Q_\nu(\eta)$$

and

$$R(r) = C_\nu r^\nu + D_\nu r^{-(\nu+1)} \qquad \text{for } \nu \ne -\frac{1}{2}$$

$$= Er^{-1/2} + Fr^{-1/2}\log(r) \qquad \text{for } \nu = -\frac{1}{2}$$

The temperature must be bounded at $r = 0$, and $\eta = \pm 1$, thus:

$$\nu = \text{integer} = n, \qquad n = 0, 1, 2 \ldots$$

$$B_n = 0, \qquad \text{and } D_n = 0, \qquad n = 0, 1, 2, \ldots$$

Thus, the eigenfunctions satisfying Laplace's equation and bounded inside the sphere has the form:

$$T_n(r, \eta) = r^n P_n(\eta)$$

and the general solution can be written as Generalized Fourier series in terms of all possible eigenfunctions:

$$T(r, \eta) = \sum_{n=0}^{\infty} E_n T_n(r, \eta)$$

Satisfying the remaining non-homogeneous boundary condition at the surface $r = a$, one obtains:

$$T(a, \eta) = g(\eta) = \sum_{n=0}^{\infty} E_n a^n P_n(\eta)$$

where:

$$g(\eta) = f(\cos^{-1} \eta)$$

Using the orthogonality of the eigenfunctions, the Fourier coefficients are given by:

$$E_n = \frac{2n+1}{2a^n} \int_{-1}^{1} g(\eta) P_n(\eta) d\eta$$

and

$$T = \sum_{n=0}^{\infty} \frac{2n+1}{2} \left(\frac{r}{a}\right)^n P_n(\eta) \left[\int_{-1}^{1} g(\eta) P_n(\eta) d\eta \right]$$

If $f(\theta) = T_o$ = constant, then:

$$\int_{-1}^{1} P_n(\eta) d\eta = 2 \qquad n = 0$$

$$= 0 \qquad n = 1, 2, 3, \ldots$$

Thus, the solution inside the solid sphere with constant temperature on its surface is constant throughout, i.e.:

$$T(r, \theta) = T_o \qquad \text{everywhere.}$$

Example 6.4 *Steady State Temperature Distribution in a Solid Cylinder*

Obtain the temperature distribution in a cylinder of length, L, and radius, c, such that the temperature at its surfaces are prescribed as follows:

$$T = T(r,\theta,z)$$

$$T(c,\theta,z) = f(\theta,z)$$

$$T(r,\theta,0) = 0$$

$$T(r,\theta,L) = 0$$

The differential equation satisfied by the temperature, T, becomes, Appendix C:

$$\frac{\partial^2 T}{\partial r^2} + \frac{1}{r}\frac{\partial T}{\partial r} + \frac{1}{r^2}\frac{\partial^2 T}{\partial \theta^2} + \frac{\partial^2 T}{\partial z^2} = 0$$

Let $T = R(r)\, U(\theta)\, Z(z)$, then the equation can be put in the form:

$$\frac{R''}{R} + \frac{1}{r}\frac{R'}{R} + \frac{1}{r^2}\frac{U''}{U} + \frac{Z''}{Z} = 0$$

Letting:

$$\frac{Z''}{Z} = -a^2 \qquad \text{and} \qquad \frac{U''}{U} = -b^2$$

then the partial differential equation separates into the following three ordinary differential equations:

$$r^2 R'' + r R' - (a^2 r^2 + b^2) R = 0$$

$$U'' + b^2 U = 0$$

$$Z'' + a^2 Z = 0$$

The choice of the sign for a^2 and b^2 are again guided by the boundary conditions. Since one of the boundary conditions in the r-coordinate is not homogeneous, then one needs to specify the sign of $a^2 > 0$ and $b^2 > 0$ to assure that the solutions in the z and θ coordinates are oscillatory functions.

There are four distinct solutions to the above equations, depending on the value of a and b:

(1) If $a \neq 0$ and $b \neq 0$, then the solutions become:

$$R = A_b\, I_b(ar) + B_b\, K_b(ar)$$

$$Z = C_a \sin(az) + D_a \cos(az)$$

$$U = E_b \sin(b\theta) + F_b \cos(b\theta)$$

where I_b and K_b are the modified Bessel Functions of the first and second kind of order b. For single-valuedness of the solution, $U(\theta) = U(\theta + 2\pi)$, requires that:

b is an integer = n = 1, 2, 3, ...

For boundedness of the solution at $r = 0$, one must set $B_n = 0$. The boundary conditions are satisfied next:

$$T(r,\theta,0) = 0 \qquad\qquad D_a = 0$$

$T(r,\theta,L) = 0 \qquad \sin(aL) = 0$, then $a_M L = m\pi \qquad$ where $m = 1, 2, 3, \ldots$

Thus, the eigenfunctions satisfying homogeneous boundary conditions are:

$$T_{nm} = \left(G_{nm}\sin(n\theta) + H_{nm}\cos(n\theta)\right) I_n\left(\frac{m\pi}{L}r\right)\sin\left(\frac{m\pi}{L}z\right) \qquad m, n = 1, 2, 3, \ldots$$

(2) If $a \neq 0$, $b = 0$, then the solutions become:

$$R = A_o I_o(ar) + B_o K_o(ar)$$

$$Z = C_a \sin(az) + D_a \cos(az)$$

$$U = E_o\theta + F_o$$

Again, single-valuedness requires that $E_o = 0$, and boundedness at $r = 0$ requires that $B_o = 0$, and:

$T(r,\theta,0) = 0 \qquad D_a = 0$

$T(r,\theta,L) = 0 \qquad \sin(aL) = 0 \qquad a_m L = m\pi, \qquad m\ 1, 2, 3, \ldots$

The solutions for this case are:

$$T_{om} = I_o\left(\frac{m\pi}{L}r\right)\sin\left(\frac{m\pi}{L}z\right)$$

(3) If $a = 0$, $b \neq 0$, then the solutions become:

$$R = A_b r^b + B_b r^{-b}$$

$$Z = C_o z + D_o$$

$$U = E_b \sin(b\theta) + F_b \cos(b\theta)$$

Single-valuedness requires that b = integer = $n = 1, 2, 3, \ldots$ and boundedness requires that $B_n = 0$. Therefore, the boundary conditions imply:

$T(r,\theta,0) = 0 \qquad D_o = 0$

$T(r,\theta,L) = 0 \qquad C_o = 0$

which results in a trivial solution:

$$T_{no} = 0$$

(4) If $a = 0$, $b = 0$, then the solutions become:

$$R = A_o \log r + B_o$$

$$Z = C_o z + D_o$$

$$U = E_o\,\theta + F_o$$

Single-valuedness requires that $E_o = 0$, and boundedness requires that $A_o = 0$:

$$T(r,\theta,0) = 0 \qquad D_o = 0$$

$$T(r,\theta,L) = 0 \qquad C_o = 0$$

which results in trivial solution:

$$T_{0,0} = 0$$

Finally, the solutions of the problem can be written as:

$$T_{nm} = I_n\left(\frac{m\pi}{L}r\right)\sin\left(\frac{m\pi}{L}z\right)\begin{bmatrix}\sin(n\theta)\\ \cos(n\theta)\end{bmatrix} \qquad n = 0, 1, 2, 3, \ldots \qquad m = 1, 2, 3, \ldots$$

The solutions T_{nm} contains orthogonal eigenfucntions in z and θ. The general solution can then be written as a Generalized Fourier series in terms of the general solutions T_{nm} as follows:

$$T = \sum_{n=0}^{\infty}\sum_{m=1}^{\infty} I_n\left(\frac{m\pi}{L}r\right)\sin\left(\frac{m\pi}{L}z\right)(G_{nm}\sin(n\theta) + H_{nm}\cos(n\theta))$$

Satisfying the remaining non-homogeneous boundary condition at r = c results in:

$$T(c,\theta,z) = f(\theta,z) = \sum_{n=0}^{\infty}\sum_{m=1}^{\infty} I_n\left(\frac{m\pi}{L}c\right)\sin\left(\frac{m\pi}{L}z\right)(G_{nm}\sin(n\theta) + H_{nm}\cos(n\theta))$$

Using the orthogonality of the Fourier sine and cosine series, one can evaluate the Fourier coefficients:

$$G_{nm} = \frac{2}{\pi L I_n(\frac{m\pi}{L}c)}\int_0^{2\pi}\int_0^{L} f(\theta,z)\sin(n\theta)\sin(\frac{m\pi}{L}z)\,dz\,d\theta \quad m, n = 1, 2, 3, \ldots$$

and

$$H_{nm} = \frac{\varepsilon_n}{\pi L I_n(\frac{m\pi}{L}c)}\int_0^{2\pi}\int_0^{L} f(\theta,z)\cos(n\theta)\sin(\frac{m\pi}{L}z)\,dz\,d\theta \qquad m = 1, 2, 3, \ldots$$

$$n = 0, 1, 2, \ldots$$

where ε_n is the Neumann factor.

Example 6.5 *Ideal Fluid Flow Around an Infinite Cylinder*

Obtain the particle velocity of an ideal fluid flowing around an infinite rigid impenetrable cylinder of radius a. The fluid has a velocity = V_o for r >> a. Since the cylinder is infinite and the fluid velocity at infinity is independent of z, then the velocity potential is also independent of z. The velocity potential ϕ satisfies Laplace's equation:

$$\nabla^2\phi = 0 \qquad \text{where} \qquad \phi = \phi\,(r,\theta)$$

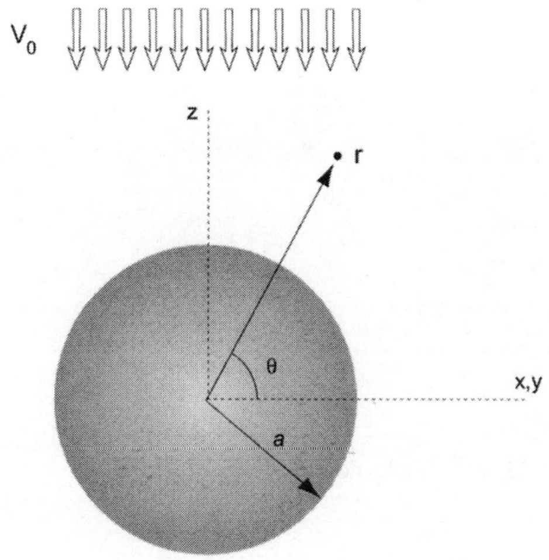

and the particle velocity is defined by:

$$\vec{V} = -\nabla\phi$$

The boundary conditions r = a requires that the normal velocity vanishes at r = a, i.e.:

$$\vec{n}\bullet\vec{V} = V_n = -\frac{\partial\phi(a,\theta)}{\partial r} = 0$$

Let $\phi = R(r)\,U(\theta)$, then:

$$r^2 R'' + r R' - k^2 R = 0$$

$$U'' + k^2 U = 0$$

If k = 0, then the solutions are:

$$U = A_o\theta + B_o$$

$$R = C_o \log r + D_o$$

If $k \neq 0$:

$$U = A_k \sin(k\theta) + B_k \cos(k\theta)$$

$$R = C_k r^k + D_k r^{-k}$$

The velocity components in the r and θ directions are the radial velocity, $V_r = -\dfrac{\partial\phi}{\partial r}$ and the angular velocity, $V_\theta = -\dfrac{1}{r}\dfrac{\partial\phi}{\partial\theta}$. Both of these components must be single valued. The velocity field for k = 0 is:

$$V_\theta = -\frac{1}{r}(C_o \log r + D_o)A_o \qquad \text{and} \qquad V_r = -\frac{C_o}{r}(A_o\theta + B_o)$$

Single-valuedness of the velocity field requires that $A_o = 0$, since $V_r(\theta) = V_r(\theta + 2\pi)$. For $k \neq 0$, the velocity field is:

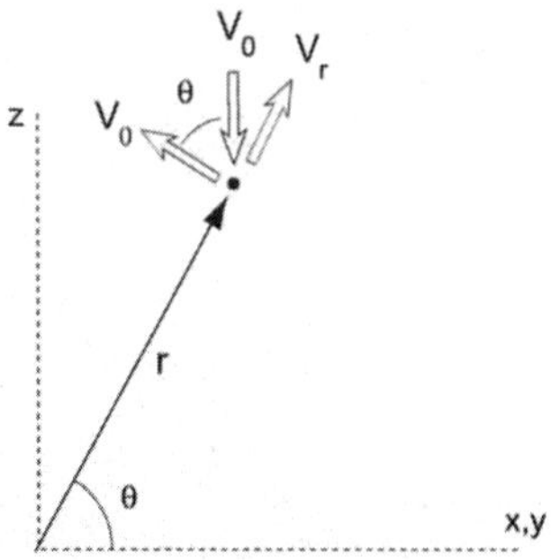

$$V_\theta = -k\,(C_k\, r^{k-1} + D_k\, r^{-(k+1)})\,(A_k \cos(k\theta) - B_k \sin(k\theta))$$

and

$$V_r = -\,k\,(C_k r^{k-1} - D_k r^{-(k+1)})\,(A_k \sin(k\theta) + B_k \cos(k\theta))$$

Requiring that $V_r(\theta) = V_r(\theta + 2\pi)$ or $V_\theta(\theta) = V_\theta(\theta + 2\pi)$ dictates that k is an integer = n. Thus, the velocity potential becomes:

$$\phi_0 = B_0\,(C_0 \log r + Do)$$

and

$$\phi_n = (A_n \sin(n\theta) + B_n \cos(n\theta))\,(C_n\, r^n + D_n\, r^{-n}) \qquad n = 1, 2, 3, \ldots$$

Furthermore, the velocity field must be bounded as $r \to \infty$. Examining the expressions for V_r and V_θ for $r >> a$ and $k = n \geq 1$, then boundedness as $r \to \infty$ requires that $C_n = 0$ for $n \geq 2$. The boundary condition must be satisfied at r = a:

$$V_r(a,\theta) = 0$$

for k = 0:

$$V_r(a,\theta) = -\frac{C_o}{a} B_o = 0 \qquad \text{or} \qquad C_o B_o = 0$$

for k = n:

$$V_r(a,\theta) = -n(C_n r^{n-1} - D_n r^{-(n+1)})(A_n \sin(n\theta) + B_n \cos(n\theta))\Big|_{r=a} = 0$$

or

$$D_n = a^{2n}\, C_n$$

and, hence:

$$D_n = 0 \qquad \text{for } n \geq 2$$

Thus, the general solution for the velocity potential becomes:

$$\phi = E_o + (E_1 \cos(\theta) + F_1 \sin(\theta))(r + a^2 r^{-1})$$

The radial and angular velocities become:

$$V_r = -(E_1 \cos(\theta) + F_1 \sin(\theta))(1 - \frac{a^2}{r^2})$$

$$V_\theta = +(E_1 \cos(\theta) - F_1 \sin(\theta))(1 + \frac{a^2}{r^2})$$

The radial and angular velocities must approach the given velocity V_o in the far-field of the cylinder, i.e.:

$$V_r \to -V_o \sin(\theta) \qquad \text{and} \qquad V_\theta \to -V_o \cos(\theta)$$

which, when compared to the expressions for V_r and V_θ, gives:

$$E_1 = 0 \qquad \text{and} \qquad F_1 = V_o$$

Thus, the solution for the velocity field takes the final form:

$$V_r = -V_o (1 - \frac{a^2}{r^2}) \sin(\theta)$$

$$V_\theta = -V_o (1 + \frac{a^2}{r^2}) \cos(\theta)$$

and

$$\phi = E_o + V_o (r + \frac{a^2}{r}) \sin(\theta)$$

Note that the velocity potential is unique to within a constant, due to the Neumann boundary condition, and unbounded. However, all the physical quantities (V_r and V_θ) are unique, single-valued and bounded.

Example 6.6 *Electrostatic Field Within a Sphere*

Obtain the electric field strength produced in two metal hemispheres, radius r = a, separated by a narrow gap, the surface of the upper half has a constant potential ϕ_o, the surface of the lower half is being kept at zero potential, i.e.:

$$\phi(a,\theta) = f(\theta) = \begin{cases} \phi_o & 0 \le \theta < \pi/2 \\ 0 & \pi/2 < \theta \le \pi \end{cases}$$

Since the sphere's shape and the boundary condition are independent of the polar angle, then the solutions can be assumed to be independent of the polar angle, i.e. axisymmetric. The equation satisfied by the electric potential ϕ in a spherical coordinates, for axisymmetric distribution, is given by (Appendix C):

$$\nabla^2\phi = \frac{\partial^2\phi}{\partial r^2} + \frac{2}{r}\frac{\partial\phi}{\partial r} + \frac{1}{r^2}\frac{\partial^2\phi}{\partial\theta^2} + \frac{\cot(\theta)}{r^2}\frac{\partial\phi}{\partial\theta} = 0$$

Let $\phi(r,\theta) = R(r)\, U(\theta)$, then the solution as given in Example 6.3 becomes:

$$\phi_k = [A_k P_k(\eta) + B_k Q_k(\eta)][C_k r^k + D_k r^{-k-1}]$$

where $\eta = \cos\theta$

Boundedness of the voltages E_r and E_θ at r = 0 and $\eta = \pm 1$ requires that k = integer = n, $B_n = 0$ and $D_n = 0$. Thus, the solution which satisfies Laplace's equation is:

$$\phi_n(r,\eta) = r^n P_n(\eta) \qquad n = 0, 1, 2, \ldots$$

and the general solution can be written as a Generalized Fourier series:

$$\phi = \sum_{n=0}^{\infty} F_n\, r^n\, P_n(\eta)$$

Satisfying the boundary condition at r = a:

$$\phi\,(a,\eta) = \sum_{n=0}^{\infty} F_n\, r^n\, P_n(\eta) = g(\eta)$$

where $g(\eta) = f(\cos^{-1}\eta)$. Thus, using the orthogonality of the eignefunctions $P_n(\eta)$, results in the following expression for the Fourier coefficients:

$$F_n = \frac{2n+1}{2a^n} \bullet \int_{-1}^{+1} g(\eta) P_n(\eta)\, d\eta = \frac{2n+1}{2a^n}\phi_o \int_{0}^{+1} P_n(\eta)\, d\eta$$

The first few Fourier constants become:

$$F_o = \frac{\phi_o}{2} \qquad F_1 = \frac{3}{4a}\phi_o \qquad F_3 = -\frac{7}{16a^3}\phi_o \qquad F_5 = \frac{11}{32a^5}\phi_o$$

$$F_{2n} = 0 \qquad n = 1,2,3,4,...$$

Therefore, the potential can be written as:

$$\phi\,(r,\eta) = \frac{\phi_o}{2}\left\{1 + \frac{3}{2}\left(\frac{r}{a}\right) P_1(\eta) - \frac{7}{8}\left(\frac{r}{a}\right)^3 P_3(\eta) + \frac{11}{16}\left(\frac{r}{a}\right)^5 P_5(\eta) - \ldots\right\}$$

The electric field strength $\vec{E} = E_r\,\vec{e}_r + E_\theta\,\vec{e}_\theta = \nabla\phi$ can be evaluated as follows:

$$E_r(r,\eta) = -\frac{\partial\phi}{\partial r} = -\frac{\phi_o}{2a}\left\{\frac{3}{2}P_1(\eta) - \frac{21}{8}\left(\frac{r}{a}\right)^2 P_3(\eta) + \frac{55}{16}\left(\frac{r}{a}\right)^4 P_5(\eta) - \ldots\right\}$$

$$E_\theta(r,\eta) = -\frac{1}{r}\frac{\partial\phi}{\partial\theta} = -\frac{\phi_o}{2a}\sqrt{1-\eta^2}\left\{\frac{3}{2}P_1'(\eta) - \frac{7}{8}\left(\frac{r}{a}\right)^2 P_3'(\eta) + \frac{11}{16}\left(\frac{r}{a}\right)^4 P_5'(\eta) - \ldots\right\}$$

6.10 The Poisson Equation

Solution of Poisson's equation may be obtained in terms of eigenfunctions. Two distinct types of problems involving Poisson's equation will be discussed; those with homogeneous boundary conditions and those with non-homogeneous ones.

In problems involving homogeneous boundary conditions, one may attempt to construct an orthogonal eigenfunction set first, which is then used to expand the source function in Poissons equation.

Start with the following system:

$$\nabla^2\phi = f(P) \qquad \text{P in V} \tag{6.49}$$

together with homogeneous boundary conditions of the Dirichlet, Neumann or Robin type, written in general form:

$$U_i(\phi(P)) = 0 \qquad P \text{ on } S \tag{6.50}$$

Starting with the Helmholtz equation

$$\nabla^2\psi + \lambda\psi = 0 \tag{6.51}$$

whose solution must satisfy the same homogeneous boundary conditions that ϕ satisfies, i.e.:

$$U_i(\psi(P)) = 0 \qquad P \text{ on } S$$

The homogeneous Helmholtz system in eqs. (6.50) and (6.51) would generate an orthogonal eigenfunction set $\{\psi_M(P)\}$, M being one, two, or three dimensional integer, such that:

$$\nabla^2\psi_M + \lambda\psi_M = 0 \tag{6.52}$$

where each eigenfunction satisfies $U_i(\psi_M(P)) = 0$. The eigenfunction set is orthogonal where the orthogonality integral is defined by:

$$\int_V \psi_M \psi_K \, dV = 0 \qquad M \neq K$$

$$= N_M \qquad M = K \tag{6.53}$$

Expanding the solution in Generalized Fourier series in terms of the orthogonal eigenfunctions:

$$\phi = \sum_M E_M \, \psi_M(P) \tag{6.54}$$

and substituting eq. (6.54) in Poisson's equation (6.49) and eq. (6.52):

$$\nabla^2\phi = \sum_M E_M \nabla^2\psi_M(P) = -\sum_M \lambda_M E_M \psi_M(P) = f(P)$$

One can use the orthogonality integral in eq. (6.53) to obtain an expression for the Fourier coefficients E_M as:

$$E_M = \frac{-1}{N_M \lambda_M} \int_V \psi_M(P) f(P) \, dV \tag{6.55}$$

If the system is completely inhomogeneous, in other words if the equation is of the Poisson's type and the boundary conditions are inhomogeneous, one can use the linearity of the problem and linear superposition to obtain the solution. Thus, for the following system:

$$\nabla^2\phi = f(P) \qquad P \text{ in } V \tag{6.49}$$

subject to the general form of boundary condition:

$$k\frac{\partial\phi(P)}{\partial n} + h\phi(P) = g(P) \qquad P \text{ on } S \qquad k, h \geq 0 \tag{6.56}$$

where k and h may or may not be zero. Let the solution be a linear combination of two solutions:

$$\phi = \phi_1 + \phi_2$$

such that ϕ_1 and ϕ_2 satisfy the following systems:

$$\nabla^2\phi_1 = 0 \qquad\qquad \nabla^2\phi_2 = f(P)$$

$$k\frac{\partial\phi_1(P)}{\partial n} + h\phi_1(P) = g(P) \qquad k\frac{\partial\phi_2(P)}{\partial n} + h\phi_2(P) = 0 \qquad P \text{ on } S \tag{6.57}$$

Thus, ϕ_1 satisfies a Laplacian system and ϕ_2 satisfies a Poisson's system with homogeneous boundary conditions.

Example 6.7 *Heat Distribution in an Annular Sheet*

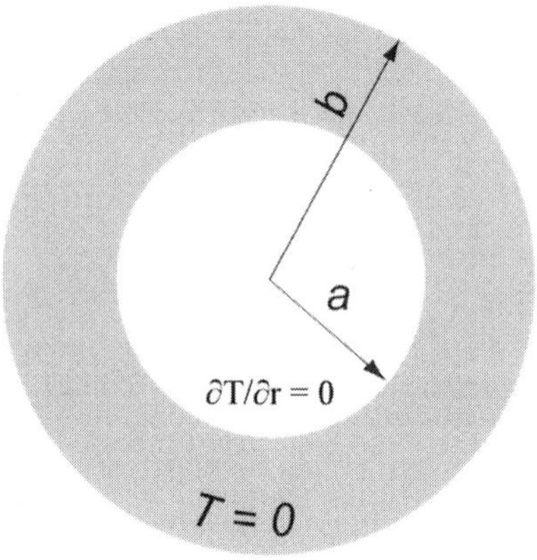

Obtain the temperature distribution in an annular sheet with heat source distribution q, such that the temperature satisfies:

$$\nabla^2 T = -\, q(r,\theta)$$

The outer boundary of the sheet, at r = b, is kept at zero temperature, while the inner boundary, at r = a, is insulated, i.e. for $T = T(r,\theta)$:

$$T(b,\theta) = 0 \qquad \text{and} \qquad \frac{\partial T(a,\theta)}{\partial r} = 0$$

The system, from which one can obtain an eigenfunction set, can be written in the form of the Helmholtz equation satisfying the same homogeneous boundary conditions, i.e.:

$$\nabla^2\psi + l^2\psi = 0 \qquad l^2 \text{ undertermined}$$

$$\psi(b,\theta) = 0 \qquad \frac{\partial\psi}{\partial r}(a,\theta) = 0$$

Let $\psi(r,\theta) = R(r)\,F(\theta)$, then the equation becomes:

$$\frac{r^2R'' + rR'}{R} + \frac{F''}{F} + l^2r^2 = 0$$

or

$$F'' + k^2 F = 0$$

$$r^2 R'' + r R' + (l^2 r^2 - k^2) R = 0$$

where the sign of the separation constant k^2 is chosen to give oscillating functions in the r and θ coordinates, since the boundary conditions are homogeneous in both variables. The solutions of the two ordinary differential equations become for $l \neq 0$:

$$F = A \sin(k\theta) + B \cos(k\theta)$$

$$R = C\, J_k\,(l\, r) + D\, Y_k\,(l\, r)$$

Single-valuedness requires that k is an integer = n, where n = 0, 1, 2, ... The two homogeneous boundary conditions are satisfied as follows:

$$C\, J_n\,(l\, b) + D\, Y_n\,(l\, b) = 0$$

$$C\, J'_n\,(l\, a) + D\, Y'_n\,(l\, a) = 0$$

which results in the following characteristic equation:

$$J_n(l\, b)\, Y'_n(l\, a) - J'_n(l\, a)\, Y_n(l\, b) = 0 \qquad l \neq 0$$

The characteristic equation can be written in terms of the ratio of the radii, c = b/a, i.e.:

$$J_n(c\, l\, a)\, Y'_n(l\, a) - J'_n(l\, a)\, Y_n(c\, l\, a) = 0$$

which has an infinite number of roots for each equation whose index is n:

$$l_{nm}\, a = \mu_{nm} \qquad m = 1, 2, 3, \ldots \qquad n = 0, 1, 2, \ldots$$

where μ_{nm} represents the mth root of the nth characteristic equation. The ratio of the constatns D/C is given by:

$$\frac{D}{C} = -\frac{J_n(c\mu_{nm})}{Y(c\mu_{nm})}$$

which can be substituted into the expression for R(r). Thus, the eigenfunctions ψ_{nm} can be written as follows:

$$\psi_{nm} = R_{nm}(r) \begin{bmatrix} \sin(n\theta) \\ \cos(n\theta) \end{bmatrix} \qquad n = 0, 1, 2, \ldots \qquad m = 1, 2, 3, \ldots$$

where:

$$R_{nm}(r) = J_n(\mu_{nm} \frac{r}{a}) - \left[\frac{J_n(c\,\mu_{nm})}{Y_n(c\,\mu_{nm})} Y_n(\mu_{nm} \frac{r}{a})\right]$$

It should be noted that angular eigenfunctions as well as the radial eignefunctions, R_{nm}, are orthogonal, i.e.:

$$\int_a^b r\, R_{nm}\, R_{nq}\, dr = \begin{cases} 0 & \text{if} \quad m \neq q \\ N_{nm} & \text{if} \quad m = q \end{cases}$$

Expanding the temperature T in a General Fourier series in terms of the eigenfunctions $\psi_{nm}(r,\theta)$ as follows:

$$T(r,\theta) = \sum_{n=0}^{\infty} \sum_{m=1}^{\infty} R_{nm}(r)\left[A_{nm}\sin(n\theta) + B_{nm}\cos(n\theta)\right]$$

The solution for the Fourier coefficients A_{nm} and B_{nm} can be obtained in the form given in eq. (6.55), using the orthogonality of R_{nm} and the Fourier sine and cosine series as:

$$A_{nm} = \frac{a^2}{\pi \mu_{nm}^2 N_{nm}} \int_a^b \int_0^{2\pi} r\, q(r,\theta)\, R_{nm}(r) \sin(n\theta)\, d\theta\, dr \qquad n, m = 1, 2, 3, \ldots$$

$$B_{nm} = \frac{a^2 \varepsilon_n}{2\pi \mu_{nm}^2 N_{nm}} \int_a^b \int_0^{2\pi} r\, q(r,\theta)\, R_{nm}(r) \cos(n\theta)\, d\theta\, dr \qquad n = 0, 1, 2, \ldots$$

$$m = 1, 2, 3, \ldots$$

where ε_n is the Neumann factor.

6.11 The Helmholtz Equation

The solution of homogeneous and non-homogeneous Helmholtz equation is outlined in this section. Consider the Helmholtz equation (6.24):

$$\nabla^2\phi + \lambda\phi = f(P) \qquad P \text{ in } V \tag{6.24}$$

subject to homogeneous boundary conditions (6.50):

$$U_i(\phi(P)) = 0 \qquad P \text{ on } S \tag{6.50}$$

The homogeneous eigenvalue system given in eqs. (6.51) and (6.52) generate an eigenfunction set that is orthogonal as defined in eq. (6.53). The eigenfunctions $\phi_M(P)$ satisfy Helmholtz equation when $\lambda = \lambda_M$, i.e. (6.52) :

$$\nabla^2\phi_M + \lambda_M \phi_M = f(P) \tag{6.52}$$

One can show that the eigenvalues are non-negative. Multiplying the Helmholtz equation on ϕ_M by ϕ_M and integrating on V, one obtains:

$$\int_V \phi_M[\nabla^2\phi_M + \lambda_M\phi_M]dV = -\int_V (\nabla\phi_M)\bullet(\nabla\phi_M)dV + \lambda_M \int_V \phi_M^2 dV$$
$$+ \int_S \phi_M \frac{\partial\phi_M}{\partial n} dS = 0$$

which can be rewritten as:

$$\int_V |\nabla\phi_M|^2\, dV - \lambda_M \int_V \phi_M^2\, dV = \int_S \phi_M \frac{\partial\phi_M}{\partial n} dS = 0 \tag{6.58}$$

Now one can solve for λ_M for the given boundary condition:

(a) Dirichlet: $\phi_M(P) = 0$ P on S, then:

$$\lambda_M = \frac{\int\limits_V |\nabla\phi_M|^2 \, dV}{\int\limits_V \phi_M^2 \, dV} > 0 \tag{6.59}$$

(b) Neumann: $\dfrac{\partial\phi_M(P)}{\partial n} = 0$ P on S, then:

the same conclusions about λ_M in eq. (6.59) are made.

(c) Robin: $\dfrac{\partial\phi_M(P)}{\partial n} + h\phi_M(P) = 0$ P on S and $h > 0$ then:

$$\lambda_M = \frac{\int\limits_V |\nabla\phi_M|^2 \, dV + h\int\limits_S \phi_M^2 \, dS}{\int\limits_V \phi_M^2 \, dV} > 0 \tag{6.60}$$

Thus, the eigenvalues corresponding to these boundary conditions are **real and non-negative**.

One can show that the eigenfunctions are also orthogonal. Let ϕ_M and ϕ_K be two eigenfunctions satisfying eq. (6.38) corresponding to eigenvalues λ_M and λ_K, with $\lambda_M \neq \lambda_K$, i.e.:

$$\begin{aligned} \nabla^2\phi_M + \lambda_M\phi_M &= 0 \\ \nabla^2\phi_K + \lambda_K\phi_K &= 0 \end{aligned} \tag{6.61}$$

Multiplying the first equation in eq. (6.61) by ϕ_K, and the second in eq. (6.61) by ϕ_M, subtracting the resulting equalities and integrating over V, one obtains:

$$\int\limits_V [\phi_K\nabla^2\phi_M - \phi_M\nabla^2\phi_K] dV + (\lambda_M - \lambda_K)\int\limits_V \phi_M\phi_K \, dV \tag{6.62}$$

From vector calculus, it can be shown that:

$$\int\limits_V f\,\nabla^2 g \, dV = -\int\limits_V (\nabla g)\bullet(\nabla g) dV + \int\limits_S f\frac{\partial g}{\partial n} dS$$

Thus, eq. (6.62) becomes:

$$\int\limits_S [\phi_K\frac{\partial\phi_M}{\partial n} - \phi_M\frac{\partial\phi_K}{\partial n}] dS = (\lambda_K - \lambda_M)\int\limits_V \phi_M\phi_K \, dV \tag{6.63}$$

If the eigenfunctions ϕ_K and ϕ_M satisfy one of the boundary conditions [eq. (6.50)], then the left side of eq. (6.63) vanishes resulting in:

$$\int\limits_V \phi_M\,\phi_K \, dV = 0 \qquad M \neq K$$

To solve the non-homogeneous system, expand the solution ϕ in Generalized Fourier series in terms of the eigenfunctions $\psi_M(P)$ of the corresponding homogeneous system (6.54) as follows:

$$\phi = \sum_M E_M \phi_M(P) \tag{6.54}$$

Substituting the solution in (6.54) into eq. (6.24) and eq. (6.52) one obtains:

$$\begin{aligned}\nabla^2\phi + \lambda\phi &= \nabla^2 \sum_M E_M \phi_M(P) + \lambda \sum_M E_M \phi_M(P) \\ &= -\sum_M \lambda_M E_M \phi_M(P) + \lambda \sum_M E_M \phi_M(P) \\ &= \sum_M (\lambda - \lambda_M) E_M \phi_M(P) = F(P)\end{aligned} \tag{6.64}$$

Multiplying eq. (6.64) by $\phi_K(P)$ and integrating over V, one obtains, after using the orthogonality integral (6.53):

$$E_K = \frac{1}{(\lambda - \lambda_K) N_K} \int_V F(P) \phi_K \, dV \tag{6.65}$$

One notes that if $\lambda = 0$, one retrieves the solution of Poisson's equation.

A few examples of systems satisfying Helmholtz equation in the field of vibration and harmonic waves will be given below.

Example 6.8 *Forced Vibration of a Square Membrane*

Obtain the steady state response of a stretched square membrane, whose sides are fixed and have a length = L, which is being excited by distributed forces q(P,t) having the following distribution:

$$q(x,y,z,t) = q_o \sin(\omega t) \qquad q_o = \text{constant}$$

Since the forces are harmonic in time, one can assume a steady state solution for the forced vibration. Let the displacement, w(x,y,t) satisfying (equation 6.7):

$$\frac{\partial^2 w}{\partial x^2} + \frac{\partial^2 w}{\partial y^2} = \frac{1}{c^2}\frac{\partial^2 w}{\partial t^2} - \frac{q_o}{S}\sin(\omega t) \qquad c^2 = \frac{S}{\rho}$$

have the following time dependence:

$$w(x,y,t) = W(x,y)\sin(\omega t)$$

then, the amplitude of vibration W(x,y) satisfies the Helmholtz equation:

$$\nabla^2 W + k^2 W = -q_o/S \qquad k = \omega / c$$

One must find the set of orthogonal eigenfunctions of the system, such that the solution W can be expanded in them. Thus, consider the solution to the associated homogeneous Helmholtz system on $\overline{W}$:

$$\nabla^2 \overline{W} + b^2 \overline{W} = 0 \qquad \text{b undetermined constant}$$

that satisfies the following boundary condition: $\overline{W}(P) = 0$, for P on C, the contour boundary of the membrane.

Let:

$$\overline{W}(P) = X(x)\,Y(y)$$

Substituting $\overline{W}(P)$ into the Helmholtz equation results in two homogeneous ordinary differential equations:

$$X'' + (b^2 - a^2)X = 0 \qquad \begin{array}{ll} a \neq b & X = A\sin(ux) + B\cos(ux) \\ a = b & X = Ax + B \end{array}$$

$$Y'' + a^2 Y = 0 \qquad \begin{array}{ll} a \neq 0 & Y = C\sin(ay) + D\cos(ay) \\ a = 0 & Y = Cy + D \end{array}$$

where $u = \sqrt{b^2 - a^2}$. One can now solve for the separation constants, a and b, given the boundary conditions. At the boundaries: y = 0, and y = L:

$$\overline{W}(x,0) = 0 \qquad D = 0$$

$$\overline{W}(x,L) = 0 \qquad \sin(aL) = 0 \qquad a_m L = m\pi \qquad m = 1,2,3,\ldots$$

If a = 0, then C = 0, which results in a trivial solution. At the boundaries x = 0, and x = L:

$$\overline{W}(0,y) = 0 \qquad B = 0$$

$$\overline{W}(L,y) = 0 \qquad \sin(uL) = 0 \qquad u_n L = n\pi \qquad n = 1, 2, 3, \ldots$$

if a = b, A = 0, which results in a trivial solution. The eignevalues b_{nm} are thus determined by:

$$u_n = \frac{n\pi}{L} = \sqrt{b^2 - a_m{}^2}$$

$$b_{nm} = \frac{\pi}{L}\sqrt{m^2 + n^2}$$

Thus, the eigenfunctions of the system can be written as:

$$\overline{W}_{mn}(x,y) = \sin\left(\frac{n\pi}{L}x\right)\sin\left(\frac{m\pi}{L}y\right)$$

It should be noted that non-trivial solutions (Mode Shapes) exist when:

$$k_{nm} = b_{nm} = \frac{\omega_{nm}}{c}$$

so that the natural frequencies of the membrane are given by:

$$\omega_{nm} = \frac{c\pi}{L}\sqrt{m^2 + n^2}$$

Expanding the solution W in a Generalized Fourier series of the eigenfunctions:

$$W(x,y) = \sum_{n=1}^{\infty}\sum_{m=1}^{\infty} E_{nm}\sin\left(\frac{n\pi}{L}x\right)\sin\left(\frac{m\pi}{L}y\right)$$

then the Fourier coefficients E_{nm} of the double Fourier series can be obtained from eq. (6.65) in an integral form:

$$E_{nm} = \frac{-4}{L^2(k^2 - k_{nm}^2)} \int_0^L \int_0^L \frac{q_o}{S} \sin\left(\frac{n\pi}{L}x\right) \sin\left(\frac{m\pi}{L}y\right) dx\, dy$$

$$= \frac{-16}{m\, n\, \pi^2 (k^2 - k_{nm}^2)} \frac{q_o}{S} \qquad \text{if m and n are both odd}$$

$$= 0 \qquad \text{if either m or n is even}$$

Finally, the response of the membrane to a uniform dynamic load is:

$$w(x, y, t) = \frac{-16 q_o}{\pi^2 S} \sin(\omega t) \sum_{n=1}^{\infty} \sum_{m=1}^{\infty} \frac{\sin\left(\frac{n\pi}{L}x\right) \sin\left(\frac{m\pi}{L}y\right)}{m\, n (k^2 - k_{mn}^2)} \qquad \text{for m, n odd}$$

Example 6.9 *Free Vibration of a Circular Plate*

Obtain the axisymmetric mode shapes and natural frequencies of a free, vibrating plate, having a radius = a, and whose perimeter is fixed. Let the displacement of the plate w can be written as follows:

$$w(r,t) = W(r)\, e^{i\omega t}$$

then the equation of motion satisfied by W (see equation 6.11) becomes:

$$-\nabla^4 W + k^4 W = 0 \qquad k^4 = \frac{\rho h}{D} \omega^2$$

The equation can be separated as follows:

$$(\nabla^2 - k^2)(\nabla^2 + k^2) W = 0$$

whose solution can be sought to the following equations for $k \neq 0$:

$$(\nabla^2 + k^2) W = 0 \qquad W = A\, J_o(kr) + B\, Y_o(kr)$$

$$(\nabla^2 - k^2) W = 0 \qquad W = C\, I_o(kr) + D K_o(kr)$$

where J_o and Y_o are Bessel functions of first and second kind and I_o and K_o are modified Bessel functions of the first and second kind respectively, all of them are of order zero.

Boundedness of the solution at r = 0 requires that B = D = 0, so the total solution can be written as follows:

$$W(r) = A\, J_o(kr) + C\, I_o(kr)$$

For a fixed plate the boundary conditions are w = 0 and $\partial w/\partial r = 0 = 0$ at r = a, and are satisfied by:

$$W(a) = A\, J_o(ka) + C\, I_o(ka) = 0$$

$$\frac{\partial W}{\partial r}(a) = k\left[A\, J_o'(ka) + C\, I_o'(ka)\right] = 0$$

which gives the characteristic equation:

$$J_0(ka)I'_0(ka) - I_0(ka)J'_0(ka) = 0$$

Let the roots α_n, where α = ka, of the characteristic equation be designated as:

$$\alpha_n = k_n a \qquad n = 1, 2, 3, \ldots$$

where it can be shown that there is no zero root. The eigenvalues are $\lambda_n = k_n^4 = \alpha_n^4/a^4$. The eigenfunctions can be evaluated by finding the ratio C/A = $-J_0(\alpha_n)/I_0(\alpha_n)$ and substituting that ratio into the solution. Since $J_0(\alpha_n) \neq 0$, then one may factor it out. Thus, the natural frequencies ω_n and the mode shapes W_n are then found to be:

$$\omega_n = \sqrt{\frac{D}{\rho h}\frac{\alpha_n^2}{a^2}}$$

$$W_n = \frac{J_0(\frac{\alpha_n r}{a})}{J_0(\alpha_n)} - \frac{I_0(\frac{\alpha_n r}{a})}{I_0(\alpha_n)} \qquad n = 1, 2, 3, \ldots$$

Example 6.10 *Free Vibration of Gas Inside a Rigid Spherical Enclosure*

Obtain the mode shapes and the corresponding natural frequencies of a gas vibrating inside a rigid spherical enclosure whose radius is a.

The velocity potential $\psi(r,\theta,\phi,t)$ of a vibrating gas inside a rigid sphere is assumed to have harmonic time dependence, such that:

$$\psi(r,\theta,\phi,t) = W(r,\theta,\phi)\, e^{i\omega t}$$

where W satisfies the Helmholtz equation. Assuming that W can be written as:

$$W(r,\theta,\phi) = R(r)\, S(\theta)\, M(\phi)$$

then the Helmholtz equation becomes:

$$\frac{R''}{R} + \frac{2}{r}\frac{R'}{R} + \frac{1}{r^2}\left[\frac{S''}{S} + \cos\theta\frac{S'}{S}\right] + \frac{1}{r^2 \sin^2\theta}\frac{M''}{M} + k^2 = 0$$

which separates into three ordinary differential equations:

$$r^2 R'' + 2r\, R' + [k^2 r^2 - \nu(\nu+1)]R = 0$$

$$M'' + \alpha^2 M = 0$$

$$S'' + \cot\theta\, S' + [\nu(\nu+1) - \frac{\alpha^2}{\sin^2\theta}]\, S = 0$$

the last of which transforms to the following equation if one substitutes $\eta = \cos\theta$:

$$\frac{d}{d\eta}[(1-\eta^2)\frac{dS}{d\eta}] + [\nu(\nu+1) - \frac{\alpha^2}{1-\eta^2}]S = 0$$

The separation constants ν and α^2 must be positive or zreo to give oscillating solutions of the three ordinary differential equations. The solution of these equations can be written as follows:

$$R = A\, j_\nu(kr) + B\, y_\nu(kr) \qquad k \neq 0$$

$$S = C\, P_\nu^\alpha(\eta) + D\, Q_\nu^\alpha(\eta) \qquad \alpha \neq 0$$

$$M = E \sin(\alpha\theta) + F \cos(\alpha\theta)$$

where j_ν and y_ν are the spherical Bessel Functions of the first and second kind of order ν, P_ν^α and Q_ν^α are the associated Legendre functions of the first and second kind, degree ν and order α. Single-valuedness requires that α be an integer = m = 0, 1, 2, ... and boundedness at r = 0 and $\eta = \pm 1$ requires that:

$$B = D = 0 \qquad \text{and} \qquad \nu \text{ is an integer} = n = 0, 1, 2, \ldots$$

The boundary condition at r = a requires that the normal (radial) velocity must vanish, i.e.:

$$V_r = -\left.\frac{\partial R}{\partial r}\right|_{r=a} = 0 \qquad \text{or } j_n'(ka) = j_n'(\mu) = 0 \qquad \text{where } \mu = ka$$

Let μ_{nl} designate the lth root of the nth equation. It can be shown that the roots $\mu_{nl} \neq 0$. The mode shapes and natural frequencies of a vibrating gas inside a spherical enclosure become:

$$W_{mnl} = j_n(\mu_{nl} \frac{r}{a})\; P_n^m(\cos\theta) \begin{bmatrix} \sin(n\phi) \\ \cos(n\phi) \end{bmatrix}$$

and

$$\omega_{nl} = \frac{c}{a} \mu_{nl} \qquad m, n = 0, 1, 2, 3, \ldots \qquad l = 1, 2, 3, \ldots$$

6.12 The Diffusion Equation

The most general system governed by the diffusion equation takes the form of a non-homogeneous partial differential equation, boundary and initial conditions, having the form

$$\nabla^2\phi = \frac{1}{K}\frac{\partial\phi}{\partial t} + F(P,t) \qquad P \text{ in } V,\ t > 0 \tag{6.66}$$

where $\phi = \phi(P,t)$ is the dependent variable satisfying time-independent non-homogeneous boundary conditions of Dirichlet, Neumann or Robin type, i.e., they are only spatially dependent:

$$U(\phi(P,t)) = l(P) \qquad P \text{ on } S,\ t > 0 \tag{6.67}$$

and the initial conditions:

$$\phi(P,0^+) = g(P) \qquad P \text{ in } V \tag{6.68}$$

and F(P,t) is a time and space dependent source. The restriction on only spatially dependent boundary conditions is due to the goal of obtaining solutions in terms of eigenfunction expansions, such restrictions will be removed in Chapter 7.

Since the non-homogeneous boundary conditions are only spatially dependent, one can split the solution ϕ into two components one being transient (time dependent), and the other steady state (time independent).

Let:

$$\phi = \phi_1(P,t) + \phi_2(P) \tag{6.69}$$

where the first component satisfies the following system:

$$\begin{aligned} &\nabla^2\phi_1 = \frac{1}{K}\frac{\partial\phi_1}{\partial t} + F(P,t) \\ &U(\phi_1) = 0 \\ &\phi_1(P,0^+) = g(P) - \phi_2(P) = h(P) \end{aligned} \tag{6.70}$$

and the second component satisfies Laplace's system:

$$\begin{aligned} &\nabla^2\phi_2 = 0 \\ &U(\phi_2) = l(P) \end{aligned} \tag{6.71}$$

The two systems in eqs. (6.70) and (6.71) add up to the original system defined in eqs. (6.66) through (6.68). The system in (6.71) is a Laplace system, which was explored in Section (6.10). Once the system in (6.71) is solved, then the initial condition of the system (6.70) is determined. To obtain a solution of the system defined by eqs. (6.70), one needs to obtain an eigenfunction set from a homogeneous Helmholtz equation with the boundary conditions specified as in (6.70), i.e.:

$$\nabla^2\phi_M + \lambda_M\,\phi_M = 0 \tag{6.72}$$

subject to the same homogeneous boundary conditions in (6.70)

$$U(\phi_M) = 0$$

so that the resulting eigenfunctions are orthogonal, satisfying the orthogonality integral:

$$\begin{aligned} \int_V \phi_M\,\phi_K\,dV &= 0 \qquad M \neq K \\ &= N_M \quad M = K \end{aligned}$$

The solution of the system in (6.70) involves the expansion of the function $\phi(P,t)$ in a Generalized Fourier series in terms of the spatially dependent eigenfunctions, but with time dependent Fourier coefficients:

$$\phi_1 = \sum_M E_M(t)\,\phi_M(P) \tag{6.73}$$

The solution ϕ_1 satisfies the boundary conditions of (6.70), i.e.:

$$U(\phi_1) = U(\sum_M E_M\phi_M) = \sum_M E_M\,U(\phi_M) = 0$$

Substituting the solution (6.73) into the differential equation of (6.70) results in:

$$\begin{aligned} \nabla^2\phi_1 = \sum_M E_M(t)\,\nabla^2\phi_M &= -\sum_M \lambda_M E_M(t)\phi_M(P) \\ &= \frac{1}{K}\sum_M E'_M(t)\phi_M(P) + F(P,t) \end{aligned} \tag{6.74}$$

which uses eq. (6.72). Rearranging eq. (6.74) results in a more compact form:

$$\sum_M \left(E'_M(t) + \lambda_M K E_M(t)\right)\phi_M = -K\,F(P,t) \tag{6.75}$$

Multiplying eq. (6.75) by $\phi_N(P)$ and integrating over the volume results in a first order ordinary differential equation on the Fourier coefficients:

$$E'_M(t) + \lambda_M K E_M(t) = -K \frac{\int_V \phi_M(P)F(P,t)\,dV}{\int_V \phi_M^2(P)\,dV} = F_M(t) \tag{6.76}$$

The solution of the non-homogeneous first-order differential equation (6.76) is obtained in the form, given in Section 1.2:

$$E_M(t) = C_M\, e^{-K\lambda_M t} + \int_0^t F_M(\eta)e^{-\lambda_M K(t-\eta)}d\eta \tag{6.77}$$

One can use the initial condition at t = 0 to determine the unknown constant C_M

$$\phi_1(P,0^+) = \sum_M E_M(0)\,\phi_M(P) = h(P) = \sum_M C_M\,\phi_M(P) \tag{6.78}$$

since $E_M(0) = C_M$. Thus, using the orthogonality of the eigenfunctions, the constants C_M become:

$$C_M = \frac{1}{N_M}\int_V h(P)\,\phi_M(P)\,dV \tag{6.79}$$

The evaluation of C_M concludes the determination of the Fourier coefficients $E_M(t)$. The solution in (6.77) is a linear combination of two parts, one dependant on the initial condition, C_M, and the other dependant on the source component, $F_M(t)$. If the heat source is not time dependent, i.e. if F(P,t) = Q(P) only, then $F_M = Q_M$, a constant, and the solution for $E_M(t)$ simplifies to:

$$E_M(t) = C_M e^{-K\lambda_M t} + \frac{Q_M}{\lambda_M K}[1 - e^{-K\lambda_M t}] \tag{6.80}$$

and C_M is defined by eq. (6.79).

Example 6.11 *Heat Flow in a Finite Thin Rod*

Obtain the heat flow in a finite rod of length L, whose ends are kept at constant temperature a and b. The rod is heated initially to a temperature f(x) and has a distributed, time-independent heat source, Q(x), such that, for T = T(x,t):

$$\frac{\partial^2 T}{\partial x^2} = \frac{1}{K}\frac{\partial T}{\partial t} - \frac{Q(x)}{K\rho c}$$

and

$T(0,t) = a = \text{constant}$ $\qquad$ $T(L,t) = b = \text{constant}$ $\qquad$ $T(x,0^+) = f(x)$

Let $T = T_1(x,t) + T_2(x)$ such that:

$$\frac{\partial^2 T_1}{\partial x^2} = \frac{1}{K}\frac{\partial T_1}{\partial t} - \frac{Q(x)}{K\rho c} \qquad \frac{\partial^2 T_2}{\partial x^2} = 0$$

$$T_1(0,t) = 0 \qquad T_2(0,t) = a$$

$$T_1(L,t) = 0 \qquad T_2(L,t) = b$$

$$T_1(x,0^+) = f(x) - T_2(x) = h(x)$$

The solution for $T_2(x)$ can be readily found as:

$$T_2(x) = \frac{b-a}{L}x + a$$

To solve for $T_1(x,t)$ one must develop an eigenfunction set satisfying the boundary conditions:

$$X'' + k^2 X = 0 \qquad X = A\sin(kx) + B\cos(kx)$$

which satisfies the following boundary conditions:

$$X(0) = 0 \qquad B = 0$$

$$X(L) = 0 \qquad \sin(kL) = 0 \quad \text{or} \quad k_n = \frac{n\pi}{L} \qquad n = 1, 2, 3, \ldots$$

Thus, the eigenfunctions and eigenvalues of the system become:

$$X_n = \sin\left(\frac{n\pi}{L}x\right) \qquad n = 1, 2, 3, \ldots$$

$$\lambda_n = n^2\pi^2 / L^2$$

Expanding T_1 in terms of time-dependent Fourier coefficients, E_n, and the associated eigenfunctions, X_n produces:

$$T_1(x,t) = \sum_{n=1}^{\infty} E_n(t)\sin\left(\frac{n\pi}{L}x\right)$$

subject to the initial condition:

$$T_1(x,0^+) = f(x) - T_2(x) = h(x)$$

Following the development in eq. (6.79), the constants C_n are given by:

$$C_n = \frac{2}{L}\int_0^L [f(x) - T_2(x)]\sin\left(\frac{n\pi}{L}x\right)dx \qquad n = 1, 2, 3, \ldots$$

Following the development for a time-independent heat source, eq. (6.76) gives:

$$Q_n = \frac{2}{\rho c L}\int_0^L Q(x)\sin\left(\frac{n\pi}{L}x\right)dx \qquad n = 1, 2, 3, \ldots$$

so that the final solution for $E_n(t)$, eq. (6.80), is given by:

$$E_n(t) = C_n e^{-Kn^2\pi^2 t/L^2} + \frac{Q_n L^2}{n^2\pi^2}[1 - e^{-Kn^2\pi^2 t/L^2}]$$

It should be noted that as $t \to \infty$, a steady state temperature distribution is given by Q_n only:

$$E_n(t) \to \frac{Q_n L^2}{n^2\pi^2} \quad \text{as} \quad t \to \infty$$

Example 6.12 *Heat Flow in a Circular Sheet*

Obtain the heat flow in a solid sheet whose radius is a and whose perimeter is kept at zero temperature. The sheet is initially heated and has an explosive point heat source applied at the center of the sheet so that the temperature T(r,t) satisfies the following system:

$$\nabla^2 T = \frac{1}{K}\frac{\partial T}{\partial t} - \delta(r)\frac{Q_0 e^{-\alpha t}}{2\pi K\rho c r} \qquad 0 \le r \le a \qquad t > 0 \qquad \alpha > 0$$

$$T(a,t) = 0$$

$$T(r,0^+) = T_o\,(1 - r^2/a^2)$$

when $\delta(r)$ is the Dirac delta function (Appendix D). Since the boundary conditions are homogeneous, then $T_2 = 0$, and $T(r,t) = T_1(r,t)$. To find the eigenfunctions of the system in cylindrical coordinates, one solves the Helmholtz system:

$$\nabla^2 R + \lambda R = \frac{d^2R}{dr^2} + \frac{1}{r}\frac{dR}{dr} + k^2 R = 0 \quad \text{where} \quad \lambda = k^2$$

which has a solution of the form:

$$R(r) = A\,J_0(kr) + B\,Y_0(kr)$$

Since the temperature is bounded at the origin, $r = 0$, let $B = 0$. Satisfying the boundary condition at $r = a$, $R(a) = A\,J_0(ka) = 0$. Letting $\mu = ka$, then $J_0(\mu) = 0$ has an infinite number of non-zero roots: $\mu_n = k_n a$, $n = 1, 2, 3, \ldots$, and the eigenfunctions and eigenvalues become:

$$R_n(r) = J_0(\mu_n \frac{r}{a}) \qquad \lambda_n = \frac{\mu_n^2}{a^2} \qquad n = 1, 2, 3, \ldots$$

and the orthogonality condition is (4.86):

$$\int_0^a r\,J_0(\mu_n \frac{r}{a})\,J_0(\mu_m \frac{r}{a})\,dr = 0 \qquad n \ne m$$

$$= N_n = \frac{a^2}{2} J_1^2(\mu_n) \qquad n = m$$

Expanding the temperature T(r,t) into an infinite series of the eigenfunctions:

$$T(r,t) = \sum_{n=1}^{\infty} E_n(t) R_n(r)$$

then one can follow the development of the solution through eqs. (6.70) through (6.79). The Fourier series of the source term of eq. (6.76) is given by:

$$F_n(t) = \frac{Q_o e^{-\alpha t}}{2\pi\rho c N_n} \int_0^a r \frac{\delta(r)}{r} J_o(\mu_n \frac{r}{a})\, dr = \frac{Q_o e^{-\alpha t}}{a^2 \pi \rho c J_1^2(\mu_n)}$$

The integral part of the solution for $E_n(t)$ in eq. (6.77) due to the point source is evaluated separately from the initial condition, yielding:

$$\frac{Q_o e^{-Kt\mu_n^2/a^2}}{a^2\pi\rho c J_1^2(\mu_n)} \int_0^t e^{-\alpha\eta} e^{-K\eta\mu_n^2/a^2}\, d\eta = \frac{Q_o[e^{-\alpha t} - e^{-Kt\mu_n^2/a^2}]}{a^2\pi\rho c[K\mu_n^2/a^2 - \alpha] J_1^2(\mu_n)}$$

The constant C_n of eq. (6.79) due to the initial condition is also obtained through eqs. (3.103) and (3.105):

$$C_n = \frac{T_o}{N_n} \int_0^a r(1 - \frac{r^2}{a^2}) J_o(\mu_n \frac{r}{a}) dr = \frac{4T_o J_1(\mu_n)}{\mu_n^3 N_n} = \frac{8T_o}{\mu_n^3 J_1(\mu_n)}$$

Finally, the solution for the Fourier coefficient $E_n(t)$ is given by:

$$E_n(t) = \frac{8T_o e^{-Kt\mu_n^2/a^2}}{\mu_n^3 J_1(\mu_n)} + \frac{Q_o[e^{-\alpha t} - e^{-Kt\mu_n^2/a^2}]}{a^2\pi\rho c[K\mu_n^2/a^2 - \alpha] J_1^2(\mu_n)}$$

One can clearly see that the temperature tends to zero as $t \to \infty$, since the source itself also vanishes as $t \to \infty$.

Example 6.13 *Heat Flow in a Finite Cylinder*

Obtain the heat flow in a cylinder of length L and radius a whose surface is being kept at zero temperature, which has an initial temperature distribution. Thus, if $T = T(r,\theta,z,t)$, then:

$$T(a,\theta,z,t) = 0, \quad T(r,\theta,0,t) = 0, \quad T(r,\theta,L,t) = 0, \quad T(r,\theta,z,0^+) = f(r,\theta,z)$$

Since the boundary conditions are homogeneous, then there is no steady state component, and the temperature satisfies the homogeneous heat flow equation in cylindrical coordinates as follows:

$$\frac{\partial^2 T}{\partial r^2} + \frac{1}{r}\frac{\partial T}{\partial r} + \frac{1}{r^2}\frac{\partial^2 T}{\partial \theta^2} + \frac{\partial^2 T}{\partial z^2} = \frac{1}{k}\frac{\partial T}{\partial t}$$

The eigenfunction of the Helmholtz equation can be obtained by letting:

$$\phi = R(r)\, F(\theta)\, Z(z)$$

then the partial differential equations can be satisfied by three ordinary differential equations:

$R''+\frac{1}{r}R'+(k^2-\frac{b^2}{r^2})R=0$	$k \neq 0$	$R = A J_b(kr)+BY_b(kr)$
	$k=0,\ b\neq 0$	$R = Ar^b + Br^{-b}$
	$k=0,\ b=0$	$R = A + B\log(r)$
$F''+b^2F=0$	$b \neq 0$	$F = C\sin(b\theta)+D\cos(b\theta)$
	$b=0$	$F = C_o\theta + D_o$
$Z''+c^2Z=0$	$c \neq 0$	$Z = G\sin(cz)+H\cos(cz)$
	$c=0$	$Z = Gz+H$

where the signs of the separation constants k^2, b^2, and c^2 were chosen to result in oscillating functions.

Single-valuedness of $F(\theta)$ requires that b = integer = n = 1, 2, 3, ... and $C_o = 0$. Boundedness at r = 0 requires that B = 0. Satisfying the boundary condition at r = a for R(r), one obtains for $k \neq 0$:

$$J_n(ka) = J_n(\mu) = 0 \qquad \mu_{nl} = k_{nl}a \qquad l = 1, 2, 3, \ldots \qquad n = 0, 1, 2, \ldots$$

where μ_{nl} is the lth root for the nth equation, and $\mu_{nl} \neq 0$. For k = 0, A = 0, resulting in a trivial solution for R(r). For $c \neq 0$:

$$Z(0) = 0 \qquad H = 0$$

$$Z(L) = 0 \qquad \sin(cL) = 0 \qquad c_m = \frac{m\pi}{L} \qquad m = 1, 2, 3, \ldots$$

There is only the trivial solution Z(z) for c = 0.

Thus, the eigenfunctions and eigenvalues can be written as follows:

$$\phi_{nml} = \sin(\frac{m\pi}{L}z)\,J_n(\mu_{nl}\frac{r}{a})\begin{bmatrix}\sin(n\theta)\\ \cos(n\theta)\end{bmatrix}$$

$$\lambda_{nml} = \frac{\mu_{nl}^2}{a^2} + \frac{m^2\pi^2}{L^2}$$

Since there are two different functional forms of the eigenfunctions, one must use two different time-dependent Fourier coefficients for the final solution for T. Letting:

$$T(r,\theta,z,t) = \sum_{n=0}^{\infty}\sum_{m=1}^{\infty}\sum_{l=1}^{\infty}\left[C_{nml}(t)\sin(n\theta)+D_{nml}(t)\cos(n\theta)\right]\sin(\frac{m\pi}{L}z)\,J_n(\mu_{nl}\frac{r}{a})$$

then the initial condition can be evaluated from:

$$T(r,\theta,z,0^+) = f(r,\theta,z) = \sum_{n=0}^{\infty}\sum_{m=1}^{\infty}\sum_{l=1}^{\infty}\sin(\frac{m\pi}{L}z)\,J_n(\mu_{nl}\frac{r}{a})\bullet$$

$$\bullet\left[C_{nml}(t)\sin(n\theta)+D_{nml}(t)\cos(n\theta)\right]$$

The solution for $C_{nml}(t)$ and $D_{nml}(t)$ for a source-free cylinder becomes:

$$C_{nml}(t) = \overline{C}_{nml}\exp(-\lambda_{nml}K\,t)$$

and

$$D_{nml}(t) = \overline{D}_{nml}\exp(-\lambda_{nml}K\,t)$$

Using eqs. (6.77-6.79), the constants $\overline{C}_{nml}$ and $\overline{D}_{nml}$ become:

$$\bar{C}_{nml} = \frac{4}{\pi L a^2 J_{n+1}^2(\mu_{nl})} \int_0^a \int_0^L \int_0^{2\pi} r\, f \sin(\frac{m\pi}{L} z)\ \sin(n\theta)\, J_n(\mu_{nl} \frac{r}{a}) d\theta\, dz\, dr$$

and

$$\bar{D}_{nml} = \frac{2\varepsilon_n}{\pi L a^2 J_{n+1}^2(\mu_{nl})} \int_0^a \int_0^L \int_0^{2\pi} r\, f \sin(\frac{m\pi}{L} z)\ \cos(n\theta)\, J_n(\mu_{nl} \frac{r}{a}) d\theta\, dz\, dr$$

6.13 The Vibration Equation

Solutions to the homogenous or non-homogenous vibration or wave equations can be obtained in terms of eigenfunction expansions.

The types of non-homogenous problems encountered in transient vibration or wave equation with time dependent sources and non-homogenous boundary conditions are again restricted to time-independent boundary conditions. This limitation is imposed in order to take full advantage of the eigenfunction expansion method. These limitations will be relaxed in Chapter 7. The system, composed of a non-homogeneous partial differential equation, boundary and initial conditions on the dependent variable $\phi(P,t)$ are:

$$\nabla^2 \phi = \frac{1}{c^2} \frac{\partial^2 \phi}{\partial t^2} + F(P,t) \qquad P \text{ in } V \qquad t > 0 \qquad (6.81)$$

where $F(P,t)$ is a time and space dependent source and the function $\phi(P,t)$ satisfies non-homogenous Dirichlet, Neumann or Robin type, spatially-dependent boundary conditions:

$$U(\phi(P,t)) = l(P) \qquad P \text{ on } S \qquad (6.82)$$

and non-homogenous initial conditions for P in V:

$$\phi(P,0^+) = h(P), \qquad \frac{\partial \phi}{\partial t}(P,0^+) = f(P) \qquad (6.83)$$

Due to the space dependence only of the boundary conditions, one may split the solution into a transient component and a steady state component, i.e.

$$\phi(P,t) = \phi_1(P,t) + \phi_2(P) \qquad (6.84)$$

such that $\phi_1(P,t)$ satisfies the following system:

$$\nabla^2 \phi_1 = \frac{1}{c^2} \frac{\partial^2 \phi_1}{\partial t^2} + F(P,t) \qquad P \text{ in } V, \qquad t > 0 \qquad (6.85)$$

and the homogenous form of the boundary conditions given in (6.82) and the initial conditions (6.83):

$$U(\phi_1(P,t)) = 0 \qquad P \text{ on } S, \qquad t > 0 \qquad (6.86)$$

$$\frac{\partial \phi_1}{\partial t}(P,0^+) = f(P) \qquad P \text{ in } V \qquad (6.87)$$

$$\phi_1(P,0^+) = h(P) - \phi_2(P) = g(P) \qquad P \text{ in } V \qquad (6.88)$$

The second steady state part $\phi_2(P)$ satisfies the system:

$$\nabla^2\phi_2 = 0 \qquad \text{P in V} \qquad t > 0 \tag{6.89}$$

$$U(\phi_2(P,t)) = l(P) \qquad \text{P on S} \tag{6.90}$$

The steady state component ϕ_2 satisfies a non-homogenous Laplace system, see Section 6.10.

To solve the system (6.85) to (6.88), one starts out by developing the eigenfunctions from the associated Helmholtz system, as was discussed in Section 6.13, eq. (6.72). Expanding the solution $\phi_1(P,t)$ in the eigenfunction of the homogenous Helmholtz system with time dependent Fourier coefficients:

$$\phi_1(P,t) = \sum_M E_M(t)\phi_M(P) \tag{6.91}$$

and substituting the solution in (6.91) into eq. (6.85) we get:

$$\begin{aligned}\nabla^2\phi_1(P,t) &= \sum_M E_M(t)\nabla^2\phi_M(P) = -\sum_M \lambda_M E_M(t)\phi_M(P)\\ &= \frac{1}{c^2}\sum_M E''_M(t)\phi_M(P) + F(P,t)\end{aligned} \tag{6.92}$$

The above equation can be rewritten in compact form as:

$$\sum_M [E''_M(t) + c^2\lambda_M E_M(t)]\phi_M(P) = -c^2 F(P,t) \tag{6.93}$$

Multiplying the series by $\phi_K(P)$, integrating over the volume, and using the orthogonality integrals (6.53), one obtains a second order ordinary differential equation on $E_M(t)$ as:

$$E''_K(t) + c^2\lambda_M E_K(t) = -\frac{c^2}{N_K}\int_V F(P,t)\phi_K(P)dV = F_K(t) \tag{6.94}$$

The general solution of eq. (6.94) can be written as:

$$\begin{aligned}E_K(t) &= A_K \sin(ct\sqrt{\lambda_K}) + B_K\cos(ct\sqrt{\lambda_K})\\ &+ \frac{1}{c\sqrt{\lambda_K}}\int_0^t F_K(\eta)\sin(c\sqrt{\lambda_K}(t-\eta))\,d\eta\end{aligned} \tag{6.95}$$

The initial conditions of $E_K(t)$ are:

$$E_K(0) = B_K \qquad \text{and} \qquad E'_K(0) = c\sqrt{\lambda_K}A_K \tag{6.96}$$

where the constants A_K and B_K are obtained from the initial conditions (6.87) and (6.88) as follows:

$$\phi_1(P,0^+) = g(P) = \sum_K E_K(0)\phi_K(P)$$

$$\frac{\partial\phi_1}{\partial t}(P,0^+) = f(P) = \sum_K E'_K(0)\phi_K(P)$$

which, upon use of the orthogonality integrals (6.53), results an integral form for the constants A_K and B_K as:

$$E_K(0) = B_K = \frac{\int_V g(P)\phi_K(P)dV}{N_K} \tag{6.97}$$

and

$$A_K = \frac{\int_V f(P)\phi_K(P)\,dV}{c\sqrt{\lambda_K}N_K} \tag{6.98}$$

The evaluation of the constants A_K and B_K concludes the evaluation of the time-dependent Fourier coefficient $E_K(t)$ and hence results in the total solution $\phi(P,t)$.

Example 6.14 *Transient Motion of a Square Plate*

Obtain the transient motion of a square plate, whose sides of length L are simply supported (hinged). The plate is initially displaced from rest, such that, if $w = w(x,y,t)$, then:

$$\nabla^4 w + \frac{\rho h}{D}\frac{\partial^2 w}{\partial t^2} = 0 \qquad 0 \le x, \quad y \le L, \quad t > 0$$

The boundary and initial conditions become (see Section 6.3.3):

$$w(P,t) = 0 \qquad P \text{ on } C$$

$$\frac{\partial^2 w}{\partial n^2}(P,t) + \nu\frac{\partial^2 w}{\partial s^2}(P,t) = 0 \qquad P \text{ on } C$$

$$w(x,y,0^+) = f(x,y) \qquad \frac{\partial w}{\partial t}(x,y,0^+) = 0$$

Thus:

$$w(0,y,t) = 0 \quad \text{and} \quad \frac{\partial^2 w}{\partial x^2}(0,y,t) = 0 \quad \text{since} \quad \frac{\partial w}{\partial y}(0,y,t) = 0$$

$$w(L,y,t) = 0 \quad \text{and} \quad \frac{\partial^2 w}{\partial x^2}(L,y,t) = 0 \quad \text{since} \quad \frac{\partial w}{\partial y}(L,y,t) = 0$$

$$w(x,0,t) = 0 \quad \text{and} \quad \frac{\partial^2 w}{\partial y^2}(x,0,t) = 0 \quad \text{since} \quad \frac{\partial w}{\partial x}(x,0,t) = 0$$

$$w(x,L,t) = 0 \quad \text{and} \quad \frac{\partial^2 w}{\partial y^2}(x,L,t) = 0 \quad \text{since} \quad \frac{\partial w}{\partial x}(x,L,t) = 0$$

Since the problem does not involve sources or non-homogenous boundary conditions, then only the transient component of eq. (6.84) remains. Starting with the associated Helmholtz equation:

$$-\nabla^4 W + b^4 W = 0$$

One can split the fourth order operator as a commutable product of operators:

$$(\nabla^2 - b^2)(\nabla^2 + b^2) W = 0$$

such that if $W = W_1 + W_2$, then the solution to W can be obtained from the following pair of differential equations:

$$(\nabla^2 - b^2)\, W_1 = 0$$

and

$$(\nabla^2 + b^2)\, W_2 = 0$$

Letting $W_{1,2} = X(x)\, Y(y)$, one obtains:

$$X'' + c^2 X = 0 \qquad\qquad X = A \sin(cx) + B \cos(cx)$$

$$Y'' - (c^2 + b^2) Y = 0 \qquad\qquad Y = C \sinh(ey) + D \cosh(ey)$$

where $e^2 = c^2 + b^2$ and:

$$X'' + d^2 X = 0 \qquad\qquad X = E \sin(dx) + F \cos(dx)$$

$$Y'' + (b^2 - d^2) Y = 0 \qquad\qquad Y = G \sin(fy) + H \cos(fy)$$

where $f^2 = b^2 - d^2$. Each of these solutions must satisfy the boundary conditions, which results in:

$$B = D = C = F = H = 0$$

and

$$\sin(dL) = 0 \qquad d_n = \frac{n\pi}{L} \qquad n = 1, 2, 3, \ldots$$

$$\sin(fL) = 0 \qquad f_n = \frac{m\pi}{L} \qquad m = 1, 2, 3, \ldots$$

Thus, the eigenfunctions and eigenvalues become:

$$W_{nm} = \sin\left(\frac{n\pi}{L} x\right) \sin\left(\frac{m\pi}{L} y\right)$$

$$b_{nm}^4 = \left(\frac{n^2\pi^2}{L^2} + \frac{m^2\pi^2}{L^2}\right)^2$$

and the resonence frequencies of a free plate, k_{nm}, are given by:

$$k_{nm} = \sqrt{\frac{D}{\rho h}}\, b_{nm}^2 = \sqrt{\frac{D}{\rho h}}\, \frac{\pi^2}{L^2}(n^2 + m^2)$$

Expanding the solution into the eigenfunctions of the problem, gives:

$$w(x, y, t) = \sum_{n=1}^{\infty} \sum_{m=1}^{\infty} E_{nm}(t)\, W_{nm}(x, y)$$

where the Fourier coefficients do not contain a source component:

$$E_{nm}(t) = A_{nm}\sin(k_{nm}t) + B_{nm}\cos(k_{nm}t)$$

The initial conditions as given in eqs. (6.97) and (6.98) results in:

$$B_{nm} = \frac{4}{L^2}\int_0^L\int_0^L f(x,y)\sin(\frac{n\pi}{L}x)\sin(\frac{m\pi}{L}y)\,dx\,dy$$

and

$$A_{nm} = 0$$

Thus, the final solution for the response of the plate is given by:

$$w = \sum_{n=1}^{\infty}\sum_{n=1}^{\infty} B_{nm}\sin(\frac{n\pi}{L}x)\sin(\frac{m\pi}{L}y)\cos(k_{nm}t)$$

Example 6.15 *Forced Vibration of a Circular Membrane*

Obtain the transient motion of a circular membrane, whose radius is a, in response to transverse time-varying forces q(r,t). The membrane is initially deformed to a displacement f(r) and released from the rest.

Since the shape of the membrane, the boundary conditions, and the source term are not dependent on θ, then the motion of the membrane will be independent of θ, i.e.: axi-symmetric. The equation of motion satisfied by an axi-symmetric displacement w(r,t) can be written as follows:

$$\frac{\partial^2 w}{\partial r^2} + \frac{1}{r}\frac{\partial w}{\partial r} = \frac{1}{c^2}\frac{\partial^2 w}{\partial t^2} - \frac{q(r,t)}{S} \qquad 0 \le r \le a \qquad t > 0$$

with boundary and initial conditions given as:

$$w(a,t) = 0, \qquad w(r,0^+) = f(r), \qquad \text{and} \qquad \frac{\partial w}{\partial t}(r,0^+) = 0$$

Since the boundary conditions are homogenous, then the steady state part of the solution vanishes and w(r,t) becomes the transient solution.

The eigenfunctions of the system can be obtained by solving the associated Helmholtz eq.:

$$r^2 R'' + r R' + k^2 r^2 R = 0 \qquad R = A\,J_0(kr) + B\,Y_0(kr)$$

Boundedness at $r = 0$ requires that $B = 0$, and the boundary condition $R(a) = 0$ gives the characteristic equation: $J_0(ka) = J_0(\mu) = 0$, where $\mu = ka$. Let μ_n be the nth root of the characteristic equation (where n = 1, 2, 3, ...) then the eigenfunctions become:

$$R_n(r) = J_0(\mu_n \frac{r}{a})$$

There is no zero root of the characteristic equation. Writing out the solution in terms of the eigenfunctions with time-dependent Fourier coefficients:

$$w(r,t) = \sum_{n=1}^{\infty} E_n(t) R_n(r)$$

then, the solution for the first component of $E_n(t)$, given in eq. (6.95) that is due to the initial conditions only, results in:

$$E_n(t) = A_n \sin(\frac{\mu_n}{a} ct) + B_n \cos(\frac{\mu_n}{a} ct)$$

with

$$B_n = \frac{2}{a^2 [J_1(\mu_n)]^2} \int_0^a r f(r) J_0(\mu_n \frac{r}{a})\, dr$$

and

$$A_n = 0$$

The second component of $E_n(t)$ that depends on the source term requires that one first evaluates $F_n(t)$ as:

$$F_n(t) = \frac{2c^2}{a^2 [J_1(\mu_n)]^2} \int_0^a r \frac{q(r,t)}{S} J_0(\mu_n \frac{r}{a})\, dr$$

which gives the component of $E_n(t)$ due to the source as:

$$E_n(t) = \frac{a}{c\mu_n} \int_0^t \sin(\frac{c\mu_n}{a}(t-\eta))\, F_n(\eta)\, d\eta$$

Thus, the two parts of $E_n(t)$ were found and the transient solution of the response of the plate evaluated.

If the applied load on the membrane takes the form of an impulsive point force of the form:

$$q(r,t) = \frac{P_o}{2\pi} \delta(r) \delta(t - t_o)$$

where δ is the Dirac delta function and represents a point force of magnitude P_o applied implulsively at $t = t_o$. Using the properties of the Dirac delta function (Appendix D) one obtains:

$$F_n(t) = \frac{P_o \delta(t-t_o) c^2}{\pi a^2 [J_1(\mu_n)]^2} \int_0^a r \frac{\delta(r)}{r} J_0(\mu_n \frac{r}{a})\, dr = \frac{P_o \delta(t-t_o) c^2}{\pi a^2 [J_1(\mu_n)]^2}$$

which when substituted in the integral for $E_n(t)$ for the source component results in:

$$E_n(t) = \frac{P_o c}{\pi a \mu_n [J_1(\mu_n)]^2} \int_0^t \sin(\frac{c\mu_n}{a}(t-\eta))\, \delta(t-t_o)\, d\eta$$

$$= \frac{P_o c}{\pi a \mu_n [J_1(\mu_n)]^2} \sin(\frac{c\mu_n}{a}(t-t_o))\, H(t-t_o)$$

where H(x) is the Heaviside unit step function (Appendix D).

6.14 The Wave Equation

The solutions of the scalar wave equation, both in transient as well as steady-state cases, will be discussed in this section.

6.14.1 Wave Propagation in an Infinite, One-Dimensional Medium

Wave propagation in an infinite one-dimensional medium is governed by the following system:

$$y = y(x,t)$$

$$\frac{\partial^2 y}{\partial x^2} = \frac{1}{c^2}\frac{\partial^2 y}{\partial t^2} \qquad -\infty < x < \infty \qquad t > 0$$

$$y(x,0^+) = f(x)$$

$$\frac{\partial y}{\partial t}(x,0^+) = g(x)$$

Letting u = x - ct and v = x + ct, then the wave equation transforms to:

$$\frac{\partial^2 y}{\partial u \partial v} = 0$$

whose solution can be shown to have the form:

$$y = F(u) + G(v)$$

$$= F(x - ct) + G(x + ct)$$

The solution must satisfy the initial conditions:

$$y(x,0^+) = f(x) = F(x) + G(x)$$

$$\frac{\partial y}{\partial t}(x,0^+) = g(x) = -c\left[\frac{dF(x)}{dx} - \frac{dG(x)}{dx}\right]$$

Differentiating the first equation with respect to x, one obtains:

$$F'(x) + G'(x) = f'(x)$$

and rewriting the second initial condition as:

$$F'(x) + G'(x) = -\frac{g(x)}{c}$$

then, one can obtain explicit expression for F and G, upon integration:

$$F(x) = \frac{f(x)}{2} - \frac{1}{2c}\int_0^x g(\eta)\,d\eta + C$$

and

$$G(x) = \frac{f(x)}{2} + \frac{1}{2c}\int_0^x g(\eta)\,d\eta - C$$

and hence, substituting for independent variables x by u or v, one gets:

$$F(u) = \frac{f(u)}{2} - \frac{1}{2c}\int_0^u g(\eta)\,d\eta + C = \frac{f(x-ct)}{2} + \frac{1}{2c}\int_{x-ct}^0 g(\eta)\,d\eta + C$$

and

$$G(v) = \frac{f(v)}{2} + \frac{1}{2c}\int_0^v g(\eta)\,d\eta - C = \frac{f(x+ct)}{2} + \frac{1}{2c}\int_0^{x+ct} g(\eta)\,d\eta - C$$

which results in the final solution in an infinite one-dimensional continuum as:

$$y(x,t) = \frac{1}{2}[f(x-ct) + f(x+ct)] + \frac{1}{2c}\int_{x-ct}^{x+ct} g(\eta)\,d\eta$$

where f(x-ct) and f(x+ct) represent the propagation in the positive and the negative directions of x, having the form f(x) and traveling at a constant speed of c.

Example 6.16 *Transient Wave Propagation in a Stretched String*

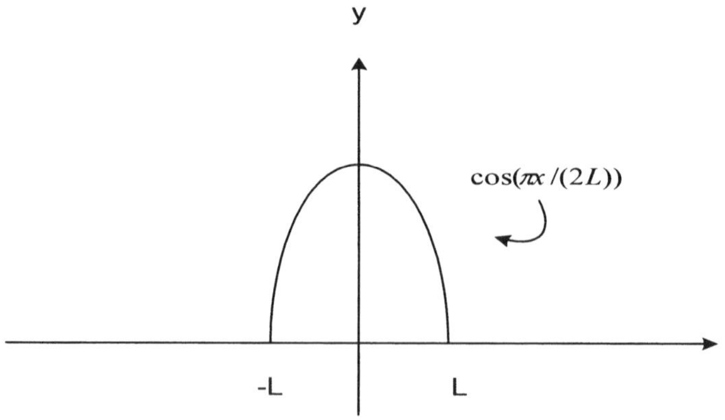

Obtain the transient displacement in an infinite stretched string, such that:

$$y(x,0^+) = f(x) = \begin{cases} 0 & x \le -L \\ \cos\left(\dfrac{\pi x}{2L}\right) & -L \le x \le L \\ 0 & x \ge L \end{cases}$$

and

$$\frac{\partial y}{\partial t}(x, 0^+) = 0$$

Note that the initial displacement can be written as:

$$y(x, 0^+) = \cos\left(\frac{\pi x}{2L}\right)\{H(x + L) - H(x - L)\}$$

where the Heaviside function $H(\eta)$ is defined in Appendix D:

The wave solution for the displacement then becomes:

$$y(x, t) = \frac{1}{2}\cos\left(\frac{\pi(x - ct)}{2L}\right)\{H((x - ct) + L) - H((x - ct) - L)\}$$

$$+\frac{1}{2}\cos\left(\frac{\pi(x + ct)}{2L}\right)\{H((x + ct) + L) - H((x + ct) - L)\}$$

which represents two half-cosine shaped waves traveling along the positive and negative x-axis at a constant speed of c.

6.14.2 Spherically Symmetric Wave Propagation in an Infinite Medium

Spherically symmetric wave propagation in an infinite medium is governed by the following system:

$$y = y(r,t)$$

$$\frac{\partial^2 y}{\partial r^2} + \frac{2}{r}\frac{\partial y}{\partial r} = \frac{1}{c^2}\frac{\partial^2 y}{\partial t^2} \qquad r \geq 0 \qquad t > 0$$

$$y(r,0^+) = f(r), \qquad \frac{\partial y}{\partial t}(r,0^+) = g(r)$$

Let $z(r,t) = r\, y(r,t)$, then the system transforms to:

$$\frac{\partial^2 z}{\partial r^2} = \frac{1}{c^2}\frac{\partial^2 z}{\partial t^2} \qquad r \geq 0 \qquad t > 0$$

$$z(r,0^+) = r\, f(r)$$

$$\frac{\partial z}{\partial t}(r,0^+) = g(r)$$

which has the following solution as developed in 6.15.1 above:

$$z(r, t) = \frac{1}{2}[(r - ct)\, f(r - ct) + (r + ct)\, f(r + ct)] + \frac{1}{2c}\int_{r-ct}^{r+ct} \eta\, g(\eta)\, d\eta$$

which becomes after transformation:

$$y(r, t) = \frac{1}{2r}[(r - ct) f(r - ct) + (r + ct)\, f(r + ct)] + \frac{1}{2cr}\int_{r-ct}^{r+ct} \eta\, g(\eta)\, d\eta$$

6.14.3 Plane Harmonic Waves

Plane harmonic wave propagation in continuous media is governed by the following Helmholtz equation:

$$\phi(P,t) = F(P)\, e^{i\omega t} \qquad P = P(x,y,z)$$

$$\frac{\partial^2 F}{\partial x^2} + \frac{\partial^2 F}{\partial y^2} + \frac{\partial^2 F}{\partial z^2} + k^2 F = 0 \qquad \text{where } k = \frac{\omega}{c}$$

Let F = X(x) Y(y) Z(z), then:

$$X'' + a^2 X = 0 \qquad X = A\, e^{iax} + B\, e^{-iax}$$

$$Y'' + b^2 Y = 0 \qquad Y = C\, e^{iby} + D\, e^{-iby}$$

$$Z'' + c^2 Z = 0 \qquad Z = E\, e^{icz} + F\, e^{-icz}$$

where $k^2 = a^2 + b^2 + d^2$. Letting $a = kl$, $b = km$, and $d = kn$, then the solution of the wave equation (wave functions) in cartesian coordinates becomes:

$$\phi(x,y,z,t) = \exp\,[ik\,(\pm lx \pm my \pm nz + ct)]$$

where

$$l^2 + m^2 + n^2 = 1$$

If one lets $l = \cos(\nu,x)$, $m = \cos(\nu,y)$, and $n = \cos(\nu, z)$, where ν represents the unit normal to the plane wave front, then the requirement that $l^2 + m^2 + n^2 = 1$ is satisfied. The solution developed in this section is the general solution for the scalar plane wave propagation in three dimensional space.

Example 6.17 *Reflection of Acoustic Waves from a Pressure Release Plane Surface*

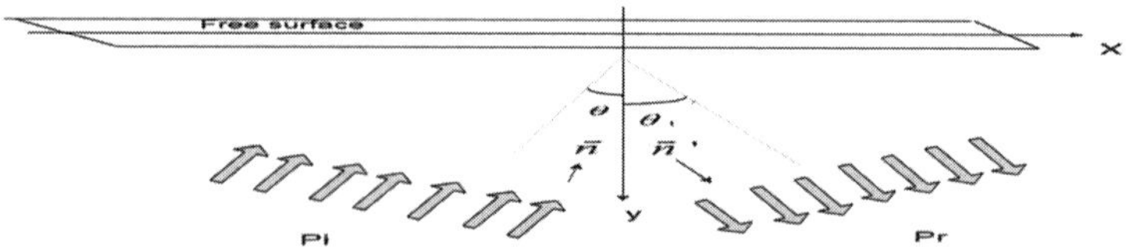

An incident plane pressure wave p_i, $p_i = p_o \exp\,[ik(l_1 x + m_1 y + ct)]$ where $l_1 = \cos(x,n)$ and $m_1 = \cos(y,n)$, impinges on a pressure-release plane surface as shown in the accompanying figure. Since the acoustic pressure satisfies the wave equation:

$$\nabla^2 p = \frac{1}{c^2}\frac{\partial^2 p}{\partial t^2}$$

then let the reflected wave p_r be a plane wave solution of the wave equation where the normal is n′:

$$p_r = A \exp [i\alpha\, (l_2 x + m_2 y + ct)]$$

where $l_2 = \cos(n',x)$ and $m_2 = \cos(n', y)$. At the pressure-release surface, the total pressure must vanish, such that:

$$p_i(y{=}0) + p_r(y{=}0) = 0$$

or

$$p_o \exp [ik(l_1 x + ct)] + A \exp [i\alpha\, (l_2 x + ct)] = 0$$

In order for the equation to be satisfied identically for all x and t, then:

$$kc = +\alpha c \qquad \text{or} \qquad \alpha = +k$$

and

$$k l_1 = \alpha l_2 \qquad \text{or} \qquad l_2 = +l_1$$

Since:

$$l_1 = \cos\left(\frac{3\pi}{2} + \theta\right) = +\sin(\theta)$$

$$l_2 = \cos\left(\frac{\pi}{2} - \theta'\right) = +\sin(\theta')$$

then $\sin\theta' = \sin\theta$ and $\theta = \theta'$, and $A = -\,p_o$. Finally, since $m_1 = \cos(\pi + \theta) = -\cos(\theta)$ then $m_2 = \cos(\theta') = \cos\theta = -\,m_1$. Thus, the reflected wave becomes:

$$p_r = -\,p_o \exp [ik\, (l_1 x - m_1 y + ct)]$$

The reflected wave has an amplitude of opposite sign to the incident wave, equal incident and reflected angles, and the same frequency ω as in the incident wave.

Example 6.18 *Reflection and Refraction of Plane Waves at an Interface*

Consider an incident plane acoustic pressure wave p_i:

$$p_i = p_o \exp [ik\, (lx + my + c_1 t)$$

existing in medium 1, (see accompanying figure) is incident at the interface between medium 1, and medium 2 and $k = \omega/c_1$. Let ρ_1 and c_1 be the density and sound speed in medium 1 and ρ_2 and c_2 be the corresponding ones for medium 2. Since the plane reflected wave p_1 is a solution of the wave equation in medium 1, let:

$$p_1 = A \exp [i\alpha_1\, (l_1 x + m_1 y + c_1 t)$$

Since the refracted wave p_2 is a solution of the wave equation in medium 2, let:

$$p_2 = B \exp [i\alpha_2\, (l_2 x + m_2 y + c_2 t)$$

Continuity of the pressure and the normal particle velocity at the interface y = 0 requires, respectivly, that:

$$p_i(x,0) + p_1(x,0) = p_2(x,0)$$

and

$$(v_i(x,0))_n + (v_1(x,0))_n = (v_2(x,0))_n$$

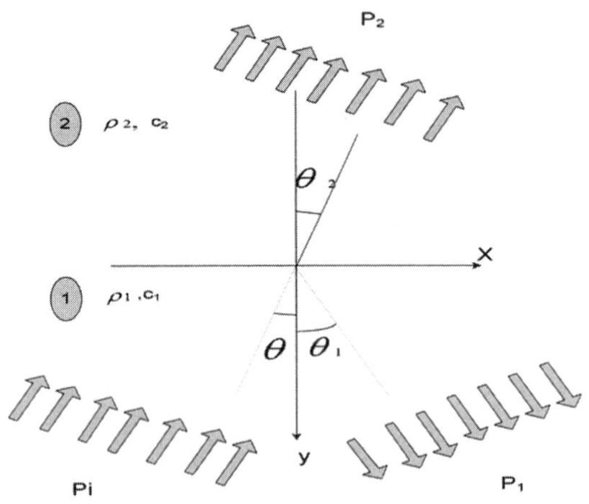

Thus, substituting the expressions for p_i, p_1 and p_2:

$$p_o \exp[ik(lx + c_1t)] + A \exp[i\alpha_1(l_1x + c_1t)] = B \exp[i\alpha_2(l_2x + c_2t)$$

which can be satisfied iff:

$$kl = \alpha_1 l_1 = \alpha_2 l_2$$

$$k c_1 = \alpha_1 c_1 = \alpha_2 c_2$$

and

$$B - A = p_o$$

Thus, these relationships require that:

$$\alpha_1 = k \qquad\qquad l_1 = l$$

and

$$\alpha_2 = k\frac{c_1}{c_2} = \frac{\omega}{c_2} \qquad\qquad l_2 = k\frac{l}{\alpha_2} = l\frac{c_2}{c_1}$$

Expressing the direction cosines in terms of θ, θ_1 and θ_2:

$$l = \cos(\frac{3\pi}{2} + \theta) = \sin(\theta)$$

$$l_1 = \cos(\frac{\pi}{2} - \theta_1) = \sin(\theta_1)$$

$$l_2 = \cos(\frac{3\pi}{2} + \theta_2) = \sin(\theta_2)$$

results in the following relationships:

$$\theta_1 = \theta$$

and

$$\sin(\theta_2) = \frac{c_2}{c_1}\sin(\theta) \qquad \text{(Snell's Law)}$$

If $c_2 < c_1$, then the maximum value of the refraction angle θ_2 occurs when $\theta = \pi/2$, called the critical angle θ_c:

$$\theta_2 = \sin^{-1}\left(\frac{c_2}{c_1}\right)$$

If $c_2 > c_1$, then θ_1 has a maximum value when $\theta_2 = \pi/2$:

$$\theta_c = \theta_1 = \sin^{-1}\left(\frac{c_1}{c_2}\right)$$

If $\theta > \theta_c$, then all of the wave reflects off the surface and none of the wave refracts into the other material at the boundary.

We can now solve for the amplitude of the transmitted and reflected wave. Since the normal velocity of the fluid at y = 0 interface is the component v_y, defined through the velocity potential ϕ:

$$v_y = -\frac{\partial\phi}{\partial y} \qquad \text{and} \qquad p = \rho\frac{\partial\phi}{\partial t} = i\omega\,\rho\phi$$

so that the velocity can be expressed in terms of the acoustic pressure:

$$v_y = \frac{i}{\omega\rho}\frac{\partial p}{\partial y}$$

Thus, substituting the expression for p for all three waves in the equation on the normal velocity:

$$\frac{km}{\omega\rho_1}p_0 + \frac{\alpha_1 m_1}{\omega\rho_1}A = \frac{\alpha_2 m_2}{\omega\rho_2}B$$

where:

$$m = \cos(\pi + \theta) = -\cos\theta$$

$$m_1 = \cos(-\theta_1) = \cos\theta = -m$$

and

$$m_2 = \cos(\pi + \theta_2) = -\cos\theta_2 = -[1-(c_2/c_1)^2 m^2]^{1/2}$$

Thus:

$$p_0 = A + \gamma B \qquad \text{where } \gamma = \frac{\cos(\theta_2)}{\cos(\theta)}\frac{\rho_1 c_1}{\rho_2 c_2}$$

Also, $p_o + A = B$. Solving for A and B, one obtains:

$$A = \frac{1-\gamma}{1+\gamma}p_o$$

and

$$B = \frac{2}{1+\gamma} p_o$$

Note that if $\gamma = 1$, then $A = 0$ and $B = p_o$, which means there is a complete penetration of the incident wave due to impedance matching at the boundary.

6.14.4 Cylindrical Harmonic Waves

Harmonic waves in the right circular-cylindrical coordinate system in an infinite medium is governed by the following Helmholtz equation:

$$\phi(P,t) = F(P)\, e^{i\omega t} \qquad P = P\,(r,\theta,z)$$

$$\frac{\partial^2 F}{\partial r^2} + \frac{1}{r}\frac{\partial F}{\partial r} + \frac{1}{r^2}\frac{\partial^2 F}{\partial \theta^2} + \frac{\partial^2 F}{\partial z^2} + k^2 F = 0 \qquad k = \frac{\omega}{c}$$

$$r \geq 0 \qquad 0 \leq \theta \leq 2\pi \qquad -\infty \leq z \leq \infty$$

Let $F = R(r)\, E(\theta)\; Z(z)$, then the equation separates into the following three ordinary differential equations:

$$r^2 R'' + rR' + (a^2 r^2 - b^2) R = 0 \qquad R = A\, H_b^{(1)}(ar) + B\, H_b^{(2)}(ar)$$

$$E'' + b^2 E = 0 \qquad E = C \sin(b\theta) + D \cos(b\theta)$$

$$Z'' + d^2 Z = 0 \qquad Z = G \exp(idz) + H \exp(-idz)$$

where $k^2 = a^2 + d^2$.

Single-valuedness of the solution requires that $E(\theta) = E(\theta + 2\pi)$ which results in:

$$b = \text{integer} = n \qquad n = 0, 1, 2, \ldots$$

Letting $a = kl$ and $d = km$, then the cylindrical wave functions become:

$$\phi = \begin{Bmatrix} H_n^{(1)}(lkr) \\ H_n^{(2)}(lkr) \end{Bmatrix} \bullet \begin{Bmatrix} \sin(n\theta) \\ \cos(n\theta) \end{Bmatrix} \bullet \begin{Bmatrix} e^{ikmz} \\ e^{-ikmz} \end{Bmatrix} e^{i\omega t}$$

For kr >> 1, the Hankel Functions approach the following asymptotic values:

$$H_n^{(1)}(krl) \approx \sqrt{\frac{2}{\pi krl}}\; e^{i(krl - n\pi/2 - \pi/4)}$$

and

$$H_n^{(2)}(krl) \approx \sqrt{\frac{2}{\pi krl}}\; e^{i(krl - n\pi/2 - \pi/4)}$$

Thus, multiplying by the time harmonic function gives:

$$H_n^{(1)}(krl)e^{i\omega t} \approx \sqrt{\frac{2}{\pi krl}}\; e^{ik(rl+ct)}\; e^{-i(n\pi/2 + \pi/4)}$$

and

$$H_n^{(2)}(krl)\, e^{i\omega t} \approx \sqrt{\frac{2}{\pi krl}}\; e^{-ik(rl-ct)}\; e^{i(n\pi/2 + \pi/4)}$$

which denotes that $H_n^{(1)}$ and $H_n^{(2)}$ represent incoming and outgoing waves, respectively.

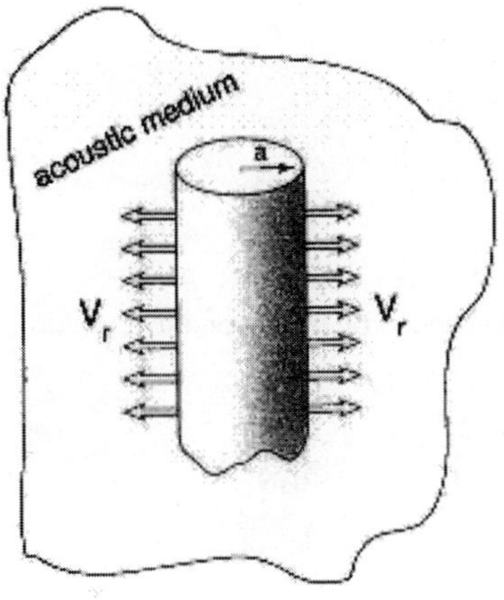

Example 6.19 *Acoustic Radiation from an Infinite Cylinder*

An infinite pulsating cylinder is submerged in an infinite acoustic medium. If the surface of the cylinder has the following normal velocity:

$$V_r(a,\theta) = f(\theta)\cos(\omega t)$$

obtain the pressure field in the acoustic medium.

Since the velocity potential $\phi(r,\theta,t)$ satisfies the axisymmetric wave equation in cylindrical coordinates, then the solution can be written as an infinite sum of all possible wave functions:

$$\phi(r,\theta,t) = \sum_{n=0}^{\infty} [A_n H_n^{(1)}(kr) + B_n H_n^{(2)}(kr)][C_n \sin(n\theta) + D_n \cos(n\theta)]e^{i\omega t}$$

One can write the boundary condition in complex form and then take the real part of the solution. Thus, letting

$$V_r(a,\theta) = f(\theta)\, e^{i\omega t}$$

then since the acoustic radiation is obviously outgoing, one must set $A_n = 0$ and $B_n = 1$. The radial component of the velocity is then given by:

$$V_r(a,\theta) = -\frac{\partial\phi}{\partial r}(a,\theta,t) = -k\sum_{n=0}^{\infty} H_n'^{(2)}(ka)[C_n \sin(n\theta) + D_n \cos(n\theta)]e^{i\omega t}$$

$$= f(\theta)e^{i\omega t}$$

which are integrated to give the Fourier coefficients of the expansion:

$$C_o = 0$$

$$C_n = \frac{-1}{\pi k H_n'^{(2)}(ka)} \int_0^{2\pi} f(\theta)\sin(n\theta)\,d\theta \qquad n = 1, 2, 3, \ldots$$

$$D_n = \frac{-\varepsilon_n}{2\pi k H_n'^{(2)}(ka)} \int_0^{2\pi} f(\theta)\cos(n\theta)\,d\theta \qquad n = 1, 2, 3, \ldots$$

where ε_n is the Neumann factor.

The velocity potential and the acoustic pressure can be developed by combining the two integrals as:

$$\phi(r,\theta,t) = -\sum_{n=0}^{\infty} \frac{\varepsilon_n H_n^{(2)}(kr)}{2\pi k H_n'^{(2)}(ka)} \left\{ \int_0^{2\pi} f(\eta)\cos(n[\theta - \eta])\,d\eta \right\} e^{i\omega t}$$

$$p(r,\theta,t) = \rho \frac{\partial \phi}{\partial t} = -i \frac{\rho c}{2\pi} \sum_{n=0}^{\infty} \frac{\varepsilon_n H_n^{(2)}(kr)}{H_n'^{(2)}(ka)} \left\{ \int_0^{2\pi} f(\eta)\cos(n[\theta - \eta])\,d\eta \right\} e^{i\omega t}$$

Thus, the acoustic pressure is the real part of the above expression:

$$p(r,\theta,t) = \frac{\rho c}{2\pi} \sum_{n=0}^{\infty} \frac{\varepsilon_n \left[\Theta_n \sin(\omega t) + \Gamma_n \cos(\omega t)\right]}{J_n'^2(ka) + Y_n'^2(ka)} \left\{ \int_0^{2\pi} f(\eta)\cos(n[\theta - \eta])\,d\eta \right\} e^{i\omega t}$$

where:

$$\Theta_n = J_n(kr)J_n'(ka) + Y_n(kr)Y_n'(ka) \quad \text{and} \quad \Gamma_n = J_n(kr)Y_n'(ka) - Y_n(kr)J_n'(ka)$$

6.14.5 Spherical Harmonic Waves

Spherical harmonic waves obey the following Helmholtz equation:

$$\frac{\partial^2 F}{\partial r^2} + \frac{2}{r}\frac{\partial F}{\partial r} + \frac{1}{r^2 \sin(\theta)}\frac{\partial}{\partial \theta}\left(\sin(\theta)\frac{\partial F}{\partial \theta}\right) + \frac{1}{r^2 \sin^2(\theta)}\frac{\partial^2 F}{\partial \phi^2} + k^2 F = 0$$

Letting $F = R(r)\,S(\theta)\,M(\phi)$, then the equation separates into three equations (Example 6.10), with the following wave solutions:

$$F = \begin{Bmatrix} h_n^{(1)}(kr) \\ h_n^{(2)}(kr) \end{Bmatrix} P_n^m(\cos\theta) \begin{Bmatrix} \sin(m\phi) \\ \cos(m\phi) \end{Bmatrix}$$

where $h_n^{(1)}$ and $h_n^{(2)}$ are spherical Hankel functions representing incoming and outgoing radial waves, respectively, and P_n^m are the associated Legendre functions.

Example 6.20 *Scattering of a Plane Wave from a Rigid Sphere*

An incident plane pressure wave is incident on a rigid sphere whose radius is a in an infinite acoustic medium. Obtain the scattered acustic pressure field.

Let the incident pressure wave p_i to have the following form:

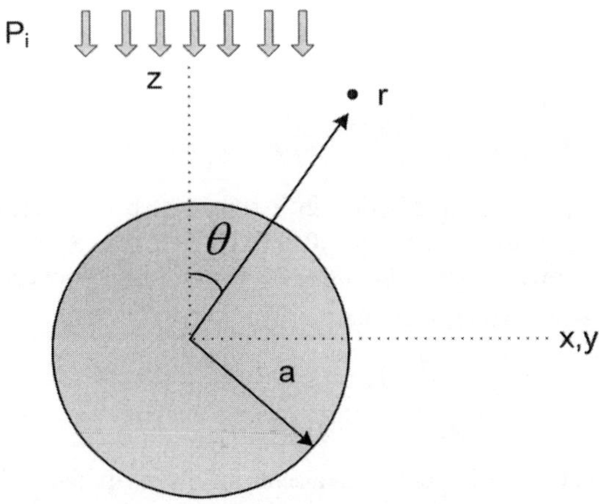

$$p_i = p_o\, e^{ikz}\, e^{i\omega t} = p_o\, e^{ikr\cos\theta}\, e^{i\omega t}$$

Expanding the plane wave in terms of axisymmetric spherical wave functions with m = 0, one obtains:

$$p_i = p_o e^{ikr\cos\theta}\, e^{i\omega t} = p_o \sum_{n=0}^{\infty} i^n (2n+1)\, j_n(kr)\, P_n(\cos\theta) e^{i\omega t}$$

The scattered pressure field p_s can also be written in terms of outgoing axisymmetric spherical wave functions as follows:

$$p_s = \sum_{n=0}^{\infty} E_n\, h_n^{(2)}(kr)\, P_n(\cos\theta)\, e^{i\omega t}$$

The total pressure field p in the infinite acoustic medium is then the sum of the incident and scattered fields, i.e.:

$$p = p_i + p_s$$

The normal component of the particle velocity at the surface of the rigid sphere must vanish, resulting in:

$$V_r(a,\theta) = \frac{i}{\omega\rho}\frac{\partial p}{\partial r}(a,\theta) = \frac{i}{\omega\rho}\left[\frac{\partial p_i}{\partial r}(a,\theta) + \frac{\partial p_s}{\partial r}(a,\theta)\right] = 0$$

which, upon substitution for the series for the incident and scattered fields, yields:

$$p_o\, i^n (2n+1) j_n'(ka) + E_n h_n'^{(2)}(ka) = 0$$

Or

$$E_n = -p_o \frac{i^n (2n+1) j_n'(ka)}{h_n'^{(2)}(ka)} \qquad n = 0, 1, 2, \ldots$$

Thus, the scattered pressure field p_s is given by the sum of spherical wave functions in the form:

$$p_s = -p_o \sum_{n=0}^{\infty} \frac{i^n (2n+1)\, j_n'(ka)}{h_n'^{(2)}(ka)}\, h_n^{(2)}(kr) P_n(\cos\theta) e^{i\omega t}$$

PROBLEMS

Section 6.9

1. Obtain the steady state temperature distribution in a square slab of sidelength = L defined by $0 \le x, y \le L$. The faces x = 0 and y = 0 are kept at a zero temperature, the face y = L is kept at a temperature T_o and the face x = L has a heat convection to an ambient medium with zero temperature, such that:

$$\frac{\partial T}{\partial x} + bT = 0 \qquad \text{at} \qquad x = L$$

2. Obtain the steady state temperature distribution in a semi-infinite strip, defined by 0 $\le x \le L$ and $y \ge 0$. The surfaces x = 0 and x = L are kept at zero temperature and the surface y = 0 has a temperature distribution:

$$T(x,0) = T_o\, f(x)$$

6 Obtain the steady state temperature in a semi-infinite slab, defined by $0 \le x \le L$, y ≥ 0. The surface x = 0 is insulated, the surface x = L has heat convection to an ambiend medium with zero temperature, such that:

$$\frac{\partial T}{\partial x} + bT = 0 \qquad \text{at} \qquad x = L$$

and the surface y = 0 is kept at a temperature $T(x,0) = T_o\, f(x)$.

4. Obtain the steady state temperature distribution in a square plate of sidelength = L defined by $0 \le x, y \le L$. The faces x = 0 and x = L are kept at zero temperature, its face y = 0 is insulated and its face y = L has a temperature distribution $T(x,L) = T_o\, f(x)$.

5. Obtain the steady state temperature distribution in a semi-circular sheet having a radius = a defined by $0 \le r \le a$ and $0 \le \theta \le \pi$. The straight face is kept at zero temperature and the cylindrical face, r = a, is kept at a constant temperature T_o.

6. Obtain the temperature distribution in a circular sector whose radius is "a" which subtends an angle b defined by $0 \le r \le a$ and $0 \le \theta \le b$. The straight faces are kept at zero temperature while the surface r = a is kept at a temperature:

$$T(a,\theta) = T_o\, f(\theta)$$

7. Obtain the temperature distribution in an infinite sheet having a circular cavity of radius "c" defined by $r \ge c$ and $0 \le \theta \le 2\pi$. The temperature on the circular boundary is kept at temperature:

$$T = T_o\, f(\theta)$$

8. Obtain the steady state temperature distribution in a right parallelopiped having the dimensions a, b, and c aligned with the x, y and z axes, respectively. The surfaces x = 0, x = a, y = 0, y = b, and z = 0 are kept at zero temperature, while the surface z = c is kept at temperature:

$$T(x,y,c) = T_0\, f(x,y)$$

9. Obtain the steady state temperature distribution in a finite cylinder of length L and radius "a" defined by $0 \le r \le a$ and $0 \le z \le L$. The cylinder is kept at zero temperature at z = 0, while the surface z = L is kept at a temperature:

$$T(r,L) = T_0\, f(r)$$

The surface at r = a dissipates heat to an outside medium having a zero temperature, such that:

$$\frac{\partial T}{\partial r}(a,z) + b\, T(a,z) = 0$$

10. Obtain the steady state temperature distribution in a hollow finite cylinder of length L, of outside and inside radii "b" and "a", respectively. The cylinder is kept at zero temperature on the surfaces r = a, r = b and z = 0, while the surface z = L is kept at a temperature:

$$T(r,L) = T_0\, f(r)$$

I. Obtain the steady state temperature distribution in a finite cylinder of length L and radius a defined by $0 \le r \le a$ and $0 \le z \le L$. The cylinder is kept at zero temperature at surfaces z = 0 and r = a, while the surface z = L is kept at a temperature:

$$T(r,L) = T_0\, f(r)$$

12. Obtain the steady state temperature distribution in a cylinder of length L and radius a defined by $0 \le r \le a$ and $0 \le z \le L$. The cylinder is kept at zero temperature at surfaces z = 0 and z = L, while the surface r = a is kept at a temperature:

$$T(a,z) = T_0\, f(z)$$

13. Obtain the steady state temperature distribution of a sphere of radius = a defined by $0 \le r \le a$ and $0 \le \theta \le \pi$. The surface of the sphere is heated to a temperature:

$$T(a,\theta) = T_0\, f(\cos\theta)$$

Also obtain the solution for f = 1.

14. Obtain the steady state temperature distribution in an infinite solid having a spherical cavity of radius = a defined by $r \ge a$, $0 \le \theta \le \pi$. The temperature at the surface of the cavity is kept at:

$$T(a,\theta) = T_0\, f(\cos\theta)$$

Also obtain the solution for f = 1.

15. Determine the steady state temperature distribution in a black metallic sphere (radius equals a), defined by $0 \le r \le a$ and $0 \le \theta \le \pi$, which is being heated by the sun's rays. The heating, by convection, of the sphere at its surface satisfies:

$$\frac{\partial T}{\partial r}(a,\theta) + bT(a,\theta) = b\, f(\theta)$$

where

$$f(\theta) = \begin{cases} T_0 \cos(\theta) & 0 \le \theta \le \pi/2 \\ 0 & \pi/2 \le \theta \le \pi \end{cases}$$

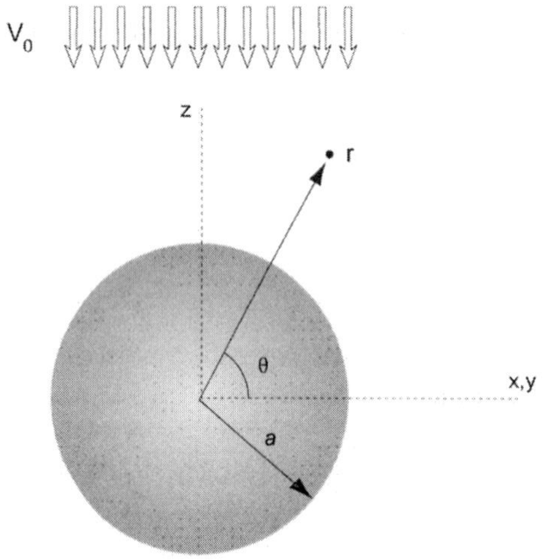

16. Determine the particle velocity of an ideal incompressible irrotational fluid flowing around a rigid sphere whose radius = a. The fluid has a velocity at infinity:

$$V_z = -V_0 \qquad z \gg a$$

17. Determine the steady state temperature distribution in a solid hemisphere whose radius is "a" defined by $0 \le r \le a$, $0 \le \theta \le \pi/2$. The hemisphere's convex surface is kept at constant temperature T_0 and its base is kept at zero temperature.

18. Obtain the steady state temperature distribution in a hollow metallic sphere whose inner and outer radii are a and b, respectively. The temperature at the outer surface is kept at zero temperature, while the temperature on the inner surface is kept at:

$$T(a,\theta) = T_0\, f(\cos\theta)$$

19. Determine the temperature distribution in a semi-infinite cylinder whose radius = a defined by $0 \le r \le a$, $0 \le \theta \le 2\pi$ and $z \ge 0$. The temperature of the surface r = a is kept at zero temperature and the temperature of the base is:

$$T(r,\theta,0) = T_0\, f(r,\theta)$$

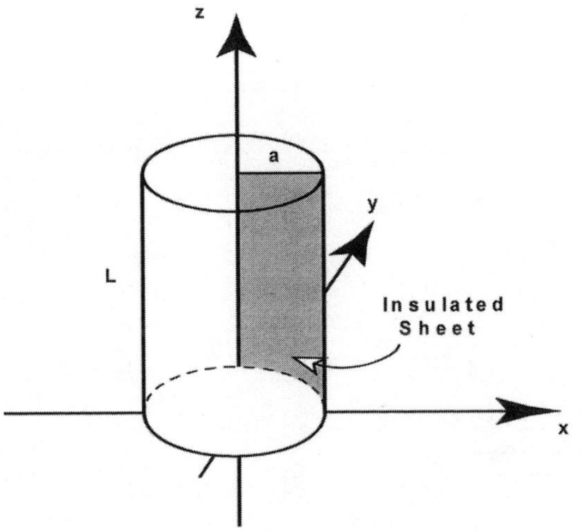

20. Determine the steady state temperature distribution in a solid finite cylinder of length = L and radius = a defined by $0 \le r \le a$, $0 \le \theta \le 2\pi$ and $0 \le z \le L$. The cylinder has an insulated surface at $\theta = 0$, see the accompanying figure, extending from its axis to the outer surface. The cylinder is kept at zero temperature at its two ends ($z = 0$ and $z = L$), and is heated at its convex surface to a temperature:

    $$T(a,z,\theta) = T_0\ f(z,\theta)$$

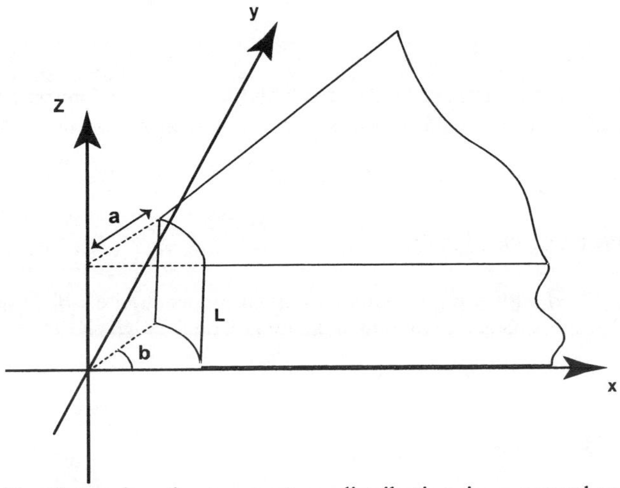

21. Determine the temperature distribution in a curved wedge occupying the region $a \le r \le \infty$, $o \le z \le L$ and $0 \le \theta \le b$, see the accompanying figure. The surfaces $z = 0$ and $z = L$ are kept at zero temperature, the surface $\theta = 0$ and $\theta = b$ are insulated and the cylindrical surface $r = a$ is kept at a temperature:

    $$T(a,z,\theta) = T_0\ f(z,\theta)$$

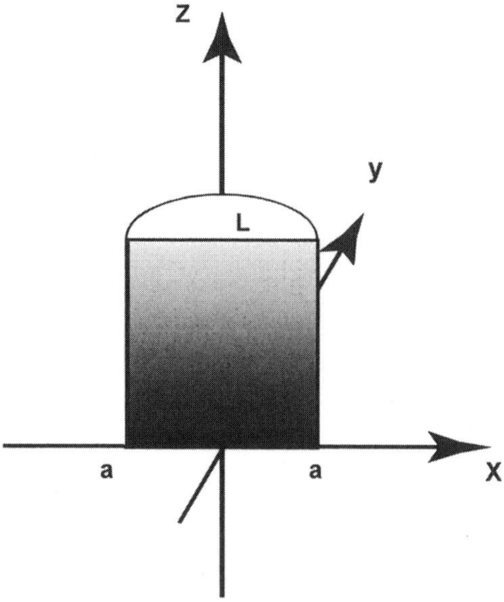

22. Determine the temperature distribution in a hemi-cylinder of length = L and radius = a defined by $0 \le r \le a$, $0 \le \theta < \pi$ and $0 \le z \le L$. The convex surface at r = a, the two plane surfaces at $\theta = 0$ and π and the lower base at z = 0 are kept at zero temperature, while the upper base at z = L is kept at a temperature:

$$T(r, \theta, L) = T_o\, f(r,\theta)$$

Section 6.10

(a) A metallic sphere of radius a and defined by $0 \le r \le a$ and $0 \le \theta \le \pi$ is kept at zero temperature at its surface. A heat source is located in a spherical region inside the sphere, such that:

$$\nabla^2 T = -q(r, \cos\theta) \qquad 0 \le r < b$$

find the steady state temperature distribution.

24. A spherical container is filled with a liquid whose walls are impenetrable. If a point sink of magnitude Q exists at its center so that the velocity potential satisfies:

$$\nabla^2 \psi = \frac{Q_0\, \delta(r)}{4\pi\, r^2} \qquad 0 \le r \le a$$

find the velocity field inside the sphere.

25. A finite circular cylindrical container with impenetrable walls is full with an incompressible liquid, occupying the space $0 \le r \le a$ and $0 \le z \le L$. A point source and a point sink of magnitudes Q_o are located on the axis of the cylinder at z = L/4 and 3L/4, respectively, such that the velocity potential $\psi(r,z)$ satisfies:

$$\nabla^2 \psi = -\frac{Q_0\, \delta(r)}{2\pi\, r}[\delta(z - L/4) - \delta(z - 3L/4)]$$

Find the velocity field inside the container.

Section 6.11

26. Determine the Eigenfunctions (modes) and Eigenvalues (natural frequencies) of membranes having the following shapes and boundaries:
    (a) Semi-circular membrane, radius = a, fixed on all its boundaries.
    (b) Annular membrane, radii b, and a (b > a), fixed on all its boundaries.
    (b) Annular membrane, fixed on the outer boundary r = b and free at its inner boundary r = a.
    (d) A circular sector, radius = a, subtending an angle = c, fixed on all its boundaries.
    (e) A circular sector membrane, radius = a, subtending an angle = c, fixed on its straight edges and free at its circular boundary.
    (f) An annular sector membrane, radii b, and a (b > a), subtending an angle = c, fixed on all its boundaries.
    (g) An annular sector, radii b and a (b > a), subtending an angle = c, fixed on its straight boundaries and free on its circular boundaries.
    (h) A rectangular membrane, of dimension a and b, with sides whose length = a are fixed and sides whose length = b are free.

27. Determine the mode shapes and natural frequencies of a vibrating gas in a rigid cylindrical tube of length = L and radius = a. The tube is closed by two rigid plates at its ends. Let the velocity potential be:

    $\phi = \phi\ (r,\theta,z)$

28. Determine the mode shapes for the tube in problem 27, where the ends of the tube are open (pressure release).

29. Determine the mode shapes and natural frequencies of a vibrating gas entrapped in the space between a rigid sphere of radius = a and a concentric rigid spherical shell of radius = b (b > a).

30. Determine the response of a rectangular membrane, measuring a,b, under the influence of sinusoidal time varying force field, i.e.:

    $q(x,y,t) = q_o\ f(x,y)\ \sin(\omega t)$

31. Determine the response of a circular membrane of radius = a, fixed on its perimeter, and acted upon by distributed forces:

    $q(r,\theta,t) = q_o\ f(r,\theta)\ \sin(\omega t)$

32. Obtain the mode shapes and natural frequencies of a circular plate, radius = a, whose boundary is simply supported, such that at the boundary r = a

$$w(a,\theta) = 0$$

and

$$\frac{\partial^2 w}{\partial r^2}(a,\theta) + \frac{\nu}{r}\frac{\partial w}{\partial r}(a,\theta) = 0$$

33. Determine the responses of a square plate, sidelength = L, whose sides are simply supported. The plate is excited by a distributed force:

$$q(x,y,t) = q_o\, f(x,y)\, \sin(\omega t)$$

34. A rectangularly shaped membrane is being excited to harmonic motion such that:

$$w = w(x,y)$$

and

$$\nabla^2 w + k^2 w = \frac{F_o}{S}\,\delta(x - \frac{a}{2})\,\delta(y - \frac{b}{2}) \qquad 0 \le x \le a \qquad 0 \le y \le b$$

$$w(x,0) = 0 \qquad w(x,b) = 0 \qquad \frac{\partial w}{\partial x}(0,y) = 0 \qquad \frac{\partial w}{\partial x}(a,y) = 0$$

Obtain the solution w(x, y).

Section 6.12

35. Determine the temperature distribution in a rod of length = L and whose ends are kept at zero temperature. The rod was heated initially, such that:

$$T(x,0^+) = \begin{cases} T_o & 0 \le x < L/2 \\ 0 & L/2 < x \le L \end{cases}$$

36. Determine the temperature distribution in a rod of length = L, where there is heat convection to an outside medium at both ends. The temperature of the outside medium is kept at zero temperature. The temperature of the rod was initially raised to:

$$T(x,0^+) = T_o\, f(x)$$

37. Determine the temperature distribution in a rectangular sheet occupying the region $0 \le x \le a$ and $0 \le y \le b$. The sides of the plate are kept at zero temperature, while the sheet was initially raised to a temperature

$$T(x,y,0^+) = T_o\, f(x,y)$$

and the sheet is heated by a source Q:

$$Q = Q_o\, \delta(x - a/2)\, \delta(y - b/2)\, e^{-\alpha t}, \qquad \alpha > 0$$

38. Determine the axisymmetric temperature distribution in a circular slab of radius = a, whose perimeter is kept at a zero temperature. The slab is initially heated to a temperature:

$$T(r,0^+) = T_0\, f(r)$$

39. Determine the axisymmetric temperature distribution for a circular slab of radius = a, such that the slab conducts heat through its perimeter to an outside medium whose temperature is kept at zero temperature. The slab is initially heated to a temperature:

$$T(r,0^+) = T_0\, f(r)$$

with an impulsive heat point source at its center:

$$Q = Q_0 \frac{\delta(r)}{2\pi r}\delta(t - t_0)$$

40. Determine the temperature distribution in a circular slab, radius = a, whose perimeter is kept at zero temperature. The slab is heated initially to a temperature:

$$T(r,\theta,0^+) = T_0\, f(r,\theta)$$

and has an impulsive heat point source at (r_0,θ_0):

$$Q = Q_0 \frac{\delta(r - r_0)}{2\pi r}\delta(t - t_0)\delta(\theta - \theta_0)$$

41. Determine the temperature distribution in a solid sphere of radius = a, whose surface is kept at zero temperature and is heated initially to a constant temperature = T_0.

42. Determine the temperature distribution in a solid sphere of radius = a, whose surface conducts heat to an outside medium that is being kept at zero temperature. The sphere is heated initially to a temperature:

$$T(r,0^+) = T_0\, f(r)$$

43. Determine the temperature distribution in a cube having a sidelength = L. The cubes' surfaces are kept at zero temperature and the cube is initially heated to a temperature:

$$T(x,y,z,0^+) = T_0\, f(x,y,z)$$

44. Determine the temperature distribution in a sphere having a radius = a, whose surface is kept at zero temperature. The sphere is initially heated such that:

$$T(r,\theta,\phi,0^+) = T_0\, f(r,\cos\theta,\phi)$$

45. A rectangular sheet is immersed in a zero temperature bath on two of its sides, and is kept at zero temperature at the other two. The sheet is heated by a point source at its center. The sheet is initially kept at zero temperature, such that

$$\nabla^2 T = \frac{1}{K}\frac{\partial T}{\partial t} - \frac{Q_o}{k}\delta(x - \frac{a}{2})\delta(y - \frac{b}{2})\sin(\omega t)$$

$$0 \le x \le a0 \le y \le bt > 0 \qquad T = T(x,y,t)$$

with boundary conditions of:

$$T(x,0,t) = 0 \quad T(x,b,t) = 0 \quad T(x,y,0) = 0$$

$$\frac{\partial T}{\partial x}(0,y,t) - \gamma\, T(0,y,t) = 0 \qquad \frac{\partial T}{\partial x}(a,y,t) + \gamma\, T(a,y,t) = 0$$

Obtain the temperature distribution T(x,y,t) in the sheet for $t > 0$.

46. A semi-circular metal sheet is heated by a point source. The sheet is initially kept at zero temperature, such that:

$$\nabla^2 T = \frac{1}{K}\frac{\partial T}{\partial t} - \frac{Q_o}{k}\frac{\delta(r - r_o)}{r}\delta(\theta - \pi/4)\,\delta(t - t_o)$$

$$0 \le r \le a\, 0 \le \theta \le \pi \qquad t, t_o > 0 \qquad T = T(r,\theta,t)$$

with the following boundary conditions:

$$T(r,0,t) = 0 \quad T(r,\pi,t) = 0 \quad \frac{\partial T}{\partial r}(a,\theta,t) = 0 \quad \frac{\partial T}{\partial r}(r,\theta,0) = 0$$

Obtain the solution for the transient temperature $T(r,\theta,t)$.

47. Obtain the temperature distribution in a rod of length L with a heat sink Q. The end $x = 0$ is insulated and the end $x = L$ is connected to a zero temperature ambient liquid bath. Find the temperature $T = T(x,t)$ satisfying:

$$\frac{\partial^2 T}{\partial x^2} = \frac{1}{K}\frac{\partial T}{\partial t} + \frac{Q_o}{k}\delta(x - x_o)e^{-at} \qquad a > 0 \qquad x_o > 0$$

Subject to the boundary and initial conditions:

$$T(x,0^+) = 0 \quad \frac{\partial T}{\partial x}(0,t) = 0 \quad \frac{\partial T}{\partial x}(L,t) + b\,T(L,t) = 0$$

48. A rectangular sheet is heated by a point source at its center. The sheet is initially kept at zero temperature, such that:

$$\nabla^2 T = \frac{1}{K}\frac{\partial T}{\partial t} - \frac{Q_o}{k}\delta(x - \frac{a}{2})\delta(y - \frac{b}{2})e^{-ct} \qquad c > 0$$

$$0 \le x \le a \quad 0 \le y \le b \quad t > 0 \qquad T = T(x,y,t)$$

Subject to the boundary and initial conditions:

$$T(x,0,t) = T(x,b,t) = 0$$

$$\frac{\partial T}{\partial x}(0,y,t) = 0 \qquad \frac{\partial T}{\partial x}(a,y,t) = 0 \qquad T(x,y,0) = 0$$

Obtain the temperature distribution in the sheet for $t > 0$

49. A completely insulated hemi-cylinder is heated such that its temperature $T(r,\theta,z,t)$ satisfies:

$$\nabla^2 T = \frac{1}{K}\frac{\partial T}{\partial t} + Q_o\frac{\delta(r - r_o)}{2\pi r}\delta(\theta - \pi/2)\delta(z - z_o)\delta(t - t_o)$$

$$0 \le r \le a \quad 0 \le z \le L \qquad 0 \le \theta \le \pi \qquad t, t_o > 0$$

Subject to the boundary and initial conditions:

$$\frac{\partial T}{\partial z}(r,\theta,0,t) = 0 \qquad \frac{\partial T}{\partial z}(r,\theta,L,t) = 0 \qquad \frac{1}{r}\frac{\partial T}{\partial \theta}(r,0,z,t) = 0$$

$$\frac{1}{r}\frac{\partial T}{\partial \theta}(r,\pi,z,t) = 0 \qquad \frac{\partial T}{\partial r}(a,\theta,z,t) = 0 \qquad T(r,\theta,z,0^+) = 0$$

Obtain the temperature in the cylinder for $t > 0$.

50. Obtain the temperature distribution in a solid sheet of length L with a heat sink Q_o. The two ends of the sheet $x = 0$ and $x = L$ are immersed in an ambient fluid whose temperature is constant at T_o. If the temperature $T = T(x,t)$ satisfies:

$$\frac{\partial^2 T}{\partial x^2} = \frac{1}{K}\frac{\partial T}{\partial t} + \frac{Q_o}{k}\delta(x - \frac{L}{2})\delta(t - t_o) \qquad 0 \le x \le L \qquad t, t_o > 0$$

Subject to the boundary and initial conditions, for $a > 0$:

$$T(x,0^+) = T_1 = \text{constant} \quad T(0,t) - a\frac{\partial T}{\partial x}(0,t) = T_o \quad T(L,t) + a\frac{\partial T}{\partial x}(a,t) = T_o$$

Obtain the temperature distribution as a function of time.

Section 6.13

51. Determine the vibration response of a string, having a length = L, fixed at both ends. The string was initially displaced such that

$$y(x,0^+) = f(x) \qquad \frac{\partial y}{\partial t}(x,0^+) = g(x)$$

52. If the string in problem 51 is plucked from rest, such that:

$$f(x) = \begin{cases} W_0 x / a & 0 \le x \le a \\ W_0 (L-x)/(L-a) & a \le x \le L \end{cases}$$

$$g(x) = 0$$

obtain an expression for the subsequent motion of the string.

53. Determine the longitudinal displacement of a rod, having a length = L, which is fixed at x = 0 and is free at x = L. The rod is initially displaced, such that:

$$u(x,0^+) = f(x) \qquad \frac{\partial u}{\partial t}(x,0^+) = g(x)$$

54. A stretched string of length = L is fixed at x = 0 and is elastically supported at x = L, such that:

$$\frac{\partial y}{\partial x}(L,t) + \gamma\, y(L,t) = 0$$

The string is initially displaced from rest, such that:

$$y(x,0^+) = y_0 x \qquad \frac{\partial y}{\partial t}(x,0^+) = 0$$

Determine the subsequent vibration response of the string.

55. A string, having a length = L, is struck by hammer at its center, such that the initial velocity imparted to the string is described by:

$$\frac{\partial y}{\partial t}(x,0^+) = \frac{I}{2\varepsilon\rho}\begin{cases} 0 & 0 \le x < L/2-\varepsilon \\ 1 & L/2-\varepsilon < x < L/2+\varepsilon \\ 0 & L/2+\varepsilon < x \le L \end{cases}$$

$$y(x,0^+) = 0$$

where I represents the total impulse of the hammer and ρ is the density per unit length of the string. The string is fixed at both ends.

(a) Obtain the subsequent displacement of the string.

(b) If $\varepsilon \to 0$, obtain an expression for the subsequent motion.

56. A string, length = L, fixed at both ends and initially at rest is acted upon by a distributed force f(x,t) per unit length. Obtain an expression for the forced motion of the string

57. If the distributed force in problem 56 is taken to be an impulsive concentrated force, such that:

$$f(x,t) = P_0 \delta(x - L/2)\, \delta(t)$$

where δ is the Dirac delta function, determine the subsequent motion of the string.

58. Determine the motion of a rectangular membrane, occupying the region $0 \le x \le a$, $0 \le y \le b$, where the membrane is initially displaced and set in motion such that:

$$W(x,y,0^+) = f(x,y) \qquad \frac{\partial W}{\partial t}(x,y,0^+) = g(x,y)$$

The membrane is fixed along its perimeter.

59. Determine the free vibration of a circular membrane, radius = a, whose perimeter is fixed. The membrane is initially set in motion, such that:

$$W(r,\theta,0^+) = f(r,\theta) \qquad \frac{\partial W}{\partial t}(r,\theta,0^+) = g(r,\theta)$$

60. An annular shaped membrane is set into motion by initially displacing it from rest, i.e.:

$$W(r,\theta,0^+) = f(r,\theta) \qquad \frac{\partial W}{\partial t}(r,\theta,0^+) = 0$$

The membrane has outer and inner radii b and a respectively. Determine the subsequent free vibration of the membrane.

61. Determine the response of a circular membrane, radius = a, when acted upon by a concentrated impulsive force described by:

$$f(r,t) = P_0 \frac{\delta(r)}{2\pi r}\, \delta(t)$$

The boundary of the membrane is fixed, and the membrane is initially undeformed and at rest.

62. Determine the response of a square membrane initially at rest, side length = L, when acted upon by an impulsive force located at x_0, y_0, described by:

$$f(x,y,t) = P_0 \delta(x - x_0)\, \delta(y - y_0)\delta(t)$$

where P_0 is total force. The sides of the membrane are fixed.

63. Determine the response of a circular membrane radius = a, initially at rest and undeformed, when acted upon by a concentrated impulsive force located at r_o , θ_o described by:

$$f(r,\theta,t) = P_0 \frac{\delta(r-r_o)}{r} \delta(\theta-\theta_o)\delta(t)$$

The membrane is fixed on its boundary.

64. A bar of length L is connected to a spring at one end and the other end is free. The bar is being excited by a point force such that, u = u(x,t) and:

$$\frac{\partial^2 u}{\partial x^2} = \frac{1}{c^2}\frac{\partial^2 u}{\partial t^2} - \frac{F_o}{AE}\delta(x)\delta(t-t_o) \qquad 0 \le x \le L \qquad t, t_o > 0$$

Subject to boundary and initial conditions:

$$\frac{\partial u}{\partial x}(0,t) = 0 \quad \frac{\partial u}{\partial x}(L,t) + \frac{\gamma}{AE}u(L,t) =$$

$$u(x,0) = 0 \quad \frac{\partial u}{\partial t}(x,0) = 0$$

Obtain the transient response of the string u(x,t).

65. A pie-shaped stretched membrane is excited to motion from rest by a mechanical point force, such that its displacement w = w(r,θ,t) satisfies:

$$\nabla^2 w = \frac{1}{c^2}\frac{\partial^2 w}{\partial t^2} - \frac{P_o}{S}\frac{\delta(r-r_o)}{r}\delta(\theta-\theta_o)\delta(t-t_o)$$

$0 \le r \le a \qquad r_o > 0 \quad 0 \le \theta \le b \qquad t, t_o > 0$

w = 0 on the boundary

$$w(r,\theta,0^+) = 0 \qquad \frac{\partial w}{\partial t}(r,\theta,0^+) = 0$$

Obtain the solution to the transient vibration of the membrane w(r,θ,t).

66. A rectangular stretched membrane is acted on by a time dependent point force such that its displacement w (x, y, t) is governed by:

$$\nabla^2 w = \frac{1}{c^2}\frac{\partial^2 w}{\partial t^2} - \frac{P_o}{S}\delta(x-x_o)\delta(y-y_o)\delta(t-t_o)$$

$0 \le x \le a0 \le y \le bt, t_o > 0$

where δ is the Dirac function and the boundary conditions are:

$$w(x,0,t) = w(x,b,t) = 0 \qquad \frac{\partial w}{\partial x}(0,y,t) = \frac{\partial w}{\partial x}(a,y,t) = 0$$

If the membrane was initially at rest, and was initially deformed such that:

$$w(x,y,0^+) = w_o \sin(\frac{\pi}{b}y) \qquad \frac{\partial w}{\partial t}(x,y,0^+) = 0$$

obtain an expression for the displacement w(x,y,t).

67. A semi-circular stretched membrane is excited to motion by a point force, such that:

$$W = W(r,\theta,t)$$

$$\nabla^2 W = \frac{1}{c^2}\frac{\partial^2 W}{\partial t^2} - \frac{P_o}{S}\frac{\delta(r-r_o)}{r}\delta(\theta-\frac{\pi}{2})\delta(t-t_o)$$

$$0 \le r \le a \quad 0 \le \theta \le \pi \qquad t, t_o > 0$$

where δ is the Dirac function and the initial boundary conditions are:

$$W(a,\theta,t) = 0 \qquad \frac{\partial W}{\partial \theta}(r,0,t) = \frac{\partial W}{\partial \theta}(r,\pi,t) = 0$$

$$W(r,\theta,0^+) = 0 \qquad \frac{\partial W}{\partial t}(r,\theta,0^+) = 0$$

Obtain the solution for the transient vibration $W(r,\theta,t)$

68. A semi-circular annular stretched membrane fixed on its perimeter, is excited to motion by a point force. Obtain the solution for the transient vibration $W(r,\theta,t)$ satisfying:

$$\nabla^2 W = \frac{1}{c^2}\frac{\partial^2 W}{\partial t^2} - \frac{P_o}{S}\frac{\delta(r-r_o)}{r}\delta(\theta-\frac{\pi}{2})\delta(t-t_o)$$

$$0 \le r \le a \quad 0 \le \theta \le \pi \qquad t, t_o > 0$$

where δ is the Dirac function and the initial conditions are:

$$W(r,\theta,0^+) = 0 \qquad \frac{\partial W}{\partial t}(r,\theta,0^+) = 0$$

69. A bar of length L is connected to springs at both ends. The bar is being excited by a point force at the center such that $u = u(x,t)$ satisfies:

$$\frac{\partial^2 u}{\partial x^2} = \frac{1}{c^2}\frac{\partial^2 u}{\partial t^2} - \frac{F_o}{AE}\delta(x-\frac{L}{2})\delta(t-t_o) \qquad 0 \le x \le L \qquad t, t_o > 0$$

With boundary and initial conditions:

$$\frac{\partial u}{\partial x}(0,t) - \frac{\zeta}{AE}u(0,t) = 0 \qquad \frac{\partial u}{\partial x}(L,t) + \frac{\gamma}{AE}u(L,t) = 0$$

$$u(x,0^+) = 0 \qquad \frac{\partial u}{\partial t}(x,0^+) = 0$$

Obtain the transient response of the bar $u(x,t)$.

70. An annular circular membrane, initially at rest and undeformed, is excited to transient forced vibration such that the displacement $W(r,\theta,t)$ satisfies:

$$\nabla^2 W = \frac{1}{c^2}\frac{\partial^2 W}{\partial t^2} - \frac{P_0}{S}\frac{\delta(r - r_0)}{r}\delta(\theta - \frac{\pi}{2})\delta(t - t_0)$$

$a \leq r \leq b \qquad t, t_0 > 0$

$W(a,\theta,t) = W(b,\theta,t) = 0, \qquad W(r,\theta,0^+) = 0 \qquad \frac{\partial W}{\partial t}(r,\theta,0^+) = 0$

Obtain the solution for the transient vibration $W(r, \theta, t)$.

71. An acoustic medium is contained inside a rigid spherical container of radius = a. If the medium is initially disturbed, such that the velocity potential $\phi(r,t)$ satisfies the following initial conditions:

$\phi(r,0^+) = f(r) \qquad \frac{\partial \phi}{\partial t}(r,0^+) = g(r)$

determine the radial particle velocity v_r of the entrapped medium.

72. An acoustic medium occupies an infinite cylinder, radius = a. If the medium is initially disturbed, such that the velocity potential $\phi(r,t)$ satisfies the following initial conditions:

$\phi(r,0^+) = f(r) \qquad \frac{\partial \phi}{\partial t}(r,0^+) = g(r)$

determine the radial particle velocity of the entrapped medium.

Section 6.14

73. A semi-infinite stretched string is set into motion by initially displacing it such that:

$y(x,0^+) = f(x) \qquad \frac{\partial y}{\partial t}(x,0^+) = g(x)$

The string is fixed at x = 0. Obtain the solution y (x,t).

74. A semi-infinite stretched string initially at rest, is set into motion by giving the end x = 0 the following displacement:

$y(0,t) = Y_0 \sin(\omega t)$

Obtain the solution for the subsequent motion.

75. A sphere, radius = a, oscillates in an infinite acoustic medium, such that its radial velocity V_r at the surface is given by:

$V_r(a,t) = V_0\, e^{-i\omega t}$

Obtain the acoustic pressure everywhere in the medium.

76. A sphere, radius = a, is oscillating in an infinite acoustic medium, such that its radial velocity V_r at its surface is given by:

$$V_r(a,\theta,t) = V_o\ f(\cos\theta)\ e^{-i\omega t}$$

Obtain the acoustic pressure everywhere in the medium.

77. A plane acoustic wave impinges on an infinite cylindrical air bubble (pressure release surfaces) of radius = a. If the incident wave is described by:

$$p_i = p_o\ e^{ikz}\ e^{i\omega t} \qquad k = \omega/c$$

obtain the scattered acoustic pressure.

78. A plane acoustic wave impinges on an infinite rigid cylinder of radius = a. If the incident wave is described by:

$$p_i = p_o\ e^{ikz}\ e^{i\omega t} \qquad k = \omega/c$$

obtain the scattered acoustic pressure.

79. A plane acoustic wave impinges on a spherical air bubble (pressure release surface) of radius = a. If the incident plane wave is described by:

$$p_i = p_o\ e^{ikz}\ e^{i\omega t} \qquad k = \omega/c$$

obtain the scattered acoustic pressure.

80. A plane acoustic wave travelling in an acoustic medium (density ρ_1, velocity c_1) impinges on a spherical acoustic body (density ρ_2, velocity c_2) of radius = a. If the incident wave is described by:

$$p_i = p_o e^{ik_1 z}\ e^{i\omega t} \qquad k_1 = \omega/c_1$$

obtain the scattered acoustic pressure in the outer medium.

81. A hemi-spherical speaker, radius = a, is set in an infinite plane rigid baffle and is in contact with a semi-infinite acoustic medium. If the radial surface velocity V_r is given by:

$$V_r(a,t) = V_o\ e^{i\omega t}$$

obtain the pressure field in the acoustic medium.

82. Obtain the pressure field in the acoustic medium of problem 81, where the baffle is a pressure-release baffle.

83. If the velocity field in Example 6.19 is given by:

$$f(\sigma) = \begin{cases} 1 & -\alpha < \sigma < \alpha \\ 0 & \alpha < \sigma < 2\pi - \alpha \end{cases}$$

(a) Obtain the pressure field in the acoustic medium

(b) If $v_o = \dfrac{Q_o}{2a\alpha}$ where Q_o is the strength of the volume flow of the line source, obtain the pressure field when $\alpha \to 0$.

84. A semi-infinite duct of rectangular crossection has rigid walls and is filled with an acoustic medium. The duct occupies the region $0 \le x \le a, \quad 0 \le y \le b, \quad z \ge 0$.If a rectangular piston, located at $z = 0$, is vibrating with an axial velocity V_z described by:

$$V_z = V_o \; f(x,y) \; e^{i\omega t}$$

(a) obtain the pressure field inside the duct.

(b) Show that only the plane wave solution, propagating along the duct, exists if:

$f(x,y) = 1$

85. A semi-finite cylindrical duct has rigid walls and is filled with an acoustic medium. The duct occupies the region $0 \le r \le a$, $0 \le \theta \le 2\pi$, and $z > 0$. If a piston, located at $z = 0$, is vibrating with an axial velocity V_z described by:

$$V_z = V_o \; f(r,\theta) \; e^{i\omega t}$$

obtain the pressure field inside the duct.

7

INTEGRAL TRANSFORMS

7.1 Fourier Integral Theorem

If f(x) is a bounded function in $-\infty < x < \infty$, and has at most only a finite number of ordinary discontinuities, and if the integral:

$$\int_{-\infty}^{\infty} |f(x)| dx$$

is absolutely convergent, then at every point x where there exists a left and right-hand derivative, f(x) can be represented by the following integral:

$$\frac{1}{2}[f(x+0)+f(x-0)] = \frac{1}{\pi}\int_0^{\infty}\int_{-\infty}^{\infty} f(\xi)\cos(u(\xi - x))d\xi\, du$$

The function f(x), $-L \le x \le L$, can be represented by a Fourier series as follows:

$$\frac{1}{2}[f(x+0)+f(x-0)] = a_0 + \sum_{n=1}^{\infty}[a_n \cos(\frac{n\pi}{L}x) + b_n \sin(\frac{n\pi}{L}x)]$$

where:

$$a_0 = \frac{1}{2L}\int_{-L}^{L} f(\xi)d\xi$$

$$a_n = \frac{1}{L}\int_{-L}^{L} f(\xi)\cos(\frac{n\pi}{L}\xi)d\xi$$

and

$$b_n = \frac{1}{L}\int_{-L}^{L} f(\xi)\sin(\frac{n\pi}{L}\xi)d\xi$$

Thus, adding the two integrals for a_n and b_n gives:

$$\frac{1}{2}[f(x+0)+f(x-0)] = \frac{1}{2L}\int_{-L}^{L} f(\xi)\, d\xi + \sum_{n=1}^{\infty}\frac{1}{L}\int_{-L}^{L} f(\xi)\cos(\frac{n\pi}{L}(\xi - x))\, d\xi$$

Let:

$$u_n = \frac{n\pi}{L} \qquad \text{and} \qquad \Delta u_n = u_{n+1} - u_n = \frac{\pi}{L}$$

then the integrals can be rewritten as:

$$f(x) = \frac{1}{2L}\int_{-L}^{+L} f(\xi)\,d\xi + \frac{1}{\pi}\sum_{n=1}^{\infty} \Delta u_n \int_{-L}^{+L} f(\xi)\cos(u_n(\xi - x))\,d\xi$$

Define the integral to equal $F(u_n)$, i.e.:

$$F(u_n) = \int_{-L}^{+L} f(\xi)\cos(u_n(\xi - x))\,d\xi$$

then the series converges to an integral in the limit $L \to \infty$ and $\Delta u_n \to 0$ as follows:

$$\lim_{\Delta u_n \to 0} \sum_{n=1}^{\infty} F(u_n)\Delta u_n \to \int_0^{\infty} F(u)\,du$$

Since the function is absolutely integrable, then the first term vanishes because:

$$\lim_{L\to\infty} \frac{1}{2L}\int_{-L}^{+L} f(x)\,dx \to 0 \quad \text{and } F(u) \text{ converges}$$

Thus, the representation of the function f(x) by a double integral becomes:

$$\frac{1}{2}[f(x+0) + f(x-0)] = \frac{1}{\pi}\int_0^{\infty}\int_{-\infty}^{\infty} f(\xi)\cos(u(\xi - x))\,d\xi\,du$$

$$= \frac{1}{\pi}\int_0^{\infty}\left[\int_{-\infty}^{\infty} f(\xi)\cos(u_n\xi)\,d\xi\right]\cos(ux)\,du$$

$$+ \frac{1}{\pi}\int_0^{\infty}\left[\int_{-\infty}^{\infty} f(\xi)\sin(u_n\xi)\,d\xi\right]\sin(ux)\,du \tag{7.1}$$

7.2 Fourier Cosine Transform

If $f(x) = f(-x)$ for $-\infty < x < \infty$ or, if $f(x) = 0$ for the range $-\infty < x < 0$ where one can choose $f(x) = f(-x)$ for the range $-\infty < x < 0$, then the second integral in eq. (7.1) vanishes and the integral representation can be rewritten as:

$$\frac{1}{2}[f(x+0) + f(x-0)] = \frac{2}{\pi}\int_0^{\infty}\left[\int_0^{\infty} f(\xi)\cos(u_n\xi)\,d\xi\right]\cos(ux)\,du \qquad x \geq 0$$

Define the **Fourier cosine transform** as:

$$F_c(u) = \int_0^{\infty} f(\xi)\cos(u\xi)\,d\xi \tag{7.2}$$

then, the **inverse Fourier cosine transforms** becomes:

$$f(x) = \frac{2}{\pi}\int_0^{\infty} F_c(u)\cos(ux)du \qquad x \geq 0$$

where $F_c(u)$ is an even function of u and cos (ux) is known as the kernel of the Fourier cosine transform.

7.3 Fourier Sine Transform

If $f(x) = -f(-x)$ in $-\infty < x < \infty$ or if $f(x) = 0$ in the range $-\infty < x < 0$, where one can choose $f(x) = -f(-x)$ in the range $-\infty < x < 0$, then the first integral of eq. (7.1) vanishes and:

$$\frac{1}{2}[f(x+0)+f(x-0)] = \frac{2}{\pi}\int_0^{\infty}\left[\int_0^{\infty} f(\xi)\sin(u\xi)\,d\xi\right]\sin(ux)\,du \qquad x \geq 0$$

Define the **Fourier sine transform** as:

$$F_s(u) = \int_0^{\infty} f(\xi)\sin(u\xi)\,d\xi \tag{7.3}$$

then the **inverse Fourier sine transform** becomes:

$$f(x) = \frac{2}{\pi}\int_0^{\infty} F_s(u)\sin(ux)du \qquad x \geq 0$$

where $F_s(u)$ is an odd function of u and sin (ux) is the kernel of the Fourier sine transform.

7.4 Complex Fourier Transform

The integral representation in eq. (7.1) can be used to develop a new transform. Define the function G_1 as the inner integral of eq. (7.1):

$$G_1(u) = \int_{-\infty}^{+\infty} f(\xi)\cos(u(\xi - x))\,d\xi$$

then the function $G_1(u)$ is an even function in u.

Define the function:

$$G_2(u) = \int_{-\infty}^{+\infty} f(\xi)\sin(u(\xi - x))\, d\xi$$

then the function $G_2(u)$ is an odd function in u. Thus, the integral of $G_2(u)$ vanishes over $[-\infty, \infty]$, i.e.:

$$\int_{-\infty}^{\infty} G_2(u)\, du = 0$$

If one adds this integral to that of eq. (7.1), a new representation of f(x) results:

$$f(x) = \frac{1}{2\pi}\int_{-\infty}^{\infty} G_1(u)\, du + \frac{i}{2\pi}\int_{-\infty}^{\infty} G_2(u)\, du = \frac{1}{2\pi}\int_{-\infty}^{\infty}\int_{-\infty}^{\infty} f(\xi) e^{iu(\xi - x)}\, d\xi\, du$$

Define the **complex Fourier transform** as:

$$F(u) = \frac{1}{2\pi}\int_{-\infty}^{\infty} f(\xi) e^{iu\xi}\, d\xi \qquad (7.4a)$$

then, the **inverse complex Fourier transform** becomes:

$$f(x) = \frac{1}{2\pi}\int_{-\infty}^{\infty} f(u)\, e^{-iux}\, du \qquad (7.4b)$$

7.5 Multiple Fourier Transform

Functions of two independent variables can be transformed by a double Fourier Complex transform. Let f(x,y) be defined in $-\infty < x < \infty$ and $-\infty < y < \infty$, such that:

$$\int_{-\infty}^{\infty}\int_{-\infty}^{\infty} |f(x, y)|\, dx\, dy \text{ exists.}$$

Thus, letting the Fourier Complex transform from x and y to u and v, then the transformation is done by successive integration:

$$\bar{f}(u, y) = \int_{-\infty}^{\infty} f(x, y) e^{iux}\, dx$$

$$\bar{F}(u, v) = \int_{-\infty}^{\infty} \bar{f}(u, y) e^{ivy}\, dy = \int_{-\infty}^{\infty}\int_{-\infty}^{\infty} f(x, y) e^{i(ux + vy)}\, dx\, dy$$

then the inverse Fourier Complex transforms from u and v to x and y can also be done by successive integrations:

$$\bar{f}(u,y) = \frac{1}{2\pi}\int_{-\infty}^{\infty} F(u,v)e^{-ivy}\,dv$$

$$f(x,y) = \frac{1}{2\pi}\int_{-\infty}^{\infty} \bar{f}(u,y)e^{-iux}\,du = \frac{1}{(2\pi)^2}\int_{-\infty}^{\infty}\int_{-\infty}^{\infty} F(u,v)e^{-i(ux+vy)}\,du\,dv$$

If the function f is a function of n independent variables, $f = f(x_1, x_2,..., x_n)$, then one can define a **multiple complex Fourier transform** as follows:

$$F(u_1, u_2, ..., u_n) =$$

$$= \int_{-\infty}^{\infty}\int_{-\infty}^{\infty} ... \int_{-\infty}^{\infty} f(x_1, x_2, ..., x_n)e^{i(u_1x_1+u_2x_2+...+u_nx_n)}dx_1\,dx_2...dx_n$$

then the **inverse multiple complex Fourier transform** becomes

$$f(x_1, x_2, ..., x_n) =$$

$$= \frac{1}{(2\pi)^n}\int_{-\infty}^{\infty}\int_{-\infty}^{\infty} ... \int_{-\infty}^{\infty} F(u_1, u_2, ..., u_n)e^{-i(u_1x_1+u_2x_2+...+u_nx_n)}du_1\,du_2...du_n$$

The transforms can be rewritten symbolically by using **x** and **u** as vectors in n-dimensional space, thus:

$$F(\mathbf{u}) = \int_{R_n(\mathbf{x})} f(\mathbf{x})e^{i\mathbf{u}\bullet\mathbf{x}}d\mathbf{x} \qquad (7.5a)$$

and

$$f(\mathbf{x}) = \int_{R_n(\mathbf{u})} F(\mathbf{u})e^{-i\mathbf{u}\bullet\mathbf{x}}d\mathbf{u} \qquad (7.5b)$$

where R_n represents the integration over the entire volume in n-dimensional space, and **x** and **u** are n-dimensional vectors.

7.6 Hankel Transform of Order Zero

If the function f(x,y) depends on x and y in the following form:

$$f(x,y) = f(\sqrt{x^2 + y^2})$$

then the Fourier Complex transform becomes:

$$F(u,v) = \int_{-\infty}^{\infty}\int_{-\infty}^{\infty} f(\sqrt{x^2 + y^2})e^{i(ux+vy)}dx\,dy$$

Transforming the integral to cylindrical coordinates:

$x = r\cos(\theta)$ $\qquad$ $y = r\sin(\theta)$ $\qquad$ and $\qquad$ $dA = r\,dr\,d\theta$

$$u = \rho \cos(\theta) \qquad v = \rho \sin(\theta) \qquad \text{and} \qquad dA = \rho \, d\rho \, d\theta$$

then the double integral transforms to:

$$F_1(\rho,\phi) = \int_0^\infty \int_0^{2\pi} r\, f(r) e^{ir\rho\cos(\theta-\phi)} dr\, d\theta$$

Integrating the inner integrand on θ, one obtains:

$$\int_0^{2\pi} e^{ir\rho\cos(\theta-\phi)}\, d\theta = \int_{-\phi}^{2\pi-\phi} e^{ir\rho\cos\theta_1}\, d\theta_1 = \left[\int_{-\phi}^{0} + \int_0^{2\pi} - \int_{2\pi-\phi}^{2\pi} \right] \left\{ e^{ir\rho\cos\theta_1}\, d\theta_1 \right\}$$

where $\theta_1 = \theta - \phi$. The first integral above becomes; with $\theta_2 = \theta_1 + 2\pi$:

$$\int_{-\phi}^{0} e^{ir\rho\cos\theta_1}\, d\theta_1 = \int_{2\pi-\phi}^{2\pi} e^{ir\rho\cos(\theta_2-2\pi)}\, d\theta_2 = \int_{2\pi-\phi}^{2\pi} e^{ir\rho\cos\theta_2}\, d\theta_2$$

thus, the first and third integrals cancel out, leaving the second integral which can be evaluated in closed form as:

$$\int_0^{2\pi} e^{ir\rho\cos(\theta-\phi)}\, d\theta = \int_0^{2\pi} e^{ir\rho\cos\theta_2}\, d\theta_2 = 2\pi J_o(r\rho)$$

where use of the integral representation of Bessel functions was made, see eq. (3.101). Thus, the integral transform becomes:

$$F_1(\rho) = 2\pi \int_0^\infty r\, f(r) J_o(r\rho)\, dr$$

and the inverse transform takes the form:

$$f(x,y) = f(\sqrt{x^2+y^2}) = \frac{1}{(2\pi)^2} \int_{-\infty}^{\infty} \int_{-\infty}^{\infty} F(u,v) e^{-i(ux+vy)} du\, dv$$

$$f(r) = \frac{1}{(2\pi)^2} \int_0^\infty \int_0^{2\pi} F_1(\rho)\, e^{ir\rho\cos(\theta-\phi)} \rho\, d\rho\, d\phi$$

The integral over $F_1(\rho)$ can be evaluated in a similar manner to the first integral so that:

$$f(r) = \frac{1}{(2\pi)^2} \int_0^\infty F_1(\rho) 2\pi J_o(r\rho)\, \rho\, d\rho$$

Therefore, the integral representation of f(r) becomes:

$$f(r) = \int_0^\infty \left\{ \int_0^\infty f(t) J_o(\rho t)\, t\, dt \right\} J_o(r\rho) \rho\, d\rho$$

Define the **Hankel transform** $F(\rho)$ as:

$$F(\rho) = \int_0^{\infty} r\, f(r) J_0(r\rho)\, dr \tag{7.6a}$$

then the **inverse Hankel transform** is given by:

$$f(r) = \int_0^{\infty} \rho F(\rho) J_0(r\rho)\, d\rho \tag{7.6b}$$

7.7 Hankel Transform of Order ν

A treatment of Hankel transform of order ν similar to Hankel transform of order zero is given in Reference [1]. Let $f = f(x_1, x_2, \ldots, x_n)$, then the Fourier transform and its inverse were defined in Section 7.5. If the function f depends on $x_1, x_2, \ldots, x_n$ as follows:

$$f = f\left(\sqrt{x_1^2 + x_2^2 + \ldots + x_n^2}\right)$$

then:

$$F(u_1, u_2, \ldots, u_n) = \int_{-\infty}^{\infty}\int_{-\infty}^{\infty} \cdots \int_{-\infty}^{\infty} f\left(\sqrt{x_1^2 + x_2^2 + \ldots + x_n^2}\right) e^{i\sum_{k=1}^{n} x_k u_k}\, dx_1 dx_2 \ldots dx_n$$

Performing a similar coordinate transformation as was done for the Hankel transform, define:

$$r^2 = x_1^2 + x_2^2 + \ldots + x_n^2$$

and

$$\rho^2 = u_1^2 + u_2^2 + \ldots + u_n^2$$

with the following coordinate transformation:

$$u_k = \rho\, a_{1k} \qquad k = 1, 2, \ldots, n$$

and

$$y_j = \sum_{k=1}^{n} a_{jk} x_k \qquad j = 1, 2, \ldots, n$$

In matrix notation, the transformation can be represented by:

$$[y_j] = [a_{jk}][x_k]$$

such that the coefficients a_{jk}, $j \neq 1$, are chosen to make the vector transformation orthogonal, i.e., the matrix:

$$[a_{jk}] = [a_{jk}]^{-1} \quad \text{or} \qquad \sum_{k=1}^{n} a_{jk} a_{ki} = \sum_{k=1}^{n} a_{jk} a_{ik} = \delta_{ji} = \begin{cases} 1 & i = j \\ 0 & i \neq j \end{cases}$$

where δ_{ij} is the Kronecher delta. Thus, the coordinates x_k are given by:

$$[x_k] = [a_{jk}]^{-1}[y_j] = [a_{kj}]^T[y_j]$$

and

$$r^2 = \sum_{k=1}^{n} x_k^2 = \sum_{k=1}^{n}\sum_{j=1}^{n}\sum_{l=1}^{n}[a_{kj}][y_j][a_{kl}][y_l] = \sum_{j=1}^{n}\sum_{l=1}^{n}[y_j]\delta_{jl}[y_l] = \sum_{l=1}^{n} y_l^2$$

The volume element becomes:

$$dx_1\, dx_2 \ldots dx_n = \{[a_{1j}][dy_j]\}\,\{[a_{2k}][dy_k]\} \ldots \{[a_{nl}][dy_l]\} = dy_1\, dy_2 \ldots dy_n$$

and

$$\sum_{k=1}^{n} u_k x_k = \sum_{k=1}^{n}\sum_{j=1}^{n} u_k a_{kj} y_j = \rho \sum_{k=1}^{n}\sum_{j=1}^{n} a_{1k} a_{kj} y_j = \rho \sum_{k=1}^{n}\sum_{j=1}^{n} \delta_{1j} y_j = \rho y_1$$

Thus:

$$F(u_1, u_2, \ldots, u_n) = \int_{-\infty}^{\infty}\int_{-\infty}^{\infty} \ldots \int_{-\infty}^{\infty} f\left(\sqrt{y_1^2 + z^2}\right) e^{i\rho y_1}\, dy_1 dy_2 \ldots dy_n$$

where:

$$z^2 = y_2^2 + y_3^2 + \ldots + y_n^2$$

One must find a function R, such that:

$$dy_2\, dy_3 \ldots dy_n = R\, dz$$

where R is the surface area of a sphere in n-dimensional space:

$$F(u_1, u_2, \ldots, u_n) = \int_{-\infty}^{\infty}\int_{-\infty}^{\infty} \ldots \int_{-\infty}^{\infty} f\left(\sqrt{y_1^2 + z^2}\right) e^{i\rho y_1} R\, dz\, dy_1$$

To evaluate the form of R, start with the following integral:

$$\int_{-\infty}^{\infty}\int_{-\infty}^{\infty} \ldots \int_{-\infty}^{\infty} F\left(\sqrt{y_2^2 + y_3^2 + \ldots y_n^2}\right) dy_2 dy_3 \ldots dy_n = \int_{0}^{\infty} F(z) R\, dz$$

Since the volume element $dy_2\, dy_3 \ldots dy_n$ represents (n-1) dimensional space, let:

$$R = S\, z^{n-2}$$

Choose $F(z) = \exp[-z^2]$, then:

$$\int_{-\infty}^{\infty}\int_{-\infty}^{\infty} \ldots \int_{-\infty}^{\infty} \exp[-(y_2^2 + y_3^2 + \ldots y_n^2)] dy_2 dy_3 \ldots dy_n = \pi^{(n-1)/2}$$

where the following integral was used:

$$\int_{-\infty}^{\infty} \exp[-x^2] dx = \sqrt{\pi}$$

The integral for dz can be evaluated:

$$\int_0^{\infty} \exp[-z^2] S z^{n-2} dx = \frac{S}{2} \Gamma(\frac{n-1}{2}) \qquad n \geq 2$$

Thus, the surface of a unit sphere in n-dimensional space is:

$$S = \frac{2\pi^{(n-1)/2}}{\Gamma(\frac{n-1}{2})}$$

so that the surface element is given by:

$$dy_2\, dy_3 \ldots dy_n = \frac{2\pi^{(n-1)/2}}{\Gamma(\frac{n-1}{2})} z^{n-2} dz$$

Hence:

$$F(u_1, u_2, \ldots, u_n) = \frac{2\pi^{(n-1)/2}}{\Gamma(\frac{n-1}{2})} \int_{-\infty}^{\infty} \int_0^{\infty} f(\sqrt{y_1^2 + z^2})\, e^{i\rho y_1}\, z^{n-2}\, dz\, dy_1$$

Let $z = r \sin\theta$, $y_1 = r\cos\theta$. Then $dz\, dy_1 = r\, dr\, d\theta$, and the above equation becomes:

$$F(u_1, u_2, \ldots, u_n) = \frac{2\pi^{(n-1)/2}}{\Gamma(\frac{n-1}{2})} \int_0^{\infty} r^{n-1} f(r) \left\{ \int_0^{\pi} e^{ir\rho\cos\theta} (\sin\theta)^{n-2}\, d\theta \right\} dr = F(\rho)$$

The inner integral becomes, (see equation 3.101):

$$\int_0^{\pi} e^{ir\rho\cos\theta} (\sin\theta)^{n-2}\, d\theta = \frac{\Gamma(\frac{1}{2})\Gamma(\frac{n-1}{2})}{\left(\frac{\rho r}{2}\right)^{\nu}} J_\nu(r\rho)$$

where $\nu = \dfrac{n-2}{2}$, and $n \geq 1$. Thus:

$$F(\rho) = \frac{(2\pi)^{n/2}}{\rho^{\nu}} \int_0^{\infty} r^{n/2} f(r)\, J_\nu(r\rho)\, dr, \qquad n \geq 1 \tag{7.7}$$

The inversion can be worked out in a similar manner:

$$f(x_1, x_2, \ldots, x_n) = \frac{1}{(2\pi)^n} \int_{-\infty}^{\infty} \int_{-\infty}^{\infty} \ldots \int_{-\infty}^{\infty} F(\rho)\, e^{i\sum_{k=1}^{n} x_k u_k}\, du_1 du_2 \ldots du_n$$

which can be shown to be equal to:

$$f(r) = \frac{1}{(2\pi)^{n/2} r^{\nu}} \int_0^{\infty} \rho^{n/2} F(\rho)\, J_\nu(r\rho)\, d\rho \tag{7.8}$$

Thus, combining eqs. (7.7) and (7.8), one obtains:

$$f(r) = \frac{1}{r^{\nu}} \int_0^{\infty} \rho J_{\nu}(r\rho) \left\{ \int_0^{\infty} r^{n/2} f(r)\, J_{\nu}(r\rho)\, dr \right\} d\rho \tag{7.9}$$

If one defines:

$$\bar{F}(\rho) = \frac{\rho^{\nu} F(\rho)}{(2\pi)^{n/2}} = \int_0^{\infty} f(r) r^{\nu} J_{\nu}(r\rho)\, r\, dr$$

then the inverse integral takes the form:

$$\bar{f}(r) = r^{\nu} f(r) = \int_0^{\infty} \bar{F}(\rho) J_{\nu}(r\rho) \rho\, d\rho$$

Redefining the functions f(r) and F(ρ) by:

$$g(r) = r^{\nu} f(r)$$

and

$$G(\rho) = \rho^{\nu} F(\rho)$$

Then g(r) and G(ρ) are defined by the following integrals:

$$G(\rho) = \int_0^{\infty} g(r) J_{\nu}(r\rho)\, r\, dr$$

$$g(r) = \int_0^{\infty} G(\rho) J_{\nu}(r\rho) \rho\, d\rho \tag{7.10}$$

valid for $\nu \geq 0$. G(ρ) is known as the **Hankel transform of order** ν of g(r) and g(r) is known as the **inverse Hankel transform of order** ν. Thus:

$$g(r) = \int_0^{\infty} \left\{ \int_0^{\infty} g(\xi) J_{\nu}(\rho\xi) \xi\, d\xi \right\} J_{\nu}(r\rho)\, \rho\, d\rho \tag{7.11}$$

Multiplying eq. (7.11) by $\sqrt{r}$, and defining h(r) = $\sqrt{r}$ g(r) then:

$$h(r) = \int_0^{\infty} \left\{ \int_0^{\infty} \sqrt{\rho\xi}\, h(\xi) J_{\nu}(\rho\xi)\, d\xi \right\} \sqrt{\rho r}\, J_{\nu}(r\rho)\, d\rho \tag{7.12}$$

This is known as the **Hankel Integral Theorem**.

7.8 General Remarks about Transforms Derived from the Fourier Integral Theorem

Since the transforms derived in Section 7.2 to 7.7 were derived from the Fourier Integral theorem, then the functions they are applied to must satisfy the following conditions and limitations:

(1) The function f(x) must be bounded and piecewise continuous.

(2) The function f(x) must have a left-handed and a right-handed derivative at every point of ordinary discontinuity.

(3) The function must have a finite number of maxima and minima.

(4) The function must be absolutely integrable, i.e., f(x) must necessarily decay as $|x| >> 1$.

These restrictions rule out a wide range of functions when applied in Engineering and Physics. It should also be noted that the transform and its inverse involve integrations on the real axis.

7.9 Generalized Fourier Transform

Let f(x), $-\infty < x < \infty$, be a function that is not absolutely integrable, that is:

$$\int_{-\infty}^{\infty} |f(x)| dx$$

does not converge, but it could increase at most at an exponential rate, i.e.:

$$|f(x)| < Ae^{ax} \qquad \text{for} \qquad x > 0$$

and

$$|f(x)| < Be^{bx} \qquad \text{for} \qquad x < 0$$

where a and b are real numbers. Thus, one can choose an exponential e^{cx} such that f(x) e^{cx} is absolutely integrable, e.g.:

$$\left| \int_0^{\infty} f(x) e^{cx} dx \right| \le A \int_0^{\infty} e^{ax} e^{cx} dx = \frac{A}{a+c} e^{(a+c)x} \Big|_0^{\infty} = -\frac{A}{a+c}$$

provided that $c < -a$, and

$$\left| \int_{-\infty}^{0} f(x) e^{cx} dx \right| \le B \int_{-\infty}^{0} e^{bx} e^{cx} dx = \frac{B}{b+c} e^{(b+c)x} \Big|_{-\infty}^{0} = \frac{B}{b+c}$$

provided that $c > -b$.

The complex Fourier transform was defined, for absolutely integrable functions:

$$F(u) = \int_{-\infty}^{\infty} f(\xi) e^{iu\xi} d\xi = \int_{-\infty}^{0} f(\xi) e^{iu\xi} d\xi + \int_{0}^{\infty} f(\xi) e^{iu\xi} d\xi$$

Define the following one-sided function:

$$g_1(x) = \begin{cases} e^{-v_1 x} f(x) & x > 0 \\ \frac{1}{2} f(0^+) & x = 0 \\ 0 & x < 0 \end{cases} \qquad v_1 > a \tag{7.13}$$

where $g_1(x)$ is absolutely integrable on $[0,\infty]$, then the Fourier transform of $g_1(x)$ becomes:

$$F_+(u, v_1) = \int_{-\infty}^{\infty} g_1(\xi) e^{iu\xi} d\xi = \int_{0}^{\infty} g_1(\xi) e^{iu\xi} d\xi = \int_{0}^{\infty} f(\xi) e^{i(u+iv_1)\xi} d\xi \tag{7.14}$$

Define the following one-sided function:

$$g_2(x) = \begin{cases} 0 & x > 0 \\ \frac{1}{2} f(0^+) & x = 0 \\ e^{-v_2 x} f(x) & x < 0 \end{cases} \qquad v_2 > b \tag{7.15}$$

where $g_2(x)$ is absolutely integrable over $[-\infty,0]$, then the Fourier transform of $g_2(x)$ becomes:

$$F_-(u, v_2) = \int_{-\infty}^{\infty} g_2(\xi) e^{iu\xi} d\xi = \int_{-\infty}^{0} g_2(\xi) e^{iu\xi} d\xi = \int_{-\infty}^{0} f(\xi) e^{i(u+iv_2)\xi} d\xi \tag{7.16}$$

The Fourier inverse transforms of F_+ and F_- become:

$$g_1(x) = \frac{1}{2\pi} \int_{-\infty}^{\infty} F_+(u, v_1) e^{-iux} du \tag{7.17}$$

and

$$g_2(x) = \frac{1}{2\pi} \int_{-\infty}^{\infty} F_-(u, v_2) e^{-iux} du \tag{7.18}$$

Multiplying eq. (7.17) by $\exp[v_1 x]$ and eq. (7.18) by $\exp[v_2 x]$, one obtains:

$$e^{v_1 x} g_1(x) = \frac{1}{2\pi} \int_{-\infty}^{\infty} F_+(u, v_1) e^{-i(u+iv_1)x} du = f(x) \qquad x > 0 \tag{7.19}$$

and

$$e^{v_2 x} g_2(x) = \frac{1}{2\pi} \int_{-\infty}^{\infty} F_-(u, v_2) e^{-i(u+iv_2)x} du = f(x) \qquad x < 0 \tag{7.20}$$

Combining eqs. (7.19) and (7.20), one can reconstruct f(x) again as defined in eqs. (7.13) and (7.15):

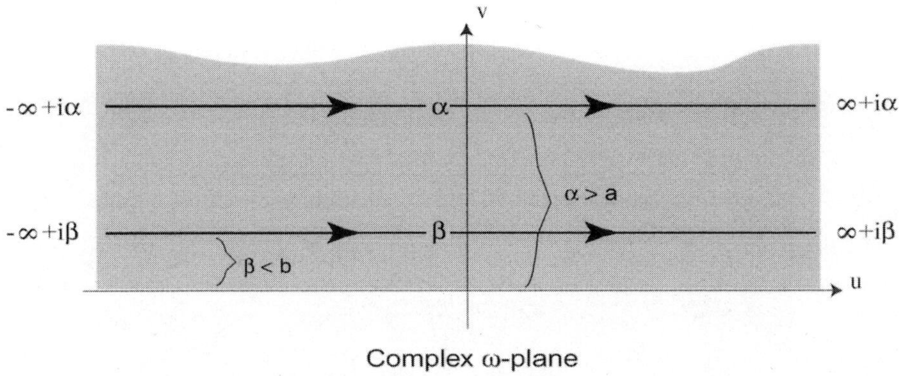

Fig. 7.1 *Paths for Generalized Fourier Transform*

$$f(x)=\frac{1}{2\pi}\left\{\int_{-\infty}^{\infty} F_+(u,v_1)e^{-i(u+iv_1)x}du+\int_{-\infty}^{\infty} F_-(u,v_2)e^{-i(u+iv_2)x}du\right\}$$

where $v_1 > a$ and $v_2 < b$. Using the transformation:

$$\omega = u + iv_{1,2} \qquad d\omega = du$$

then the new limits become:

$$u = -\infty \qquad \omega = -\infty + iv_{1,2}$$

$$u = \infty \qquad \omega = \infty + iv_{1,2}$$

one can rewrite the integral as follows:

$$f(x)=\frac{1}{2\pi}\left\{\int_{-\infty+i\alpha}^{\infty+i\alpha} F_+(\omega)e^{-i\omega x}d\omega+\int_{-\infty+i\beta}^{\infty+i\beta} F_-(\omega)e^{-i\omega x}d\omega\right\} \qquad \begin{matrix}\alpha > a\\ \beta < b\end{matrix} \tag{7.21}$$

where the functions $F_+(\omega)$ and $F_-(\omega)$ are defined by:

$$\begin{aligned} F_+(\omega) &= \int_0^{\infty} f(\xi)e^{i\omega\xi}d\xi \qquad Im(\omega) = v > a \\ F_-(\omega) &= \int_{-\infty}^{0} f(\xi)e^{i\omega\xi}d\xi \qquad Im(\omega) = v < b \end{aligned} \tag{7.22}$$

Equation (7.22) defines the **Generalized Fourier transform** and eq. (7.21) defines the **inverse Generalized Fourier transform**. It should be noted that the transform variable ω is complex, that the transform integrals are real, but the inverse transform is an integral in the complex plane ω. The paths of integration for the inverse transforms are shown in Fig. 7.1.

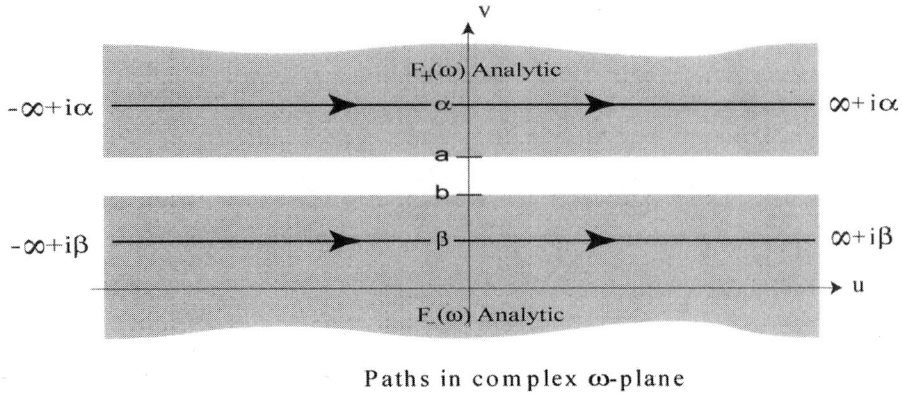

Fig. 7.2 *Paths for Generalized Fourier Transform*

Since the transforms $F_+(\omega)$ and $F_-(\omega)$ are functions of a complex variable ω, the region of analyticity of these complex functions must be examined. The function $F_+(\omega)$, as defined in eq. (7.22), is an absolutely convergent integral, provided that $\text{Im}(\omega) = v > a$. Let $\omega = u + iv$, $F_+(\omega) = U_+(u,v) + iV_+(u,v)$, then U_+ and V_+ must necessarily satisfy the Cauchy-Riemann conditions given in eq. (5.5), where:

$$U_+(u,v) = \int_0^\infty f(\xi)\, e^{-v\xi} \cos(u\xi)\, d\xi$$

and

$$V_+(u,v) = \int_0^\infty f(\xi)\, e^{-v\xi} \sin(u\xi)\, d\xi$$

The partial derivatives of U_+ and V_+ can be obtained by differentiating the integrands, since the integrals are absolutely convergent:

$$\frac{\partial U_+}{\partial u} = \frac{\partial V_+}{\partial v} = -\int_0^\infty \xi\, f(\xi)\, e^{-v\xi} \sin(u\xi)\, d\xi$$

and

$$\frac{\partial U_+}{\partial v} = -\frac{\partial V_+}{\partial u} = -\int_0^\infty \xi\, f(\xi)\, e^{-v\xi} \cos(u\xi)\, d\xi$$

which satisfy the Cauchy-Riemann conditions. Thus, the necessary and sufficient conditions for analyticity are satisfied, provided the partial derivatives are continuous and convergent, which is true in this case, since the function:

$$\left|x e^{-vx} f(x)\right| < A e^{-(v-a)x} \qquad \text{for} \qquad Im\,(\omega) = v > a$$

Thus, $F_+(\omega)$ is analytic in the upper half plane of ω above the line $v = a$, as shown in Fig. 7.2.

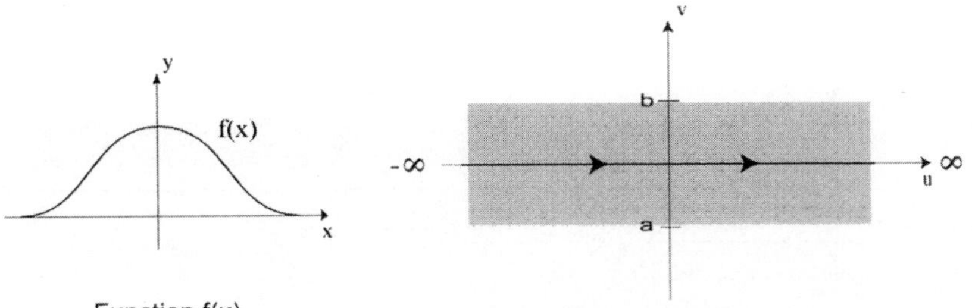

Function f(x) — Path in complex ω-plane

Fig. 7.3 *Path for Absolutely Convergent Fourier Transform*

Similarly, $F_-(\omega)$ is analytic in the lower half plane of ω, below the line v = b, as shown in Fig. 7.2. the contour integration for the inverse transformation must then be taken in those shaded regions shown in Fig. 7.2.

The contour integrals of the inverse transforms depend on the rate at which f(x) becomes exponentially unbounded. Some special cases, which reflect the relative values of a and b are enumerated below:

(i) $a < 0$ and $b > 0$.

The function f(x) vanishes as $x \to \pm\infty$. Then there exists a region of analyticity that is common to both transforms. Any common line contour, where $a < v < b$, can be used for the inverse transform, hence one may choose $\alpha = \beta = 0$, as shown in Fig. 7.3. Then, the two transforms F_+ and F_- become:

$$F_+(\omega)\big|_{v=0} = F_+(u) = \int_0^\infty f(x)e^{iux}dx$$

$$F_-(\omega)\big|_{v=0} = F_-(u) = \int_{-\infty}^0 f(x)e^{iux}dx$$

so that the two integrals can be combined into one integral over the real axis:

$$F(u) = F_+(u) + F_-(u) = \int_{-\infty}^\infty f(x)e^{iux}dx$$

The inverse transform becomes, with v = 0:

$$f(x) = \frac{1}{2\pi}\left\{\int_{-\infty}^\infty F_+(u, v_1)e^{-iux}du + \int_{-\infty}^\infty F_-(u, v_2)e^{-iux}du\right\} = \frac{1}{2\pi}\int_{-\infty}^\infty F(u)\, e^{-iux}du$$

which is the complex Fourier transform and its inverse as defined in eq. (7.4).

(ii) $a < b$

In this case, f(x) is not in general absolutely integrable, as shown in Fig. 7.4, but there is a common region of analyticity for the transform as shown in the shaded section in Fig. 7.4. Thus, it is convenient to choose a common line-contour for the inverse

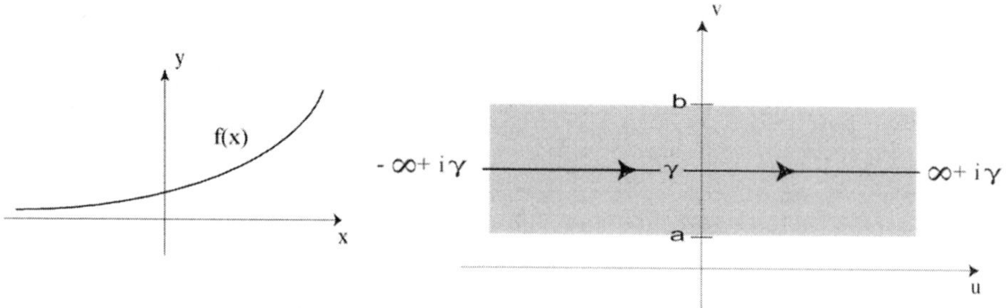

Function f(x)

Path in complex ω-plane

Fig. 7.4 *Path for One-Sided Fourier Transform*

transform $\alpha = \beta = \gamma$. Hence, the Fourier transforms F_+ and F_- are defined in the same manner as given in eq. (7.22), while the inverse transform is taken on a common line, where $a < \gamma < b$:

$$f(x)=\frac{1}{2\pi}\left\{\int_{-\infty+i\gamma}^{\infty+i\gamma} F_+(\omega)e^{-i\omega x}d\omega + \int_{-\infty+i\gamma}^{\infty+i\gamma} F_-(\omega)e^{-i\omega x}d\omega\right\} \tag{7.23}$$

Further discussion can be carried out for the possible signs of a and b:

(a) If $a > 0$ then $b > 0$, then f(x) is a function that vanishes as $x \to -\infty$ and becomes unbounded as $x \to \infty$.

(b) If $b > 0$, then $a < 0$, then f(x) is a function that vanishes as $x \to \infty$ and becomes unbounded as $x \to -\infty$.

In either case, since the function is unbounded on only one side of the real axis, then one can choose a common value for v such that:

$$a < v = Im\,(\omega) < b$$

and

$$F(\omega) = F_+(\omega) + F_-(\omega) = \int_{-\infty}^{0} f(x)e^{i\omega x}dx + \int_{0}^{\infty} f(x)e^{i\omega x}dx = \int_{-\infty}^{\infty} f(x)e^{i\omega x}dx \tag{7.24}$$

where F(ω) is analytic. The inverse transform becomes.

$$f(x)=\frac{1}{2\pi}\int_{-\infty+i\gamma}^{\infty+i\gamma} F(\omega)e^{-i\omega x}d\omega \qquad \text{where} \qquad a < \gamma < b \tag{7.25}$$

It should be noted that the function $F_+(\omega)$ and $F_-(\omega)$ may have poles in the complex plane ω.

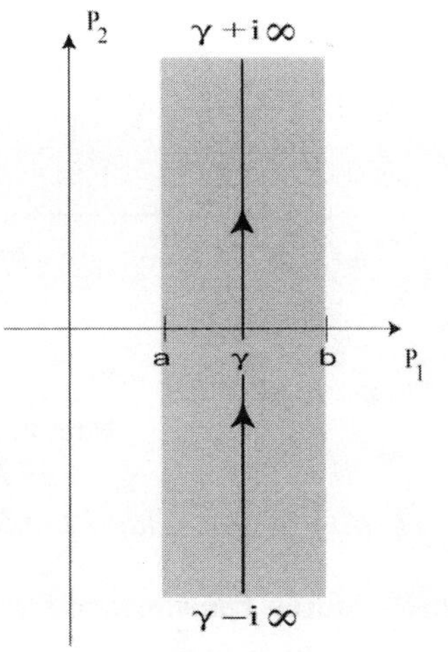

Path in complex p-plane

Fig. 7.5 *Path for Two-Sided Laplace Transform*

7.10 Two-Sided Laplace Transform

If one makes the transformation

$p = -i\omega = p_1 + ip_2 = v - iu$

and if a < b, then one can define the **two-sided Laplace transform**:

$$F_{LII}(p) = \int_{-\infty}^{\infty} f(x) e^{-px} dx \qquad a < p_1 = Re\,(p) < b \tag{7.26}$$

and the **inverse two-sided Laplace transform** is then defined by:

$$f(x) = \frac{1}{2\pi i} \int_{\gamma - i\infty}^{\gamma + i\infty} F_{LII}(p) e^{px} dp \tag{7.27}$$

where γ is any line contour in the region of analyticity of $F_{LII}(p)$, shown as the shaded area in Fig. 7.5.

7.11 One-Sided Generalized Fourier Transform

If the function f(x) is defined so that:

$f(x) = 0 \qquad x < 0$

and

$|f(x)| < Ae^{ax} \qquad x > 0$

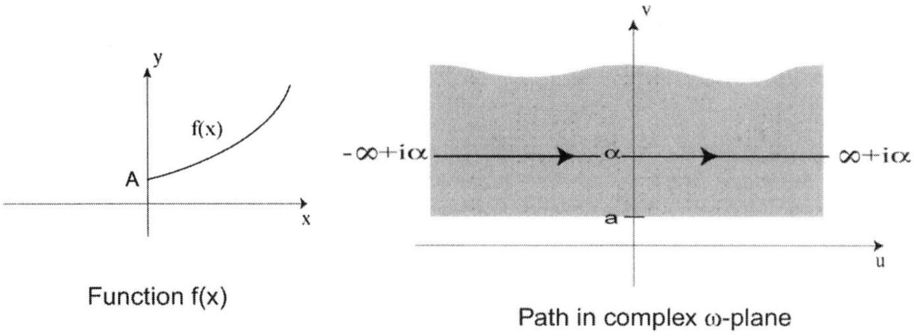

Fig. 7.6 *Path for One-Sided Laplace Transform*

then the **one-sided Generalized Fourier transform** of f(x) can be written as:

$$F_I(\omega)=\int_0^{\infty} f(x)e^{-i\omega x}dx \qquad Im(\omega)=v>b$$

and the **inverse one-sided Generalized Fourier transform** is then defined by the integral:

$$f(x)=\frac{1}{2\pi}\int_{-\infty-i\alpha}^{\infty+i\alpha} F_I(\omega)e^{-i\omega x}d\omega \qquad \alpha>a \tag{7.28}$$

The transform $F_I(\omega)$ is analytic above the line v = a, hence the inverse transformation is performed along a line $v = \alpha > a$. (See Fig. 7.6) Thus, let the line $v = \alpha$ be above all the singularities of $F_I(\omega)$.

7.12 Laplace Transform

If the function f(x) is once again defined as:

$$f(x)=0 \qquad x<0$$

and

$$|f(x)|<Ae^{ax} \qquad x>0$$

then the **Laplace transform** of f(x) becomes:

$$F(p)=\int_0^{\infty} f(x)e^{-px}dx \qquad Re(p)-p_1>a \tag{7.29}$$

and the **inverse Laplace transform** is then defined by:

$$f(x)=\frac{1}{2\pi i}\int_{\gamma-i\infty}^{\gamma+i\infty} F(p)e^{px}dp \qquad \gamma<a$$

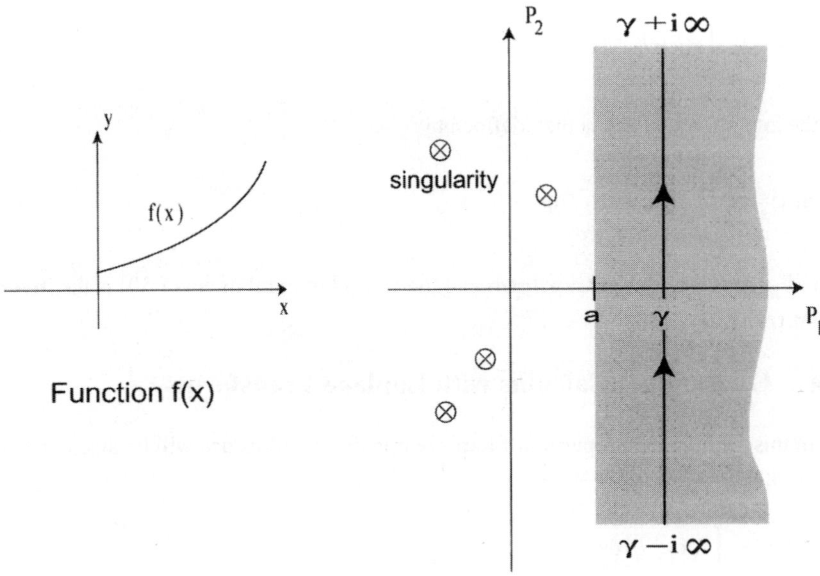

Path in complex p-plane

Fig. 7.7 *Path for Laplace Transform*

The transform F(p) is analytic to the right of $p_1 = a$, so that one may, choose $p_1 = \gamma$ such that all the singularities of F(p) are located to the left of the line $p_1 = \gamma$, (see Fig. 7.7).

7.13 Mellin Transform

For the case of a < b, the two sided Laplace transform can be altered by making the following transformation on the independent variable x:

$$x = -\log\eta \quad \text{or} \quad \eta = e^{-x} \qquad dx = -\frac{d\eta}{\eta}$$

then the two sided Laplace transform takes the form:

$$F_{LII}(p) = \int_{-\infty}^{\infty} f(x)e^{-px}dx = \int_{\infty}^{0} f(-\log\eta)e^{p\log\eta}\frac{d\eta}{\eta} = \int_{0}^{\infty} f(-\log\eta)\eta^{p-1}\,d\eta$$

and the inverse Laplace transform is then defined by:

$$f(x) = f(-\log\eta) = \frac{1}{2\pi i}\int_{\gamma - i\infty}^{\gamma + i\infty} F_{LII}(p)e^{-p\log\eta}dp = \frac{1}{2\pi i}\int_{\gamma - i\infty}^{\gamma + i\infty} F_{LII}(p)\eta^{-p}dp$$

To redefine these transform integrals, let:

$$f(-\log x) = g(x) \qquad 0 < x < \infty$$

then the Laplace transform becomes:

$$F_m(p) = \int_0^\infty g(x) x^{p-1} dx$$

and the integral transform is then defined by:

$$g(x) = \frac{1}{2\pi i} \int_{\gamma - i\infty}^{\gamma + i\infty} F_m(p) x^{-p} dp \tag{7.30}$$

where $F_m(p)$ is the **Mellin transform** and the second integral of eq. (7.30) is the **inverse Mellin transform**.

7.14 Operational Calculus with Laplace Transforms

In this section, the properties of Laplace transform and its use will be discussed. The following notations will be used:

$$L\, f(x) = \int_0^\infty f(x) e^{-px} dx$$

$$f(x) = L^{-1}\, F(p) = \frac{1}{2\pi i} \int_{\gamma - i\infty}^{\gamma + i\infty} F(p) e^{px} dp$$

7.14.1 The Transform Function

The transform function F(p) of a function f(x) can be shown to vanish as $p \to \infty$:

$$|F(p)| = \left| \int_0^\infty f(x) e^{-px} dx \right| \le A \int_0^\infty e^{-(p-a)x} dx = \frac{A}{p-a}$$

Thus, F(p) vanishes as p goes to infinity. Similarly, one can show that the transform of the functions $x^n e^{ax}$ vanish as $p \to 0$:

$$\lim_{p \to \infty} \int_0^\infty x^n f(x) e^{-px} dx = 0$$

This proves that the Laplace integral is uniformly convergent if $p > a$. This property allows the differentiation of F(p) with respect to p, i.e.:

$$F(p)^{(n)} = \frac{d^n}{dp^n} \int_0^\infty f(x) e^{-px} dx = \int_0^\infty (-1)^n x^n f(x) e^{-px} dx$$

which also proves that the derivatives of F(p) also vanish as $p \to \infty$, i.e.:

$$\lim_{p \to \infty} F(p)^{(n)} = 0 \qquad n = 0, 1, 2, \ldots$$

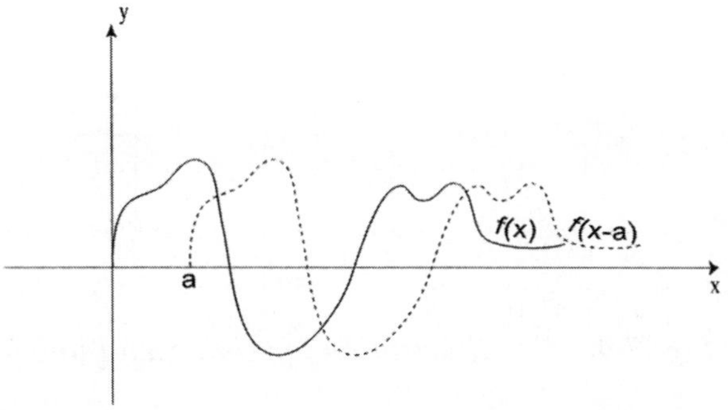

Fig. 7.8 *Illustration for the Shift Theorem*

7.14.2 Shift Theorem

If a function is shifted by an offset = a, as shown in Fig. 7.8, then let

$$g(x) = f(x\text{-}a)\, H(x\text{-}a) \qquad x > 0$$

where H(x-a) is the Heaviside step function, see Appendix D, so that its Laplace transform is:

$$L\,g(x) = G(p) = \int_0^\infty f(x-a) H(x-a) e^{-px} dx = \int_a^\infty f(x-a)\, e^{-px} dx$$

$$= \int_0^\infty f(u) e^{-p(a+u)} du = e^{-pa} F(p) \qquad (7.31)$$

7.14.3 Convolution (Faltung) Theorems

Convolution theorems give the inversion of products of transformed functions in the form of definite integrals, whose integrands are products of the inversion of the individual transforms, known as **Convolution Integrals**. Let the functions G(p) and K(p) be Laplace transforms of g(x) and k(x), respectively, and

$$F(p) = G(p)\, K(p)$$

where the Laplace transforms of k(x) and g(x) are defined as:

$$K(p) = L\,k(x) = \int_0^\infty k(x) e^{-px} dx$$

and

$$G(p) = L\,g(x) = \int_0^\infty g(x) e^{-px} dx$$

Thus, the product of these transforms, after suitable substitutions of the independent variables, can be written as:

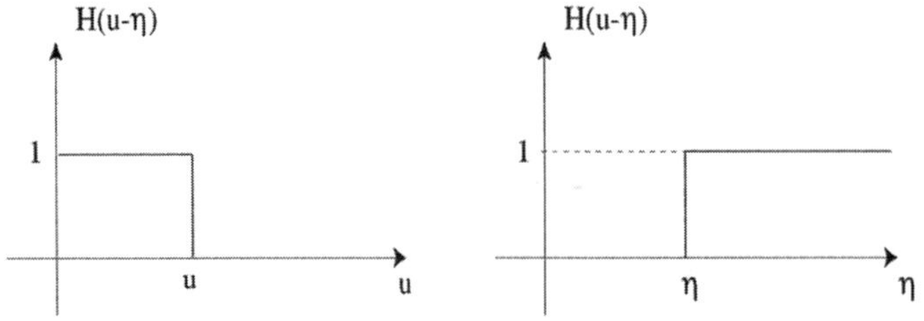

Fig. 7.9 Illustration for Unit Step Function

$$F(p) = G(p)K(p) = \int_0^\infty\int_0^\infty k(\xi)g(\eta)e^{-p(\xi+\eta)}d\xi\, d\eta = \int_0^\infty\left[\int_0^\infty k(\xi)e^{-p(\xi+\eta)}d\xi\right]g(\eta)\, d\eta$$

Let $u = \xi + \eta$ in the inner integral, then $d\xi = du$, and the integral can be transformed to:

$$f(p) = \int_0^\infty\left[\int_\eta^\infty k(u-\eta)e^{-pu}du\right]g(\eta)d\eta = \int_0^\infty\left[\int_0^\infty k(u-\eta)H(u-\eta)e^{-pu}du\right]g(\eta)d\eta$$

$$= \int_0^\infty\left[\int_0^\infty g(\eta)k(u-\eta)H(u-\eta)d\eta\right]e^{-pu}du$$

where $H(u-\eta)$ is the Heaviside function (see figure 7.9).

Thus, using the definition of F(p), and comparing it with the inner integral, one obtains:

$$f(x) = \int_0^\infty g(\eta)k(x-\eta)H(x-\eta)d\eta = \int_0^x g(\eta)k(x-\eta)d\eta \tag{7.32}$$

Similarly, one could also show that:

$$f(x) = \int_0^x k(\eta)g(x-\eta)\, d\eta$$

Convolution theorems for a larger number of products of transformed functions can be obtained in a similar manner, e.g. if F(p) is the product of three transform functions:

$$F(p) = G(p)\, K(p)\, M(p)$$

then the convolution integral for f(x) is given in many forms, two of which are given below:

$$f(x) = \int_0^x\int_0^\xi g(x-\xi)\, k(\xi-\eta)m(\eta)d\eta\, d\xi$$

and

$$f(x) = \int_0^x \int_0^{x-\xi} g(x-\xi-\eta)k(\xi)\, m(\eta)\, d\eta\, d\xi \tag{7.33}$$

7.14.4 Laplace Transform of Derivatives

The Laplace transform of the derivatives of f(x) can be obtained in terms of the Laplace transform of the function f(x). Starting with the first derivative of f(x):

$$L\frac{\partial f}{\partial x} = \int_0^\infty \frac{\partial f}{\partial x} e^{-px} dx = f(x)e^{-px}\Big|_0^\infty + p\int_0^\infty f(x)e^{-px} dx = p\,F(p) - f(0^+)$$

$$L\frac{\partial^2 f}{\partial x^2} = \int_0^\infty \frac{\partial^2 f}{\partial x^2} e^{-px} dx = \frac{\partial f}{\partial x} e^{-px}\Big|_0^\infty + p\int_0^\infty \frac{\partial f}{\partial x} e^{-px} dx =$$

$$= p^2 F(p) - pf(0^+) - \frac{\partial f}{\partial x}(0^+)$$

Similarly:

$$L\frac{\partial^n f}{\partial x^n} = p^n F(p) - \sum_{k=0}^{n-1} p^{n-k-1}\frac{\partial^k f}{\partial x^k}(0^+) \tag{7.34}$$

7.14.5 Laplace Transform of Integrals

Define the indefinite integral g(x) as:

$$g(x) = \int_0^x f(y)\, dy$$

then its Laplace transform can be evaluated using the definition:

$$Lg(x) = G(p) = \int_0^\infty g(x)e^{-px} dx = -g(x)\frac{e^{-px}}{p}\Big|_0^\infty + \frac{1}{p}\int_0^\infty \frac{dg}{dx} e^{-px} dx$$

$$= \frac{1}{p}\int_0^\infty f(x)\, e^{-px} dx = \frac{F(p)}{p} \tag{7.35}$$

because $g(0) = 0$, and $dg/dx = f(x)$.

7.14.6 Laplace Transform of Elementary Functions

The Laplace transform for few elementary functions are as follows:

$$L[e^{ax} f(x)] = \int_0^x e^{ax} f(x)\, e^{-px}\, dx = \int_0^x f(x)\, e^{-(p-a)x}\, dx = F(p-a) \tag{7.36}$$

$$L\,[x\, f(x)] = \int_0^x x\, f(x)\, e^{-px}\, dx = -\frac{d}{dp}\int_0^x f(x) e^{-px}\, dx = -\frac{dF(p)}{dp}$$

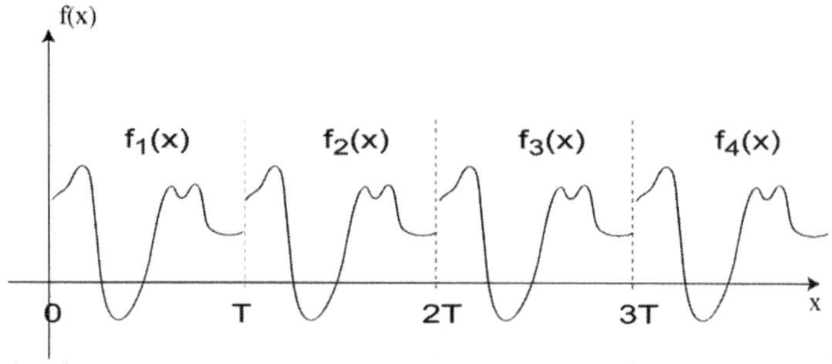

Fig. 7.10 *Illustration for Periodic Function*

$$L[x^2 f(x)] = \int_0^X x^2 f(x) e^{-px} \, dx = \frac{d^2}{dp^2} \int_0^X f(x) e^{-px} \, dx = \frac{d^2 F}{dp^2}$$

and, in general:

$$L[x^n f(x)] = (-1)^n \frac{d^n F}{dp^n} \qquad n \geq 0 \tag{7.37}$$

The Laplace transform of the Heaviside function H(x) is:

$$L[1] = L[H(x)] = 1/p$$

and that of a shifted Heaviside function H(x-a) is:

$$L[H(x-a)] = e^{-pa} L[H(x)] = \frac{e^{-pa}}{p} \tag{7.38}$$

where eq. (7.31) was used. The Laplace transform of a power of x is then derived from eq. (7.37) as:

$$L[x^n] = L[x^n H(x)] = (-1)^n \frac{d^n}{dp^n}\left(\frac{1}{p}\right) = \frac{n!}{p^{n+1}} \tag{7.39}$$

The Laplace transform of the Dirac Delta Function $\delta(x)$ is (see Appendix D):

$$L[\delta(x)] = \int_0^X \delta(x) e^{-px} \, dx = e^{-px}\Big|_{x=0} = 1$$

One should note that F(p) does not vanish as $p \to \infty$ because the function is a point-function and does not conform to the requirements on f(x).

The Laplace transform of a shifted Dirac function:

$$L[\delta(x-a)] = e^{-pa} \tag{7.40}$$

7.14.7 Laplace Transform of Periodic Functions

Let f(x) be a periodic function, with a periodicity = T, as shown in Figure 7.10 i.e.:

$f(x) = f(x+T)$

Define the functions $f_n(x)$:

$$f_1(x) = \begin{cases} f(x) & 0 \le x \le T \\ 0 & x < 0,\ x > T \end{cases}$$

$$f_2(x) = \begin{cases} f(x) & T \le x \le 2T \\ 0 & x < T, x > 2T \end{cases}$$
$$= f_1(x - T)$$

$$f_3(x) = \begin{cases} f(x) & 2T \le x \le 3T \\ 0 & x < 2T,\ x > 3T \end{cases}$$
$$= f_1(x - 2T)$$

$$f_{n+1}(x) = \begin{cases} f(x) & nT \le x \le (n+1)T \\ 0 & x < nT, x > (n+1)T \end{cases}$$
$$= f_1(x - nT)$$

Thus, the function f(x) is the sum of an infinite number of the functions $f_n(x)$:

$$f(x) = \sum_{n=1}^{\infty} f_n(x) = \sum_{n=0}^{\infty} f_{n+1}(x) = \sum_{n=0}^{\infty} f_1(x - nT)$$

Using the shift theorem eq. (7.31) on the shifted functions, one obtains:

$$L f_1(x - nT) = e^{-npT} F_1(p)$$

where $F_1(p)$ is the Laplace transform of $f_1(x)$. The Laplace transform of f(x) as a sum of shifted functions becomes the sum of the Laplace transform at the shifted functions:

$$L\,f(x) = \sum_{n=0}^{\infty} F_1(p)e^{-npT} = F_1(p) \sum_{n=0}^{\infty} \left(e^{-pT}\right)^n$$

which can be summed up using the geometric series summation formula, resulting in:

$$L\,f(x) = \frac{F_1(p)}{1 - e^{-pT}} \tag{7.41}$$

where:

$$F_1(p) = L\,f_1(x) = \int_0^T f_1(x)e^{-px}dx$$

7.14.8 Heaviside Expansion Theorem

If the transform F(p) is a rational function of two polynomials, i.e.,

$$F(p) = \frac{N(p)}{D(p)}$$

where D(p) is a polynomial of degree n and N(p) is a polynomial of degree m ≤ n, then one can obtain an inverse transform of F(p) by the method of partial fractions. Let the n roots of D(p) be labeled p_1, p_2 ..., p_n and assume that none of these roots are roots of N(p). The denominator D(p) can then be factored out in terms of its roots, i.e.:

$$D(p) = (p-p_1)\,(p-p_2)\,\ldots\,(p-p_n)$$

The factorization depends on whether all of the roots p_j, j = 1, 2, .., are distinct or some are repeated:

(i) If all the roots of the denominator D(p) are distinct, then one can expand F(p) as follows:

$$F(p) = \frac{N(p)}{D(p)} = \frac{A_1}{p-p_1} + \frac{A_2}{p-p_2} + \ldots + \frac{A_n}{p-p_n} = \sum_{j=1}^{n} \frac{A_j}{p-p_j}$$

where, the unknown coefficients A_1, A_2, ..., A_n can be obtained as follows:

$$A_j = \lim_{p \to p_j} \left[(p-p_j)\frac{N(p)}{D(p)} \right] = \frac{N(p_j)}{D'(p_j)} \qquad j = 1, 2, 3, \ldots n$$

The inverse transform F(p) can be readily obtained as the sum of the inverse of each of these terms:

$$f(x) = \sum_{j=1}^{n} A_j\, e^{p_j x} \tag{7.42}$$

(ii) If only one root is repeated k times, then, taking that root to be p_1, one can obtain the partial fractions as follows:

$$F(p) = \frac{N(p)}{D(p)} = \frac{A_1}{p-p_1} + \frac{A_2}{(p-p_1)^2} + \ldots \frac{A_{k-1}}{(p-p_1)^{k-1}} + \frac{A_k}{(p-p_1)^k} + Q(p)$$

where Q(p) has poles at points other that p_1, i.e., simple poles at p_{k+1}, p_{k+2}, ..., p_n. The function Q(p) can be factored out as:

$$Q(p) = \frac{A_{k+1}}{p-p_{k+1}} + \frac{A_{k+2}}{p-p_{k+2}} + \ldots + \frac{A_n}{p-p_n}$$

which can be treated in the same manner as was outlined in Item (i) above.

Letting G(p) = $(p-p_1)^k$ F(p), then the constants A_1 to A_k can be evaluated as follows:

$$A_k = \lim_{p \to p_1} [G(p)] = G(p_1) \qquad A_{k-1} = \frac{1}{1!}\frac{dG(p_1)}{dp}$$

$$A_{k-2} = \frac{1}{2!}\frac{d^2G(p_1)}{dp^2}, \ldots, A_1 = \frac{1}{(k-1)!}\frac{d^{(k-1)}G(p_1)}{dp^{(k-1)}}$$

i.e.:

$$A_j = \frac{1}{(k-j)!}\frac{d^{(k-j)}G(p_1)}{dp^{(k-j)}} \qquad j = 1, 2, \ldots, k$$

To evaluate the contacts A_{k+1} to A_n, one uses the same formulae in (i). Thus, the inverse transform of the part of the function F(p) corresponding to the repeated root p_1 takes the form:

$$f(x) = e^{p_1 x} \sum_{j=1}^{k} \frac{x^{j-1}}{(j-1)!(k-j)!} \frac{d^{(k-j)} G(p_1)}{dp^{(k-j)}} + q(x) \tag{7.43}$$

where eqs. (7.39) and (7.36) were used. The remainder function q(x) is the same as given in eq. (7.42) with the index ranging from j = k+1 to n.

7.14.9 The Addition Theorem

If an infinite series of functions $f_n(x)$ representing a function f(x):

$$f(x) = \sum_{n=0}^{\infty} f_n(x)$$

is uniformly convergent on [0,∞], and if either the integral of |f(x)|:

$$\int_0^{\infty} e^{-px} |f(x)| \, dx$$

or the sum of the integrals of $|f_n(x)|$:

$$\sum_{n=0}^{\infty} \int_0^{\infty} e^{-px} |f_n(x)| \, dx$$

converges, then:

$$L f(x) = F(p) = L \sum_{n=0}^{\infty} f_n(x) = \sum_{n=0}^{\infty} L f_n(x) = \sum_{n=0}^{\infty} F_n(p) \tag{7.44}$$

Example 7.1

Various examples of the Leplace transform, which illustrate the various theorems discussed above, are given below:

(i) sin (ax)H(x):

First, rewrite sin (ax) as a sum of exponentials:

$$\sin(ax) = \frac{1}{2i}(e^{iax} - e^{-iax}),$$

then, using eq. (7.36):

$$L[e^{iax} H(x)] = \frac{1}{p - ia}$$

The Leplace transform of sin(ax) is found to be:

$$L[\sin(ax)H(x)] = \frac{1}{2i}\left[\frac{1}{p - ia} - \frac{1}{p + ia}\right] = \frac{a}{p^2 + a^2}$$

(ii) e^{bx} sin(ax)

Since the Laplace transform of sin(ax) is now known, one can use eq. (7.36) to evaluate the product, i.e.:

$$L\,[e^{bx}\sin(ax)] = F(p-b) = \frac{a}{(p-b)^2 + a^2}$$

(iii) sin (ax) H(x) (periodic function):

Since the function sin (ax) is periodic with periodicity T = 2π/a, then:

$$F_1(p) = \int_0^{2\pi/a} \sin(ax) e^{-px}\,dx = \frac{a}{p^2+a^2}(1 - e^{-2p\pi/a})$$

then:

$$F(p) = L\,\sin(ax) = \frac{F_1(p)}{1-e^{-pT}} = \frac{a}{p^2+a^2}(1-e^{-2p\pi/a})\frac{1}{(1-e^{-2p\pi/a})} = \frac{a}{p^2+a^2}$$

(iv) Find the inverse transform of $F(p) = \dfrac{a}{p^2 - a^2}$:

F(p) can be written as the product of two functions:

$$F(p) = \frac{a}{p^2-a^2} = \frac{a}{p+a} \bullet \frac{1}{p-a}$$

then the inverse transform of the product can be obtained by the convolution theorem.
Letting:

$$G(p) = \frac{a}{p+a} \qquad \text{and} \qquad K(p) = \frac{1}{p-a}$$

then the inverse transform of G(p) and K(p) are known to be (see eq. 7.53):

$$g(x) = ae^{-ax} \qquad \text{and} \qquad k(x) = e^{ax}$$

so that the inverse transform of F(p) can be obtained in the form of a convolution integral:

$$f(x) = \int_0^x \left(ae^{-a\eta}\right)\left(e^{a(x-\eta)}\right)d\eta = -\frac{e^{ax}}{2}[e^{-2ax} - 1] = \sinh(ax)$$

Alternatively, since the function F(p) has two simple poles whose denominator has two roots, p = ± a, then one can use the Heaviside theorem to obtain an inverse transform:

$$F(p) = \frac{A_1}{p-a} + \frac{A_2}{p+a}$$

Since the roots are distinct then:

$$A_1 = \left.\frac{a}{2p}\right|_{p=a} = \frac{1}{2} \qquad \text{and} \quad A_2 = \left.\frac{a}{2p}\right|_{p=-a} = -\frac{1}{2}$$

so that:

$$F(p) = \frac{1}{2}\left[\frac{1}{p-a} - \frac{1}{p+a}\right]$$

and

$$f(x) = \frac{1}{2}\left[e^{ax} - e^{-ax}\right] = \sinh(ax)$$

(v) Find the inverse transform of F(p), defined as:

$$F(p) = \frac{p+a}{(p+b)(p+c)^2} \qquad b \neq c$$

The function F(p) has a simple pole at p = -b and a pole of order 2 at p = -c.
Let:

$$F(p) = \frac{A_1}{p+c} + \frac{A_2}{(p+c)^2} + \frac{A_3}{(p+b)}$$

then the coefficients A_j are found from the partial fraction theorem:

$$G(p) = (p+c)^2 F(p) = \frac{p+a}{p+b}$$

$$A_1 = \frac{dG}{dp}(-c) = \left.\frac{b-a}{(p+b)^2}\right|_{p=-c} = \frac{b-a}{(b-c)^2}$$

$$A_2 = G(p_1) = G(-c) = \frac{a-c}{b-c}$$

$$A_3 = (p+b)\,F(p)\big|_{p=b} = \frac{a-b}{(c-b)^2}$$

Thus, the inverse transform of F(p) is given by:

$$f(x) = \frac{1}{b-c}\left\{e^{-cx}\left[\frac{b-a}{b-c} + (a-c)x\right] + e^{-bx}\frac{a-b}{b-c}\right\}$$

where eqs. (7.42) and (7.43) were used.

7.15 Solution of Ordinary and Partial Differential Equations by Laplace Transforms

One may use Laplace transform to solve ordinary and partial differential equations for semi-infinite independent variables. For use of the Laplace transform on time, where $t > 0$, one would require initial conditions at $t = 0$. In this case, application of Laplace on time for the first or second derivations in time requires the specification of one or two initial conditions, respectively, as required by the uniqueness theorem. Use of Laplace transform on space is more problematic. Use of Laplace on x for the second derivative $\partial^2 y(x,t)/\partial x^2$ would require the specification of y(0,t) and $\partial y(0,t)/\partial x$. However, uniqueness theorem requires that only one of these two boundary conditions *can be specified* at the origin. Hence, one must assume that the unknown boundary condition is a given function. For example, if y(0,t) = f(t), a specified function, then one must assume that $\partial y(0,t)/\partial x = g(t)$; an unknown function. The function g(t) must be solved for eventually after finding y(x,t) in terms of g(t). The reverse would also be true: if $\partial y(0,t)/\partial x = f(t)$, then y(0,t) = g(t); an unknown function. This indicates that the Laplace transform is more suited to use on time rather than space.

In this section, the Laplace transform will be applied on various ordinary or partial differential equations in the following examples.

Example 7.2

Obtain the solution y(t) of the following initial value problem:

$$\frac{d^2y}{dt^2} + a^2 y = f(t) \qquad t > 0$$

with the initial conditions of:

$$y(0) = C_1 \qquad \frac{dy}{dt}(0) = C_2$$

Applying the Laplace transform on the variable t to the ordinary differential equation, the system transforms to an algebraic equation as follows:

$$p^2 \, Y(p) - p \, y(0) - \frac{dy}{dt}(0) + a^2 Y(p) = F(p)$$

where $Y(p) = L \, y(t)$. After inserting the initial conditions, one can find the solution in the transform plane p:

$$Y(p) = \frac{F(p)}{p^2 + a^2} + \frac{pC_1 + C_2}{p^2 + a^2}$$

To obtain the inverse transforms of the first term, one needs to use the convolution theorem since f(t) was not explicitly specified:

$$L^{-1}\left[\frac{1}{p^2 + a^2}\right] = \frac{\sin(at)}{a} \qquad L^{-1}\left[\frac{p}{p^2 + a^2}\right] = \cos(at)$$

Thus, using the Convolution theorem:

$$y(t) = \int_0^t f(t - \eta) \frac{\sin(a\eta)}{a} d\eta + C_1 \cos(at) + \frac{C_2}{a} \sin(at)$$

Example 7.3

Obtain the solution to the following integro-differential equation by Laplace transform:

$$\frac{dy}{dt} + ay = f(t) + \int_0^t g(t - \eta) y(\eta) d\eta$$

with the initial condition $y(0) = 0$.

Applying the Laplace transform on the equation, and using the Convolution theorem, one obtains:

$$pY(p) - y(0) + AY(p) = F(p) + G(p) \, Y(p)$$

which can be solved for Y(p):

$$Y(p) = \frac{F(p)}{p + a - G(p)} = F(p) K(p)$$

where $K(p) = \dfrac{1}{p + a - G(p)}$. Then:

$$y(t) = \int_0^t f(t-\eta)\,k(\eta)\,d\eta$$

Example 7.4

Obtain the solution to the following initial value problem by use of Laplace transform:

$$\frac{d^2y}{dt^2} + t\frac{dy}{dt} - 2y = 1 \qquad t \geq 0$$

with the initial conditions of:

$$y(0) = 0 \qquad \frac{dy}{dt}(0) = 0$$

Applying the Laplace transform to the equation, and noting that the equation has non-constant coefficients, the Laplace transform for $[t\, y'(t)]$ becomes:

$$L\left[t\frac{dy}{dt}\right] = -\frac{d}{dp}\left[L\frac{dy}{dt}\right] = -\frac{d}{dp}[pY(p) - y(0)] = -p\frac{dY}{dp} - Y$$

then:

$$p^2Y - py(0) - \frac{dy}{dt}(0) - p\frac{dY}{dp} - Y - 2Y = \frac{1}{p}$$

or:

$$\frac{dY}{dp} + \left(\frac{3}{p} - p\right)Y = -\frac{1}{p^2}$$

The homogeneous solution Y_h becomes:

$$Y_h = C\exp\left[-\int\left(\frac{3}{p} - p\right)dp\right] = C\frac{e^{p^2/2}}{p^3}$$

and the particular solution Y_{par} is found to be:

$$Y_{par} = \frac{1}{p^3}$$

Thus, the total solution can be written as follows:

$$Y = C\frac{e^{p^2/2}}{p^3} + \frac{1}{p^3}$$

Since the limit of Y(p) goes to zero as p goes to infinity, then C = 0 and $Y(p) = 1/p^3$. The inverse transform gives (see eq. 7.39):

$$y(t) = \frac{t^2}{2}$$

Example 7.5 Forced Vibration of a Stretched Semi-Infinite String

A semi-infinite free stretched string, initially undisturbed, is being excited at its end x = 0, such that, for y = y(x,t):

$$\frac{\partial^2 y}{\partial x^2} = \frac{1}{c^2}\frac{\partial^2 y}{\partial t^2} \qquad x > 0 \qquad t > 0$$

together with the initial and boundary conditions:

$$y(0,t) = f(t) \qquad y(x,0^+) = 0 \qquad \frac{\partial y}{\partial t}(x,0^+) = 0$$

The differential equation satisfied by the string is first transformed on the time variable, that is:

$$L_t\left[\frac{\partial^2 y}{\partial x^2}\right] = \frac{1}{c^2} L_t\left[\frac{\partial^2 y}{\partial t^2}\right]$$

where the symbol L_t signifies Laplace transformation on the variable t. Let:

$$Y(x,p) = \int_0^\infty y(x,t)\, e^{-pt} dt$$

then the transform of the partial derivatives on the spatial variable x is:

$$L_t\left[\frac{\partial^2 y}{\partial x^2}\right] = \frac{d^2 Y}{dx^2}(x,p)$$

and the transform of the partial derivative on the time variable is:

$$L_t\left[\frac{\partial^2 y}{\partial t^2}\right] = p^2 Y - [py(x,0^+) + \frac{\partial y}{\partial t}(x,0^+)] = p^2 Y$$

Transforming the boundary condition at x = 0:

$$L_t\, y(0,t) = Y(0,p) = L_t\, f(t) = F(p)$$

Thus, the system transforms to following boundary value problem:

$$\frac{d^2 Y}{dx^2} - \frac{p^2}{c^2} Y = 0$$

$$Y(0,p) = F(p)$$

The solution of the differential equation can be shown to be:

$$Y = Ae^{-px/c} + Be^{px/c}$$

The solution Y must vanish as $x \to \infty$, which require that B = 0. The boundary condition at x = 0 is satisfied next:

$$Y(0,p) = A = F(p)$$

so that the solution in the transform plane is finally found to be:

$$Y = F(p)\, e^{-px/c}$$

The inverse transform is given by:

$$y(x,t) = L_t^{-1}[F(p)e^{-px/c}] = \begin{cases} f(t - \frac{x}{c}) & t > x/c \\ 0 & t < x/c \end{cases}$$

which can be written in terms of the Heaviside function:

$$y(x,t) = f(t - \frac{x}{c})\, H(t - \frac{x}{c})$$

where the shift theorem (eq. 7.31) was used. The solution exhibits the physical property that any disturbance at $x = 0$ arrives at a station x at a time $t = x/c$ having the same time dependence as the original disturbance.

Example 7.6 Heat Flow in a Semi-Infinite Rod

Obtain the heat flow in a semi-infinite rod, where its end is heated, such that:

$T = T(x,t)$

$$\frac{\partial^2 T}{\partial x^2} = \frac{1}{K}\frac{\partial T}{\partial t} \qquad x > 0 \qquad t > 0$$

subject to the following initial and boundary conditions:

$$T(0,t) = f(t) \qquad T(x,0^+) = 0$$

Applying the Laplace transform on t on the equation, and defining:

$$\bar{T}(x,p) = \int_0^\infty T(x,t)e^{-pt}\,dt$$

then the equation and the boundary condition transform to:

$$\frac{d^2\bar{T}}{dx^2} = \frac{1}{K}[p\bar{T} - T(x,0)] = \frac{p}{K}\bar{T}$$

and

$$\bar{T}(0,p) = F(p)$$

The differential equation on the transform temperature T becomes:

$$\frac{d^2\bar{T}}{dx^2} - \frac{p}{K}\bar{T} = 0$$

and has the two solutions:

$$\bar{T} = Ae^{-\sqrt{p/K}\,x} + Be^{+\sqrt{p/K}\,x}$$

Boundedness of T as $x \to \infty$ requires that $B = 0$. Satisfying the boundary condition:

$$\bar{T}(0,p) = F(p) = A$$

then, the solution in the complex plane p is given by:

$$\bar{T} = F(p)e^{-\sqrt{p/K}\,x}$$

The inverse transform of $\exp[-\sqrt{p/K}\,x]$ [from Laplace Transform Tables] is given as follows:

$$L^{-1}\left[e^{-a\sqrt{p}}\right] = \frac{a}{2\sqrt{\pi t^3}}\, e^{-a^2/4t} \quad \text{where} \quad a = \sqrt{\frac{x}{K}}$$

Thus, using the convolution theorem (eq. 7.32):

$$T(x,t) = \frac{x}{2\sqrt{\pi K}} \int_0^t f(t-\eta)\eta^{-3/2}\, e^{-x^2/(4K\eta)} d\eta$$

Let:

$$\xi = \frac{x}{2\sqrt{K\eta}} \qquad d\xi = -\frac{x}{4\sqrt{K\eta^3}} d\eta \qquad \text{and} \qquad \chi = \frac{x}{2\sqrt{Kt}}$$

then the integral representation of the temperature becomes:

$$T(x,t) = \frac{2}{\sqrt{\pi}} \int_\chi^\infty f(t - \frac{x^2}{4K\xi^2}) e^{-\xi^2} d\xi$$

$$= \frac{2}{\sqrt{\pi}} \left\{ \int_0^\infty f(t - \frac{x^2}{4K\xi^2}) e^{-\xi^2} d\xi - \int_0^\chi f(t - \frac{x^2}{4K\xi^2}) e^{-\xi^2} d\xi \right\}$$

If $f(t) = T_0$ = constant, then the integral can be solved:

$$T(x,t) = \frac{2}{\sqrt{\pi}} T_0 \left\{ \int_0^\infty e^{-\xi^2} d\xi - \int_0^\chi e^{-\xi^2} d\xi \right\} = T_0[1 - \text{erf}(\chi)] = T_0 \,\text{erfc}(\chi)$$

where erf(y) is the error function as defined in B5.1 (Appendix B) and erfc(y) = 1 - erf(y) and erf (∞) = 1.

Example 7.7 Vibration of a Finite Bar

A finite bar, initially at rest, is induced to vibration by a force f(t) applied at its end $x = L$ for $t > 0$. The bar's displacement y(x,t) satisfies the following system:

$$\frac{\partial^2 y}{\partial x^2} = \frac{1}{c^2} \frac{\partial^2 y}{\partial t^2}$$

$$y(0,t) = 0 \qquad y(x,0^+) = 0$$

$$\frac{\partial y}{\partial t}(x,0^+) = 0 \qquad AE\frac{\partial y}{\partial x}(L,t) = f(t)\, H(t)$$

Applying Laplace transform on the time variable, the equation transforms to:

$$\frac{d^2 Y}{dx^2}(x,p) = \frac{1}{c^2}[p^2 Y(x,p) - p y(x,0^+) - \frac{\partial y}{\partial t}(x,0^+)] = \frac{p^2}{c^2} Y(x,p)$$

where $Y(x, p) = L_t\, y(x,t)$. Transforming the boundary conditions:

$$Y(0,p) = 0 \quad \text{and} \quad \frac{dY}{dx}(L,p) = \frac{F(p)}{AE}$$

The solution of the differential equation on the transformed variable Y can be written as follows:

$$Y(x,p) = De^{-px/c} + Be^{+px/c}$$

which is substituted in the two boundary conditions:

$$Y(0,p) = D + B = 0$$

$$\frac{dY}{dx}(L,p) = \frac{p}{c}[-De^{-pL/c} + Be^{+pL/c}] = \frac{F(p)}{AE}$$

The unknown coefficients are readily evaluated:

$$B = -D = \frac{c}{p}\frac{F(p)}{AE}\frac{1}{e^{-pL/c} + e^{+pL/c}}$$

and the transformed solution has the form:

$$Y(x,p) = \frac{c}{p}\frac{F(p)}{AE}\frac{e^{+px/c} - e^{-px/c}}{e^{+pL/c} + e^{-pL/c}}$$

Separating the solution into two parts:

$$Y(x,p) = \frac{c}{AE}F(p)\,G(x,p)$$

where G(p) is defined as:

$$G(x,p) = \frac{1}{p}\frac{e^{+px/c} - e^{-px/c}}{e^{+pL/c} + e^{-pL/c}} = \frac{G_1(x,p)}{1 - e^{-4pL/c}}$$

where $G_1(x,p)$ represents the transform of the first part of a periodic function whose periodicity is T = 4L/c and is given by:

$$G_1(x,p) = \frac{1}{p}[e^{-p(L-x)/c} - e^{-p(x+L)/c} - e^{-p(3L-x)/c} + e^{-p(3L+x)/c}]$$

The inverse transform of e^{-ap}/p is H(t - a) which results in a inverse of $G_1(x, p)$:

$$\begin{aligned} g_1(x,t) &= H[t-(L-x)/c] - H[t-(L+x)/c] \\ &\quad - H[t-(3L-x)/c] + H[t-(3L+x)/c] \end{aligned}$$

The inverse transform of G(x,p) is given by g(x, t) where:

$$g(x,t) = g_1(x,t) \qquad 0 \le t \le 4L/c$$

and g(x, t) is a periodic function with period T = 4L/c, i.e.:

$$g(x,t) = g(x,t+4L/c)$$

so that the periodic function can be written as:

$$g(x,t) = \sum_{n=0}^{\infty} g_1(x, t - 4nL/c)$$

The final solution to the displacement y(x, t) requires the use of the convolution integral:

$$y(x,t) = \frac{c}{AE}\int_0^t g(x,u)f(t-u)\,du$$

If $f(t) = F_0$ = constant, then $F(p) = F_0/p$, and:

$$Y(x,p) = \frac{c}{p^2}\frac{F_0}{AE}\frac{e^{+px/c} - e^{-px/c}}{e^{+pL/c} + e^{-pL/c}}$$

The transform of the deformation at the end x = L then becomes:

$$Y(L,p)=\frac{c}{p^2}\frac{F_o}{AE}\tanh(pL/c)$$

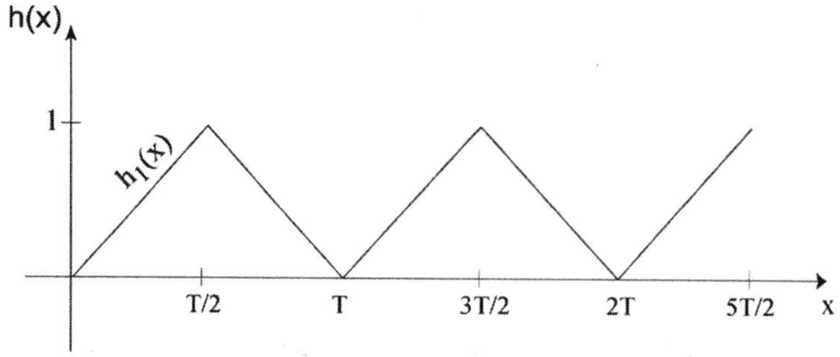

Fig. 7.11 *Saw-Tooth Function*

The transform of a saw-tooth (triangular) wave h(x) defined by $h_1(x)$, $0 \le x \le T$, (as shown in figure 7.11) is defined as:

$$h_1(x)=\begin{cases} 2x/T & 0 \le x \le T/2 \\ 2(1-x/T) & T/2 \le x \le T \end{cases}$$

and

$$\mathcal{L}\, h(x)=\frac{2}{Tp^2}\tanh(\frac{pT}{4})$$

Thus, the inverse transform of the deformation becomes, with T = 4L/c:

$$\frac{y(L,t)}{y_o} = 2\,h(t) = \text{dynamic deflection/static deflection}$$

where y_o is the static deflection defined by:

$$y_o = \frac{F_o L}{AE}$$

The maximum value y(L, t) attains is $2y_o$ at t = 2L/c, 6L/c, ... The deflection at any other point x can be developed in an infinite series form:

$$Y(x,p)=\frac{c}{p^2}\frac{F_o}{AE}\frac{e^{-p(L-x)/c}-e^{-p(L+x)/c}}{1+e^{-2pL/c}}$$

$$\frac{Y(x,p)}{y_o}=\frac{c}{L}U(x,p)=\frac{c}{L}\frac{U_1(x,p)}{1-e^{-4pL/c}}$$

$$U_1(x,p)=\frac{1}{p^2}[e^{-p(L-x)/c}-e^{-p(x+L)/c}-e^{-p(3L-x)/c}+e^{-p(x+3L)/c}]$$

where U(x,p) represent a periodic function, u(x,t) = u(x,t + 4L/c) with U_1 being the transform of the function u(x,t) within the first period $0 \le t \le 4L/c$. Noting that from equation 7.31:

$$\mathcal{L}^{-1}[\frac{1}{p^2}e^{-ap}]=(t-a)\,H[t-a]$$

then, the solution y(x,t) is given by the periodic function u(x,t) = u(x, t + 4L/c):

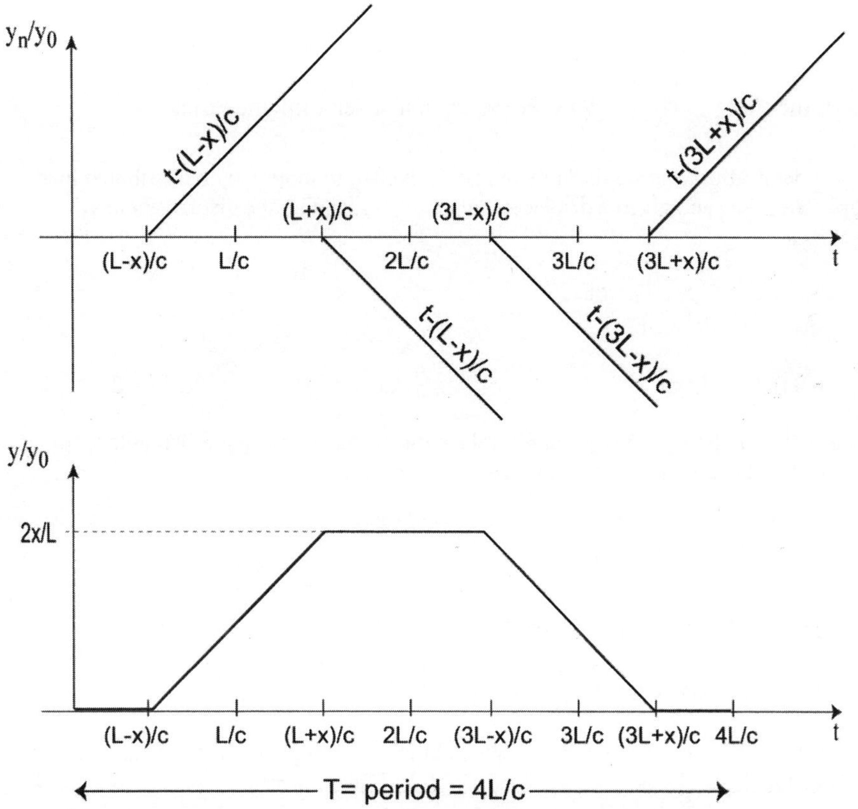

Fig. 7.12 *Construction of the Propagating Wave*

$$\frac{y(x,p)}{y_0} = \frac{c}{L} u(x,p)$$

The inverse transform of $U_1(x, p)$ is then found as:

$$u_1(x,t) = [(t - \frac{L-x}{c})H(t - \frac{L-x}{c}) - (t - \frac{L+x}{c})H(t - \frac{L+x}{c})$$
$$- (t - \frac{3L-x}{c})H(t - \frac{3L-x}{c}) + (t - \frac{3L+x}{c})H(t - \frac{3L+x}{c})]$$

for $0 \le t \le 4L/c$.

The solution $u_1(x,t)$ for the first period $0 \le t \le 4L/c$ is made up of the first arrival of the wave at $t = (L - x)/c$ which is then followed by three reflections, two at $x = 0$ and one at $x = L$. This solution is shown graphically for the first period $t = 4L/c$ in the accompanying plot, see Fig. 7.12. Note that from that time on, the displacement is periodic with a period of $T = 4L/c$.

Use of the Laplace transform on the time variable t requires that two initial values be given, which are required for uniqueness. However, use of the Laplace transform on the spatial variable x, requires two boundary conditions at $x = 0$, of which only one is prescribed. One can solve such problems by assuming the unknown boundary condition and then solve for it, by satisfying the remaining boundary condition.

Example 7.8 Wave Propagation in a Semi-Infinite String

A semi-infinite string, initially at rest, is excited to motion by a distributed load applied at $t = t_o$ and given a displacement at $x = 0$ such that the displacement $y(x,t)$ satisfies:

$$\frac{\partial^2 y}{\partial x^2} = \frac{1}{c^2}\frac{\partial^2 y}{\partial t^2} + \frac{P_o}{T_o} e^{-bx}\,\delta(t - t_o) \qquad x > 0 \qquad t, t_o > 0$$

$$y(0,t) = y_o\, H(t) \qquad y(x,0^+) = 0 \qquad \frac{\partial y}{\partial t}(x,0^+) = 0$$

Obtain the solution $y(x, t)$ by using Laplace transform on the spatial variable x.

Define the Laplace transform on x:

$$\mathcal{L}_x[y(x,t)] = Y(p,t) = \int_0^\infty y(x,t) e^{-px}\,dx$$

Applying the Laplace transform on the differential equation:

$$p^2 Y(x,p) - p y(0,t) - \frac{\partial y(0,t)}{\partial x} = \frac{1}{c^2}\frac{d^2 Y(x,p)}{dt^2} + \frac{P_o}{T_o}\frac{\delta(t - t_o)}{p + b}$$

Since the displacement at $x = 0$ was given, but not the slope, $\partial y/\partial x$ is not known, then assume that:

$$\frac{\partial y}{\partial x}(0,t) = f(t)$$

so that the differential equation takes the form:

$$\frac{d^2 Y}{dt^2} - c^2 p^2 Y = -c^2 p^2 y_o H(t) - c^2 f(t) - c^2 \frac{P_o}{T_o}\frac{\delta(t - t_o)}{p + b} = Q(t)$$

The homogeneous and particular solutions are given by:

$$Y_h = A \sinh(cpt) + B \cosh(cpt)$$

$$Y_p = \frac{1}{cp}\int_0^t Q(u) \sin[cp(t - u)]\,du$$

$$= \frac{y_o}{p}(1 - \cosh(cpt)) - \frac{c}{p(p + b)}\frac{P_o}{T_o} \sinh[cp(t - u)]\, H[t - t_o]$$

$$- \frac{c}{p}\int_0^t f(u) \sinh[cp(t - u)]\,du$$

Using initial conditions:

$$Y(p,0) = 0 = B$$

$$\frac{dY(p,0)}{dt} = pcA = 0$$

so that Y(p,t) = Y_p(p,t). The inverse transform of Y(p,t) is then given by:

$$y(x,t) = \frac{y_0}{2} H[ct-x] - \frac{c}{2}\int_0^t f(u)\big(H[x+c(t-u)] - H[x-c(t-u)]\big)du$$

$$-\frac{cP_0}{2bT_0} H[t-t_0][1-e^{-b[x+c(t-t_0)]}]\,H[x+c(t-t_0)]$$

$$+\frac{cP_0}{2bT_0} H[t-t_0][1-e^{-b[x-c(t-t_0)]}]\,H[x-c(t-t_0)]$$

The solution for y(x,t) still contains the unknown boundary condition f(t). Differentiating y partially with x and setting x = 0 one obtains:

$$\frac{\partial y}{\partial x}(0,t) = f(t) = -\frac{y_0}{2}\delta(ct) - \frac{c}{2}\int_0^t f(u)\big(\delta[c(t-u)] - \delta[c(u-t)]\big)du$$

$$-\frac{cP_0}{2bT_0} H[t-t_0]\Big(be^{-bc(t-t_0)} + 2\sinh[bc(t-t_0)]\delta[c(t-t_0)]\Big)$$

where δ(u) = δ(-u) and δ(cu) = δ(u)/c were used (Appendix D).

The integral in the last expression can be shown to equal f(t)/2, so that f(t) is finally obtained as:

$$f(t) = -\frac{y_0}{2}\delta(t) - \frac{cP_0}{bT_0} H[t-t_0]\Big(be^{-bc(t-t_0)} + 2\sinh[bc(t-t_0)]\delta[c(t-t_0)]\Big)$$

Substituting f(t) into the integral term of y(x, t) results in the following expression:

$$\frac{y_0}{2} H[ct-x] - \frac{cP_0}{2bT_0} H[c(t-t_0)-x]H[t-t_0](e^{-b[x-c(t-t_0)]} - 1)$$

Substituting the last expression into that for y(x, t) gives a final solution:

$$y(x,t) = y_0 H[ct-x] - \frac{cP_0}{bT_0} H[t-t_0]e^{-bx}\sinh[bc(t-t_0)]$$

$$+\frac{cP_0}{bT_0}\sinh[bc(t-t_0)-bx]H[(t-t_0)-x/c]$$

where H(-u) = 1 - H(u) was used in the expression.

7.16 Operational Calculus with Fourier Cosine Transform

The Fourier cosine transform of a function f(x) was defined in 7.2 as follows:

$$F_c[f(x)] = F_c(u) = \int_0^\infty f(x)\cos(ux)\,dx$$

then:

$$f(x)=\frac{2}{\pi}\int_0^\infty F_c(u)\cos(ux)\,du$$

7.16.1 Fourier Cosine Transform of Derivatives

The Fourier transform of the derivative of f(x) is derived as:

$$F_c[\frac{\partial f(x)}{\partial x}] = \int_0^\infty \frac{\partial f(x)}{\partial x}\cos(ux)dx = f(x)\cos(ux)|_0^\infty + u\int_0^\infty f(x)\sin(ux)\,dx$$

$$= uF_s(u) - f(0^+)$$

The transform of the second derivative of f(x):

$$F_c[\frac{\partial^2 f(x)}{\partial x^2}] = \int_0^\infty \frac{\partial^2 f(x)}{\partial x^2}\cos(ux)dx = \frac{\partial f}{\partial x}\cos(ux)\Big|_0^\infty + u\int_0^\infty \frac{\partial f}{\partial x}\sin(ux)\,dx$$

$$= -\frac{\partial f}{\partial x}(0^+) + u\, f(x)\sin(ux)|_0^\infty - u^2\int_0^\infty f(x)\cos(ux)dx$$

$$= -u^2F_c(u) - \frac{\partial f}{\partial x}(0^+)$$

In general, the Fourier cosine transform of even and odd derivatives are:

$$F_c[\frac{\partial^{2n} f}{\partial x^{2n}}] = (-1)^n u^{2n}F_c(u) - \sum_{m=0}^{n-1} (-1)^m u^{2m}\frac{\partial^{2n-2m-1} f(0^+)}{\partial x^{2n-2m-1}} \qquad n \geq 1$$

$$\text{provided that } \left|\frac{\partial^m f}{\partial x^m}\right| \to 0 \text{ as } x \to \infty \text{ for } m \leq (2n-1) \tag{7.45}$$

and

$$F_c[\frac{\partial^{2n+1} f}{\partial x^{2n+1}}] = (-1)^n u^{2n+1}F_s(u) - \sum_{m=0}^{n} (-1)^m u^{2m}\frac{\partial^{2n-2m} f(0^+)}{\partial x^{2n-2m}} \qquad n \geq 0$$

$$\text{provided that } \left|\frac{\partial^m f}{\partial x^m}\right| \to 0 \text{ as } x \to \infty \text{ for } m \leq 2n \tag{7.46}$$

It should be noted the Fourier cosine transform of even derivatives of a function gives the Fourier cosine transform of the function, and requires initial conditions of odd derivatives. However, the Fourier cosine transform of odd derivatives leads to the Fourier sine transform of the function, and hence not conductive to solving problems.

7.16.2 Convolution Theorem

The convolution theorem for Fourier cosine transform can be developed for products of transformed functions. Let $H_c(u)$ and $G_c(u)$ be the Fourier cosine transforms of h(x) and g(x), respectively. Then:

$$F_c^{-1}[H_c(u)G_c(u)] = \frac{2}{\pi}\int_0^\infty H_c(u)G_c(u)\cos(ux)du$$

$$= \frac{2}{\pi}\int_0^\infty H_c(u)\left\{\int_0^\infty g(\xi)\cos(u\xi)d\xi\right\}\cos(ux)\,du$$

$$= \int_0^\infty g(\xi)\left\{\frac{2}{\pi}\int_0^\infty H_c(u)\cos(u\xi)\cos(ux)du\right\}d\xi$$

$$= \int_0^\infty g(\xi)\left\{\frac{1}{\pi}\int_0^\infty H_c(u)[\cos(u(x+\xi))+\cos(u(x-\xi))]du\right\}d\xi$$

$$= \frac{1}{2}\int_0^\infty g(\xi)[h(x+\xi)+h(|x-\xi|)]d\xi \tag{7.47}$$

7.16.3 Parseval Formula

If one sets x = 0 in eq. (7.47), one obtains:

$$\frac{2}{\pi}\int_0^\infty H_c(u)G_c(u)\,du = \int_0^\infty g(\xi)h(\xi)d\xi \tag{7.48}$$

If $G_c(u) = H_c(u)$, an integral known as the **Parseval formula for the Fourier cosine transform** is obtained:

$$\frac{2}{\pi}\int_0^\infty H_c^2(u)\,du = \int_0^\infty h^2(\xi)d\xi \tag{7.49}$$

The Fourier cosine transform can be used to evaluate definite improper integrals.

Example 7.9

The Fourier cosine transform of the following exponentials:

$$h(x) = \frac{e^{-ax}}{a} \quad a > 0 \qquad g(x) = \frac{e^{-bx}}{b} \quad b > 0$$

becomes:

$$H_c(u) = \frac{1}{u^2+a^2} \qquad G_c(u) = \frac{1}{u^2+b^2}$$

Hence, one can evaluate the following integral:

$$\int_0^\infty \frac{du}{(u^2+a^2)(u^2+a^2)} = \frac{\pi}{2}\int_0^\infty \frac{e^{-ax}}{a}\frac{e^{-bx}}{b}dx = \frac{\pi}{2ab(a+b)}$$

by use of eq. (7.48).

Example 7.10 Heat Flow in a Semi-Infinite Rod

Obtain the heat flow in a semi-infinite rod, initially at zero temperature, where the heat flux at its end x = 0 is prescribed, such that the temperature T = T(x,t) satisfies the following system:

$$\frac{\partial^2 T}{\partial x^2} = \frac{1}{K}\frac{\partial T}{\partial t} \qquad x > 0 \qquad t > 0$$

$$\frac{\partial T}{\partial x}(0,t) = -\frac{l(t)}{k} \qquad T(x,0^+) = 0 \qquad \lim_{x\to\infty} T(x,t) \to 0$$

Since the Fourier cosine transform requires odd-derivative boundary conditions, see Equation 7.45, it is well suited for application to the present problem. Defining the transform of the temperature:

$$\bar{T}(u,t) = \int_0^\infty T(x,t)\cos(ux)\,dx$$

then the application of Fourier cosine transform to the differential equation and initial condition results in:

$$F_c\left[\frac{\partial^2 T}{\partial x^2}\right] = -u^2\bar{T} - \frac{\partial T}{\partial x}(0,t) = -u^2\bar{T} + \frac{l(t)}{k} = \frac{1}{K}F_c\left[\frac{\partial T}{\partial x}\right] = \frac{1}{K}\frac{d\bar{T}}{dt}$$

$$F_c[T(x,0^+)] = \bar{T}(u,0^+) = 0$$

Thus, the equation governing the transform of the temperature:

$$\frac{d\bar{T}}{dt} + u^2 K\bar{T} = K\frac{l(t)}{k}$$

can be written as an integral, eq. (1.9):

$$\bar{T}(u,t) = Ce^{-u^2Kt} + \frac{K}{k}\int_0^t l(t-\eta)e^{-ku^2\eta}\,d\eta$$

which must satisfy the initial condition:

$$\bar{T}(u,0^+) = C = 0$$

Thus, the solution is found in the form of an integral:

$$\bar{T}(u,t) = \frac{K}{k}\int_0^t l(t-\eta)e^{-ku^2\eta}\,d\eta$$

Applying the inverse transformation on the exponential function within the integrand:

$$T(x,t) = \frac{2}{\pi}\frac{K}{k}\int_0^t l(t-\eta)\left\{\int_0^\infty e^{-Ku^2\eta}\cos(ux)du\right\}d\eta$$

Using integral or transform tables:

$$I(x) = \int_0^\infty e^{-au^2}\cos(ux)du = \frac{1}{2}\sqrt{\frac{\pi}{a}}\,e^{-x^2/4a}$$

one finally obtains the solution:

$$T(x,t) = \sqrt{\frac{K}{\pi k^2}}\int_0^t \frac{l(t-\eta)}{\sqrt{\eta}}\,e^{-x^2/(4K\eta)}\,d\eta$$

7.17 Operational Calculus with Fourier Sine Transform

The Fourier sine transform, as defined in 7.3, will be discussed in this section. Let the Fourier sine transform of a function f(x) be defined as:

$$F_S[f(x)] = F_S(u) = \int_0^\infty f(x)\sin(ux)dx$$

then:

$$f(|x|)\,\mathrm{sgn}\,x = \frac{2}{\pi}\int_0^\infty F_s(u)\sin(ux)\,du$$

where the signum functions sgn is defined by:

$$\mathrm{sgn}(x) = \begin{cases} x/|x| & x \neq 0 \\ 0 & x = 0 \end{cases}$$

7.17.1 Fourier Sine Transform of Derivatives

The Fourier sine transform of the derivative of f(x) can be derived as:

$$F_S[\frac{\partial f(x)}{\partial x}] = \int_0^\infty \frac{\partial f(x)}{\partial x}\sin(ux)dx = f(x)\sin(ux)\Big|_0^\infty - u\int_0^\infty f(x)\cos(ux)\,dx = -uF_c(u)$$

The transform of the second derivative of f(x):

$$F_S[\frac{\partial^2 f(x)}{\partial x^2}] = \int_0^\infty \frac{\partial^2 f(x)}{\partial x^2}\sin(ux)dx = \frac{\partial f}{\partial x}\sin(ux)\Big|_0^\infty - u\int_0^\infty \frac{\partial f}{\partial x}\cos(ux)dx$$

$$= 0 - u\left\{f(x)\cos(ux)\Big|_0^\infty + u\int_0^\infty f(x)\sin(ux)\,dx\right\} = -u^2F_S(u) + uf(0^+)$$

and in general:

$$F_s[\frac{\partial^{2n} f}{\partial x^{2n}}] = (-1)^n u^{2n} F_s(u) + \sum_{m=1}^{n} (-1)^{m+1} u^{2m-1} \frac{\partial^{2n-2m} f(0^+)}{\partial x^{2n-2m}} \qquad n \geq 1$$

$$\text{provided that } \left|\frac{\partial^m f}{\partial x^m}\right| \to 0 \text{ as } x \to \infty \text{ for } m \leq (2n-1) \tag{7.50}$$

and

$$F_s[\frac{\partial^{2n+1} f}{\partial x^{2n+1}}] = (-1)^{n+1} u^{2n+1} F_c(u) + \sum_{m=1}^{n} (-1)^{m+1} u^{2m-1} \frac{\partial^{2n-2m+1} f(0^+)}{\partial x^{2n-2m+1}} \qquad n \geq 1$$

$$\text{provided that } \left|\frac{\partial^m f}{\partial x^m}\right| \to 0 \text{ as } x \to \infty \text{ for } m \leq 2n \tag{7.51}$$

It should be noted that the Fourier sine transform of even derivatives of a function give the Fourier sine transform of the function, and requires initial conditions of even derivatives. The Fourier sine transform of odd derivatives give the Fourier cosine transform of the function, and thus cannot be used to solve problems.

7.17.2 Convolution Theorem

It can be shown that there is no convolution theorem for the Fourier sine transform. Let H_s (u) and G_s (u) be the Fourier sine transforms of h(x) and g(x) respectively. Then:

$$F_s^{-1}[H_s(u)G_s(u)] = \frac{2}{\pi}\int_0^\infty H_s(u)G_s(u)\sin(ux)\,du$$

$$= \frac{2}{\pi}\int_0^\infty H_s(u)\left\{\int_0^\infty g(\xi)\sin(u\xi)d\xi\right\}\sin(ux)\,du$$

$$= \int_0^\infty g(\xi)\left\{\frac{2}{\pi}\int_0^\infty H_s(u)\sin(u\xi)\sin(ux)du\right\}d\xi$$

$$= \int_0^\infty g(\xi)\left\{\frac{1}{\pi}\int_0^\infty H_s(u)[\cos(u|x-\xi|) - \cos(u(x+\xi))]\,du\right\}d\xi$$

which cannot be put in a convolution form, since the integrals are cosine and not sine transforms.

If H_s (u) and G_c (u) are the Fourier sine transform of h(x) and the Fourier cosine transform of g(x), respectively, then the inverse sine transform of this product becomes:

$$F_s^{-1}[H_s(u)G_c(u)] = \frac{2}{\pi}\int_0^\infty H_s(u)G_c(u)\sin(ux)\,du$$

$$= \frac{2}{\pi}\int_0^\infty G_c(u)\left\{\int_0^\infty h(\xi)\sin(u\xi)d\xi\right\}\sin(ux)\,du$$

$$= \int_0^\infty h(\xi)\left\{\frac{2}{\pi}\int_0^\infty G_c(u)\sin(u\xi)\sin(ux)\,du\right\}d\xi$$

$$= \int_0^\infty h(\xi)\left\{\frac{1}{\pi}\int_0^\infty G_c(u)[\cos(u(x-\xi)) - \cos(u(x+\xi))]\,du\right\}d\xi$$

$$= \frac{1}{2}\int_0^\infty h(\xi)[g(|x-\xi|) - g(x+\xi)]\,d\xi \tag{7.52}$$

This means that if there is a product of two functions, $F_1(u)\cdot F_2(u)$, then call $F_1(u) = H_s(u)$, and $F_2(u) = G_c(u)$. To use the convolution theorem use the inverse transform of $h(x) = F_s^{-1}(H_s(u))$, and that of $g(x) = F_c^{-1}(G_c(u))$, to obtain $h(x)$ and $g(x)$.

7.17.3 Parseval Formula

Consider the following integral:

$$\frac{2}{\pi}\int_0^\infty H_s(u)G_s(u)\cos(ux)du = \frac{2}{\pi}\int_0^\infty H_s(u)\left\{\int_0^\infty g(\xi)\sin(u\xi)\,d\xi\right\}\cos(ux)du$$

$$= \int_0^\infty g(\xi)\left\{\frac{2}{\pi}\int_0^\infty H_s(u)\sin(u\xi)\cos(ux)\,du\right\}d\xi$$

$$= \int_0^\infty g(\xi)\left\{\frac{1}{\pi}\int_0^\infty H_s(u)[\sin(u(x+\xi)) + \sin(u(\xi-x))]du\right\}d\xi$$

$$= \frac{1}{2}\int_0^\infty g(\xi)[h(x+\xi) + h(|x-\xi|)\mathrm{Sgn}(\xi-x)]\,d\xi \tag{7.53}$$

If x is set to zero in eq. (7.53), one obtains:

$$\frac{2}{\pi}\int_0^\infty H_s(u)G_s(u)\,du = \int_0^\infty g(\xi)h(\xi)\,d\xi \tag{7.54}$$

and if $H_s(u) = G_s(u)$, then:

$$\frac{2}{\pi}\int_0^\infty H_s^2(u)\,du = \int_0^\infty g^2(\xi)\,d\xi \tag{7.55}$$

which is the **Parseval formula for the Fourier sine transform.**

Example 7.11 Heat Flow in a Semi-Infinite Rod

Obtain the heat flow in a rod, initially at zero temperature, where the temperature is prescribed at its end x = 0, such that T(x,t) satisfies the following system:

$$\frac{\partial^2 T}{\partial x^2} = \frac{1}{K}\frac{\partial T}{\partial t} \qquad x > 0 \qquad t > 0$$

$$T(0,t) = f(t) \qquad T(x,0^+) = 0$$

Since the Fourier sine transform requires even derivative boundary conditions, see eq. (7.50), it is well suited for application to the present problem. Define:

$$\bar{T}(u,t) = \int_0^\infty T(x,t)\sin(ux)\,dx$$

$$\frac{d\bar{T}}{dt} + u^2K\bar{T} = Kuf(t)$$

Thus, the solution for the transform of T is given by eq. (1.9):

$$\bar{T}(u,t) = Ce^{-u^2Kt} + Ku\int_0^t f(t-\eta)e^{-Ku^2\eta}\,d\eta$$

Satisfying the initial condition:

$$\bar{T}(u,0^+) = C = 0$$

then the solution of the transform of T becomes:

$$\bar{T}(u,t) = Ku\int_0^t f(t-\eta)e^{-Ku^2\eta}\,d\eta$$

The inverse transform integral is then defined by:

$$T(x,t) = \frac{2K}{\pi}\int_0^t f(t-\eta)\left\{\int_0^\infty ue^{-Ku^2\eta}\sin(ux)\,du\right\}d\eta$$

To evaluate the inner integral, one can use the integral tables:

$$I(x) = \int_0^\infty e^{-au^2}\cos(ux)\,du = \frac{1}{2}\sqrt{\frac{\pi}{a}}\,e^{-x^2/4a}$$

Then, differentiating I(x) with x, one can find the inverse transform of the solution:

$$\frac{dI(x)}{dx} = -\frac{x}{4a}\sqrt{\frac{\pi}{a}}\, e^{-x^2/4a} = -\int_0^\infty u e^{-au^2} \sin(ux)\, du$$

so that the solution of the temperature is given by:

$$T(x,t) = \frac{x}{2\sqrt{\pi K}} \int_0^t f(t-\eta)\, \eta^{-3/2} e^{-x^2/(4K\eta)}\, d\eta$$

Compare this result with the result of Example 7.6.

Example 7.12 Free Vibration of a Stretched Semi-Infinite String

Obtain the amplitude of vibration in a stretched, free, semi-infinite string, such that, y = y(x,t) satisfies the following system:

$$\frac{\partial^2 y}{\partial x^2} = \frac{1}{c^2}\frac{\partial^2 y}{\partial t^2} \qquad x > 0 \qquad t > 0$$

$$y(0,t) = 0 \qquad \lim_{x\to\infty} y(x,t) \to 0$$

$$y(x,0^+) = f(x) \qquad \frac{\partial y}{\partial t}(x,0^+) = g(x)$$

Since the boundary condition is an even derivative, then apply Fourier sine transform to the system. Defining Y(u,t) as the transform of y(x,t), then application of Fourier sine transform to the differential equation and the initial conditions results in:

$$F_s\left[\frac{\partial^2 y}{\partial x^2}\right] = -u^2 Y + u y(0,t) = -u^2 Y = \frac{1}{c^2} F_s\left[\frac{\partial^2 y}{\partial t^2}\right] = \frac{1}{c^2}\frac{d^2 Y}{dt^2}$$

$$F_s[y(x,0^+)] = Y(u,0^+) = F_s(f(x)) = F(u)$$

$$F_s\left[\frac{\partial y}{\partial t}(x,0^+)\right] = \frac{dY}{dt}(u,0^+) = F_s[g(x)] = G(u)$$

Thus, the transformed system of differential equation and initial conditions:

$$\frac{d^2 Y}{dt^2} + c^2 u^2 Y = 0, \quad Y(u,0^+) = F(u), \quad \frac{dY}{dt}(u,0^+) = G(u)$$

$$Y = A \sin(uct) + B \cos(uct)$$

Satisfying the two initial conditions yields the final transformed solution:

$$Y(u,t) = \frac{G(u)}{uc}\sin(uct) + F(u)\cos(uct)$$

and the solution y(x, t) can now be written in terms of two inverse transform integrals:

$$y(u,t) = \frac{2}{\pi}\int_0^\infty \frac{G(u)}{uc}\sin(uct)\sin(ux)\,du + \frac{2}{\pi}\int_0^\infty F(u)\cos(uct)\sin(ux)\,du$$

The second integral can be evaluated readily:

$$\frac{2}{\pi}\int_0^\infty F(u)\cos(uct)\sin(ux)\,du = \frac{1}{\pi}\int_0^\infty F(u)[\sin(u(x+ct))+\sin(u(x-ct))]\,du$$
$$= \frac{1}{2}[f(x+ct)+f(|x-ct|)\,\mathrm{Sng}(x-ct)]$$

The first integral can be evaluated as follows:

$$\frac{1}{\pi c}\int_0^\infty \frac{G(u)}{u}[\cos(u(x-ct))-\cos(u(x+ct))]\,du$$

Since:

$$g(|v|)\mathrm{Sgn}\,v = \frac{2}{\pi}\int_0^\infty g(u)\sin(uv)\,du$$

then:

$$\int_0^v g(|\eta|)\mathrm{Sgn}\,\eta\,d\eta = \frac{2}{\pi}\int_0^\infty G(u)\left\{\int_0^v \sin(u\eta)\,d\eta\right\}du = -\frac{2}{\pi}\int_0^\infty \frac{G(u)}{u}\cos(uv)\,du + F$$

where $F = \frac{2}{\pi}\int_0^\infty \frac{G(u)}{u}\,du$.

Thus:

$$\frac{2}{\pi}\int_0^\infty \frac{G(u)}{u}\cos(uv)\,du = -\int_0^v g(|\eta|)\mathrm{Sgn}\,\eta\,d\eta + F = \int_{|v|}^0 g(|\eta|)\,d\eta + F$$

The first integral then becomes:

$$\frac{1}{\pi c}\int_0^\infty \frac{G(u)}{u}[\cos(u(x-ct))-\cos(u(x+ct))]\,du$$
$$= \frac{1}{2c}\left(\int_{|x-ct|}^0 g(\eta)\,d\eta + F\right) - \frac{1}{2c}\left(\int_{x+ct}^0 g(\eta)\,d\eta + F\right)$$
$$= \frac{1}{2c}\int_{|x-ct|}^{x+ct} g(\eta)\,d\eta$$

Thus, the total solution becomes:

$$y(x,t) = \frac{1}{2}[f(x+ct)+f(|x-ct|)\,\mathrm{Sng}(x-ct)] + \frac{1}{2c}\int_{|x-ct|}^{x+ct} g(\eta)\,d\eta$$

7.18 Operational Calculus with Complex Fourier Transform

The complex Fourier transform was defined in eq. (7.4). Let F(u) represent the complex Fourier transform of f(x), defined as follows:

$$F(f(x)) = F(u) = \int_{-\infty}^{\infty} f(x)e^{iux}\,dx$$

then:

$$f(x) = \frac{1}{2\pi}\int_{-\infty}^{\infty} F(u)e^{-iux}\,du$$

7.18.1 Complex Fourier Transform of Derivatives

The complex Fourier transform of the first derivative is easily calculated:

$$F[\frac{\partial f}{\partial x}] = \int_{-\infty}^{\infty} \frac{\partial f}{\partial x} e^{iux}\,dx = f(x)e^{iux}\Big|_{-\infty}^{\infty} - iu\int_{-\infty}^{\infty} f(x)e^{iux}\,dx = (-iu)\,F(u)$$

The transform of the second derivative of f(x) is:

$$F[\frac{\partial^2 f}{\partial x^2}] = \int_{-\infty}^{\infty} \frac{\partial^2 f}{\partial x^2} e^{iux}\,dx = \frac{\partial f}{\partial x} e^{iux}\Big|_{-\infty}^{\infty} - iu\int_{-\infty}^{\infty} \frac{\partial f}{\partial x} e^{iux}\,dx$$

$$= -iu\left\{ fe^{iux}\Big|_{-\infty}^{\infty} - iu\int_{-\infty}^{\infty} fe^{iux}\,dx \right\} = (iu)^2 F(u)$$

In general:

$$F[\frac{\partial^n f}{\partial x^n}] = (-iu)^n\,F(u) \qquad n \geq 0$$

$$\text{provided that } \left|\frac{\partial^m f}{\partial x^m}\right| \to 0 \text{ as } x \to \infty \text{ for } m \leq (n-1) \tag{7.56}$$

7.18.2 Convolution Theorem

The Convolution theorem for the complex Fourier transform for a product of transforms is developed in this section. Let F(u) and G(u) represent the complex Fourier transform of f(x) and g(x), respectively. Then, the inverse transform of the product is defined as:

$$F^{-1}[F(u)G(u)] = \frac{1}{2\pi}\int_{-\infty}^{\infty} F(u)G(u)\,e^{-iux}\,du$$

$$= \frac{1}{2\pi}\int_{-\infty}^{\infty} F(u)\left\{\int_{0}^{\infty} g(\xi)e^{iu\xi}\,d\xi\right\} e^{-iux}\,du$$

$$= \int_{-\infty}^{\infty} g(\xi)\left\{\frac{1}{2\pi}\int_{0}^{\infty} F(u)e^{-iu(x-\xi)}\,du\right\} d\xi$$

$$= \int_{-\infty}^{\infty} g(\xi)f(x-\xi)\,d\xi$$

Similarly, it can be shown that the last integral can also be written in the form:

$$\int_{-\infty}^{\infty} f(\xi)g(x-\xi)\,d\xi \tag{7.57}$$

7.18.3 Parseval Formula

If one sets x = 0 in eq. (7.57) one obtains:

$$\frac{1}{2\pi}\int_{-\infty}^{\infty} F(u)G(u)du = \int_{-\infty}^{\infty} g(\xi)\,f(-\xi)\,d\xi = \int_{-\infty}^{\infty} g(-\xi)f(\xi)\,d\xi \tag{7.58}$$

which does not lead to a Parseval formula. However, if one defines the complex conjugate of G(u) as follows:

$$G^*(u) = \int_{-\infty}^{\infty} g(x)e^{-iux}\,dx$$

then:

$$\frac{1}{2\pi}\int_{-\infty}^{\infty} F(u)G^*(u)e^{-iux}du = \frac{1}{2\pi}\int_{-\infty}^{\infty} F(u)\left\{\int_{-\infty}^{\infty} g(\xi)\,e^{-iu\xi}d\xi\right\} e^{-iux}du$$

$$= \int_{-\infty}^{\infty} g(\xi)\left\{\frac{1}{2\pi}\int_{-\infty}^{\infty} F(u)e^{-iu(\xi+x)}du\right\} d\xi = \int_{-\infty}^{\infty} g(\xi)f(\xi+x)\,d\xi \tag{7.59}$$

If one again sets x = 0 in eq. (7.59), one obtains:

$$\frac{1}{2\pi}\int_{-\infty}^{\infty} F(u)G^*(u)\,du = \int_{-\infty}^{\infty} g(\xi)f(\xi)\,d\xi \tag{7.60}$$

If g(x) = f(x), then one obtains the **Parseval formula for complex Fourier transforms**:

$$\frac{1}{2\pi}\int_{-\infty}^{\infty} F(u)F^*(u)\,du = \frac{1}{2\pi}\int_{-\infty}^{\infty} |F(u)|^2\,du = \int_{-\infty}^{\infty} f^2(\xi)\,d\xi \tag{7.61}$$

Example 7.13 Vibration of a Free Infinite Stretched String

A free infinite string is induced to motion by imparting it with an initial displacement and velocity. Let the displacement y = y(x,t), then the equation of motion and initial conditions are:

$$\frac{\partial^2 y}{\partial x^2} = \frac{1}{c^2}\frac{\partial^2 y}{\partial t^2} \qquad -\infty < x < \infty \qquad t > 0$$

$$y(x,0^+) = f(x) \qquad \frac{\partial y}{\partial t}(x,0^+) = g(x)$$

Using the complex Fourier transform on x, one obtains, with Y(u,t) being the transform of y(x,t):

$$-u^2 Y = \frac{1}{c^2}\frac{d^2Y}{dt^2}$$

$$Y(u,0^+) = F(u) \qquad \frac{dY}{dt}(u,0^+) = G(u)$$

The solution of the differential equation is readily obtained as:

$$Y(u,t) = A\sin(uct) + B\cos(uct)$$

which, after satisfying the transformed initial conditions gives the final solution:

$$Y(u,t) = \frac{G(u)}{uc}\sin(uct) + F(u)\cos(uct)$$

The inversion of the transformed solution can be evaluated in two parts:

$$F^{-1}[F(u)\cos(uct)] = \frac{1}{2\pi}\int_{-\infty}^{\infty} F(u)\frac{e^{iuct}+e^{-iuct}}{2}e^{-iux}\,du$$

$$= \frac{1}{2\pi}\int_{-\infty}^{\infty} \frac{1}{2}F(u)(e^{-iu(x-ct)} + e^{-iu(x+ct)})\,du$$

$$= \frac{1}{2}[f(x-ct) + f(x+ct)]$$

and

$$F^{-1}[\frac{G(u)}{uc}\sin(uct)] = \frac{1}{2\pi}\int_{-\infty}^{\infty} G(u)\frac{e^{iuct}-e^{-iuct}}{2iuc}e^{-iux}\,du$$

$$= \frac{1}{4\pi}\int_{-\infty}^{\infty} \frac{G(u)}{iuc}(e^{-iu(x-ct)} - e^{-iu(x+ct)})\,du$$

Since the integral definition of the inverse transform is:

$$g(\eta) = \frac{1}{2\pi}\int_{-\infty}^{\infty} G(u)e^{-iu\eta}\,du$$

then integrating this again results in the following relationship:

$$\frac{1}{c}\int_{0}^{v} g(\eta)\,d\eta = -\frac{1}{2\pi c}\int_{-\infty}^{\infty} \frac{G(u)}{iu}[e^{-iuv} - 1]\,du$$

Using this form, the two integrals in the inverse transform of G(u)/cu become:

$$\frac{1}{2\pi c}\int_{-\infty}^{\infty} \frac{G(u)}{iu} e^{-iu(x-ct)}\,du = -\frac{1}{c}\int_{0}^{x-ct} g(\eta)\,d\eta + \frac{1}{2\pi c}\int_{-\infty}^{\infty} \frac{G(u)}{iu}\,du$$

and

$$\frac{1}{2\pi c}\int_{-\infty}^{\infty} \frac{G(u)}{iu} e^{-iu(x+ct)}\,du = -\frac{1}{c}\int_{0}^{x+ct} g(\eta)\,d\eta + \frac{1}{2\pi c}\int_{-\infty}^{\infty} \frac{G(u)}{iu}\,du$$

Finally, adding the two expressions, one obtains:

$$F^{-1}[\frac{G(u)}{uc}\sin(uct)] = \frac{1}{2c}\int_{x-ct}^{x+ct} g(\eta)\,d\eta$$

The total solution y(x,t) is recovered by adding the two parts:

$$y(x,t) = \frac{1}{2}[f(x+ct) + f(x-ct)] + \frac{1}{2c}\int_{x-ct}^{x+ct} g(\eta)\,d\eta$$

The solution given above is the well-known solution for wave propagation in an infinite one-dimensional medium.

Example 7.14 Heat Flow in an Infinite Rod

Obtain the temperature in a given infinite rod, with a given initial temperature distribution. Let T = T(x,t), then the temperature T satisfies the system:

$$\frac{\partial^2 T}{\partial x^2} = \frac{1}{K}\frac{\partial T}{\partial t} \qquad -\infty < x < \infty \qquad t > 0$$

and

$$T(x,0^+) = f(x)$$

Applying the complex Fourier transform on the space variable x, the differential equation and the initial condition are transformed to:

$$-u^2 T^* = \frac{1}{K}\frac{dT^*}{dt}$$

and

$$T^*(x,0^+) = F(x)$$

where T*(u, t) is the transform of T(x, t). The solution to the first-order equation is given by eq. (1.9):

$$T^*(u,t) = C e^{-u^2 Kt}$$

which, upon satisfaction of the initial condition, results in the final transformed solution:

$$T^*(u,t) = F(u) e^{-u^2 Kt}$$

The inversion of the solution can be written in terms of convolution integrals. Starting with the inverse of the exponential term:

$$F^{-1}[e^{-u^2Kt}] = \frac{1}{2\pi}\int_{-\infty}^{\infty} e^{-u^2Kt} e^{-iux}\,du = \frac{1}{\pi}\int_{0}^{\infty} e^{-u^2Kt}\cos(ux)\,du =$$

$$= \frac{1}{\sqrt{4\pi Kt}}\, e^{-x^2/(4Kt)}$$

Thus, using the convolution theorem in eq. (7.57), one obtains:

$$T(x,t) = \frac{1}{2\pi}\int_{-\infty}^{\infty} F(u)e^{-u^2Kt} e^{-iux}\,du = \frac{1}{\sqrt{4\pi Kt}}\int_{-\infty}^{\infty} f(x-\xi)e^{-\xi^2/(4Kt)}\,d\xi$$

7.19 Operational Calculus with Multiple Fourier Transform

Multiple Fourier transforms were discussed in Section 7.5, and given in eq. (7.5). Let:

$f = f(x,y) \qquad -\infty < x < \infty \qquad -\infty < y < \infty$

be an absolutely integrable function, then define:

$$F_{xy}[f(x,y)] = F(u,v) = \int_{-\infty}^{\infty}\int_{-\infty}^{\infty} f(x,y)\, e^{i(ux+vy)}\,dx\,dy$$

and

$$f(x,y) = \frac{1}{4\pi^2}\int_{-\infty}^{\infty}\int_{-\infty}^{\infty} F(u,v)\, e^{-i(ux+vy)}\,du\,dv$$

7.19.1 Multiple Transform of Partial Derivatives

The multiple transform of partial derivatives are defined as follows:

$$F_{xy}\left[\frac{\partial^{n+m} f}{\partial x^n \partial y^m}\right] = (-iu)^n (-iv)^m F(u,v) \tag{7.62}$$

$$F_{xy}\left[\nabla^2 f\right] = F_{xy}\left[\frac{\partial^2 f}{\partial x^2} + \frac{\partial^2 f}{\partial y^2}\right] = (-iu)^2 F(u,v) + (-iv)^2 F(u,v)$$
$$= -(u^2 + v^2)F(u,v) \tag{7.63}$$

$$F_{xy}\left[\nabla^4 f\right] = F_{xy}\left[\frac{\partial^4 f}{\partial x^4} + 2\frac{\partial^4 f}{\partial x^2 \partial y^2} + \frac{\partial^4 f}{\partial y^4}\right] = (u^2 + v^2)^2 F(u,v) \tag{7.64}$$

7.19.2 Convolution Theorem

The convolution theorem for multiple transforms can be treated in the same manner as single transforms. Let F(u,v) and G(u,v) be the Fourier multiple transform of the functions f(x,y) and g(x,y), respectively.

Then:

$$\frac{1}{4\pi^2}\int_{-\infty}^{\infty}\int_{-\infty}^{\infty} F(u,v)G(u,v)\,e^{-i(ux+vy)}\,du\,dv$$

$$=\frac{1}{4\pi^2}\int_{-\infty}^{\infty}\int_{-\infty}^{\infty} F(u,v)\left\{\int_{-\infty}^{\infty}\int_{-\infty}^{\infty} g(\xi,\eta)e^{i(u\xi+v\eta)}\,d\xi\,d\eta\right\}e^{-i(ux+vy)}\,du\,dv$$

$$=\int_{-\infty}^{\infty}\int_{-\infty}^{\infty} g(\xi,\eta)\left\{\frac{1}{4\pi^2}\int_{-\infty}^{\infty}\int_{-\infty}^{\infty} F(u,v)e^{-i[u(x-\xi)+v(y-\eta)]}\,du\,dv\right\}d\xi\,d\eta$$

$$=\int_{-\infty}^{\infty}\int_{-\infty}^{\infty} g(\xi,\eta)\,f(x-\xi,y-\eta)\,d\xi\,d\eta \qquad (7.65)$$

Example 7.15 Wave Propagation in Infinite Plates

A free, infinite plate is induced to vibration by initially displacing it from equilibrium, and releasing it from rest. Let w = w(x,y,t), then the equation of motion and the initial conditions are:

$$\nabla^4 w+\beta^4\frac{\partial^2 w}{\partial t^2}=0 \qquad |x|<\infty \qquad |y|<\infty \qquad t>0$$

where $\beta^4 = \rho h/D$, and

$$w(x,y,0^+) = f(x,y) \qquad\qquad \frac{\partial w}{\partial t}(x,y,0^+) = 0$$

Applying the multiple Fourier transforms on the space variables x and y:

$$F_{xy}\left[\nabla^4 w\right]=(u^2+v^2)^2\,W$$

$$F_{xy}\left[\frac{\partial^2 w}{\partial t^2}\right]=\frac{d^2 W}{dt^2}$$

where:

$$W(u,v,t)=\int_{-\infty}^{\infty}\int_{-\infty}^{\infty} w(x,y,t)e^{i(ux+vy)}\,dx\,dy$$

The equation of motion and the initial condition transform to the following system:

$$(u^2+v^2)^2 W+\beta^4\frac{d^2W}{dt^2}=0$$

and

$$W(u,v,0^+) = f(x,y) \qquad \frac{dW}{dt}(u,v,0^+) = 0$$

The solution for the transform W becomes:

$$W = A\sin\left(\frac{u^2+v^2}{\beta^2}t\right) + B\cos\left(\frac{u^2+v^2}{\beta^2}t\right)$$

which results in the following solution upon satisfaction of the two initial conditions:

$$W = F(u,v)\cos\left(\frac{u^2+v^2}{\beta^2}t\right)$$

Since $\cos\left((u^2+v^2)t/\beta^2\right)$ is not absolutely integrable, then one cannot obtain its multiple complex inverse readily. This can be rectified by adding a diminishingly small damping by defining G(u) as:

$$G(u) = e^{-\varepsilon u^2} e^{iau^2} \qquad \varepsilon > 0$$

which reverts to the function $\exp(iau^2)$ when $\varepsilon \to 0$. The inverse transform of G(u) is defined by:

$$g(x) = \frac{1}{2\pi}\int_{-\infty}^{\infty} e^{-(\varepsilon - ia)u^2} e^{-iux}\, du = \frac{1}{\pi}\int_{0}^{\infty} e^{-(\varepsilon - ia)u^2}\cos(ux)\, du$$

$$= \frac{1}{2\pi}\sqrt{\frac{\pi}{\varepsilon - ia}}\, e^{-x^2/(4(\varepsilon - ia))}$$

Taking the limit $\varepsilon \to 0$ in the integral, one can readily obtain the inverse:

$$\frac{1}{2\pi}\int_{-\infty}^{\infty}\cos(au^2)\, e^{-iux} du = \frac{1}{\sqrt{8\pi a}}\left[\cos\left(\frac{x^2}{4a}\right) + \sin\left(\frac{x^2}{4a}\right)\right]$$

and

$$\frac{1}{2\pi}\int_{-\infty}^{\infty}\sin(au^2)\, e^{-iux} du = \frac{1}{\sqrt{8\pi a}}\left[\cos\left(\frac{x^2}{4a}\right) - \sin\left(\frac{x^2}{4a}\right)\right] \qquad (7.66)$$

In a similar manner, one can use the limiting process on the double integral where one defines G(u,v) as:

$$G(u,v) = e^{-\varepsilon(u^2+v^2)} e^{ia(u^2+v^2)} \qquad \varepsilon > 0$$

then:

$$g(x,y) = \frac{1}{4\pi^2}\int_{-\infty}^{\infty}\int_{-\infty}^{\infty} e^{-\varepsilon(u^2+v^2)} e^{ia(u^2+v^2)} e^{-i(ux+vy)}\, du\, dv$$

$$\to \frac{-i}{4\pi a} e^{-i(x^2+y^2)/4a} \quad \text{as} \quad \varepsilon \to 0$$

Hence:

$$\frac{1}{4\pi^2}\int_{-\infty}^{\infty}\int_{-\infty}^{\infty}\cos[a(u^2+v^2)]\,e^{-i(ux+vy)}\,du\,dv=\frac{1}{4\pi a}\sin\left(\frac{x^2+y^2}{4a}\right) \tag{7.67}$$

and

$$\frac{1}{4\pi^2}\int_{-\infty}^{\infty}\int_{-\infty}^{\infty}\sin[a(u^2+v^2)]\,e^{-i(ux+vy)}\,du\,dv=\frac{1}{4\pi a}\cos\left(\frac{x^2+y^2}{4a}\right)$$

Once the inverse transform of $\cos\left((u^2+v^2)t/\beta^2\right)$ is found, one then substitutes this into the convolution theorem, eq. (7.65), giving the final solution:

$$w(x,y,t)=\frac{\beta^2}{4\pi t}\int_{-\infty}^{\infty}\int_{-\infty}^{\infty}f(x-\xi,y-\eta)\sin\left(\beta^2\frac{\xi^2+\eta^2}{4t}\right)d\xi\,d\eta$$

7.20 Operational Calculus with Hankel Transform

The Hankel transform of order zero was discussed in Section 7.6 and was defined in eq. (7.6) and Hankel transform of order ν was discussed in Section 7.7 and was given in eq. (7.10).

Define the Hankel transform of order ν as:

$$H_\nu[f(r)]=F_\nu(\rho)=\int_0^{\infty}r\,f(r)\,J_\nu(r\rho)\,dr \qquad \nu\geq-\frac{1}{2} \tag{7.68}$$

7.20.1 Hankel Transform of Derivatives

$$H_\nu[\frac{\partial f}{\partial r}]=\int_0^{\infty}\frac{\partial f}{\partial r}J_\nu(r\rho)\,r\,dr=f(r)\,J_\nu(r\rho)\,r\Big|_0^{\infty}-\int_0^{\infty}f(r)\frac{\partial}{\partial r}\left(r\,J_\nu(r\rho)\right)dr$$

Using the identity, see equation 3.13:

$$\frac{d}{dr}\left(r\,J_\nu(r\rho)\right)=J_\nu(r\rho)+r\frac{dJ_\nu(r\rho)}{dr}=\rho r J_{\nu-1}(r\rho)-(\nu-1)J_\nu(r\rho)$$

then the integral becomes:

$$-\int_0^{\infty}f(r)\left[\rho r J_{\nu-1}(r\rho)-(\nu-1)J_\nu(r\rho)\right]dr=-\rho\,F_{\nu-1}(\rho)+(\nu-1)\int_0^{\infty}f(r)J_\nu(r\rho)dr$$

Using the identity given in eq. (3.16), the last equation becomes:

$$-\rho F_{\nu-1}(\rho) + \rho\frac{\nu-1}{2\nu}\left[\int_0^\infty r\, f(r) J_{\nu+1}(r\rho)dr + \int_0^\infty r\, f(r) J_{\nu-1}(r\rho)dr\right]$$

$$= -\rho F_{\nu-1}(\rho) + \frac{\nu-1}{2\nu}\rho\left[F_{\nu+1}(\rho) + F_{\nu-1}(\rho)\right]$$

$$= -\frac{\rho}{2\nu}[(\nu+1)F_{\nu-1}(\rho) - (\nu-1)F_{\nu+1}(\rho)]$$

Finally, the Hankel transform of the first derivative becomes:

$$H_\nu\left(\frac{\partial f}{\partial r}\right) = \frac{\rho}{2\nu}[(\nu-1)F_{\nu+1}(\rho) - (\nu+1)F_{\nu-1}(\rho)] \tag{7.69}$$

provided that:

$$\lim_{r\to 0} r^{\nu+1} f(r) \to 0 \quad \text{and} \quad \lim_{r\to\infty} \sqrt{r}\, f(r) \to 0$$

Similarly, using eq. (7.69):

$$H_\nu\left(\frac{\partial^2 f}{\partial r^2}\right) = H_\nu\left(\frac{\partial}{\partial r}\frac{\partial f}{\partial r}\right) = \frac{\rho}{2\nu}\left[(\nu-1)H_{\nu+1}\left(\frac{\partial f}{\partial r}\right) - (\nu+1)H_{\nu-1}\left(\frac{\partial f}{\partial r}\right)\right]$$

$$= \frac{\rho^2}{4\nu}\left\{\frac{\nu+1}{\nu-1}[\nu F_{\nu-2} - (\nu-2)F_\nu] - \frac{\nu-1}{\nu+1}[(\nu+2)F_\nu - \nu F_{\nu+2}]\right\}$$

$$= \frac{\rho^2}{4}\left\{\frac{\nu+1}{\nu-1}F_{\nu-2}(\rho) - 2\frac{\nu^2-3}{\nu^2-1}F_\nu(\rho) + \frac{\nu-1}{\nu+1}F_{\nu+2}(\rho)\right\} \tag{7.70}$$

provided that:

$$\lim_{r\to 0} r^{\nu+1} f'(r) \to 0 \quad \text{and} \quad \lim_{r\to\infty} \sqrt{r}\, f'(r) \to 0$$

as well as the limit requirements on f(r) in eq. (7.69).

The transform of the two-dimensional Laplacian in cylindrical coordinates defined as:

$$\nabla^2 f = \frac{d^2 f}{dr^2} + \frac{1}{r}\frac{df}{dr} \qquad \text{with} \qquad f = f(r)$$

can be obtained as follows:

$$H_\nu(\nabla^2 f) = \int_0^\infty r\left\{\frac{d^2 f}{dr^2} + \frac{1}{r}\frac{df}{dr}\right\} J_\nu(r\rho)dr = \int_0^\infty \frac{d}{dr}\left(r\frac{df}{dr}\right) J_\nu(r\rho)\,dr$$

$$= r\frac{df}{dr}J_\nu(r\rho)\Big|_0^\infty - \int_0^\infty r\frac{df}{dr}\frac{dJ_\nu(r\rho)}{dr}dr$$

$$= -r\,f(r)\frac{df}{dr}J_\nu(r\rho)\Big|_0^\infty + \int_0^\infty f(r)\frac{d}{dr}\left[r\frac{dJ_\nu(r\rho)}{dr}\right]dr$$

$$= \int_0^\infty f(r)\left[\frac{\nu^2}{r^2} - \rho^2\right] r\, J_\nu(r\rho)\, dr$$

where Bessel's equation in eq. (3.161) was used, and provided that:

$$\lim_{r\to 0}(-r^{\nu+2} + \nu r^\nu) f(r) \to 0 \qquad \lim_{r\to\infty} \sqrt{r}\, f(r) \to 0$$

and

$$\lim_{r\to 0} r^{\nu+1} f'(r) \to 0 \qquad \lim_{r\to\infty} \sqrt{r}\, f'(r) \to 0$$

Thus, the Hankel transform of order ν of the νth Laplacian becomes:

$$H_\nu\left(\frac{d^2 f}{dr^2} + \frac{1}{r}\frac{df}{dr} - \frac{\nu^2}{r^2} f\right) = -\rho^2 H_\nu(f(r)) = -\rho^2 F_\nu(\rho) \tag{7.71}$$

and for $\nu = 0$, the Hankel transform of order zero of the axisymmetric Laplacian becomes:

$$H_0\left(\frac{d^2 f}{dr^2} + \frac{1}{r}\frac{df}{dr}\right) = -\rho^2 F_0(\rho) \tag{7.72}$$

7.20.2 Convolution Theorem

It can be shown that there is no closed form convolution theorem for the Hankel transforms. Let $F_\nu(\rho)$ and $G_\nu(\rho)$ be the Hankel transform of order ν of f(r) and g(r), respectively. Then:

$$\int_0^\infty F_\nu(\rho) G_\nu(\rho) J_\nu(r\rho)\rho\, d\rho = \int_0^\infty F_\nu(\rho)\left\{\int_0^\infty g(\eta) J_\nu(\eta\rho)\,\eta\, d\eta\right\} J_\nu(r\rho)\,\rho\, d\rho$$

$$= \int_0^\infty g(\eta)\left\{\int_0^\infty F_\nu(\rho) J_\nu(r\rho) J_\nu(\eta\rho)\,\rho\, d\rho\right\}\eta\, d\eta$$

The inner integral contains a product of $J_\nu(r\rho)\, J_\nu(\eta\rho)$, which cannot be written in an additive form in a simple manner.

7.20.3 Parseval Formula

Let $F_\nu(\rho)$ and $G_\nu(\rho)$ be the Hankel transforms of order ν of the functions f(r) and g(r), respectively, then:

$$\int_0^\infty F_\nu(\rho) G_\nu(\rho)\rho\, d\rho = \int_0^\infty F_\nu(\rho)\left\{\int_0^\infty g(r) J_\nu(r\rho)\, r\, dr\right\}\rho\, d\rho$$

$$= \int_0^\infty g(r)\left\{\int_0^\infty F_\nu(\rho) J_\nu(r\rho)\,\rho\, d\rho\right\} r\, dr = \int_0^\infty g(r) f(r)\, r\, dr \tag{7.73}$$

Also, for f(r) = g(r) results in a Parseval Formula for Hankel transform:

$$\int_0^\infty F_\nu^2(\rho)\rho\, d\rho = \int_0^\infty f^2(r)\, r\, dr$$

Example 7.16 Axisymmetric Wave Propagation in an Infinite Membrane

A stretched infinite membrane is initially deformed such that the axisymmetric displacement w(r,t) satisfies the following equation and initial conditions:

$$\nabla^2 w = \frac{1}{c^2}\frac{\partial^2 w}{\partial t^2} \qquad r \geq 0 \qquad t > 0$$

$$w(r,0^+) = f(r) \qquad \frac{\partial w}{\partial r}(r,0^+) = g(r)$$

Since the problem is axisymmetric, without dependence on the rotational angle θ, a Hankel transform of order zero is appropriate. Applying the Hankel transform of order zero to the differential equation and initial conditions one obtains:

$$H_0(\nabla^2 w) = H_0(\frac{d^2 w}{dr^2} + \frac{1}{r}\frac{dw}{dr}) = -\rho^2 W(\rho,t) = \frac{1}{c^2} H_0(\frac{\partial^2 w}{\partial t^2}) = \frac{1}{c^2}\frac{d^2 W}{dt^2}$$

$$H_0(w(r,0^+)) = W(\rho,0^+) = F_0(\rho)$$

$$H_0(\frac{\partial w}{\partial t}(r,0^+)) = \frac{dW}{dt}(\rho,0^+) = G_0(\rho)$$

where:

$$W(\rho,t) = \int_0^\infty r\, w(r,t)\, J_0(r\rho)\, dr$$

Then, the equation of motion transforms to:

$$\frac{d^2 W}{dt^2} + \rho^2 c^2 W = 0$$

whose solution, satisfying the two initial conditions becomes:

$$W(\rho,t) = F_0(\rho)\cos(\rho ct) + \frac{G_0(\rho)}{\rho c}\sin(\rho ct)$$

Since there is no convolution theorem, one must invert the total solution, which can only be done if f(r) and g(r) are given explicitly, e.g., if the initial displacement f(r) is given by:

$$f(r) = w_0 \frac{a}{\sqrt{a^2 + r^2}}$$

and the initial velocity g(r) = 0, then the transform (from transform tables) of f(r) becomes:

$$\int_0^\infty \frac{r\, J_0(r\rho)}{\sqrt{r^2+k^2}}\, dr = \frac{e^{-k\rho}}{\rho}$$

Thus, the transform of the initial displacement field is given by:

$$F_0(\rho) = w_0 \frac{a}{\rho} e^{-a\rho} \quad \text{and} \quad G_0(\rho) = 0$$

and, the transform of the displacement w(r,t) is given by the expression:

$$W(\rho,t) = w_0 \frac{a}{\rho} e^{-a\rho} \cos(\rho ct) = w_0 \frac{a}{\rho} e^{-a\rho} \mathit{Re}\,[e^{-i\rho ct}]$$

Letting H(ρ,t) represent the complex function in W(ρ,t):

$$H(\rho,t) = w_0 \frac{a}{\rho} e^{-\rho(a+ict)}$$

then its inverse Hankel transform can be written in an integral form:

$$h(r,t) = H_0^{-1}\left[H(\rho,t)\right] = w_0 a \int_0^\infty e^{-\rho(a+ict)} J_0(r\rho)\, d\rho$$

Noting that the inversion of the Hankel transform of exp[kρ]/ρ is given by:

$$\int_0^\infty \frac{e^{-k\rho}}{\rho} J_0(r\rho)\rho\, d\rho = \frac{1}{\sqrt{r^2+k^2}}$$

then the inverse transform of H(ρ,t) becomes:

$$h(r,t) = w_0 a \frac{1}{\sqrt{(a+ict)^2 + r^2}}$$

and the solution can be obtained explicitly:

$$w(r,t) = \mathit{Re}\,[h(r,t)] = \frac{w_0 a}{\sqrt{2R}} \left[1 + \frac{r^2 + a^2 - c^2t^2}{r^2}\right]^{1/2}$$

where:

$$R = \left(r^2 + a^2 - c^2t^2\right)^2 + 4a^2c^2t^2$$

PROBLEMS

Section 7.14

1. Find the Laplace transform of the following functions using the various theorems in Section 7.14 and without resorting to integrations:

   (a) $\cos(at)$ (b) $t \sin(at)$

   (c) $e^{at}\cos(bt)$ (d) $\sin(at)\sinh(at)$

   (e) $t^n e^{-at}$ (f) $\cos(at)\sinh(at)$

2. Obtain the Laplace transform of the following functions:

   (a) $f(t/a)$ (b) $e^{bt} f(t/a)$

   (c) $\dfrac{d}{dt}[e^{-at}f(t)]$ (d) $\dfrac{d}{dt}[t^2 f(t)])$

   (e) $t\,e^{-at}f(t)$ (f) $e^t \dfrac{d^2f}{dt^2}$

   (g) $t^n f(t/a)$ (h) $\dfrac{d}{dt}[e^{at}\dfrac{df}{dt}]$

   (i) $\dfrac{d}{dt}[t f(t)]$ (j) $\sinh(at) f(t)$

   (k) $\displaystyle\int_0^t x f(x)\,dx$ (l) $\displaystyle\int_0^t f(t-x)\,dx$

3. Obtain the Laplace transform of the following periodic functions; where $f(t+T) = f(t)$ and $f_1(t)$ represents the function defined over the first period:

   (a) $f_1(t) = t \qquad 0 \le t \le T$

   (b) $f(t) = |\sin(at)|$

   (c) $f_1(t) = \begin{cases} +1 & 0 \le t < T/2 \\ -1 & T/2 < t \le T \end{cases}$

   (d) $f_1(t) = t(\pi - t) \qquad T = \pi$

   (e) $f_1(t) = \begin{cases} 1 & 0 \le t < T/2 \\ 0 & T/2 < t \le T \end{cases}$

   (f) $f_1(t) = \begin{cases} 0 & 0 \le t < T/4 \\ 1 & T/4 < t < 3T/4 \\ 0 & 3T/4 < t \le T \end{cases}$

(g) $|\cos(\omega t)|$

4. Obtain the inverse Laplace transform of the following transforms by using the theorems in Section 7.14:

(a) $\dfrac{1}{(p-a)(p-b)}$ (b) $\dfrac{a^2}{p(p^2+a^2)}$

(c) $\dfrac{a^3}{p^2(p^2+a^2)}$ (d) $\dfrac{2a^3}{(p^2+a^2)^2}$

(e) $\dfrac{2ap}{(p^2+a^2)^2}$ (f) $\dfrac{4a^3}{p^4+4a^4}$

(g) $\dfrac{2ap^2}{(p^2+a^2)^2}$ (h) $\dfrac{2a^3}{p^4-a^4}$

Section 7.15

5. Obtain the solution to the following ordinary differential equations subject to the stated initial conditions by the use of Laplace transform on y(t):

(a) $y'' + k^2 y = f(t)$ $y(0) = A$ $y'(0) = B$

(b) $y'' - k^2 y = f(t)$ $y(0) = A$ $y'(0) = B$

(c) $y^{(iv)} - a^4 y = 0$ $y(0) = 0$ $y'(0) = 0$ $y''(0) = A$ $y'''(0) = B$

(d) $y^{(iv)} - a^4 y = f(t)$ $y(0) = y'(0) = y''(0) = y'''(0) = 0$

(e) $y''' + 6y'' + 11y' + 6y = f(t)$ $y(0) = y'(0) = y''(0) = 0$

(f) $y''' + 5y'' + 8y' + 4y = f(t)$ $y(0) = y'(0) = y''(0) = 0$

(g) $y^{(iv)} + 4y''' + 6y'' + 4y' + y = f(t)$ $y(0) = y'(0) = y''(0) = y'''(0) = 0$

(h) $y'' + 2y' + y = f(t)$ $y(0) = y'(0) = 0$

(i) $y'' + 4y' + 4y = A\,t\,\delta(t-t_o)$ $y(0) = 1$ $y'(0) = 0$ $t_o > 0$

(j) $y'' + y' - 2y = 1 - 2t$ $y(0) = 0$ $y'(0) = 4$

(k) $y'' - 5y' + 6y = A\,\delta(t-t_o)$ $y(0) = 0$ $y'(0) = B$ $t_o > 0$

6. Obtain the solution to the following integro-differential equation subject to the stated initial conditions by use of the Laplace transform on y(t):

(a) $y'' + 3.5\,y' + 2y = 2\int_0^t y(x)\,dx + A\,\delta(t-t_o)$ $y(0) = y'(0) = 0$ $t_o > 0$

(b) $y' + \int_0^t y(x)\cosh(t-x)\,dx = 0$ $\qquad y(0) = A$

(c) $y' - \int_0^t y(x)dx = 2$ $\qquad y(0) = 1$

(d) $y'' + k^2 y = f(t) + \int_0^t g(t-x)\,y(x)\,dx$ $\qquad y(0) = y'(0) = 0$

(e) $y' + 3ay + a^2 \int_0^t y(x)\,e^{-a(t-x)}\,dx = \delta(t)$ $\qquad y(0) = 1$

(f) $y' + 5y + 4\int_0^t y(x)dx = f(t)$ $\qquad y(0) = 0$

(g) $y' + 3y + 2\int_0^t y(x)dx = A\,e^{-at}$ $\qquad y(0) = B$

(h) $y' + \int_0^t y(x)dx = f(t)$ $\qquad y(0) = 0$

(i) $y' - ay + \int_0^t y(x)e^{a(t-x)}\,dx = f(t)$ $\qquad y(0) = 0$

(j) $y'' + 3y' + 3y + \int_0^t y(x)dx = f(t)$ $\qquad y(0) = y'(0) = 0$

7. Solve the following coupled ordinary differential equations subject to the stated initial conditions by the use of Laplace transform, where x = x (t) and y = y (t):

(a) $y'' - a^2 x = U$ $\qquad$ U and V are constants
$\quad x'' + a^2 y = V$ $\qquad x(0) = x'(0) = y(0) = y'(0) = 0$

(b) $y'' + 2x' + y = 0$ $\qquad x(0) = y(0) = 1$
$\quad x'' + 2y' + x = 0$ $\qquad x'(0) = y'(0) = 0$

(c) $y'' - 3x' + x = 0$ $\qquad x(0) = 1 \qquad y(0) = 3$
$\quad x'' - 3y' + y = 0$ $\qquad x'(0) = 2$

(d) $x' + x + y' + 2y = f(t)$ $\qquad f(0) = g(0) = 0$
$\quad x' + 2x + y' + y = g(t)$ $\qquad x(0) = y(0) = 0$

(e) $x'' + y = g(t)$ $\quad$ $x(0) = y(0) = 0$
$y'' + x = f(t)$ $\quad$ $x'(0) = y'(0) = 0$

(f) $x' + y = g(t)$ $\quad$ $x(0) = y(0) = 0$
$y' + x = f(t)$

(g) $y'' + 2x + 3y' = A = \text{constant}$ $\quad$ $x(0) = x_0$ $\quad$ $y(0) = y_0$
$x'' + 2y + 3x' = B = \text{constant}$ $\quad$ $x'(0) = y'$ $\quad$ $(0) = 0$

8. The wave equation for a one-dimensional medium under a distributed pulsed load is given by:

$$\frac{\partial^2 y}{\partial x^2} = \frac{1}{c^2}\frac{\partial^2 y}{\partial t^2} + Ae^{-bx}\delta(t - t_0) \qquad b > 0 \qquad t, t_o > 0$$

$$y(0,t) = 0 \qquad y(x,0) = 0 \qquad \frac{\partial y}{\partial t}(x,0) = 0$$

Obtain the solution y(x, t) by use of the Laplace transform.

9. The following system obeys the diffusion equation with a time decaying source:

$$\frac{\partial^2 u}{\partial x^2} = \frac{1}{K}\frac{\partial u}{\partial t} - Q_0 e^{-bt} \qquad x > 0 \qquad t > 0 \qquad a > 0 \qquad b > 0$$

$$u(0,t) = T_0 e^{-at} \qquad u(x,0^+) = 0 \qquad Q = \text{constant}$$

Obtain the solution u(x, t) by the use of the Laplace transform.

10. The wave equation for a semi-infinite rod under the influence of a point force is given by:

$$\frac{\partial^2 y}{\partial x^2} = \frac{1}{c^2}\frac{\partial^2 y}{\partial t^2} - y_0\delta(x - x_o)\delta(t - t_o) \quad x > 0 \qquad t, t_o > 0$$

where:

$$y = y(x,t) \qquad \frac{\partial y}{\partial x}(0,t) = 0 \quad y(x,0^+) = 0 \qquad \frac{\partial y}{\partial t}(x,0^+) = 0$$

and δ is the Dirac delta function. Obtain the solution y(x,t) explicitly by use of the Laplace transform.

11. The temperature distribution in a semi-infinite rod obeys the diffusion equation such that:

$$\frac{\partial^2 T}{\partial x^2} = \frac{1}{K}\frac{\partial T}{\partial t} - Q_0\delta(x - x_o)\delta(t - t_o) \quad x > 0 \qquad t, t_o > 0$$

where:

$$T = T(x,t) \qquad T(0,t) = 0 \qquad T(x,0^+) = 0$$

Obtain explicitly the temperature distribution in the rod by use of Laplace transform.

12. A finite string is excited to motion such that its deflection y(x,t) is governed by the wave equation:

$$\frac{\partial^2 y}{\partial x^2} = \frac{1}{c^2}\frac{\partial^2 y}{\partial t^2} \qquad 0 \le x \le L \qquad t > 0$$

$$\frac{\partial y}{\partial x}(0,t) = 0 \qquad \frac{\partial y}{\partial x}(L,t) = 0 \qquad \frac{\partial y}{\partial t}(x,0^+) = 0$$

$$y(x,0^+) = y_0(x - \frac{L}{2})^2$$

Obtain an explicit expression for the displacement y(x,t) by use of Laplace transform on time.

13. The displacement y(x, t) in a semi-infinite rod is governed by:

$$\frac{\partial^2 y}{\partial x^2} = \frac{1}{c^2}\frac{\partial^2 y}{\partial t^2} \qquad x > 0 \qquad t > 0$$

$$y(0,t) = V_0 t \qquad \frac{\partial y}{\partial t}(x,0^+) = -V_0 \qquad y(x,0^+) = 0$$

Obtain the solution y(x, t) explicitly by Laplace transform.

14. A finite rod is undergoing a displacement y(x, t) such that:

$$\frac{\partial^2 y}{\partial x^2} = \frac{1}{c^2}\frac{\partial^2 y}{\partial t^2} \qquad 0 \le x \le L \qquad t > 0$$

$$y(0,t) = y_0 H(t) \qquad y(L,t) = -y_0 H(t)$$

$$\frac{\partial y}{\partial t}(x,0^+) = 0 \qquad y(x,0^+) = 0$$

Obtain an expression for the displacement y (x,t) explicitly by Laplace transform. Sketch the displacement y (L/4,t), using at least the first four terms in the solution, in their order of the arrival times.

15. A stretched semi-infinite string is excited to vibration such that y = y(x,t):

$$\frac{\partial^2 y}{\partial x^2} = \frac{1}{c^2}\frac{\partial^2 y}{\partial t^2} + \frac{P_0}{T_0}e^{-bx}\,\delta(t - t_0) \qquad x > 0 \qquad t, t_o > 0$$

$$y(0,t) = y_0 H(t) \qquad y(x,0^+) = 0 \qquad \frac{\partial y}{\partial t}(x,0^+) = 0$$

where δ is the Dirac delta function. Obtain the solution y(x, t) explicitly by use of Laplace transforms.

16. A semi-infinite rod is heated such that the temperature $y = y(x, t)$ satisfies the following system:

$$\frac{\partial^2 y}{\partial x^2} = \frac{1}{K}\frac{\partial y}{\partial t} \qquad x > 0 \qquad t, t_o > 0$$

$$y(0,t) = T_0\,\delta(t-t_0) \qquad y(x,0^+) = T_0$$

where δ is the Dirac delta function. Obtain the temperature distribution $y(x,t)$ explicitly by use of Laplace transforms.

17. A semi-infinite rod is heated such that:

$$\frac{\partial^2 T}{\partial x^2} = \frac{1}{K}\frac{\partial T}{\partial t} \qquad x > 0 \qquad t > 0$$

$$T = T(x,t) \qquad T(0,t) = 0 \qquad T(x,0^+) = T_0 e^{-bx}$$

Obtain the solution $T(x, t)$ explicitly, using the Laplace transform.

18. A stretched semi-infinite string is excited to motion such that:

$$\frac{\partial^2 y}{\partial x^2} = \frac{1}{c^2}\frac{\partial^2 y}{\partial t^2} \qquad y = y(x,t) \qquad x > 0 \qquad t > 0$$

$$y(x,0^+) = 0 \qquad \frac{\partial y}{\partial t}(x,0^+) = 0$$

$$\frac{\partial y}{\partial x}(0,t) - \gamma\, y(0,t) = H(t)$$

where γ is the spring constant. Find the displacement $y(x,t)$ explicitly, using Laplace transform.

19. Find the displacement $y(x,t)$ explicitly by use of Laplace transforms:

$$\frac{\partial^2 y}{\partial x^2} = \frac{1}{c^2}\frac{\partial^2 y}{\partial t^2} \qquad x \geq 0 \qquad t \geq 0$$

$$y(x,0^+) = 0 \qquad \frac{\partial y}{\partial t}(x,0^+) = V_0 \qquad y(0,t) = -\frac{1}{2}a_0 t^2$$

20. Find the temperature distribution $T(x, t)$ by use of Laplace transforms:

$$\frac{\partial^2 T}{\partial x^2} = \frac{1}{K}\frac{\partial T}{\partial t} - Q_0\,\delta(t - t_0) \qquad b > 0 \qquad x \geq 0 \qquad t, t_0 > 0$$

$$T(x,0^+) = 0 \qquad \frac{\partial T}{\partial x}(0,t) - b\,T(0,t) = -bT_0$$

21. The temperature in a semi-infinite bar is governed by the following equation. Obtain the solution by use of the Laplace transform.

$$\frac{\partial^2 T}{\partial x^2} = \frac{1}{K}\frac{\partial T}{\partial t} - \frac{Q_0}{k} e^{-at} \qquad x > 0 \qquad t > 0$$

$$T(x,0^+) = 0 \quad T(0,t) = T_0\, H(t)$$

22. A finite bar, initially at rest and fixed at both ends, is induced to vibration such that the displacement y(x,t) is governed by:

$$\frac{\partial^2 y}{\partial x^2} = \frac{1}{c^2}\frac{\partial^2 y}{\partial t^2} - \frac{F_0}{AE}\delta(x - x_0)\,\delta(t - t_0) \qquad 0 \le x, x_0 \le L \quad t, t_0 \ge 0$$

$$y(x,0^+) = \frac{\partial y}{\partial t}(x,0^+) = 0 \qquad y(0,t) = y(L,t) = 0$$

Obtain the solution y(x,t) by use of Laplace transforms.

23. A finite bar, initially at rest, is induced to vibration such that the displacement y(x,t) is governed by:

$$\frac{\partial^2 y}{\partial x^2} = \frac{1}{c^2}\frac{\partial^2 y}{\partial t^2} - \frac{F_0}{AE}\sin(at) \qquad 0 \le x \le L \qquad t \ge 0$$

$$y(x,0^+) = \frac{\partial y}{\partial t}(x,0^+) = 0 \qquad y(0,t) = y(L,t) = 0$$

Obtain the solution y(x,t) by use of Laplace transforms.

24. The temperature in a semi-infinite bar is governed by the following system. Obtain the solution by use of the Laplace transform:

$$\frac{\partial^2 T}{\partial x^2} = \frac{1}{K}\frac{\partial T}{\partial t} - \frac{Q_0}{k}\delta(t - t_0) \qquad x > 0 \qquad t, t_0 > 0$$

$$T(x,0^+) = 0 \qquad T(0,t) = T_0 \frac{t}{a}\begin{cases} 1 & t \le a \\ 0 & t > a \end{cases}$$

25. A semi-infinite stretched string is induced to vibration such that y = y (x,t):

$$\frac{\partial^2 y}{\partial x^2} = \frac{1}{c^2}\frac{\partial^2 y}{\partial t^2} + \frac{P_0}{T_0} e^{-bx}\sin(at) \qquad x > 0, \quad t > 0, \quad b > 0$$

$$y(x,0^+) = 0 \qquad \frac{\partial y}{\partial t}(x,0^+) = 0 \qquad y(0,t) = 0$$

Obtain the solution y(x,t) by use of Laplace transform.

26. A semi-infinite rod is heated such that the temperature, T(x,t), satisfies:

$$\frac{\partial^2 T}{\partial x^2} = \frac{1}{K}\frac{\partial T}{\partial t} + Q \qquad x > 0 \qquad t > 0$$

$$T(0,t) = T_0\, \delta(t - t_0) \qquad T(x,0^+) = 0$$

where Q is a constant. Obtain the solution by use of the Laplace transform.

27. Find the displacement y(x, t) explicitly by use of Laplace transforms:

$$\frac{\partial^2 y}{\partial x^2} = \frac{1}{c^2}\frac{\partial^2 y}{\partial t^2} - \frac{F_0}{AE} e^{-at} \qquad x > 0, \quad t > 0, \quad a > 0$$

$$\frac{\partial y}{\partial t}(x,0^+) = y(x,0^+) = 0 \qquad y(0,t) = y_0 \cos(bt)$$

28. The temperature, T(x,t), in a semi-infinite bar is governed by the following equation. Obtain the solution by use of the Laplace transform:

$$\frac{\partial^2 T}{\partial x^2} = \frac{1}{K}\frac{\partial T}{\partial t} - \frac{Q_0}{k}\delta(t - t_0) \qquad x > 0 \qquad t, t_0 > 0$$

$$T(x,0^+) = 0 \qquad \frac{\partial T}{\partial x}(0,x) = F\, t e^{-at} \qquad a > 0$$

29. A semi-infinite stretched string is induced to vibration such that the displacement, y(x,t) satisfies:

$$\frac{\partial^2 y}{\partial x^2} = \frac{1}{c^2}\frac{\partial^2 y}{\partial t^2} \qquad x > 0 \qquad t > 0 \qquad b > 0$$

$$y(x,0^+) = y_0 e^{-bx} \qquad \frac{\partial y}{\partial t}(x,0^+) = 0 \qquad y(0,t) = A\, H(t)$$

Obtain the solution y(x,t) by use of the Laplace transform.

30. A semi-infinite rod is heated such that the temperature satisfies:

$$\frac{\partial^2 T}{\partial x^2} = \frac{1}{K}\frac{\partial T}{\partial t} + Q_0 \sin(at) \qquad x > 0 \qquad t > 0 \qquad a > 0$$

$$T = T(x,t) \qquad T(0,t) = T_0\, t \qquad T(x,0^+) = 0$$

Obtain the solution by use of the Laplace transform.

Section 7.16

31. Do Problem 8 by Fourier cosine transform.

32. Do Problem 9 by Fourier cosine transform.

33. Do Problem 11 by Fourier cosine transform.

34. Do Problem 14 by Fourier cosine transform.

35. Do Problem 15 by Fourier cosine transform.

36. Do Problem 16 by Fourier cosine transform.

37. Do Problem 17 by Fourier cosine transform.

38. Do Problem 19 by Fourier cosine transform.

39. Do Problem 21 by Fourier cosine transform.

40. Do Problem 24 by Fourier cosine transform.

41. Do Problem 26 by Fourier cosine transform.

42. Do Problem 26 by Fourier cosine transform.

43. Do Problem 27 by Fourier cosine transform.

44. Do Problem 29 by Fourier cosine transform.

45. Do Problem 30 by Fourier cosine transform.

Section 7.17

46. Do Problem 10 by Fourier sine transform.

47. Do Problem 18 by Fourier sine transform.

48. Do Problem 20 by Fourier sine transform.

49. Do Problem 28 by Fourier sine transform.

Section 7.18

50. Obtain the response of an infinite vibrating bar under distributed load by use of complex Fourier transform:

$$\frac{\partial^2 y}{\partial x^2} = \frac{1}{c^2}\frac{\partial^2 y}{\partial t^2} - A\, e^{-b|x|} H(t) \qquad -\infty < x < \infty, \quad t > 0, \qquad b > 0$$

$$y\,(x,0^+) = 0 \qquad \frac{\partial y}{\partial t}(x,0^+) = 0$$

51. Obtain the response of an infinite string under distributed loads by use of complex Fourier transform:

$$\frac{\partial^2 y}{\partial x^2} = \frac{1}{c^2}\frac{\partial^2 y}{\partial t^2} - \frac{q_0}{\tau_0}\, e^{-b|x|} \sin(\omega t) \quad -\infty < x < \infty, \quad t > 0, \qquad b > 0$$

$$y\,(x,0^+) = 0 \qquad \frac{\partial y}{\partial t}(x,0^+) = 0$$

52. Obtain the response of an infinite string subject to a point load by use of complex Fourier transform:

$$\frac{\partial^2 y}{\partial x^2} = \frac{1}{c^2}\frac{\partial^2 y}{\partial t^2} - \frac{P_o}{T_o}\,\delta\,(x - x_o)\sin(\omega t) \qquad -\infty < x < \infty, \quad t > 0$$

$$y\,(x,0^+) = 0 \qquad \frac{\partial y}{\partial t}(x,0^+) = 0$$

53. Obtain the temperature distribution, T(x,t), in an infinite rod, by use of complex Fourier transform:

$$\frac{\partial^2 T}{\partial x^2} = \frac{1}{K}\frac{\partial T}{\partial t} - \frac{Q_o}{k}\,e^{-b|x|} \qquad -\infty < x < \infty, \quad t > 0, \qquad b > 0$$

$$T(x,0^+) = 0$$

54. Obtain the temperature distribution, T(x,t), in an infinite rod, by use of complex Fourier transform:

$$\frac{\partial^2 T}{\partial x^2} = \frac{1}{K}\frac{\partial T}{\partial t} - \frac{Q_o}{k}\,\delta\,(x - x_o)\sin(\omega t) \qquad -\infty < x < \infty, \quad t > 0$$

$$T(x,0^+) = 0$$

55. Obtain the temperature distribution, T(x,t), in an infinite rod, by use of complex Fourier transform:

$$\frac{\partial^2 T}{\partial x^2} = \frac{1}{K}\frac{\partial T}{\partial t} - \frac{Q_o}{k}\,e^{-b|x|}\,\delta\,(t - t_o) \qquad -\infty < x < \infty, \quad t > 0, \qquad b > 0$$

$$T(x,0^+) = 0$$

8

GREEN'S FUNCTIONS

8.1 Introduction

In this chapter, the solution of non-homogeneous ordinary and partial differential equations is obtained by an integral technique known as Green's function method. In essence, the system's response is sought for a point source, known as Green's function, so that the solution for a distributed source is obtained as an integral of this function over the source strength region.

8.2 Green's Function for Ordinary Differential Boundary Value Problems

Consider the following ordinary linear boundary value problem:

$$\mathbf{L}\, y = \begin{cases} f(x) & a < x < b \\ 0 & x < a \text{ or } x > b \end{cases} \tag{8.1}$$

$$\mathbf{U}_i\,(y) = \gamma_i \qquad i = 1, 2, \ldots, n \tag{8.2}$$

where **L** is an nth order ordinary, linear, differential operator with non-constant coefficients, given in (4.27) and $\mathbf{U}_i$ are the non-homogeneous boundary conditions in (4.35).

Define the Green's function $g(x|\xi)$:

$$\mathbf{L}\, g(x|\xi) = \delta(x-\xi) \tag{8.3}$$

$$\mathbf{U}_i(g) = 0 \qquad i = 1, 2, \ldots, n \tag{8.4}$$

where $\delta(x)$ is the Dirac delta function (Appendix D). The solution $g(x|\xi)$ is then the solution of the system due to a point source located at $x = \xi$, satisfying homogeneous boundary conditions. The solution of eqs. (8.3) and (8.4) gives the Green's function for the problem. It should be noted that, in general, $g(x|\xi)$ is not symmetric in (x,ξ). Rewriting eq. (8.1) and substituting eq. (8.3) for the operator **L:**

$$\mathbf{L}y = f(x) = \int_a^b f(\xi)\,\delta(x-\xi)\,d\xi = \int_a^b \mathbf{L}\,g(x|\xi)\,f(\xi)\,d\xi = \mathbf{L}\int_a^b g(x|\xi)\,f(\xi)\,d\xi$$

Hence, the particular solution of the system in eq. (8.1) $y_p(x)$ is given by:

$$y_p = \int_a^b f(\xi)\,g(x|\xi)\,d\xi \tag{8.5}$$

Substituting the particular solution y_p in eq. (8.5) in the boundary conditions, one finds that they satisfy homogeneous conditions; since the Green's function $g(x|\xi)$ satisfies the same:

$$U_i\left(y_p(x)\right) = U_i\left\{\int_a^b f(\xi)\, g(x|\xi)\, d\xi\right\} = \int_a^b f(\xi)\, U_i\left(g(x\,|\,\xi)\right) d\xi = 0$$

Thus, the total solution for the boundary value problem posed in eqs. (8.1) and (8.2) is:

$$y = y_h(x) + y_p(x)$$

where $y_h(x)$ is the homogeneous solution of the differential equations $\mathbf{L}y = 0$, and y_p is the particular solution that satisfies the non-homogeneous equation with homogeneous boundary condition. It follows that the homogeneous solutions, with n independent solutions $\{\mathbf{y}_i(x)\}$ satisfies the non-homogeneous boundary conditions eq. (8.2).

Example 8.1

Obtain the total solution for the following system:

$$\mathbf{L}\, y = x^2 y'' - 2xy' + 2y = 1 \qquad 1 < x < 2$$

$$y(1) = 3 \qquad y'(2) = 2$$

The homogeneous equation $\mathbf{L}y=0$ yields the following two independent solutions:

$$y_1(x) = x^2 \qquad y_2(x) = x$$

To obtain the Green's function for this system, $g(x|\xi)$ satisfies:

$$\mathbf{L}\, g(x|\xi) = x^2 \frac{d^2 g(x|\xi)}{dx^2} - 2x \frac{dg(x|\xi)}{dx} + 2g(x|\xi) = \delta(x - \xi)$$

$$g(1|\xi) = 0 \qquad \frac{dg}{dx}(2|\xi) = 0$$

To evaluate the Green's function, let:

$$g(x|\xi) = g_h(x|\xi) + g_p(x|\xi)$$

where:

$$\mathbf{L}\, g_h(x|\xi) = 0$$

and

$$\mathbf{L}\, g_p(x|\xi) = \delta(x\text{-}\xi)$$

so that:

$$g_h(x|\xi) = Ax^2 + Bx$$

To obtain the particular solution, one needs to resort to the method of variation of the parameters (Section 1.7), i.e.:

$$g_p(x|\xi) = v_1(x)x^2 + v_2(x)x$$

so that the solution for a second-order differential equation is given by (1.26) as:

$$g_p(x|\xi) = \int_1^x \frac{\eta^2 x - \eta x^2}{(-\eta^2)} \frac{\delta(\eta-\xi)}{\eta^2} d\eta = \left[\frac{x^2}{\xi^3} - \frac{x}{\xi^2}\right] H(x-\xi)$$

The total Green's function becomes:

$$g(x|\xi) = Ax^2 + Bx + \left[\frac{x^2}{\xi^3} - \frac{x}{\xi^2}\right] H(x-\xi)$$

Satisfying the boundary condition on $g(x|\xi)$ results in:

$$A = -B = \frac{1}{3}\left[\frac{1}{\xi^2} - \frac{4}{\xi^3}\right]$$

and the Green's function for this problem is given by:

$$g(x|\xi) = \frac{1}{3}\left(\frac{1}{\xi^2} - \frac{4}{\xi^3}\right)(x^2 - x) + \left(\frac{x^2}{\xi^3} - \frac{x}{\xi^2}\right) H(x-\xi)$$

It should be noted that this Green's function is not symmetric, i.e., $g(x|\xi) \neq g(\xi|x)$. Using the Green's function, the particular solution $y_p(x)$ is:

$$y_p(x) = \int_1^2 g(x|\xi)\, f(\xi)\, d\xi = \frac{1}{6}(x^2 - 4x + 3)$$

Note that:

$$y_p(1) = 0, \quad \text{and } y_p'(2) = 0$$

Thus, the total solution becomes:

$$y = y_h + y_p \qquad y_h = c_1 x^2 + c_2 x$$

which upon satisfying the non-homogeneous boundary gives:

$$y_h = (10x - x^2) / 3$$

and

$$y(x) = (-x^2 + 16x + 3) / 6$$

8.3 Green's Function for an Adjoint System

One can develop a Green's function for the adjoint system to a given boundary value problem. For the boundary value problem in eqs. (8.1) and (8.2), there exists an adjoint differential operator **K** given in (4.28) and the associated adjoint boundary condition $\mathbf{V}_i(y) = 0$ in (4.36). Let the Green's function for the adjoint system be $g^*(x|\xi)$ satisfy:

$$\mathbf{K}\, g^*(x|\xi) = \delta(x-\xi) \tag{8.6}$$

and satisfy the adjoint boundary conditions:

$$\mathbf{V}_i(g^*(x|\xi)) = 0 \tag{8.7}$$

The resulting adjoint Green's function $g^*(x|\xi)$ is, in general, not symmetric in (x,ξ). Multiplying eq. (8.6) by $y_p(x)$ and eq. (8.1) by $g^*(x|\xi)$ and after subtracting the two equations and integrating over the range (a,b), one obtains:

$$\int_a^b \left(g^* \mathbf{L} y_p - y_p \mathbf{K} g^*\right) dx = \int_a^b \left[g^*(x|\xi)\, f(x) - y_p(x)\, \delta(x-\xi)\right] dx \qquad (8.8)$$

The left-hand side of eq. (8.8) vanishes due to the definition of an adjoint system (see Section 4.12). The right-hand side then gives:

$$y_p(x) = \int_a^b f(\xi)\, g^*(\xi|x)\, d\xi \qquad (8.9)$$

Thus, the particular solution can also be obtained as an integral over the source distribution f(x) and the adjoint Green's function $g^*(x|\xi)$.

Example 8.2

For the system given in Example 8.1, obtain the adjoint Green's function $g^*(x|\xi)$. The adjoint operator **K:**

$$\mathbf{K}\, g^*(x|\xi) = x^2 g^{*''} + 6x g^{*'} + 6 g^* = \delta(x-\xi)$$

and the adjoint boundary conditions become:

$$g^*(1|\xi) = 0 \qquad\qquad g^{*'}(2|\xi) + 2g^*(2|\xi) = 0$$

Following a similar method of solution, one obtains the Green's function $g^*(x|\xi)$ as:

$$g^*(x|\xi) = \frac{1}{3}\left(\xi^2 - 4\xi\right)\left(\frac{1}{x^2} - \frac{1}{x^3}\right) + \left(\frac{\xi}{x^2} - \frac{\xi^2}{\xi^3}\right) H(x-\xi)$$

It should be noted that $g^*(x|\xi)$ is not symmetric in (x,ξ).

8.4 Symmetry of the Green's Functions and Reciprocity

In general, both Green's functions are not symmetric in (x,ξ). However, the Green's function $g(x|\xi)$ and its adjoint form $g^*(x|\xi)$ are related. Rewriting the two ordinary differential equations (8.3) and (8.6) as:

$$\mathbf{L}\, g(x|\xi) = \delta(x-\xi) \qquad (8.3)$$

$$\mathbf{K}\, g^*(x|\eta) = \delta(x-\eta) \qquad (8.6)$$

multiplying eq. (8.3) by $g^*(x|\eta)$ and eq. (8.6) by $g(x|\xi)$, subtracting and integrating the resulting two equalities one obtains:

$$\int_a^b \left[g^* \mathbf{L} g - g \mathbf{K} g^*\right] dx = 0 = \int_a^b \left[g^*(x|\eta)\, \delta(x-\xi) - g(x|\xi)\, \delta(x-\eta)\right] dx$$

The left-hand side vanishes and the right-hand side gives:

$$g^*(\xi|\eta) = g(\eta|\xi) \qquad (8.10)$$

This means that while the two Green's functions are not symmetric, they are symmetric with each other. This can be seen in Examples 8.1 and 8.2.

If the operator **L** is self-adjoint (see (4.34)), then **L** = **K** and $\mathbf{U}_i(y) = \mathbf{V}_i(y)$. This means that $g^*(x|\xi) = g(x|\xi)$ and hence:

$$g(x|\xi) = g(\xi|x) \tag{8.11}$$

which means that Green's function is symmetric in (x,ξ). This symmetry is known as the "Reciprocity" principle in physical systems. It indicates that the response of a system at x due to a point source at ξ is equal to the response at ξ due to a point source at x.

Example 8.3

If one rewrites the operator in Example 8.1 into a self-adjoint form $\bar{\mathbf{L}}$, (see Section 4.11), one obtains:

$$\bar{\mathbf{L}}y = \frac{d}{dx}\left(\frac{1}{x^2}\frac{dy}{dx}\right) + \frac{2}{x^4}y = \frac{1}{x^4} \qquad 1 < x < 2$$

Note that the source function becomes $f(x)=x^{-4}$. Defining $\bar{g}(x|\xi)$ as the Green's function for the self-adjoint operator $\bar{\mathbf{L}}$:

$$\bar{\mathbf{L}}\bar{g}(x|\xi) = \frac{d}{dx}\left(\frac{1}{x^2}\frac{d\bar{g}}{dx}\right) + \frac{2}{x^4}\bar{g} = \delta(x-\xi)$$

$$\bar{g}(x\,|\,\xi) = 0 \qquad \frac{d\bar{g}}{dx}(2|\xi) = 0$$

Following the method used to find the Green's function in Examples 8.1 and 8.2 results in:

$$\bar{g}(x|\xi) = \frac{1}{3}\left(\xi^2 - 4\xi\right)\left(x^2 - x\right) + \left(x^2\xi - x\xi^2\right)H(x-\xi)$$

Note that:

$$\bar{g}(x|\xi) = \bar{g}(\xi|x)$$

The particular solution is now given by:

$$y_p(x) = \int_1^2 \bar{g}(x|\xi)\frac{1}{\xi^4}\,d\xi$$

which is the same as in Example 8.1.

8.5 Green's Function for Equations with Constant Coefficients

If the operator **L** is one with constant coefficients, one can show that:

$$g_p(x|\xi) = g_p(x-\xi) \tag{8.12}$$

This can be done by making the transformation, $\eta = x-\xi$, so that:

$$\mathbf{L}_x = \mathbf{L}_\eta$$

and

$$\mathbf{L}_\eta \, \bar{g}_p(\eta|0) = \delta(\eta)$$

resulting in:

$$g_p = g_p(\eta) \quad \text{or} \quad g_p = g_p(x-\xi)$$

One still has to add the homogeneous solution g_h, so that the total Green's function satisfies the boundary condition. The resulting Green's function then, would not be dependent on x - ξ.

Example 8.4

For static longitudinal deformation of a bar under a distributed force field:

$$\frac{d^2u}{dx^2} = f(x) = x \qquad 0 < x < L$$

$$u(0) = 0 \qquad u(L) = 0$$

To construct the Green's function, let:

$$\frac{d^2g}{dx^2} = \delta(x-\xi)$$

$$g(0|\xi) = 0 \qquad g(L|\xi) = 0$$

Since the equation is one with constant coefficients, then one can solve for $g(x|0)$:

$$\frac{d^2g}{dx^2} = \delta(x)$$

To obtain the solution by direct integration:

$$\frac{dg_p}{dx} = \int_0^x \delta(x)\,dx = H(x), \qquad g_p(x) = \int_0^x H(x)\,dx = x\,H(x)$$

$$g_p(x|\xi) = g(x-\xi) = (x-\xi)H(x-\xi)$$

$$g_h = C_1 x + C_2, \qquad g(0|\xi) = C_2 = 0$$

$$g(L|\xi) = (L-\xi) + C_1 L = 0 \qquad C_1 = \frac{\xi - L}{L}$$

$$g(x|\xi) = (x-\xi)\,H(x-\xi) + x\frac{\xi - L}{L}$$

$$u(x) = \int_0^L g(x|\xi) f(\xi)\,d\xi = \frac{x}{6}(x^2 - L^2)$$

8.6 Green's Functions for Higher-Ordered Sources

If the source field of a system is a distributed field of higher-order than a simple source, one can show that the Green's function for such a system is obtainable from that for a simple source. For example, if the Green's function for a dipole source or a mechanical couple is desired then:

$$L\, g_1(x|\xi) = \delta_1(x-\xi) = -\frac{d\delta(x-\xi)}{dx} \tag{8.13}$$

where $\delta_1(x-\xi)$ represents a positive unit couple or dipole, see Section D.2, and $g_1(x|\xi)$ is the Green's function for a dipole/couple source. Starting with the definition of $g(x|\xi)$ for a point source:

$$\mathbf{L}\, g(x|\xi) = \delta(x-\xi) \tag{8.3}$$

and differentiating eq. (8.3) once partially with ξ, one gets:

$$\mathbf{L}\frac{\partial g}{\partial \xi}(x|\xi) = \frac{\partial \delta(x-\xi)}{\partial \xi} = -\frac{\partial \delta(x-\xi)}{\partial x} = \delta_1(x-\xi)$$

where the last equality is the identity (D.49). Thus:

$$g_1(x|\xi) = \frac{\partial g}{\partial \xi}(x|\xi) \tag{8.14}$$

In a similar fashion, one can obtain the Green's function for distributed source fields of higher-ordered sources (quadrupoles, octopoles, etc.):

$$\mathbf{L}\, g_N(x|\xi) = \delta_N(x-\xi) = (-1)^N \frac{\partial^N \delta(x-\xi)}{\partial x^N} \tag{8.15}$$

where $\delta_N(x)$ is the Nth order Dirac delta function, see Section D.3, then one can show that:

$$g_N(x|\xi) = \frac{\partial^N g(x|\xi)}{\partial \xi^N} \tag{8.16}$$

8.7 Green's Function for Eigenvalue Problems

Consider a non-homogeneous eigenvalue problem obeying the Sturm-Liouville system of 2nd order (see Section 4.15), i.e.:

$$\frac{d}{dx}\left(p\frac{dy}{dx}\right) + (q+\lambda r)y = f(x) \qquad a < x < b \tag{8.17}$$

$$U_i(y) = 0 \qquad i = 1, 2 \tag{8.18}$$

where p(x), q(x) and r(x) are defined for a positive-definite system and the boundary conditions in eq. (8.18) are any pair allowed in this system and detailed in Section 4.15. Since the operator is self-adjoint, the resulting Green's function is symmetric in (x,ξ). The Green's function depends on (x,ξ) and the parameter λ. Thus $g = g(x|\xi,\lambda)$ satisfies the following:

$$\frac{d}{dx}\left(p\frac{dg}{dx}\right) + (q+\lambda r)g = \delta(x-\xi) \tag{8.19}$$

$$U_i(g) = 0 \qquad\qquad i = 1, 2 \tag{8.20}$$

The total solution for the system of eqs. (8.17) and (8.18) then becomes:

$$y(x) = \int_a^b f(\xi)\, g(x|\xi,\lambda)\, d\xi \tag{8.21}$$

Example 8.5 Green's Function for the Vibration of a Finite String

Consider the forced vibration of a stretched string of length L under a distributed time harmonic source f(x). The equation of motion for the string is:

$$\frac{d^2y}{dx^2} + \frac{\omega^2}{c^2} y = -\frac{f(x)}{T_0} \qquad 0 < x < L$$

$$y(0) = 0 \qquad\qquad y(L) = 0$$

where ω is the frequency of the source field, T_0 is the tension in the string and c is the sound speed, see Section 4.10.

The Green's function satisfies:

$$\frac{d^2g}{dx^2} + k^2 g = \delta(x-\xi) \qquad 0 < x, \xi < L$$

$$g(0|\xi,k) = 0 \qquad\qquad g(L|\xi,k) = 0$$

where:

$$k = \omega/c$$

The method used is to obtain the homogeneous and particular parts of the Green's function.

$$g = g_h + g_p$$

$$g_h = A \sin(kx) + B\cos(kx)$$

$$g_p = v_1(x)\sin(kx) + v_2(x)\cos(kx)$$

The particular solution of g becomes, using the results of the solution of problem 7(a) in Chapter 1:

$$g_p(x|\xi,k) = \frac{1}{k}\sin(k(x-\xi))\, H(x-\xi)$$

$$g = A\sin(kx) + B\cos(kx) + \frac{1}{k}\sin(k(x-\xi))\, H(x-\xi)$$

satisfying both boundary conditions:

$$g(0|\xi,k) = 0 \qquad g(L|\xi,k) = 0$$

results in:

$$g(x|\xi,k) = \frac{1}{k\sin(kL)}\left[\sin(kL)\sin(k(x-\xi))\ H(x-\xi) - \sin(kx)\sin(k(L-\xi))\right]$$

This is a closed form Green's function. Note that if sin (kL) = 0 or $k_n = n\pi/L$, the Green's function becomes unbounded. These are the resonance frequencies of the stretched string.

In general, one can do the same for the general eigenvalue problems in Section 4.13. Let the non-homogeneous eigenvalue problem be defined as in Section 4.13 as:

$$\mathbf{L}\,y + \lambda\,\mathbf{M}\,y = f(x) \qquad a < x < b \tag{8.22}$$

$$U_i(y) = 0 \qquad i = 1, 2, ..., 2n$$

where **L** is $2n^{th}$ and **M** is $2m^{th}$ self-adjoint ordinary differential operators, with n > m. Define Green's function to satisfy the following equation and boundary conditions:

$$\mathbf{L}g + \lambda\mathbf{M}g = \delta(x-\xi) \tag{8.23}$$

$$U_i(g) = 0 \qquad i = 1, 2, ..., 2n$$

The solution for the Green's function above is obtainable in a closed form. One can also derive the Green's function in terms of the eigenfunction of the system defined by eq. (8.22). Let the eigenfunction $\phi_\ell(x)$ and eigenvalue λ_ℓ be the solution of:

$$\mathbf{L}\phi_\ell + \lambda_\ell\,\mathbf{M}\phi_\ell = 0$$

$$U_i(\phi_\ell) = 0 \qquad i = 1, 2, ..., 2n$$

Since **L** and **M** are self-adjoint, the eigenfunctions are orthogonal, with the orthogonality defined in (4.45). Expanding the Green's function in a series of the eigenfunctions:

$$g(x|\xi,\lambda) = \sum_\ell E_\ell\phi_\ell(x) \tag{8.24}$$

then the expansion constants E_ℓ are given by (4.73) as:

$$E_\ell = \frac{1}{(\lambda-\lambda_\ell)N_\ell}\int_a^b \delta(x-\xi)\phi_\ell(x)\,dx = \frac{\phi_\ell(\xi)}{(\lambda-\lambda_\ell)N_\ell} \tag{8.25}$$

where N_ℓ is the normalization constant (4.45). The resulting Green's function becomes:

$$g(x|\xi,\lambda) = \sum_\ell \frac{\phi_\ell(\xi)\,\phi_\ell(x)}{(\lambda-\lambda_\ell)N_\ell} \tag{8.26}$$

It should be noted that the Green's function in eq. (8.26) is a symmetric function in (x,ξ). The total solution is in the form of an infinite series resulting from the substitution of eq. (8.26) in the integral eq. (8.21).

Example 8.6 Green's Function for the Vibration of a Finite String

Following Example 8.5, one can obtain the Green's function for the stretched string using eigenfunction expansion. Starting with the homogeneous equation:

$$u'' + \lambda u = 0$$

$$u(0) = 0 \qquad u(L) = 0$$

one can show that the eigenfunctions and eigenvalues are:

$$\phi_n(x) = \sin(n\pi x / L) \qquad n = 1, 2, \ldots$$

$$\lambda_n = n^2 \pi^2 / L^2 \qquad n = 1, 2, \ldots$$

The Green's function then becomes:

$$g(x|\xi, \lambda) = \frac{2}{L} \sum_{n=1}^{\infty} \frac{\sin(n\pi x/L)\sin(n\pi \xi/L)}{(\lambda - \lambda_n)}$$

8.8 Green's Function for Semi-Infinite One-Dimensional Media

The Green's function for semi-infinite media cannot be obtained through the methods outlined in the previous sections. Essentially, the dependent variables y(x) must be absolutely integrable over the semi-infinite region. Furthermore, boundary value problems in a semi-infinite region have boundary conditions on one end only. In such problems, use of integral transforms such as Fourier transforms becomes necessary. There is no general method of solution, as each problem requires the use of a specific transform tailored for that problem.

Example 8.7 Green's Function for the Longitudinal Vibration of a Semi-Infinite Bar

Obtain the response of a semi-infinite bar vibrating in longitudinal mode. The bar is excited to vibration by a distributed harmonic force $f(x)e^{-i\omega t}$, where f(x) is bounded and absolutely integrable. The longitudinal displacement of the bar $y(x)e^{-i\omega t}$ obeys the following equation (see section 4.3):

$$-\frac{d^2y}{dx^2} - k^2 y = \frac{f(x)}{AE} \qquad x > 0$$

$$y(0) = 0 \qquad k^2 = \omega^2/c^2 \qquad c = \sqrt{E/\rho}$$

where A is the cross-sectional area and E is the Young's modulus. The Green's function then satisfies the following system:

$$-\frac{d^2g}{dx^2} - k^2 g = \delta(x - \xi), \qquad g(0|\xi) = 0$$

The Dirichlet boundary condition at x=0 requires the use of Fourier sine transform, as it requires even-derivative boundary conditions, see section 7.16. Applying the Fourier sine transform on the differential equation on the Green's function, eq. (7.50):

$$u^2\bar{g}(u) - k^2\bar{g} + ug(0) = \int_0^{\infty} \delta(x - \xi)\sin(ux)\,dx = \sin(u\xi)$$

where $\bar{g}(u)$ is the Fourier sine transform of $g(x|\xi)$ and u is the transform variable. The transform of g(x) is obtained from above as:

$$\bar{g}(u) = \frac{\sin(u\xi)}{u^2 - k^2}$$

The inverse Fourier sine transform of $\bar{g}(u)$ is thus given by:

$$g(x|\xi) = \frac{2}{\pi}\int_0^\infty \frac{\sin(u\xi)\,\sin(ux)}{u^2 - k^2}\,du$$

In the inverse transformation, care must be taken to insure that waves propagate outward in the farfield and no waves are reflected from the farfield, i.e., no incoming waves in the farfield. To insure this, one would assume that the medium has material absorption that would insure that outgoing waves decay and hence no incoming (i.e., reflected) waves could possibly originate from the farfield. This can be accomplished by making the material constant complex. Letting the Young's modulus become complex:

$$E^* = E(1 - i\eta) \qquad \eta << 1$$

then:

$$c^* = \sqrt{\frac{E^*}{\rho}} \cong c(1 - i\,\eta/2)$$

so that:

$$k^* = \frac{\omega}{c^*} \cong k(1 + i\eta/2)$$

is a complex number.

Rewriting the inverse transform with complex k* results in:

$$g(x|\xi) = \frac{1}{\pi}\int_0^\infty \frac{\cos(u(x-\xi)) - \cos(u(x+\xi))}{u^2 - k^{*2}}\,du$$

The integrals can be evaluated using integration in the complex plane. The first integral becomes:

$$\int_0^\infty \frac{\cos(u(x+\xi))}{u^2 - k^{*2}}\,du = \frac{1}{2}\int_{-\infty}^\infty \frac{\cos(u(x+\xi))}{u^2 - k^{*2}}\,du$$

$$= \frac{1}{4}\int_{-\infty}^\infty \frac{e^{iu(x+\xi)}}{u^2 - k^{*2}}\,du + \frac{1}{4}\int_{-\infty}^\infty \frac{e^{-iu(x+\xi)}}{u^2 - k^{*2}}\,du$$

To evaluate the first integral, one can close the real axis path with a semi-circular contour of radius R in upper-half plane. Using the residue theorem for the simple pole at u = k*:

$$\int_{-\infty}^\infty + \int_{C_R} = 2i\pi\, r(k^*) = \frac{2i\pi}{4}\left.\frac{e^{iu(x+\xi)}}{2u}\right|_{u=k^*} = \frac{i\pi}{4}\frac{e^{ik^*(x+\xi)}}{k^*}$$

The integral on C_R vanishes as the radius $R \to \infty$. Similarly, the second integral can be evaluated by closing the real axis with a semi-circular contour of radius R in the lower-half plane.

$$\int_{-\infty}^{\infty} + \int_{C_R} = -\frac{2i\pi}{4} r(-k^*) = -\frac{i\pi}{2}\frac{e^{-iu(x+\xi)}}{2u}\Bigg|_{u=k^*} = \frac{i\pi}{4}\frac{e^{ik^*(x+\xi)}}{k^*}$$

The sum of the two integrals then becomes:

$$\frac{i\pi}{2}\frac{e^{ik^*(x+\xi)}}{k^*}$$

The second integral can be evaluated by similar methods:

$$\int_0^{\infty}\frac{\cos u(x-\xi)}{u^2-k^{*2}}\,du = \frac{1}{2}\int_{-\infty}^{\infty}\frac{\cos u(x-\xi)}{u^2-k^{*2}} = \frac{1}{4}\int_{-\infty}^{\infty}\frac{e^{iu(x-\xi)}+e^{-iu(x-\xi)}}{u^2-k^{*2}}\,du$$

Again, these integrals will be evaluated by closing them in the complex plane. However, since the sign of $x - \xi$ could change depending on $x > \xi$ or $x < \xi$, it also would change whether one closes the contours in the upper or lower half-planes of the complex u-plane.

<u>For $x > \xi$</u>

Since $x - \xi > 0$, then one can use the results of the first integral, giving the integral as:

$$i\frac{\pi}{2}\frac{e^{ik^*(x-\xi)}}{k^*}$$

<u>For $x < \xi$</u>

Since $\xi - x > 0$, then rewrite the integral as:

$$\int_{-\infty}^{\infty}\frac{e^{iu(x-\xi)}}{u^2-k^{*2}}\,du = \int_{-\infty}^{\infty}\frac{e^{-iu(\xi-x)}}{u^2-k^{*2}}\,du$$

resulting in the integral as:

$$= \frac{i\pi}{2}\frac{e^{ik^*(\xi-x)}}{k^*}$$

Finally, assembling the two integrals, one obtains the Green's function for $x > \xi$ or $x < \xi$. Letting $\eta \to 0$, $k^* \to k$, results in the final solution for $g(x|\xi)$:

$$g(x|\xi) = \begin{cases} \dfrac{i}{2k}\left[e^{ik(\xi+x)} - e^{ik(\xi-x)}\right] = +\dfrac{e^{ik\xi}}{k}\sin(kx) & x < \xi \\ \dfrac{i}{2k}\left[e^{ik(\xi+x)} - e^{ik(x-\xi)}\right] = +\dfrac{e^{ikx}}{k}\sin(k\xi) & x > \xi \end{cases}$$

Note that the Green's function $g(x|\xi)\, e^{-i\omega t}$ represents only outgoing waves in the farfield, $x > \xi$ but has a standing wave for $0 < x < \xi$. The response of the bar to a distributed load is the integral of the Green's function convolved with the source term, i.e.:

$$y(x) = \int_0^\infty g(x|\xi) \frac{f(\xi)}{AE} d\xi$$

$$= +\frac{1}{AEk}\left\{ e^{ikx} \int_0^x f(\xi) \sin(k\xi)\, d\xi + \sin(kx) \int_x^\infty e^{ik\xi} f(\xi) d\xi \right\}$$

8.9 Green's Function for Infinite One-Dimensional Media

For infinite media, one must apply Fourier Complex transform. In this case, the dependent variable and all its derivatives up to (2n-1) must decay at some rate. Furthermore, the source distribution must be absolutely integrable.

Example 8.8 Green's Function for the Vibration in an Infinite String

Obtain the displacement field of an infinite vibrating stretched string undergoing forced vibration due to a distributed time-harmonic load $f(x)\, e^{-i\omega t}$. Since the string is infinite in extent, there are no boundary conditions to satisfy. For this problem, Fourier Complex transform is an ideal transform. The equation of motion of the string (Section 4.2) is given as:

$$-\frac{d^2y}{dx^2} - k^2 y = \frac{f(x)}{T_0} \qquad -\infty < x < \infty$$

The Green's function satisfies the differential equation:

$$-\frac{d^2g}{dx^2} - k^2 g = \delta(x - \xi)$$

Applying the Fourier Complex transform (see Section 7.17) to the differential equation on the independent variable x, one obtains:

$$u^2 g^* - k^2 g^* = e^{-iu\xi}$$

where:

$$g^*(u) = \int_{-\infty}^{\infty} e^{-iux} g(x|\xi)\, dx$$

solving for g^*, one gets:

$$g^*(u) = \frac{e^{-iu\xi}}{u^2 - k^2}$$

The Green's function is evaluated from the inverse transform of g*(u):

$$g(x|\xi) = \frac{1}{2\pi} \int_{-\infty}^{\infty} g^*(u) e^{+iux}\, du = \frac{1}{2\pi} \int_{-\infty}^{\infty} \frac{e^{iu(x-\xi)}}{u^2 - k^2}\, du$$

Depending on whether $x > \xi$ or $x < \xi$, one may close the path on the real axis with a circular contour in the upper/lower half planes of the complex u-plane. In order to avoid the creation of reflected waves from the farfield region $x \to \pm\infty$, one must again add a limiting absorption to the material constants, as was done in Example 8.7. Thus, the Green's function written for $k = k^*$ becomes:

$$g(x|\xi) = \frac{1}{2\pi} \int_{-\infty}^{\infty} \frac{e^{iu(x-\xi)}}{u^2 - k^{*2}}\, du$$

For $x < \xi$, closure is made in the lower-half plane, resulting in Green's function, after making $k^* \to k$, as:

$$g(x|\xi) = \frac{i}{2k} e^{ik(\xi - x)} \qquad x < \xi$$

For $x > \xi$, the closure is performed in the upper-half plane, resulting in a Green's function of:

$$g(x|\xi) = \frac{i}{2k} e^{ik(x-\xi)} \qquad x > \xi$$

The Green's function for the different regions can be written in one compact form as:

$$g(x|\xi) = \frac{i}{2k} e^{ik|x-\xi|}$$

Note that the Green's function represents outgoing waves in the farfield $x \to \infty$ or $-\infty$. The displacement field due to a source distribution f(x) as:

$$y(x) = \frac{1}{2\pi} \int_{-\infty}^{\infty} g(x|\xi) \frac{f(\xi)}{T_0}\, d\xi = \frac{i}{2\pi k T_0} \left\{ e^{ikx} \int_{-\infty}^{x} e^{-ik\xi} f(\xi)\, d\xi + e^{-ikx} \int_{x}^{\infty} e^{ik\xi} f(\xi)\, d\xi \right\}$$

8.10 Green's Function for Partial Differential Equations

The use of Green's function for partial differential equations parallels the treatment given to ordinary differential equations. There are, however, some differences that need to be clarified. The first is that the definition of a self-adjoint operator. Let the linear partial differential operator **L** be defined as (see section D.7):

$$\mathbf{L}\phi(\mathbf{x}) = \sum_{|\mathbf{k}| \le n} \mathbf{a_k}(\mathbf{x})\, \partial^k\, \phi(\mathbf{x}) \qquad (8.27)$$

where **x** is an mth dimensional independent variable and n is the highest-order partial derivative of the operator **L**. The adjoint operator **K** then is defined as:

$$\mathbf{K}\,\phi(\mathbf{x}) = \sum_{|\mathbf{k}| \le n} (-1)^{|\mathbf{k}|} \partial^k \left[a_k(\mathbf{x})\, \phi(\mathbf{x})\right] \qquad (8.28)$$

If **K** = **L**, then **L** is self-adjoint.

Example 8.9

The Laplacian operator ∇^2 in cartesian coordinates in three dimensions:

$$\nabla^2\psi = \frac{\partial^2\psi}{\partial^2 x} + \frac{\partial^2\psi}{\partial y^2} + \frac{\partial^2\psi}{\partial z^2}$$

is self-adjoint, since a_k are constants. The Laplacian operator in cylindrical coordinates (r,θ,z) written as:

$$\mathbf{L}\,\psi = \nabla^2\psi = \frac{\partial^2\psi}{\partial r^2} + \frac{1}{r}\frac{\partial\psi}{\partial r} + \frac{\partial^2\psi}{\partial z^2} + \frac{1}{r^2}\frac{\partial^2\psi}{\partial\theta^2}$$

is not self-adjoint, since the adjoint operator **K** is given by:

$$\mathbf{K}\,\psi = \frac{\partial^2\psi}{\partial r^2} - \frac{1}{r}\frac{\partial\psi}{\partial r} + \frac{1}{r^2}\psi + \frac{\partial^2\psi}{\partial z^2} + \frac{1}{r^2}\frac{\partial^2\psi}{\partial\theta^2}$$

and is not equal to **L**.

However, if one modifies **L** such that:

$$\overline{\mathbf{L}}\psi = r\,\mathbf{L}\,\psi = r\frac{\partial^2\psi}{\partial r^2} + \frac{\partial\psi}{\partial r} + r\frac{\partial^2\psi}{\partial z^2} + \frac{1}{r}\frac{\partial^2\psi}{\partial\theta^2}$$

then $\overline{\mathbf{K}} = \overline{\mathbf{L}}$, i.e., the operator $\overline{\mathbf{L}}$ is self-adjoint.

For the general partial differential equation, one can show that:

$$v(\mathbf{x})\,\mathbf{L}u(\mathbf{x}) - u(\mathbf{x})\,\mathbf{K}\,v(\mathbf{x}) = \nabla\cdot\vec{\mathbf{P}}(u,v) \qquad (8.29)$$

where ∇ is the gradient in n-dimensional space defined by:

$$\nabla = \vec{e}_1\frac{\partial}{\partial x_1} + \vec{e}_2\frac{\partial}{\partial x_2} + \ldots + \vec{e}_n\frac{\partial}{\partial x_n} \qquad (8.30)$$

P is a bi-linear form of u and v, and $\vec{e}_j$ are the unit base vectors. Integrating the two sides of eq. (8.29) over the volume:

$$\int_R (v\mathbf{L}u - u\mathbf{K}v)\,\mathbf{dx} = \int_R \nabla\cdot\vec{P}\,\mathbf{dx} = \int_S \vec{n}\cdot\vec{P}\,\mathbf{dS} \qquad (8.31)$$

where $\vec{n}$ is the outward normal to the surface enclosing the region R. The last integral transformation is the divergence theorem stated as:

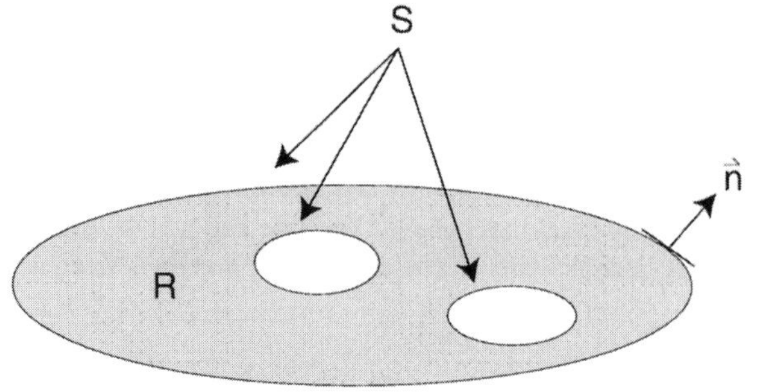

Fig. 8.1 *Illustration for the Divergence Theorem*

$$\int_R \nabla \cdot \vec{\mathbf{B}}\,\mathbf{dx} = \int_S \vec{\mathbf{n}} \cdot \vec{\mathbf{B}}\,\mathbf{dS} \qquad (8.32)$$

where $\vec{\mathbf{B}}$ is a vector function and $\nabla\cdot$ is the divergence of a vector. In this integral **dx** is a volume element in the region R (shaded region), $\vec{n}$ is a unit outward normal vector, defined positive away from the region R, and **S** is the sum of all the surfaces enclosing the region R, see Figure 8.1.

Example 8.10

For the Laplacian in cylindrical coordinates in three dimensional space given in Example 8.9:

$$\mathrm{v}\mathbf{L}\,\mathrm{u} - \mathrm{u}\mathbf{K}\,\mathrm{v} = \nabla\cdot\left\{\left[\frac{\mathrm{uv}}{\mathrm{r}} - \mathrm{u}\frac{\partial \mathrm{v}}{\partial \mathrm{r}} + \mathrm{v}\frac{\partial \mathrm{u}}{\partial \mathrm{r}}\right]\vec{\mathrm{e}}_r + \left[\frac{\mathrm{v}}{\mathrm{r}}\frac{\partial \mathrm{u}}{\partial \theta} - \frac{\mathrm{u}}{\mathrm{r}}\frac{\partial \mathrm{v}}{\partial \theta}\right]\vec{\mathrm{e}}_\theta + \left[\mathrm{v}\frac{\partial \mathrm{u}}{\partial \mathrm{z}} - \mathrm{u}\frac{\partial \mathrm{v}}{\partial \mathrm{z}}\right]\vec{\mathrm{e}}_z\right\}$$

$$= \nabla\cdot\vec{\mathbf{P}}(\mathrm{u},\mathrm{v})$$

8.11 Green's Identities for the Laplacian Operator

In this section, the derivation of the transformation given for the integrals in eq. (8.31) are performed for the Laplacian operator. Since the Laplacian operator is self-adjoint in cartesian coordinates, then $\mathbf{L} = -\nabla^2 = \mathbf{K}$.

If one lets $\vec{\mathbf{B}} = \mathrm{v}\nabla\mathrm{u}$, in eq. (8.32) where v and u are scalar functions and ∇ is the gradient, then:

$$\nabla\cdot\vec{\mathbf{B}} = \mathrm{v}\nabla^2\mathrm{u} + (\nabla\mathrm{u})\cdot(\nabla\mathrm{v}) \qquad (8.33)$$

Similarly, if one lets $\vec{\mathbf{B}} = \mathrm{u}\nabla\mathrm{v}$, in eq. (8.32) then one gets:

$$\nabla\cdot\vec{\mathbf{B}} = \mathrm{u}\nabla^2\mathrm{v} + (\nabla\mathrm{u})\cdot(\nabla\mathrm{v}) \qquad (8.34)$$

subtraction of the two identities eqs. (8.33) and (8.34) results in a new identity:

$$\mathrm{u}\nabla^2\mathrm{v} - \mathrm{v}\nabla^2\mathrm{u} = \nabla\cdot(\mathrm{u}\nabla\mathrm{v} - \mathrm{v}\nabla\mathrm{u}) = \nabla\cdot\vec{\mathbf{P}} \qquad (8.35)$$

where:

$$\vec{\mathbf{P}} = \mathrm{u}\nabla\mathrm{v} - \mathrm{v}\nabla\mathrm{u} \qquad (8.36)$$

is a bi-linear function of u and v. Integrating eq. (8.35) over the volume R:

$$\int_R \left(v\nabla^2 u - u\nabla^2 v\right) \mathbf{dx} = \int_R \nabla\cdot\vec{\mathbf{P}}\,\mathbf{dx} = \int_S \vec{n}\cdot\vec{\mathbf{P}}\,\mathbf{dS} \tag{8.37}$$

where the last integral resulted from the use of the divergence theorem.

The last integral can be simplified to:

$$\int_S \vec{n}\cdot(v\nabla u - u\nabla v)\,\mathbf{dS} = \int_S \left(v\frac{\partial u}{\partial n} - u\frac{\partial v}{\partial n}\right)\mathbf{dS}$$

resulting in the identity:

$$\int_R \left(v\nabla^2 u - u\nabla^2 v\right)\mathbf{dx} = \int_S \left(v\frac{\partial u}{\partial n} - u\frac{\partial v}{\partial n}\right)\mathbf{dS} \tag{8.38}$$

The terms in the integral over the surface S represent boundary conditions.

8.12 Green's Identity for the Helmholtz Operator

The Helmholtz equation has an operator given as:

$$\mathbf{L} = -\nabla^2 - \lambda$$

so that it is also self-adjoint, since the Laplacian is a self-adjoint operator. Substituting for ∇^2 in eq. (8.35) by **L** above then:

$$v(-\nabla^2 - \lambda)u - u(-\nabla^2 - \lambda)v = u\nabla^2 v - v\nabla^2 u = \nabla\cdot\bar{P}$$

The Green's identity for this operator becomes:

$$\int_R (u\mathbf{L}v - v\mathbf{L}u)\,\mathbf{dx} = \int_S \left(u\frac{\partial v}{\partial n} - v\frac{\partial u}{\partial n}\right)\mathbf{dS} \tag{8.39}$$

The terms in the integral over the surface S represent boundary conditions.

8.13 Green's Identity for Bi-Laplacian Operator

The bi-Laplacian operator ∇^4 is defined as:

$$\mathbf{L} = -\nabla^4 = -\nabla^2\nabla^2$$

which is a self-adjoint operator and shows up in the theory of elastic plates. In order to use the results for the Green's identity for the Laplacian, let $\nabla^2 u = U$ and $\nabla^2 v = V$, then:

$$\begin{aligned}
u\mathbf{L}v - v\mathbf{L}u &= v\nabla^4 u - u\nabla^4 v = v\nabla^2 U - u\nabla^2 V \\
&= \left[\nabla\cdot(v\nabla U) - \nabla v\cdot\nabla U\right] - \left[\nabla\cdot(u\nabla V) - \nabla u\cdot\nabla V\right] \\
&= \nabla\cdot\left[v\nabla\left(\nabla^2 u\right) - u\nabla\left(\nabla^2 v\right)\right] + \left[\nabla u\cdot\nabla\left(\nabla^2 v\right) - \nabla v\cdot\nabla\left(\nabla^2 u\right)\right]
\end{aligned}$$

Rewriting the terms in the second bracketed quantities:

$$\nabla u\cdot\nabla\left(\nabla^2 v\right) = \nabla\cdot\left[(\nabla u)\left(\nabla^2 v\right)\right] - \left(\nabla^2 u\right)\left(\nabla^2 v\right)$$

$$\nabla v \cdot \nabla\left(\nabla^2 u\right) = \nabla \cdot \left[(\nabla v)\left(\nabla^2 u\right)\right] - \left(\nabla^2 v\right)\left(\nabla^2 u\right)$$

then the Green's identity for the bi-Laplacian can be written as:

$$v\nabla^4 u - u\nabla^4 v = \nabla \cdot \left\{v\nabla(\nabla^2 u) - u\nabla(\nabla^2 v) + \nabla u(\nabla^2 v) - \nabla v(\nabla^2 u)\right\} = \nabla \cdot \vec{\mathbf{P}} \tag{8.40}$$

Integrating eq. (8.40) over the volume R:

$$\int_R \left(v\nabla^4 u - u\nabla^4 v\right)\mathbf{dx} = \int_R \nabla \cdot \vec{\mathbf{P}}\,\mathbf{dx} = \int_S \vec{n} \cdot \vec{\mathbf{P}}\,\mathbf{dS}$$

$$= \int_S \left[v\frac{\partial(\nabla^2 u)}{\partial n} - u\frac{\partial(\nabla^2 v)}{\partial n} + \nabla^2 v\frac{\partial u}{\partial n} - \nabla^2 u\frac{\partial v}{\partial n}\right]\mathbf{dS} \tag{8.41}$$

The terms in the integral over the surface S represent boundary conditions.

Similarly, if one has a bi-Laplacian Helmholtz type operator, i.e., if $L = -\nabla^4 + \lambda$, then:

$$\int_R (uLv - vLu)\,\mathbf{dx} = \int_R \nabla \cdot \vec{\mathbf{P}}\,\mathbf{dx} = \int_S \vec{n} \cdot \vec{\mathbf{P}}\,\mathbf{dS}$$

$$= \int_S \left[v\frac{\partial(\nabla^2 u)}{\partial n} - u\frac{\partial(\nabla^2 v)}{\partial n} + \nabla^2 v\frac{\partial u}{\partial n} - \nabla^2 u\frac{\partial v}{\partial n}\right]\mathbf{dS} \tag{8.42}$$

8.14 Green's Identity for the Diffusion Operator

For the diffusion equation, the operator and its adjoint are defined as:

$$\mathbf{L} = \frac{\partial}{\partial t} - \kappa\nabla^2 \qquad \text{and} \qquad \mathbf{K} = -\frac{\partial}{\partial t} - \kappa\nabla^2 \tag{8.43}$$

Thus, these operators give the following identity:

$$v\mathbf{L}u - u\mathbf{K}v = \left(v\frac{\partial u}{\partial t} + u\frac{\partial v}{\partial t}\right) - \kappa\left(v\nabla^2 u - u\nabla^2 v\right)$$

$$= \frac{\partial}{\partial t}uv - \kappa\nabla \cdot (v\nabla u - u\nabla v) \tag{8.44}$$

Here we are dealing in four-dimensional space, i.e., (x,y,z, and t). In this space one defines a new gradient $\vec{\vec{\nabla}}$ as:

$$\vec{\vec{\nabla}} = \frac{\partial}{\partial t}\vec{e}_t + \vec{\nabla} \tag{8.45}$$

where $\vec{\nabla}$ is the spatial gradient defined in eq. (8.30) and $\vec{e}_t$ is a unit temporal base vector in time, orthogonal to the spatial unit base vectors. Using the new gradient, one can rewrite eq. (8.44) as:

$$v\mathbf{L}u - u\mathbf{K}v = \vec{\vec{\nabla}} \cdot \vec{\mathbf{P}}(u, v) \tag{8.46}$$

where $\vec{\mathbf{P}}(u, v)$ is defined as:

$$\vec{\mathbf{P}} = uv\,\vec{e}_t - \kappa\left(v\vec{\nabla}u - u\vec{\nabla}v\right) \tag{8.47}$$

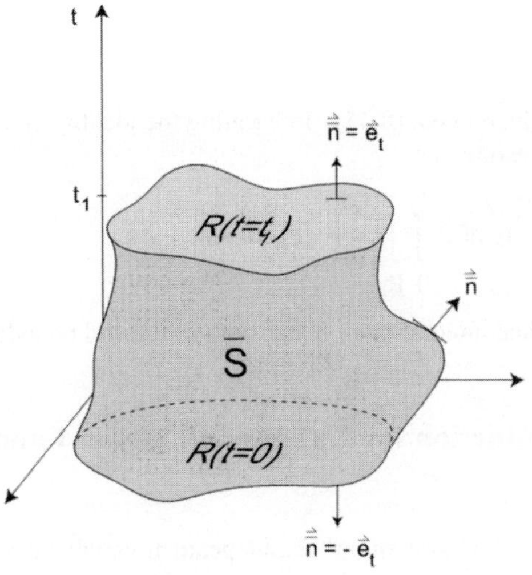

Figure 8.2 *Integration in the Space-Time Region*

Integrating the identity in eq. (8.46), one obtains:

$$\int_0^t \int_R (v\mathbf{L}u - u\mathbf{K}v)\,dx\,dt = \int_{\bar{S}} \bar{\vec{n}} \cdot \vec{\mathbf{P}}\,d\bar{S} \tag{8.48}$$

where $\bar{\vec{n}}$ is a unit normal to surface $\bar{S}$ enclosing the region (R,t) in (x,y,z,t) space, see Figure 8.2. The unit normal $\bar{\vec{n}}$ to the surface $\bar{S}$ can be resolved into temporal and unit spatial base vectors as:

$$\bar{\vec{n}} = n_t \vec{e}_t + \vec{n}$$

with $\vec{n}$ being the unit normal to the surface S(x,y,z,t) enclosing the region R(x,y,z,t). Note that $\bar{\vec{n}}$ on the surface $\bar{S}$(x,y,z,t=0) is $-\vec{e}_t$ and that on $\bar{S}$ (x,y,z,t) is $+\vec{e}_t$. The terms in the surface integral $\bar{S}$ represent boundary and initial conditions.

8.15 Green's Identity for the Wave Operator

The scalar wave equation operator can be defined as:

$$\mathbf{L} = \frac{\partial^2}{\partial t^2} - c^2\nabla^2 \tag{8.49}$$

which is a self-adjoint operator, so that the identity can be developed by:

$$v\mathbf{L}u - u\mathbf{L}v = \left[v\frac{\partial^2 u}{\partial t^2} - u\frac{\partial^2 v}{\partial t^2}\right] - c^2\left(v\nabla^2 u - u\nabla^2 v\right)$$

$$= \frac{\partial}{\partial t}\left[v\frac{\partial u}{\partial t} - u\frac{\partial v}{\partial t}\right] - c^2\nabla\cdot(v\nabla u - u\nabla v) = \bar{\vec{\nabla}}\cdot\vec{\mathbf{P}}(u,v) \tag{8.50}$$

with $\vec{\mathbf{P}}(u,v)$ defined as:

$$\vec{\mathbf{P}}(u,v) = \vec{e}_t\left[v\frac{\partial u}{\partial t} - u\frac{\partial v}{\partial t}\right] - c^2\left[v\nabla u - u\nabla v\right] \tag{8.51}$$

and the gradient $\vec{\vec{\nabla}}$ given in eq. (8.45). Integrating the identity eq. (8.50) over the spatial region R and time, one obtains:

$$\int_0^t \int_R (v\mathbf{L}u - u\mathbf{L}v)\,d\mathbf{x}\,dt = \int_0^t \int_R \vec{\vec{\nabla}} \cdot \vec{\mathbf{P}}\,d\mathbf{x}\,dt = \int_{\bar{S}} \vec{\bar{n}} \cdot \vec{\mathbf{P}}\,d\bar{\mathbf{S}} \tag{8.52}$$

The terms in the surface integral over $\bar{S}$ has both initial and boundary conditions.

8.16 Green's Function for Unbounded Media-Fundamental Solution

Consider the following system on the independent variable u(**x**):

$$\mathbf{L}u(\mathbf{x}) = f(\mathbf{x}) \qquad \mathbf{x} \text{ in } R_n \tag{8.53}$$

where **L** is an operator in n-dimensional space and f(**x**) is the source term that is absolutely integrable over the unbounded region R_n. Define the Green's function $g(\mathbf{x}|\xi)$ for the unbounded region, known as the **Fundamental solution**, and the Green's function $g^*(\mathbf{x}|\xi)$ for the adjoint operator **K** to satisfy:

$$\mathbf{L}\,g(\mathbf{x}|\xi) = \delta(\mathbf{x} - \xi) \tag{8.54}$$

$$\mathbf{K}\,g^*(\mathbf{x}|\xi) = \delta(\mathbf{x} - \xi) \tag{8.55}$$

where $g(\mathbf{x}|\xi)$ and $g^*(\mathbf{x}|\xi)$ must decay in the farfield at a prescribed manner. It should be noted that $g^*(\mathbf{x}|\xi) = g(\xi|\mathbf{x})$. Multiplying eq. (8.53) by $g^*(\mathbf{x}|\xi)$ and eq. (8.55) by u(**x**) and integrating over the unbounded region, one obtains:

$$\int_{R_n} \left(u\mathbf{K}g^* - g^*\mathbf{L}u\right)d\mathbf{x} = \int_{R_n} \left[u(\mathbf{x})\,\delta(\mathbf{x} - \xi) - g^*(\mathbf{x}|\xi)\,f(\mathbf{x})\right]d\mathbf{x}$$

$$= u(\xi) - \int_{R_n} g^*(\mathbf{x}|\xi)\,f(\mathbf{x})\,d\mathbf{x}$$

The integral on the left-hand side can be written as a surface integral, see eq. (8.31), over the surface S_n. The surface S_n of an unbounded medium could be taken as a large spherical surface with a radius $R \to \infty$. The integrand then must decay with R at a rate that would make the integral vanish. The condition on $g(\mathbf{x}|\xi)$ would also require that it decays at a prescribed rate as $R \to \infty$. Thus, if the left-hand side of eq. (8.31) vanishes, then:

$$u(\xi) = \int_{R_n} g^*(\mathbf{x}|\xi)\,f(\mathbf{x})\,d\mathbf{x} = \int_{R_n} g(\xi|\mathbf{x})\,f(\mathbf{x})\,d\mathbf{x}$$

Changing **x** by ξ and vice versa, one can write the solution for u(**x**) as:

$$u(\mathbf{x}) = \int_{R_n} g(\mathbf{x}|\xi)\,f(\xi)\,d\xi \tag{8.56}$$

If the operator **L** is self-adjoint, then the Green's function is symmetric, i.e., $g(\mathbf{x}|\xi) = g(\xi|\mathbf{x})$. Furthermore, if the operator **L** is one with constant coefficients, then:

$$g(\mathbf{x}|\xi) = g(\mathbf{x} - \xi) \tag{8.57}$$

8.17 Fundamental Solution for the Laplacian

Consider the Poisson equation in cartesian coordinates:

$$-\nabla^2 u = f(\mathbf{x}) \qquad \mathbf{x} \text{ in } R_n \tag{8.58}$$

where the Laplacian is a self-adjoint operator with constant coefficients. The solution $u(\mathbf{x})$ can be obtained as an integral over the Green's function and the source $f(\mathbf{x})$ given in eq. (8.56).

8.17.1 Three-Dimensional Space

Define the Fundamental solution $g(\mathbf{x}|\xi)$ to satisfy:

$$-\nabla^2 g(\mathbf{x}|\xi) = \delta(\mathbf{x} - \xi) = \delta(x_1 - \xi_1)\,\delta(x_2 - \xi_2)\,\delta(x_3 - \xi_3)$$

Since the Laplacian has constant coefficients, then one solves for the Green's function with $\xi = 0$, i.e., the point source is transferred to the origin:

$$-\nabla^2 g = \delta(x_1)\,\delta(x_2)\,\delta(x_3) \tag{8.59}$$

Since the source is at the origin, one can transform the cartesian coordinates to spherical coordinates for a spherically symmetric source, with the point source defined in spherical coordinates as:

$$-\nabla^2 g(r) = \frac{\delta(r)}{4\pi r^2} \tag{8.60}$$

To ascertain the rate of decay of $g(r)$ with r, integrate eq. (8.59) over R_n, resulting

$$\int_{R_n} \nabla^2 g\,\mathbf{dx} = \int_{S_n} \left.\frac{\partial g}{\partial n}\right|_{\text{on } S_n} \mathbf{dS} = -1 \tag{8.61}$$

Since g depends on r only, then $\frac{\partial g}{\partial n}$ on S, the spherical surface whose radius is R, becomes $\frac{dg}{dr}(R)$ which is a constant in the farfield R since $g = g(r)$ only. The last integral in eq. (8.61) then becomes:

$$\frac{dg}{dR} \cdot 4\pi R^2 = -1$$

or

$$\frac{dg}{dR} \cong -\frac{1}{4\pi R^2} \qquad \text{for } R >> 1$$

or

$$g(R) \cong \frac{1}{4\pi R} \qquad \text{for } R >> 1 \tag{8.62}$$

This means that the Green's function for an unbounded three-dimensional region must decay as 1/R. Returning to eq. (8.60), one can integrate the differential equation directly by writing the Laplacian in spherical coordinates in r only, i.e.:

$$-\frac{1}{r^2}\frac{d}{dr}\left(r^2\frac{dg}{dr}\right) = \frac{\delta(r)}{4\pi r^2} \tag{8.63}$$

Direct integration results in:

$$g(r) = \frac{1}{4\pi r} + C$$

since g(R) for R >> 1 must decay as 1/R, then $C \equiv 0$, giving:

$$g(r) = \frac{1}{4\pi r} = \frac{1}{4\pi\left[x_1^2 + x_2^2 + x_3^2\right]^{1/2}}$$

$$g(\mathbf{x} - \boldsymbol{\xi}) = \frac{1}{4\pi\left[(x_1 - \xi_1)^2 + (x_2 - \xi_2)^2 + (x_3 - \xi_3)^2\right]^{1/2}} = \frac{1}{4\pi|\mathbf{x} - \boldsymbol{\xi}|} \tag{8.64}$$

8.17.2 Two-Dimensional Space

In two dimensional space, the Green's function satisfies:

$$-\nabla^2 g(\mathbf{x}|\boldsymbol{\xi}) = \delta(\mathbf{x} - \boldsymbol{\xi}) = \delta(x_1 - \xi_1)\delta(x_2 - \xi_2) \tag{8.65}$$

As the two-dimensional Laplacian has constant coefficients, one can shift the source to the origin:

$$-\nabla^2 g = \delta(x_1)\delta(x_2) \tag{8.66}$$

Since the source is at the origin, then one can transform eq. (8.66) to cylindrical coordinate in two-dimensional space and the Green's function becomes g(r):

$$-\nabla^2 g = -\frac{1}{r}\frac{d}{dr}\left(r\frac{dg}{dr}\right) = \frac{\delta(r)}{2\pi r} \tag{8.67}$$

Again, to define the behavior of g(r) as $r \to \infty$, one can integrate eq. (8.66) over the unbounded region R_n:

$$\int_{R_n} \nabla^2 g \, d\mathbf{x} = \int_{S_n} \frac{\partial g}{\partial n}\bigg|_{\text{on } S_n} d\mathbf{S} \cong \frac{dg}{dr}(R) \cdot 2\pi R = -1$$

so that:

$$\frac{dg}{dR} \cong -\frac{1}{2\pi R}$$

or:

$$g(R) \cong -\frac{1}{2\pi}\log R \qquad \text{for } R >> 1 \tag{8.68}$$

Integrating eq. (8.67) directly, one obtains:

$$g(r) = -\frac{1}{2\pi}\log r + C$$

Again C must be neglected in order that g(r) behaves as in eq. (8.68), giving:

$$g(r) = -\frac{1}{2\pi}\log r = -\frac{1}{2\pi}\log\left[x_1^2 + x_2^2\right]^{1/2}$$

$$g(\mathbf{x}-\boldsymbol{\xi}) = -\frac{1}{2\pi}\log\left[(x_1-\xi_1)^2 + (x_2-\xi_2)^2\right]^{1/2} = -\frac{1}{2\pi}\log|\mathbf{x}-\boldsymbol{\xi}| \tag{8.69}$$

8.17.3 One-Dimensional Space

For the one-dimensional case, g satisfies:

$$-\frac{d^2g}{dx^2} = \delta(x-\xi)$$

Direct integration yields the fundamental solution of:

$$g(x|\xi) = -\frac{1}{2}|x-\xi| \tag{8.70}$$

8.17.4 Development by Construction

One can derive the Green's function by construction, which is yet another method for development of the Green's function. First, enclose the source region at the origin by an infinitesimal sphere R_ε of radius ε. Starting with the definition of Green's function:

$$-\nabla^2 g = \delta(\mathbf{x}) \qquad \mathbf{x} \text{ in } R_\varepsilon$$

then, since the source region is confined to the origin:

$$\nabla^2 g = 0 \qquad \text{outside } R_\varepsilon$$

Rewriting the Laplacian in terms of spherical coordinates, then:

$$\nabla^2 g = \frac{1}{r^2}\frac{d}{dr}\left[r^2\frac{dg}{dr}\right] = 0 \quad \text{outside } R_\varepsilon$$

By directly integrating the differential equation above, one obtains:

$$g = \frac{C_1}{r} + C_2$$

Since g decays as $r \to \infty$, then $C_2 = 0$. Integrating the equation over the infinitesimal sphere R_ε:

$$-\int_{R_\varepsilon} \nabla^2 g \, \mathbf{dx} = \int_{R_\varepsilon} \delta(\mathbf{x}) \, \mathbf{dx} = 1$$

$$= -\int_{R_\varepsilon} \nabla\cdot\nabla g \, \mathbf{dx} = -\int_{S_\varepsilon} \left.\frac{\partial g}{\partial n}\right|_{\text{on}S_\varepsilon} \mathbf{dS}$$

$$= -\int_{S_\varepsilon} \left.\frac{\partial g}{\partial r}\right|_{r=\varepsilon} \mathbf{dS} = \frac{C_1}{\varepsilon^2}\int_{S_\varepsilon} \mathbf{dS} = 4\pi C_1$$

where the normal derivative $\frac{\partial}{\partial n} = \frac{\partial}{\partial r}$. The constant becomes $C_1 = \frac{1}{4\pi}$ and the Green's function $g = \frac{1}{4\pi r}$.

8.17.5 Behavior for Large R

The behavior of u(**x**) as $R \to \infty$ can be postulated from eq. (8.55). The integration over the surface of an infinitely large sphere of radius R must vanish. Thus:

$$\int_{R_n} (uLg - gLu)\,\mathbf{dx} = \int_{S_n(R\to\infty)} \left(u\frac{\partial g}{\partial n} - g\frac{\partial u}{\partial n}\right) \mathbf{dS} \to 0$$

where $\vec{n}$ is the unit outward normal and $\frac{\partial}{\partial n} = \frac{\partial}{\partial R}$. For three-dimensional space, $g \cong \frac{1}{R}$, and $\frac{\partial g}{\partial n} \cong -\frac{1}{R^2}$ so that the surface integral above becomes:

$$\lim_{R\to\infty} \left[\frac{u(R)}{R^2} + \frac{\partial u(R)}{\partial R} \cdot \frac{1}{R}\right] 4\pi R^2 \to 0$$

This requires that the function u and its derivative behave as:

$$u(R) + R\frac{\partial u(R)}{\partial R} \approx \frac{1}{R^p} \qquad p > 0 \tag{8.71}$$

For two-dimensional media, $g \cong \log r$, $\frac{\partial g}{\partial n} \cong \frac{1}{r}$, so that the surface integral above becomes:

$$\lim_{R\to\infty} \left[\frac{u(R)}{R} + \frac{\partial u(R)}{\partial R} \cdot (\log R)\right] 2\pi R \to 0$$

This requires that the function u and its derivative behave as:

$$u(R) + R\log R\frac{\partial u(R)}{\partial R} \approx \frac{1}{R^p} \qquad p > 0 \tag{8.72}$$

8.18 Fundamental Solution for the Bi-Laplacian

Consider the bi-Laplacian in two-dimensional space:

$$-\nabla^4 g = \delta(\mathbf{x} - \xi) \qquad \mathbf{x} \text{ in } R_n \tag{8.73}$$

then one can again shift the source to the origin, since the bi-Laplacian has constant coefficients.

$$-\nabla^4 g = \delta(\mathbf{x}) = \delta(x_1)\delta(x_2) \tag{8.74}$$

Rewriting this equation in two-dimensional cylindrical coordinates:

$$-\nabla^4 g = -\left(\nabla^2\right)^2 g = -\left[\frac{1}{r}\frac{d}{dr}\left(r\frac{d}{dr}\right)\right]^2 g = \frac{\delta(r)}{2\pi r} \tag{8.75}$$

Direct integration of eq. (8.75) results in:

$$g = -\frac{r^2}{8\pi}[\log r - 1] + C_1 r^2 \log r + C_2 r^2 + C_3$$

Integrating eq. (8.74) over large circular area R_n with $R >> 1$, one obtains the condition that as $R \to \infty$:

$$\nabla^2 g \cong -\frac{1}{2\pi}\log R \qquad R >> 1$$

This requires that all the arbitrary constants C_1, C_2 and C_3 vanish, giving g as:

$$g = \frac{r^2}{8\pi}[1 - \log r] \tag{8.76}$$

8.19 Fundamental Solution for the Helmholtz Operator

Consider the Helmholtz equation:

$$-\nabla^2 u - \lambda u = f(\mathbf{x}) \qquad \mathbf{x} \text{ in } R_n \tag{8.77}$$

where $L = -\nabla^2 - \lambda$ is a self-adjoint operator. The Green's function satisfies the Helmholtz equation:

$$-\nabla^2 g - \lambda g = \delta(\mathbf{x} - \xi) \tag{8.78}$$

Since the operator has constant coefficients, then once again, the source could be transformed to the origin. The solution for $u(\mathbf{x})$ can be obtained as an integral over the source term $f(\mathbf{x})$ and the Green's function, as in eq. (8.56).

8.19.1 Three-Dimensional Space

To develop the Green's function by construction, enclose the source at the origin by an infinitesimal sphere R_ε, such that:

$$-\nabla^2 g - \lambda g = 0 \qquad \text{outside } R_\varepsilon$$

Replacing λ by k^2 and writing the equation in spherical coordinates one obtains a homogeneous equation:

$$\frac{1}{r^2}\frac{d}{dr}\left[r^2 \frac{dg}{dr}\right] + k^2 g = 0 \qquad \text{outside } R_\varepsilon$$

which has the solution:

$$g = C_1 \frac{e^{ikr}}{r} + C_2 \frac{e^{-ikr}}{r} \tag{8.79}$$

with $e^{-i\omega t}$ assumed for the time dependence leading to the Helmholtz equation, the two solutions represent outgoing and incoming waves, respectively. For outgoing waves, let $C_2 = 0$. Integrating eq. (8.78) with the source at the origin over R_ε:

$$-\int_{R_\varepsilon} \nabla^2 g \, d\mathbf{x} - k^2 \int_{R_\varepsilon} g \, d\mathbf{x} = 1 \tag{8.80}$$

The first integral can be transformed to a surface integral over the infinitesimal sphere:

$$\int_{R_\varepsilon} \nabla^2 g \, \mathbf{dx} = \int_{S_\varepsilon} \left.\frac{\partial g}{\partial n}\right|_{r=\varepsilon} \mathbf{dS} = 4\pi\varepsilon^2 C_1\left(\frac{ik}{\varepsilon} - \frac{1}{\varepsilon^2}\right) e^{ik\varepsilon} \tag{8.81}$$

Taking the limit of eq. (8.81) as $\varepsilon \to 0$, the integral approaches $-4\pi C_1$. The second integral in eq. (8.80) can also be shown to vanish in the limit as $\varepsilon \to 0$:

$$\left|\int_{R_\varepsilon} g \, \mathbf{dx}\right| = \left|\int_0^\varepsilon \frac{e^{ikr}}{r} \cdot 4\pi r^2 dr\right| \le 4\pi \left|\int_0^\varepsilon r \, dr\right| = 2\pi\varepsilon^2$$

which vanishes as $\varepsilon \to 0$. This results in the evaluation of $C_1 = \frac{1}{4\pi}$, so that the Green's function becomes:

$$g = \frac{e^{ikr}}{4\pi r} \tag{8.82}$$

The Green's function for a general source location ξ:

$$g(\mathbf{x}-\xi) = \frac{e^{ik|\mathbf{x}-\xi|}}{4\pi|\mathbf{x}-\xi|} \tag{8.83}$$

8.19.2 Two-Dimensional Space

Following the same procedure for the development of the Green's function in three dimensional space, the two-dimensional analog can be written as:

$$-\nabla^2 g - \lambda g = -\frac{1}{r}\frac{d}{dr}\left[r\frac{dg}{dr}\right] - k^2 g = \delta(\mathbf{x}) \tag{8.84}$$

For the solution outside a small circular area R_ε whose radius is ε, the homogeneous solution of eq. (8.84) is given by:

$$g = C_1 \, H_0^{(1)}(kr) + C_2 \, H_0^{(2)}(kr) \tag{8.85}$$

where $H_0^{(1)}$ and $H_0^{(2)}$ are the Hankel functions of the first and second kind, respectively. For outgoing waves in R_n, let $C_2 = 0$. Integrating eq. (8.84) over a small circular area R_ε, one can evaluate the first integral as:

$$\int_{R_\varepsilon} \nabla^2 g \, \mathbf{dx} = \int_{S_\varepsilon} \left.\frac{\partial g}{\partial n}\right|_{r=\varepsilon} \mathbf{dS} = C_1 \, k \, H_0^{(1)'}(k\varepsilon) \cdot 2\pi\varepsilon = -2\pi\varepsilon \, C_1 \, k \, H_1^{(1)}(k\varepsilon)$$

Taking the limit as $\varepsilon \to 0$, the integral approaches $4iC_1$. In a similar manner to the three-dimensional Green's function, the second integral can be shown to vanish as $\varepsilon \to 0$. Finally the Green's function can be written as:

$$g = \frac{i}{4} H_0^{(1)}(kr) \tag{8.86}$$

which, when the source location is transferred from the origin to ξ, gives:

$$g(\mathbf{x}-\xi) = \frac{i}{4} H_0^{(1)}(k|\mathbf{x}-\xi|) \tag{8.87}$$

8.19.3 One-Dimensional Space

The Green's function for the Helmholtz operator was worked out in Example 8.8 as:

$$g(x|\xi) = \frac{i}{2k} e^{ik|x-\xi|} \tag{8.88}$$

8.19.4 Behavior for Large R

The behavior of u(**x**) for the Helmholtz operator as r → ∞ can be postulated from eq. (8.55). The integration over the surface of an infinitely large sphere of radius R must vanish. Thus:

$$\int_{R_n} (uLg - gLu)\,\mathbf{dx} = \int_{S_n(R\to\infty)} \left(u\frac{\partial g}{\partial n} - g\frac{\partial u}{\partial n}\right)\mathbf{dS} \to 0 \tag{8.39}$$

where $\vec{n}$ is the unit outward normal and $\frac{\partial}{\partial n} = \frac{\partial}{\partial R}$. For three-dimensional space, $g \cong \frac{e^{ikR}}{R}$, and $\frac{\partial g}{\partial n} \cong \frac{e^{ikR}}{R^2}(ikR-1)$, so that the surface integral in eq. (8.70) becomes:

$$\lim_{R\to\infty}\left[ikR\frac{u(R)}{R^2} - \frac{\partial u(R)}{\partial R}\cdot\frac{1}{R}\right]e^{ikR}R^2 \to 0$$

This is known as the Sommerfeld Radiation Condition for three-dimensional space. This requires that the function u and its derivative behave as:

$$ik\,u(R) - \frac{\partial u(R)}{\partial R} \approx \frac{1}{R^p} \qquad p > 1 \tag{8.89}$$

For two-dimensional media, $g \cong H_0^{(1)}(kr)$, and $\frac{\partial g}{\partial n} \cong -kH_1^{(1)}(kr)$, so that the surface integral in eq. (8.70) becomes:

$$\lim_{R\to\infty}\left[-k\,H_1^{(1)}(kR)u(R) - H_0^{(1)}(kR)\frac{\partial u(R)}{\partial R}\right]2\pi R \to 0$$

This is known as the Sommerfeld Radiation Condition for two-dimensional space. This requires that the function u and its derivative behave as:

$$ik\,u(R) - \frac{\partial u(R)}{\partial R} \approx \frac{1}{R^p} \qquad p > 1/2 \tag{8.90}$$

8.20 Fundamental Solution for the Operator, $-\nabla^2 + \mu^2$

There is another operator that is related to the Helmholtz operator, defined as:

$$\mathbf{L}\,u(\mathbf{x}) = \left(-\nabla^2 + \mu^2\right)u(\mathbf{x}) = f(\mathbf{x}) \qquad \mathbf{x} \text{ in } R_n \tag{8.91}$$

One can see that this operator is related to the Helmholtz operator by making $\lambda = -\mu^2$ or $\mu = -ik = -i\sqrt{\lambda}$. The substitution of μ in the final results for the Green's function for the Helmholtz operator:

$$\left(-\nabla^2+\mu^2\right)g(\mathbf{x}|\xi)=\delta(\mathbf{x}-\xi) \tag{8.92}$$

results in the following Green's function.

8.20.1 Three-Dimensional Space

Substitution of μ in eq. (8.82) gives:

$$g(r)=\frac{e^{-\mu r}}{4\pi r} \tag{8.93}$$

8.20.2 Two-Dimensional Space

Substitution of μ in eq. (8.86) results in:

$$g=\frac{i}{4}H_0^{(1)}(i\mu r)=\frac{1}{2\pi}K_0(\mu r) \tag{8.94}$$

where K_0 is the modified Bessel function of the first kind.

8.20.3 One-Dimensional Space

Substitution of μ in eq. (8.88) results in the Green's function for one-dimensional media as:

$$g=\frac{e^{-\mu|x|}}{2\mu} \tag{8.95}$$

8.21 Causal Fundamental Solution for the Diffusion Operator

For the diffusion operator:

$$\frac{\partial u}{\partial t}-\kappa\nabla^2 u=f(\mathbf{x},t) \qquad \mathbf{x} \text{ in } R_n, \qquad t>0 \tag{8.96}$$

the Green's function satisfies the following system:

$$\frac{\partial g}{\partial t}-\kappa\nabla^2 g=\delta(\mathbf{x}-\xi)\delta(t-\tau) \tag{8.97}$$

where:

$$g=g(\mathbf{x},t|\xi,\tau)$$

satisfying the initial condition $g(\mathbf{x},0^+|\xi,\tau)=0$, and satisfies the causality condition:

$$g=0 \qquad \text{for } t<\tau$$

which states that the solution is null until $t=\tau$. Since the diffusion operator has constant coefficients, one can shift the ξ and τ to the origin

$$g=g(\mathbf{x},t|0,0)=g(\mathbf{x},t)$$

satisfying the diffusion operator:

$$\frac{\partial g}{\partial t} - \kappa \nabla^2 g = \delta(\mathbf{x})\,\delta(t) \tag{8.98}$$

with the causality condition is now given by:

$$g = 0 \qquad t < 0$$

In order to obtain the Fundamental solution, one can apply the Laplace transform on time. Using the definitions and operations of Laplace transform in section (7.14) and defining the Laplace transform:

$$L\,g(\mathbf{x},t) = \bar{g}(\mathbf{x},p)$$

the differential equation (8.98) transforms to:

$$-\nabla^2 \bar{g} + \frac{p}{\kappa}\bar{g} = \frac{\delta(\mathbf{x})}{\kappa} \tag{8.99}$$

Equation (8.99) resembles eq. (8.92) with solutions for three- and two-dimensional media with $\mu^2 = \frac{p}{\kappa}$.

8.21.1 Three-Dimensional Space

With $\mu = \sqrt{\frac{p}{\kappa}}$, the transform of the Green's function for three-dimensional space in eq. (8.93) gives:

$$\bar{g}(r,p) = \frac{e^{-r\sqrt{p/\kappa}}}{4\pi\kappa\, r}$$

whose Laplace inverse transform gives:

$$g(r,t) = \frac{e^{-r^2/(4\kappa t)}}{(4\pi\kappa t)^{3/2}}\, H(t) \tag{8.100}$$

Rewriting eq. (8.100) to revert to the source space and time ξ and τ gives:

$$g(\mathbf{x}-\xi, t-\tau) = \frac{e^{-|\mathbf{x}-\xi|^2/[4\kappa(t-\tau)]}}{[4\pi\kappa(t-\tau)]^{3/2}}\, H(t-\tau) \tag{8.101}$$

note that the resulting expression for g is causal.

8.21.2 Two-Dimensional Space

The transform of the Green's function in two-dimensional space given in eq. (8.92), with $\mu = \sqrt{\frac{p}{\kappa}}$ is:

$$\bar{g}(r,p) = \frac{1}{2\pi\kappa} K_0\left(r\sqrt{p/\kappa}\right)$$

has an inverse Laplace transform of:

$$g(r,t) = \frac{1}{4\pi\kappa t} e^{-r^2/(4\kappa t)}\, H(t) \tag{8.102}$$

which, upon transforming the coordinates to the source location and time results in the following expression:

$$g(\mathbf{x}-\xi,t-\tau)=\frac{1}{[4\pi\kappa(t-\tau)]}e^{-|\mathbf{x}-\xi|^2/[4\kappa(t-\tau)]}H(t-\tau) \tag{8.103}$$

8.21.3 One-Dimensional Space

with $\mu=\sqrt{\frac{p}{\kappa}}$, the transform of the Green's function for one dimensional medium can be written as (see (8.95)):

$$\bar{g}(x,p)=\frac{e^{-x\sqrt{p/\kappa}}}{2\sqrt{p\kappa}} \tag{8.104}$$

The inverse Laplace transform of (8.104) can be shown to have the form:

$$g(x,t)=\frac{1}{[4\pi\kappa t]^{1/2}}e^{-x^2/(4\kappa t)}H(t) \tag{8.105}$$

and transforming the origin to the actual location:

$$g(x-\xi,t-\tau)=\frac{1}{[4\pi\kappa(t-\tau)]^{1/2}}e^{-(x-\xi)^2/[4\kappa(t-\tau)]}H(t-\tau) \tag{8.106}$$

Defining the Green's function $g^*(x|\xi)$ for the adjoint operator **K** as:

$$\mathbf{K}\,g^*=-\frac{\partial g^*}{\partial t}-\kappa\nabla^2 g^*=\delta(\mathbf{x}-\xi)\delta(t-\tau)$$

and using the form eq. (8.56), one can write down the solution for u(**x**,t) as:

$$u(\mathbf{x},t)=\int_0^\infty\int_{R_n}g(\mathbf{x},t|\xi,\tau)f(\xi,\tau)\,d\xi\,d\tau \tag{8.107}$$

8.22 Causal Fundamental Solution for the Wave Operator

For the wave operator, the solution u(**x**) satisfies:

$$\left(\frac{\partial^2}{\partial t^2}-c^2\nabla^2\right)u(\mathbf{x},t)=f(\mathbf{x},t) \qquad \mathbf{x}\text{ in }R_n, \qquad t>0 \tag{8.108}$$

and the Green's function then satisfies:

$$\left(\frac{\partial^2}{\partial t^2}-c^2\nabla^2\right)g(\mathbf{x},t|\xi,\tau)=\delta(\mathbf{x}-\xi)\,\delta(t-\tau) \tag{8.109}$$

satisfying the homogeneous initial conditions:

$$g(\mathbf{x},0^+|\xi,\tau)=0 \qquad \text{and} \qquad \frac{\partial g}{\partial t}(\mathbf{x},0^+|\xi,\tau)=0$$

with the causality condition:

$$g = 0 \qquad \text{and} \qquad \frac{\partial g}{\partial t} = 0 \qquad t < \tau \tag{8.110}$$

since the wave operator has constant coefficients, the source location is transferred to the origin, such that eqs. (8.109) and (8.110) become:

$$\left(\frac{\partial^2}{\partial t^2} - c^2\nabla^2\right) g(\mathbf{x}, t) = \delta(\mathbf{x})\delta(t) \tag{8.111}$$

and

$$g = 0 \quad \text{and} \quad \frac{\partial g}{\partial t} = 0 \qquad t < 0$$

Applying Laplace transform on time, one obtains the equation on the transform of the Green's function as:

$$\left(-\nabla^2 + \frac{p^2}{c^2}\right)\bar{g}(\mathbf{x}, p) = \frac{\delta(\mathbf{x})}{c^2} \tag{8.112}$$

The solution of eq. (8.112) can be developed from eqs. (8.93) to (8.95) with $\mu^2 = \frac{p^2}{c^2}$.

8.22.1 Three-Dimensional Space

The solution for the transform $\bar{g}$ can be obtained from eq. (8.93) with $\mu = \frac{p}{c}$:

$$\bar{g}(r, p) = \frac{e^{-pr/c}}{4\pi c^2 r} \tag{8.113}$$

The inverse transform of $\bar{g}(r,t)$ then becomes:

$$g(r, t) = \frac{\delta\,(t - \frac{r}{c})}{4\pi c^2 r} = \frac{\delta(ct - r)}{4\pi cr} \tag{8.114}$$

Note that the Green's function is a spherical shell source at r = ct of decreasing strength with $\frac{1}{r}$. Transferring back to the location of the source:

$$g(\mathbf{x} - \xi, t - \tau) = \frac{\delta\left[t - \tau - |\mathbf{x} - \xi|/c\right]}{4\pi c^2 |\mathbf{x} - \xi|} \tag{8.115}$$

8.22.2 Two-Dimensional Space

Here the Laplace transformed solution is given by:

$$\bar{g}(r, p) = \frac{1}{2\pi c^2} K_0\left(\frac{p}{c} r\right) \tag{8.116}$$

The inverse Laplace transform of eq. (8.116) can be shown to be:

$$g(r, t) = \frac{H(ct - r)}{2\pi c\left[c^2 t^2 - r^2\right]^{1/2}} H(t) \tag{8.117}$$

Note that the Green's function has a trail that decays with ct at any fixed position r with a sharp wavefront at ct = r. The Green's function can be transferred to the location of the source to give the Green's function as:

$$g(\mathbf{x}-\xi, t-\tau) = \frac{H\left[c(t-\tau)-|\mathbf{x}-\xi|\right]}{2\pi c\left[c^2(t-\tau)^2-|\mathbf{x}-\xi|^2\right]^{1/2}} H(t-\tau) \tag{8.118}$$

8.22.3 One-Dimensional Space

For the one-dimensional medium:

$$\bar{g}(x, p) = \frac{e^{-p|x|/c}}{2pc} \tag{8.119}$$

which may be inverted by Laplace transform to give:

$$g(x,t) = \frac{H\left[t-|x|/c\right]}{2c} H(t) \tag{8.120}$$

Note that the Green's function is constant, but has two sharp wavefronts at x = ±ct. Upon transferring to the location of the source, one obtains:

$$g(x-\xi, t-\tau) = \frac{H\left[t-\tau-|x-\xi|/c\right]}{2c} H(t-\tau) \tag{8.121}$$

Since the wave operator is self-adjoint, then the solution u(**x**,t) can be written from eq. (8.56) as:

$$u(\mathbf{x},t) = \int_0^\infty \int_{R_n} g(\mathbf{x}, t|\xi, \tau) f(\xi, \tau)\, d\xi\, d\tau$$

8.23 Fundamental Solutions for the Bi-Laplacian Helmholtz Operator

The fundamental solution for the bi-Laplacian Helmholtz operator applies to the vibration of elastic plates. Since the plate is a two-dimensional medium, then the Fundamental solution satisfies the following equation:

$$\left(-\nabla^4 + k^4\right) g(\mathbf{x}|\xi) = \delta(\mathbf{x}-\xi) \qquad \mathbf{x} \text{ in } R_n \tag{8.122}$$

Since the operator has constant coefficients, then one can transform the source location to the origin and write out the operator in cylindrical coordinates in the radial distance:

$$\left(-\nabla^4 + k^4\right) g(r) = \frac{\delta(r)}{2\pi r} \tag{8.123}$$

To obtain the solution for g(r), one can apply the Hankel transform on r, see Section 7.19, where the Hankel transform of g(r) is $\bar{g}(\rho)$:

$$\left(\rho^4 - k^4\right)\bar{g}(\rho) = -\frac{1}{2\pi}$$

resulting in the solution:

$$\bar{g}(\rho) = \frac{-1}{2\pi\left(\rho^4 - k^4\right)}$$

The inverse Hankel transform of $\bar{g}(\rho)$ can be shown to be:

$$g(r) = \frac{-1}{2\pi}\int_0^\infty \frac{\rho\, J_0(r\rho)}{\rho^4 - k^4}\, d\rho \tag{8.124}$$

In order to perform the integration in the complex plane, one needs to extend the integration on ρ to (-∞). Using the identities (3.38) and (3.39), one can substitute for J_0 by $H_0^{(1)}$ and $H_0^{(2)}$ as:

$$g(r) = -\frac{1}{4\pi}\int_0^\infty \frac{\rho\left[H_0^{(1)}(r\rho) + H_0^{(2)}(r\rho)\right]}{\rho^4 - k^4}\, d\rho$$

Since $H_0^{(2)}(r\rho) = -H_0^{(1)}(-r\rho)$, the integral in eq. (8.125) can be extended to -∞ giving:

$$g(r) = -\frac{1}{4\pi}\int_{-\infty}^\infty \frac{\rho\, H_0^{(1)}(r\rho)}{\rho^4 - k^4}\, d\rho \tag{8.125}$$

Since $H_0^{(1)}(x)$ behaves as $e^{ix}/\sqrt{x}$ for x >> 1, then one can close the contour in the upper half-plane. The integrand has four simple poles, two real and two imaginary. Using the principle of limiting absorption then the four simple poles would rotate counterclockwise by an angle equal to the infinitesimal damping coefficient η, such that $k^* = k(1 + i\eta)$.

The two simple poles that fall in the upper-half plane are k^* and ik^*. The final solution for g(r) becomes the sum of two residues after letting $k^* \rightarrow k$:

$$g(r) = -\frac{i}{8k^2}\left[H_0^{(1)}(kr) - H_0^{(1)}(ikr)\right] \tag{8.126}$$

The Hankel function of an imaginary argument can be replaced by $-2iK_0(kr)/\pi$, so that the final expression for the Fundamental solution is written as:

$$g(r) = -\frac{1}{8k^2}\left[iH_0^{(1)}(kr) - \frac{2}{\pi}K_0(kr)\right] \tag{8.127}$$

8.24 Green's Function for the Laplacian Operator for Bounded Media

In this section, the Green's function is developed for bounded media for the Laplacian operator. This is accomplished through the surface integrals that were developed when the Green's identities were derived earlier. For the Laplacian operator, start with the Green's identity in eq. (8.38) and the differential equation (8.58). Let v=g(**x**|ξ) in eq. (8.38) and u(**x**) from eq. (8.58), one obtains the following:

$$\int_R \left[-g(x|\xi)f(x) + u\delta(x-\xi)\right]\mathbf{dx} = \int_S \left[g\frac{\partial u}{\partial n} - u\frac{\partial g}{\partial n}\right]_S \mathbf{dS}$$

which, upon rearrangement gives:

$$u(\xi) = \int_R g(x|\xi)f(x)\,\mathbf{dx} + \int_S \left[g(x|\xi)\frac{\partial u(\mathbf{x})}{\partial n_x} - u(x)\frac{\partial g(x|\xi)}{\partial n_x}\right]\mathbf{dS_x}$$

Since the Laplacian is a self-adjoint operator, then one can change the independent variable $\mathbf{x}$ to ξ and vice versa, giving:

$$u(\mathbf{x}) = \int_R g(x|\xi)f(\xi)\,d\xi + \int_{S_\xi} \left[g(x|\xi)\frac{\partial u(\xi)}{\partial n_\xi} - u(\xi)\frac{\partial g(x|\xi)}{\partial n_\xi}\right]_{S_\xi} dS_\xi \qquad (8.128)$$

This solution is composed of two integrals. The first is a volume integral over the volume source distribution. The second is a surface integral that requires the specification of the function u($\mathbf{x}$) and the normal derivative $\partial u(\mathbf{x})|\partial n$ at every point on the surface. Those requirements would over-specify the boundary conditions. Only one boundary condition can be prescribed at every point of the surface for a unique solution. To adjust the surface integrals so that only one boundary condition needs to be specified at each point of the surface, an auxiliary function $\bar{g}$ is defined such that:

$$-\nabla^2\bar{g}(x|\xi) = 0 \qquad \mathbf{x} \text{ in } R \qquad (8.129)$$

Substituting $v = \bar{g}(x|\xi)$ in eq. (8.38) one obtains:

$$-\int_R \bar{g}(x|\xi)f(x)\,\mathbf{dx} = \int_{S_x} \left[\bar{g}\frac{\partial u}{\partial n} - u\frac{\partial \bar{g}}{\partial n}\right]_{S_x} dS_x$$

Again switching $\mathbf{x}$ to ξ and vice versa, one obtains a new identity on the auxiliary function:

$$-\int_R \bar{g}(x|\xi)f(\xi)\,d\xi = \int_{S_\xi} \left[\bar{g}(x|\xi)\frac{\partial u(\xi)}{\partial n_\xi} - u(\xi)\frac{\partial \bar{g}(x|\xi)}{\partial n_\xi}\right]_{S_\xi} dS_\xi \qquad (8.130)$$

Defining $G(x|\xi) = g - \bar{g}$ and subtracting eq. (8.130) and eq. (8.128) results in a new identity:

$$u(\mathbf{x}) = \int_R G(x|\xi)f(\xi)\,d\xi - \int_{S_\xi} \left[u(\xi)\frac{\partial G(x|\xi)}{\partial n_\xi} - G(x|\xi)\frac{\partial u(\xi)}{\partial n_\xi}\right]_{S_\xi} dS_\xi \qquad (8.131)$$

Depending on the prescribed boundary condition, one can eliminate one of the two surface integrals.

8.24.1 Dirichlet Boundary Condition

If the function is prescribed on the boundary:

$$u(\mathbf{x}) = h(\mathbf{x}) \qquad \mathbf{x} \text{ on } S$$

then one needs to drop the second integral in eq. (8.131) by requiring that:

$$G(\mathbf{x}|\xi)\Big|_{S_\xi} = 0 \qquad \xi \text{ on } S_\xi$$

or, due to the symmetry of the Green's function:

$$G(\mathbf{x}|\xi)\Big|_{S_x} = 0 \qquad \mathbf{x} \text{ on } S_x$$

The function G must satisfy either of these conditions. Substituting this condition on G into eq. (8.131) results in the final form of the solution:

$$u(\mathbf{x}) = \int_R G(\mathbf{x}|\xi) f(\xi)\, d\xi - \int_{S_\xi} h(\xi) \frac{\partial G}{\partial n_\xi}\Big|_{S_\xi} dS_\xi \qquad (8.132)$$

8.24.2 Neumann Boundary Condition

If the normal derivative is specified on the surface, i.e.:

$$\frac{\partial u(\mathbf{x})}{\partial n} = h(\mathbf{x}) \qquad \mathbf{x} \text{ on } S$$

then one needs to eliminate the first integral of eq. (8.125) by letting:

$$\frac{\partial G(\mathbf{x}|\xi)}{\partial n_\xi}\Big|_{S_\xi} = 0 \qquad \xi \text{ on } S_\xi$$

or, due to the symmetry of the Green's function:

$$\frac{\partial G(\mathbf{x}|\xi)}{\partial n_x}\Big|_{S_x} = 0 \qquad \mathbf{x} \text{ on } S_x$$

Again, the function G must satisfy either of these two conditions. Substituting this condition into eq. (8.131) results in the final solution expressed as:

$$u(\mathbf{x}) = \int_R G(\mathbf{x}|\xi) f(\xi)\, d\xi + \int_{S_\xi} G(\mathbf{x}|\xi)\Big|_{S_\xi} h(\xi)\, dS_\xi \qquad (8.133)$$

8.24.3 Robin Boundary Condition

For impedance-type Robin boundary condition expressed as:

$$\frac{\partial u(\mathbf{x})}{\partial n} + \gamma u(\mathbf{x}) = h(\mathbf{x}) \qquad \mathbf{x} \text{ on } S \qquad (8.134)$$

substituting eq. (8.134) in eq. (8.131) and rearranging the terms

$$u(\mathbf{x}) = \int_R G(\mathbf{x}|\xi) f(\xi)\, d\xi - \int_{S_\xi} \left\{ u(\xi) \frac{\partial G(\mathbf{x}|\xi)}{\partial n_\xi} - G(\mathbf{x}|\xi)\left[h(\xi) - \gamma u(\xi)\right] \right\}_{S_\xi} dS_\xi$$

which, since the boundary condition is given by h(**x**), let:

$$\frac{\partial G}{\partial n_\xi}(\mathbf{x}|\xi)+\gamma G(\mathbf{x}|\xi)\Big|_{S_\xi}=0 \qquad \xi \text{ on } S_\xi$$

Again, due to the symmetry of the Green's function, G must also satisfy:

$$\frac{\partial G}{\partial n_x}(\mathbf{x}|\xi)+\gamma G(\mathbf{x}|\xi)\Big|_{S_x}=0 \qquad \mathbf{x} \text{ on } S_x$$

Thus, one ends up with the final solution:

$$u(\mathbf{x})=\int_R G(\mathbf{x}|\xi)f(\xi)\,d\xi+\int_{S_\xi} G(\mathbf{x}|\xi)\Big|_{S_\xi} h(\xi)\,dS_\xi \tag{8.135}$$

8.25 Construction of the Auxiliary Function-Method of Images

To construct the function G, composed of the difference between an auxiliary function and the fundamental solution, one needs to find an auxiliary function $\bar{g}(\mathbf{x}|\xi)$. The most important property of $\bar{g}(\mathbf{x}|\xi)$ is that its Laplacian is zero, i.e., the location of the source of the function $\bar{g}(\mathbf{x}|\xi)$ at ξ must not be within the region R. Then the function $\bar{g}$ can be constructed as a Fundamental solution $\bar{g}(\mathbf{x}|\bar{\xi})$ whose source $\bar{\xi}$ is located outside the region R. This can be achieved by assuming that $\bar{g}(\mathbf{x}|\bar{\xi})$ has a source location $\bar{\xi}$ located at the mirror image of the source ξ, so that when ξ approaches the surface, its image $\bar{\xi}$ also reaches the surface.

8.26 Green's Function for the Laplacian for Half-Space

Consider a three-dimensional half-space, see Figure 8.3, such that:

$$-\nabla^2 u(x,y,z) = f(x,y,z) \qquad -\infty < x,\ y < \infty,\ z > 0$$

then one can set up the geometry of the problem with the observation point P(x,y,z) and the source point Q(ξ,η,ζ). The image of Q at Q′ is located at $\bar{\xi}=(\xi,\eta,-\zeta)$ so that Q has an image behind a mirror placed at the surface $\zeta = z = 0$.

Define:

$$|\mathbf{x}|^2 = r^2 = x^2 + y^2 + z^2$$

$$|\xi|^2 = \rho^2 = \xi^2 + \eta^2 + \zeta^2$$

$$r_1^2 = |\mathbf{x}-\xi|^2 = (x-\xi)^2 + (y-\eta)^2 + (z-\zeta)^2$$

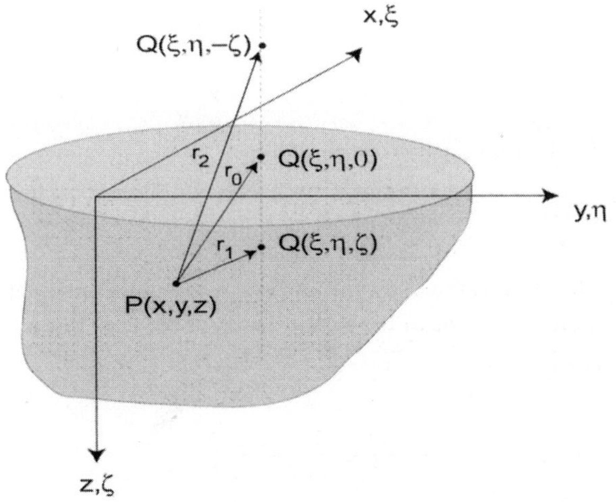

Fig. 8.3 *Half-Space*

$$r_2^2 = \left|\mathbf{x} - \bar{\boldsymbol{\xi}}\right|^2 = (x-\xi)^2 + (y-\eta)^2 + (z+\zeta)^2$$
$$r_0^2 = (x-\xi)^2 + (y-\eta)^2 + z^2$$

The Fundamental Green's function for this problem is given in eq. (8.64):

$$g = \frac{1}{4\pi r_1} \tag{8.136}$$

where r_1 is the distance between the source at Q and the field point at P. Since the Green's function depends on the particular boundary condition, the two simple boundary conditions, Dirichlet and Neumann, will be discussed first.

8.26.1 Dirichlet Boundary Condition

Here the boundary condition is given as:

$u(x,y,0) = h(x,y)$

This type of boundary condition requires that G=0 on the boundary. Let the location of the source $\bar{\xi}$ for Q' to be located at $\bar{\xi} = (\xi, \eta, -\zeta)$, then the auxiliary function $\bar{g}$ can take the form:

$$\bar{g} = \frac{C}{4\pi r_2} \tag{8.137}$$

where r_2 is the distance between the image of the source at Q' and the field point at P. Since the location of the image is always in the region $\zeta < 0$ for all ζ, then it satisfies the homogeneous Laplacian. The constant C is thus needed to satisfy the condition $G = 0$ on $\zeta = 0$ or $z = 0$. On the surface $\zeta = 0$, the two radii r_1 and r_2 are equal, i.e., $r_1 = r_2 = r_0$. Similarly, when $z = 0$, the two radii r_1 and r_2 are also equal. Thus, for $G = 0$ on the surface, C must be equal to one, i.e.:

$$\bar{g} = \frac{1}{4\pi r_2}$$

and the Green's function for the half-space becomes:

$$G = \frac{1}{4\pi}\left(\frac{1}{r_1} - \frac{1}{r_2}\right) \tag{8.138}$$

Once the Green's function is found for the half-space, one needs to obtain the normal derivative at the surface in terms of the ξ coordinates as required in eq. (8.132). Noting that $\vec{n} = -\vec{e}_\zeta$, then $\frac{\partial}{\partial n} = -\frac{\partial}{\partial \zeta}$:

$$\left.\frac{\partial G}{\partial n}\right|_{\zeta=0} = -\left.\frac{\partial G}{\partial \zeta}\right|_{\zeta=0} = -\frac{1}{4\pi}\frac{2z}{r_0^3} = \frac{1}{4\pi}\frac{\partial}{\partial z}\left(\frac{2}{r_0}\right)$$

The final solution is then written in terms of two integrals:

$$4\pi\, u(x,y,z) = \int_{-\infty}^{\infty}\int_{-\infty}^{\infty}\int_{0}^{\infty}\left(\frac{1}{r_1} - \frac{1}{r_2}\right) f(\xi,\eta,\zeta)\, d\xi\, d\eta\, d\zeta + 2z\int_{-\infty}^{\infty}\int_{-\infty}^{\infty}\frac{h(\xi,\eta)}{r_0^3}\, d\xi\, d\eta$$

The last surface integral can also be replaced by:

$$-2\frac{\partial}{\partial z}\int_{-\infty}^{\infty}\int_{-\infty}^{\infty}\frac{h(\xi,\eta)}{r_0}\, d\xi\, d\eta$$

8.26.2 Neumann Boundary Condition

Here the boundary condition is given on the normal derivative as:

$$\frac{\partial u}{\partial n} = -\frac{\partial u}{\partial z}(x,y,0) = h(x,y)$$

For Neumann type boundary condition, one must show that $\partial G/\partial n = -\partial G/\partial \zeta = 0$ on the boundary $\zeta = 0$ or $-\partial G/\partial z = 0$ on the boundary $z = 0$. On S_ζ this condition becomes:

$$\left.\frac{\partial \bar{g}}{\partial \zeta}\right|_{\zeta=0} = \left.\frac{\partial g}{\partial \zeta}\right|_{\zeta=0} = \frac{1}{4\pi}\frac{z}{r_0^3}$$

Again, using the auxiliary function $\bar{g}$ given in eq. (8.137) results in:

$$\left.\frac{\partial \bar{g}}{\partial \zeta}\right|_{\zeta=0} = -\frac{C}{4\pi}\frac{z}{r_0^3}$$

For $\partial G/\partial \zeta = 0$ this requires that C is set to -1, giving the final form for G for half-space as:

$$G = \frac{1}{4\pi}\left(\frac{1}{r_1} + \frac{1}{r_2}\right) \tag{8.139}$$

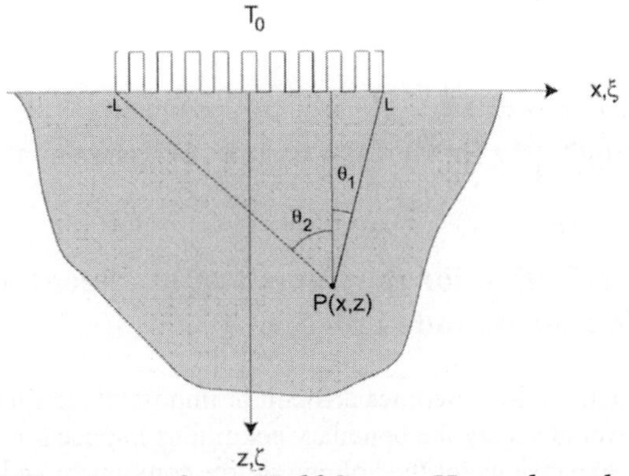

***Fig. 8.4** Geometry for a Half-Space Heated on the Boundary*

Similarly, C = -1 also satisfies the condition $\partial G/\partial z = 0$ on z = 0. The surface integral eq. (8.133) requires the evaluation of G on the surface $\zeta = 0$:

$G|_{\zeta=0} = 1/(2\pi r_0)$

Finally, the solution for the half-space is given by:

$$4\pi\, u(x,y,z) = \int_{-\infty}^{\infty}\int_{-\infty}^{\infty}\int_{0}^{\infty}\left(\frac{1}{r_1}+\frac{1}{r_2}\right) f(\xi,\eta,\zeta)\, d\xi\, d\eta\, d\zeta + 2\int_{-\infty}^{\infty}\int_{-\infty}^{\infty}\frac{h(\xi,\eta)}{r_0}\, d\xi\, d\eta$$

Example 8.11 Temperature Distribution in a Semi-Infinite Sheet

Evaluate the temperature distribution T(x,z) in a two-dimensional solid sheet, where there is no heat source and the temperature on the boundary is given by:

$$T(x,0) = T_0 \qquad |x| < L$$
$$= 0 \qquad |x| > L$$

Here $\nabla^2 T = 0$ so that the temperature T(x,z) is given in terms of a surface integral. For two-dimensional media, it can be shown that the temperature is given by:

$$4\pi T(x,z) = 4z\int_{-\infty}^{\infty}\frac{h(\xi)}{r_0^2}\, d\xi = 4zT_0\int_{-L}^{+L}(x-\xi)^2 + z^2 d\xi$$

$$4\pi T(x,z) = 4z\int_{-\infty}^{\infty}\frac{h(\xi)}{r_0^2}\, d\xi = 4zT_0\int_{-L}^{+L}(x-\xi)^2 + z^2 d\xi$$

$$= 4T_0 \arctan\left(\frac{\xi - x}{z}\right)\Bigg|_{-L}^{+L} = 4T_0\left\{\arctan\left(\frac{L-x}{z}\right) + \arctan\left(\frac{L+x}{z}\right)\right\}$$

$$= 4T_0(\theta_1 + \theta_2)$$

where $\theta_1 + \theta_2$ is the total angle subtended by the two dotted lines. Therefore, the temperature is:

$$T(x,z) = \frac{T_0}{\pi}(\theta_1 + \theta_2)$$

Note that if $|x| < L$ and $z \to 0$, then $\theta_1 + \theta_2 \to \pi$, which satisfies the boundary condition. Furthermore, for $|x|$ finite, if $z/L >> 1$, $\theta_1 \to 0$ and $\theta_2 \to 0$ and thus the temperature vanishes as $z \to \infty$.

8.27 Green's Function for the Laplacian by Eigenfunction Expansion for Bounded Media

For bounded media, it is sometimes difficult or impossible to find a closed form solution for G that would satisfy the boundary conditions imposed in Section 8.24. Thus, one must use series expansions for the volume source component and another for the non-homogeneous boundary condition component.

Let $G = G_1$ or G_2 such that:

$$-\nabla^2 G_1 = \delta(\mathbf{x} - \xi) \qquad \nabla^2 G_2 = 0$$

$$U_i(G_1) = 0 \qquad U_i(G_2) = \delta(\mathbf{x} - \xi) \qquad \mathbf{x} \text{ on } S_x$$

where the boundary forms $U_i(G)$ takes the form:

Dirichlet: $$G(\mathbf{x}|\xi)\big|_{S_x} = \delta(\mathbf{x} - \xi) \qquad \mathbf{x} \text{ on } S_x$$

Neumann: $$\frac{\partial G}{\partial n_x}(\mathbf{x}|\xi)\Big|_{S_x} = \delta(\mathbf{x} - \xi) \qquad \mathbf{x} \text{ on } S_x$$

Robin: $$\frac{\partial G}{\partial n_x}(\mathbf{x}|\xi) + \gamma G(\mathbf{x}|\xi)\Big|_{S_x} = \delta(\mathbf{x} - \xi) \qquad \mathbf{x} \text{ on } S_x$$

For the volume source Green's function G_1, one would expand it in terms of the eigenfunctions of the interior problem, see section 6.11. For the non-homogeneous boundary condition Green's function G_2, one must resort to the solution of the Laplace's equation, see section 6.10.

For G_1, the eigenfunctions of the problem $\Psi_M(\mathbf{x})$ must then satisfy the homogeneous boundary conditions. Expanding G_1 in terms of the eigenfunction set $\Psi_M(\mathbf{x})$ given in (6.54), then the generalized Fourier series coefficient is given by (6.55) with $f(P) = -\delta(\mathbf{x}-\xi)$, resulting in:

$$E_M = \frac{\Psi_M(\xi)}{\lambda_M N_M}$$

Finally, the Green's function G_1 for the volume source is given by:

$$G_1(\mathbf{x},\xi) = \sum_M \frac{\Psi_M(\mathbf{x})\,\Psi_M(\xi)}{\lambda_M N_M} \tag{8.140}$$

Note that G_1 is symmetric in $\mathbf{x}$ and ξ. The solution for the Green's function G_2 for the non-homogeneous boundary conditions depends on the geometry of the problem. The component G_1 is substituted in the volume source integral and the component G_2 as the kernel in the appropriate surface integral of eqs. (8.131), (8.133), or (8.135).

8.28 Green's Function for a Circular Area for the Laplacian

For a two-dimensional region having a circular boundary there are two Green's functions, one for the interior region and another for the exterior region.

The Laplacian operator in two-dimensional space in cylindrical coordinates is given by:

$$-\nabla^2 u(r,\theta) = f(r,\theta) \qquad \text{interior} \qquad r \le a,\ 0 \le \theta \le 2\pi$$

$$= 0 \qquad \text{exterior} \qquad r \ge a,\ 0 \le \theta \le 2\pi$$

Let the field point be designated $P(r,\theta)$, the source point $Q(\rho,\phi)$ and the image point $\bar{Q}(\bar{\rho},\phi)$, see Figures 8.5 and 8.6. The source point Q and its image $\bar{Q}$ are located on the same radial ray, i.e., the angle ϕ defines both the source and its image angle. For the image point $\bar{Q}$, it can be shown that its radial distance $\bar{\rho}$ is given by $\bar{\rho} = a^2/\rho$.

The distances r_1, r_2 and r_0 can be defined as:

$$r_1^2 = r^2 + \rho^2 - 2r\rho\cos(\theta-\phi)$$

$$r_2^2 = r^2 + \bar{\rho}^2 - 2r\bar{\rho}\cos(\theta-\phi)$$

$$r_0^2 = r^2 + a^2 - 2ar\cos(\theta-\phi)$$

Note that when the source point Q approaches the boundary, $\rho \to a$, $\bar{Q}$ also approaches the boundary, i.e., $\bar{\rho} \to a$. If one would let the field point P reach the boundary, then the resulting two similar triangles OPQ and $OP\bar{Q}$ has the following proportionality for their sides:

$$\frac{a}{\rho} = \frac{\bar{\rho}}{a} = \frac{r_2}{r_1}$$

so that:

$$\left.\frac{1}{r_1}\right|_{\text{on C}} = \left.\frac{a}{\rho}\frac{1}{r_2}\right|_{\text{on C}}$$

The Fundamental Green's function is given by eq. (8.69):

$$g = -\frac{1}{2\pi}\log r_1 = \frac{1}{4\pi}\log(r_1^{-2})$$

8.28.1 Interior Problem

(a) Dirichlet boundary condition

For Dirichlet boundary condition given as:

$$u(a,\theta) = h(\theta) \qquad 0 \le \theta \le 2\pi$$

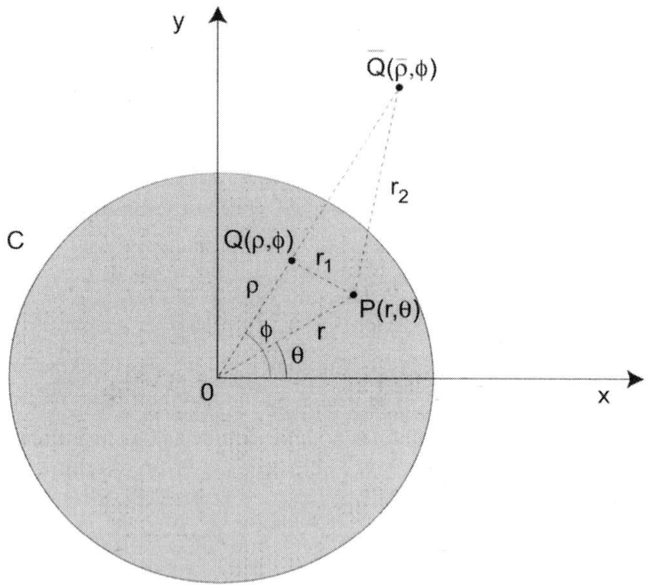

***Fig. 8.5** Geometry for the Interior Circular Region*

For the interior problem, see Figure 8.5, let us use the equality in eq. (8.141) to guide the choice of the auxiliary function $\bar{g}$, i.e., let:

$$\bar{g} = \frac{C}{4\pi}\log\left(\frac{\rho}{a}r_2\right)^{-2} \tag{8.141}$$

The choice of the auxiliary function with a constant multiplier ρ/a is dictated by the equality given above. It should be noted that since the factor ρ/a is constant in (r, θ) coordinates, then $\nabla^2\bar{g} = 0$ for $r < a$.

With the definition $G = g - \bar{g}$, then G (at $\rho = a$ or $r = a$) = 0. Note that on S_ξ, $\rho = a$, then $\bar{\rho} = a$, and hence $r_1 = r_2 = r_0$, then the constant C must be set to one. Similarly, G = 0 is also satisfied on S_x at $r = a$, where $C = 1$. Thus, the function G becomes:

$$G = \frac{1}{4\pi}\left\{\log\left(\frac{\rho}{a}r_2\right)^2 - \log(r_1^2)\right\} = \frac{1}{4\pi}\log\left(\frac{\rho^2 r_2^2}{a^2 r_1^2}\right) \tag{8.142}$$

Since $\partial/\partial n = \partial/\partial\rho$ on C, differentiating eq. (8.142) and evaluating the gradient at the surface $\rho = \bar{\rho} = a$, results in the expression:

$$\left.\frac{\partial G}{\partial n}\right|_C = -\frac{1}{2\pi a}\frac{a^2 - r^2}{a^2 + r^2 - 2ar\cos(\theta - \phi)}$$

The final solution for $u(r,\theta)$ can be expressed by area and contour integrals as, see eq. (8.132):

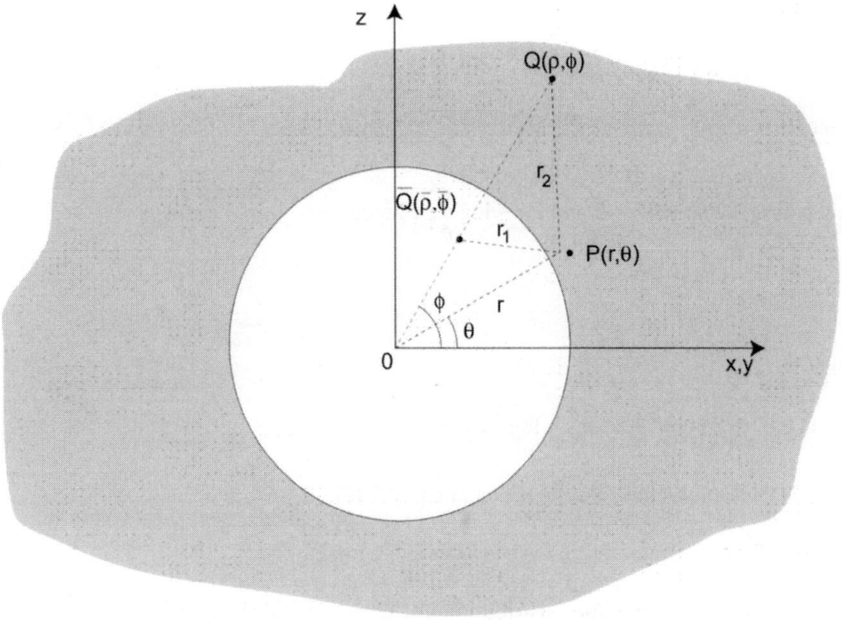

***Fig. 8.6** Geometry for the Exterior Circular Region*

$$4\pi\, u(r,\theta) = -\int_0^a \int_0^{2\pi} \log\left(\frac{a}{\rho}\frac{r_1}{r_2}\right)^2 f(\rho,\phi)\rho\, d\rho\, d\phi$$

$$+\frac{2(a^2-r^2)}{a}\int_0^{2\pi} \frac{h(\phi)\, a\, d\phi}{r^2+a^2-2ar\cos(\theta-\phi)} \qquad (8.143)$$

The same Green's function can be obtained by the use of image sources and by requiring the Green's function satisfy the boundary conditions. Starting with:

$$\bar{g} = \frac{C}{4\pi}\log r_2^2 + \frac{D(r,\rho)}{4\pi}$$

where $\nabla^2 D = 0$, and solving the homogeneous equation on D results in: $D = E(\rho)\log r + F(\rho)$. The constant C as well as the functions $E(\rho)$ and $F(\rho)$ will be evaluated through the satisfaction of the boundary conditions at $r = a$ or $\rho = a$.

$$G = g - \bar{g} = -\frac{1}{4\pi}\left\{\log r_1^2 + C\log r_2^2 + E(\rho)\log r + F(\rho)\right\}$$

$$G\big|_{\rho=\bar{\rho}=a} = 0 = -\frac{1}{4\pi}\left\{\log r_0^2 + C\log r_0^2 + E(a)\log r + F(a)\right\} = 0$$

Thus, $C = -1$, $E(a) = 0$ and $F(a) = 0$.

$$G\big|_{r=a} = 0 = -\frac{1}{4\pi}\left\{\log r_1^2(a) - \log r_2^2(a) + E(\rho)\log a + F(\rho)\right\} = 0$$

Therefore:

$$r_1^2(r=a) = a^2 + \rho^2 - 2a\rho\cos(\theta-\phi)$$

$$r_2^2(r=a) = a^2 + \bar{\rho}^2 - 2a\bar{\rho}\cos(\theta-\phi) = \frac{a^2}{\rho^2}\,r_1^2(a)$$

This indicates that $E(\rho) = 0$ and $F_0 = \log\frac{a^2}{\rho^2}$, giving:

$$D(r,\rho) = \log\frac{a^2}{\rho^2}$$

Thus:

$$G = -\frac{1}{4\pi}\left\{\log r_1^2 - \log r_2^2 + \log\frac{a^2}{\rho^2}\right\}$$

which is the same as the solution given in eq. (8.142).

Example 8.12 Temperature Distribution in a Circular Sheet

Calculate the temperature distribution in a circular solid sheet of radius = a, with no sources and the temperature on the boundary is a constant T_0. Here $f(r,\theta) = 0$ and $h(\theta) = T_0$. Thus:

$$4\pi\,T(r,\theta) = 2(a^2 - r^2)T_0 \int_0^{2\pi} \frac{d\phi}{a^2 + r^2 - 2ar\cos(\theta-\phi)}$$

Since the integral is symmetric with θ, one can let $\theta = 0$, resulting in the solution:

$$4\pi T = 4T_0 \arctan\left[\frac{a^2 - r^2}{(a-r)^2}\tan\phi\right]_{-\pi/2}^{\pi/2} = 4T_0\pi$$

or:

$$T(r,\theta) = T_0$$

This shows that the temperature is constant throughout the circular region.

(b) Neumann boundary condition

For the Neumann boundary condition given by:

$$\frac{\partial u}{\partial r}(a,\theta) = h(\theta) \qquad 0 \le \theta \le 2\pi$$

the gradient $\partial G/\partial n = 0$ on C.

For Neuman boundary conditions, one again can obtain the Green's function by the method of images. However, one must again adjust the image source by a function that guarantees the Neumann boundary condition, i.e.:

$$\left.\frac{\partial G}{\partial \rho}\right|_{\rho=a} = 0 \qquad \text{and} \qquad \left.\frac{\partial G}{\partial r}\right|_{r=a} = 0$$

Starting as above, let:

$$G = g - \bar{g} = -\frac{1}{4\pi}\left\{\log r_1^2 + C\log r_2^2 + D(r,\rho)\right\}$$

then $D(r,\rho)$ must satisfy $\nabla^2 D = 0$ or $D = E(\rho)\log r + F(\rho)$.

$$\left.\frac{\partial G}{\partial \rho}\right|_{\rho=a} = -\frac{1}{2\pi}\left\{\frac{a - r\cos(\theta-\phi)}{r_0^2} - C\frac{a - r\cos(\theta-\phi)}{r_0^2} + \left.\frac{\partial D}{\partial \rho}\right|_{\rho=a}\right\} = 0$$

This indicates that $C = +1$ and $\left.\frac{\partial D}{\partial \rho}\right|_{\rho=a} = 0$.

Therefore $E(\rho) = \text{constant} = E_0$ and $F(\rho) = \text{constant} = F_0$. Also:

$$\left.\frac{\partial G}{\partial r}\right|_{r=a} = -\frac{1}{2\pi}\left\{\frac{a - \rho\cos(\theta-\phi)}{r_1^2(a)} + \frac{\rho(\rho - a\cos(\theta-\phi))}{a r_1^2(a)} + \frac{1}{2}\left.\frac{\partial D}{\partial r}\right|_{r=a}\right\} = 0$$

where

$$r_1^2(a) = a^2 + \rho^2 - 2a\rho\cos(\theta-\phi)$$

The term dependant on (θ,ϕ) must vanish indicating that $E_0 = -2$. The constant F_0 can be adjusted to give a non-dimensional argument for the logarithmic function in G, i.e., let $F_0 = -\log a^2$. Thus, the source term represented by D is located at the center and gives out the correct flux at r = a to nullify the flux of G at r = a. Therefore:

$$D = -2\log r - 2\log a = -\log(a^2 r^2)$$

and

$$G = -\frac{1}{4\pi}\left\{\log r_1^2 + \log r_2^2 - \log(a^2 r^2)\right\} = -\frac{1}{4\pi}\log\left(\frac{r_1^2 r_2^2}{a^2 r^2}\right) \tag{8.144}$$

The final solution for $U(r,\theta)$ can be expressed as area and contour integrals, i.e.:

$$4\pi\, u(r,\theta) = -\int_0^a\int_0^{2\pi} \log\left(\frac{r_1^2 r_2^2}{a^2 r^2}\right) f(\rho,\phi)\rho\, d\rho\, d\phi + 2a\int_0^{2\pi}\log\left(\frac{r_0^2}{a^2}\right) h(\phi)\, d\phi \tag{8.145}$$

One may also obtain the Green's function in terms of eigenfunctions by attempting to split $G = G_1$ or G_2, with G_1 in terms of an eigenfunction expansion eq. (8.140). The eigenfunctions for this problem:

$$\Psi_{nm}(r,\theta) = J_n\left(\mu_{nm}\frac{r}{a}\right)\begin{Bmatrix}\sin n\theta\\ \cos n\theta\end{Bmatrix} \qquad n = 1,2,3\ldots \quad m = 1,2,3\ldots$$

and the eigenvalues are:

$$\lambda_{nm} = \mu_{nm}^2 / a^2$$

which are the roots of $J_n'(\mu_{nm}) = 0$.

Expanding G_1 in terms of these eigenfunctions:

$$G_1 = \sum_{n=1}^{\infty}\sum_{m=1}^{\infty} E_{nm}\, J_n\!\left(\mu_{nm}\frac{r}{a}\right)\cos n\theta + \sum_{n=1}^{\infty}\sum_{m=1}^{\infty} H_{nm}\, J_n\!\left(\mu_{nm}\frac{r}{a}\right)\sin n\theta$$

and using the point source representation in two-dimensional space for cylindrical coordinates given by:

$$\delta(\mathbf{x}-\boldsymbol{\xi}) = \frac{\delta(r-\rho)\delta(\theta-\phi)}{r} \qquad 0 \le r, \rho \le a, \qquad 0 \le \theta, \phi \le 2\pi$$

gives an expression for the Fourier constants:

$$\begin{matrix} E_{nm} \\ H_{nm} \end{matrix} = \frac{J_n\!\left(\mu_{nm}\frac{\rho}{a}\right)}{\lambda_{nm}\, N_{nm}} \begin{Bmatrix} \sin n\phi \\ \cos n\phi \end{Bmatrix}$$

where:

$$N_{nm} = \frac{\pi a^2}{2\mu_{nm}^2}\left(\mu_{nm}^2 - n^2\right) J_n^2\left(\mu_{nm}\right)$$

$$\begin{matrix} E_{nm} \\ H_{nm} \end{matrix} = \frac{2\, J_n\!\left(\mu_{nm}\frac{\rho}{a}\right)\begin{Bmatrix} \sin n\phi \\ \cos n\phi \end{Bmatrix}}{\pi a^2\left(\mu_{nm}^2 - n^2\right) J_n^2\left(\mu_{nm}\right)}$$

The final form for the Green's function G_1 is given in the form

$$G_1(r,\theta|\rho,\phi) = \frac{2}{\pi a^2}\sum_{n=1}^{\infty}\sum_{m=1}^{\infty} \frac{J_n\!\left(\mu_{nm}\frac{\rho}{a}\right) J_n\!\left(\mu_{nm}\frac{r}{a}\right)\cos(n(\theta-\phi))}{\left(\mu_{nm}^2 - n^2\right) J_n^2\left(\mu_{nm}\right)} \tag{8.146}$$

For the second component G_2, which satisfies Laplace's equation with non-homogeneous boundary conditions, one can show that G_2 can be expanded in the form:

$$G_2(r,\theta\,|\,\phi) = \sum_{n=0}^{\infty} A_n r^n \cos(n\theta) + \sum_{n=1}^{\infty} B_n r^n \sin(n\theta)$$

Substituting

$$\left.\frac{\partial G_2(r,\theta\,|\,\phi)}{\partial r}\right|_{r=a} = \frac{\delta(\theta-\phi)}{a}$$

results in the Green's function G_2 as:

$$G_2(r,\theta|\phi) = \sum_{n=1}^{\infty}\frac{1}{n\pi}\left(\frac{r}{a}\right)^n \cos(n(\theta-\phi)) \tag{8.147}$$

These components are included in the integrand of eq. (8.133). It should be noted that the final solution is unique to within a constant.

8.28.2 Exterior Problem

For the exterior problem, see Figure 8.6, the field and the source points are located outside the circle r = a and the image point $\bar{Q}$ is located within the circle.

(a) Dirichlet boundary condition

For the Dirichlet boundary condition one can use the same function G as in eq. (8.142)

$$4\pi\, u(r,\theta) = -\int_a^{\infty}\int_0^{2\pi} \log\left(\frac{\rho}{a}\frac{r_1}{r_2}\right)^2 f(\rho,\phi)\rho\, d\rho\, d\phi - 2\left(a^2 - r^2\right)\int_0^{2\pi} \frac{h(\phi)}{r_0^2}\, d\phi \qquad (8.148)$$

(b) Neumann boundary condition

For the Neumann boundary condition

$$\frac{\partial u}{\partial n} = -\frac{\partial u}{\partial r}(a,\theta) = h(\theta) \qquad 0 \le \theta \le 2\pi$$

Here again one may use the same Green's function given in eq. (8.144), such that the final solutions given by:

$$4\pi\, u(r,\theta) = -\int_a^{\infty}\int_0^{2\pi} \log\left(\frac{r_1^2\, r_2^2}{a^2 r^2}\right) f(\rho,\phi)\rho\, d\rho\, d\phi + 2a\int_0^{2\pi} \log\left(\frac{r_0^2}{a^2}\right) h(\phi)\, d\phi \qquad (8.149)$$

One may also find the Green's function by eigenfunction expansion in terms of the angular coordinate θ. Following the proceeding treatment for the interior problem, one can split $G = G_1$ or G_2. Since this is an exterior problem, there is no eigenfunction set for the component G_1 in the radial coordinate r. Starting with the differential equation G_1 satisfies:

$$-\nabla^2 G_1 = \delta(\mathbf{x} - \xi) = \frac{\delta(r-\rho)\delta(\theta-\phi)}{r} \qquad \rho, r \ge a \qquad 0 \le \theta, \phi \le 2\pi$$

Expanding G_1:

$$G_1 = E_0^{(2)}(r) + \sum_{n=1}^{\infty}\left(E_n^{(1)}(r)\sin n\theta + E_n^{(2)}(r)\cos n\theta\right)$$

and using the orthogonality of the circular functions, one obtains for $n \ge 1$:

$$\frac{d^2 E_n^{(1)(2)}}{dr^2} + \frac{1}{r}\frac{d^2 E_n^{(1)(2)}}{dr} - \frac{n^2}{r^2} E_n^{(1)(2)} = -\frac{\delta(r-\rho)}{\pi r}\begin{Bmatrix}\sin n\phi \\ \cos n\phi\end{Bmatrix}$$

Applying Hankel transform on $E_n^{(1)(2)}(r)$ (see Section 7.7) and letting $\bar{E}_n^{(1)(2)}(u)$ be the Hankel transform of $E_n^{(1)(2)}(r)$, eq. (7.11), then for $n \ge 1$:

$$-u^2\, \bar{E}_n^{(1)(2)}(u) = -\int_a^{\infty} \frac{\delta(r-\rho)}{r}\, r\, J_n(ur)\, dr \begin{Bmatrix}\sin n\phi \\ \cos n\phi\end{Bmatrix}$$

$$E_n^{(1)(2)}(u)=\frac{J_n(\rho u)}{\pi u^2}\begin{Bmatrix}\sin(n\phi)\\ \cos(n\phi)\end{Bmatrix} \qquad n\geq 1$$

The inverse Hankel transform of nth order gives:

$$E_n^{(1)(2)}(r)=\frac{1}{2\pi n}\begin{Bmatrix}\sin(n\phi)\\ \cos(n\phi)\end{Bmatrix}\begin{Bmatrix}(r/\rho)^n & r<\rho\\ (\rho/r)^n & r>\rho\end{Bmatrix}$$

For $E_0^{(2)}(r)$, see eq. (8.69):

$$E_0^{(2)}(r)=-\frac{1}{2\pi}\log(r_1)=-\frac{1}{\pi}\left[r^2+\rho^2-2r\rho\cos(\theta-\phi)\right]$$

Finally, substituting these expressions into the series for G_1:

$$G_1(r,\theta|\rho,\phi)=-\frac{1}{\pi}\log\left[r^2+\rho^2-2r\rho\cos(\theta-\phi)\right]+\frac{1}{2\pi}\sum_{n=1}^{\infty}\frac{r^n}{n\rho^n}\cos(n(\theta-\phi)) \qquad \text{for } r<\rho \tag{8.150}$$

$$G_1(r,\theta|\rho,\phi)=-\frac{1}{\pi}\log\left[r^2+\rho^2-2r\rho\cos(\theta-\phi)\right]+\frac{1}{2\pi}\sum_{n=1}^{\infty}\frac{\rho^n}{nr^n}\cos(n(\theta-\phi)) \qquad \text{for } r>\rho \tag{8.151}$$

For the solution to the second component G_2, one can obtain the solution by use of the solution of Laplace's equation with Neumann boundary condition:

$$G_2(r,\theta|\phi)=\frac{1}{\pi}\sum_{n=1}^{\infty}\left(\frac{a}{r}\right)^n\cos(n(\theta-\phi)) \tag{8.152}$$

8.29 Green's Function for Spherical Geometry for the Laplacian

For a three-dimensional region having a spherical boundary, there are two Green's functions, one for the interior and one for the exterior of the spherical surface at $r = a$, see Figures 8.7 and 8.8.

The Laplacian operator in three-dimensional space in spherical coordinates can be written as:

$$-\nabla^2 u(r,\theta,\phi)=f(r,\theta,\phi) \qquad \text{interior } r\leq a,\ 0\leq\theta\leq\pi,\ 0\leq\phi\leq 2\pi$$

$$\text{exterior } r\geq a,\ 0\leq\theta\leq\pi,\ 0\leq\phi\leq 2\pi$$

The source point $Q(\rho,\bar\theta,\bar\phi)$ has an image at $\bar Q(\bar\rho,\bar\theta,\bar\phi)$ such that $\bar\rho = a^2/\rho$. The distances r_1, r_2 and r_0 are given by:

$$r_1^2=r^2+\rho^2-2r\rho\cos\theta_0$$

$$r_2^2 = r^2 + \bar{\rho}^2 - 2r\bar{\rho}\cos\theta_0$$

where:

$$\cos\theta_0 = \cos\theta\cos\bar{\theta} + \sin\theta\sin\bar{\theta}\cos(\phi - \bar{\phi})$$

The Fundamental solution for three-dimensional space is given by, with r_1 replacing r:

$$g = \frac{1}{4\pi r_1} \tag{8.136}$$

8.29.1 Interior Problem

<u>(a) Dirichlet boundary condition</u>

For the Dirichlet boundary condition:

$$u(a,\theta,\phi) = h(\theta,\phi)$$

the choice of the auxiliary function $\bar{g}$ follows the same development for a circular area, i.e., the equality (8.141). This leads to the choice of auxiliary function as:

$$\bar{g} = \frac{C}{4\pi}\frac{a}{\rho}\frac{1}{r_2}$$

so that for G to vanish at the spherical surface $\rho = a$, the constant $C = 1$, results in an expression for G as:

$$G = \frac{1}{4\pi}\left(\frac{1}{r_1} - \frac{a}{\rho}\frac{1}{r_2}\right) \tag{8.153}$$

The normal gradient $\partial|\partial n = \partial|\partial\rho$ is needed, which can be shown to give:

$$\left.\frac{\partial G}{\partial n}\right|_{\rho = a} = -\frac{a^2 - r^2}{4\pi a r_0^3}$$

The final solution for u can be written in terms of a volume integral and a surface integral:

$$4\pi u(r,\theta,\phi) = \int_0^a\int_0^\pi\int_0^{2\pi}\left(\frac{1}{r_1} - \frac{a}{\rho}\frac{1}{r_2}\right) f(\rho,\bar{\theta},\bar{\phi})\rho^2 \sin\bar{\theta}\, d\bar{\theta}\, d\bar{\phi}\, d\rho$$

$$+(a^2 - r^2)a\int_0^\pi\int_0^{2\pi}\frac{h(\bar{\theta},\bar{\phi})}{r_0^3}\sin\bar{\theta}\, d\bar{\theta}\, d\bar{\phi} \tag{8.154}$$

<u>(b) Neumann boundary condition</u>

For Neumann boundary condition:

$$\frac{\partial u}{\partial r}(a,\theta,\phi) = h(\theta,\phi)$$

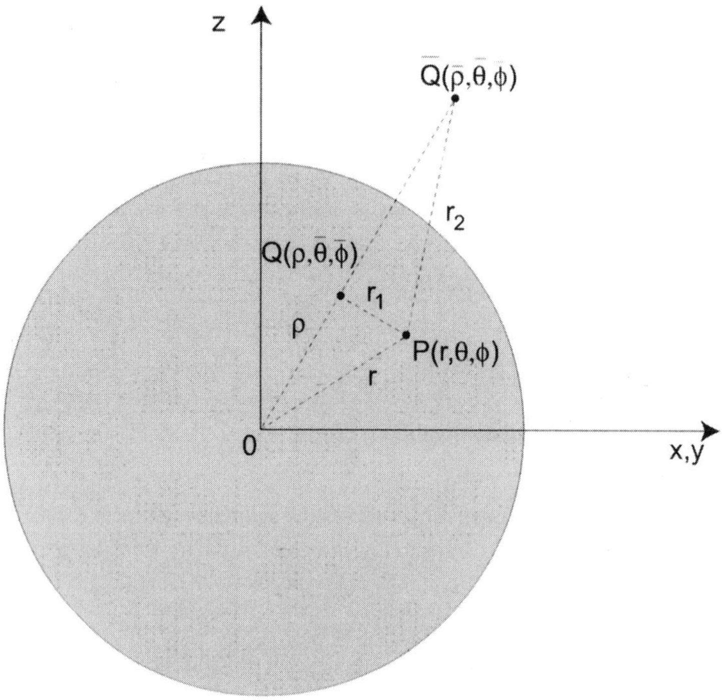

***Fig. 8.7** Geometry for the Interior Spherical Region*

the auxiliary function $\bar{g}$ cannot be found in a closed form, as was the case for the cylindrical problem. Here again, one needs to split the Green's function $G = G_1$ or G_2 where G_1 is obtained for the point source for the volume source distribution and G_2 for the non-homogeneous Neumann boundary condition as was done in section 8.27.

8.29.2 Exterior Problem

Development of the Green's function for the exterior spherical problem closely follows that of the circular region.

(a) Dirichlet boundary condition

Here let the Green's function be the same as in eq. (8.151), so the normal gradient of G is needed. The normal gradient then is $\partial G/\partial n = -\partial G/\partial \rho$.

(b) Neumann boundary condition

For Neumann boundary condition, one must follow the analysis of the exterior problem of the exterior cylindrical problem by letting $G = G_1$ or G_2 as was done in Section 8.27.

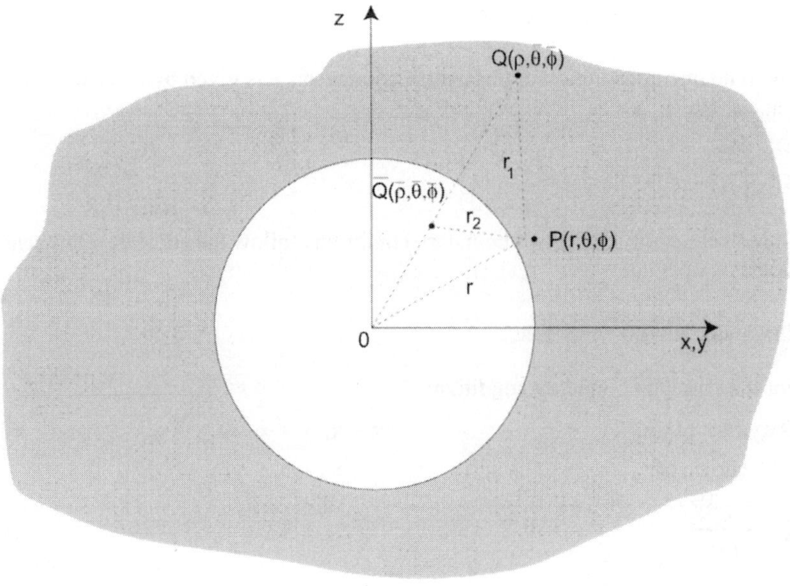

Fig. 8.8 *Geometry for the Exterior Spherical Region*

8.30 Green's Function for the Helmholtz Operator for Bounded Media

Consider the Helmholtz operator in Section 8.12. Substituting for u from eq. (8.77) and v = g from eq. (8.78) into the equality in eq. (8.39) results in the same expression given in (8.128):

$$u(\mathbf{x}) = \int_R g(\mathbf{x}|\xi) f(\xi)\, d\xi + \int_{S_\xi} \left[g(\mathbf{x}|\xi) \frac{\partial u(\xi)}{\partial n_\xi} - u(\xi) \frac{\partial g(\mathbf{x}|\xi)}{\partial n_\xi} \right]_{S_\xi} dS_\xi \qquad (8.128)$$

Following the analysis undertaken for the Laplacian, let the auxiliary function $\bar{g}(x|\xi)$ satisfy:

$$-\nabla^2 \bar{g}(x|\xi) - \lambda \bar{g}(x|\xi) = 0 \qquad \mathbf{x} \text{ in } R \qquad (8.155)$$

Letting the Green's function G for the bounded media be defined as $G = g - \bar{g}$, then the final solution for the non-homogeneous problem is the same as the Laplacian's, eqs. (8.132) to (8.153).

8.31 Green's Function for the Helmholtz Operator for Half-Space

Refer to the geometry of three- or two-dimensional half-spaces in section 8.26. For two-dimensional space, delete the coordinate y from three-dimensional system, such that $-\infty < x < \infty$, $z > 0$.

8.31.1 Three-Dimensional Half-Space

The fundamental solution in three-dimensional space is given by 8.82, with r_1 replacing r:

$$g = \frac{e^{ikr_1}}{4\pi r_1} \tag{8.82}$$

The Green's function for the two boundary conditions follow the same development for the Laplacian operator.

(a) Dirichlet boundary condition

For the Dirichlet boundary condition:

$$u(x,y,0) = h(x,y) \qquad -\infty < x,\ y < \infty \tag{8.156}$$

Here, the choice of:

$$\bar{g} = \frac{C\, e^{ikr_2}}{4\pi r_2} \tag{8.157}$$

requires that C = 1 to make $G(\zeta=0) = 0$ or $G(z=0) = 0$. The Green's function then is:

$$G = \frac{1}{4\pi}\left(\frac{e^{ikr_1}}{r_1} - \frac{e^{ikr_2}}{r_2}\right) \tag{8.158}$$

The Dirichlet boundary condition eq. (8.155) requires the evaluation of the normal gradient of G on the surface, given by:

$$4\pi \left.\frac{\partial G}{\partial n_\xi}\right|_{\zeta=0} = 2\left(ik - \frac{1}{r_0}\right)\frac{z}{r_0^2}\, e^{ikr_0} \tag{8.159}$$

The final solution for u(**x**) can be shown to be:

$$4\pi\, u(x,y,z) = \int_0^\infty \int_{-\infty}^\infty \int_{-\infty}^\infty \left(\frac{e^{ikr_1}}{r_1} - \frac{e^{ikr_2}}{r_2}\right) f(\xi,\eta,\zeta)\, d\xi\, d\eta\, d\zeta$$

$$+ 2z \int_{-\infty}^\infty \int_{-\infty}^\infty \left(\frac{1}{r_0} - ik\right)\frac{e^{ikr_0}}{r_0^2}\, h(\xi,\eta)\, d\xi\, d\eta \tag{8.160}$$

(b) Neumann boundary condition

For the Neumann boundary condition:

$$\frac{\partial u}{\partial n} = -\left.\frac{\partial u}{\partial z}\right|_{z=0} = h(x,y) \tag{8.161}$$

the Green's function must satisfy $\partial G/\partial n = -\partial G/\partial \zeta = 0$ on the surface $\zeta = 0$ or $-\partial G/\partial z = 0$ on the surface z = 0. It can be shown that the constant C = -1, giving the Green's function as:

$$G = \frac{1}{4\pi}\left(\frac{e^{ikr_1}}{r_1} + \frac{e^{ikr_2}}{r_2}\right) \tag{8.162}$$

The final solution for u(x,y,z) can be written as:

$$4\pi\, u(x,y) = \int_0^{\infty}\int_{-\infty}^{\infty}\int_{-\infty}^{\infty}\left(\frac{e^{ikr_1}}{r_1} + \frac{e^{ikr_2}}{r_2}\right) f(\xi,\eta,\zeta)\, d\xi\, d\eta\, d\zeta$$

$$+\,2\int_{-\infty}^{\infty}\int_{-\infty}^{\infty}\frac{e^{ikr_0}}{r_0}\, h(\xi,\eta)\, d\xi\, d\eta \tag{8.163}$$

8.31.2 Two-Dimensional Half-Space

The fundamental Green's function for two-dimensional space is given by, with r_1 replacing r:

$$g = \frac{i}{4} H_0^{(1)}(kr_1) \tag{8.86}$$

<u>(a) Dirichlet boundary condition</u>

For the boundary condition, one must satisfy G = 0 on $\zeta = 0$ or z = 0, such that:

$$\bar{g} = \frac{iC}{4} H_0^{(1)}(kr_2) \tag{8.164}$$

so that C = 1 resulting in the Green's function as:

$$G = \frac{i}{4}\left[H_0^{(1)}(kr_1) - H_0^{(2)}(kr_2)\right] \tag{8.165}$$

so that:

$$\left.\frac{\partial G}{\partial n}\right|_{\zeta=0} = \left.\frac{\partial G}{\partial \zeta}\right|_{\zeta=0} = -\frac{ikz}{2r_0} H_1^{(1)}(kr_0)$$

and the final solution can be shown to have the form:

$$4i\, u(x,z) = -\int_0^{\infty}\int_{-\infty}^{\infty}\left[H_0^{(1)}(kr_1) - H_0^{(1)}(kr_2)\right] f(\xi,\zeta)\, d\xi\, d\zeta$$

$$-2kz\int_{-\infty}^{\infty}\frac{h(\xi)}{r_0} H_1^{(1)}(kr_0)\, d\xi \tag{8.166}$$

<u>(b) Neumann boundary condition</u>

For the Neumann boundary condition, the normal gradient must vanish on the surface $\zeta = 0$ or z = 0, requiring that C = -1, giving G as:

$$G = \frac{i}{4}\left[H_0^{(1)}(kr_1) + H_0^{(2)}(kr_2)\right] \tag{8.167}$$

The final solution u(x,z) is given by:

$$4iu(x,z) = -\int_0^{\infty}\int_{-\infty}^{\infty}\left[H_0^{(1)}(kr_1) + H_0^{(2)}(kr_2)\right]f(\xi,\zeta)\,d\xi\,d\zeta$$

$$-2\int_{-\infty}^{\infty} H_0^{(1)}(kr_0)\,h(\xi)\,d\xi \tag{8.168}$$

8.31.3 One-Dimensional Half-Space

The fundamental Green's function for one dimensional half-space is given by eq. (8.88):

$$g(x|\xi) = \frac{i}{2k}e^{ik|x-\xi|} \tag{8.88}$$

(a) Dirichlet boundary condition

For the Dirichlet boundary condition G = 0 on $\xi = 0$ or x = 0, such that:

$$G = \frac{i}{2k}\left[e^{ik|x-\xi|} - e^{ik(x+\xi)}\right] \tag{8.169}$$

(b) Neumann boundary condition

For the Neumann boundary condition $\partial G/\partial\xi = 0$ on $\xi = 0$ or $\partial G/\partial x = 0$ on x = 0, such that:

$$G = \frac{i}{2k}\left[e^{ik|x-\xi|} + e^{ik(x+\xi)}\right] \tag{8.170}$$

(c) Robin boundary condition

For the Robin boundary condition, G must satisfy, $-\partial G/\partial\xi + \gamma G = 0$ on $\xi = 0$. In this case, it is not a simple matter to readily enforce this condition. For this boundary condition, a less direct method is needed to obtain G. With $G = g - \bar{g}$, define a new function w(x) as:

$$w(x) = \frac{dG}{dx} - \gamma G \tag{8.171}$$

then:

$$w(0) = 0$$

Substituting w(x) into the Helmholtz equation:

$$\left(-\frac{d^2}{dx^2} - k^2\right) w(x) = \left(-\frac{d^2}{dx^2} - k^2\right)\left(\frac{\partial G}{\partial x} - \gamma G\right) = \left(-\frac{d^2}{dx^2} - k^2\right)\left(\frac{\partial g}{\partial x} - \gamma g\right)$$

$$= \delta'(x-\xi) - \gamma\,\delta(x-\xi)$$

With w(x) satisfying the Dirichlet boundary condition w(0) = 0, one can use the results of eq. (8.169) for the final solution for w(x) with the source term given above:

$$w(x) = \frac{i}{2k}\int_0^\infty \left[\frac{\partial\delta(\eta-\xi)}{\partial\eta} - \gamma\delta(\eta-\xi)\right]\left[e^{ik|x-\eta|} - e^{ik(x+\eta)}\right] d\eta$$

Integrating the above expression, one can show that:

$$w(x) = \left[\frac{i\gamma}{2k} - \frac{1}{2}\right] e^{ik(x+\xi)} - \left[\frac{i\gamma}{2k} + \frac{1}{2}\,\text{sgn}(x-\xi)\right] e^{ik|x-\xi|}$$

where the signum function sgn(x) = +1 for x > 0, and = -1 for x < 0. Note that w(0) = 0. Returning to the first order ordinary differential equation on the function G with w(x) being the non-homogeneity:

$$\frac{dG}{dx} - \gamma G = w(x)$$

then the solution for G in terms of w(x) is given in (1.9) as:

$$G = -\,e^{\gamma x}\int_x^\infty w(\eta)\, e^{-\gamma\eta}\, d\eta \qquad (8.172)$$

The integration in eq. (8.172) is straightforward. However, the integration for the second part of eq. (8.171) requires that separate integrals must be performed for $x > \xi$ and $x < \xi$.

The final solution for $G(x|\xi)$ becomes:

$$G(x|\xi) = \frac{i}{2k}\left[\frac{ik+\gamma}{ik-\gamma}\, e^{ik(x+\xi)} + e^{ik|x-\xi|}\right] \qquad (8.173)$$

Note that if $\gamma = 0$, one recovers the Neumann boundary condition solution in eq. (8.170) and if $\gamma \to \infty$, one recovers the Dirichlet boundary condition solution in eq. (8.171).

8.32 Green's Function for a Helmholtz Operator in Quarter-Space

Consider the field in a three-dimensional quarter-space, see Figure 8.9. The quarter-space is defined in the region $0 < x, z < \infty$, $-\infty < y < \infty$. Let the field point be P(x,y,z) and the source point be $Q(\xi,\eta,\zeta)$. There is an image of Q at $Q_1(\xi,\eta,-\zeta)$ about the x-y plane, another image of Q about the y-z plane at $Q_2(-\xi,\eta,\zeta)$. There is an image of Q_1 about the

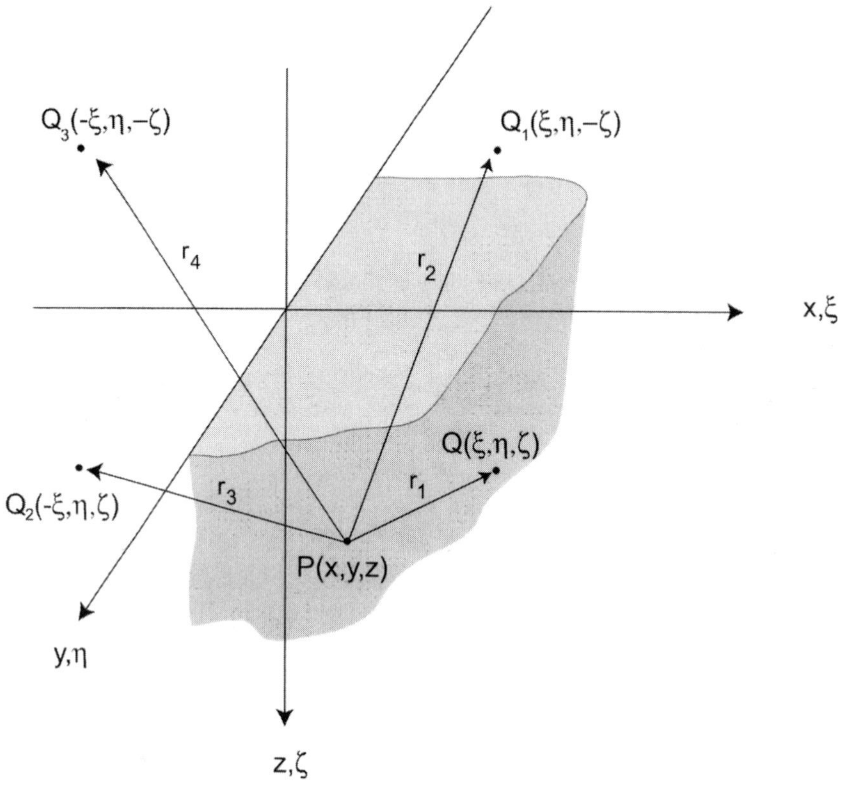

Fig. 8.9 *Geometry for a Three Dimensional Quarter-Space*

y-z plane and an image of Q_2 about the x-y plane, both coinciding at $Q_3(-\xi,\eta,-\zeta)$. Define the radii for the problem as:

$$r^2 = x^2 + y^2 + z^2 \qquad \rho^2 = \xi^2 + \eta^2 + \zeta^2$$

$$r_1^2 = (x-\xi)^2 + (y-\eta)^2 + (z-\zeta)^2 \qquad r_2^2 = (x-\xi)^2 + (y-\eta)^2 + (z+\zeta)^2$$

$$r_3^2 = (x+\xi)^2 + (y-\eta)^2 + (z-\zeta)^2 \qquad r_4^2 = (x+\xi)^2 + (y-\eta)^2 + (z+\zeta)^2$$

$$r_{01}^2 = (x-\xi)^2 + (y-\eta)^2 + z^2 \qquad r_{02}^2 = x^2 + (y-\eta)^2 + (z-\zeta)^2$$

$$r_{03}^2 = (x+\xi)^2 + (y-\eta)^2 + z^2 \qquad r_{04}^2 = x^2 + (y-\eta)^2 + (z+\zeta)^2$$

Consider the following problem:

$$(-\nabla^2 - \lambda)\, u = f(x,y,z) \qquad x,\ z \ge 0,\ -\infty < y < \infty$$

$$u(x,y,0) = h_1(x,y) \tag{8.174}$$

$$\frac{\partial u}{\partial n} = -\frac{\partial u}{\partial x}(0, y, z) = h_2(y, z)$$

The fundamental solution in three-dimensional space is given by:

$$g = \frac{e^{ikr_1}}{4\pi r_1} \tag{8.82}$$

Since the images Q_1, Q_2 and Q_3 are located outside the quarter-space, one can choose three auxiliary functions as:

$$\bar{g} = \frac{1}{4\pi}\left\{C_1\frac{e^{ikr_2}}{r_2} + C_2\frac{e^{ikr_3}}{r_3} + C_3\frac{e^{ikr_4}}{r_4}\right\} \tag{8.175}$$

With the definition $G = g - \bar{g}$, then the Green's function must satisfy the following boundary conditions:
on the surface S_1:

$$G\big|_{\zeta=0} = 0 \qquad G\big|_{z=0} = 0$$

on the surface S_2:

$$\frac{\partial G}{\partial n} = -\frac{\partial G}{\partial \xi}\bigg|_{\xi=0} = 0 \qquad -\frac{\partial G}{\partial x}\bigg|_{x=0} = 0$$

When Q approaches the surface S_1, $r_1 = r_2 = r_{01}$ and $r_3 = r_4 = r_{03}$. Thus:

$$4\pi G\big|_{\zeta=0} = \frac{e^{ikr_{01}}}{r_{01}} - C_1\frac{e^{ikr_{01}}}{r_{01}} - C_2\frac{e^{ikr_{03}}}{r_{03}} - C_3\frac{e^{ikr_{03}}}{r_{03}} = 0$$

This requires that $C_1 = 1$ and $C_2 = -C_3$. When Q approaches the surface S_2, $r_1 = r_3 = r_{02}$ and $r_2 = r_4 = r_{04}$. Thus:

$$-4\pi\frac{\partial G}{\partial \xi}\bigg|_{\xi=0} = x\left(ik - \frac{1}{r_{02}}\right)\frac{e^{ikr_{02}}}{r_{02}^2}(1 + C_2) + x\left(ik - \frac{1}{r_{04}}\right)\frac{e^{ikr_{04}}}{r_{04}^2}(-C_1 + C_3) = 0$$

This requires that $C_2 = -1$ and $C_1 = C_3$. Finally, the constants carry the value $C_1 = 1$, $C_2 = -1$ and $C_3 = 1$ so that the Green's function takes the final form:

$$4\pi G = \frac{e^{ikr_1}}{r_1} - \left[\frac{e^{ikr_2}}{r_2} - \frac{e^{ikr_3}}{r_3} + \frac{e^{ikr_4}}{r_4}\right]$$

If one would want to establish an algorithm for determining the signs of the images, i.e., C_1, C_2 and C_3, one can follow the subsequent rules:

(1) The sign of the constant is the same as the source for a Dirichlet boundary condition, if the image is reflected over the actual boundary.

(2) The sign is reversed if the image is reflected over an extension of the Dirichlet boundary.

(3) The sign is a reverse of the source for the Neumann boundary condition, if the image is reflected of the actual boundary.

(4) The sign is the same as the source if the image is reflected over an extension of the Neumann boundary.

With this construct, the sign for $C_1 = 1$, the sign of $C_2 = -1$, the sign of C_3 should be the same as C_1 because of the reflection about a Neumann boundary extension and should be the opposite of C_2 because it is a reflection about the Dirichlet boundary extension, i.e., $C_3 = C_1 = -C_2 = +1$.

8.33 Causal Green's Function for the Wave Operator in Bounded Media

Consider the wave operator:

$$\left(\frac{\partial^2}{\partial t^2} - c^2\nabla^2\right) u(\mathbf{x},t) = f(\mathbf{x},t) \qquad \mathbf{x} \text{ in R}, \; t > 0 \quad (8.108)$$

together with the initial and boundary conditions:

$$u(\mathbf{x},0) = f_1(\mathbf{x}) \frac{\partial u}{\partial t}(\mathbf{x},0) = f_2(\mathbf{x})$$

(a) Dirichlet: $u(\mathbf{x},t)\big|_S = h(\mathbf{x},t)$

(b) Neumann: $\dfrac{\partial u}{\partial n}(\mathbf{x},t)\Big|_S = h(\mathbf{x},t)$

(c) Robin: $\dfrac{\partial u}{\partial n}(\mathbf{x},t) + \gamma u(\mathbf{x},t)\Big|_S = h(\mathbf{x},t)$

The causal fundamental Green's function $g(\mathbf{x},t|\xi,\tau)$ was defined in eq. (8.109). Consider the adjoint casual fundamental Green's function $g(\xi,\tau|\mathbf{x},t)$ which satisfies:

$$\left(\frac{\partial^2}{\partial t^2} - c^2\nabla^2\right) g(\xi,\tau|\mathbf{x},t) = \delta(\xi - \mathbf{x})\,\delta(\tau - t)$$

$$g(\xi,\tau|\mathbf{x},t) = 0 \qquad \tau < t$$

It should be noted that since the wave operator is self-adjoint, then:

$$g(\mathbf{x},t|\xi,\tau) = g(\xi,\tau|\mathbf{x},t) \qquad (8.176)$$

Consider the special case of a time-independent region R and surface S. Substitute $u(\mathbf{x},t)$ from above, eq. (8.108) and the adjoint causal Green's function $v = g(\xi,\tau|\mathbf{x},t)$ into Green's identity for the wave operator eq. (8.52). Since the region R and its surface S do not change in time, the surface $\bar{S}$ takes a cylindrical surface form shown in Figure 8.10. On the cylindrical surface $\bar{S}$, $\bar{\bar{n}} = \vec{n}$, while on the surface $t = 0$ and $t = T$, the normal $\bar{\bar{n}} = -\vec{e}_t$ and $\vec{e}_t$, respectively.

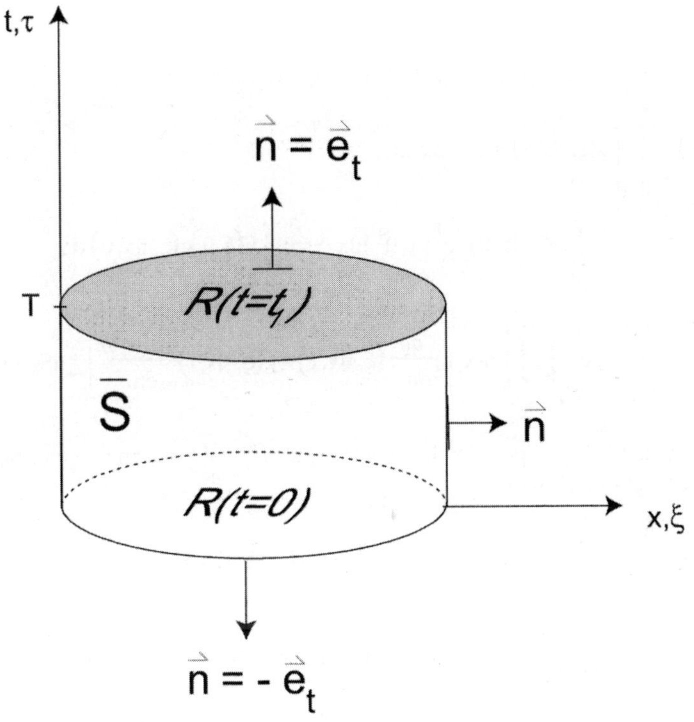

Fig. 8.10 *Geometry for the Time-Space Wave Equation.*

Thus, the Green's identity results in the following integrals:

$$\int_0^T \int_R g(\boldsymbol{\xi},\tau|\boldsymbol{x},t) f(\boldsymbol{x},t)\,\boldsymbol{dx}dt - \int_0^T \int_R u(\boldsymbol{x},t)\delta(\boldsymbol{\xi}-\boldsymbol{x})\delta(\tau-t)\,\boldsymbol{dx}dt$$

$$= -\int_R \left[g(\xi,\tau|\mathbf{x},0)\frac{\partial u}{\partial t}(\mathbf{x},0) - u(\mathbf{x},0)\frac{\partial g(\xi,\tau|\mathbf{x},0)}{\partial t} \right] \mathbf{dx}$$

$$+ \int_R \left[g(\xi,\tau|\mathbf{x},T)\frac{\partial u}{\partial t}(\mathbf{x},T) - u(\mathbf{x},T)\frac{\partial g(\xi,\tau|\mathbf{x},T)}{\partial t} \right] \mathbf{dx}$$

$$+ c^2 \int_0^T \int_{S_x} \left[u(\mathbf{x},t)\frac{\partial g}{\partial n_x}(\xi,\tau|\mathbf{x},t) - g(\xi,\tau|\mathbf{x},t)\frac{\partial u(\mathbf{x},t)}{\partial n_x} \right]_{S_x} dS_x\, dt \tag{8.177}$$

If one takes T large enough to exceed $t = \tau$, then the causality of g will make the upper limit $t = \tau$ and the third integrand evaluated at $t = T$ vanishes, since $g = 0$, $t = T > \tau$.

Since $\partial g|\partial t = -\partial g|\partial \tau$, then eq. (8.177) can be simplified to:

$$u(\xi,\tau) = \int_0^\tau \int_R g(\xi,\tau|\mathbf{x},t) f(\mathbf{x},t)\, d\mathbf{x}\, dt$$

$$+ \int_R g(\xi,\tau|\mathbf{x},0) \frac{\partial u}{\partial t}(\mathbf{x},0^+)\, d\mathbf{x} + \frac{\partial}{\partial \tau} \int_R u(\xi,0) g(\xi,\tau|\mathbf{x},0)\, d\mathbf{x}$$

$$- c^2 \int_0^\tau \int_{S_x} \left[u(\mathbf{x},t) \frac{\partial g}{\partial n_x}(\xi,\tau|\mathbf{x},t) - g(\xi,\tau|\mathbf{x},t) \frac{\partial u(\mathbf{x},t)}{\partial n_x} \right]_{S_x} dS_x\, dt \qquad (8.178)$$

One can rewrite the last expression by switching $\mathbf{x}$ to ξ and t to τ and vice versa after noting that $g(\xi,\tau|\mathbf{x},t)=g(\mathbf{x},t|\xi,\tau)$ giving:

$$u(\mathbf{x},t) = \int_0^t \int_R g(\mathbf{x},t|\xi,\tau) f(\xi,\tau)\, d\xi\, d\tau + \int_R g(\mathbf{x},t|\xi,0) f_2(\xi)\, d\xi$$

$$+ \frac{\partial}{\partial t} \int_R g(\mathbf{x},t|\xi,0) f_1(\xi)\, d\xi - c^2 \int_0^t \int_{S_\xi} \left[u \frac{\partial g}{\partial n_\xi} - g \frac{\partial u}{\partial n_\xi} \right]_{S_\xi} dS_\xi\, d\tau \qquad (8.179)$$

The expression shows that the response depends linearly on the initial conditions.

Example 8.13 Transient Vibration of an Infinite String

Obtain the transient response of an infinite stretched string under a distributed load q(x,t), which is initially set in motion, such that:

$$\frac{\partial^2 u}{\partial t^2} - c^2 \frac{\partial^2 u}{\partial x^2} = \frac{q(x,t)}{T_0} \qquad -\infty < x < \infty$$

$$u(x,0) = f_1(x) \qquad \frac{\partial u}{\partial t}(x,0) = f_2(x)$$

For the one-dimensional problem, see eq. (8.121):

$$g(x,t|\xi,\tau) = \frac{1}{2c} H\left[(t-\tau) - |x-\xi|/c\right] H(t-\tau)$$

The Heaviside function can be replaced by:

$$H(a-|b|) = H(a-b) + H(a+b) - 1 \quad \text{for } a > 0$$

Thus, the function g can be rewritten as:

$$g(x,t|\xi,\tau) = \frac{1}{2c} \left\{ H\left[(t-\tau) - (x-\xi)/c\right] + H\left[(t-\tau) + (x-\xi)/c\right] - 1 \right\}$$

For an infinite string, the boundary condition integrals in eq. (8.179) vanish, leaving integrals on the source and the two initial conditions in eq. (8.179). The first integral on the source term can be written by:

$$\frac{1}{2c}\int_0^t \frac{q(\xi,\tau)}{T_0} H\left[(t-\tau)-|x-\xi|/c\right] d\xi \, d\tau = \frac{1}{2cT_0}\int_0^t \int_{x-c(t-\tau)}^{x+c(t-\tau)} q(\xi,\tau)\, d\xi \, d\tau$$

The second integral can be written as:

$$\frac{1}{2c}\int_{-\infty}^{\infty} f_2(\xi) H\left[\tau - |x-\xi|/c\right] d\xi = \frac{1}{2c}\int_{x-ct}^{x+ct} f_2(\xi)\, d\xi$$

The third integral on the initial condition requires the time derivative of g:

$$\frac{\partial g}{\partial t}\left(x, t|\xi, 0^+\right) = \frac{1}{2c}\left\{\delta\left[t-(x-\xi)/c\right] + \delta\left[t+(x-\xi)/c\right]\right\}$$

so that the third integral becomes:

$$= \frac{1}{2c}\int_{-\infty}^{\infty} f_1(\xi)\left\{\delta\left[t-(x-\xi)/c\right] + \delta\left[t+(x-\xi)/c\right]\right\} d\xi = \frac{1}{2}\left[f_1(x+ct) + f_1(x-ct)\right]$$

The final solution for u(x,t) becomes:

$$u(x,t) = \frac{1}{2c}\int_0^t \int_{x-c(t-\tau)}^{x+c(t-\tau)} \frac{q(\xi,\tau)}{T_0} d\xi\, d\tau + \frac{1}{2}\left[f_1(x+ct) + f_1(x-ct)\right] + \frac{1}{2c}\int_{x-ct}^{x+ct} f_2(\xi)\, d\xi$$

For a bounded medium, the requirement to specify u and $\partial u/\partial n$ on the surface makes the problem overspecified and the solution non-unique. Let the auxiliary causal function $\bar{g}$ to satisfy:

$$\frac{\partial^2 \bar{g}}{\partial t^2} - c^2 \nabla^2 \bar{g} = 0 \qquad \mathbf{x} \text{ in R, } t > 0 \tag{8.180}$$

$$\bar{g} = 0 \qquad t < \tau$$

Following the development of eq. (8.179) for g, one obtains:

$$0 = \int_0^t \int_R \bar{g}(\mathbf{x}, t|\xi, \tau) f(\xi, \tau)\, d\xi\, d\tau + \int_R \bar{g}(\mathbf{x}, t|\xi, 0) \frac{\partial u}{\partial t}(\xi, 0)\, d\xi$$

$$+ \frac{\partial}{\partial t}\int_R \bar{g}(\mathbf{x}, t|\xi, 0) u(\xi, 0)\, d\xi - c^2 \int_0^t \int_{S_\xi} \left[u \frac{\partial \bar{g}}{\partial n_\xi} - \bar{g}\frac{\partial u}{\partial n_\xi}\right]_{S_\xi} dS_\xi\, d\tau \tag{8.181}$$

Subtraction of the two equations (8.179) and (8.181), together with the definition $G = g - \bar{g}$ results in the final solution:

$$u(\mathbf{x},t) = \int_0^t \int_R G(\mathbf{x}, t|\xi, \tau) f(\xi, \tau)\, d\xi\, d\tau + \int_R G(\mathbf{x}, t|\xi, 0) f_2(\xi)\, d\xi$$

$$+\frac{\partial}{\partial t}\int_R G(x,t|\xi,0)f_1(\xi)d\xi - c^2\int_0^t\int_{S_\xi}\left[u\frac{\partial G}{\partial n_\xi} - G\frac{\partial u}{\partial n_\xi}\right]_{S_\xi} dS_\xi\, d\tau \qquad (8.182)$$

For the following boundary conditions, one must set conditions on the function G as:

(a) Dirichlet: $G|_{S_\xi} = 0$

(b) Neumann: $\left.\frac{\partial G}{\partial n_\xi}\right|_{S_\xi} = 0$

(c) Robin: $\left.\frac{\partial G}{\partial n_\xi} + \gamma G\right|_{S_\xi} = 0$

The boundary integrals in weq. (8.18) take the forms in eqs (8.132) to (8.135).

Example 8.14 Transient Longitudinal Vibration in an Infinite Bar

Obtain the transient displacement field of a semi-infinite bar at rest, which is set in motion by displacing the bar at the boundary x = 0. The system satisfies the following equation:

$$\frac{\partial^2 u}{\partial t^2} - c^2\frac{\partial^2 u}{\partial x^2} = 0 \qquad u(x,0) = 0 \qquad \frac{\partial u}{\partial t}(x,0) = 0 \qquad u(0,t) = h(t)$$

Using the method of images, let the image of the source at ξ be located at $-\xi$, giving:

$$G = \frac{1}{2c}\left\{H\left[(t-\tau) - |x-\xi|/c\right] - CH\left[(t-\tau) - |x+\xi|/c\right]\right\}$$

The Green's function satisfies $G|_{x=0} = 0$ if C = 1. Rewriting G in a more convenient form using:

H(a-|b|) = H(a-b) + H(a+b) - 1 for a > 0

$$G = \frac{1}{2c}\left\{H\left[(t-\tau) - (x-\xi)/c\right] + H\left[(t-\tau) + (x-\xi)/c\right]\right.$$
$$\left. - H\left[(t-\tau) - (x+\xi)/c\right] - H\left[(t-\tau) + (x+\xi)/c\right]\right\} H(t-\tau)$$

$$\left.\frac{\partial G}{\partial n_\xi}\right|_{\xi=0} = -\left.\frac{\partial G}{\partial \xi}\right|_{\xi=0} = -\frac{1}{c^2}\delta[t-\tau-x/c]$$

giving the final solution:

$$u(x,t) = \int_0^t h(\tau)\delta[t-\tau-x/c]\,d\tau = h(t-x/c)H(ct-x)$$

8.34 Causal Green's Function for the Diffusion Operator for Bounded Media

Consider a system undergoing diffusion, such that:

$$\frac{\partial u}{\partial t} - \kappa\nabla^2 u = f(\mathbf{x},t) \qquad \mathbf{x} \text{ in R}, \ t > 0 \tag{8.96}$$

together with initial conditions:

$$u(\mathbf{x},0) = f_1(\mathbf{x}) \qquad \mathbf{x} \text{ in R}$$

and the boundary conditions:

(a) Dirichlet: $u(\mathbf{x},t)\big|_S = h(\mathbf{x},t)$ $\quad$ $\mathbf{x}$ on S

(b) Neumann: $\dfrac{\partial u}{\partial n}(\mathbf{x},t)\Big|_S = h(\mathbf{x},t)$ $\quad$ $\mathbf{x}$ on S

(c) Robin: $\dfrac{\partial u}{\partial n}(\mathbf{x},t) + \gamma u(\mathbf{x},t)\Big|_S = h(\mathbf{x},t)$ $\quad$ $\mathbf{x}$ on S

The causal fundamental solution $g(\mathbf{x},t|\xi,\tau)$ satisfies:

$$\frac{\partial g}{\partial t} - \kappa\nabla^2 g = \delta(\mathbf{x}-\xi)\delta(t-\tau) \tag{8.97}$$

$$g = 0 \qquad t < \tau$$

$$g(x,0^+|\xi,\tau) = 0$$

Let the adjoint causal fundamental solution $g^*(\mathbf{x},t|\xi,\tau)$ satisfy:

$$-\frac{\partial g^*}{\partial t} - \kappa\nabla^2 g^* = \delta(\mathbf{x}-\xi)\delta(t-\tau) \tag{8.183}$$

$$g^*(\xi,\tau|\mathbf{x},0^+) = 0$$

$$g^* \equiv 0 \qquad t > \tau$$

The two causal Green's functions are related by the symmetry conditions:

$$g(\mathbf{x},t|\xi,\tau) = g^*(\xi,\tau|\mathbf{x},t) \tag{8.184}$$

$$g(\xi,\tau|\mathbf{x},t) = g^*(\mathbf{x},t|\xi,\tau)$$

Using $v = g^*(\mathbf{x},t|\xi,\tau)$ and u(**x**,t) into the Green's identity eq. (8.48), with the surfaces shown in Figure 8.10:

$$\int_0^T \int_R g^*(\mathbf{x},t|\xi,\tau)\, f(\mathbf{x},t)\, d\mathbf{x}\, dt - u(\xi,\tau) =$$

$$= -\kappa \int_0^T \int_{S_x} \left[g^* \frac{\partial u}{\partial n_x} - u \frac{\partial g}{\partial n_x} \right]_{S_x} dS_x \, dt$$

$$- \int_R u(\mathbf{x},0)\, g^*(\mathbf{x},0|\xi,\tau)\, \mathbf{dx} + \int_R u(\mathbf{x},T)\, g^*(\mathbf{x},T|\xi,\tau)\, \mathbf{dx}$$

Again since g^* is causal, let T be taken large enough so that $g = 0$ for $t = T > \tau$. Rearranging the terms gives:

$$u(\xi,\tau) = \int_0^\tau \int_R g^*(\mathbf{x},t|\xi,\tau)\, f(\mathbf{x},t)\, \mathbf{dx}\, dt$$

$$+ \int_R u(\mathbf{x},0)\, g^*(\mathbf{x},0|\xi,\tau)\, \mathbf{dx} + \kappa \int_0^\tau \int_{S_x} \left[g^* \frac{\partial u}{\partial n_x} - u \frac{\partial g^*}{\partial n_x} \right]_{S_x} dS_x \, dt \qquad (8.185)$$

For a bounded medium let the auxiliary causal function $\bar{g}$ satisfy:

$$\frac{\partial \bar{g}}{\partial t} - \kappa \nabla^2 \bar{g} = 0$$

$$\bar{g} = 0 \qquad t < \tau$$

and the adjoint causal auxiliary function $\bar{g}^*$ satisfies:

$$-\frac{\partial \bar{g}^*}{\partial t} - \kappa \nabla^2 \bar{g}^* = 0$$

$$\bar{g}^* = 0 \qquad t > \tau$$

Using $\bar{g}(\xi,\tau|\mathbf{x},t) = \bar{g}^*(\mathbf{x},t|\xi,\tau)$ into the Green's identity eq. (8.48) results in an equation similar to eq. (8.185):

$$0 = \int_0^\tau \int_R \bar{g}^*(\mathbf{x},t|\xi,\tau)\, f(\mathbf{x},t)\, \mathbf{dx}\, dt$$

$$+ \int_R u(x,0)\bar{g}^*(x,0\,|\,\xi,\tau)dx + k \int_{S_x} \left[\bar{g}^* \frac{\partial u}{\partial n_x} - u \frac{\partial g^*}{\partial n_x} \right]_{S_x} dS_x dt \qquad (8.186)$$

Subtraction of eq. (8.186) from eq. (8.185) and using the definition $G^* = g^* - \bar{g}^*$ results in:

$$u(\xi,\tau) = \int_0^\tau \int_R G^*(\mathbf{x},t|\xi,\tau)\, f(\mathbf{x},t)\, \mathbf{dx}\, dt$$

$$+\int_R u(\mathbf{x},0)\,G^*(\mathbf{x},0|\xi,\tau)\,\mathbf{dx}+\kappa\int_0^{\tau}\int_{S_x}\left[G^*\frac{\partial u}{\partial n_x}-u\frac{\partial G^*}{\partial n_x}\right]_{S_x} dS_x\,dt$$

Switching $\mathbf{x}$ to ξ and t to τ and vice versa and recalling that:

$$G^*(\xi,\tau|\mathbf{x},t)=G(\mathbf{x},t|\xi,\tau)$$

one can rewrite the last expression to:

$$u(\mathbf{x},t)=\int_0^t\int_R G(\mathbf{x},t|\xi,\tau)\,f(\xi,\tau)\,\mathbf{d\xi}\,d\tau$$

$$+\int_R f_1(\xi)\,G(\mathbf{x},t|\xi,0)\,\mathbf{d\xi}+\kappa\int_0^t\int_{S_\xi}\left[G\frac{\partial u}{\partial n_\xi}-u\frac{\partial G}{\partial n_\xi}\right]_{S_\xi} dS_\xi\,d\tau \tag{8.187}$$

where $G=g-\bar{g}$. Thus, for the different types of boundary conditions:

(a) Dirichlet: $G|_{S_\xi}=0$

(b) Neumann: $\left.\dfrac{\partial G}{\partial n_\xi}\right|_{S_\xi}=0$

(c) Robin: $\left.\dfrac{\partial G}{\partial n_\xi}+\gamma G\right|_{S_\xi}=0$

The boundary integrals follow the same forms as in eqs. (8.132) to (8.135).

Example 8.15 Heat Flow in a Semi-Infinite Bar

Consider a source-free semi-infinite bar being heated at it's boundary, such that:

$$\frac{\partial u}{\partial t}-\kappa\nabla^2 u=0 \qquad x\geq 0,\ t>0$$

$$u(x,0^+)=0 \qquad u(0,t)=u_o\,h(t)$$

To construct $G(x,t|\xi,\tau)$ for a Dirichlet boundary condition, then both conditions must be satisfied; $G(0,t|\xi,\tau)=0$ and $G(x,t|0,\tau)=0$.

The fundamental Green's function $g(x,t|\xi,\tau)$ is given in eq. (8.106). To construct the auxiliary function $\bar{g}$, let the image source be located at ($-\xi$) such that:

$$\bar{g}(x,t\,|-\xi,\tau)=C\frac{e^{-(x+\xi)^2/[4\kappa(t-\tau)]}}{[4\pi\kappa(t-\tau)]^{1/2}}H(t-\tau) \tag{8.101}$$

then to make $G(0,t|\xi,\tau)=0$ requires that $C=-1$, and G becomes:

$$G = \frac{H(t-\tau)}{[4\pi\kappa(t-\tau)]^{1/2}} \left\{ e^{-(x-\xi)^2/[4\kappa(t-\tau)]} - e^{-(x+\xi)^2/[4\kappa(t-\tau)]} \right\} \tag{8.101}$$

The final solution requires the evaluation of $\partial G/\partial n_\xi$:

$$\left.\frac{\partial G}{\partial n_\xi}\right|_{\xi=0} = -\left.\frac{\partial G}{\partial \xi}\right|_{\xi=0} = -\frac{x\,H(t-\tau)}{\sqrt{4\pi}\,[\kappa(t-\tau)]^{3/2}}\, e^{-x^2/[4\kappa(t-\tau)]}$$

Therefore, the temperature distribution in the bar due to the non-homogenous boundary condition is given by:

$$u(x,t) = \frac{u_o x}{\sqrt{4\pi\kappa}} \int_0^t \frac{h(\tau)}{(t-\tau)^{3/2}}\, e^{-x^2/[4\kappa(t-\tau)]}\, d\tau$$

Example 8.16 Temperature Distribution in a Semi-Infinite Bar

Find the temperature distribution in a source-free, semi-infinite solid bar with Newton's law of cooling at the boundary where the external ambient temperature is $u_o h(t)$, such that the temperature u(x,t) satisfies:

$$\frac{\partial u}{\partial t} - \kappa \frac{\partial^2 u}{\partial x^2} = 0 \qquad x \geq 0,\ t > 0$$

$$u(x,0) = 0 \qquad -\frac{\partial u}{\partial x}(0,t) + \gamma\, u(0,t) = \gamma\, u_o\, h(t)$$

Here the boundary condition is the Robin condition such that:

$$\left.\left(\frac{\partial G}{\partial n_\xi} + \gamma G\right)\right|_{\xi=0} = \left.\left(-\frac{\partial G}{\partial \xi} + \gamma G\right)\right|_{\xi=0} = 0$$

or

$$\left.\left(-\frac{\partial G}{\partial x} + \gamma G\right)\right|_{x=0} = 0$$

Let the function $w(x,t|\xi,\tau)$ be defined by:

$$w(x,t|\xi,\tau) = \frac{\partial G}{\partial x} - \gamma G$$

Substituting w into the diffusion equation and recalling that $G = g - \bar{g}$, then:

$$\left(\frac{\partial}{\partial t} - \kappa\frac{\partial^2}{\partial x^2}\right) w = \left(\frac{\partial}{\partial x} - \gamma\right)\left(\frac{\partial g}{\partial t} - \kappa\frac{\partial^2 g}{\partial x^2}\right) - \left(\frac{\partial}{\partial x} - \gamma\right)\left(\frac{\partial \bar{g}}{\partial t} - \kappa\frac{\partial^2 \bar{g}}{\partial x^2}\right)$$

$$= \left(\frac{\partial}{\partial x} - \gamma\right)\left[\delta(x-\xi)\,\delta(t-\tau)\right]$$

$$= \left[\frac{\partial \delta}{\partial x}(x-\xi) - \gamma\,\delta(x-\xi)\right]\delta(t-\tau)$$

together with the boundary condition on w, $w(0,t|\xi,\tau) = 0$. This shows that the function w satisfies the Dirichlet boundary with the above prescribed source term. The Green's function for a Dirichlet boundary condition is given in Example (8.15):

$$w(x,t|\xi,\tau) = \int_0^t \int_0^\infty G(x,t\,|\,\eta,\zeta)\left[\frac{\partial\delta}{\partial\eta}(\eta-\xi) - \gamma\,\delta(\eta-\xi)\right]\delta(\zeta-\tau)\,d\eta\,d\zeta$$

$$= \frac{H(t-\tau)}{\sqrt{4\pi\kappa(t-\tau)}}\left\{\gamma\left[e^{-(x+\xi)^2/[4\kappa(t-\tau)]} - e^{-(x-\xi)^2/[4\kappa(t-\tau)]}\right] + \frac{\partial}{\partial x}\left[e^{-(x+\xi)^2/[4\kappa(t-\tau)]} + e^{-(x-\xi)^2/[4\kappa(t-\tau)]}\right]\right\}$$

Integrating the equation for G, one obtains:

$$G = -e^{\gamma x}\int_x^\infty w(u,t\,|\,\xi,\tau)\,e^{-\gamma u}\,du$$

Integrating by parts the second bracketed quantity in w, results in the following expression for G:

$$G(x,t\,|\,\xi,\tau) = \frac{H(t-\tau)}{\sqrt{4\pi\kappa(t-\tau)}}\left[e^{-(x-\xi)^2/[4\kappa(t-\tau)]} - e^{-(x+\xi)^2/[4\kappa(t-\tau)]}\right]$$

$$-2\gamma\,e^{\gamma x}\left[\int_x^\infty e^{-\gamma u}e^{-(u+\xi)^2/[4\kappa(t-\tau)]}\,du\right]H(t-\tau)$$

The last expression in the integral form can be shown to result in:

$$-\gamma\,H(t-\tau)\,e^{\gamma[x+\xi+\kappa\gamma(t-\tau)]}\,\text{erfc}\left[\frac{x+\xi}{\sqrt{4\kappa(t-\tau)}} + \gamma\sqrt{\kappa(t-\tau)}\right]$$

Note that if $\gamma = 0$, one retrieves the Green's function for Neumann boundary condition. If the limit is taken as $\gamma \rightarrow \infty$, then one obtains the Green's function for Dirichlet boundary condition, matching that given in Example (8.15).

The final solution is given by:

$$u(x,t) = \kappa\,\gamma\,u_0\int_0^t G(x,t\,|\,0,\tau)\,h(\tau)\,d\tau$$

8.35 Method of Summation of Series Solutions in Two-Dimensional Media

The Green's function can also be obtained for two-dimensional media for Poisson's and Helmholts eqs. in closed form by summing the series solutions. This method was developed by Melnikov. Since the fundamental Green's function is logarithmic, then all the Green's functions will involve logarithmic solutions as well. This method depends on the following expansion for a logarithmic function:

$$\log\sqrt{1-2u\cos\phi+u^2} = -\sum_{n=1}^{\infty}\frac{u^n}{n}\cos(n\phi) \tag{8.188}$$

provided that $|u| < 1$, and $0 \le \phi \le 2\pi$

The method of finding the Green's function depends on the geometry of the problem and the boundary conditions.

8.35.1 Laplace's Equation in Cartesian Coordinates

In order to show how this method may be applied it is best to work out an example.

Example 8.17 *Green's Function for a Semi-Infinite Strip*

Consider the semi-infinite strip $0 \le x \le L$, $y \ge 0$ for the function u(x,y) satisfying:

$$-\nabla^2 u = f(x,y) \qquad 0 \le x \le L,\ y \ge 0$$

Subject to the Dirichlet boundary condition on all three sides, i.e.:

$$u(0,y) = 0 \qquad u(h,y) = 0 \qquad u(x,0) = 0$$

and $u(x,\infty)$ is bounded.

One may obtain the solution in an infinite series of eigenfunctions in the x-coordinates, since the two boundary conditions on x = 0, L are homogenous. These eigenfunctions are given by sin $(n\pi x/L)$ [see Chapter 6, problem 2], which satisfies the two homogenous boundary conditions.

Let the final solution be expanded in these eigenfunctions as:

$$u = \sum_{n=1}^{\infty} u_n(y)\sin(\frac{n\pi}{L}x)$$

Substituting into Laplace's equation and using the orthogonality of the eigenfunctions one obtains:

$$\frac{d^2u_n}{dy^2} - (\frac{n^2\pi^2}{L^2})u_n = -f_n(y) \qquad n = 1, 2, 3, \ldots$$

where

$$f_n(y) = \frac{2}{L}\int_0^L f(x,y)\sin(\frac{n\pi}{L}x)\,dx$$

The solution for the non-homogenous differential equation is given by the solution to Chapter 1, problem 7d:

$$u_n(y) = A_n\sinh(\frac{n\pi}{L}y) + B_n\cosh(\frac{n\pi}{L}y) + \frac{L}{n\pi}\int_0^y f_n(\eta)\sinh(\frac{n\pi}{L}(y-\eta))\,d\eta$$

Since $u(0,x) = 0$, then $B_n = 0$ and:

$$u_n(y) = A_n \sinh(\frac{n\pi}{L} y) + \frac{L}{n\pi} \int_0^y f_n(\eta) \sinh(\frac{n\pi}{L}(y-\eta))\, d\eta$$

Also since $u(x,\infty)$ is bounded then:

$$A_n = -\frac{L}{n\pi} \int_0^\infty f_n(\eta) e^{-n\pi\eta/L}\, d\eta$$

The final solution is then:

$$u_n(y) = \frac{-L}{2n\pi} \int_0^\infty f_n(\eta) \left[e^{n\pi(y-\eta)/L} H(\eta - y) - e^{-n\pi(y+\eta)/L} + e^{-n\pi(y-\eta)/L} H(\eta - y) \right] d\eta$$

Thus, the Green's function for the y-component is given by:

$$G_n(y \mid \eta) = \frac{L}{2n\pi} \begin{cases} e^{-n\pi(y+\eta)/L} - e^{n\pi(y-\eta)/L} & y \le \eta \\ e^{-n\pi(y+\eta)/L} - e^{-n\pi(y-\eta)/L} & y \ge \eta \end{cases}$$

Note that $G_n(0|\eta) = 0$.

Thus, the solution for $u_n(y)$ is given by:

$$u_n(y) = \int_0^\infty G_n(y \mid \eta) f_n(\eta) d\eta = \frac{2}{L} \int_0^\infty \int_0^L G_n(y \mid \eta)\, f(\xi, \eta) \sin(\frac{n\pi}{L} \xi)\, d\xi\, d\eta$$

$$u(x,y) = \frac{2}{L} \sum_{n=1}^\infty \int_0^L \int_0^\infty G_n(y \mid \eta) \sin(\frac{n\pi}{L} \xi) \sin(\frac{n\pi}{L} x)\, f(\xi, \eta)\, d\xi\, d\eta$$

$$= \int_0^L \int_0^\infty \left\{ \frac{2}{L} \sum_{n=1}^\infty G_n(y \mid \eta) \sin(\frac{n\pi}{L} \xi) \sin(\frac{n\pi}{L} x) \right\} f(\xi, \eta)\, d\xi\, d\eta$$

Thus, the Green's function $G(x,y|\xi,\eta)$ is given by:

$$G(x, y \mid \xi, \eta) = \frac{2}{L} \sum_{n=1}^\infty G_n(y \mid \eta) \sin(\frac{n\pi}{L} \xi) \sin(\frac{n\pi}{L} x)$$

$$= \frac{1}{L} \sum_{n=1}^\infty G_n(y \mid \eta) \left[\cos(\frac{n\pi}{L}(x - \xi)) - \cos(\frac{n\pi}{L}(x + \xi)) \right]$$

For the region $y \ge \eta$:

$$G(x,y\,|\,\xi,\eta)=\frac{1}{2\pi}\sum_{n=1}^{\infty}\frac{1}{n}\left[\exp(-\frac{n\pi}{L}(y+\eta))-\exp(-\frac{n\pi}{L}(y-\eta))\right]\cdot$$

$$\cdot\left[\cos(\frac{n\pi}{L}(x-\xi))-\cos(\frac{n\pi}{L}(x+\xi))\right]$$

Using the summation formula eq. (8.188), on the first of four terms, one gets the following closed form:

$$\frac{1}{L}\sum_{n=1}^{\infty}\exp(-\frac{n\pi}{L}(y+\eta))\cos(\frac{n\pi}{L}(x-\xi))$$

Here:

$$u=\exp(-\frac{\pi}{L}(y+\eta))\qquad \phi=\frac{\pi}{L}(x-\xi)$$

$$=\frac{-1}{2\pi}\log\sqrt{1-2\exp(-\frac{\pi}{L}(y+\eta))\cos(\frac{\pi}{L}(x-\xi))+\exp(-\frac{2\pi}{L}(y+\eta))}$$

Similarly, one obtains the closed form for each of the remaining three series, resulting in the final form:

$$G(x,y\,|\,\xi,\eta)=\frac{1}{4\pi}\log(\frac{AB}{CD}) \tag{8.189}$$

where:

$$A=1-2\exp(\frac{\pi}{L}(y+\eta))\cos(\frac{\pi}{L}(x-\xi))+\exp(\frac{2\pi}{L}(y+\eta))$$

$$B=1-2\exp(\frac{\pi}{L}(y-\eta))\cos(\frac{\pi}{L}(x+\xi))+\exp(\frac{2\pi}{L}(y-\eta))$$

$$C=1-2\exp(\frac{\pi}{L}(y-\eta))\cos(\frac{\pi}{L}(x-\xi))+\exp(\frac{2\pi}{L}(y-\eta))$$

$$D=1-2\exp(\frac{\pi}{L}(y+\eta))\cos(\frac{\pi}{L}(x+\xi))+\exp(\frac{2\pi}{L}(y+\eta))$$

The method of images would have resulted in an infinite number of images.

8.35.2 Laplace's Equation in Polar Coordinates

The use of the summation for obtaining closed form solutions for circular regions in two dimensions can be best illustrated by examples.

Example 8.18 Green's Function for the Interior/Exterior of a Circular Region with Dirichlet Boundary Conditions

First, consider the solution in the interior circular region $r\le a$ with Dirichlet boundary condition governed by Poisson's equation, such that:

$$-\nabla^2 u=f(r,\theta)\qquad u(a,\theta)=0\qquad r\le a,\;\; 0\le\theta\le 2\pi$$

The eigenfunctions in angular coordinates are:

$$\sin(n\theta) \qquad n = 1, 2, 3, \ldots$$

$$\cos(n\theta) \qquad n = 0, 1, 2, \ldots$$

Expanding the solution in terms of these eigenfunctions:

$$u = u_0(r) + \sum_{n=1}^{\infty} u_n^c(r)\cos(n\theta) + \sum_{n=1}^{\infty} u_n^s(r)\sin(n\theta)$$

so that the functions u_n satisfy:

$$\frac{1}{r}\frac{d}{dr}\left(r\frac{du_o}{dr}\right) = -f_o(r)$$

$$\frac{1}{r}\frac{d}{dr}\left(r\frac{du_n^{c,s}}{dr}\right) - \frac{n^2}{r^2}u_n^{c,s} = -f_n^{c,s}(r)$$

where:

$$f_o(r) = \frac{1}{2\pi}\int_0^{2\pi} f(r,\theta)\, d\theta$$

and

$$f_n^{c,s}(r) = \frac{1}{\pi}\int_0^{2\pi} f(r,\theta) \begin{matrix}\cos(n\theta)\\ \sin(n\theta)\end{matrix}\, d\theta$$

Integrating the differential equation for $u_o(r)$ gives:

$$u_o(r) = A_o \log r + B_o + \int_0^r \log\left(\frac{\rho}{r}\right) f_o(\rho)\rho\, d\rho$$

The condition that $u_o(0)$ is bounded requires that $A_o = 0$ and the boundary condition at $r = a$ results in the expression for B_o:

$$B_o = -\int_0^a \log\left(\frac{\rho}{a}\right) f_o(\rho)\rho\, d\rho$$

so that:

$$u_o(r) = \int_0^r \log\left(\frac{\rho}{r}\right) f_o(\rho)\rho\, d\rho - \int_0^a \log\left(\frac{\rho}{a}\right) f_o(\rho)\rho\, d\rho$$

$$= \int_0^a G_o(r\,|\,\rho)\, f_o(\rho)\rho\, d\rho$$

where:

$$G_0(r\,|\,\rho) = \begin{cases} -\log(\frac{\rho}{a}) & r \le \rho \\ -\log(\frac{r}{a}) & r \ge \rho \end{cases}$$

It should be noted that $G_0(a|\rho) = G_0(r|a) = 0$ as required by the Dirichlet boundary condition.

Integrating the differential equation for $u_n^{c,s}(r)$ results in the following solution:

$$u_n^{c,s}(r) = A_n^{c,s} r^{-n} + B_n^{c,s} r^n + \frac{1}{2n} \int_0^r \left[\left(\frac{\rho}{r}\right)^n - \left(\frac{r}{\rho}\right)^n\right] f_n^{c,s}(\rho)\rho\, d\rho$$

Again $u_n^{c,s}(0)$ is bounded requires that $A_n = 0$ and the boundary on r = a requires that:

$$B_n^{c,s} = -\frac{1}{2n\, a^n} \int_0^a \left[\left(\frac{\rho}{a}\right)^n - \left(\frac{a}{\rho}\right)^n\right] f_n^{c,s}(\rho)\,\rho\, d\rho$$

Thus, the final solution for $u_n^{c,s}(r)$ becomes:

$$u_n^{c,s}(r) = \frac{1}{2n} \int_0^r \left[\left(\frac{\rho}{r}\right)^n - \left(\frac{r}{\rho}\right)^n\right] f_n^{c,s}(\rho)\rho\, d\rho - \frac{1}{2n}\left(\frac{r}{a}\right)^n \int_0^a \left[\left(\frac{\rho}{a}\right)^n - \left(\frac{a}{\rho}\right)^n\right] f_n^{c,s}(\rho)\,\rho\, d\rho$$

which can be written as:

$$u_n^{c,s}(r) = \frac{1}{2n} \int_0^a \left[\left(\frac{\rho}{r}\right)^n - \left(\frac{r}{\rho}\right)^n\right] f_n^{c,s}(\rho)\rho\, H(r-\rho)\, d\rho$$

$$- \frac{1}{2n}\left(\frac{r}{a}\right)^n \int_0^a \left[\left(\frac{\rho}{a}\right)^n - \left(\frac{a}{\rho}\right)^n\right] f_n^{c,s}(\rho)\rho\, d\rho$$

$$= \int_0^a G_n(r\,|\,\rho)\; f_n^{c,s}(\rho)\rho\, d\rho$$

where

$$G_n(r\,|\,\rho) = \begin{cases} \frac{1}{2n}\left[\left(\frac{r}{\rho}\right)^n - \left(\frac{r\rho}{a^2}\right)^n\right] & \text{for} \quad r \le \rho \\ \frac{1}{2n}\left[\left(\frac{\rho}{r}\right)^n - \left(\frac{r\rho}{a^2}\right)^n\right] & \text{for} \quad r \ge \rho \end{cases}$$

Note that $G_n(a|\rho) = G_n(r|a) = 0$ as required by the Dirichlet boundary condition. Finally, the solution for $u(r,\theta)$ is obtained by solutions into the original eigenfunction expansion:

$$u(r,\theta)=\int_0^a G_0(r\,|\,\rho)\; f_0(\rho)\rho\, d\rho$$

$$+\sum_{n=1}^{\infty}\int_0^a G_n(r\,|\,\rho)\;\left[f_n^c(\rho)\cos(n\theta)+f_n^s(\rho)\sin(n\theta)\right]\rho\, d\rho$$

$$=\frac{1}{2\pi}\int_0^{2\pi}\int_0^a G_0(r\,|\,\rho)\; f(\rho,\phi)\,\rho\, d\rho\, d\phi$$

$$+\frac{1}{\pi}\sum_{n=1}^{\infty}\int_0^{2\pi}\int_0^a G_n(r\,|\,\rho)\;\left[\cos(n\varphi)\cos(n\theta)+\sin(n\phi)\sin(n\theta)\right]f(\rho,\phi)\rho\, d\rho\, d\phi$$

$$=\frac{1}{2\pi}\int_0^{2\pi}\int_0^a G_0(r\,|\,\rho)\; f(\rho,\phi)\,\rho\, d\rho\, d\phi$$

$$+\frac{1}{\pi}\sum_{n=1}^{\infty}\int_0^{2\pi}\int_0^a G_n(r\,|\,\rho)\cos(n(\theta-\varphi))\, f(\rho,\phi)\,\rho\, d\rho\, d\phi$$

Thus, the Green's function becomes:

$$G(r,\theta\,|\,\rho,\phi)=\frac{1}{2\pi}\left[G_0(r\,|\,\rho)+2\sum_{n=1}^{\infty} G_n(r\,|\,\rho)\cos(n(\theta-\varphi))\right]$$

The series can be summed, eq. (8.188) as:

$$2\sum_{n=1}^{\infty} G_n(r\,|\,\rho)\cos(n(\theta-\varphi))=\sum_{n=1}^{\infty}\frac{1}{n}\left[\left(\frac{r}{\rho}\right)^n-\left(\frac{r\rho}{a^2}\right)^n\right]\cos(n(\theta-\varphi))$$

$$=\frac{1}{2}\log\left(\frac{1-2(\frac{r\rho}{a^2})\cos(\theta-\varphi)+(\frac{r\rho}{a^2})^2}{1-2(\frac{r}{\rho})\cos(\theta-\varphi)+(\frac{r}{\rho})^2}\right)$$

$$=\frac{1}{2}\log\left(\frac{1-2(\frac{r}{\bar\rho})\cos(\theta-\varphi)+(\frac{r}{\bar\rho})^2}{1-2(\frac{r}{\rho})\cos(\theta-\varphi)+(\frac{r}{\rho})^2}\right)$$

where $\bar\rho = a^2/\rho$ is the location of the image of the source at ρ. Therefore:

$$G(r,\theta\,|\,\rho,\phi)=\frac{1}{4\pi}\left\{-\log\left(\frac{\rho}{a}\right)^2+\log\left[\left(\frac{\bar\rho^2-2r\bar\rho\cos(\theta-\varphi)+r^2}{\rho^2-2r\rho\cos(\theta-\varphi)+r^2}\right)\left(\frac{\rho}{\bar\rho}\right)^2\right]\right\}$$

$$=\frac{1}{4\pi}\log\left(\frac{\rho^2}{a^2}\frac{r_2^2}{r_1^2}\right) \tag{8.190}$$

the notation for r_2 and r_1 are given in Section (8.28). Note that the last answer is the same as the one given in eq. (8.142).

In the exterior region $r \geq a$, one can use the same Green's function, with the notation that $\rho > a$ is the source location and hence $\bar{\rho} = a^2 / \rho < a$.

Example 8.19 Green's Function for the Interior Region of a Circular Region with Neumann Boundary

Consider the solution in the interior region of a circle $r \leq a$ with Neumann boundary condition, governed by Poisson's equation, such that:

$$-\nabla^2 u = f(r,\theta) \quad \text{and} \quad \frac{\partial u}{\partial r}(a,\theta) = 0$$

Following Example 8.17, then u_o is given by:

$$u_o(r) = A_o \log r + B_o + \int_0^r \log(\frac{\rho}{r})\, f_o(\rho)\rho\, d\rho$$

Requiring that $u_o(0)$ is bounded and satisfying the Neumann boundary condition results in:

$$A_o = \int_0^a f_o(\rho)\,\rho\, d\rho$$

and

$$u_o(r) = \int_0^a \log(\rho)\, f_o(\rho)\rho\, d\rho + \int_r^a \log(\frac{r}{\rho})\, f_o(\rho)\rho\, d\rho$$

$$= \int_0^a G_o(r\,|\,\rho)\, f_o(\rho)\rho\, d\rho$$

where

$$G_o(r\,|\,\rho) = \begin{cases} \log(\frac{\rho}{a}) & \rho \leq r \\ \log(\frac{r}{a}) & \rho \geq r \end{cases}$$

For the functions $u_n^{c,s}(r)$:

$$u_n^{c,s}(r) = A_n^{c,s} r^{-n} + B_n^{c,s} r^n + \frac{1}{2n}\int_0^r \left[\left(\frac{\rho}{r}\right)^n - \left(\frac{r}{\rho}\right)^n\right] f_n^{c,s}(\rho)\rho\, d\rho$$

Requiring that $u_n^{c,s}(0)$ is bounded and $\frac{\partial u_n}{\partial r}(a) = 0$ results in:

$$B_n^{c,s} = \frac{1}{2n\, a^n}\int_0^a \left[\left(\frac{\rho}{a}\right)^n + \left(\frac{a}{\rho}\right)^n\right] f_n^{c,s}(\rho)\rho\, d\rho$$

Finally, the function $u_n^{c,s}(r)$ can be written in compact form as:

$$u_n^{c,s}(r) = \int_0^a G_n(r \mid \rho)\, f_n^{c,s}(\rho)\rho\, d\rho$$

where:

$$G_n(r \mid \rho) = \begin{cases} \dfrac{1}{2n}\left[\left(\dfrac{r}{\rho}\right)^n + \left(\dfrac{r\rho}{a^2}\right)^n\right] & \text{for} \quad r \le \rho \\[2ex] \dfrac{1}{2n}\left[\left(\dfrac{\rho}{r}\right)^n + \left(\dfrac{r\rho}{a^2}\right)^n\right] & \text{for} \quad r \ge \rho \end{cases}$$

Substituting $G_o(r|\rho)$ and $G_n(r|\rho)$ into the solutions for $u_n^{c,s}$ and those in turn into she series for $u(r,\theta)$ results in the solution given by:

$$u(r,\theta) = \frac{1}{2\pi}\int_0^{2\pi}\int_0^a \left\{ G_o(r \mid \rho) + 2\sum_{n=1}^{\infty} G_n(r \mid \rho)\cos(n(\theta - \varphi)) \right\} f(\rho,\phi)\rho\, d\rho\, d\phi$$

Summing the series results in the form given in eq. (8.188):

$$\begin{aligned} G(r,\theta \mid \rho,\phi) &= \frac{1}{4\pi}\left\{ \log\left(\frac{r}{a}\right)^2 - \log\left(1 - 2\left(\frac{r\rho}{a^2}\right)\cos(\theta - \phi) + \left(\frac{r\rho}{a^2}\right)^2\right) \right. \\ &\qquad \left. - \log\left(1 - 2\left(\frac{r}{\rho}\right)\cos(\theta - \phi) + \left(\frac{r}{\rho}\right)^2\right)\right\} \\ &= \frac{1}{4\pi}\left\{ \log\left(\frac{r}{a}\right)^2 + \log \rho^2\bar{\rho}^2 - \log(r_1^2 r_2^2)\right\} \\ &= -\frac{1}{4\pi}\log\left(\frac{r_1^2 r_2^2}{a^2 r^2}\right) \end{aligned} \qquad (8.191)$$

The last expression is written in the notation of Section (8.28) and matches the solution given in eq. (8.144).

The Greens function is symmetric in (r,ρ) and satisfies:

$$\frac{\partial G}{\partial r}(a,\theta \mid \rho,\phi) = \frac{\partial G}{\partial \rho}(r,\theta \mid a,\phi) = 0$$

In the exterior region, one may use the form in eq. (8.190) with the notation at the source $\rho > a$ and its image $\bar{\rho} = a^2/\rho < a$.

PROBLEMS

Section 8.1

Obtain the Green's function for the following boundary value problems:

1. $\dfrac{d^2y}{dx^2} + y = x,$ $\quad 0 \le x \le 1 y(0) = 1 \quad y'(1) = 0$

   also obtain y(x)

2. $\dfrac{d}{dx}\left(x\dfrac{dy}{dx}\right) - \dfrac{n^2}{x}y = f(x),$ $\quad 0 \le x \le 1 y(0)$ finite $\quad y(1) = 0$

3. $x^2\dfrac{d^2y}{dx^2} + x\dfrac{dy}{dx} - n^2y = f(x),$ $\quad 0 \le x \le 1 y(0)$ finite $\quad y(1) = 0$

4. $\dfrac{d^2y}{dx^2} - k^2y = f(x),$ $\quad 0 \le x \le L \quad y(0) = 0 \quad y(L) = 0$

5. $-\dfrac{d^4y}{dx^4} = f(x),$ $\quad 0 \le x \le L \quad y(0) = y'(0) = 0, y''(L) = y'''(L) = 0$

6. $-\dfrac{d^4y}{dx^4} = f(x),$ $\quad 0 \le x \le L \quad y(0) = y'(0) = 0, y(L) = y'(L) = 0$

7. $-\dfrac{d^4y}{dx^4} = f(x)$ $\quad 0 \le x \le L \quad y(0) = y''(0) = 0, y(L) = y''(L) = 0$

Section 8.7

Obtain the Green's function for the following eigenvalue problems by:

(a) Direct integration (b) Eigenfunction expansion

8. $\dfrac{d^2y}{dx^2} + k^2y = f(x)$ $\quad 0 \le x \le L \quad y(0) = 0 \quad y(L) = 0$

9. $-\dfrac{d^4y}{dx^4} + \beta^4y = f(x)$ $\quad 0 \le x \le L$

   (i) $y(0) = 0,\ y''(0) = 0,\ y(L) = 0,\ y''(L) = 0$

   (ii) $y(0) = 0,\ y'(0) = 0,\ y(L) = 0,\ y'(L) = 0$

10. $\frac{d}{dx}\left(x\frac{dy}{dx}\right)-\frac{n^2}{x}y+k^2xy=f(x)$ $\quad 0 \le x \le 1 \quad y(0)$ finite $\quad y(1)=0$

11. $x^2\frac{d^2y}{dx^2}+2x\frac{dy}{dx}+k^2x^2y=f(x)$ $\quad 0 \le x \le 1 \quad y(0)$ finite $\quad y(1)=0$

Section 8.8

12. Find the Green's function for a beam on an elastic foundation having a spring constant γ^4:

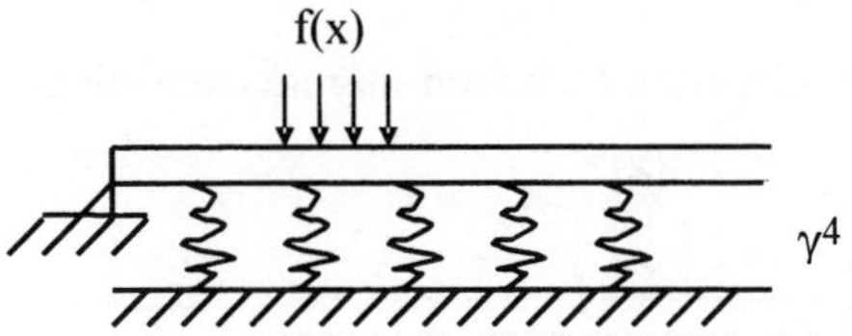

$$-\frac{d^4y}{dx^4}-\gamma^4y=f(x) \qquad x \ge 0 \qquad y(0)=0 \qquad y''(0)=0$$

13. Find the Green's function for a vibrating string under tension and resting on an elastic foundation whose spring constant is γ:

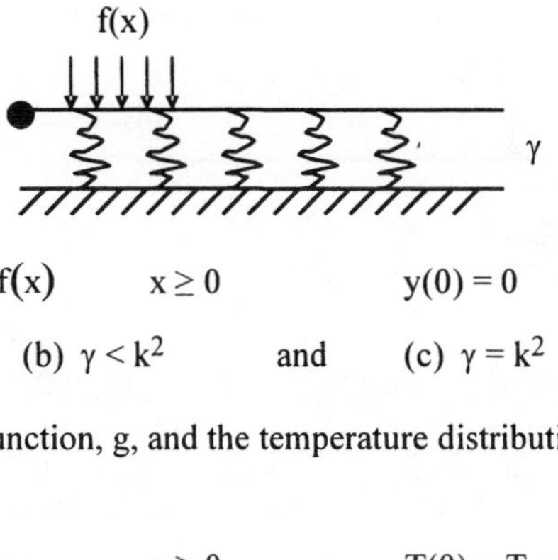

$$-\frac{d^2y}{dx^2}+\gamma y-k^2y=f(x) \qquad x \ge 0 \qquad y(0)=0$$

for: (a) $\gamma > k^2$ (b) $\gamma < k^2$ and (c) $\gamma = k^2$

14. Obtain the Green's function, g, and the temperature distribution, T, in a semi-infinite bar, such that:

$$-\frac{d^2T}{dx^2}=f(x) \qquad x \ge 0 \qquad T(0)=T_1=\text{const}$$

15. Find the Green's function for a semi-infinite, simply supported vibrating beam:

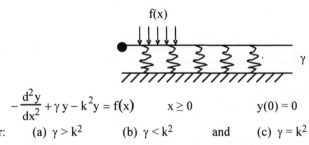

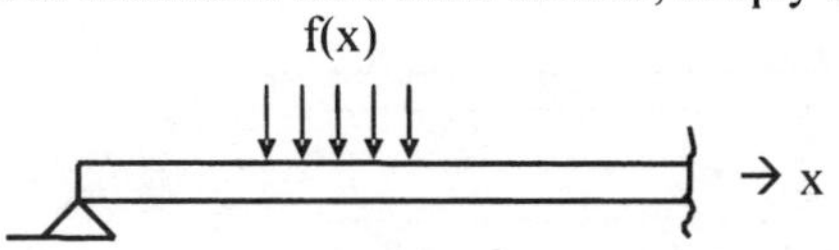

$$-\frac{d^4y}{dx^4}+\beta^4y=f(x) \qquad x \ge 0 \qquad y(0)=0 \qquad y''(0)=0$$

16. Find the Green's function for the semi-infinite fixed-free vibrating beam:

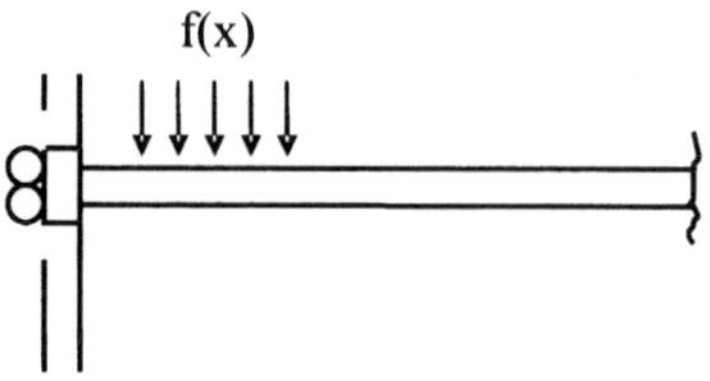

$$-\frac{d^4y}{dx^4}+\beta^4 y = f(x) \qquad x \ge 0 \qquad y'(0)=0 \qquad y'''(0)=0$$

17. Find the Green's function for a semi-infinite fixed vibrating beam such that:

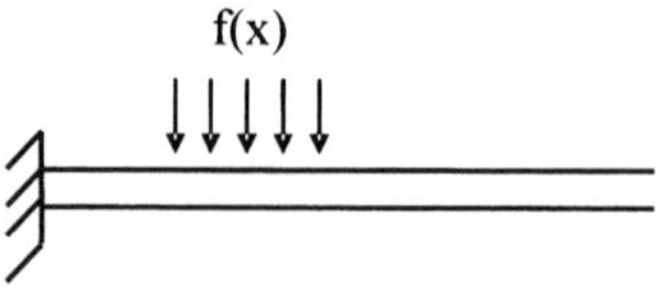

$$-\frac{d^4y}{dx^4}+\beta^4 y = f(x) \qquad x \ge 0 \qquad y(0)=0 \qquad y'(0)=0$$

18. Find the Green's function for a vibrating semi-infinite, simply supported beam resting on an elastic foundation, whose elastic constant is γ^4, such that:

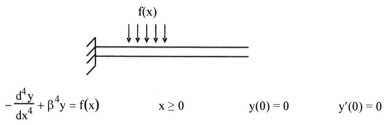

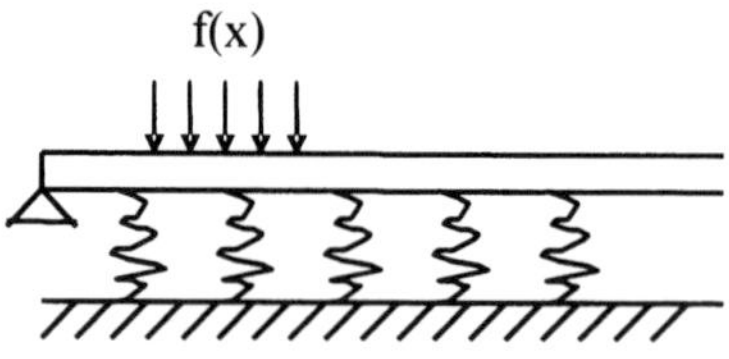

$$-\frac{d^4y}{dx^4}-\gamma^4 y+\beta^4 y = f(x) \qquad x \ge 0 \qquad y(0)=0 \qquad y''(0)=0$$

for (a) $\gamma > \beta$ (b) $\gamma < \beta$ and (c) $\gamma = \beta$

19. Find the Green's function for a vibrating semi-infinite fixed-free beam resting on an elastic foundation, whose elastic constant is γ^4, such that:

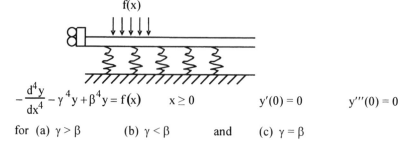

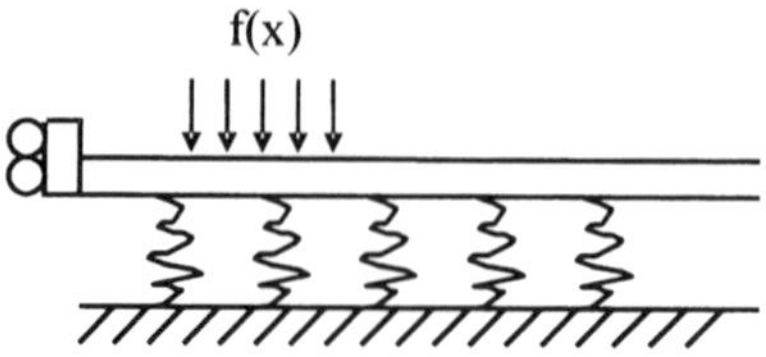

$$-\frac{d^4y}{dx^4}-\gamma^4 y+\beta^4 y = f(x) \qquad x \ge 0 \qquad y'(0)=0 \qquad y'''(0)=0$$

for (a) $\gamma > \beta$ (b) $\gamma < \beta$ and (c) $\gamma = \beta$

Section 8.9

20. Find the Green's function for an infinite beam on an elastic foundation:

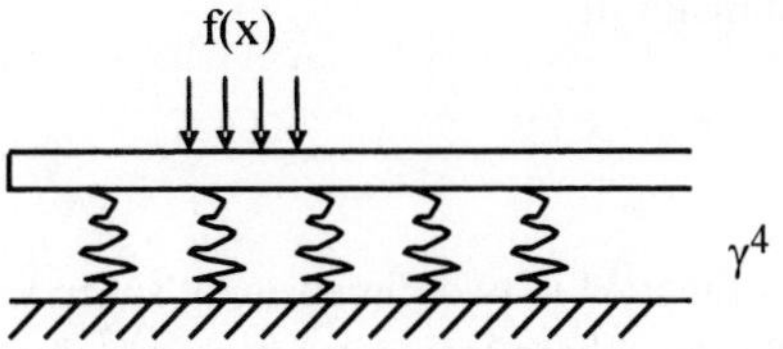

$$-\frac{d^4y}{dx^4} - \gamma^4 y = f(x) \qquad -\infty < x < \infty$$

21. Find the Green's function for a vibrating string under tension and resting on an elastic foundation, whose elastic constant is γ:

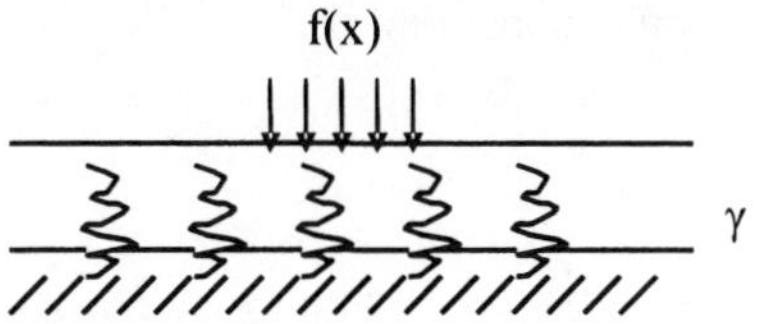

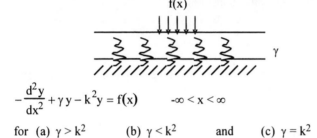

$$-\frac{d^2y}{dx^2} + \gamma y - k^2 y = f(x) \qquad -\infty < x < \infty$$

for (a) $\gamma > k^2$ (b) $\gamma < k^2$ and (c) $\gamma = k^2$

22. Find the Green's function for the temperature distribution in an infinite solid rod:

$$-\frac{d^2T}{dx^2} = f(x) \qquad -\infty < x < \infty$$

23. Find the Green's function for an infinite vibrating beam:

$$-\frac{d^4y}{dx^4} + \beta^4 y = f(x) \qquad -\infty < x < \infty$$

24. Find the Green's function for an infinite vibrating beam resting on an elastic foundation, whose elastic constant is γ^4:

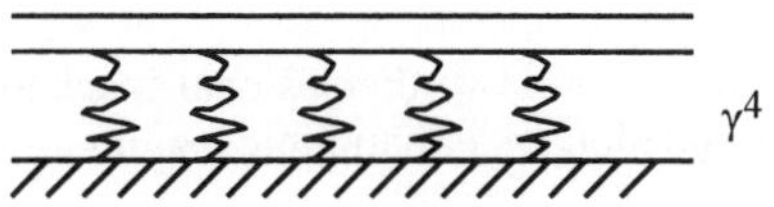

$$-\frac{d^4y}{dx^4} - \gamma^4 y + \beta^4 y = f(x) \qquad -\infty < x < \infty$$

for (a) $\gamma > \beta$ (b) $\gamma < \beta$ and (c) $\gamma = \beta$

Sections 8.17 - 8.20

25. Find the Fundamental Green's function in two-dimensional space for a stretched membrane by use of Hankel transform:

    $-\nabla^2 g = \delta(\mathbf{x} - \xi)$

26. Find the Fundamental Green's function in two-dimensional space for a stretched membrane on an elastic foundation, whose spring constant is γ^2, by use of Hankel transform:

    $(-\nabla^2 + \gamma^2)g = \delta(\mathbf{x} - \xi)$

27. Find the Fundamental Green's function for a vibrating membrane in two-dimensional space by use of Hankel transform:

    $(-\nabla^2 - k^2)g = \delta(\mathbf{x} - \xi)$

28. Find the Fundamental Green's function for a vibrating stretched membrane resting on an elastic foundation, such that:

    $-\nabla^2 g + (\gamma - \kappa^2) g = \delta(\mathbf{x} - \xi)$

    (a) $\gamma > \kappa^2$ (b) $\gamma < \kappa^2$

29. Find the Fundamental Green's function in two-dimensional space for an elastic plate by use of Hankel transform:

    $-\nabla^4 g = \delta(\mathbf{x} - \xi)$

30. Find the Fundamental Green's function in two-dimensional space for a plate on elastic foundation (γ^4 being the elastic spring constant) such that:

    $-\nabla^4 g - \gamma^4 g = \delta(\mathbf{x} - \xi)$

    (a) by Hankel or (b) by construction

31. Find the Fundamental Green's function in two-dimensional space for a vibrating plate supported on an elastic foundation under harmonic loading, by use of Hankel transform, such that:

    $-\nabla^4 g + k^4 g - \gamma^4 g = \delta(\mathbf{x} - \xi)$

    for (a) $k > \gamma$(b) $k < \gamma$

    where γ^4 represents the spring constant per unit area and k^4 represents the frequency parameter.

Sections 8.21 - 8.23

For the following problems, obtain the Fundamental Green's function by (a) Hankel transform only, (b) simultaneous application of Hankel on space and Laplace transform on time, or (c) construction after Laplace transform on time. For Laplace transform on time, let $\delta(t)$ be replaced by $\delta(t-\varepsilon)$, so that the source term is not confused with the initial condition. Let $\varepsilon \to 0$ in the final solution.

32. Find the Fundamental Green's function for the diffusion equation in two-dimensional space $g(x,t)$, such that:

$$\frac{\partial g}{\partial t} - \kappa \nabla^2 g = \delta(\mathbf{x} - \boldsymbol{\xi})\,\delta(t - \tau) \qquad g(\mathbf{x},0|\boldsymbol{\xi},\tau) = 0$$

33. Do Problem 32 for three-dimensional space.

34. Find the Fundamental Green's function for the wave equation in two-dimensional space for wave propagation in a stretched membrane:

$$\frac{\partial^2 g}{\partial t^2} - c^2 \nabla^2 g = \delta(\mathbf{x} - \boldsymbol{\xi})\,\delta(t - \tau) \qquad g(\mathbf{x},0|\boldsymbol{\xi},\tau) = 0 \qquad \frac{\partial g}{\partial t}(\mathbf{x},0|\boldsymbol{\xi},\tau) = 0$$

35. Do Problem 34 in three-dimensional space.

36. Find the Fundamental Green's function for wave propagation in an infinite elastic beam such that:

$$-c^2 \frac{\partial^4 g}{\partial x^4} - \frac{\partial^2 g}{\partial t^2} = \delta(\mathbf{x} - \boldsymbol{\xi})\,\delta(t - \tau) \qquad g(\mathbf{x},0|\boldsymbol{\xi},\tau) = 0 \qquad \frac{\partial g}{\partial t}(\mathbf{x},0|\boldsymbol{\xi},\tau) = 0$$

37. Find the Fundamental Green's function in two-dimensional space for wave propagation in an elastic plate such that:

$$-c^2 \nabla^4 g - \frac{\partial^2 g}{\partial t^2} = \delta(\mathbf{x} - \boldsymbol{\xi})\,\delta(t - \tau) \qquad g(\mathbf{x},0|\boldsymbol{\xi},\tau) = 0 \qquad \frac{\partial g}{\partial t}(\mathbf{x},0|\boldsymbol{\xi},\tau) = 0$$

38. Find the Fundamental Green's function in two-dimensional space for a stretched membrane on an elastic foundation with a spring constant γ, such that:

$$\frac{\partial^2 g}{\partial t^2} + \gamma\, g - c^2 \nabla^2 g = \delta(\mathbf{x} - \boldsymbol{\xi})\,\delta(t - \tau) \qquad g(\mathbf{x},0|\boldsymbol{\xi},\tau) = 0 \qquad \frac{\partial g}{\partial t}(\mathbf{x},0|\boldsymbol{\xi},\tau) = 0$$

39. Obtain the solution for Poisson's equation in one-dimensional space for a semi-infinite medium:

$$-\frac{d^2u(x)}{dx^2} = f(x) \qquad\qquad x \geq 0$$

with Robin boundary condition:

$$-\frac{du(0)}{dx} + \gamma\, u(0) = h$$

40. Obtain the solution for Poisson's equation in two-dimensional space for half space:

$$-\nabla^2 u = f(x, z) \qquad\qquad -\infty < x < \infty, \qquad z \geq 0$$

with (a) Dirichlet or (b) Neumann boundary contiditions.

Sections 8.24 - 8.34

Obtain the Green's functions G for the following bounded media and systems, with D and N designating Dirichlet and Neumann boundary conditions, respectively.

41. Poisson's Equation in two-dimensional space in quarter space:

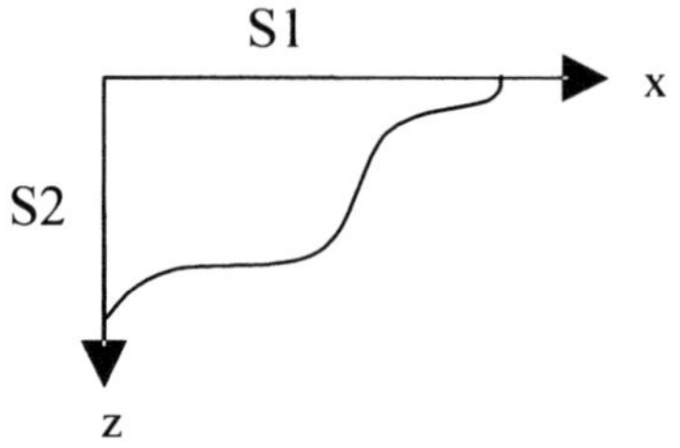

$$-\nabla^2 u = f(x,z) \qquad\qquad z, x \geq 0$$

The boundary conditions are specified in order S1, S2

(a) N, N (b) D, D (c) N, D(d) D, N

42. Do Problem 41 in three dimensions in quarter space:

$$-\nabla^2 u = f(x,y,z) \qquad x, z \geq 0 \qquad -\infty < y < \infty$$

43. Helmholtz Equation in two dimensions in quarter space:

$$-\nabla^2 u - k^2 u = f(x,z) \qquad x, z \geq 0$$

same boundary condition pairs as in Problem 41.

44. Do Problem 43 in three dimensions, same boundary conditions as in Problem 41, where:

$$-\nabla^2 u - k^2 u = f(x,y,z) \quad x, z \geq 0 \qquad -\infty < y < \infty$$

45. Poisson's Equation for eighth space:

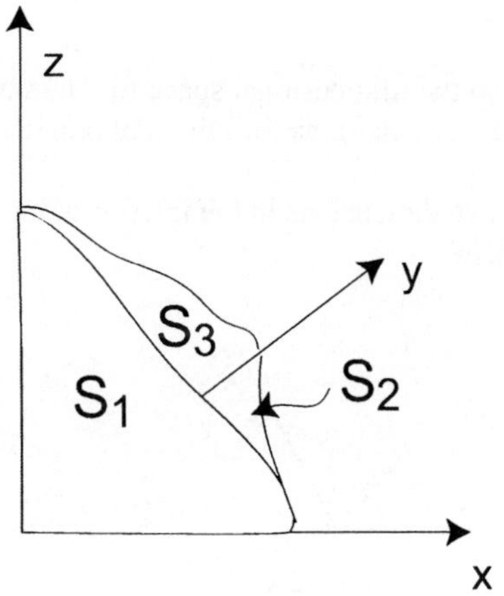

$-\nabla^2 u = f(x,y,z) \qquad x, y, z \geq 0$

with boundary conditions on surface:

S1 (xz plane), S2 (xy plane) and S3 (yz plane) given in order S1, S2, S3

(a) D, D, D (b) N, N, N (c) D, D, N (d) D, N, N

46. Do Problem 45 for the Helmholtz Equation:

$-\nabla^2 u - k^2 u = f(x,y,z) \qquad x, y, z \geq 0$

47. Poisson's Equation in two dimensions in a two-dimensional infinite strip

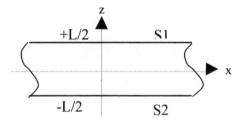

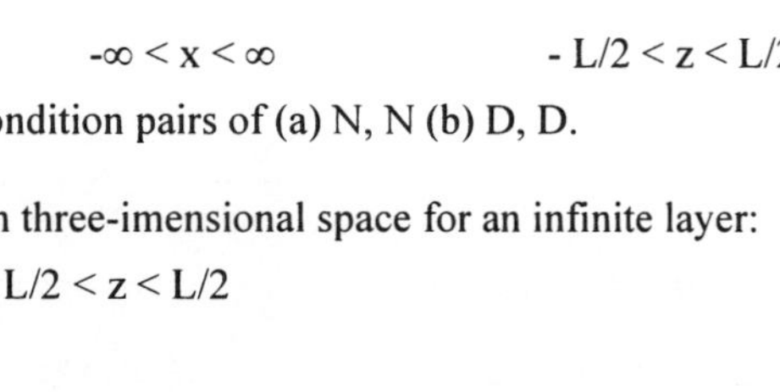

$-\nabla^2 u = f(x,z) \qquad -\infty < x < \infty \qquad -L/2 < z < L/2$

with boundary condition pairs of (a) N, N (b) D, D.

48. Do Problem 47 in three-imensional space for an infinite layer:

$-\infty < x,y < \infty, \quad -L/2 < z < L/2$

49. Helmholtz Equation in two-dimensional space in an infinite strip, same boundary conditions pairs as in Problem 47.

50. Do Problem 49 for three-dimensional space in an infinite layer.

    $-\infty < x,y < \infty, \quad -L/2 < z < L/2$

51. Find Green's function in two-dimensional space for Helmholtz equation in the interior and exterior of a circular area for Dirichlet boundary condition.

52. Poisson's Equation in two dimensions in the interior of a two-dimensional wedge, whose angle is $\pi/3$ where:

    $r \geq 0, \quad 0 \leq \theta \leq \pi/3$

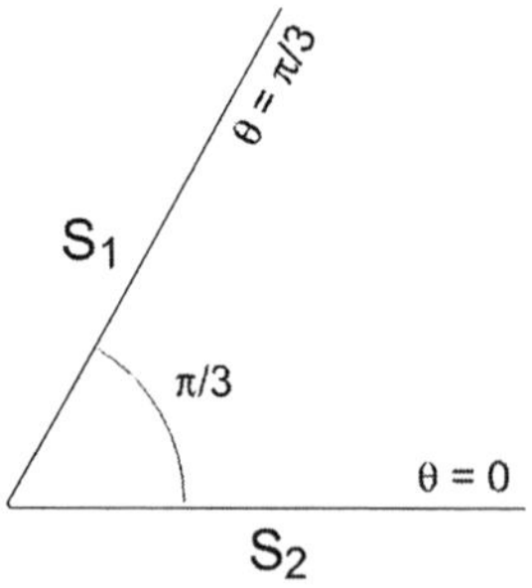

with boundary condition pairs of (a) N-N, (b) D-D

53. Helmholtz Equation for the geometry in Problem 52.

9

ASYMPTOTIC METHODS

9.1 Introduction

In this chapter on asymptotic methods, the emphasis is placed on asymptotic evaluation of integrals and asymptotic solution of ordinary differential equations. The general form of the integrals involves an integrand that is a real or complex function multiplied by an exponential. If the exponential function has an argument that can become large, then it is possible to get an asymptotic value of the integral by one of a few methods. In the following sections, a few of these methods are outlined.

9.2 Method of Integration by Parts

In this method, repeated use is made of integrations by part to create a series with descending powers of a larger parameter.

Example 9.1

Consider the integral I(a):

$$I(a) = \int_u^\infty x^n e^{-ax}\, dx$$

integration by parts results in:

$$I(a) = -\frac{x^n}{a} e^{-ax}\Big|_u^\infty - \frac{n}{a}\int_u^\infty x^{n-1} e^{-ax}\, dx$$

$$= \frac{u^n}{a} e^{-au} - \frac{n}{a}\int_u^\infty x^{n-1} e^{-ax}\, dx$$

Repeated integration of the integral above results in:

$$I(a) = e^{-au}\sum_{k=0}^{n} \frac{u^{n-k}}{a^{k+1}} \frac{n!}{(n-k)!}$$

9.3 Laplace's Integral

Integrals of the Laplace's type can be evaluated asymptotically by use of Taylor series expansion about the origin and integrating the resulting series term by term. Let the integral be given by:

$$f(\rho) = \int_0^\infty e^{-\rho t}\, F(t)\, dt \qquad (9.1)$$

Expanding F(t) in a Taylor series about t = 0, F(t) can be written as a sum, i.e.:

$$F(t) = \sum_{n=0}^{\infty} \frac{F^{(n)}(0)}{n!}\, t^n$$

where $F^{(n)}$ is the nth derivative. Integrating each term in eq. (9.1) results in an asymptotic series for f(ρ):

$$f(\rho) = \sum_{n=0}^{\infty} \frac{F^{(n)}(0)}{\rho^{n+1}} \qquad (9.2)$$

where the Watson's Lemma was used:

$$\int_0^\infty t^\nu\, e^{-\rho t}\, dt = \frac{\Gamma(\nu+1)}{\rho^{\nu+1}} \qquad (9.3)$$

and where Γ(x) is the Gamma function, see Appendix B1.

Example 9.2

Consider the following integral, which is known to have a closed form:

$$I(s) = \int_0^\infty \frac{e^{-st}}{\sqrt{1+t}}\, dt = \sqrt{\frac{\pi}{s}}\, e^s\, \mathrm{erfc}\left(\sqrt{s}\right)$$

The term $(1+t)^{-1/2}$ can be expanded in a Taylor series:

$$(1+t)^{-1/2} = 1 - \frac{t}{2} + \frac{1\cdot 3}{2!\, 2^2}\, t^2 - \frac{1\cdot 3\cdot 5}{3!\, 2^3}\, t^3 + \ldots$$

which, upon integration via eq. (9.3) results in:

$$I(s) = \frac{1}{s} - \frac{1}{2\, s^2} + \frac{1\cdot 3}{2^2\, s^3} - \frac{3\cdot 5}{2^3\, s^4} + \ldots$$

Equating this expression to the $\mathrm{erfc}(\sqrt{s})$ one obtains an asymptotic series for the erfc(z):

$$\mathrm{erfc}\left(\sqrt{s}\right) \sim \sqrt{\frac{s}{\pi}}\, e^{-s} \left\{ \frac{1}{s} - \frac{1}{2\, s^2} + \frac{1\cdot 3}{2^2\, s^3} - \frac{3\cdot 5}{2^3\, s^4} + \ldots \right\}$$

$$\mathrm{erfc}(z) \sim \frac{z}{\sqrt{\pi}} e^{-z^2} \left\{ \frac{1}{z^2} - \frac{1}{2\,z^4} + \frac{1 \cdot 3}{2^3\,z^6} - \frac{3 \cdot 5}{2^4\,z^8} + \ldots \right\}$$

9.4 Steepest Descent Method

Consider an integral of the form:

$$I_C = \int_C e^{\rho\, f(z)}\, F(z)\, dz \qquad (9.4)$$

where C is a path of integration in the complex plane, $z = x + iy$, f(z) and F(z) are analytic functions and ρ is a real constant. It is desired to find an asymptotic value of this integral for large ρ. The **Steepest Descent Method (SDM)** involves finding a point, called the **Saddle Point (SP)**, and a path through the point, called the **Steepest Descent Path (SDP)**, so that the integrand decays exponentially along that path and the integral can be approximately evaluated for a large argument ρ. Letting the analytic function f(z) be defined as:

$$f(z) = u(x,y) + i\, v(x,y) \qquad (9.5)$$

then the path of integration is chosen such that the real part of f(z) = u(x,y) has a maximum value at some point z_0. This would maximize the real part of the exponential function, especially when $\rho >> 1$. To locate the point z_0 where u(x,y) is maximized, the extremum point(s) are found by finding the point(s) where the partial derivatives with respect to x and y vanish, i.e.:

$$\frac{\partial u}{\partial x} = 0, \quad \frac{\partial u}{\partial y} = 0 \qquad (9.6)$$

Since f(z) is an analytic function, then u and v are harmonic functions, i.e., $\nabla^2 u = 0$, which indicates that u(x,y) cannot have points of absolute maxima or minima in the entire z-plane. Hence, the points where eq. (9.6) is satisfied are stationary points, $z_0 = x_0 + iy_0$. The topography near z_0 for u(x,y) = constant would be a surface that resembles a saddle, i.e., paths originating from z_0 either descend, stay at the same level, or ascend, see Figure 9.1. To choose a path through the saddle point z_0, one obviously must choose paths where u(x,y) has a relative maximum at z_0, so that u(x,y) decreases on the path(s) away from z_0, i.e., a path of decent from the point z_0. This would mean that the exponential function has a maximum value at z_0 and decays exponentially away from the SP z_0. This would result in an integral that would converge. On the other hand, if one chooses a path starting from z_0 where u(x,y) has a relative *minimum* at z_0, i.e., u(x,y) increases along C' , then the exponential function increases exponentially away from the saddle point at z_0. This would result in an integral that will diverge along that path.

Since $\partial u/\partial x = 0$ and $\partial u/\partial y = 0$ at the SP z_0, and f(z) is analytic at z_0, then the partial derivatives $\partial v/\partial x = 0$ and $\partial v/\partial y = 0$ due to the Cauchy-Riemann conditions. This indicates that:

$$\left.\frac{df}{dz}\right|_{z_0} = f'(z_0) = 0 \tag{9.7}$$

The roots of eq. (9.7) are thus the saddle points of f(z).

One must choose a path C′ originating from the SP, z_0, i.e., a path of decent from z_0 so that the real part of the exponential function decreases along C′. This would lead to a convergent integral along C′ as ρ becomes very large. In order to improve the convergence of the integral, especially with a large argument ρ, one needs to find the steepest of all the descent paths C′. This means that one must find the path C′ so that the function u(x,y) decreases at a maximum rate as z traverses along the path C′ away from z_0. To find such a path, defined by a distance parameter "s" where u decreases at the fastest rate, the absolute value of the rate of change of u(x,y) along the path "s" must be maximized, i.e., $|\partial u/\partial s|$ is maximum along C′. Let the angle θ be the angle between the tangent to the path C′ at z_0 and the x-axis, then the slope along the path C′ is given by:

$$\frac{\partial u}{\partial s} = \frac{\partial u}{\partial x}\frac{\partial x}{\partial s} + \frac{\partial u}{\partial y}\frac{\partial y}{\partial s} = \frac{\partial u}{\partial x}\cos\theta + \frac{\partial u}{\partial y}\sin\theta$$

To find the orientation θ where $\partial u / \partial s$ is maximized, then one obtains the extremum of the slope as a function of the local orientation angle θ of C′ with x, i.e.:

$$\frac{\partial}{\partial \theta}\left(\frac{\partial u}{\partial s}\right) = -\frac{\partial u}{\partial x}\sin\theta + \frac{\partial u}{\partial y}\cos\theta = 0$$

Using the Cauchy-Riemann conditions:

$$\frac{\partial u}{\partial x} = \frac{\partial v}{\partial y} \qquad \text{and} \qquad \frac{\partial u}{\partial y} = -\frac{\partial v}{\partial x}$$

then the equation above becomes:

$$-\frac{\partial v}{\partial y}\sin\theta - \frac{\partial v}{\partial x}\cos\theta = -\frac{\partial v}{\partial s} = 0 \tag{9.8}$$

Integrating eq. (9.8) with respect to the distance along C′, s, results in v = constant along C′. Thus, the function u(x,y) changes most rapidly on path C′ defined by v = constant. Since the path must pass through the SP at z_0, then the equation of the path is defined by:

$$v(x,y) = v(x_0,y_0) = v_0 \tag{9.9}$$

Equation (9.9) defines path(s) C′ from z_0 having the most rapid change in the slope. Thus, eq. (9.9) defines a path(s) where u(x,y) increases or decreases most rapidly. It is imperative that one finds *the path(s)* where the function u(x,y) *decreases* most rapidly and this path is to be called **Steepest Descent Path (SDP)**.

To identify which of the paths are SDP, it is sufficient to examine the topography near z_0. Since f(z) is an analytic function at z_0, then one can expand the function f(z) in a Taylor series about z_0, giving:

$$f(z) = a_0 + a_1(z - z_0) + a_2(z - z_0)^2 + a_3(z - z_0)^3 + \ldots$$

where:

$$a_n = \frac{f^{(n)}(z_0)}{n!}$$

Due to the definition of the SP at z_0, then the second term vanishes, since:

$$a_1 = f'(z_0) = 0$$

If, in addition, $a_2(z_0) = a_3(z_0) = ... = a_m(z_0) = 0$ also, so that the first non-vanishing coefficient is a_{m+1}, then, in the neighborhood of z_0, f(z) can be approximated by the first two non-vanishing terms of the Taylor series about z_0, i.e.:

$$f(z) \approx f(z_0) + (z - z_0)^{m+1} \frac{f^{(m+1)}(z_0)}{(m+1)!}$$

where terms of degree higher than (m+1) were neglected in comparison with the $(m+1)^{st}$ term. Defining:

$$\frac{f^{(m+1)}(z_0)}{(m+1)!} = ae^{ib}$$

and the local topography near the SP z_0 by:

$$z - z_0 = r\, e^{i\theta}$$

then the function f(z) in the neighborhood of the SP can be described by:

$$f(z) \approx f(z_0) + ae^{ib}(re^{i\theta})^{m+1} = u_0 + iv_0 + r^{m+1}\, ae^{i[(m+1)\theta+b]}$$

where:

$$u_0 = u(x_0,y_0) \qquad\qquad v_0 = v(x_0,y_0)$$

Hence, the real and imaginary parts of f(z) in the neighborhood of z_0 are, respectively:

$$u = u_0 + ar^{m+1}\cos[(m+1)\theta+b]$$

and

$$v = v_0 + ar^{m+1}\sin[(m+1)\theta+b]$$

The steepest descent and steepest ascent paths are given by $v = v_0 =$ constant, or:

$$\sin[(m+1)\theta+b] = 0$$

The various paths of steepest ascent or descent have local orientation angles θ with the x-axis given by:

$$\theta = \frac{n\pi}{m+1} - \frac{b}{m+1} \qquad\qquad n = 0, 1, 2, ..., (2m+1)$$

Substitution of θ in the expression for u(θ) above and noting that, for steepest descent paths, u_0 has a local maximum at z_0 on C′ and hence, $u - u_0 < 0$ for any point (x,y) on C′, then $\cos(n\pi) < 0$, indicating that n must be odd. The number of steepest descent paths are thus (m + 1), and are defined by:

$$\theta_{SDP} = \frac{2n+1}{m+1}\pi - \frac{b}{m+1} \qquad\qquad n = 0, 1, 2, ..., m$$

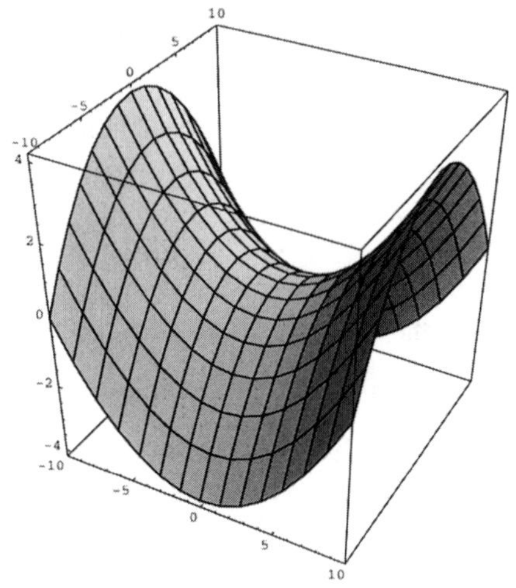

Fig. 9.1 *Surface in the Neighborhood of a Saddle Point*

To evaluate the integral over C in eq. (9.4), the original path C must be closed with any two of the 2m SDP paths C′, call them C_1' and C_2', each originating from z_0. Invoking the Cauchy Residue theorem for the closed path $C + C_1' + C_2'$ let:

$$w = f(z_0) - f(z) = (u_0 + iv_0) - (u + iv)$$

The preceding equality can be used to obtain a conformal transformation w = w(z), on each of the two paths C_1' and C_2' which can be inverted to give z = z(w). This transformation from the z-plane to the w-plane transforms the original path C as well as the paths $C_{1,2}'$ to new paths in the w-plane. It should be noted that this conformal transformation is usually not easily invertible.

Since $v = v_0$ on $C_{1,2}'$, then the function w is real on the two SDP $C_{1,2}'$, i.e.:

$$w\Big|_{C_1', C_2'} = u_0 - u$$

When $z = z_0$, then w = 0 and when |z| on $C_{1,2}' \to \infty$, $w \to \infty$, so that the integrals on $C_{1,2}'$ are performed over the real axis of the w-plane, i.e.:

$$I_{C_{1,2}'} = \int_0^\infty e^{\rho[f(z_0)-w]}\bar{F}(w)\left(\frac{dz}{dw}\right) dw = e^{\rho f(z_0)} \int_0^\infty e^{-\rho w} \frac{\bar{F}(w)}{(dw/dz)} dw \qquad (9.10)$$

where $\bar{F}(w) = F(z(w))$ and (dw/dz) are complex function in the w-plane, since the conformal transformation z = z(w) is complex.

Expanding $\dfrac{\bar{F}(w)}{(dw/dz)}$ in a Taylor series in w about w = 0, then:

$$\frac{\bar{F}(w)}{(dw/dz)} = \sum_{n=0}^{\infty} \bar{F}_n \, w^{n+v} \qquad (9.11)$$

where ν is a non-integer constant, resulting from the derivative dw/dz.

It should be noted that the slope dw/dz has a different value on C_1' and C_2'. Substituting eq. (9.11) into eq. (9.10), integrating the resulting series term by term, and using Watson's Lemma in eq. (9.3), the integral in eq. (9.10) becomes:

$$I_{C'_{1,2}} \sim e^{\rho f(z_0)} \sum_{n=0}^{\infty} \bar{F}_n \frac{\Gamma(n+\nu+1)}{\rho^{n+\nu+1}} \tag{9.12}$$

Note that $I_{C_1'}$ and $I_{C_2'}$ have different series based on the path taken. Thus, if C is an infinite path, and one must close it with an infinite path, then two paths $C'_{1,2}$ must be joined to C, resulting in:

$$I_C = I_{C_1'} - I_{C_2'} \pm 2\pi i \text{ [sum of residues of the poles between } C + C_1' + C_2' \text{]} \tag{9.13}$$

The sign for the residues depends on the sense of the path(s) of closure between C, C_1', and C_2', which may be clockwise for some poles and counterclockwise for other poles. The paths C_1' and C_2' start from w = 0 and end in w = ∞ along each path, so that the sign assigned for C_2' is negative.

9.5 Debye's First-Order Approximation

There are first order approximations to the integrals in eq. (9.10). Principally, these approximations assume that the major contribution to the integral comes from the section of the path near the saddle point, especially when ρ is very large. This means that the first term in eq. (9.12) would suffice if ρ is sufficiently large. To obtain the first order approximation, one can neglect higher order terms in $\bar{F}(w)$ and (dw/dz) in such a way that a closed form expression can be obtained for the first order term. Thus, an approximate value for w can be obtained by neglecting higher order terms in w:

$$w = f(z_0) - f(z) \approx -(z - z_0)^{m+1} \frac{f^{(m+1)}(z_0)}{(m+1)!} \tag{9.14}$$

Thus, for z near z_0, the conformal transformation between w and z can be obtained explicitly in a closed form by the approximation:

$$(z - z_0) \approx \left[-\frac{(m+1)!}{f^{(m+1)}(z_0)} \right]^{1/(m+1)} w^{1/(m+1)} = [cw]^{1/(m+1)} \tag{9.15}$$

where the complex constant c is given by:

$$c = -\frac{(m+1)!}{f^{(m+1)}(z_0)}$$

Note that the (m+1) roots have different values along the different paths C_m'. Differentiating this approximation for z with respect to w results in:

$$\frac{dz}{dw} \approx \frac{c^{1/(m+1)} w^{-m/(m+1)}}{m+1}$$

Similarly, the function F(z) can be approximated by its value at z_0:

$$F(z) \approx F(z_0)$$

Thus, the integrals $I_{C_1', C_2'}$ become:

$$I_{C_1', C_2'} \sim \frac{c^{1/(m+1)} F(z_0) e^{\rho f(z_0)}}{m+1} \int_0^{\infty} e^{-\rho w} w^{-m/(m+1)} dw$$

$$I_{C_1', C_2'} \sim \frac{c^{1/(m+1)} \Gamma(1/(m+1))}{m+1} \frac{F(z_0) e^{\rho f(z_0)}}{\rho^{1/(m+1)}} \tag{9.16}$$

The first-order approximation to the integrals ineq. (9.4) is thus given by:

$$I_C \approx I_{C_1'} - I_{C_2'}$$

$$= \frac{\Gamma((m+1)^{-1}) F(z_0) e^{\rho f(z_0)}}{(m+1)\rho^{(m+1)^{-1}}} \left\{ c^{(m+1)^{-1}} \Big|_{on C_1'} - c^{(m+1)^{-1}} \Big|_{on C_2'} \right\} \tag{9.17}$$

where the residues of the poles were neglected. Equation (9.17) represents the leading term in the approximation of the asymptotic series. Note for m = 1, the two roots of c are opposite in signs and hence the expression in the bracket is simply double the first term in the bracket, i.e.:

$$I_c \approx F(z_0) \frac{e^{\rho f(z_0)}}{\rho^{1/2}} \sqrt{2\pi/(-f''(z_0))} \qquad (m = 1) \tag{9.18}$$

Example 9.3

Obtain the Debye's approximation for the factorial of a large number, known as Sterling's Formula. The Gamma function is given as an integral:

$$\Gamma(k+1) = \int_0^{\infty} t^k e^{-t} dt$$

When k is an integer n, $\Gamma(n+1) = n!$. To obtain a Debye's approximation for the asymptotic value for a large k, the integrand must be slowly varying. This is not the case here as the function t^k becomes unbounded for k large. Furthermore, the exponential term does not have the parameter k in the exponent. Let t = kz, then:

$$\Gamma(k+1) = k^{k+1} \int_0^{\infty} e^{-kz} z^k dz = k^{k+1} \int_0^{\infty} e^{k(\log z - z)} dz$$

For the last integral, F(z) = 1 and:

$$f(z) = \log(z) - z$$

The saddle point z_0 is derived from $f'(z_0) = z_0^{-1} - 1 = 0$, so that the saddle point is located at z_0=+1. Evaluating the function in eq. (9.18) gives:

$$f(z_0) = f(1) = -1, \qquad f''(z_0) = -1 = 1 \cdot e^{i\pi}$$

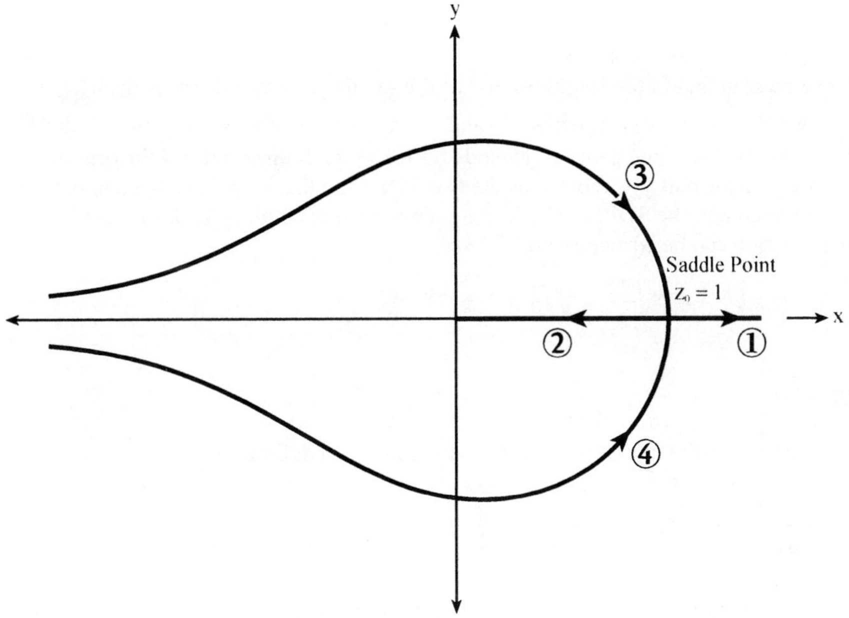

Figure 9.2 *Steepest Descent and Ascent Paths for Example 9.3*

therefore:

$$a = 1 \qquad \text{and} \qquad b = \pi.$$

Since $f''(z_0) \neq 0$, then the saddle point is of rank one ($m = 1$) and hence the SDP in the neighborhood of z_0 make tangent angles given by:

$$\theta_{SDP} = \frac{2n+1}{2}\pi - \frac{\pi}{2} \qquad n = 0, 1$$

$$= 0, \pi$$

The SDP equation is given by $v = v_0 =$ constant. The function $f(z) = \log(z) - z$ can be written in terms of cylindrical coordinates. Let $z = re^{i\theta}$, then:

$$f(z) = \log(r) + i\theta - re^{i\theta} = \log(r) - r\cos\theta + i(\theta - r\sin\theta)$$

Here:

$$u = \log(r) - r\cos\theta$$

$$v = \theta - r\sin\theta$$

The saddle point $z_0 = 1$ has $r = 1$, $\theta = 0$ and thus $v_0 = 0$. The equation of the SDP becomes:

$$v = \theta - r\sin\theta = v_0 = 0$$

or:

$$r = \frac{\theta}{\sin\theta}$$

The four paths are shown in Figure 9.2.

It can be seen that in the neighborhood of the saddle point $z_0 = 1$, the paths $\theta_{SDP} = 0$, π are paths "1" and "2", so that paths "3" and "4" are the steepest ascent paths. Path "2" extends from z = 1 to 0 and path "1" extends from 1 to ∞. It turns out that the original path on the positive real axis represents the two SDP's, so that there is no need to deform the original path into the SDP's. The leading term of the asymptotic series for the Gamma function can be written as eq. (9.18):

$$\Gamma(k+1) = k^{k+1}\, e^{-k}\sqrt{\frac{2\pi}{k}} = \sqrt{2\pi}\, e^{-k}\, k^{k+1/2}$$

Example 9.4

Find the first-order approximation for Airy's function defined as:

$$A_i(z) = \frac{1}{\pi}\int_0^{\infty}\cos\left(s^3/3 + sz\right)ds = \frac{1}{2\pi}\int_{-\infty}^{\infty}\exp\left[i\left(s^3/3 + sz\right)\right]ds$$

To obtain an asymptotic approximation for large z, the first exponential terms is also not a slowly varying function. To merge the first exponential with the second, let $s = \sqrt{z}\, t$:

$$A_i(z) = \frac{\sqrt{z}}{2\pi}\int_{-\infty}^{\infty}\exp\left[iz^{3/2}\left(t^3/3 + t\right)\right]dt$$

Letting $x = z^{3/2}$ one can write out the integral as:

$$A_i(x^{2/3}) = \frac{x^{1/3}}{2\pi}\int_{-\infty}^{\infty}\exp\left[ix\left(t^3/3 + t\right)\right]dt$$

One can evaluate the first-order approximation for large x. In this integral F(t) = 1 and

$f(t) = i\,(t^3/3 + t)$

The saddle points are given by $f'(t_0) = i\,(t_0^2 + 1) = 0$ resulting in two saddle points, $t_0 = \pm i$. To map the SDP:

$$f(\pm i) = \mp\frac{2}{3}$$

$$f''(\pm i) = \mp 2 = 2\begin{cases} e^{i\pi} \\ e^{i0} \end{cases}$$

Here b = 2 and $\theta = \pi$ for $t_0 = +i$ and $\theta = 0$ for $t_0 = -i$. It should be noted that since $f''(t_0) \neq 0$, m = 1 for both saddle points. Letting $t = \xi + i\eta$, then the SDP path equations for both saddle points are given by:

$$v(\xi,\eta) = \text{Im } f(t) = \xi^3/3 - \xi\eta^2 + \xi = v_0(\xi_0,\eta_0) = v_0(0,\pm 1) = 0$$

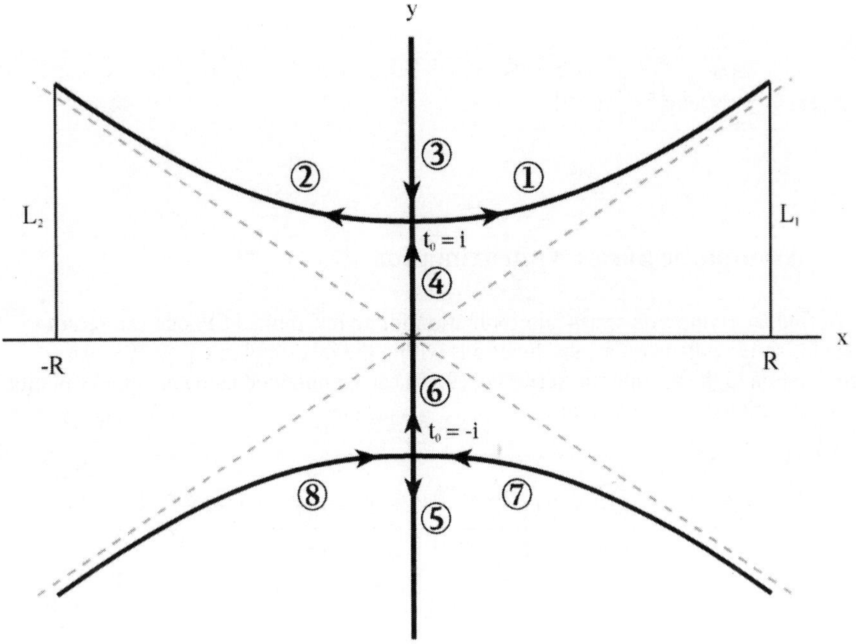

Figure 9.3 : *Steepest Descent and Ascent Paths for Example 9.4*

The paths of steepest ascent or descent are plotted for $t_0 = +i$ (paths 1-4) and for $t_0 = -i$ (paths 5-8), see Figure 9.3.

For the SP at $t_0 = +i$, path "3" extends from i to i∞ and path "4" extends from i to –i∞. For the SP at $t_0 = -i$, the path "5" extends from -i to -i∞ and path "6" extends from -i to i∞. It should be noted that path "4" partially overlaps path "5" and path "3" partially overlaps path "6". For the SP at $t_0 = +i$, $f''(+i) = 2e^{i\pi}$, so that the steepest descent paths near $t_0 = +i$ make tangent angles given by:

$$\theta_{SDP} = \frac{2n+1}{2}\pi - \frac{\pi}{2} = 0, \pi$$

Thus, the SDP's for $t_0 = +i$ are paths "1" and "2" having tangent angles 0 and π, while the paths "3" and "4" are steepest ascent paths. For $t_0 = -i$, $f''(-i) = +2$, so that the SDP make tangent angles π/2 and 3π/2 near the saddle point $t_0 = -i$.

Since there are two saddle points, one can connect the original path (-∞,∞) to either paths "1" and "2" through $t_0 = +i$ or "5" and "6" through $t_0 = -i$. Considering the second choice, the closure with the original path with "6" and "5" through $t_0 = -i$, requires going through $t_0 = i$ along paths "3" and "4" which were steepest ascent paths for $t_0 = +i$. Thus, this will result in the integrals becoming unbounded. Thus, the only choice left is to close that original path (-∞,∞) through $t_0 = +i$ by connecting to the paths "1" and "2" by line segments L_1 and L_2. To obtain a first-order approximation, then:

$$A_i(x) \approx \frac{x^{1/3}}{2\pi} \cdot 1 \cdot e^{x(-2/3)} \sqrt{\frac{2\pi}{-(-2x)}} = \frac{x^{-1/6}}{2\sqrt{\pi}}\, e^{-2x/3}$$

so that:

$$A_i(z) \approx \frac{z^{-1/4}}{2\sqrt{\pi}} \exp\left(-\frac{2}{3} z^{2/3}\right)$$

9.6 Asymptotic Series Approximation

To find an asymptotic series approximation for an mth ranked SP, one can return to the Taylor series expansion for the functions within the integrand in eq. (9.10). An approximation to the asymptotic series eq. (9.10) can be obtained using an approximation for the derivative dz/dw. Letting:

$$w = f(z_0) - f(z)$$

$$= \frac{(z-z_0)^{m+1}}{c}\left[1 + \frac{(z-z_0)}{m+2}\frac{f(z_0)^{(m+2)}}{f(z_0)^{(m+1)}} + \frac{(z-z_0)^2}{(m+2)(m+3)}\frac{f(z_0)^{(m+3)}}{f(z_0)^{(m+1)}} + \ldots\right] \tag{9.19}$$

then:

$$\frac{dw}{dz} = \frac{(m+1)}{c}(z-z_0)^m\left[1 + \frac{(z-z_0)}{m+1}\frac{f(z_0)^{(m+2)}}{f(z_0)^{(m+1)}} + \frac{(z-z_0)^2}{(m+1)(m+2)}\frac{f(z_0)^{(m+3)}}{f(z_0)^{(m+1)}} + \ldots\right] \tag{9.20}$$

In the neighborhood of $z = z_0$, then, using the expression for $z - z_0$ in eq. (9.15), on obtains:

$$\frac{dw}{dz} \approx \frac{m+1}{c^{1/(m+1)}}\, w^{m/(m+1)}\left[b_0 + b_1 w^{1/(m+1)} + b_2 w^{2/(m+1)} + \ldots\right]$$

$$= \frac{(m+1)}{c^{1/(m+1)}}\, w^{m/(m+1)} \sum_{n=0}^{\infty} b_n w^{n/(m+1)}$$

where:

$$b_0 = 1$$

$$b_1 = -\frac{c^{(m+2)/(m+1)}}{(m+1)(m+1)!} f(z_0)^{(m+2)}$$

$$b_2 = -\frac{c^{(m+3)/(m+1)}}{(m+1)(m+2)(m+1)!} f(z_0)^{(m+3)}$$

$$b_3 = -\frac{c^{(m+4)/(m+1)}}{(m+1)(m+2)(m+3)(m+1)!} f(z_0)^{(m+4)}$$

Also, the function F(z) can also be expanded in a Taylor series as follows:

$$F(z)=\sum_{0}^{\infty}\frac{F^{(n)}(z_0)}{n!}(z-z_0)^n \approx \sum_{0}^{\infty}\frac{F^{(n)}(z_0)}{n!}\,c^{n/(m+1)}\,w^{n/(m+1)}$$

so that the integrand of eq. (9.10) becomes:

$$\frac{F(z)}{(dw/dz)} \approx \frac{c^{1/(m+1)}}{m+1}\,w^{-m/(m+1)}\,\frac{\sum\limits_{n=0}^{\infty}\frac{F^{(n)}(z_0)}{n!}\,c^{n/(m+1)}\,w^{n/(m+1)}}{\sum\limits_{n=0}^{\infty} b_n\,w^{n/(m+1)}} \tag{9.21}$$

$$=\frac{c^{1/(m+1)}}{m+1}\,w^{-m/(m+1)}\sum_{n=0}^{\infty} d_n\,w^{n/(m+1)}$$

where:

$$d_0 = F(z_0)$$

$$d_1 = -b_1\,F(z_0) + c^{1/(m+1)}\,F_1'(z_0)$$

$$d_2 = \left(b_1^2 - b_2\right)F(z_0) - b_1\,F'(z_0)\,c^{1/(m+1)} + \frac{F''(z_0)}{2!}\,c^{2/(m+1)}$$

Substituting eq. (9.21) into eq. (9.10) one obtains:

$$I_{C_1',C_2'} \approx e^{\rho f(z_o)}\,\frac{c^{1/(m+1)}}{m+1}\sum_{n=0}^{\infty} d_n \int_0^{\infty} e^{-\rho w}\,w^{(n-m)/(m+1)}\,dw$$

$$= e^{\rho f(z_o)}\,\frac{c^{1/(m+1)}}{m+1}\sum_{n=0}^{\infty} d_n\,\frac{\Gamma\!\left(\frac{n+1}{m+1}\right)}{\rho^{(n+1)/(m+1)}} \tag{9.22}$$

It should be noted that the first term in the asymptotic series in eq. (9.22) is the same one given in eq. (9.16). The expression in eq. (9.22) is useful when a simple relationship $z = z(w)$ cannot be found, and thus the expansion in eq. (9.11) is not possible.

Another transformation that could be used to make the integrands even that would eliminate the odd terms in the Taylor series expansion is given by:

$$\frac{1}{2}y^2 = f(z_o) - f(z) = u_0 - u \qquad \text{real on } C' \tag{9.23}$$

In addition, the integration over the two paths C′ could be substituted by one integral over (-∞ to +∞). Thus:

$$\bar{I}_{C'} = 2I_{C'} = e^{\rho f(z_o)}\int_{-\infty}^{\infty} e^{-\rho y^2/2}\,\bar{F}(y)\,\frac{dy}{(dy/dz)} \tag{9.24}$$

Expanding the integrand in a Taylor series, and retaining only the even terms since the odd terms will vanish gives:

$$\frac{\bar{F}(y)}{(dy/dz)} = \sum_{n=0}^{\infty} \bar{F}_{2n}\, y^{2n+2\nu} \tag{9.25}$$

where ν is a non-integer constant.

Thus, the integral over the entire length of the steepest descent path $\left(2\bar{I}_{C'}\right)$ can be obtained as follows:

$$\bar{I}_{C'} = 2\bar{I}_{C'} = e^{\rho f(z_0)} \sum_{n=0}^{\infty} \bar{F}_{2n} \int_{-\infty}^{\infty} e^{-\rho y^2/2}\, y^{2n+2\nu}\, dy$$

$$= \sqrt{\frac{2\pi}{\rho}}\, e^{\rho f(z_0)} \sum_{n=0}^{\infty} \frac{\Gamma(2n+2\nu+1)}{\rho^{n+\nu+1}\Gamma(n+\nu+1)}\, \bar{F}_{2n} \tag{9.26}$$

Example 9.5

Obtain the asymptotic series for Airy's function of Example 9.3. Starting with the integral given in Example 9.3 then the transformation about the saddle point at t_0=+i is given by:

$$w = f(t_0) - f(t) = -2/3 - i\,(t^3/3 + t) = (t - i)^2 - i\,(t - i)^3/3$$

The preceding conformal transformation between t and w can be inverted exactly, since the formula is a cubic equation. However, this would result in a complicated transformation $t = t(w)$. Instead, one can try to find a good approximation valid near the SP at $t_0 = i$.

To obtain a transformation from t to w, we can obtain, approximately, an inverse formula. Let the term (t – i) be represented by:

$$t - i = \frac{\pm\sqrt{w}}{\left[1 - i(t-i)/3\right]^{1/2}}$$

Again, since the integral has the greatest contribution near the saddle point, then one may approximate the term (t – i) by:

$$t - i \approx \pm\sqrt{w}$$

Substituting this approximation for (t – i) in the denominator of the formula above, one can obtain the approximate conformal transformation from t to w:

$$t - i \approx \frac{\pm\sqrt{w}}{\left[1 - \left(\pm i\sqrt{w}/3\right)\right]^{1/2}}$$

The +/- signs represent the transformation formula for the paths "1" and "2" of Figure 9.3. Expanding the denominator in an infinite series about w = 0, one obtains:

$$t - i \approx \sum_{n=1}^{\infty} \frac{(\pm 1)^n\, i^{n-1}\, \Gamma(3n/2 - 1)\, w^{n/2}}{n!\,\Gamma(n/2)\, 3^{n-1}}$$

The derivatives dt/dw can be obtained readily:

$$\frac{dt}{dw} \approx \frac{3}{2}\sum_{n=1}^{\infty} \frac{(\pm 1)^n\, i^{n-1}\, \Gamma(3n/2-1)\, w^{n/2-1}}{(n-1)!\,\Gamma(n/2)\, 3^n}$$

The product of F(w) = 1 and dt/dw can be substituted in the integral in eq. (9.10). The integrals require the evaluation of the following:

$$\int_0^{\infty} e^{-xw} w^{n/2-1}\, dw = \frac{\Gamma(n/2)}{x^{n/2}}$$

Thus, the two integrals on paths "1" and "2" are given by:

$$I_{C_1',C_2'} \approx \frac{3x^{1/3}}{4\pi} e^{-2x/3} \sum_{n=1}^{\infty} \frac{(\pm 1)^n\, \Gamma(3n/2-1)}{(n-1)!\, 3^n\, x^{n/2}}$$

$$= \frac{x^{-1/6}}{4\pi} e^{-2x/3} \sum_{n=0}^{\infty} \frac{(\pm 1)^{n+1}\, i^n\, \Gamma[(3n+1)/2]}{n!\,(9x)^{n/2}}$$

Therefore:

$$I_C = I_{C_1'} - I_{C_2'} = \frac{x^{-1/6}\, e^{-2x/3}}{4\pi} \sum_{n=0}^{\infty} \frac{\left[1-(-1)^{n+1}\right] i^n\, \Gamma[(3n+1)/2]}{n!\,(9x)^{n/2}}$$

Rewriting the final results in terms of z and simplifying the final expression gives:

$$A_i(z) \approx \frac{z^{-1/4} \exp\left[-2z^{3/2}/3\right]}{2\pi} \sum_{m=0}^{\infty} \frac{(-1)^m\, \Gamma(3m+1/2)}{(2m)!\,(9z^{3/2})^m}$$

It should be noted that the first-order approximation of $A_i(z)$, i.e., the $m = 0$ term results

$$A_i(z) \approx \frac{Z^{-1/4} exp[-2Z^{3/2}/3]}{2\pi} \Gamma(\frac{1}{2})$$

which matches the expression derived in Example 9.4. The resulting asymptotic series is conditionally convergent.

9.7 Method of Stationary Phase

The **Stationary Phase** method is analogous to the Steepest Descent method, although the approach and reasoning for the approximation is different. Performing the integration in the complex plane results in the two methods having identical outcomes. Consider the integral:

$$I(\rho) = \int_C F(z)\, e^{i\rho f(z)}\, dz \tag{9.27}$$

where f(z) is an analytic function and F(z) is a slowly varying function. Thus, as ρ becomes larger, the exponential term oscillates in increasing frequency. Since the exponential can be written in terms of circular functions, then as ρ increases, the frequency of the circular functions increases, so much so that these circular functions oscillate rapidly between +1 and -1. This then tends to cancel out the integral of F(z) when ρ becomes very large for sufficiently large z. The major contribution to the integral then occurs when f(z) has a minimum so that the exponential function oscillates the least. This occurs when:

$$f'(z_0) = 0$$

where z_0 (x_0, y_0) is called the **Stationary Phase Point (SPP)**. Letting f(z) = u + iv, then:

$$e^{i\rho f(z)} = e^{-\rho v}\, e^{i\rho u}$$

If F(z) is a slowly varying function, then most of the contribution to the integral comes from near the SPP z_0, where the exponential oscillates the least. Expanding the function f(z) about the SPP z_0:

$$f(z) = f(z_0) + 1/2\ f''(z_0)\,(z-z_0)^2 + \ldots$$

and defining:

$$w = f(z) - f(z_0) = -\,1/2\ f''(z_0)\,(z-z_0)^2 - \ldots$$

then the integral becomes:

$$I(\rho) = e^{i\rho f(z_0)} \int_{C'} \frac{F(z(w))}{(dw/dz)}\, e^{-i\rho w}\, dw \tag{9.28}$$

where C′ is the Stationary Phase path defined by v = constant = v_0 and $v_0 = v(x_0,y_0)$. This is the same path defined for the Steepest Descent Path. For an equivalent Debye's first-order approximation for m = 1, let:

$$w \approx -\frac{1}{2} f''(z_0)(z - z_0)^2$$

$$dw/dz \approx -f''(z_0)(z - z_0) \approx \sqrt{2w}$$

$$F(z_0) \approx F(z(w{=}0))$$

then the integral in eq. (9.28) becomes:

$$I(\rho) \approx e^{i\rho f(z_o)}\, F(z_o) \int\limits_{-\infty}^{\infty} \frac{e^{-i\rho w}}{\sqrt{2w}}\, dw = e^{i\rho f(z_o)}\, F(z_o) \sqrt{\frac{2\pi}{\rho f''(z_o)}}\, e^{i\pi/4} \tag{9.29}$$

9.8 Steepest Descent Method in Two Dimensions

If the integral to be evaluated asymptotically is a double integral of the form:

$$I = \int\limits_{-\infty}^{\infty} \int\limits_{-\infty}^{\infty} F(u,v)\, e^{\rho f(u,v)}\, du\, dv \tag{9.30}$$

then one can follow a similar approach to Section 9.4. The saddle point in the double-complex space is given by:

$$\frac{\partial f}{\partial u} = 0 \quad \text{and} \quad \frac{\partial f}{\partial v} = 0$$

which defines the location of saddle point(s) (u_s,v_s) in the double complex space.

Expanding the function f(u,v) about the saddle point u_s,v_s by a Taylor series, and neglecting terms higher than quadratic terms, one obtains:

$$f(u,v) \approx f(u_s,v_s) + \frac{1}{2}\left[a_{11}(u-u_s)^2 + 2\,a_{12}(u-u_s)(v-v_s) + a_{22}(v-v_s)^2\right] + \ldots$$

Making a transformation about (u_s,v_s) such that:

$$\frac{1}{2}\left[b_1 x^2 + b_2 y^2\right] = f(u_s,v_s) - f(u,v)$$

results in the transformation:

$$a_{11}(u-u_s)^2 + 2\,a_{12}(u-u_s)(v-v_s) + a_{22}(v-v_s)^2 = -b_1 x^2 - b_2 y^2$$

which is made possible by finding the transformation:

$$u - u_s = r_{11}x + r_{12}y$$

$$v - v_s = r_{21}x + r_{22}y$$

where the matrix r_{ij} is a rotation matrix, with $r_{12} = -r_{21}$. Thus:

$$I = \int\limits_{-\infty}^{\infty}\!\!\int e^{\rho\left[f(u_s,v_s) - b_1 x^2/2 - b_2 y^2/2\right]}\, \bar{F}(x,y) \frac{dx\, dy}{(dx/du)(dy/dv)} \tag{9.31}$$

Expanding the integrand into a double Taylor series:

$$\frac{\bar{F}(x,y)}{(dx/du)(dy/dv)} = \sum_{m=0}^{\infty} \sum_{n=0}^{\infty} F_{nm}\, x^{2n+2\nu}\, y^{2m+2\gamma}$$

where ν and γ result from the derivative transformations, then one can integrate the series term by term, resulting in the asymptotic series:

$$I = e^{\rho f(u_s,v_s)} \sum_{m=0}^{\infty} \sum_{n=0}^{\infty} F_{nm} \int_{-\infty}^{\infty} x^{2n+2\nu} e^{-\rho b_1 x^2/2} dx \int_{-\infty}^{\infty} y^{2m+2\gamma} e^{-\rho b_2 y^2/2} dy$$

$$= e^{\rho f(u_s,v_s)} \frac{2\pi}{\sqrt{b_1 b_2}} \sum_{n=0}^{\infty} \sum_{m=0}^{\infty} F_{nm} \frac{\Gamma(2n+2\nu+1)\Gamma(2m+2\gamma+1)}{\rho^{n+m+\nu+\gamma+1} b_1^n b_2^m \Gamma(n+\nu)\Gamma(m+\gamma)} \tag{9.32}$$

9.9 Modified Saddle Point Method: Subtraction of a Simple Pole

The expansion of a function by a Taylor series about a point has a radius of convergence equal to the distance between that point and the closest singularity in the complex plane. This is generally true for the transformations of the type given in eq. (9.10) and primarily due to the factor (dw/dz). Thus, the series expansion given in eq. (9.11) or eq. (9.21) about the saddle point would not be valid for an infinite extent, so that the integrations in eqs. (9.10), (9.12) and (9.16) cannot be carried out to $\pm\infty$. The closer the singularity comes to the saddle point, the shorter the radius of convergence and, hence, the larger value of ρ for which the asymptotic series can be evaluated. To alleviate this problem, few methods were devised to account for the singularity in the function F(z) and hence extend the region of applicability of the asymptotic series.

One method would subtract the pole of the singular function F(z) and expand the remainder of the function in a Taylor series. Letting the function $\frac{\overline{F}(y)}{dy/dz} = G(y)$ in eq. (9.25), then the integral in (9.24) becomes:

$$I = e^{\rho f(z_0)} \int_{-\infty}^{\infty} G(y)\, e^{-\rho y^2/2} dy \tag{9.33}$$

Let the function F(z) have a simple pole at $z = z_1$, then the function G(y) have a simple pole at y = b corresponding to the simple pole at $z = z_1$. The Laurent's series for G(y) can then be written as:

$$G(y) = \frac{a}{y-b} + g(y)$$

where the location of the pole at $z = z_1$ or y = b is given by:

$$b = \sqrt{2}\sqrt{f(z_0) - f(z_1)}$$

and

$$a = \lim_{y \to b} (y-b) G(y) \tag{9.34}$$

is the residue of G(y) at y = b. The function g(y) is analytic at y = 0 and at y = b, so that a Taylor series expansion is possible, whose radius of convergence extends from zero to the closest singularity to y = 0 farther than that at y = b. Thus, the range of validity has now been improved by extending the radius of convergence to the next and farther singularity. Of course, if no other singularity exists, g(y) has an infinite radius of convergence. Expanding the function g(y) in a Taylor series in y, the integral in eq. (9.33) becomes:

$$I = e^{\rho f(z_0)}\, a \int_{-\infty}^{\infty} \frac{e^{-\rho y^2/2}}{y-b}\, dy + e^{\rho f(z_0)} \sum_{n=0}^{\infty} g_{2n} \int_{-\infty}^{\infty} y^{2n} e^{-\rho y^2/2} dy \qquad (9.35)$$

where the odd terms of the Taylor series were dropped because their integral is zero and:

$$g_{2n} = \frac{1}{(2n)!} \frac{d^{2n} g(0)}{dy^{2n}}$$

The second integral in eq. (9.35) gives the same series as in eq. (9.26) with g_{2n} substituting for $\bar{F}_{2n}$ and $\nu = 0$. The first integral can be evaluated by letting:

$$A(\rho, b) = a \int_{-\infty}^{\infty} \frac{e^{-\rho y^2/2}}{y-b}\, dy = ab \int_{-\infty}^{\infty} \frac{e^{-\rho y^2/2}}{y^2 - b^2}\, dy \qquad (9.36)$$

The above expression resulted from splitting the integrand as follows:

$$\frac{a}{y-b} = \frac{a(y+b)}{y^2 - b^2} = \frac{ay}{y^2 - b^2} + \frac{ab}{y^2 - b^2}$$

whose first term integral, being odd, vanishes. Differentiating eq. (9.36) with ρ:

$$\frac{dA}{d\rho} = -\frac{ab}{2} \int_{-\infty}^{\infty} \frac{y^2}{y^2 - b^2} e^{-\rho y^2/2} dy = -\frac{ab}{2} \int_{-\infty}^{\infty} \left(1 + \frac{b^2}{y^2 - b^2}\right) e^{-\rho y^2/2} dy$$

$$= -\frac{b^2}{2} A(\rho, b) - \frac{ab}{2} \int_{-\infty}^{\infty} e^{-\rho y^2/2} dy = -\frac{b^2}{2} A(\rho, b) - \frac{ab}{2} \sqrt{\frac{2\pi}{\rho}}$$

Thus, a differential equation on A(ρ,b) results, i.e.:

$$\frac{dA}{d\rho} + \frac{b^2}{2} A = -ab\sqrt{\frac{\pi}{2}}\, \rho^{-1/2} \qquad (9.37)$$

Letting:

$$A(b, \rho) = e^{-\rho b^2/2} B(b, \rho) \qquad (9.38)$$

then B(ρ,b) satisfies the following differential equation:

$$\frac{dB}{d\rho} = -ab\sqrt{\frac{\pi}{2}} \frac{e^{\rho b^2/2}}{\sqrt{\rho}}$$

There are two methods that can be employed to obtain an expression for B(b,ρ). Following Baños, the function B(b,ρ) becomes:

$$B(b, \rho) = B(b, 0) - ab\sqrt{\frac{\pi}{2}} \int_0^{\rho} \frac{e^{b^2 t/2}}{\sqrt{t}}\, dt = B(b, 0) - i\pi a\, \mathrm{erf}\left(-ib\sqrt{\rho/2}\right) \qquad (9.39)$$

provided that *Re* $b^2 > 0$, or equivalently $-\pi/4 < \arg b < \pi/4$, and:

$$A(b,\rho) = e^{-\rho b^2/2}\left[B(b,0) - i\pi a \operatorname{erf}\left(-ib\sqrt{\rho/2}\right)\right] \tag{9.40}$$

To find B(b,0), let $\rho = 0$ in eqs. (9.36) and (9.40), so that:

$$A(b,0) = B(b,0) = ab\int_{-\infty}^{\infty}\frac{dy}{y^2-b^2} = a\int_0^{\infty}\left(\frac{1}{y-b}-\frac{1}{y+b}\right)dy$$

$$= a\log\left(\frac{y-b}{y+b}\right)\Bigg|_0^{\infty} = -a\lim_{y\to 0}\log\left(\frac{y-b}{y+b}\right)$$

$$= \begin{cases} i\pi a & 0 < \arg b < \pi/4 \quad \text{or} \quad 0 < \arg b^2 < \pi/2 \\ -i\pi a & -\pi/4 < \arg b < 0 \quad \text{or} \quad -\pi/2 < \arg b^2 < 0 \end{cases}$$

Thus:

$$A(b,\rho) = \frac{1}{2}P\left[1-\operatorname{erf}\left(-ib\sqrt{\rho/2}\right)\right] \qquad \text{for} \qquad 0 < \arg b < \pi/4$$

$$A(b,\rho) = -\frac{1}{2}P\left[1+\operatorname{erf}\left(-ib\sqrt{\rho/2}\right)\right] \qquad \text{for} \qquad -\pi/4 < \arg b < 0 \tag{9.41}$$

where P is the residue of the function A at y - b in eq. (9.36) given by:

$$P = 2\pi i a\, e^{-\rho b^2/2} \tag{9.42}$$

The two expressions given in eq. (9.41) can be written in one form as:

$$A(b,\rho) = \frac{P}{2}\operatorname{erfc}\left(-ib\sqrt{\rho/2}\right) - P\,H\left(-\arg b^2\right) \tag{9.43}$$

where $\arg b^2$ was substituted for arg b, since both are equivalent. Thus, the asymptotic series given by eq. (9.35) is given in full by:

$$I \sim e^{\rho f(z_0)}\left\{\frac{P}{2}\operatorname{erfc}\left(-ib\sqrt{\rho/2}\right) - P\,H\left(-\arg b^2\right) + \sqrt{2\pi}\sum_{n=0}^{\infty} g_{2n}\frac{(2n)!}{\rho^{n+1/2}\,n!}\right\} \tag{9.44}$$

If $|b|\sqrt{\rho} \gg 1$, then the first-order approximation of the asymptotic value of eq. (9.44) becomes:

$$I \to \sqrt{\frac{2\pi}{\rho}}\, e^{\rho f(z_o)}\left(-\frac{a}{b}+g_0\right) = \sqrt{\frac{2\pi}{\rho}}\, e^{\rho f(z_o)}\, G(0) \quad \text{for} \quad |b|\sqrt{\rho} \gg 1 \tag{9.45}$$

Felsen and Marcuvitz present a different method of evaluation of the integral for $B(b,\rho)$ in eq. (9.38). Starting with eqs. (9.36) and (9.28):

$$B(b,\rho) = e^{\rho b^2/2}A(b,\rho) = ab\int_{-\infty}^{\infty}\frac{e^{-\rho(y^2-b^2)/2}}{y^2-b^2}\,dy$$

then one can express the denominator as an integral as:

$$= ab \int_{-\infty}^{\infty} \left[\int_{\rho}^{\infty} e^{-\eta(y^2 - b^2)/2} \, d\eta \right] dy$$

where the condition for existence of the integral is:

$Re\ b^2 < 0$

Separating the integrals above, results in:

$$B = ab \int_{\rho}^{\infty} e^{+b^2\eta/2} \left[\int_{-\infty}^{\infty} e^{-\eta y^2/2} dy \right] d\eta = ab\sqrt{\frac{\pi}{2}} \int_{\rho}^{\infty} \frac{e^{\eta b^2/2}}{\sqrt{\eta}} \, d\eta \tag{9.46}$$

The integral in eq. (9.46) becomes:

$$B(b, \rho) = a\pi \frac{b}{\pm ib} \operatorname{erfc}\left(\pm ib\sqrt{\rho/2}\right) \tag{9.47}$$

where the sign is chosen so that the complementary error function converges, i.e.:

$\operatorname{Re}(\mp ib) > 0$

Thus, the positive sign is chosen when $Im\ b < 0$ and the negative sign is chosen when $Im\ b > 0$. This results in:

$$B(b,\rho) = \pm i\pi a \operatorname{erfc}\left(\mp ib\sqrt{\rho/2}\right) \qquad Im\ b \gtrless 0 \tag{9.48}$$

Since erfc (x) = 2 – erfc (-x), then:

$$B(b, \rho) = i\pi a \operatorname{erfc}\left(-ib\sqrt{\rho/2}\right) \qquad Im\ b > 0$$

and

$$B(b, \rho) = i\pi a \operatorname{erfc}\left(-ib\sqrt{\rho/2}\right) - 2i\pi a \qquad Im\ b < 0 \tag{9.49}$$

Finally, the resulting expressions for A can be written as one:

$$A(b, \rho) = \frac{P}{2} \operatorname{erfc}\left(-ib\sqrt{\rho/2}\right) - PH(Im\ b) \tag{9.50}$$

The condition that $Im\ b \gtrless 0$ is equivalent to the condition $\arg b \gtrless 0$ or $\arg b^2 \gtrless 0$.

Another method suggested by Ott for the evaluation of integrals asymptotically when the saddle point is close to a simple pole is the factorization method. Essentially, the integrand in eq. (9.33), G(y), is factored as an analytic function h(y) divided by (y - b), i.e.:

$$I = e^{\rho f(z_0)} \int_{-\infty}^{\infty} \frac{h(y)}{y - b} e^{-\rho y^2/2} dy \tag{9.51}$$

Expanding the analytic function h(y) in a Taylor series about y = b, i.e.:

$$h(y) = \sum_{n=0}^{\infty} h_n (y - b)^n$$

then, the integral becomes:

$$I \sim e^{\rho f(z_0)} \int_{-\infty}^{\infty} \frac{h_0}{y-b} e^{-\rho y^2/2} dy + e^{\rho f(z_0)} \sum_{n=0}^{\infty} h_{n+1} \int_{-\infty}^{\infty} (y-b)^n e^{-\rho y^2/2} dy$$

The first integral was developed earlier in eq. (9.43). The integrals in the series can be integrated term by term. The final form of the asymptotic series becomes:

$$I \sim e^{\rho f(z_0)} \begin{Bmatrix} \frac{P}{2} \operatorname{erfc}\left(-ib\sqrt{\rho/2}\right) - P H\left(-\arg b^2\right) \\ +\sqrt{2\pi} \sum_{n=0}^{\infty} (-1)^n (n!) h_{n+1} \left[\sum_{k=0}^{E(n/2)} \frac{b^{n-2k}}{(n-2k)!\,k!\,2^k \rho^{k+1/2}} \right] \end{Bmatrix} \tag{9.52}$$

where $P = 2\pi i\, h(b)$ and the symbol $E(n/2)$ denotes the largest even integer less than n/2. The expression in eq. (9.52) has a complementary error function just as that given in eq. (9.44). However, the asymptotic series in eq. (9.44) depends on the large parameter ρ only, while the series in eq. (9.52) depends further on the location of the pole with respect to the saddle point. This is not usually desirable, because the radius of convergence of the series in eq. (9.52) depends on the pole location given by "b".

9.10 Modified Saddle Point Method: Subtraction of Pole of Order N

If the function G(y) in eq. (9.33) has a pole of order N, then one can expand the function G(y) in a Laurent's series as follows:

$$G(y) = \frac{a_{-N}}{(y-b)^N} + \frac{a_{-N+1}}{(y-b)^{N-1}} + \ldots + \frac{a_{-1}}{(y-b)} + g(y) \tag{9.53}$$

where g(y) is an analytic function at y = b. Define:

$$A_{-k}(\rho, b) = \int_{-\infty}^{\infty} \frac{e^{-\rho y^2/2}}{(y-b)^k} dy = \frac{1}{k-1} \frac{d}{db} A_{-k+1}(\rho, b) \qquad k = 2, 3, \ldots \tag{9.54}$$

Recalling the expressions in eqs. (9.38) and (9.47) one obtains:

$$A_{-1} = \int_{-\infty}^{\infty} \frac{e^{-\rho y^2/2}}{(y-b)} dy = \pm i\pi e^{-\rho b^2/2} \operatorname{erfc}\left(\mp ib\sqrt{\rho/2}\right)$$

then:

$$A_{-2} = \frac{d}{db} A_{-1} = \pm i\pi e^{-\rho b^2/2} \left\{ -\rho b \operatorname{erfc}\left(\mp ib\sqrt{\rho/2}\right) + \frac{d}{db} \operatorname{erfc}\left(\mp ib\sqrt{\rho/2}\right) \right\}$$

Since:

$$\operatorname{erfc}(x) = \frac{2}{\sqrt{\pi}} \int_{x}^{\infty} e^{-y^2} dy$$

then:

$$\frac{d}{dx}(\mathrm{erfc}(x)) = -\frac{2}{\sqrt{\pi}} e^{-x^2}$$

$$A_{-2} = -\sqrt{2\pi\rho} \mp i\pi b\rho e^{-\rho b^2/2} \mathrm{erfc}\left(\mp ib\sqrt{\rho/2}\right) = -\sqrt{2\pi\rho} - \rho b\, A_{-1} \tag{9.55}$$

Likewise, A_{-3}, A_{-4}, etc. can be computed by a similar procedure. It should be noted that if $|b|\sqrt{\rho} >> 1$, then the asymptotic value of erfc (x) gives:

$$A_{-2} \to \sqrt{\frac{2\pi}{\rho}} \frac{1}{b^2}$$

which is of the same order as A_{-1} given in eq. (9.45).

9.11 Solution of Ordinary Differential Equations for Large Arguments

In Chapter 2, the solution of ordinary differential equations for small arguments was presented by use of ascending power series: the Taylor series for an expansion about a regular point or the Frobenius series for an expansion about a regular singular point. Both of these series solution converge fast if the series is evaluated near the expansion point. To obtain solutions of ordinary differential equations for large arguments, one needs to obtain solutions in a descending power series. To accomplish this, a transformation of the independent variable $\xi = 1/x$ is performed on the differential equation and a series solution in ascending power of ξ.

9.12 Classification of Points at Infinity

To classify points at infinity, one can transform the independent variable x to ξ, so that $x = \infty$ maps into $\xi = 0$. Letting $\xi = 1/x$, the differential equation (2.4) transforms to:

$$\frac{d^2y}{d\xi^2} + \frac{[2\xi - a_1(1/\xi)]}{\xi^2}\frac{dy}{d\xi} + \frac{a_2(1/\xi)}{\xi^4} y = 0 \tag{9.56}$$

Classification of the point $\xi = 0$ depends on the functions $a_1(x)$ and $a_2(x)$:

(i) $\xi = 0$ is a Regular point if:

$$a_1(x) = 2x^{-1} + p_{-2}x^{-2} + p_{-3}x^{-3} + \ldots$$

and

$$a_2(x) = q_{-4}x^{-4} + q_{-5}x^{-5} + q_{-6}x^{-6} + \ldots$$

The solution for a regular point then becomes a Taylor solution:

$$y(\xi) = \sum_{n=0}^{\infty} a_n \xi^n \qquad \text{or} \qquad y(x) = \sum_{n=0}^{\infty} a_n x^{-n}$$

which is a descending power series valid for large x.

(ii) $\xi = 0$ is a regular singular point if:

$$a_1(x) = p_{-1}x^{-1} + p_{-2}x^{-2} + p_{-3}x^{-3} + \ldots \qquad (p_{-1} \neq 0)$$

and

$$a_2(x) = q_{-2}x^{-2} + q_{-3}x^{-3} + q_{-4}x^{-4} + \ldots \qquad (q_{-2} \neq 0)$$

The solution for a regular singular point takes the form:

$$y(\xi) = \sum_{n=0}^{\infty} a_n \xi^{n+\sigma} \quad \text{or} \quad y(x) = \sum_{n=0}^{\infty} a_n x^{-n-\sigma}$$

Again the solution is in descending powers of x valid for large x.

(iii) $\xi = 0$ is an irregular singular point if:

$$a_1(x) = p_0 + p_{-1}x^{-1} + p_{-2}x^{-2} + \ldots \qquad (p_0 \neq 0)$$

and

$$a_2(x) = q_0 + q_{-1}x^{-1} + q_{-2}x^{-2} + \ldots \qquad (q_0 \neq 0)$$

While solutions for finite irregular singular points do not exist, an asymptotic solution of the following type exists:

$$y(x) \sim e^{\alpha x} \sum_{n=0}^{\infty} a_n x^{-n-\sigma}$$

The asymptotic solution approaches the solution for large x.

(iv) $\xi = 0$ is an irregular singular point of rank k, if

$$a_1(x) = p_{k-1}x^{k-1} + p_{k-2}x^{k-2} + \ldots \qquad k \geq 1$$

and

$$a_2(x) = q_{2k-2}x^{2k-2} + q_{2k-3}x^{2k-3} + \ldots \qquad k \geq 1$$

where k is the smallest integer that equals or exceeds 3/2.

For asymptotic solutions about an irregular singular point of order $k \geq 2$:

$$y(x) \sim e^{\omega(x)} \sum_{n=0}^{\infty} a_n x^{-n-\sigma}$$

where:

$$\omega(x) = \sum_{j=1}^{s} \omega_j x^j \qquad s \leq k$$

Example 9.6 Classify the Point $\xi = 0$ for the Following Differential Equations

(i) Legnedre's equation

$$(1-x^2)y''-2x\,y'+n(n+1)y=0$$

$$a_1(x)=\frac{-2x}{1-x^2} \qquad a_2(x)=\frac{n(n+1)}{1-x^2}$$

$$a_1(x)=\frac{2}{x}\frac{1}{1-1/x^2}=\frac{2}{x}\sum_{n=0}^{\infty}\left(\frac{1}{x^2}\right)^n=\frac{2}{x}+\frac{2}{x^3}+----$$

$$a_2(x)=-\frac{n(n+1)}{x^2(1-1/x^2)}=-\frac{n(n+1)}{x^2}\sum_{n=0}^{\infty}\left(\frac{1}{x^2}\right)^n=-\frac{n(n+1)}{x^2}-\frac{n(n+1)}{x^4}----$$

This means that $\xi = 0$ is a regular singular point

(ii) Bessel's equation

$$x^2y''+xy'+(x^2-p^2)y=0$$

$$a_1(x)=\frac{1}{x} \qquad a_2(x)=1-\frac{p^2}{x^2}$$

This indicates the point $\xi = 0$ is an irregular singular point of rank $k = 1$.

9.13 Solutions of Ordinary Differential Equations with Regular Singular Points

If the point $\xi = 0$ is a regular singular point, then one may substitute the Frobenius solution having the form:

$$y(\xi)=\sum_{n=0}^{\infty}a_n\xi^{n+\sigma}$$

$$y(x)=\sum_{n=0}^{\infty}a_n x^{-n-\sigma}$$

Example 9.7

Obtain the solution for large arguments of Legendre's equation:

$$(1-x^2)y''-2x\,y'+n(n+1)\,y=0$$

The point $\xi = 0$ is RSP, then assuming a Frobenius solution, one obtains

$$-a_0(\sigma+n)(\sigma-n-1)x^{-\sigma} - a_1[(\sigma+n+1)(\sigma-n)]x^{-\sigma-1}$$

$$+\sum_{m=0}^{\infty}\left[-(\sigma+m+n+2)(\sigma+m-n+1)a_{m+2}+(\sigma+m)(\sigma+m+1)a_m\right]x^{-m-\sigma-2}=0$$

For $a_0 \neq 0 \qquad \sigma_1 = -n \qquad \sigma_2 = n+1$

$a_1 = 0$

$$a_{m+2} = \frac{(\sigma+m)(\sigma+m+1)}{(\sigma+m+n+2)(\sigma+m-n+1)}\,a_m \qquad m = 0, 1, 2...$$

For σ_1 = -n, the first solution's coefficients are:

$$a_{m+2} = \frac{(m-n)(m-n+1)}{(m+2)(m-2n+1)}\,a_m \qquad m = 0, 1, 2, ...$$

$$a_2 = -\frac{n(n-1)}{2(2n-1)}\,a_0 \qquad a_4 = \frac{n(n-1)(n-2)(n-3)a_0}{2^2 2!\,(2n-1)(2n-3)}$$

$$a_6 = -\frac{n(n-1)\cdot ... \cdot(n-4)}{2^3 3!(2n-1)(2n-3)(2n-5)}\,a_0$$

when m = n, a_{n+2} = 0 and hence $a_{n+4} = a_{n+6} = \ldots = 0$. Therefore:

$$y_1 = a_0 x^{+n}\left[1 - \frac{n(n-1)}{2(2n-1)}x^{-2} + \frac{n(n-1)(n-2)(n-3)}{2^2 2!\,(2n-1)(2n-3)}x^{-4} + ... + (\;)x^{-n}\right] \qquad x \geq 1$$

It can be shown that y_1 is a polynomial of degree n, which is also identical to $P_n(x)$. Hence, it is valid for all x.

For σ_2 = n + 1, the second solution's coefficients are:

$$a_{m+2} = \frac{(m+n+1)(m+n+2)}{(n+2)(m+2n+3)}\,a_m \qquad m = 0, 1, 2, ...$$

$$a_2 = \frac{(n+1)(n+2)}{2(2n+3)}\,a_0 \qquad a_4 = \frac{(n+1)(n+2)(n+3)(n+4)}{2^2 2!\,(2n+3)(2n+5)}\,a_0$$

$$a_6 = -\frac{(n+1)\cdot ... \cdot(n+6)}{2^3 3!(2n+3)(2n+5)(2n+7)}\,a_0$$

The second solution can thus be written as

$$y_2 = a_0 x^{-n-1}\left[1 + \frac{(n+1)(n+2)}{2(2n+3)}x^{-2} + \frac{(n+1)\cdot ... \cdot(n+4)}{2^2 2!\,(2n+3)(2n+5)}x^{-4} + ...\right] \qquad x \geq 1$$

The second solution should be the representation for $Q_n(x)$ for $|x| > 1$. Letting:

$$a_0 = \frac{n!}{1\cdot 3\cdot 5\cdot ... \cdot(2n+1)}$$

results in a descending power series solution for $Q_n(x)$ for $x > 1$:

$$Q_n(x) = \frac{n!\,x^{-n-2}}{1\cdot 3\cdot \ldots \cdot (2n+1)}\left\{1 + \frac{(n+1)(n+2)}{2(2n+3)}x^{-2} + \frac{(n+1)\cdot \ldots \cdot (n+4)}{2^2\,2!\,(2n+3)(2n+5)}x^{-4} + \ldots\right\}$$

9.14 Asymptotic Solutions of Ordinary Differential Equations with Irregular Singular Points of Rank One

If the ordinary differential equation has an irregular singular point at $x = \infty$ of order $k = 1$, then an asymptotic solution can be found in a descending power series. Starting out with form of the ordinary differential equation:

$$y'' + p(x)y' + q(x)y = 0 \tag{9.57}$$

For $k = 1$, then:

$$q(x) = q_0 + \frac{q_1}{x} + \frac{q_2}{x^2} + \ldots \qquad q_0 \neq 0$$

$$p(x) = p_0 + \frac{p_1}{x} + \frac{p_2}{x^2} + \ldots \qquad p_0 \neq 0$$

one can transform this ordinary differential equation to a simpler more manageable equation by transforming the dependent variable y(x):

$$y(x) = u(x)\exp\left(-\frac{1}{2}\int p\,dx\right)$$

which transforms eq. (9.57) to:

$$u''(x) + Q(x)\,u(x) = 0 \tag{9.58}$$

where:

$$Q(x) = q(x) - \frac{1}{2}p'(x) - \frac{p^2(x)}{4}$$

Thus Q(x) has the form for $k = 1$ as:

$$Q(x) = \left(q_0 - \frac{1}{4}p_0^2\right) + \left(q_1 - \frac{p_0 p_1}{2}\right)x^{-1} + \left(q_2 + \frac{p_1}{2} - \frac{p_1^{\,2} + p_0 p_2}{4}\right)x^{-2} + \ldots$$

$$= Q_0 + Q_1 x^{-1} + Q_2 x^{-2} + \ldots = \sum_{n=0}^{\infty} Q_n x^{-n}$$

9.14.1 Normal Solutions

For $k = 1$, try an asymptotic solution with an exponential function being linear in x, i.e.:

$$u(x) \sim e^{\omega x}\sum_{n=0}^{\infty} a_n\,x^{-n-\sigma} \tag{9.59}$$

Substituting into eq. (9.58) results in a recurrence formula:

$$\left(\omega^2+Q_o\right)a_n+\left[Q_1-2\omega(\sigma+n-1)\right]a_{n-1}+\left[Q_2+\sigma+n-2\right]a_{n-2}+\sum_{k=3}^{k=n}Q_k\,a_{n-k}=0$$

If $Q_0 \neq 0$, then for n = 0:

$$\left(\omega^2+Q_o\right)a_0=0$$

since $a_0 \neq 0$, then:

$$\omega^2+Q_o=0 \qquad \omega_1=i\sqrt{Q_o} \qquad \omega_2=-i\sqrt{Q_o} \tag{9.60a}$$

This means that the first term of eq. (9.59) vanishes for all n when ω is equal to ω_1 or ω_2. For n = 1:

$$\left[Q_1-2\omega\sigma\right]a_0=0$$

which results in the value for σ since $a_0 \neq 0$

$$\sigma=\frac{Q_1}{2\omega} \quad \text{or} \quad \sigma_1=\frac{Q_1}{2\omega_1} \quad \text{and} \quad \sigma_2=\frac{Q_1}{2\omega_2} \tag{9.60b}$$

For n = 2:

$$a_1=\frac{\left(Q_2+\sigma_{1,2}\right)}{2\omega_{1,2}}a_0$$

For n≥ 3: with σ_1, σ_2, ω_1, ω_2 given above, the recurrence formula becomes:

$$2\omega_{1,2}(n-1)\,a_{n-1}=\left[Q_2+\sigma_{1,2}+n-2\right]a_{n-2}+\sum_{k=3}^{n}Q_k\,a_{n-k} \qquad n\geq 3 \tag{9.61}$$

It should be noted that both normal solutions are called **Formal Solutions**, i.e., they satisfy the differential equation, but the resulting series in general diverge. However, these solutions represent the asymptotic solutions for large argument x.

Example 9.8 Asymptotic Solutions of Bessel's Equation

Obtain the asymptotic solutions for Bessel's equation of zero order satisfying:

$$y''+\frac{1}{x}y'+y=0$$

This equation was shown to have an irregular singular point of order k = 1. Transforming y(x) to u(x) the ordinary differential equation becomes:

$$y=x^{-\frac{1}{2}}u(x)$$

$$u''+\left(1+\frac{1}{4x^2}\right)u=0$$

Here $Q_0=1$, $Q_1=0$, $Q_2=1/4$, and $Q_3=Q_4=\ldots=0$.

Thus:

$$\omega^2 = -1 \qquad \omega_1 = +i \qquad \omega_2 = -i \qquad \sigma_1 = 0 \qquad \sigma_2 = 0$$

$$a_1 = \frac{\frac{1}{4}}{2\omega} = \frac{1}{8\omega}, \text{ and } a_{n-1} = \frac{\left(\frac{1}{4} + n - 2\right) a_{n-2}}{2\omega(n-1)} \qquad n \geq 3$$

Therefore, the succeeding coefficients become:

$$a_2 = \frac{(1 \cdot 3)^2}{2!\, 8^2\, \omega^2} a_0$$

$$a_3 = \frac{(1 \cdot 3 \cdot 5)^2}{3!\, 8^3 \omega^3} a_0 \ldots$$

and by induction

$$a_n = \frac{[1 \cdot 3 \cdot 5 \cdot \cdots \cdot (2n-1)]^2}{8^n\, n!\, \omega^n} a_0 \qquad n = 1, 2, \ldots.$$

$$= \frac{\left[\frac{1}{2} \cdot \frac{3}{2} \cdot \frac{5}{2} \cdot \ldots \cdot \frac{2n-1}{2}\right]^2}{2^n\, n!\, \omega^n} a_0 = \frac{\left[\Gamma\left(n + \frac{1}{2}\right)\right]^2}{\Gamma^2\left(\frac{1}{2}\right) 2^n\, n!\, \omega^n} a_0$$

$$= \frac{\Gamma^2\left(n + \frac{1}{2}\right)}{\pi\, 2^n\, \omega^n\, n!} a_0$$

The two asymptotic solutions of Bessel's equation are:

$$y_{1,2} \sim \frac{e^{\pm ix}}{\pi\sqrt{x}} a_0 \sum_{n=0}^{\infty} (\mp i)^n \frac{\Gamma^2\left(n + \frac{1}{2}\right)}{2^n\, n!} x^{-n}$$

Choosing $a_0 = \sqrt{2\pi}\, e^{\mp i\pi/4}$, then the asymptotic solutions are those for $H_0^{(1)}(x)$ and $H_0^{(2)}(x)$, i.e.:

$$H_o^{(1)}(x) \sim \sqrt{\frac{2}{\pi x}}\, e^{i(x - \pi/4)} \sum_{n=0}^{\infty} \left(\frac{-i}{2x}\right)^2 \frac{\Gamma^2\left(n + \frac{1}{2}\right)}{n!}$$

$$H_o^{(2)}(x) \sim \sqrt{\frac{2}{\pi x}}\, e^{-i(x - \pi/4)} \sum_{n=0}^{\infty} \left(\frac{i}{2x}\right)^n \frac{\Gamma^2\left(n + \frac{1}{2}\right)}{n!}$$

Examination of the asymptotic series for $H_0^{(1)}(x)$ and $H_0^{(2)}(x)$ shows that the series should be summed up to N terms, provided that $x > N/2$.

9.14.2 Subnormal Solutions

If the series for $Q_n(x)$ happens to have $Q_o = 0$, then $\sigma_{1,2}$ become unbounded. To overcome this problem, one can perform a transformation on the independent variable x:

Let $\xi = x^{\frac{1}{2}}, x = \xi^2$ and $\eta = \xi^{-\frac{1}{2}}\, u(\xi) = x^{-\frac{1}{4}} u(x)$ which results in a new ordinary differential equations on $\eta(\xi)$:

$$\eta'' + P(\xi)\,\eta(\xi) = 0 \tag{9.62}$$

where:

$$P(\xi) = 4\xi^2 Q(\xi^2) - \frac{3}{4\xi^2}$$

If $Q_0 = 0$, and $Q_1 \neq 0$, then:

$$Q(x) = Q_1 x^{-1} + Q_2 x^{-2} + \ldots$$

or

$$Q(\xi^2) = Q_1 \xi^{-2} + Q_2 \xi^{-4} + Q_3 \xi^{-6} + \ldots$$

so that:

$$P(\xi) = 4Q_1 + \left(4Q_2 - \frac{3}{4}\right)\xi^{-2} + Q_3 \xi^{-4} + \ldots$$

Here:

$$P_o(\xi) = 4Q_1$$
$$P_1(\xi) = 0$$
$$P_2(\xi) = 4Q_2 - \frac{3}{4}$$
$$P_3 = 0$$
$$P_4(\xi) = 4Q_3$$

Now, one can use the normal solution for an irregular point of rank one on $\eta(\xi)$, i.e., let:

$$\eta(\xi) \sim e^{\omega\xi} \sum_{n=0}^{\infty} a_n \xi^{-n-\sigma} \tag{9.63}$$

so that:

$$u(x) \sim x^{1/4} e^{\omega\sqrt{x}} \sum_{n=0}^{\infty} a_n\, x^{-(n+\sigma)/2}$$

Since $P_0 = 4\,Q_1$, then:

$$\omega^2 = P_o = -4Q_1 \qquad \omega_{1,2} = \pm\, 2i\sqrt{Q_1}$$

$$\sigma = \frac{P_1}{\omega} = 0$$

so that:

$$u(x) \sim e^{\omega\sqrt{x}} \sum_{n=0}^{\infty} a_n\, x^{1/4 - n/2}$$

Again, the subnormal solutions are **Formal Solutions** as they satisfy the differential equation, but are divergent series.

Example 9.9

Obtain the asymptotic solutions for the following ordinary differential equation:

$$xy'' - y = 0$$

where:

$$Q(x) = -x^{-1}$$

so that:

$$Q_0 = 0, \; Q_1 = -1, \text{ and } Q_2 = Q_3 = \ldots = 0$$

the differential equation transforms to one on $\eta(\xi)$:

$$\eta'' + \left(-4 - \frac{3}{4\xi^2}\right)\eta = 0$$

Here:

$$P_0 = -4, \quad P_1 = 0, \quad P_2 = -\frac{3}{4}, \quad \text{and} \quad P_3 = P_4 = \ldots = 0$$

Letting:

$$\eta(\xi) = e^{\omega\xi} \sum_{n=0}^{\infty} a_n \; \xi^{-n-\sigma}$$

then $\omega^2 = 4$, $\omega_{1,2} = \pm 2$, $\sigma = 0$, and the recurrence formula becomes:

$$a_{n+1} = \frac{\left(n+\frac{3}{2}\right)\left(n-\frac{1}{2}\right)}{2\omega(n+1)} a_n \qquad n = 0, 1, 2, ----$$

so that:

$$n = 0 \qquad a_1 = -\frac{\left(\frac{1}{2}\right)\left(\frac{3}{2}\right)}{(2\omega)} a_0 = +\frac{\Gamma\left(\frac{5}{2}\right)\Gamma\left(\frac{1}{2}\right)}{\Gamma\left(-\frac{1}{2}\right)\Gamma\left(\frac{3}{2}\right)(2\omega)} a_0$$

$$n = 1 \qquad a_2 = +\frac{\Gamma\left(\frac{1}{2}\right)\frac{1}{2}\,\Gamma\left(\frac{5}{2}\right)\frac{5}{2}}{2!\,(2\omega)^2} a_0 = +\frac{\Gamma\left(\frac{7}{2}\right)\Gamma\left(\frac{3}{2}\right)}{\Gamma\left(-\frac{1}{2}\right)\Gamma\left(\frac{3}{2}\right)2!\,(2\omega)^2} a_0$$

and by induction

$$a_n = \frac{\Gamma\left(\frac{2n-1}{2}\right)\Gamma\left(\frac{2n+3}{2}\right)}{\Gamma\left(-\frac{1}{2}\right)\Gamma\left(\frac{3}{2}\right)n!\,(2\omega)^n} a_0 \qquad n \geq 1$$

Since $\Gamma\left(-\frac{1}{2}\right)\Gamma\left(\frac{3}{2}\right) = -\pi$ (Eq. B.1.5)

Therefore:

$$a_n = -\frac{1}{\pi}\frac{\Gamma\left(\frac{2n-1}{2}\right)\Gamma\left(\frac{2n+3}{2}\right)}{n!\,(2\omega)^2} a_0 \qquad n \geq 1$$

and

$$y_{1,2}(x) = a_0 e^{\pm 2\sqrt{x}}\, x^{\frac{1}{4}} \sum_{n=0}^{\infty} (\pm 1)^n \frac{\Gamma\left(\frac{2n-1}{2}\right)\Gamma\left(\frac{2n+3}{2}\right)}{n!\,(16x)^{n/2}}$$

$$= a_0 x^{\pm 2\sqrt{x}}\, x^{\frac{1}{4}} \sum_{n=0}^{\infty} (\pm 1)^n \frac{\left(n^2 - \frac{1}{4}\right)\Gamma^2\left(\frac{2n-1}{2}\right)}{(16x)^{n/2}}$$

Letting $a_0 = -\dfrac{1}{2\sqrt{\pi^3}}$ for y_1, and $a_0 = -\dfrac{1}{2\sqrt{\pi}}$ for y_2 would result in the asymptotic solution of the equation, i.e.:

$$y_1 = x^{\frac{1}{2}}\, I_{\frac{1}{4}}\left(2x^{\frac{1}{2}}\right)$$

$$y_2 = x^{\frac{1}{2}}\, K_{\frac{1}{4}}\left(2x^{\frac{1}{2}}\right)$$

Close examination of the series for the two subnormal solutions shows that they would diverge quickly, after N terms, when the argument $2\sqrt{x} > N(N+1)/2$.

9.15 The Phase Integral and WKBJ Method for an Irregular Singular Point of Rank One

Consider the same reduced eq. (9.58)

$$u'' + Q(x)u = 0$$

with:

$$Q(x) = Q_0 + Q_1 x^{-1} + Q_2 x^{-2} + \ldots$$

Then one may obtain an asymptotic solution by successive iterations. This is known as the WKBJ solution after Wentzel, Kramers, Brillouin, and Jeffrey. Starting out with terms for $x >> 1$, then:

$$u'' + Q_0 u = 0 \qquad\qquad x >> 1$$

giving:

$$u \sim A\, e^{ix\sqrt{Q_0}} + B\, e^{-ix\sqrt{Q_0}}$$

where A and B are the amplitudes and the exponential terms represent the phase of the asymptotic solutions. Thus, let the solution be written as:

$$u(x) \sim e^{ih(x)}$$

then the derivative of $h(x)$ is approximately equal to $\sqrt{Q}$ and $h'(x) \sim \sqrt{Q_0} + \ldots$ so that:

$$u'(x) \sim ih'\, e^{ih}$$

$$u''(x) \sim -e^{ih}\, (ih'' - (h')^2)$$

which when substituted in the ordinary differential equation, results in:

$$ih'' - (h')^2 = -Q(x)$$

This is a non-linear equation on h(x). To obtain a solution, one may resort to iterative methods. Let:

$$h'(x) = \sqrt{Q(x) + ih''}$$

As a first approximation, one may use $h'(x) = \sqrt{Q}$, then one can use iteration to evaluate h'(x), so that:

$$h_j'(x) = \sqrt{Q(x) + ih_{j-1}''} \qquad j = 1, 2, \ldots$$

with $h_{-1}(x) = 0$.

Starting with j = 0:

$$h_0' = \sqrt{Q(x)} \quad \text{and} \quad h_0'' = \frac{Q'}{2\sqrt{Q}}$$

then for the second iteration, j = 1:

$$h_1' = \sqrt{Q + i\frac{Q'}{2\sqrt{Q}}} = \sqrt{Q}\sqrt{1 + \frac{iQ'}{2Q^{3/2}}} \cong \sqrt{Q}\left\{1 + \frac{iQ'}{4Q^{3/2}}\right\} \approx \sqrt{Q} + \frac{iQ'}{4Q}$$

If one would stop at this iteration, then:

$$h_1 = \int \sqrt{Q(t)}\, dt + \frac{i}{4}\log Q$$

and

$$u \sim Q^{-1/4}(x) e^{\pm i \int^x \sqrt{Q'(t)}\, dt} \tag{9.64}$$

This is a first-order approximation. Continuing this process, one can get higher-ordered approximations to h(x). Using this series expression of Q(x), one can obtain an asymptotic series.

Thus, $h_0' = \sqrt{Q_o}$, $h_0'' = 0$, $h_0 = \sqrt{Q_o}\, x$, and:

$$h_1' = \sqrt{Q_o}\sqrt{1 + \frac{Q_1}{Q_o}\frac{1}{x} + \frac{Q_2}{Q_o}\frac{1}{x^2} + \ldots}$$

$$\approx \sqrt{Q_o}\left[1 + \frac{1}{2}\frac{Q_1}{Q_o}\frac{1}{x} + \frac{1}{2}\frac{Q_2}{Q_o}\frac{1}{x^2} + \ldots\right]$$

So that:

$$h_1 \approx \sqrt{Q_o}\, x + \frac{Q_1}{2Q_o}\log x - \frac{1}{2}\frac{Q_2}{\sqrt{Q_o}}\frac{1}{x}$$

$$h_1'' \approx -\frac{1}{2}\frac{Q_1}{\sqrt{Q_o}}\frac{1}{x^2}$$

and

$$h_2 = \sqrt{Q_o + \frac{Q_1}{x} + \frac{Q_2}{x^2} - \frac{i}{2}\frac{Q_1}{\sqrt{Q_o}}\frac{1}{x^2} + \ldots}$$

$$\approx \sqrt{Q_o}\sqrt{1 + \frac{Q_1}{Q_o}\frac{1}{x} + \left(\frac{Q_2}{Q_o} - \frac{i}{2}\frac{Q_1}{Q_o^{3/2}}\right)\frac{1}{x^2} + \ldots}$$

$$\approx \sqrt{Q_o}\left\{1 + \frac{1}{2}\left(\frac{Q_1}{Q_o}\right)\frac{1}{x} + \frac{1}{8Q_o}\left[4Q_2 - 2i\frac{Q_1}{\sqrt{Q_o}} - \frac{Q_1^2}{Q_o}\right]\frac{1}{x^2} + \ldots\right\}$$

then:

$$h_2(x) \sim \sqrt{Q_o}\,x + \frac{1}{2}\left(\frac{Q_1}{\sqrt{Q_o}}\right)\log x - \frac{1}{8\sqrt{Q_o}}\left[4Q_2 - 2i\frac{Q_1}{\sqrt{Q_o}} - \frac{Q_1^2}{Q_o}\right]\frac{1}{x} \ldots.$$

So that:

$$u_1(x) \sim e^{ih(x)} \sim (x)^{iQ_1/\sqrt{Q_o}}\; e^{-i\sqrt{Q_o}\,x}\; e^{+i\sqrt{Q_o}\,x}\; e^{-iA/8x}$$

$$u_2(x) \sim (x)^{-iQ_1/\sqrt{Q_o}}\; e^{-i\sqrt{Q_o}\,x}\; e^{-iA/8x} \tag{9.65}$$

where $A = \frac{1}{\sqrt{Q_o}}\left(4Q_2 - \frac{2iQ_1}{\sqrt{Q_o}} - \frac{Q_1^2}{Q_o}\right)$

Using $e^{ia} = \sum_{n=0}^{\infty} (ia)^n$, one obtains the desired asymptotic series.

Example 9.10 Asymptotic Solutions of Bessel's Equation

Obtain the asymptotic solution of Bessel functions by the WKBJ method.

$$\frac{d^2y}{dx^2} + \frac{1}{x}\frac{dy}{dx} + \left(1 - \frac{p^2}{x^2}\right)y = 0$$

Letting $y(x) = x^{-1/2}u(x)$, then:

$$\frac{d^2u}{dx^2} + \left(1 - \frac{p^2 - \frac{1}{4}}{x^2}\right)u = 0$$

where $Q = 1 - \frac{1}{x^2}\left(p^2 - \frac{1}{4}\right)$, with:

$$Q_0 = 1, \qquad Q_1 = 0, \qquad Q_2 = -\left(p^2 - \frac{1}{4}\right), \qquad Q_3 = Q_4 = \ldots = 0$$

and

$$A = -4\left(p^2 - \frac{1}{4}\right) = 1 - 4p^2$$

$$y_1 \sim x^{-1/2} e^{ix} e^{i(1-4p^2)/(8x)} \sim x^{-1/2} e^{ix} \sum_{n=0}^{\infty} \left[\frac{i}{8x}\left(1-4p^2\right)\right]^n$$

$$y_2 \sim x^{-1/2} e^{-ix} \sum_{n=0}^{\infty} \left[\frac{-i}{8x}\left(1-4p^2\right)\right]^n$$

These solutions are asymptotic solutions to $H_p^{(1)}(x)$ and $H_p^{(2)}(x)$.

9.16 Asymptotic Solutions of Ordinary Differential Equations with Irregular Singular Points of Rank Higher than One

Starting with the reduced eq. (9.58), then

$$y''(x) + Q(x)\,y(x) = 0$$

If the rank of the irregular singular point at $x = \infty$ is larger than one, then one can obtain an asymptotic solution with the exponential term having higher powers of x than one.

However, since the rank could be fractional due to its definition in Section 9.12, i.e., when $2r = 1, 3, 5, \ldots$, then one can avoid fractional powers by transforming $x = \xi^2$, and by letting $u = \xi^{-\frac{1}{2}} y(\xi)$, so that the ordinary differential equation (9.58) becomes:

$$\frac{d^2u}{d\xi^2} + \left[4\xi^2 Q\left(\xi^2\right) - \frac{3}{4\xi^2}\right] u(\xi) = 0 \tag{9.66}$$

Letting the bracketed expression be written as:

$$\frac{d^2u}{d\xi^2} + \xi^{2r} P(\xi)\, u(\xi) = 0 \tag{9.67}$$

then $P(\xi) = P_0 + P_1\xi^{-1} + P_2\xi^{-2} + \ldots$ and the new ordinary differential equation (9.67) has an irregular singular point of order "r".

Assuming an asymptotic solution of ordinary differential equations in eq. (9.67) in the form:

$$u(\xi) = e^{\omega(\xi)} \sum_{n=0}^{\infty} a_n \xi^{-n-\sigma} \tag{9.68}$$

where $\omega(\xi) = \omega_0 \xi + \frac{\omega_1}{2}\xi^2 + \ldots + \omega_{r-1}\frac{\xi^r}{r} + \omega_r \frac{\xi^{r+1}}{r+1}$.

Substituting the form in eq. (9.68) into the ordinary differential equation (9.67) results in the following series:

$$\left[(\omega')^2 + \omega'' + \xi^{2r}P(\xi)\right]\sum_{n=0}^{\infty} a_n\xi^{-n} - 2\omega'(\xi)\sum_{n=0}^{\infty} a_n(n+\sigma)\xi^{-n-1}$$
$$+\sum_{n=0}^{\infty}(n+\sigma)(n+\sigma+1)a_n\xi^{-n-2} = 0 \qquad (9.69)$$

where:

$$\omega'(\xi) = \omega_0 + \omega_1\xi^1 + \omega_2\xi^2 + \ldots + \omega_{r-1}\xi^{r-1} + \omega_r\xi^r$$

and

$$\omega''(\xi) = \omega_1 + 2\omega_2\xi + \ldots + (r-1)\omega_{r-1}\xi^{r-2} + r\omega_r\xi^{r-1}$$

Since the bracketed expression is a polynomial of degree (2r), each multiplying the first term a_0, then for $a_0 \neq 0$, that expression must vanish for ξ^k up to $k = r$, i.e.:

$$(\omega'(\xi))^2 + \omega''(\xi) + \xi^{2r}P(\xi) = 0$$

which results in the evaluation of all the coefficients $\omega_0, \omega_1, \ldots, \omega_r$.:

$$(\omega')^2 = \sum_{k=0}^{2r}\left(\sum_{i+j=k}\omega_{r-i}\omega_{r-j}\right)\xi^{2r-k}$$

$$\omega'' = \sum_{k=0}^{r-1}(r-k)\,\omega_{r-k}\xi^{r-k-1}$$

$$\xi^{2r}P(\xi) = \sum_{k=0}^{\infty} P_k\xi^{2r-k}$$

Since ω'' has ξ raised to a maximum power of (r-1), $(\omega')^2$ has ξ raised to a maximum power of 2r, then the first r terms, with powers of ξ ranging from 2r to r-1 multiply a_0, so that first r terms satisfy:

$$\sum_{i+j=k}\omega_{r-i}\omega_{r-j} + P_k = 0 \qquad k = 0,1,2,\ldots,r \qquad (9.70)$$

This would allow the evaluation of the coefficients ω_0 to ω_r, i.e.:

$$\omega_r^2 + P_0 = 0 \qquad \omega_r = \pm\sqrt{-P_0}$$

$$2\omega_r\,\omega_{r-1} + P_1 = 0 \qquad \omega_{r-1} = -\frac{P_1}{2\omega_r} \qquad (9.71)$$

$$2\omega_r\,\omega_{r-2} + \omega_{r-1}^2 + P_2 = 0 \qquad \omega_{r-2} = -\frac{P_2 + \omega_{r-1}^2}{2\,\omega_r}$$

and

$$\sigma = +\frac{1}{2\,\omega_r}\left(P_{r+1} + r\omega_r + 2\omega_0\,\omega_{r-1} + 2\omega_1\,\omega_{r-2} + ...\right)$$

The remaining equalities in eq. (9.69) would determine the series coefficients a_1, a_2, ... in terms of a_0.

Example 9.11 Asymptotic Solutions for Airy's Functions

Obtain the asymptotic solutions for Airy's function satisfying:

$$y'' - xy = 0$$

The irregular singular point $x = \infty$ is of $r = 1/2$. Due to the fractional order, then the ordinary differential equations to:

$$\frac{d^2u}{d\xi^2} + \left[-4\xi^4 - \frac{3}{4\xi^2}\right]u(\xi) = 0$$

Here $r = 2$, and:

$$P(\xi) = -4 - \frac{3}{4\xi^6}$$

$$P_0 = -4,\quad P_1 = P_2 = P_3 = P_4 = P_5 = 0,\quad P_6 = -\frac{3}{4},\quad P_7 = P_8 = ... = 0,$$

and

$$u(\xi) = \xi^{-1/2}y(\xi)$$

Let:

$$\omega(\xi) = \omega_0\xi + \frac{\omega_1}{2}\xi^2 + \frac{\omega_2}{3}\xi^3$$

Following the procedure outlined in eq. (9.71):

$$\omega_2^2 - 4 = 0 \qquad \omega_2 = \pm 2$$

$$\omega_1 = 0 \qquad \omega_0 = 0 \qquad \sigma = 1$$

Thus:

$$\omega = \omega_2\frac{\xi^3}{3},\quad \omega' = \omega_2\xi^2,\quad \omega'' = 2\omega_2\xi$$

Substituting these in eq. (9.70) and the value of ω_2:

$$\sum_{n=0}^{\infty}\,[(2\omega_2\xi - \frac{3}{4\xi^2})a_n\xi^{-n} - 2\omega_2(n+1)\,a_n\xi^{-n+1} + (n+1)(n+2)a_n\xi^{-n-2}] = 0$$

Expanding these series, one finds that $a_1 = a_2 = 0$, and:

$$a_{m+3} = \frac{(m+1)(m+2) - \frac{3}{4}}{2(m+3)\omega_2}\,a_m \quad m = 0, 1, 2, ...$$

Using the recurrence formula, one can write the two asymptotic solutions as:

$$u(\xi) \sim e^{\pm 2\xi^3/3}\,\xi^{-1}\left\{1 \pm \frac{5}{48}\xi^{-3} + \frac{5\cdot 77}{4^2\cdot 36\cdot 4}\xi^{-6} \pm \frac{5\cdot 77\cdot 221}{4^4\cdot 24\cdot 81}\xi^{-9} + \ldots\right\}$$

This asymptotic solution can be written in terms of x:

$$y_{1,2}(x) \sim e^{\pm 2x^{3/2}/3}\,x^{-1/4}\left\{1 \pm \frac{\Gamma\left(\frac{7}{2}\right)}{1!\Gamma(3/2)x^{3/2}} + \frac{\Gamma\left(\frac{13}{2}\right)}{2!\Gamma(5/2)\,x^{3}} \pm \frac{\Gamma\left(\frac{19}{2}\right)}{3!\Gamma(7/2)x^{9/2}} + \ldots\right\}$$

or:

$$y_{1,2}(x) \sim a_0\, x^{\pm 2x^{3/2}/3}\, x^{-1/4} \sum_{k=0}^{\infty} \frac{(-1)^k\,\Gamma\left(3k+\frac{1}{2}\right)}{k!\Gamma\left(k+\frac{1}{2}\right)\,x^{3k/2}}$$

One may choose $a_0 = 2\sqrt{\pi}$, so that the above series represents the two solutions of Airy's equation.

9.17 Asymptotic Solutions of Ordinary Differential Equations with Large Parameters

It is sometimes necessary to obtain a solution of an ordinary differential equation, such as Sturn-Louisville equations, with a large parameter. The series solutions near x = 0 cannot usually be evaluated when the parameter becomes large. To obtain such asymptotic solution for a large parameter, one can resort to the same methods used in Section 9.15.

9.17.1 Formal Solution in Terms of Series in x and λ

Consider an ordinary differential equation of the type:

$$\frac{d^2y}{dx^2} + p(x,\lambda)\frac{dy}{dx} + q(x,\lambda)y = 0 \tag{9.72}$$

where λ is a parameter of the ordinary differential equation, and the function p and q are given by:

$$\begin{aligned} p(x,\lambda) &= \sum_{n=0}^{\infty} p_n(x)\lambda^{k-n} \\ q(x,\lambda) &= \sum_{n=0}^{\infty} q_n(x)\lambda^{2k-n} \end{aligned} \tag{9.73}$$

where k is a positive integer, $k \geq 1$, and either $p_o \neq 0$ or $q_o \neq 0$. One can reduce the equation to a simpler form:

$$y(x) = u(x)\, e^{-\frac{1}{2}\int p(x,\lambda)dx}$$

which reduces eq. 9.72 to:

$$\frac{d^2u}{dx^2} + Q(x,\lambda)\,u(x) = 0 \tag{9.74}$$

where $Q(x,\lambda) = q(x,\lambda) - \frac{1}{2}p'(x,\lambda) - \frac{1}{4}p^2(x,\lambda)$ and can be represented by:

$$Q(x,\lambda) = \sum_{n=0}^{\infty} Q_n x\,\lambda^{2k-n} \tag{9.75}$$

A *formal solution* of the ordinary differential equation (9.74) of the form:

$$u(x) = e^{\omega(x,\lambda)} \sum_{n=0}^{\infty} u_n(x)\,\lambda^{-n} \tag{9.76}$$

where

$$\omega(x,\lambda) = \sum_{m=0}^{k-1} \omega_m(x)\,\lambda^{k-m} \tag{9.77}$$

Substituting eqs. (9.77) and (9.76) into eq. (9.74), one obtains:

$$\begin{aligned}
&\left\{\sum_{m=0}^{k-1} \omega''_m(x)\,\lambda^{k-m} + \left(\sum_{m=0}^{k-1} \omega'_m(x)\,\lambda^{k-m}\right)^2\right\} \sum_{n=0}^{\infty} u_n(x)\,\lambda^{-n} \\
&\quad + 2\left(\sum_{m=0}^{k-1} \omega'_m(x)\,\lambda^{k-m}\right) \sum_{n=0}^{\infty} u'_n(x)\,\lambda^{-n} + \sum_{n=0}^{\infty} u''_n(x)\,\lambda^{-n} \\
&\quad + \left(\sum_{n=0}^{\infty} Q_n\,\lambda^{2k-n}\right)\left(\sum_{n=0}^{\infty} u_n\,\lambda^{-n}\right) = 0
\end{aligned} \tag{9.78}$$

The coefficient of λ^{2k-n} can be factored out, resulting in the recurrence formula:

$$\begin{aligned}
&\sum_{\ell=0}^{\infty} u_{n-\ell}(x)\left[Q_\ell(x) + \sum_{m=0}^{\infty} \omega'_m(x)\,\omega'_{\ell-m}(x)\right] \\
&\quad + \sum_{\ell=0}^{\infty} u_{n-\ell}(x)\,\omega''_{\ell-k} + 2\sum_{\ell=0}^{\infty} u'_{n-\ell}(x)\,\omega'_{\ell-k}(x) + u''_{n-2k}(x) = 0 \qquad n = 0,1,2,\ldots
\end{aligned} \tag{9.79}$$

The summation is performed with the proviso that:

$$u_q = 0 \qquad q = -1, -2, \ldots$$

and

$$\omega_q = 0 \qquad q = -1, -2\ldots, \text{ and } q = k, k+1, \ldots$$

Setting n = 0 in eq. (9.75), and since $u_q = 0$ for q = -1,- 2..., then the first (k-1) terms of the first bracketed sum of eq. (9.79) must vanish, i.e.:

$$Q_\ell + \sum_{m=0}^{\infty} \omega'_m\, \omega'_{\ell-m} = 0 \qquad \ell = 0,1,2,\ldots,k-1 \tag{9.80}$$

setting $\ell = 0$ gives:

$$Q_0 + [\omega'_0(x)]^2 = 0 \qquad \text{or} \qquad \omega'_0(x) = \pm\sqrt{-Q_0(x)}$$

$$Q_1 + 2\omega'_0\omega'_1 = 0 \qquad \text{or} \qquad \omega'_1 = -\frac{Q_1(x)}{2\omega'_0} = \mp\frac{Q_1(x)}{\sqrt{-Q_0(x)}} \tag{9.81}$$

or in more general form:

$$2\omega'_0\omega'_\ell + Q_\ell + \sum_{m=0}^{m=\ell-1} \omega'_m\omega'_{\ell-m} = 0 \qquad \ell = 1,2,\ldots,k-1$$

so that: (9.82)

$$\omega'_\ell = -\frac{Q_\ell + \sum_{m=1}^{\ell-1} \omega'_m\omega'_{\ell-m}}{2\omega'_0(x)} \qquad \ell = 1,2,\ldots,k-1$$

which gives an expression for all the unknown coefficients, i.e., $\omega'_\ell, \ell = 1,2,\ldots,k-1$ in terms of the two values of $\omega'_0(x)$.

After removing the first (k-1) from the first bracketed sum of eq. (9.79), there remains:

$$\sum_{\ell=k}^{\infty} u_{n-\ell}\left[Q_\ell + \sum_{m=0}^{\infty} \omega'_m\omega'_{\ell-m}\right] + \sum_{\ell=0}^{\infty} u_{n-\ell}\omega''_{\ell-k}$$
$$+ 2\sum_{\ell=0}^{\infty} u'_{n-\ell}\omega'_{\ell-k} + u''_{n-2k} = 0 \qquad n = 1,2,3,\ldots \tag{9.83}$$

Setting n = k in eq. (9.83), one obtains a differential equation for $u'_0(x)$, i.e.:

$$2u'_0\,\omega'_0 + \left[\omega''_0 + Q_k + \sum_{m=1}^{k-1} \omega'_m\omega'_{k-m}\right] u_0 = 0$$

resulting in a linear first-order differential equation on u_0:

$$u'_0(x) + A_0(x)\,u_0(x) = 0 \tag{9.84}$$

where $A_0(x) = \dfrac{1}{2\omega'_0}\left(\omega''_0 + Q_k + \sum_{m=1}^{k-1} \omega'_m\omega'_{k-m}\right)$.

Defining:

$$\mu(x) = e^{-\int A_o(x)dx} \tag{9.85}$$

then the solution for u_0 (eq. (1.9)) can be written as:

$$u_o(x) = C_o\ \mu(x)$$

Similarly one can find formulae for u'_n:

$$u'_n + A_o(x)u_n(x) = B_n(x)$$

where:

$$B_n(x) = -\frac{1}{2\omega'_o}\left\{\sum_{\ell=1}^{n} u_{n-\ell}\left(\omega''_\ell + Q_{k+\ell} + \sum_{m=\ell+1}^{k-1} \omega'_m\omega'_{k+\ell-m}\right) + 2u'_{n-\ell}\omega'_\ell + u''_{n-k}\right\}$$

(9.86)

whose solution is given by eq. (1.9):

$$u_n(x) = C_n\ \mu(x) + \mu(x)\int\frac{B_n(x)}{\mu(x)}\,dx \qquad (9.87)$$

Note that except for the constant C_n, the homogeneous solution for $u_n(x)$ is the same function $\mu(x)$ for $u_o(x)$. Since:

$$\omega'_o = \pm\sqrt{-Q_o(x)}$$

then eqs. (9.82) and (9.87) yield two independent solutions for ω_1, ω_2 ,...ω_k and u_0 ,u_1,....

Example 9.12 Asymptotic Solution for Bessel Functions with Large Orders

Obtain the asymptotic solution of Bessel functions for large arguments and orders. Examining the Bessel's eq.:

$$z^2\frac{d^2y}{dz^2} + z\frac{dy}{dz} + \left(z^2 - p^2\right)y = 0$$

whose solutions are $J_p(z)$ and $Y_p(z)$, and letting $z = px$, then the equation transforms to:

$$x^2\frac{d^2y}{dx^2} + x\frac{dy}{dx} + p^2\left(x^2 - 1\right)y = 0$$

These solutions can be expanded for large parameter p and large argument px.
Letting:

$$y(x) = u(x)\ e^{-\frac{1}{2}\int\frac{dx}{x}} = x^{\frac{1}{2}}u(x)$$

then the equation transforms to:

$$\frac{d^2u}{dx^2} + Q(x,p)\,u = 0$$

where:

$$Q(x) = p^2\left(1 - \frac{1}{x^2}\right) + \frac{1}{4x^2}$$

Thus, here:

$$k = 1, \qquad Q_0 = 1 - \frac{1}{x^2}, \qquad Q_1 = 0, \qquad Q_2 = \frac{1}{4x^2}, \qquad Q_3 = Q_4 = ... = 0\,,$$

and:

$$\omega(x) = p\omega_0(x)$$

Therefore:

$$\omega_0' = \pm\sqrt{\frac{1}{x^2} - 1} = \pm\frac{\sqrt{1-x^2}}{x}$$

or

$$\omega_0(x) = \pm\int\frac{\sqrt{1-x^2}}{x}\,dx = \pm\left[\sqrt{1-x^2} + \log\frac{x}{1+\sqrt{1-x^2}}\right]$$

Equation (9.84) gives:

$$A_0(x) = \frac{\omega_0''}{2\omega_0'}$$

$$\mu(x) = e^{-\int\frac{\omega_0''}{2\omega_0'}dx} = (\omega_0')^{-1/2} = \frac{x^{1/2}}{(1-x^2)^{1/4}}$$

$$u_0(x) = C_0(\omega_0')^{1/2} = C_0^{\pm}\frac{x^{1/2}}{(1-x^2)^{1/4}}$$

where $C_0^{\pm}$ are constants. For n = 1

$$B_1(x) = -\frac{1}{2\omega_0'}\left\{(1 - A_0' + A_0^2)\mu(x) + \frac{1}{4x^2}\right\}$$

$$u_1(x) = C_1\,\mu(x) + \mu(x)\int\frac{B_1(x)}{\mu(x)}\,dx$$

$$= C_1\,\mu(x) - \mu(x)\int\left[\frac{1 - A_0' + A_0^2}{2\omega_0'} + \frac{1}{8x^2\omega_0'\mu}\right]dx$$

Finding closed form solutions for $u_1(x)$ has become an arduous task, which gets more so for higher ordered expansion functions $u_n(x)$. However, one can obtain the first-order asymptotic values as:

$$y_1 \sim x^{-\frac{1}{2}}\,e^{p|\omega_0|}\,u_0(x) \sim x^{-\frac{1}{2}}\,e^{p\sqrt{1-x^2}}\left(\frac{x}{1+\sqrt{1-x^2}}\right)^p + \ldots$$

$$y_2 \sim x^{-\frac{1}{2}}\,e^{-p|\omega_0|}\,u_0(x) \sim x^{-\frac{1}{2}}\,e^{-p\sqrt{1-x^2}}\left(\frac{x}{1+\sqrt{1-x^2}}\right)^{-p} + \ldots$$

9.17.2 Formal Solutions in Exponential Form

Another *formal solution* can be obtained by writing out the solution as an exponential, i.e.:

$$u'' + Q(x,\lambda)u = 0$$

$$Q(x,\lambda)=\sum_{n=0}^{\infty} Q_n(x)\,\lambda^{2k-n}$$

by use of the formal expansion:

$$u \sim e^{\omega(x,\lambda)} \tag{9.88}$$

where:

$$\omega(x,\lambda)=\sum_{n=0}^{\infty} \omega_n(x)\lambda^{k-n} \tag{9.89}$$

Substituting eq. (9.88) into the ordinary differential equations, one has to satisfy the following equality:

$$\omega''+(\omega')^2+Q=0 \tag{9.90}$$

which results in the following recurrence formulae:

$$Q_n+\sum_{m=0}^{n} \omega'_m\omega'_{n-m}=0 \qquad n=0,1,2...k-1 \tag{9.91}$$

and

$$\omega''_{n-k}+\sum_{m=0}^{n} \beta'_m\,\beta'_{n-m}+Q_n=0 \qquad n>k \tag{9.92}$$

For n = 0 in eq. (9.91) gives a value for ω_0:

$$Q_0+(\omega'_0)^2=0 \qquad \omega'_0=\pm\sqrt{-Q_0}=\pm i\sqrt{Q_0} \tag{9.93}$$

which is the same expression as in eq. (9.81):

$$Q_1+2\omega'_0\omega'_1=0 \qquad \omega'_1=-\frac{Q_1}{2\omega'_0}$$

which is the same expression as in eq. (9.81) and in general gives:

$$\omega'_n=-\frac{1}{2\omega'_0}\left[\sum_{m=1}^{n-1}\omega'_m\omega'_{n-m}+Q_n\right] \qquad n=1,2,...,k-1 \tag{9.94}$$

For $n > k$ use eq. (9.91) to give

$$\omega'_n=-\frac{1}{2\omega'_0}\left[\sum_{m=1}^{n-1}\omega'_m\omega'_{n-m}+Q_n+\omega''_{n-k}\right] \qquad n>k \tag{9.95}$$

The two formal solutions in eqs. (9.76) and (9.88) are identical if one would expand the exponential terms in $e^{\omega(x,\lambda)}$ for $n \geq k$ into an infinite series of λ^{-n}.

9.17.3 Asymptotic Solutions of Ordinary Differential Equations with Large Parameters by the WKBJ Method

Consider the special equation of the Sturm-Liouville type:

$$\frac{d^2y}{dx^2} + \lambda^2 Qy = 0$$

Following the method of Section (9.15), then one can replace the coefficients Q_i by λQ_i. Thus, the asymptotic first-order approximation given in eq. (9.64) is:

$$u_{1,2} \sim Q^{-\frac{1}{4}}(x)\, e^{\pm i\lambda \int \sqrt{Q(x)}\, dx}$$

Example 9.13 Asymptotic Solution for Airy's Functions with Large Parameter

Obtain the asymptotic approximation for Airy's function with larger parameter, satisfying

$$\frac{d^2y}{dx^2} - \lambda^2 xy = 0$$

In this case Q(x) = -x then the first-order approximations become:

$$y_{1,2} \sim (-x)^{-1/4}\, e^{\pm i\lambda \int \sqrt{-x}\, dx}$$

$$y_{1,2} \sim x^{-1/4}\, e^{\pm i2x^{3/2}/3}$$

PROBLEMS

Sections 9.2–9.3

Obtain the asymptotic series of the following functions by (a) integration by parts, or (b) Laplace integration:

1. Incomplete Gamma Function: $\Gamma(k,x) = \int_x^\infty t^{k-1}\, e^{-t}\, dt$

2. Incomplete Gamma Function: $\Gamma(k,x) = x^k\, e^{-x} \int_0^\infty e^{-xt}\, (t+1)^{k-1}\, dt$

3. Exponential Integral: $E_1(z) = e^{-z} \int_0^\infty \frac{e^{-zt}}{t+1}\, dt$

4. Exponential Integral of order n: $E_n(z) = e^{-z} \int_0^\infty \frac{e^{-zt}}{(t+1)^n}\, dt$

5. $f(z) = \int_0^\infty \frac{e^{-zt}}{t^2+1}\, dt$

6. $g(z) = \int_0^\infty t \frac{e^{-zt}}{t^2+1}\, dt$

7. $H_0^{(1)}(z) = \frac{2}{\pi} e^{i(z-\pi/4)} \int_0^\infty \frac{e^{-zw}}{\sqrt{2w+iw^2}}\, dw$

Sections 9.5–9.7

Obtain (a) the Debye leading asymptotic term and (b) the asymptotic series for:

8. Complementary error function:

$$\operatorname{erfc}(z) = \frac{1}{\sqrt{z}} e^{-z^2} \int_0^\infty e^{-t^2/4 - zt}\, dt \qquad z \gg 1 \qquad \text{(Hint: let } t = sz\text{)}$$

9. $\operatorname{erfc}(z) = \frac{2}{\pi} e^{-z^2} \int_0^\infty \frac{e^{-z^2 t^2}}{t^2+1}\, dt \qquad z \gg 1$

10. $H_\nu^{(1)}(z) = -\frac{i}{\pi} e^{-i\nu\pi/2} \int_0^\infty e^{iz/2(t+1/t)} t^{-\nu-1} dt$ $\qquad z >> 1$

Also find $J_\nu(z)$ and $Y_\nu(z)$, where $H_\nu^{(1)} = J_\nu(z) + i\,Y_\nu(z)$

11. $H_\nu^{(1)}(z) = -\frac{i}{\pi} \int_{-\infty}^{\infty + i\pi} e^{z\sinh(t) - \nu t} dt$ $\qquad z >> 1$

12. $K_\nu(z) = \frac{\sqrt{\pi}}{\Gamma(\nu + 1/2)} \left(\frac{z}{2}\right)^\nu \int_0^\infty e^{-z\cosh(t)} [\sinh(t)]^{2\nu} dt$ $\qquad z >> 1$ for $\nu > 1/2$

13. $K_\nu(z) = \frac{1}{2}\left(\frac{z}{2}\right)^\nu \int_0^\infty e^{-t - z^2/(4t)} t^{-\nu-1} dt$ $\quad z >> 1$ for $\nu > 1/2$ (Hint: let $t = zs$)

14. $K_\nu(z) = \frac{\sqrt{\pi}}{\Gamma(\nu + 1/2)} \left(\frac{z}{2}\right)^\nu \int_1^\infty e^{-zt} (t^2 - 1)^{\nu - 1/2} dt$ $\qquad z >> 1$ for $\nu > 1/2$

15. $U(n,z) = \frac{e^{-z^2/4}}{(n-1)!} \int_0^\infty e^{-zt - t^2/2} t^{n-1} dt$ $\qquad z >> 1 \qquad n \geq 1$

16. $U(n,z) = \frac{z e^{-z^2/4}}{\Gamma(n/2)} \int_0^\infty e^{-t} t^{n/2-1} (z^2 + 2t)^{-(n+1)/2} dt$ $\qquad$ (Hint: let $t = zs$)

17. Fresnel Function: $F(z) = \frac{e^{i\pi/4}}{\sqrt{2}} - \sqrt{2}\, z e^{iz^2} \int_0^\infty \frac{e^{-iz^2 t^2}}{\sqrt{t^2 + 1}} t\, dt$

18. Probability Function: $\Phi(x) = 1 - \frac{2x}{\sqrt{\pi}} e^{-x^2} \int_0^\infty \frac{e^{-x^2 t^2}}{\sqrt{t^2 + 1}} t\, dt$

Section 9.13–9.16

Obtain the asymptotic solution for large arguments ($x >> 1$) of the following ordinary differential equations

19. $\frac{d^2y}{dx^2} + 2x\frac{dy}{dx} = 0$

20. $x^2 \frac{dy}{dx^2} + x\frac{dy}{dx} + \left(x^2 - \upsilon^2\right)y = 0$

21. $x^2 \frac{dy}{dx^2} + x\frac{dy}{dx} - \left(x^2 + \upsilon^2\right)y = 0$

Section 9.17

Obtain the asymptotic solution for large parameter of the following ordinary differential equations:

22. Problem 20 for finite x, large υ.

23. Problem 20 for x and υ large.

24. Problem 21 for x and υ large.

10

NUMERICAL METHODS

10.1 Introduction

In many applications either the geometry or material composition or non linearity do not lend themselves to an analytic solution. In such cases, numerical methods are the only method for finding a solution. These methods include methods that are based on converting the differential equations to a set of algebraic equations through discretizing the physical region or through finite element methods by discretizing the region into finite elements and then apply the physical laws to each element. With the advent of fast computers and vast storage capabilities, it has become easier to implement numerical methods to solve complex problems. Furthermore, new software enables scientists and engineers to solve numerically complex and large problems that do not require the development of individual software to solve such problems.

10.2 Roots of Non-Linear Equations

To obtain the roots of polynomials or transcendental equations, various iterative methods can be employed to solve:

$$y = f(x) = 0 \tag{10.1}$$

10.2.1 Bisection Method

In this method, one obtains two values of y corresponding to two values of $x = a_i \; and \; b_i$ such that $y_1$and $y_2$have numerically opposite signs:

$$\begin{aligned} y_1 &= f(a_i) \\ y_2 &= f(b_i) \end{aligned} \tag{10.2}$$

then choose a point c_i

$$c_i = (a_i + b_i)/2 \tag{10.3}$$

If $f(c_i)$has the same sign as y_1, then choose $a_{i+1} = c_i$ and $b_{i+1} = b_i$. If $f(c_i)$has the same sign as y_2, then choose $a_{i+1} = a_i$and $b_{i+1} = c_i$. With this stipulation, repeat eqs. (10.2) and (10.3) for the next iteration $(i + 1)$. This method is illustrated in Figure 10.1.

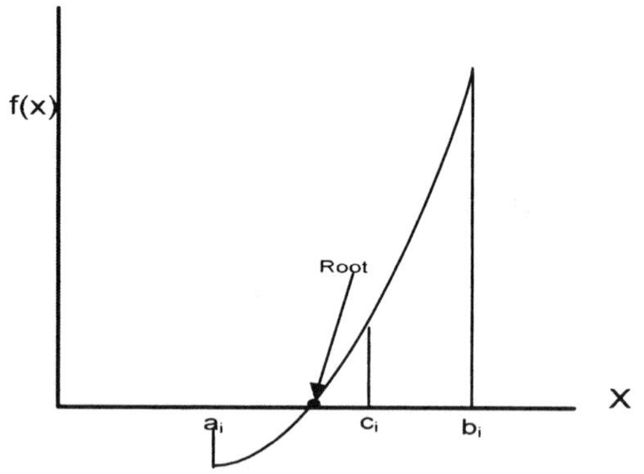

Fig. 10.1: *Scheme for Bisection Method*

Example 10.1

Obtain the smallest positive root of up to two decimal figures:

$f(x) = x^5 - 1$

Let initial $a_1 = 0$ and $b_1 = 1.5$ so that $f(a_1) < 0$ or $f(b_1) > 0$
The solution x=1.00 results in eight iterations, see Table 10.1.

Table 10.1

a_i	$f(a_i)$	b_i	$f(b_i)$	c_i	$f(c_i)$
0	-1.00	1.50	6.59	+0.75	-0.76
0.75	-0.76	1.50	6.59	1.13	+0.84
0.75	-0.76	1.13	+0.84	0.94	-0.27
0.94	-0.27	1.13	+0.84	1.04	+0.22
0.94	-0.22	1.04	0.22	0.99	-0.05
0.99	-0.05	1.04	0.22	1.02	+0.10
0.99	-0.05	1.02	+0.10	1.01	+0.05
0.99	-0.05	1.01	+0.05	1.00	0.00

10.2.2 Newton-Raphson Method

Let x_k be the approximate value of the root of eq. 10.1. Then the Taylor series of $f(x)$ about that value of x:

$$f(x) \cong f(x_k) + f'(k)(x - x_k) = 0$$

Solving for the next value of $x = x_{k+1}$

$$x_{k+1} = x_k - \frac{f(x_k)}{f'(x_k)} \tag{10.4}$$

The method is illustrated in Figure 10.2.

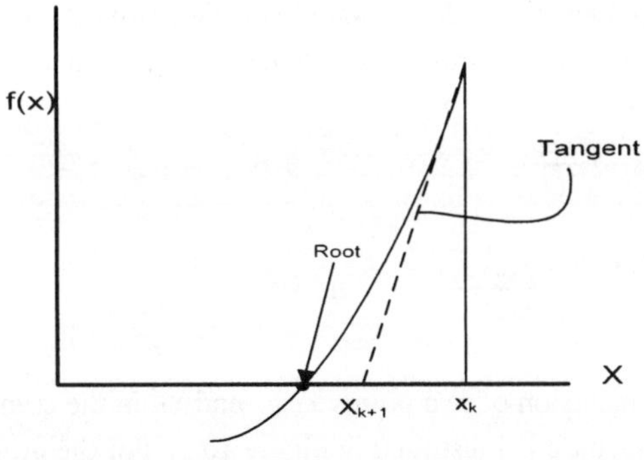

Fig. 10.2: *Scheme for Newton-Raphson Method*

Example 10.2

Find the first root of the following equation within 3 decimal figures:

$$f(x) = e^x - cosx - 1$$
$$f'(x) = e^x + sinx$$

Starting with $x_1 = 0$, the Newton-Raphson method converges in five iterations, see Table 10.2.

Table 10.2

Newton – Raphson		Von Misses ($x_0 = 0.3$)	
x_k	x_{k+1}		
0	+1.00	0	0.608
+1.00	0.669	0.608	0.598
0.669	0.604	0.598	0.603
0.604	0.604	0.603	0.601
0.601	0.301	0.601	0.602
		0.602	0.601

Sometimes, it is not easy to compute a general expression for $f'(x_k)$, so that one can use the initial value $f'(x_o)$ for $f'(x_k)$, the method known for Von-Misses. However, this method converges slowly, if it converges. Taking $x_0 = 0$, with $f'(0) = 1$, Von-Misses did not converge. Taking $x_0 = 0.3$, it converged in six steps.

10.2.3 Secant Method

Sometimes, it is not an easy task to compute the derivative $f^1(x_k)$ readily. To alleviate this problem, one can numerically calculate the derivative at x_k.

Thus $x_{k+1} = x_k - \frac{f(x_k)}{g'(x_k)}$ (10.5)

Where $g'(x_k) = \frac{f(x_k)-f(x_{k-1})}{x_k-x_{k-1}}$

This requires the inclusion of two points x_{k-1} and x_k in the computation of the first derivative. This method is illustrated in Figure 10.3. For the problem in Example (10.2), it took seven iterations to reach the desired root, as shown in Table (10.3).

Table 10.3

k	1	2	3	4	5	6	7	8
x_k	0	0.1	0.909	0.510	0.587	0.602	0.601	0.601
$f(x_k)$	-1	-0.890	0.867	-0.207	-0.034	0.002	-0.008	
$g'(x_k)$		1.100	2.172	2.692	2.247	2.400	10.000	

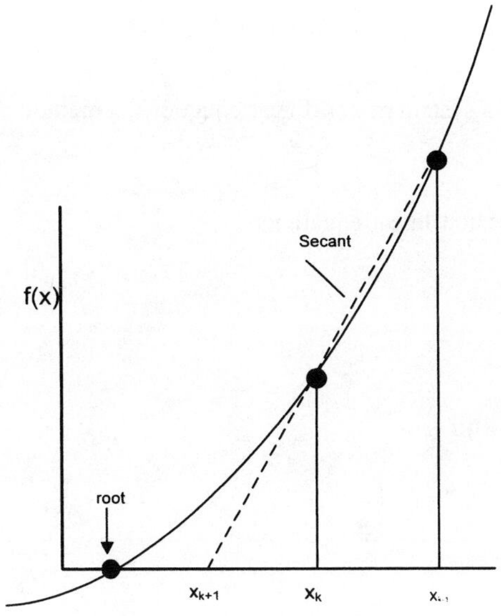

Fig. 10.3: *Scheme for Secant Method*

10.2.4 Iterative Method

If one can rewrite the eq. (10.1) as:

$$x = F(x) \tag{10.6}$$

Then one can iterate this equation as:

$$x_{k+1} = F(x_k)$$

Example 10.3

Calculate the root of $f(x) = \tanh(x) - x + 1 = 0$

Rewriting the equation such that F(x) = tanh(x)+1 then the iterative form becomes:

$$x_{k+1} = \tanh(x_k) + 1$$

Starting with $x_1 = 0.0$, then successive iterations give 1.000, 1.762, 1.943, 1.960, 1.961, and 1.961.

10.3 Roots of a System of Nonlinear Equations

To solve for the roots of a system of non-linear equations, a method based on Newton's method is used.

For the two simultaneous non-linear equations:

$$f(x,y) = 0$$
$$g(x,y) = 0 \tag{10.7}$$

10.3.1 Iterative Method

Rewriting eq. (10.6) as:

$$x = F(x,y)$$
$$y = G(x,y) \tag{10.6}$$

then one can perform simultaneous iterative as:

$$x_{k_{+1}} = F(x_k, y_k)$$
$$y_{k+1} = G(x_u, y_k) \tag{10.8}$$

Example 10.3

Solve for the root of the following simultaneous transcendental eqs.

$$x = \tanh(x+y) - 1$$

$$y = \tanh\,(y-x) + 1$$

Starting with $x_1 = 0 \;\; y_1 = 0$, the root converges to final answer in eight iterations.

10.3.2 Newton's Method

If the root is located at (x,y), the a Taylor series expansion about a neighboring point x_k, y_k is given by:

$$f(x,y) = f(x_k, y_k) + f_x(x_k, y_k)(x - x_k) + f_y(x_k, y_k)(y - y_k) = 0$$

$$g(x,y) = g(x_k, y_k) + g_x(x_k, y_k)(x - x_k) + g_y(x_k, y_k)(y - y_k) = 0$$

Where $f_x = \partial f/\partial x$ and $f_y = \partial f/\partial y$. Solving for $a_k = x - x_k$ and $b_k = y - y_k$ gives

$$a_k f_x(x_k, y_k) + b_k f_y(x_k, y_k) = -f(x_k, y_k)$$

$$a_k g_x(x_k, y_k) + b_k g_y(x_k, y_k) = -g(x_k, y_k) \qquad (10.9)$$

Solving for a_k and b_k results in the next iterative value for x and y, namely:

$a_k = x_{k+1} - x_k$ $\qquad$ $b_k = y_{k+1} - y_k$

$x_{k+1} = x_k + a_k$ $\qquad$ $y_{k+1} = y_k + b_k$

Example 10.4

Obtain the positive real root of the following non-linear equations:

$$x^2 + y^2 = 9$$

$$x^2 - y^2 = 4$$

Let $f(x,y) = x^2 + y^2 - 9$ $\quad f_x = 2x$ $\qquad f_y = 2y$

$g(x,y) = x^2 - y^2 - 4$ $\qquad g_x = 2x$ $\qquad g_y = -2y$

The iterative process is recalculated in Table 10.4.

Table 10.4

k	1	2	3	4	5
x_k	1.0	3.750	2.742	2.556	2.550
y_k	2.0	1.625	1.582	1.574	1.581
a_k	2.750	-1.008	-0.186	-0.006	-0.005
b_k	-0.375	-0.043	-0.008	0.007	0.000

It appears that the root is obtained in five iterations to within 0.001 accuracy. It should be noted that the exact root is $x_k = 2.550$ and $y_k = 1.581$.

10.4 Finite Differences

For the solution of ODE or PDE, one needs to discretize the physical region into a finite number of points and then solve for the dependent variables at those points. To this end, let us define the three types of finite differences.

10.4.1 Forward Difference

For a function $y = f(x)$, the first, second, etc., finite forward difference are defined as:

$$\Delta f_i = f_{i+1} - f_i$$

$$\Delta^2 f_i = \Delta f_{i+1} - \Delta f_i = (f_{i+2} - f_{i+1}) - (f_{i+1} - f_i) = f_{i+2} - 2f_{i+1} + f_i$$

$$\Delta^3 f_i = \Delta^2 f_{i+1} - \Delta^2 f_i = f_{i+3} - 3f_{i+2} + 3f_{i+1} - f_i$$

$$\Delta^4 f_i = \Delta^3 f_{i+1} - \Delta^3 f_i = f_{i+4} - 4f_{i+3} + 6f_{i+2} - 4f_{i+1} + f_i$$

$$\Delta^5 f_i = \Delta^4 f_{i+1} - \Delta^4 f_i = f_{i+5} - 5f_{i+4} + 10f_{i+3} - 10f_{i+2} + 5f_{i+1} - f_i \tag{10.10}$$

10.4.2 Backward Difference

The first, second, etc., finite backward differences are defined as:

$$\nabla f_i = f_i - f_{i-1}$$

$$\nabla^2 f_i = \nabla f_i - \nabla f_{i-1} = (f_i - f_{i-1}) - (f_{i-1} - f_{i-2}) = f_i - 2f_{i-1} + f_{i-2}$$

$$\nabla^3 f_i = \nabla^2 f_i - \nabla^2 f_{i-1} = f_i - 3f_{i-1} + 3f_{i-2} - f_{i-3}$$

$$\nabla^4 f_i = \nabla^3 f_i - \nabla^3 f_{i-1} = f_i - 4f_{i-1} + 6f_{i-2} - 4f_{i-3} + f_{i-4}$$

$$\nabla^5 f_i = \nabla^4 f_i - \nabla^4 f_{i-1} = f_i - 5f_{i-1} + 10f_{i-2} - 10f_{i-3} + 5f_{i-4} - f_{i-5} \tag{10.11}$$

10.4.3 Central Difference

The first, second, etc., Central Differences are defined as:

$$\delta f_{i-\frac{1}{2}} = f_i - f_{i-1}$$

$$\delta^2 f_i = \delta f_{i+\frac{1}{2}} - \delta f_{i-\frac{1}{2}} = (f_{i-1} - f_i) - (f_i - f_{i-1}) = f_{i+1} - 2f_i + f_{i-1}$$

$$\delta^3 f_{i-\frac{1}{2}} = \delta^2 f_i - \delta^2 f_{i-1} = (f_{i+1} - 2f_i + f_{i-1}) - (f_i - 2f_{i-1} + f_{i-2})$$
$$= f_{i+1} - 3f_i + 3f_{i-1} - f_{i-2}$$

$$\delta^4 f_i = \delta^3 f_{i+\frac{1}{2}} - \delta^3 f_{i-\frac{1}{2}} = (f_{i+2} - 3f_{i+1} + 3f_i - f_{i-1}) - (f_{i+1} - 3f_i + 3f_{i-1} - f_{i-2})$$
$$= f_{i+2} - 4f_{i+1} + 6f_i - 4f_{i-1} + f_{i-2}$$

$$\delta^5 f_{i-\frac{1}{2}} = \delta^4 f_i - \delta^4 f_{i-1} = f_{i+2} - 5f_{i+1} + 10f_i - 10f_{i-1} + 5f_{i-2} - f_{i-3} \tag{10.12}$$

$$\delta^6 f_i = \delta^5 f_{i+\frac{1}{2}} - \delta^5 f_{i-\frac{1}{2}} = f_{i+3} - 6f_{i+2} + 15f_{i+1} - 20f_i + 15f_{i-1} - 6f_{i-2} + f_{i-3}$$

10.5 Numerical Differentiation

To complete the differential of a function numerically, one can resort to finite difference schemes by using the Taylor series.

10.5.1 Forward Differentiation

Expanding $f(x+h)$ in terms of the function and its derivatives at x

$$f(x+h) = f(x) + f'(x)h + \frac{1}{2}f''(x)h^2 + \frac{1}{6}f'''(x)h^3 + \frac{1}{24}f^{iv}(x)h^4 + \ldots..$$

Rewriting in terms of f_i

$$f_{i+1} = f_i + f_i'h + \frac{1}{2}f_i''h^2 + \frac{1}{6}f_i'''h^3 + \frac{1}{24}f_i^{iv}h^4 + \ldots\ldots$$

Thus, the first derivative becomes

$$f_i' = (f_{i+1} - f_i)/h + R_2$$

Where $R_2(x) = -\frac{1}{2}f''h - \ldots..$

Thus $f_i' = (f_{i+1} - f_i)/h + O(h) = \Delta f_i/h + O(h)$

Expanding about $f(x + 2h)$

$$f(x + 2h) = f(x) + 2f'(x)h + 2f''(x)h^2 + \tfrac{4}{3}f'''(x)h^3 + \ldots$$

Eliminating the first derivative term from the two expansions gives:

$$f_i'' = (f_{i+2} - 2f_{i+1} + f_i)/h^2 + O(h) = \Delta^2 f_i/h^2$$

Similarly $f_i''' = \Delta^3 f_i/h^3 + O(h)$ and $f_i^{\prime v} = \Delta^4 f_i/h^4 + O(h)$.

If one needs to improve the accuracy of these computed derivatives, then one can compute it to a higher-order error. Starting with the Taylor series of f_{i+1} and f_{i+2}:

$$f_{i+1} = f_i + f_i'h + \frac{1}{2}f_i''h^2 + \frac{1}{6}f_i'''h^3 + O(h^4)$$

$$f_{i+2} = f_i + 2f_i'h + 2f_i''h^2 + \frac{4}{3} + f_i'''h^3 + O(h^4)$$

$$f_{i+3} = f_i + 3f_i'h + \frac{9}{2}f_i''h^2 + \frac{9}{2}f_i'''h^3 + O(h^4)$$

Eliminating f_i'' from the first two equations:

$$f_i' = (-f_{i+2} + 4f_{i+1} - 3f_i)/(2h) + O(h^2)$$

$$= [-\frac{1}{2}\Delta^2 f_i + \Delta f_i]/h + O(h^2)$$

This can also be achieved by substituting f_i'''in terms of finite forward differ of $O(h)$, i.e.,

$$f_{i+1} = f_i + f_i'h + \frac{1}{2}[f_{i+2} - 2f_{i+1} + f_i] + O(h^3)$$

Which when rewritten becomes:

$$f_i' = [-f_{i+2} + 4f_{i+1} - 3f_i]/(2h) + O(h^3)$$

Similarly, eliminating f_i' and f_i'' from above expressions, gives:

$$f_i'' = [-f_{i+3} + 4f_{i+2} - 5f_{i+1} + 2f_i]/h^2 + O(h^2)$$

$$= [-\Delta^3 f_i + \Delta^2 f_i]/h^2 + O(h^2)$$

In summary, one can list the expressions for the derivatives in terms of any order of error.

$$hf_i' = \Delta f_i \qquad e[O(h)] \qquad (10.13)$$

$$= \Delta f_i - \tfrac{1}{2}\Delta^2 f_i \qquad e[O(h^2)] \qquad (10.14)$$

$$= \Delta f_i - \tfrac{1}{2}\Delta^2 f_i + \tfrac{1}{3}\Delta^3 f_i \qquad e[O(h^3)] \qquad (10.15)$$

$$= \Delta f_i - \tfrac{1}{2}\Delta^2 f_i + \tfrac{1}{3}\Delta^3 f_i - \tfrac{1}{4}\Delta^4 f_i \qquad e[O(h^4)] \qquad (10.16)$$

$$h^2 f_i'' = \Delta^2 f_i \qquad e[O(h)] \qquad (10.17)$$

$$= \Delta^2 f_i - \Delta^3 f_i \qquad e[O(h^2)] \qquad (10.18)$$

$$=\Delta^2 f_i - \Delta^3 f_i + \tfrac{11}{12}\Delta^4 f_i \qquad e[O(h^3)] \qquad (10.19)$$

$$= \Delta^2 f_i - \Delta^3 f_i + \tfrac{11}{12}\Delta^4 f_i - \tfrac{5}{6}\Delta^5 f_i \qquad e[O(h^4)] \qquad (10.20)$$

$$h^3 f_i''' = \Delta^3 f_i \qquad e[O(h)] \qquad (10.21)$$

$$= \Delta^3 f_i - \tfrac{3}{2}\Delta^4 f_i \qquad e[O(h^2)] \qquad (10.22)$$

$$= \Delta^3 f_i - \tfrac{3}{2}\Delta^4 f_i + \tfrac{7}{4}\Delta^5 f_i \qquad e[O(h^3)] \qquad (10.23)$$

$$= \Delta^3 f_i - \tfrac{3}{2}\Delta^4 f_i + \tfrac{7}{4}\Delta^5 f_i - \tfrac{15}{8}\Delta^6 f_i \qquad e[O(h^4)] \qquad (10.24)$$

$$h^4 f_i^{\prime v} = \Delta^4 f_i \qquad e[O(h)] \qquad (10.25)$$

$$= \Delta^4 f_i - 2\Delta^5 f_i \qquad e[O(h^2)] \qquad (10.26)$$

$$= \Delta^4 f_i - 2\Delta^5 f_i + \tfrac{17}{6}\Delta^6 f_i \qquad e[O(h^3)] \qquad (10.27)$$

$$= \Delta^4 f_i - 2\Delta^5 f_i + \frac{17}{6}\Delta^6 f_i - \frac{7}{2}\Delta^7 f_i \qquad e[O(h^4)] \qquad (10.28)$$

One can rewrite these equations explicitly in terms of the values at i, i+1, ... for errors of order up to h^4. These are shown in Table 10.5. $h^n f_i^{(n)} = Factor\ [(\,)f_i + (\,)f_{i+1} + \cdots]$

Table 10.5
Forward Finite Difference

Derivative	Factor	i	$i+1$	$i+2$	$i+3$	$i+4$	$i+5$	$i+6$	$i+7$	$e(0h^k)$
	1	-1	+1							h
hf'_i	1/2	-3	+4	-1						h^2
	1/6	-11	+18	-9	+2					h^3
	1/12	-25	+48	-36	+16	-3				h^4
	1	1	-2	+1						h
$h^2 f''_i$	1	2	-5	+4	-1					h^2
	1/12	35	-104	+114	-56	+11				h^3
	1/12	45	-154	+214	-156	+61	-10			h^4
	1	-1	+3	-3	+1					h
$h^3 f'''_i$	1/2	-5	18	-24	+14	-3				h^2
	1/4	-17	+71	-118	+98	-41	+7			h^3
	1/8	-49	+232	-461	+496	-307	+104	-15		h^4
	1	1	-4	+6	-4	+1				h
$h^4 f^{iv}_i$	1	3	-14	+26	-24	+11	-2			h^2
	1/6	35	-186	+411	-484	+321	-114	+17		h^3
	1/6	56	-333	+852	-1219	+1056	-555	+164	-21	h^4

10.5.2 Backward Differentiation

Expanding $f(x-h)$ in terms of the function and its derivatives at:

$$f(x-h) = f(x) - f'(x)h + \frac{1}{2}f''(x)h^2 - \frac{1}{6}f'''h^3 + \frac{1}{24}f^{iv}(x)h^4 + \ldots.$$

Rewriting this in terms of f_i:

$$f_{i-1} = f_i - f_i'h + \frac{1}{2}f''h^2 - \frac{1}{6}f'''h^3 + \frac{1}{24}f^{iv}h^4 + \ldots$$

Then one can obtain an expression for f_i' as:

$$f_i' = [f_i - f_{i-1}]/h + O(h) = [\nabla f_i]/h + O(h)$$

To find high-ordered derivates, start with:

$$f(x-2h) = f(x) - 2f'(x)h + 2f''(x)h^2 - \tfrac{4}{3}f'''(x)h^3 + \ldots$$

$$f_{i-2} = f_i - 2f_i'h + 2f_i''h^2 - \tfrac{4}{3}f_i'''(x)h^3 + \ldots$$

Eliminating f_i' results in

$$f_i'' = [f_i - 2f_{i-1} + f_{i-2}]/h^2 + O(h) = [\nabla^2 \mathrm{f_i}]/\mathrm{h}^2 + \mathrm{O(h)}$$

Eliminating f_i'' results in higher-ordered approximation

$$f_i' = \frac{1}{2}[f_{i-2} - 4f_{i-1} + 3f_i]/h + O(h^2)$$

$$= [\tfrac{1}{2}\nabla^2 f_i + \nabla f_i]/h + O(h^2)$$

Following the methods used in finding expressions for forward differentiation, one can get the following list of backward differentiation formulae.

$$f_i' = [\nabla f_i]/h + O(h) \tag{10.29}$$

$$= [\nabla f_i + \tfrac{1}{2}\nabla^2 f_i]/h + O(h^2) \tag{10.30}$$

$$= [\nabla f_i + \tfrac{1}{2}\nabla^2 f_i + \tfrac{1}{3}\nabla^3 f_i]/h + 0(h^3) \tag{10.31}$$

$$= [\nabla f_i + \tfrac{1}{2}\nabla^2 f_i + \tfrac{1}{3}\nabla^3 f_i + \tfrac{1}{4}\nabla^4 f_i]/h + 0(h^4) \tag{10.32}$$

$$f_i'' = [\nabla^2 f_i]/h^2 + O(h) \tag{10.33}$$

$$= [\nabla^2 f_i + \nabla^3 f_i]/h^2 + O(h^2) \tag{10.34}$$

$$= [\nabla^2 f_i + \nabla^3 f_i + \tfrac{11}{12}\nabla^4 f_i]/h^2 + O(h^3) \tag{10.35}$$

$$= [\nabla^2 f_i + \nabla^3 f_i + \tfrac{11}{12}\nabla^4 f_i + \tfrac{5}{6}\nabla^5 f_i]/h^2 + O(h^4) \tag{10.36}$$

$$f_i''' = [\nabla^3 f_i]/h^3 + O(h) \tag{10.37}$$

$$= [\nabla^3 f_i + \tfrac{3}{2}\nabla^4 f_i]/h^3 + O(h^2) \quad (10.38)$$

$$= [\nabla^3 f_i + \tfrac{3}{2}\nabla^4 f_i + \tfrac{7}{4}\nabla^5 f_i]/h^3 + O(h^3) \quad (10.39)$$

$$= [\nabla^3 f_i + \tfrac{3}{2}\nabla^4 f_i + \tfrac{7}{4}\nabla^5 f_i + \tfrac{15}{8}\nabla^6 f_i]/h^3 + O(h^4) \quad (10.40)$$

$$f_i^{iv} = [\nabla^4 f_i]/h^4 + O(h) \quad (10.41)$$

$$= [\nabla^4 f_i + 2\nabla^5 f_i]/h^4 + O(h^2) \quad (10.42)$$

$$= [\nabla^4 f_i + 2\nabla^5 f_i + \tfrac{17}{6}\nabla^6 f_i]/h^4 + O(h^3) \quad (10.43)$$

$$= [\nabla^4 f_i + 2\nabla^5 f_i + \tfrac{17}{6}\nabla^6 f_i + \tfrac{7}{2}\nabla^7 f_i]/h^4 + O(h^4) \quad (10.44)$$

One may rewrite these expressions explicitly in terms of values at $i, i-1, i-2, \ldots$etc., for errors of order up to h^4. These are shown in Table 10.6.

$$h^n f_i^{(n)} = Factor[(..)f_i + (..)f_{i-1} + (..)f_{i-2} + \cdots]$$

Table 10.6
Backward Finite Difference

Derivative	Factor	i	$i-1$	$i-2$	$i-3$	$i-4$	$i-5$	$i-6$	$i-7$	$e(0h^k)$
hf_i'	1	+1	-1							h
	1/2	3	-4	+1						h^2
	1/6	+11	-18	+9	-2					h^3
	1/12	+25	-48	+36	-16	+3				h^4
$h^2 f_i''$	1	+1	-2	+1						h
	1	+2	-5	+4	-1					h^2
	1/12	35	-104	+114	-56	+11				h^3
	1/12	45	-154	+214	-156	+61	-10			h^4
$h^3 f_i'''$	1	1	-3	+3	-1					h
	1/2	5	-18	+24	-14	+3				h^2
	1/4	+17	-71	+118	-98	+41	-7			h^3
	1/8	+49	-232	+461	-496	+307	-104	+15		h^4
$h^4 f_i^{iv}$	1	1	-4	+6	-4	+1				h
	1	3	-14	+26	-24	+11	-2			h^2
	1/6	35	-186	+411	-484	+321	-114	+17		h^3
	1/6	56	-333	+852	-1219	+1056	-555	+164	-21	h^4

10.5.3 Central Differentiation

For this differentiation, one starts with:

$$f(x+h) = f(x) + hf'(x) + \frac{1}{2}h^2 f''(x) + \frac{1}{6}h^3 f''' + \frac{1}{24}h^4 f'^v + \ldots$$

$$f(x-h) = f(x) - hf'(x) + \frac{1}{2}h^2 f''(x) - \frac{1}{6}h^3 f''' + \frac{1}{24}h^4 f'^v + \ldots$$

Rewriting these expressions in terms of function at x_i:

$$f_{i+1} = f_i + hf_i' + \frac{1}{2}h^2 f_i'' + \frac{1}{6}h^3 f_i''' + \frac{1}{24}h^4 f_i'^v + \ldots$$

$$f_{i-1} = f_i - hf_i' + \frac{1}{2}h^2 f_i'' - \frac{1}{6}h^3 f_i''' + \frac{1}{24}h^4 f_i'^v + \ldots$$

Subtracting these expansions results in an expression for f_i':

$$f_{i+1} - f_{i-1} = 2hf_i' + \frac{1}{3}h^3 f_i''' + \ldots$$

$$f_i' = [f_{i+1} - f_{i-1}]/(2h) + O(h^2)$$

Summing these expansions gives:

$$f_{i+1} + f_{i-1} = 2f_i + h^2 f_i'' + \frac{1}{12}h^4 f_i'^v + \ldots.$$

Resulting in an expression for f_i''

$$f_i'' = [f_{i+1} - 2f_i + f_{i-1}]/h^2 + O(h^2)$$

Note that the first approximation is $O(h^2)$ instead of $O(h)$ as in forward or backaward differentials.

Expanding the expression for f_{i+2} and f_{i-2} leads to higher derivatives or higher orders. Final forms are listed below:

$$f_i' = \frac{1}{2}[\delta f_{i+\frac{1}{2}} + \delta f_{i-\frac{1}{2}}]/h + O(h^2) \quad (10.45)$$

$$= \{\frac{1}{2}[\delta f_{i+\frac{1}{2}} + \delta f_{i-\frac{1}{2}}] - \frac{1}{12}[\delta^3 f_{i+\frac{1}{2}} + \delta^3 f_{i-\frac{1}{2}}]\}/h + O(h^4) \quad (10.46)$$

$$f_i'' = [\delta^2 f_i]/h^2 + O(h^2) \tag{10.47}$$

$$= [\delta^2 f_i - \tfrac{1}{12}\delta^4 f_i]/h^2 + O(h^4) \tag{10.48}$$

$$f_i''' = \tfrac{1}{2}[\delta^3 f_{i+\frac{1}{2}} + \delta^3 f_{i-\frac{1}{2}}]/h^3 + O(h^2) \tag{10.49}$$

$$= \{\tfrac{1}{2}[\delta^3 f_{i+\frac{1}{2}} + \delta^3 f_{i-\frac{1}{2}}] - \tfrac{1}{8}[\delta^5 f_{i+\frac{1}{2}} + \delta^5 f_{i-\frac{1}{2}}]\}/h^3 + O(h^4) \tag{10.50}$$

$$f_i^{\prime v} = [\delta^4 f_i]/h^4 + O(h^2) \tag{10.51}$$

$$= [\delta^4 f_i - \tfrac{1}{6}\delta^6 f_i]/h^4 + O(h^4) \tag{10.52}$$

These equations can be rewritten in terms of values at $i, i+1, i+2$, etc., for error orders h^2 and h^4. These are shown in Table 10.7.

$$h^n f_i^{(n)} = Factor[(..)f_i + (..)f_{i-1} + (..)f_{i+1} + ...]$$

Table 10.7
Central Finite Difference

Derivative	Factor	$i-3$	$i-2$	$i-1$	i	$i+1$	$i+2$	$i+3$	$e(0h^k)$
hf_i'	1/2			-1	0	+1			h^2
	1/12		+1	-8	0	+8	-1		h^4
$h^2 f_i''$	1			+1	-2	+1			h^2
	1/12		-1	+16	-30	+16	-1		h^4
$h^3 f_i'''$	1/2		-1	+2	0	-2	+1		h^2
	1/8	1	-8	+13	0	-13	+8	-1	h^4
$h^4 f_i^{\prime v}$	1		+1	-4	+6	-4	+1		h^2
	1/6	-1	+12	-39	+56	-39	+12	-1	h^4

Example 10.4

Evaluate the first four derivatives of the following function at x_0:

$f(x) = e^x - x^2$ at $x_0 = 0.5$, step $h = 0.1$

$f'(x)$	<u>exact</u>	Eq.:	(10.13)	(10.14)	(10.15)	(10.16)
	0.6487		0.6340	0.6425	0.6482	0.6464

		Eq.:	(10.29)	(10.30)	(10.31)	(10.32)
			0.6690	0.6440	0.6493	0.6505
		Eq.:	(10.45)	(10.46)		
			0.6515	0.6487		
$f''(x)$	exact	Eq.:	(10.17)	(10.18)	(10.19)	(10.20)
	-0.3513		-0.1764	-0.3282	-0.3497	-0.3514
		Eq.:	(10.33)	(10.34)	(10.35)	(10.36)
			-0.5071	-0.3654	-0.3537	-0.3514
		Eq.:	(10.47)	(10.48)		
			-0.3499	-03518		
$f'''(x)$	exact	Eq.:	(10.21)	(10.22)	(10.23)	(10.24)
	1.6487		1.8990	1.6160	1.6525	
		Eq.:	(10.37)	(10.38)	(10.39)	(10.40)
			1.4208	1.6236	1.6462	
		Eq.:	(10.49)	(10.50)		
			1.6529	1.6487		
$f'^{v}(x)$	exact	Eq.:	(10.25)	(10.26)	(10.27)	(10.28)
	1.6487		2.0170	1.6097		
		Eq.:	(10.41)	(10.42)	(10.43)	(10.44)
			1.3521	1.6095		
		Eq.:	(10.51)	(10.52)		
			1.6514	1.6486		

10.6 Numerical Integration

In many instances, one cannot find closed form integration formulae for mathematical functions. One then resorts to integrate such functions numerically.

10.6.1 Trapezoidal Rule

To numerically integrate a function $f(x)$ over a finite region, $\int_{x_0}^{x_n} f(x)\,dx$ one can divide the integration interval $x_n - x_0$ into n equal segments with the first point being x_0 and th$(n+1)$e st point is x_n. Evaluating the functio $f(x)$n at x_i where $f_i = f(x_i)$, and connecting the points $f_0, f_1, \dots, f_n$ with straight lines, then between f_i and f_{i+1} results a trapezoid of heights f_i and f_{i+1} and a $base = h = \frac{x_n - x_0}{n}$. Summing the areas of these trapezoids results in an approximation of the areas under the curve $f(x)$:

$$\int_{x_0}^{x_n} f(x)\,dx \approx \frac{1}{2}h(f_0 + f_1) + \frac{1}{2}h(f_1 + f_2) \dots + \frac{1}{2}h(f_{n-1} + f_n)$$

$$= \frac{1}{2}h(f_0 + 2f_i + 2f_2 + \dots + 2f_{n-1} + f_n) \qquad (10.53)$$

It should be noted that as $n \gg 1$, the approximation approaches the actual exact value.

The error in the trapezoid rule is $O(h^2)$. Note that the minimum number of segments is 2.

Example 10.5

Integrate the function $f(x) = e^x$ over [0,1].

The exact value of the integral is 1.718281828

$n = 2$	$h = 0.5$	I = 1.753931092	er = 2.1%
$n = 4$	$h = 0.25$	I = 1.727221905	er = 0.5%
$n = 8$	$h = 0.125$	I = 1.720518592	er = 0.13%

10.6.2 Simpson's Rule

Dividing the range $x_n - x_0$ into n equal parts and evaluating the function $f(x)$ at $n+1$ points $x_0, x_1, \dots, x_n$ gives $n+1$ values $f_0, f_1, \dots, f_n$. Connecting the points $f_0, f_1, \dots, f_n$ by parabolic segments, one obtains an approximate value for the integral:

$$\int_{x_0}^{x_n} f(x)\,dx \approx \frac{h}{3}(f_0 + 4f_1 + 2f_2 + 4f_3 + 2f_4 + \cdots + 2f_{n-2} + 4f_{n-1} + f_n) \qquad (10.54)$$

This is known as Simpson's 1/3 Rule. The error in this integration rule is $O(h^4)$. Note that the minimum number of segments is 4.

Example 10.6

Integrate the function $f(x) = e^x$ over [0,1] using Simpson's rule.

$n = 4$ I = 1.718318842 er = 0.002%

$n = 8$ I = 1.718284155 er = 0.0001%

It can be seen that the accuracy of the Simpson's rule is higher than the trapezoidal rule for the same number of segments.

10.6.3 Romberg Integration

This method relies on the estimate of the error to improve the accuracy of the integration. This method starts by using a finite number of segments and then successively doubling the number of segments. Starting with the trapezoidal rule giving $I_h = I(h_2)$ and $I_\ell = I(h_1)$, with $h_2 = h_1/2$ the improved integral to a first-order approximation can be estimated as:

$$I^{(1)} = \frac{4\ I_h - I_\ell}{4-1} \qquad (10.55)$$

To improve the integral further, a second-order integral $I^{(2)}$:

$$I^{(2)} = \frac{4^2 I_h - I_\ell}{4^2-1} \qquad (10.56)$$

A nth order improvement on the integral results in an improved integral given by

$$I^{(n)} = \frac{4^n I_h - I_\ell}{4^n-1} \qquad (10.57)$$

Example 10.7

Using the results obtained in Example 10.5, an exact value = 1.718281828

Number of strips	**1**	**2**	**4**	**8**
	1.859140914	1.753931092	1.727221905	1.720518592
error		0.035649264	0.008940077	0.002236764
First-order approx ($n = 1$) (eq. 10.55)		1.718861151	1.718318843	1.718284152
error		0.000579323	0.00037115	0.000002324
Second-order approx ($n = 2$) (eq. 10.56)			1.718282689	1.718281839
error			0.000000861	0.000000011
Third -order approx ($n = 3$) (eq. 10.57)				1.718281826
error				0.000000002

One can see that the Romberg Scheme improved the value of the integral over the value obtained by the trapezoidal rule for the same number of strips. For example, the trapezoidal rule for 2 segments has an error of 0.035649264, while the first-order Romberg Scheme for the same two segment reduced the error to 0.000579323. The same is true for 4 segments, where the error dropped for the first-order Romberg to 0.000000861 for the second-order Romberg.

10.6.4 Gaussian Quadrature

In this scheme, a function $f(x)$ is approximated by an nth degree polynomial. In this section, we will only cover the approximation by Legendre polynomials of degree up to n.

Consider the integration formula

$$\int_{-1}^{+1} f(x)\,dx = w_0 f(x_0) + w_1 f(x_1) + \cdots + w_n f(x_n) \tag{10.58}$$

Where x_i are the zeroes of $P_{n+1}(x_i) = 0$ $i = 0,1,2,\ldots,n$. The weighting coefficients $w_0, w_1, \ldots, w_n$ are to be computed from the formula:

$$w_i = \frac{1}{P'_{n+1}(x_i)} \int_{-1}^{+1} \frac{P_{n+1}(x)}{x - x_i}\,dx \quad i = 0, 1, 2, \ldots, n$$

The weighting coefficient w_i is listed in many mathematical handbooks for any degree n [see Ref. Chap. 3, Abramowitz and Stegun].

Example 10.8

Integrate the function using four point Gaussian scheme:

$$\int_{-1}^{+1} \frac{dx}{(x+2)^2}$$

Let $n + 1 = 4$

The zeroes of $P_4(x_i) = 0$ and the corresponding weighting coefficients are:

	$\pm x_i$	w_i
	0.3399810435	0.6521451548
	0.8611363115	0.3478548451
$I =$	0.666449254	error = 0.033%

If the integration limits are not $[-1, +1]$, one can transform the integration to these limits, start with:

$$\int_a^b f(x)dx$$

Let $x = \frac{b-a}{2} z + \frac{b+a}{2}$

$$dx = \frac{b-a}{2} d_z \qquad x = a \;\; z = -1 \qquad and\; x = b \;\; z = +1$$

$$\int_a^b f(x)dx = \frac{b-a}{2} \int_{-1}^{+1} f\left(\frac{b-a}{2} z + \frac{b+a}{2}\right) d_z =$$

$$= \frac{b-a}{2} \int_{-1}^{+1} \bar{f}\,(z)dz \tag{10.59}$$

Example 10.9

Integrate $\int_0^1 e^x\, dx$ using a four-point Gaussian scheme:

$x = (z+1)/2$ $\bar{f}(z) = e^{(z+1)/2}$

$I = \frac{1}{2}\int_{-1}^{+1} e^{(z+1)/2} dz$

Same points z_i and weighting coefficients as in Example 10.8.

$I = 1.718281828$ same as the exact to 10^{-10} digits

10.7 Ordinary Differential Equations-Initial Value Problems

One of the major uses of numerical methods is the solution of ordinary (ODE) and partial (PDE) differential equations.

Initial value problems are characterized by an nth order ODE with uniqueness requiring the specification of n initial values of the dependent variable at an initial point x_0. The solution is sought for $x > x_0$.

10.7.1 Euler's Method for First-Order ODE

These problems are defined by the system

$y'(x) = f(x,y)$ $y(x_0) = y_0$ $x \geq x_0$

Using forward difference for the first derivative with $y = y(x_0)$:

$$y_{i+1} - y_i = \int_{x_i}^{x_{i+1}} f(x,y)dx = hf_i(x_i, y_i) + 0(h^2) \tag{10.60}$$

or $y_{i+1} = y_i + hf_i(x_i, y_i)$

The truncation error is $O(h^2)$, see eq. (10.13). This is known as Euler's Formula.

Example 10.10

Obtain the solution of the following system:

$y' + xy = e^{-x^2/2}$ $y(0) = 1$ $f(x,y) = e^{-x^2/2} - xy$

For $x = 0(0.1)0.4$

$$y_{i+1} = y_i + (0.1)(e^{-x_i^2/2} - x_i y_i)$$

Here $f(x,y) = e^{-x_i^2/2} - x_i y_i$

x	y	f	exact	error%
0	1	1		
0.1	1.1	0.885012	1.094514	+0.5
0.2	1.188501	0.742498	1.176238	+1.0
0.3	1.262751	0.577172	1.242797	+1.6
0.4	1.320468		1.292363	+2.2

The exact solution (Chap. 1, Problem 1)

$$y = (x+1)e^{-x^2/2}$$

For forward difference scheme, the accumulated error increases with each step.

One can also solve non-linear ODE problems numerically.

Example 10.11

Obtain the solution of the following system by Euler's Method:

$y' = xy^{\frac{1}{2}}$ $\quad\quad y(0) = 1$ $\quad\quad x = 0(0.1)0.4$

The exact solution of this problem $y = (\frac{x^2}{4} + 1)^2$

$$y_{i+1} = y_i + hx_i y_i^{\frac{1}{2}}$$

	exact	error %
$y_1 = 1 + 0 = 1.000000$	1.00500	0.5%
$y_2 = 1 + .1 * 0.1 * 1 = 1.01$	1.02010	1%

$y_3 = 1.01 + 0.1 * 0.2 * (1.01)^{\frac{1}{2}} = 1.030100$	1.045506	1.5%
$y_4 = 1.030100 + 0.1 * 0.3 * (1.0301)^{\frac{1}{2}} = 1.060548$	1.081600	1.9%

10.7.2 Euler Prediction-Corrector Method

Since Euler's Method computes the derivative value $f(x_i, y_i)$ at the starting point x_i, y_i, the error accumulates with each step. An improvement on Euler' Method involves calculating y_{i+1}, then calculating the predicted value of $f_{i+1}(x_{i+1}, y_{i+1})$ and use trapezoidal approximation between f_i and f_{i+1} to increase the accuracy of Euler's Method to $O(h^3)$.

Thus $y_{i+1} - y_i = hf_i(x_i, y_i)$

$$y_{i+1} - y_i = \frac{h}{2}[f_i(x_i, y_i) + f_{i+1}(x_{i+1}, y_{i+1})] \qquad (10.61)$$

The Predictor-Corrector method can be repeated at each point until a satisfactory value is reached.

Example 10.12

Redo Example (10.10) by Euler Predictor-Corrector Method. Let $y^{(n)}$ be the n^{th}iteration where $n = 3$.

x	$y^{(1)}$	$f^{(1)}$	$y^{(2)}$	$f^{(2)}$	$y^{(3)}$	error %
0	1.000	1.000				
0.1	1.1	0.885012	1.094251	0.885587	1.094279	-0.02%
0.2	1.182838	0.743631	1.175740	0.745119	1.175814	-0.04%
0.3	1.250326	0.580900	1.242115	0.583363	1.242238	-0.04%
0.4	1.300574	0.402887	1.291550	0.406496	1.291731	-0.05%

Example 10.13

Redo Example 10.11 by Euler Predictor-Corrector Method for up three iterations:

x	$y^{(1)}$	$f^{(1)}$	$y^{(2)}$	$f^{(2)}$	$y^{(3)}$	error %
0	1.000	0				
0.1	1.000	0.1	1.005	0.100250	1.005012	0.001%
0.2	1.015037	0.201499	1.200100	0.202000	1.020125	0.002%
0.3	1.040325	0.305989	1.045524	0.306753	1.045563	0.005%
0.4	1.076238	0.414968	1.081650	0.416010	1.081701	0.009%

It can be seen that the error quickly reduces as one takes more iterations at each point x.

10.7.3 Runge-Kutta Methods

In these methods, one avoids the successive computation of y'_{i+1} at the succeeding point x_{i+1} by matching the increment $\Delta y_{i+1} = y_{i+1} - y_i$ to a Taylor series expression of y'_i. Matching up to the second, third, and fourth derivatives results in what is known as the Runge-Kutta Method of order two, three, and four.
Second-order Runge-Kutta:

$$k_1 = hf(x_i, y_i)$$

$$k_2 = hf(x_i + h, y_i + k_1)$$

Resulting in a solution with an error of $O(h^3)$.

$$y_{i+1} = y_i + \frac{1}{2}(k_1 + k_2) \tag{10.62}$$

Third-order Runge-Kutta:

$$k_1 = hf(x_i, y_i)$$

$$k_2 = hf(x_i + h/2, y_i + k_1/2)$$

$$k_3 = hf(x_i + h, y_i - k_1 + 2k_2)$$

Resulting in a solution with an error of $O(h^4)$.

$$y_{i+1} = y_i + \frac{1}{6}(k_1 + 4k_2 + k_3) \tag{10.63}$$

Fourth-order Runge-Kutta:

$$k_1 = hf(x_i, y_i)$$

$$k_2 = hf(x_i + h/2, y_i + k_1/2)$$

$$k_3 = hf(x_i + h/2, y_i + k_2/2)$$

$$k_4 = hf(x_i + h, y_i + k_3)$$

$$y_{i+1} = y_i + \frac{1}{6}(k_1 + 2k_2 + 2k_3 + k_4) \tag{10.64}$$

Resulting in a solution with an error of $O(h^5)$.

Example 10.14

Do Example (10.11) by Runge-Kutta Method of order two, three, and four.

Runge-Kutta of Order 2:

x_i	y_i	k_1	k_2	y_{i+1}	error%
0	1.000	0	0.01	1.005	0
0.1	1.005	0.010025	0.020150	1.020087	0.001
0.2	1.020087	0.020200	0.030598	1.045486	0.002
0.3	1.045486	0.030675	0.041495	1.081571	0.003

Comparing the errors with Euler Predictor-Corrector Method shows the same approximate error. That is because both have errors of $O(h^3)$.

Runge-Kutta of Order 3:

x_i	y_i	k_1	k_2	k_3	y_{i+1}	error%
0	1.0000	0	0.005	0.010050	1.005008	0.0008
0.1	1.005008	0.010025	0.015075	0.020250	1.020104	0.0004
0.2	1.020104	0.020201	0.025375	0.030750	1.045513	0.0006
0.3	1.045513	0.030675	0.036049	0.041702	1.081609	0.0008

One can see the improvement in the solution from the second-order solution.

Runge-Kutta of Order 4:

x_i	y_i	k_1	k_2	k_3	k_4	y_{i+1}	error%
0	1.0000	0.000	0.005	0.05006	0.010025	1.005006	0.0006
0.1	1.005006	0.010025	0.015075	0.015094	0.0202	1.020100	0
0.2	1.020100	0.020201	0.025375	0.025407	0.030675	1.045507	0.00006
0.3	1.045507	0.030675	0.036049	0.036095	0.041600	1.081601	0.00002

This is a one order of magnitude improvement over the third-order solution.

10.7.4 Adams Method

Adams Method is similar to Euler Predictor-Corrector Method, but it relies on more points outside y_i and y_{i+1}. To obtain the values of y' at other points, one can approximate $f(x, y)$ by a polynomial of a certain degree such that $f(x, y)$ fits exactly at a finite number of points x.

Adams second-order formula:

$$y_{i+1} = y_i + \frac{h}{2}(-f_{i-1} + 3f_i) \tag{10.65}$$

The solution has an error $O(h^3)$.

Adams third-order formula:

$$y_{i+1} = y_i + \frac{h}{12}(5f_{i-2} - 16f_{i-1} + 23f_i) \qquad (10.66)$$

The solution has an error of $O(h^4)$.

Adams fourth-order formula:

$$y_{i+1} = y_i + \frac{h}{24}(-9f_{i-3} + 37f_{i-2} - 59f_{i-1} + 55f_i) \qquad (10.67)$$

The solution has an error of $O(h^5)$.

Since one needs more than a point to start these solutions, it is best to obtain values for these starting points by an equivalent Runge-Kutta Method.

Example 10.15

Do Example 10.11 by using initial data of Example 10.13 and Adams Method of second order, since Adams Method of second order is $O(h^3)$, same as the Euler Predictor-Corrector Method. This requires two starting points at $x = 0$ and 0.1.

x_i	y_i	f_i	y_{i+1}	error %
0	1.00	0		
0.1	1.005012	0.100250	1.020050	0.005
0.2	1.020050	0.202000	1.045337	0.016
0.3	1.045337	0.306725	1.081246	0.033
0.4	1.081246	0.415932	1.128300	0.054

Using Adams Method of third order requires three starting points at $x = 0, 0.1$ and 0.2 which will be taken from Example (10.13).

x_i	y_i	f_i	y_{i+1}	error %
0	1.00	0		
0.1	1.005012	0.100250		
0.2	1.020125	0.202002	1.045475	0.003
0.3	1.045475	0.306745	1.081511	0.008
0.4	1.081511	0.415983	1.128758	0.013

Using Adams Method of fourth order requires four starting points at $x = 0, 0.1, 0.2.$ and $0.3.$

x_i	y_i	f_i	y_{i+1}	error %
0	1.00	0		
0.1	1.005012	0.100250		
0.2	1.020050	0.202000		
0.3	1.045337	0.306725	1.081425	0.016
0.4	1.081425	0.415966	1.128730	0.016

10.7.5 System of First-Order Simultaneous ODE

Let the n dependent variables $z_1(x), z_2(x), \dots, z_n(x)$ satisfy the following n simultaneous first-order ordinary differential equations.

$$\begin{aligned}
\frac{dz_1}{dx} &= f_1(x, z_1, z_2, \dots, z_n) \qquad & z_1(x_0) &= z_{10} \\
\frac{dz_2}{dx} &= f_2(x, z_1, z_2, \dots, z_n) \qquad & z_2(x_0) &= z_{20} \qquad (10.68) \\
&\ \ \vdots & &\ \ \vdots \\
\frac{dz_n}{dx} &= f_n(x, z_1, z_2, \dots, z_n) \qquad & z_n(x_0) &= z_{n0}
\end{aligned}$$

One can solve these equations by any of the methods outlined earlier. Each equation at any point x_i is solved for $z_1, \ldots, z_n$. Then the solution is advanced by h to the next point x_{i+1}, starting with the initial values at x_0.

Example 10.16

Solve the following coupled first-order differential equations using Runge-Kutta of the second order, with $h = 0.1$:

$x = [0(0.1)0.5]$

$y' = z \qquad y(0) = 1$

$z' = y \qquad z(0) = 0$

The exact solution is $y = \cosh x \qquad z = \sinh x$

Let

$y' = f(x, y, z), \quad k_1 = hf(x_i, y_i, z_i), \quad k_2 = hf(x_i + h, y_i + k_1, z_i + m_1)$

$z' = g(x, y, z), \quad m_1 = hg(x_i, y_i, z_i), \quad m_2 = hg(x_i + h, y_i + k_1, z_i + m_1)$

x_i	y_i z_i	k_1 m_1	k_2 m_2	y_{i+1} z_{i+1}	y error% z error%
0	1.000	0.000	0.01	1.005000	0.0004
	0.000	0.1000	0.10	0.100	0.166
0.1	1.005000	0.010	0.02005	1.020025	0.004
	0.100000	0.1005	0.101500	0.201000	0.168
0.2	1.020025	0.020100	0.030300	1.045225	0.011
	0.201000	0.102003	0.104013	0.304008	0.171
0.3	1.045225	0.030401	0.040853	1.080852	0.020
	0.304008	0.104523	0.107563	0.410051	0.171

10.7.6 High-Ordered ODE

For the second-order ODE, one can convert it to a system of two simultaneous ODEs.

$$y'' = g(x, y, y') \quad y(x_o) = y_0 \quad y'(x_0) = y_1$$

Let $y'(x) = z$, *then* $y'' = z'$, so that we have

$$y' = z = f(x, y, z) \qquad y(x_o) = y_0$$

$$z' = g(x, y, z) \qquad z(x_o) = y_1$$

Similarly, one can treat a nth order ODE as a system of n first-order simultaneous ODE.

For a nth order ODE:

$$y^{(n)} = g(x, y, y', \dots, y^{(n-1)})$$

One can convert this to a n system of first-order simultaneous ODE.

Let

$$y' = z_1$$

$$y'' = z_2 = z_1'$$

$$y''' = z_3 = z_2'$$

.

.

.

$$y^{(n-1)} = z_{n-1} = z_{n-2}'$$

$$y^{(n)} = z_{n-1}' = g(x, z_1, z_2, \dots, z_{n-1})$$

Resulting in the following system of first-order ODE:

$$y' = z_1$$

$$z_1' = z_2$$

$$z_2' = z_3$$

$z'_{n-2} = z_{n-1}$
$z'_{n-1} = g(x, z_1, z_2, \dots, z_{n-1})$

10.7.7 Correction-Extrapolation of Results

The error in the finite difference schemes, eqs (10.13) to (10.52) depends on the spacing size h. One can improve the numerical results by applying the numerical scheme with different spacing h and evaluated at the same point. The error in each numerical differentiation scheme is of the order given by the remainder of the series. In the central difference schemes, the error in each formula of a differential of second order of a function $F(x)$.

$$\frac{\delta^2 F}{2} = \frac{h^2 D^2 F}{2!} + \frac{h^4 D^4 F}{4!} + \frac{h^6 D^6 F}{6!} + \frac{h^8 D^8 F}{8!} + \dots$$

So that neglecting the higher-order term of $h^4, h^6, \dots$

$$\frac{\delta^2 F}{2} = \frac{h^2 D^2 F}{2!} + h^2 e_2(x)$$

$$e_2(x) = \frac{h^2 D^4 F}{4!} + \cdots, \quad e_2(x) = O(h^2)$$

Neglecting higher-order terms of $h^6, h^8, \dots$ gives an error

$$e_2(x) = \frac{h^4}{6!} D^6 F + \cdots \quad e_2(x) = 0(h^4)$$

So that in general, the finite difference in the Central Difference Schemes can be written as:

$$e_2(x) = g_2(x)h^2 + g_4(x)h^4 + g_6(x)h^6 + \cdots.$$

Thus, formula of order h^4 is obtained by making $g_2 = 0$. The formula for order h^6 is obtained by making $g_2 = g_4 = 0$, etc. If one computes the finite difference at a point x_0, then $g_2(x_0), g_4(x_0), \dots$ are fixed constants.

$$e(x_0) = g_2(x_0)h^2 + g_4(x_0)h^4 + g_6(x_0)h^6 + \cdots$$

Let the range of x be $= L$, then dividing this length into n and m strips results:

$$h_n = \frac{L}{n} \qquad h_m = \frac{L}{m}$$

Such that the results of invoking the differentiation schemes gives I_n and I_m, and the exact value be I. Neglecting the higher-order terms if h/L is very small, gives:

$$e_n = I - I_n = \frac{g_2(x_0)L^2}{n^2}$$

$$e_m = I - I_m = \frac{g_2(x_0)L^2}{m^2}$$

Eliminating $g_2(x_0)L^2$ from both of these expressions, results in h^2- Correction Extrapolation formula for I.

$$I_{n,m} = \frac{m^2}{m^2-n^2} I_m - \frac{n^2}{m^2-n^2} I_n \tag{10.69}$$

In particular, if $m = 2n$ (doubling the number of segments):

$$I_{n,2n} = \frac{4}{3} I_{2n} - \frac{1}{3} I_n$$

If one has three values of the results for three successively larger number of strips, i.e., if we have I_n, I_m, I_k $\quad n < m < k$, then we examine the errors for the three schemes.

$$e_n = I - I_n = \frac{g_2(x_0)L^2}{n^2} + \frac{g_4(x_0)L^4}{n^4}$$

$$e_m = I - I_m = \frac{g_2(x_0)L^2}{m^2} + \frac{g_4(x_0)L^4}{m^4}$$

$$e_k = I - I_k = \frac{g_2(x_0)L^2}{k^2} + \frac{g_4(x_0)L^4}{k^4}$$

Eliminating $g_2(x_0)L^2$ and $g_4(x_0)L^4$ results in a correction formula for a better approximation in the case of three, rather than two values of the differential.

$$I_{n,m,k} = \frac{n^4}{(m^2 - n^2)(k^2 - n^2)} I_n + \frac{m^4}{(n^2 - m^2)(k^2 - m^2)} I_m$$

$$+\frac{k^4}{(k^2-n^2)(k^2-m^2)}I_k \tag{10.69}$$

In particular, if $k = 4n$ and $m = 2n$, i.e., one obtains the approximate value:

$$I_{n,2n,4n} = +0.02222\ \ I_n - 0.44444\ \ I_{2n} + 1.42222\ \ I_{4n}$$

In Example 10.5:

$I\ exact = 1.718281828$

$I_2 = 1.753931092$	$e = 2.1\%$
$I_4 = 1.727221905$	$e = 0.5\%$
$I_8 = 1.720518592$	$e = 0.13\%$
$I_{2,4} = 1.718318852$	$e = 0.0022\%\ v.s. 0.5\%$
$I_{4,8} = 1.718284157$	$e = 0.00014\%\ v.s. 0.13\%$
$I_{2,4,8} = 1.718282216$	$e = 0.000023\%\ v.s. 0.13\%$

It can be seen that the error is greatly reduced using either the two or the three extrapolation formulae.

10.8 ODE-Boundary Value Problems (BVP)

The solution of Boundary Value Problem (BVP) by finite difference methods eliminates the need for analytic integration methods. This is especially useful if the geometry of the problem is not simple or the problem has variable material properties, or has non-linear terms. The solution by finite difference includes the boundary conditions.

10.8.1 One-Dimensional BVP

A most general second-order BVP in one-dimensional can be described by (see Chapter 4).

$$y'' = f(x, y, y') \qquad a \le x \le b$$

Satisfying the following BC:

$$U_1(y) = A$$

$U_2(y) = B$

Consider a Dirichlet-Dirichlet BC:

$$y(a) = A \qquad\qquad y(b) = B$$

There are two methods for solving for $y(x)$. One method uses trial-and-error and initial value methods known as the "shooting method". The other requires the solution of simultaneous algebraic equations.

10.8.2 Shooting Method

The Shooting Method utilizes the methods of initial value problems for linear or non-linear second-order ODEs or higher-ordered equations.

To solve such problems, one starts the solution by using the known BC as one of the initial values and assume a value for the unknown second initial value. Integration of the problem to the end of the region is then performed. The resulting difference between the integration result and the given BC constitutes an error. To start the numerical procedure, one needs two initial guesses for the unknown initial condition. Then, using these two initial guesses, one can use extrapolation to start the third integration. These procedures are repeated until the desired accuracy is reached.

Let $y_t^{(0)}$ and $y_t^{(1)}$ be the first two trial guesses, resulting in error $e^{(0)}$and $e^{(1)}$, then the third trial gives $y_t^{(2)}$ is given:

$$y_t^{(2)} = \frac{e_1 y_t^{(0)} - e_0 y_t^{(1)}}{e_1 - e_0}$$

Example 10.17

Find the solution to the following by using the Shooting Method:

$$y'' + 4y^2 = 0 \qquad y(0) = 0 \qquad y(1) = 1$$

Rewriting the equation into a Forward Difference Scheme:

$$y_{i+1} = (2y_i - 4h^2 y_i^2) - y_{i-1} \qquad \text{using } h = 0.2 \quad i = 1, 2, \ldots 4$$

k	i	0	1	2	3	4	5	e
	x_i	0	0.2	0.4	0.6	0.8	1.0	

0		0	0.20	0.3936	0.5624	0.6806	0.7247	-0.2753
1		0	0.30	0.5856	0.8163	0.9404	0.9230	-0.0770
2		0	0.3388	0.6592	0.9101	1.0285	0.9777	-0.0223
3		0	0.3546	0.6891	0.9476	1.0624	0.9966	-0.0034
		0	0.3574	0.6944	0.9542	1.0683	0.9998	-0.0002

10.8.3 Equilibrium Method

To solve BVP in one dimension, one can employ Central Finite Difference equations to solve the problem. One can rewrite the ODE and the boundary condition in terms of Central Finite Difference, making sure that one uses the same order of error in terms of (h).

Example 10.18

A stretched string of length $L = 1.0$ is being loaded by a static load $f(x)/T_o = e^x$, and is supported by a distributed spring whose elastic constant =4. The string satisfies the following ODE:

$$y'' - 4y = e^x \qquad 0 \le x \le 1$$

$$y(0) = 0 \qquad y(1) = 1$$

Writing out the equation in Central Difference of $O(h^2)$, and multiplying the equation by h^2, results:

$$y_{i-1} - 2y_i + y_{i+1} - 4h^2 y_i = h^2 e^{x_i} \qquad i = 1, 2, \ldots, n-1$$

$$y_{i-1} - 2(1 + 2h^2)y_i + y_{i+1} = h^2 e^{x_i} \qquad i = 1, 2, \ldots, n-1$$

Where $n =$ number of segments.

The exact solution is given by:

$$y = 0.17978 \sinh(2x) + 0.33333(\cosh(2x) - e^x)$$

and exact value at center, $y(0.5) = 0.17606$

Starting with $n = 2$ and ending with $n = 32$, the numerical solution gives:

n	2	4	8	16	32
$y(0.5)$	0.19594	0.18141	0.17742	0.17640	0.17614
Error %	11.3	3.0	0.8	0.2	0.02

Example 10.19

A contoured beam of length 1 is acted upon by a concentrated force at $x = 1$ of unit magnitude. If the beam stiffness $EI = (1 - x^2),\ 0 \le x \le 1$, find the displacement at the end $x = 1$. The beam is fixed at $x = 0$.

The governing equation is given by:

$$(1 - x^2)\frac{d^2y}{dx^2} = M = 1(1 - x) \qquad 0 \le x \le 1$$

$$y(0) = 0 \qquad y'(0) = 0$$

The exact solution is given by:

$$y = (x + 1)\log(x + 1) - x$$

and $y(1) = 0.3862943$

Using Finite Central Difference of $O(h^2)$ and using n segments, then $h = 1/n$.

$$y_{i-1} - 2y_i + y_{i+1} = h^2(1 + x_i) \qquad i = 0, 1, 2, ...$$

For the boundary conditions $y_0 = 0$ and $y_0' = 0$, using the Central Finite Difference of $O(h^2)$ for $y_i' = y_{i-1} - y_{i+1}$ at $i = 0$, then $y_{-1} = y_{+1}$, which is introduced in the first two equations.

n	4	8	16	32
y_n	0.39405	0.38824	0.38678	0.38641
Error %	2.0	0.2	0.05	0.03
$y_{n,2n}$		0.38630	0.38629	0.38629
Error %		0.00259	0.00000	0.00000
$y_{n,2n,4n}$			0.38629	0.38629
			$e = 0.00000$	0.00000

It can be seen that the error reduces to a small percentage as the number of segments is increased.

One can also solve non-linear BVP by finite difference.

Example 10.20

The example problem 10.17 can be reworked by Central Finite Difference:

$$y'' + 4y^2 = 0 \qquad y(0) = 0 \qquad y(1) = 1$$

using n points, where $h = 1/n$

$$y_{i+1} = (4h^2 y_i^2 - 2y_i) + y_{i-1} = 0 \qquad i = 1, 2, \ldots, n-1$$

Since this resulted in a non-linear algebraic system, one can move the non-linearity to the right-hand side and solve iteratively. The resulting linear algebraic system:

$$y_{i+1}^{k+1} - 2y_i^{k+1} + y_{i-1}^{k+1} = -4h^2 (y_i^k)^2 \qquad i = 1, 2, \ldots., n-1$$

with $y_0^{(k)} = 0 \qquad y_n^{(k)} = 1.0 \qquad k = 1, 2, \ldots$

For the first iteration, $k = 1$, set $y_i^{(1)} = 0$ and solve for $y^{(2)}$. For the second iteration, $k = 2$, set $y_i^{(2)}$ from the previous iteration solution and solve for $y_i^{(3)}$. The results for $n = 5$ and $k = 10$ and 20 are shown below.

x_i / i	y_1	y_2	y_3	y_4
$k = 10$	0.3540	0.6882	0.9472	1.0637
$k = 20$	0.3574	0.6944	0.9543	1.0684

It can be seen that the solution is the same as in the Shooting Method after 20 iterations.

10.9 ODE-Eigenvalue Problems

The solution of eigenvalue Problems in one-dimension can be achieved by using Finite Difference Methods as outlined in the preceding section. Consider the self-adjoint eigenvalue problem, eq. (4.39). One can rewrite the ODE in terms of Finite Central Difference and compute the eigenvalues λ_m.

This can be achieved by segmenting the range into n parts and writing out the resulting $n * n$ algebraic equations, each having an unknown λ. One can find the n eigenvalues and corresponding eigenvectors through a matrix eigenvalue solver or do a search for the zeroes of the determinant of the $n * n$ matrix.

Example 10.21

Consider a uniform string, $\text{length} = \pi$, fixed at both ends. Find the first four eigenvalues:

ODE $\quad y'' + k^2 y = 0 \qquad 0 \leq x \leq \pi \qquad y(0) = 0 \qquad y(\pi) = 0$

The exact solution is given - Example 4.4, with $L = \pi$

$$y = C \sin mx \qquad k_m = m \qquad m = 1, 2, 3, ...$$

Rewriting the ODE in terms of Central Difference and using n segments:

$$y_{i+1} - (k^2h^2 - 2)y_i + y_{i-1} = 0 \qquad i = 1, 2, ..., n$$

$$y_0 = 0 \qquad y_n = 0 \qquad h = \pi/n$$

One should note that this leads to a system of $(n - 1) * (n - 1)$ algebraic equations.

Thus, one expects only $(n - 1)$ eigenvalues to result from this computation.

n	k_1	k_2	k_3	k_4	k_5	k_6	k_7	k_8	k_9
5	1.0	1.9	2.6	3.1					
10	1.0	2.0	2.9	3.8	4.6	5.2	5.7	6.1	6.3
15	1.0	2.0	3.0	3.9	4.8	5.7	6.4	7.1	7.8
20	1.0	2.0	3.0	4.0	4.9	5.8	6.7	7.5	8.3
5, 10	1.0	2.0	3.0	4.0					
10, 20		2.0	3.0	4.0	5.0	6.0	7.0	8.0	9.0

It can be seen that the error increases for the higher eigenvalues as n becomes larger. One can see that only the first one-half of the eigenvalues are within 5% error.

It should be noted that the 5, 10 and 10, 20 extrapolations produced errors of less than 0.1%.

Example 10.22

Do Problem 3 of Chapter 4. The problem was transformed to:

$$y'' + k^2 z^2 y = 0 \qquad 1 \le z \le 2 \qquad y(1) = 0 \qquad y(2) = 0$$

The problem has an exact $k_m = m * 2.095$ writing out the Central Finite Difference gives:

$$y_{i+1} + (k^2 z_i^2 - 2)y_i - y_{i-1} = 0 \qquad i = 1, 2, \dots, n$$

$$y_0 = 0 \qquad y_n = 0 \qquad h = 1/n \qquad z_i = 1 + ih$$

The first seven eigenvalues are listed below.

exact k_m	2.09	4.19	6.28	8.38	10.47	12.56	14.66
$n = 5$	2.03	3.88	5.34	6.90	-------	-------	------
$n = 10$	2.06	4.10	6.02	7.76	9.27	10.53	11.77
$n = 15$	2.06	4.14	6.16	8.10	9.94	11.64	13.19
$n = 20$	2.06	4.15	6.21	8.22	10.17	12.05	13.84
5, 10	2.07	4.17	6.25	8.05			
10, 20	2.06	4.17	6.27	8.37	10.47	12.56	14.53

It can be seen that the error increases for the higher eigenvalues. The error exceeds 5% for more than half of the eigenvalues.

It should be noted that the 5, 10 and 10, 20 extrapolations produced errors of less than 0.8%.

Example 10.23

Obtain the eigenvalues of a fixed-fixed beam of length $L = 1$, as in Example (4.6).

$$y^{\prime v} - k^4 y = 0 \qquad 0 \le x \le 1$$
$$y(0) = y'(0) = 0 \qquad y(1) = y'(1) = 0$$

Writing out the Finite Central Difference for n segments where $h = 1/n$:

$$y_{i+2} - 4y_{i+1} + (6 - k^4 h^4)y_i - 4y_{i-1} + y_{i-2} = 0 \qquad i = 1, \dots, n$$

$$y_0 = 0 \qquad y_n = 0$$

Also, $y_0' = 0$ $\qquad \frac{y_1 - y_{-1}}{2h} = 0 \qquad y_{-1} = y_{+1}$

$$y_n' = 0 \qquad \frac{y_{n+1} - y_{n-1}}{2h} = 0 \qquad y_{n+1} = y_{n-1}$$

The exact solution is given in Example (4.6).

Exact	4.73	7.85	10.97	14.14	17.28	20.42
$n = 8$	4.59	7.36	9.85	11.99	13.71	14.97
$n = 16$	4.69	7.72	10.68	13.52	16.23	18.78
$n = 24$	4.72	7.80	10.85	13.86	16.80	19.66
$n = 32$	4.73	7.82	10.92	13.98	17.01	19.99
8, 16	4.72	7.84	10.96	14.03	17.07	20.05
16, 32	4.74	7.85	11.00	14.13	17.27	20.39

The error drops from 5-28% in the first row, to errors below 0.15%.

Example 10.24

Obtain the buckling loads of a fixed-elastically supported beam of Example (4.8), with $r^2 = P/EI, L = 1$ and $\gamma/EI = 16.0$. starting with the original equation in eq. (4.13):

$$y^{(iv)} + r^2 y'' = 0$$

$$y(0) = 0 \qquad y'(0) = 0 \qquad y'''(1) - 16y(1) = 0 \quad y'(1) = 0$$

Writing out the Finite Central Difference for the ODE, results in the following system for n-points:

$$y_{i+2} - 4y_{i+1} + 6y_i - 4y_{i-1} + y_{i-2} + r^2h^2(y_{i+1} - 2y_i + y_{i-1}) = 0$$

$$i = 1, 2, \dots, n$$

Rewriting the preceding equation, one gets:

$$y_{i+2} + (-4 + r^2h^2)y_{i+1} + (6 - 2r^2h^2)y_i + (-4 + r^2h^2)y_{i-1} + y_{i-2} = 0$$

With $h = 1/n$ and

$y_0 = 0,$ $\quad y_0' = 0,$ $\quad y_n' = 0$ $\quad$ and $y_n''' - 16y_n = 0$

The second BC gives $\quad -y_{-1} + y_{+1} = 0$

The third BC gives $\quad -y_{n-1} + y_{n+1} = 0$

The fourth BC gives $\quad \frac{1}{2h^3}[-y_{n-2} + 2y_{n-1} - 2y_{n+1} + y_{n+2}] - 16y_n = 0$

From these BC $\quad y_{-1} = y_{+1}$

$$y_{n+1} = y_{n-1}$$

$$y_{n+2} = 32h^3 y_n + y_{n-2} - 2y_{n-1} + 2y_{n+1}$$

Substituting $y_{n+1} = y_{n-1}$ results in the last BC being $y_{n+2} = y_{n-2} + 32h^3 y_n$.

The numerical values for r are given below for $n = 8, 16, 24$ *and* 32.

Exact	4.74	6.28	9.56	12.57	15.72	18.85	22.00	25.13
$n = 8$	4.74	6.13	9.00	11.32	13.34	14.79	15.72	16.01
$n = 16$	4.75	6.25	9.39	12.25	15.11	17.78	20.31	22.63
$n = 24$	4.75	6.27	9.46	12.43	15.45	18.37	21.24	24.00
$n = 32$	4.75	6.28	9.49	12.49	15.57	18.58	21.57	24.50
8, 16	4.75	6.29	9.52	12.56	15.70	18.78	21.84	24.84
16, 32	4.75	6.29	9.52	12.57	15.72	18.85	21.99	25.12

It can be seen that improvement of the eigenvalues increase with increasing n, the error being less than 3%.

With 8, 16 and 16, 32 extrapolations, the errors are less than 0.05%.

10.10 Partial Differential Equations

One can solve PDE in a similar manner to the solution of ODE by finite difference. If one defines

$$D_x = \frac{\partial}{\partial x} \qquad D_y = \frac{\partial}{\partial y} \tag{10.70}$$

and use steps of size h in the x direction and k in the y direction, then the partial differentiation of $f(x, y)$ is accomplished through finite differences on a two-dimensional grid. Let $f(x_i, y_j) = f_{i,j}$ represent the function at point x_i, y_j, then the central finite difference representation of the partial differentials are:

$$2hD_x f_{i,j} = f_{i+1,j} - f_{i-1,j}$$

$$2kD_y f_{i,j} = f_{i,j+1} - f_{i,j-1}$$

$$h^2 D_x^2 f_{i,j} = f_{i+1,j} - 2f_{i,j} + f_{i-1,j}$$

$$k^2 D_y^2 f_{i,j} = f_{i,j+1} - 2f_{i,j} + f_{i,j-1}$$

$$h^4 D_x^4 f_{i,j} = f_{i+2,j} - 4f_{i+1,j} + 6f_{i,j} - 4f_{i-1,j} + f_{i-2,j}$$

$$k^4 D_y^4 f_{i,j} = f_{i,j+2} - 4f_{i,j+1} + 6f_{i,j} - 4f_{i,j-1} + f_{i,j-2} \tag{10.71}$$

To do a mixed derivative D_{xy} one can apply one operator D_x to an operator D_y.

$$D_y f_{i,j} = \frac{1}{2k}[(\ )_{i,j+1} - (\ )_{i,j-1}]$$

$$D_x D_y f_{i,j} = \frac{1}{2h}[\frac{1}{2k}(f_{i+1,j+1} - f_{i-1,j+1}) - \frac{1}{2k}(f_{i+1,j-1} - f_{i-1,j-1})]$$

$$D_{xy} f_{i,j} = \frac{1}{4kh}\{f_{i+1,j+1} - f_{i-1,j+1} - f_{i+1,j-1} + f_{i-1,j-1}\} \tag{10.72}$$

In a similar manner,

$$h^2k^2 D_{xxyy} f_{i,j} = f_{i+1,j+1} + f_{i-1,j+1} + f_{i+1,j-1} + f_{i-1,j-1}$$

$$-2(f_{i,j+1} + f_{i,j-1} + f_{i+1,j} + f_{i-1,j}) + 4f_{i,j} \tag{10.73}$$

Thus, the Laplacian operator:

$$\nabla^2 = \frac{\partial^2}{\partial x^2} + \frac{\partial^2}{\partial y^2} = D_x^2 + D_y^2$$

$$h^2k^2\nabla^2 f_i = k^2(f_{i+1,j} - 2f_{i,j} + f_{i-1,j}) + h^2(f_{i,j+1} - 2f_{i,j} + f_{i,j-1}) \qquad (10.74)$$

If the spacing distances are the same, i.e. $h = k$

$$h^2\nabla^2 f_i = f_{i+1,j} + f_{i-1,j} + f_{i,j+1} + f_{i,j-1} - 4f_{i,j} \qquad O(h^2) \qquad (10.75)$$

which can be illustrated in Fig. 10.4.

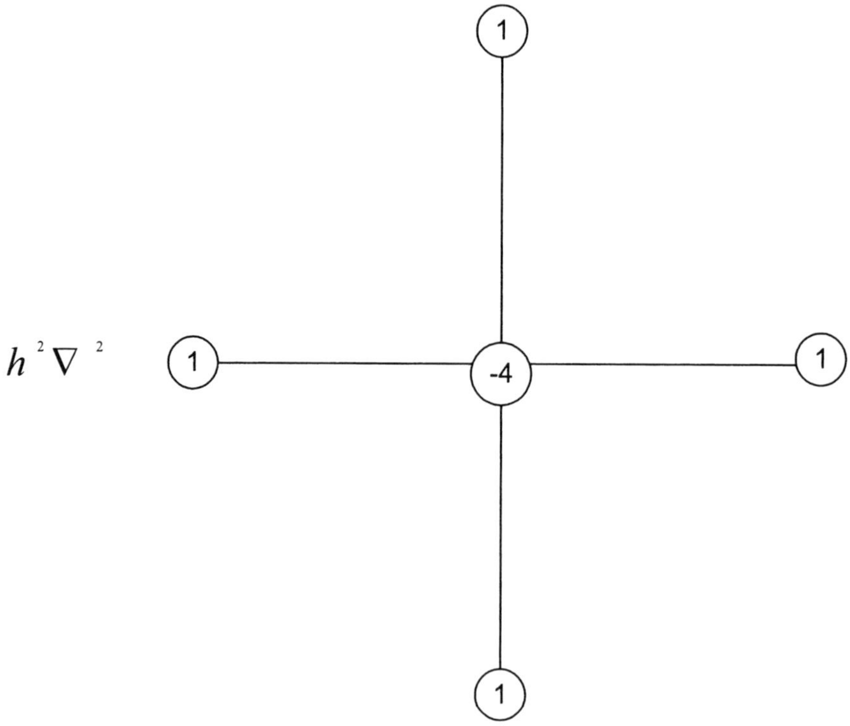

Fig. 10.4 *Cells for* ∇^2

Similarly, the biharmonic operator,

$$\nabla^4 = \frac{\partial^4}{\partial x^4} + 2\frac{\partial^4}{\partial x^2 \partial y^2} + \frac{\partial^4}{\partial y^4}$$

becomes for equal spacing $h = k$

$$h^4\nabla^4 = f_{i+2,j} + f_{i-2,j} + f_{i,j+2} + f_{i,j-2}$$
$$-8(f_{i+1,j} + f_{i-1,j} + f_{i,j+1} + f_{i,j-1})$$
$$+2(f_{i+1,j+1} + f_{i-1,j+1} + f_{i+1,j-1} + f_{i-1,j-1})$$
$$+20f_{i,j} + O(h^2) \tag{10.76}$$

Which is illustrated in Fig. 10.5.

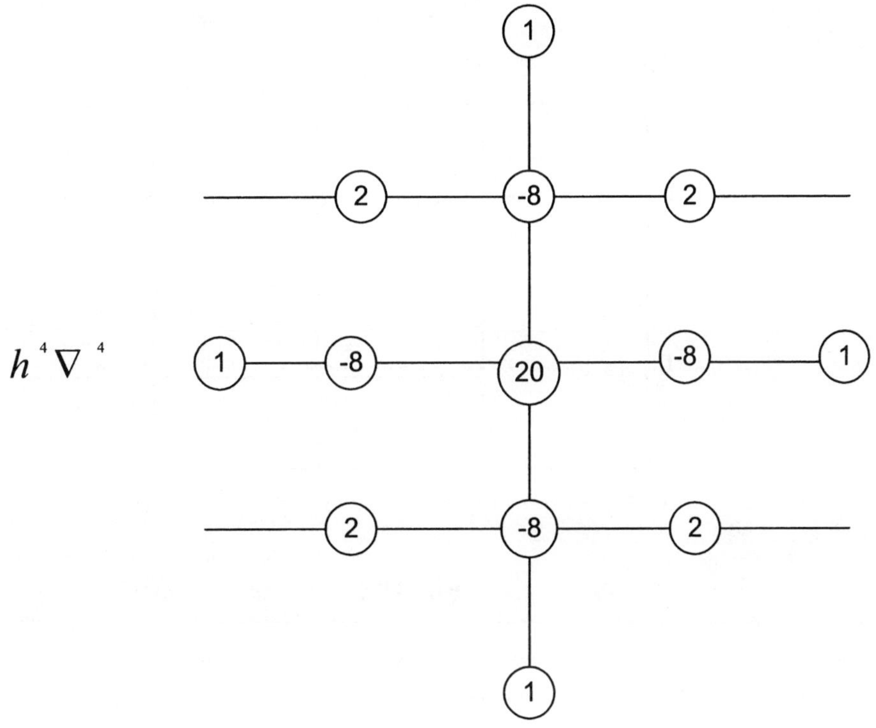

Fig. 10.5 *Cells for* ∇^4

10.10.1 Laplace Equation

For Laplace's Equation, eq. (6.28), the solution depends on the boundary conditions. These equations represent steady-state problems.

Example 10.25

Obtain the temperature distribution in a square plate, such that

$$\nabla^2 T = 0 \qquad T = T(x, y) \qquad 0 \leq x \leq 1 \qquad 0 \leq y \leq 1$$

$$T(x, 0) = 1 \qquad T(0, y) = 0 \qquad T(1, y) = 0 \qquad T(x, 1) = 0$$

Divide the length of the sides by 4, such that $h = k = 1/4$. Since at the corner points $(0,0)$ and $(1,0)$, the temperature jumps from 0 to +1 around these corners, one can average the two corner values, i.e., let:

$T(0,0) = 0.5 \qquad T(1,0) = 0.5$ as seen in Fig. 10.6.

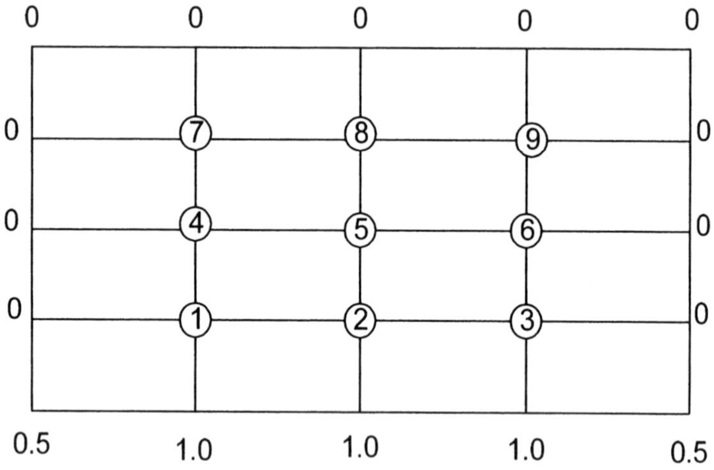

Fig. 10.6 *Square Plate*

There are nine unknown temperatures, designated $T_1, T_2, \ldots, T_9$. Due to the symmetry of the problem, (both field properties and boundary conditions), one can state that $T_1 = T_3, T_4 = T_6$ and $T_7 = T_9$, resulting in 6 unknown temperatures. Applying the scheme of Fig 10.4 at points:

Point 1 $\qquad -4T_1 + T_2 + T_4 + 1.0 = 0$

Point 2 $\qquad -4T_2 + 2T_1 + T_5 + 1.0 = 0$

Point 4 $\qquad -4T_4 + T_5 + T_7 + T_1 = 0$

Point 5 $\qquad -4T_5 + 2T_4 + T_8 + T_2 = 0$

Point 7 $\qquad -4T_7 + T_4 + T_8 = 0$

Point 8 $\qquad -4T_8 + T_5 + 2T_7 = 0$

These constitute six algebraic equations in six unknowns. Solving these equations gives for the temperature at the center as $T_5 = 0.250$.

The exact value from Example 6.1 with $L = 1, f(x) = 1$, is $T_5 = 0.2500$.

10.10.2 Poison's Equation

Solutions of problems governed by Poison's Eq. can be worked out as in Laplace's equation except for the addition of a source term at each point. In this section, one can also discuss the equivalent **Bi-Laplacian** equation.

Example 10.26

Obtain the temperature distribution in a square plate such that:

$$\nabla^2 T = -Q(x,y) \qquad 0 \le x \le 1 \qquad 0 \le y \le 1$$

with $T = 0$ on the boundary for $Q = 1.0 =$ constant. Writing the finite difference at point i,j:

$$T_{i+1,j} + T_{i-1,j} + T_{i,j+1} + T_{i,j-1} - 4T_{i,j} = -h^2 Q_{i,j}$$

The exact value at the center is $7.3669 * 10^{-2}$

a. Let $n = 2, \quad h = 1/2$

There is only one unknown, T_{11}.

$$-4T_{11} = -\frac{1}{4} * 1 \qquad T_{11} = \frac{1}{16} = 6.25 * 10^{-2} \qquad \text{(error 15.8\%)}$$

b. Let $n = 4, \quad h = 1/4$

Due to the symmetry of the geometry, boundary conditions and the source Distribution,

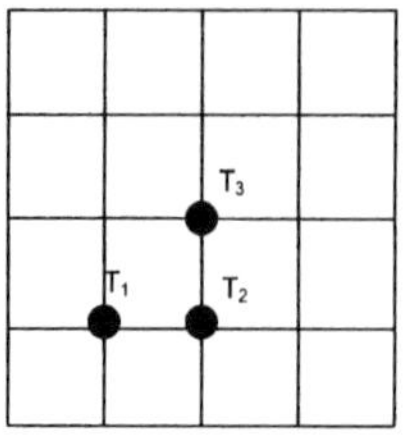

$T_{1,1} = T_{1,3} = T_{3,1} = T_{3,3}$

$T_{2,1} = T_{2,3} = T_{3,2} = T_{1,2}$

This means there are only three independent temperatures, namely $T_{1,1} = T_1$, $T_{2,1} = T_2$, and $T_{2,2} = T_3$.

Applying the ∇^2 operator at these points, one gets:

$i = 1, \quad j = 1 \qquad -4T_1 + 2T_2 = (1/4)^2 = -1/16$

$i = 2, \quad j = 1 \qquad 2T_1 - 4T_2 + T_3 = -1/16$

$i = 2, \quad j = 2 \qquad 4T_2 - 4T_3 = -1/16$

Solving for $T_2 = 7.0312 * 10^{-2}$ (error 4.6%)

c. Letting $n = 6$, $h = 1/6$, then there are only six independent temperatures, i.e., $T_{1,1} = T_1, T_{2,1} = T_2, T_{3,1} = T_3, T_{2,2} = T_4, T_{3,2} = T_5$, and $T_{3,3} = T_6$

Applying the Laplacian on the pivotal points 11, 21, 22, 31, 32 and 33 results in the following 6 algebraic equations:

$-4T_1 + 2T_2 = -1/36$

$T_1 - 4T_2 + T_3 + T_4 = -1/36$

$2T_2 - 4T_3 + T_5 = -1/36$

$2T_2 - 4T_4 + 2T_5 = -1/36$

$T_3 + 2T_4 - 4T_5 + T_6 = -1/36$

$4T_5 - 4T_6 = -1/36$

The temperature at the center, $T_6 = 7.2115 * 10^{-2}$ (error 2.1%)

Using extrapolation at the central point of the plate:

$I_{2,4} = 7.2902 * 10^{-2}$ (error 1.0%)

$I_{4,6} = 7.3576 * 10^{-2}$ (error 0.13%)

$I_{2,4,6} = 7.3637 * 10^{-2}$ (error 0.04%)

Example 10.27

Obtain the displacement field of an elastic square plate under a constant static load $q = 1.0$, eq. 6.11 with time differentials being deleted. The plate has an elastic plate stiffness $D = 1.0$, length $L = 1.0$. The plate is simply supported on all boundaries.

$\nabla^4 w = q(x, y) = 1.0 \qquad 0 \leq x \leq 1.0 \qquad 0 \leq y \leq 1.0$

$w = w(x, y)$, the boundary conditions are:

$$w(0,y) = 0 \quad \frac{\partial^2 w}{\partial x^2}(0,y) = 0 \quad w(1,y) = 0 \quad \frac{\partial^2 w}{\partial x^2}(1,y) = 0$$

$$w(x,0) = 0 \quad \frac{\partial^2 w}{\partial y^2}(x,0) = 0 \quad w(x,1) = 0 \quad \frac{\partial^2 w}{\partial y^2}(x,1) = 0$$

Dividing the region [0,1] into n segments, $h = 1/n$, one can write out the central finite difference at $x = i \quad y = j$.

$$\nabla^4 w_{i,j} = w_{i+2,j} + w_{i-2,j} + w_{i,j+2} + w_{i,j-2}$$

$$-8(w_{i+1,j} + w_{i-1,j} + w_{i,j+1} + w_{i,j-1})$$

$$-2(w_{i+1,j+1} + w_{i-1,j+1} + w_{i+1,j-1} + w_{i-1,j-1}) + 20w_{i,j} = \frac{1}{n^4}$$

$i = 1, \ldots, n-1 \qquad j = 1, \ldots, n-1$

The four BC can be written out in finite difference as:

$$w(0,y) = 0 \quad w(1,y) = 0 \quad w_{0,j} = 0 \quad w_{n,j} = 0$$

$$w(x,0) = 0 \quad w(x,1) = 0 \quad w_{i,0} = 0 \quad w_{i,n} = 0$$

For the remaining four BC:

$\frac{\partial^2 w}{\partial x^2}(0, y) = 0 \qquad\qquad w_{1,j} - 2w_{0,j} + w_{-1,j} = 0$

Since $w_{0,j} = 0$ then $w_{-1,j} = -w_{1,j}$

Similarly for $\frac{\partial^2 w}{\partial x^2}(1, y) = 0$ gives $w_{n+1,j} = -w_{n-1,j}$

For $\frac{\partial^2 w}{\partial y^2}(x, 0) = 0 \qquad w_{i,1} - 2w_{i,0} + w_{i,-1} = 0$

Since $w_{i,0} = 0$ then $w_{i,-1} = -w_{i,1}$

Similarly $w_{i,n+1} = -w_{i,n-1}$

Since the plate, the loading and plate geometry are symmetric with the x- and y-axes, and the two diagonal lines, there are fewer independent point displacement than the unknown $(n-1)^2$ of points.

The exact value of $w_{center} = 4.0608 * 10^{-3}$

<u>**$n = 2$**</u>

$i = 1, j = 1 \qquad w_{3,1} + w_{-1,1} + w_{1,3} + w_{1,-1} - 8(w_{2,1} + w_{0,1} + w_{1,2} + w_{1,0})$

$$+2(w_{2,2} + w_{0,2} + w_{2,0} + w_{0,0}) + 20w_{1,1} = \frac{1}{2^4}$$

Here $w_{-1,1} = -w_{1,1} \qquad w_{1,-1} = -w_{1,1} \qquad w_{3,1} = -w_{1,1} \qquad w_{1,3} = -w_{1,1}$

$16\, w_{1,1} = \frac{1}{2^4} = \frac{1}{16} \qquad w_{1,1} = \frac{1}{16^2} = 3.9063 * 10^{-3}$ (error = 3.8%)

<u>**$n = 4$**</u>

Here $i = 0, \dots, 4$ and $j = 0, \dots, 4$

Due to complete symmetry, there are only 3 unknowns: $w_{1,1}$, $w_{2,1}$, and $w_{2,2}$

The 3 x 3 matrix of the algebraic equations below, for $i, j = 1{,}1\ \ 2{,}1,$ and $2{,}2$:

$$\begin{vmatrix} 20 & -16 & +2 \\ -16 & +24 & -8 \\ 8 & -32 & +20 \end{vmatrix} \begin{vmatrix} w_{1,1} \\ w_{2,1} \\ w_{2,2} \end{vmatrix} = \begin{vmatrix} 1 \\ 1 \\ 1 \end{vmatrix} * \frac{1}{4^4}$$

The solution at center, $w_{1,1} = 4.028 * 10^{-3}$ (error 0.8%)

<u>$n = 6$</u>

Here $i = 0, \dots, 6$ and $j = 0, \dots, 6$

Due to complete symmetry one has only 6 unknowns : $w_{1,1}, w_{2,1}, w_{3,1}, w_{2,2}, w_{3,2},$ and w_{33}. The 6 x 6 matrix of the algebraic equation:

$$\begin{vmatrix} 18 & -16 & 2 & 2 & 0 & 0 \\ -8 & 22 & -8 & -8 & 3 & 0 \\ 2 & -16 & 19 & 4 & -8 & 1 \\ 2 & -16 & 4 & 22 & -16 & 2 \\ 0 & 6 & -8 & -16 & 25 & -8 \\ 0 & 0 & 4 & 8 & -32 & 20 \end{vmatrix} * \begin{vmatrix} w_{1,1} \\ w_{2,1} \\ w_{3,1} \\ w_{2,2} \\ w_{3,2} \\ w_{3,3} \end{vmatrix} = \begin{vmatrix} 1 \\ 1 \\ 1 \\ 1 \\ 1 \\ 1 \end{vmatrix} * \frac{1}{6^4}$$

The displacement at the center $w_{3,3} = 4.04836 * 10^{-3}$ (error 0.34%)

The extrapolation on the central displacement gives:

2,4 $4.06902*10^{-3}$ (error 0.16%)

4,6 $4.06439*10^{-3}$ (error 0.05%)

2,4,6 $4.06382*10^{-3}$ (error 0.03%)

One can see, extrapolation greatly reduces the errors.

10.10.3 The Laplacian in Cylindrical Coordinates

The Laplacian given in eq. (10.74) is derived for cartesian coordinates. To do numerical solutions in circular cylindrical coordinates, one needs to re-derive the Laplacian for these coordinate systems. Consider a two-dimensional problem with r and o being the coordinates. The Laplacian becomes (Section C.4)

$$\nabla^2 f = \frac{\partial^2 f}{\partial r^2} + \frac{1}{r}\frac{\partial f}{\partial r} + \frac{1}{r^2}\frac{\partial^2 f}{\partial \theta^2}$$

Using equal spacing in the r-and θ-direction, let

$h = \Delta r$ $\qquad$ $k = \Delta\theta$, with j and i referencing r and θ, respectively, then

$$\frac{\partial^2 f_{i,j}}{\partial r^2} = \frac{1}{h^2}(f_{i,j-1} - 2f_{i,j} + f_{i,j+1})$$

$$\frac{\partial f_{i,j}}{\partial r} = \frac{1}{2h}(f_{i,j+1} - f_{i,j-1})$$

$$\frac{\partial^2 f}{\partial \theta^2} = \frac{1}{k^2}(f_{i+1,j} - 2f_{i,j} + f_{i-1,j})$$

so that

$$h^2\nabla^2 f_{i,j} = (1 - \frac{h}{2r_i})f_{i,j-1} + (1 + \frac{h}{2r_i})f_{i,j+1} + \frac{h^2}{k^2 r_i^2} f_{i+1,j}$$

$$+ \frac{h^2}{k^2 r_i^2} f_{i-1,j} - 2(1 + \frac{h^2}{k^2 r_i^2})f_{i,j} + 0(h^2) \qquad (10.77)$$

Then $h^2\nabla^2$ operator in circular cylindrical coordinate is depicted in Fig. 10.7.

If the geometry of the problem does not depend on the angular coordinate θ, i.e., an axisymmetric problem, then

$$\nabla^2 f = \frac{d^2 f}{dr^2} + \frac{1}{r}\frac{df}{dr}$$

becomes $h^2\nabla^2 f_i = (1 - h/2r_i)\, f_{i-1} - 2f_i + (1 + h/2r_i)f_{i+1} + 0(h^2)$ (10.78)

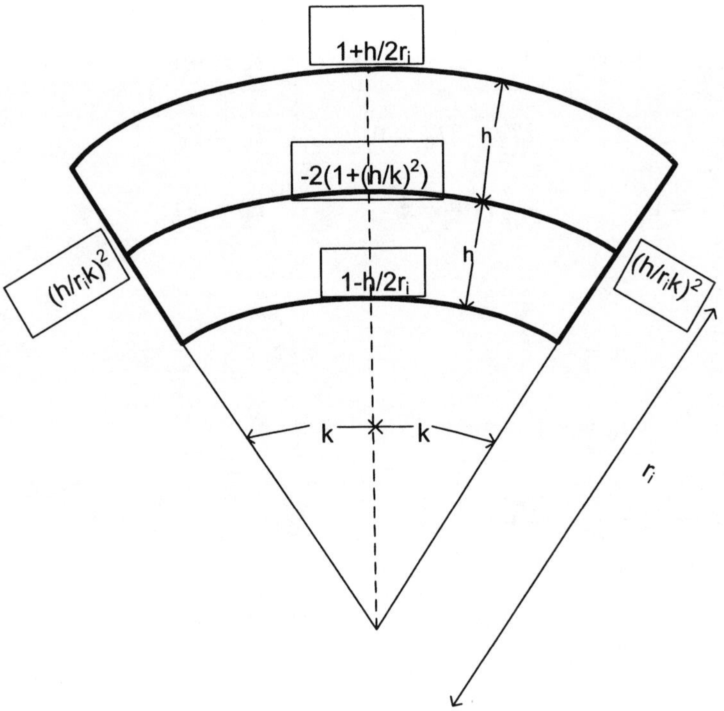

Fig. 10.7 *Cells for ∇^2 in Cylindrical Coordinates*

Example 10.28

Obtain the temperature distribution in an annular sheet, (Example 6.2) such that b = 2 and $a = 1, T = T(r)$

$\nabla^2 T = 0 \qquad 1 \le r \le 2$

$\frac{dT}{dr}(1) = 0 \qquad T(2) = 1 \qquad (T_0 = 1)$

$\frac{dT}{dr} = 0, \qquad T_{-1} - T_{+1} = 0$

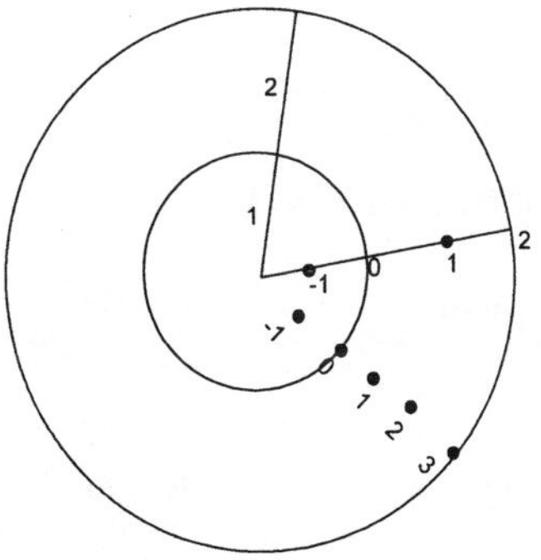

$\boldsymbol{n = 2} \qquad h = {}^1/_2$

pt 0 $\qquad r_0 = 1$

$$(1 - \frac{1}{4})T_{-1} - 2T_0 + (1 + \frac{1}{4})T_1 = 0$$

pt 1 $\qquad r_1 = 1.5$

$(1-\frac{1}{6})T_0 - 2T_1 + (1+\frac{1}{6})T_2 = 0$ $\qquad T_2 = +1$

Since $T_{-1} = T_1$ $\qquad 2T_1 - 2T_0 = 0$ $\qquad T_0 = T_1$

$\frac{5}{6}T_0 - 2T_1 + \frac{7}{6} * 1 = 0$ $\qquad -\frac{7}{6}T_1 = -\frac{7}{6}$ $\qquad T_1 = +1$

Hence, $T_0 = T_1 = 1$

n = 3 $\qquad h = \frac{1}{3}$

pt 0 $\qquad r_0 = +1$

$$(1-\frac{1}{6})T_{-1} - 2T_0 + (1+\frac{1}{6})T_1 = 0$$

Since $T_{-1} = T_1$ gives $T_1 = T_0$

pt 1 $\qquad r_1 = \frac{4}{3}$

$$(1-\frac{1}{8})T_0 - 2T_1 + (1+\frac{1}{8})T_2 = 0$$

$-\frac{9}{8}T_1 + \frac{9}{8}T_2 = 0$ $\qquad T_2 = T_1$

pt 2 $\qquad r_2 = \frac{5}{3}$

$(1-\frac{1}{10})T_1 - 2T_2 + (1+\frac{1}{10})T_3 = 0$ $\qquad T_3 = 1$

$\frac{9}{10}T_2 - 2T_2 + \frac{11}{10} = 0$ $\qquad T_2 = 1$

Hence, $T_0 = T_1 = T_2 = 1$

Exact solution gives $T = 1$ $\qquad 1 \leq r \leq 2$

Example 10.29

Solve for the temperature distribution in a semi-circular sheet of problem 6.5 with $T = T(r,\theta), a = 1, T(r,0) = T(r,\pi) = 0$

The exact solution of problem 6.5, the center point $r = 0.5,\ \theta = {}^{\pi}/_{2}$ is

$$T(\frac{1}{2},\frac{\pi}{2}) = \frac{4}{\pi}\sum_{n=0}^{\infty}(\frac{1}{2})^{2n+1}\frac{(-1)^n}{2n+1} = 0.5904$$

<u>n = 2</u>

$h = \frac{1}{2}$ $\qquad k = {}^{\pi}/_{2}$

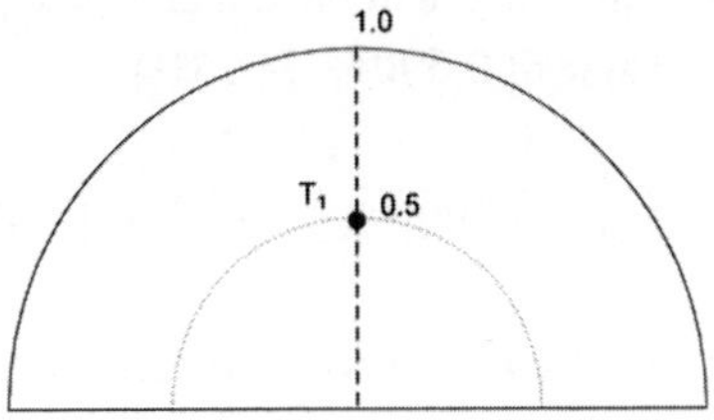

There is only one interior point $r = 0.5$

$$(1 + 1/2) * 1 - 2(1 + 4/\pi^2)T_1 = 0$$

$T_1 = T(0.5, \pi/2 = 0.5337$ (er 10%)

<u>n = 4</u>

$h = 1/4$ $\qquad k = \pi/4$

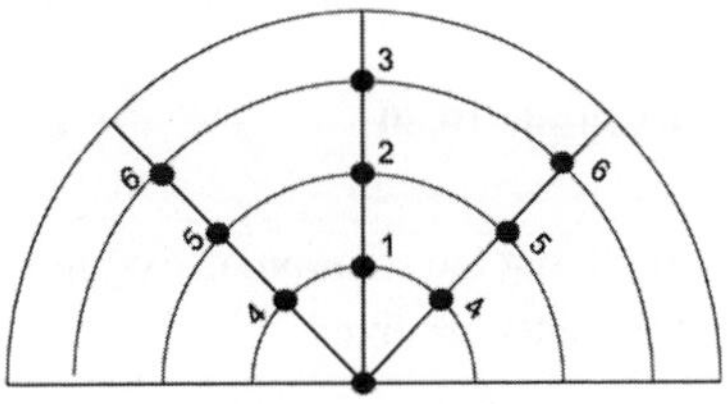

Due to the symmetry of the problem about $\theta = \pi/2$, there are only six independent temperatures.

pt 1 $\qquad r = 1/4$

$$-2(1 + 16/\pi^2)T_1 + 32\,T_4/\pi^2 + (1 + {}^{1}/_{2})T_2 = 0$$

pt 4 $\qquad r = 1/4$

$$-2(1 + 16/\pi^2)T_4 + 16\,T_1/\pi^2 + T_5(1 + {}^{1}/_{2}) = 0$$

pt 2 $\qquad r = 1/2$

$$-2(1 + 4/\pi^2)T_2 + 8\,T_5/\pi^2 + (1 + 1/4)T_3 + (1 - {}^{1}/_{4})T_1 = 0$$

pt 5 $\qquad r = 1/2$

$$-2(1 + 4/\pi^2)T_5 + 4\,T_2/\pi^2 + (1 + 1/4)T_6 + (1 - {}^{1}/_{4})T_4 = 0$$

pt 3 $r = 3/4$

$$-2(1 + 16/9\pi^2)T_3 + 32\,T_6/9\pi^2 + (1 + 1/6) * 1 + (1 - {}^{1}/_{6})T_2 = 0$$

pt 6 $r = 3/4$

$$-2(1 + 16/9\pi^2)T_6 + 16\,T_3/9\pi^2 + (1 + 1/6) * 1 + (1 - {}^{1}/_{6})T_5 = 0$$

The solution of these algebraic equations gives $T_2 = 0.5735$ compared to the exact value of 0.5904. (er. 3%)

$I_{2,4}$ =0.5868 (er. 0.6%)

10.10.4 Helmholtz Equation

Problems satisfying Helmholtz Equation comprise a wide range of scientific and engineering fields. One again converts the non-homogeneous Helmholtz Equation by writing out the Laplacian operator in terms of finite central difference.

Example 10.30

Work out the solution of Example 6.8, with the sides $L = 1$, $q_o/S = -1.0$, $k^2 = 1$. The four sides are fixed.

The exact solution at x = $y = 1/2$ is found as $w_c = w(1/2\,,1/2) = 7.795 * 10^{-2}$. Since the geometry and the loading have two-fold symmetry about the center, one can use the two-fold symmetry to reduce the size of the numerical problem.

$\nabla^2 w + k^2 w = +1.0 = q$ $\quad 0 \le x \le 1 \quad 0 \le y \le 1$

$$w(0, y) = w(1, y) = w(x, 0) = w(x, 1) = 0$$

$$w_{i+1,j} + w_{i-1,j} + w_{i,j+1} + w_{i,j-1} + (k^2h^2 - 4)w_{i,j} = h^2 q_{i,j}$$

$n = 2$ $\quad h = 1/2$ $\quad w_c = 6.667 * 10^{-2}$ $\quad$ error = 14.5%

$n = 4$ $\quad h = 1/4$ $\quad w_c = 7.457 * 10^{-2}$ $\quad$ error = 4.3%

$n = 6$ $\quad h = 1/6$ $\quad w_c = 7.639 * 10^{-2}$ $\quad$ error = 2.0%

2,4 extrapolation $\quad w_c = 7.720 * 10^{-2}$ $\quad$ error = 1.0%

4,6 extrapolation $w_c = 7.785 * 10^{-2}$ error = 0.13%

2,4,6 extrapolation $w_c = 7.793 * 10^{-2}$ error = 0.03%

Example 10.31

Find the eigenvalues of a rectangular membrane with fixed sides. The membrane occupies the region $0 \leq x \leq 2, 0 \leq y \leq 1$. Writing out the Helmholtz operator:

$$w_{i-1,j-1} + w_{i+1,j-1} + w_{i+1,j+1} + w_{i-1,j-1} + (k^2h^2 - 4)w_{i,j} = 0$$

We will use the same spacing for the x-axis and the y-axis coordinates, i.e., $h = 1/n$. It should be noted that one cannot use any symmetry conditions for the eigenvalue problems because it will eliminate all the anti-symmetric eigenfunctions.

The exact eigenfunction solution is given by (see chapter 6):

$$f_{\ell m}(x, y) = \sin\frac{\ell\pi x}{2}\sin m\pi y \qquad \ell = 1,2, ... \qquad m = 1,2, ...$$

The corresponding eigenvalues:

$$k_{\ell m} = \pi\sqrt{m^2 + \ell^2/4} \qquad \ell = 1,2, ..., m = 1,2, ...,$$

When modeling this problem numerically, one should be aware of the possible highest number modes that can be predicted, based on the number of segments *n*.

n = 2 $h = 1/2$

With this number of segments in the y-direction, there are 4 segments in the x-direction. This means that there are only 3 points at which the displacement is unknown, i.e., $w_{1,1}, w_{2,1},$ and $w_{3,1}$. This also indicates that whatever eigenvalues will be predicted, the mode numbers are limited to $m = 1$ and $\ell = 1,2, or\ 3$. Also, the resulting 3 x 3 equations predicts only 3 eigenvalues. Let the eigenvalues be written as $\lambda_{m\ell} = \pi k_{m\ell}$.

The three eigenvalues are *k*=1.024, 1.274, and 1.482, corresponding to the exact ones $k_{11} = 1.118$, $k_{12} = 1.414$ and $k_{13} = 1.803$.

n = 3 $h = 1/3$

There are ten pivotal points. This allows for modes numbers m=1 or 2 and $\ell = 1, 2, 3, 4\ or\ 5$. There are possible ten eigenvalues that could be the root of these 10 x 10 equations. The first five eigenvalues are obtained as: 1.076, 1.351, 1.654, 1.727 and 2.078.

<u>$n = 4$</u> $h = 1/4$

There are 21 pivotal points. This allows for mode numbers $m = 1, 2\ or\ 3$ and $\ell = 1, 2, 3, 4, 5, 6\ or\ 7$, for a total of 21 eigenvalues. The first five eigenvalues are found to be: 1.102, 1.402, 1.753, 1.868, and 2.048.

The exact values of the eigenvalues:

$$k_{\ell m} = \pi\sqrt{m^2 + \ell^2/4} \qquad m, \ell = 1, 2, 3, \ldots$$

$n = 2$	$n = 3$	$n = 4$	Exact	3,4 extrap.	Error
1.024	1.076	1.102	1.118	1.135	1.5%
1.274	1.351	1.402	1.414	1.468	3.8%
1.482	1.654	1.753	1.803	1.880	4.3%
	1.727	1.868	2.062	2.049	0.6%
	2.078	2.048	2.236	2.009	10.2%

It can be seen that the error, even utilizing 3,4 extrapolation increases with higher mode numbers.

Example 10.32

Obtain the eigenvalues of a rectangular plate, simply supported on all its boundaries. The plate occupies the region $0 \le x \le 2,\ \ 0 \le y \le 1$.

The plate is governed by the biharmonic Helmholtz equation (eq. 6.11). With harmonic time dependence, the biharmonic equation can be written as:

$$(\nabla^4 - \lambda^2)w = 0 \qquad 0 \le x \le 2, 0 \le y \le 1$$

$$w(0\ or\ 2, y) = 0 \qquad \frac{\partial^2 w}{\partial x^2}(0\ or\ 2, y) = 0$$

$$w(x, 0\ or\ 1) = 0 \qquad \frac{\partial^2 w}{\partial y^2}(x, 0\ or\ 1) = 0$$

The exact solution of the eigenfunctions and eigenvalues are:

$$w_{nm} = \sin\frac{\ell\pi}{2}x \sin my \qquad \ell, m = 1,2,3, \ldots.$$

$$\lambda_{\ell m}=(\frac{\ell^2}{4}+m^2)\pi^2=k_{\ell m}\pi^2$$

The two boundary conditions on each boundary are the same ones in Example 10.27, subdividing the *y*-side in *n* divisions and the *x*-side into *2n,* such that one has one value for $h=1/n$. The eigenvalue problem can be written in terms of finite difference scheme as:

$$\nabla^4 w_{i,j}-\lambda^2 w_{i,j}=w_{i+2,j}+w_{i-2,j}+w_{i,j+2}+w_{i,j-2}$$

$$-8(w_{i+1,j}+w_{i-1,j}+w_{i,j+1}+w_{i,j-1})$$

$$-2(w_{i+1,j+1}+w_{i-1,j+1}+w_{i+1,j-1}+w_{i-1,j-1})+(20-\lambda^2/n^4)w_{i,j}=0$$

$i=1,\dots,2n-1$ $\qquad j=1,\dots,n-1$

With the same boundary conditions:

$w_{0,j}=0$ $\qquad w_{2n,j}=0$ $\qquad w_{i,0}=0$ $\qquad w_{i,n}=0$

$w_{-1,j}=-w_{i,j}$ $\qquad w_{2n+1,j}=-w_{2n-1,j}$

$w_{i,-1}=-w_{i,1}$ $\qquad w_{i,n+1}=-w_{i,n-1}$

One must not assume any symmetry in the displacements at the pivotal points $w_{i,j}$, in order to find the eigenvalues for odd and even modes.

<u>n = 2</u>

There are only 3 pivotal points, $w_{1,1}=w_1$, $w_{2,1}=w_2$, and $w_{3,1}=w_3$. With these points, the eigenfunctions could be $n=1,2,$ *and* 3 and $m=1$.

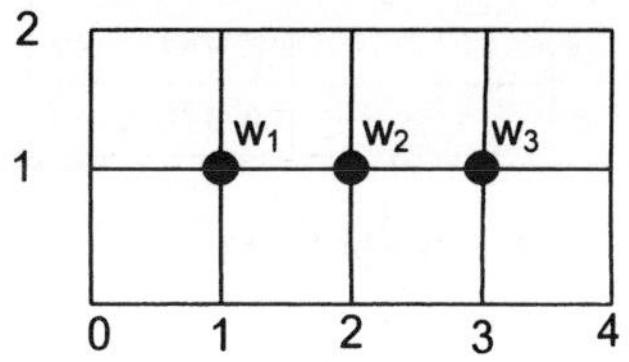

The three equations are listed below:

$i=1$ $\qquad j=1$ $\qquad (17-\lambda^2/2^4)\,w_1-8w_2+w_3=0$

$i=2$ $\qquad j=1$ $\qquad -8w_1+(18-\lambda^2/2^4)w_2-8w_3=0$

$i=3$ $\qquad j=1$ $\qquad w_1-8w_2+(17-\lambda^2/2^4)w_3=0$

There are only 3 eigenvalues:

$$k_{\ell m} = 1.048, 1.622, 2.19$$

<u>**$n = 3$**</u>

There are 10 pivotal points. The 10 x 10 algebraic equations would result in 10 eigenvalues. With these segments, one would predict $n = 1, 2, \ldots, 7$ and $m = 1, 2, and\ 3$.

The first five eigenvalues are:

$$k_{\ell m} = 1.157, 1.824, 2.736, 2.981, 4.316$$

<u>**$n = 4$**</u>

There are 21 pivotal points. The first five eigenvalues are:

$$k_{\ell m} = 1.157, 1.867, 2.933, 3.506, 4.170$$

One can improve the accuracy by using 3,4 extrapolation.

$n = 2$	$n = 3$	$n = 4$	exact	% error	3,4	% error
1.048	1.157	1.170	1.250	6.4	1.187	5.0
1.622	1.824	1.867	2.000	6.7	1.922	3.9
2.195	2.736	2.933	3.250	9.8	3.186	2.0
	2.981	3.506	4.250	17.5	4.181	1.6
	4.316	4.170	5.000	16.7	3.982	20.4

Example 10.33

Find the eigenvalues of axisymmetrically vibrating circular stretched membrane of radius = 1. The membrane is fixed at $r = 1$.

The governing equation in cylindrical coordinates

$$(\nabla^2 + k^2)w = 0 \qquad w = w(r) \qquad 0 \le r \le 1$$

From eq. (10.78), one has

$$(1 - h/2r_i)w_{i-1} - 2w_i + (1 + h/2r_i)w_{i+1} + h^2k^2w_i = 0$$

The exact solution is given by

$$w_n = J_0(k_n r)$$

Where the eigenvalues are the roots of $J_0(k_n r) = 0$, and the zeroes are given in Section 3.15:

$$k_1 = 2.405,\ k_2 = 5.520,\ k_3 = 8.654,\ \ldots$$

To use a finite central difference scheme, one can use the fact that due to symmetry, $\partial w/\partial r = 0$ at $r = 0$. Since $r \geq 0$, one cannot use central finite difference to evaluate the slope at $r = 0$, but use forward difference.

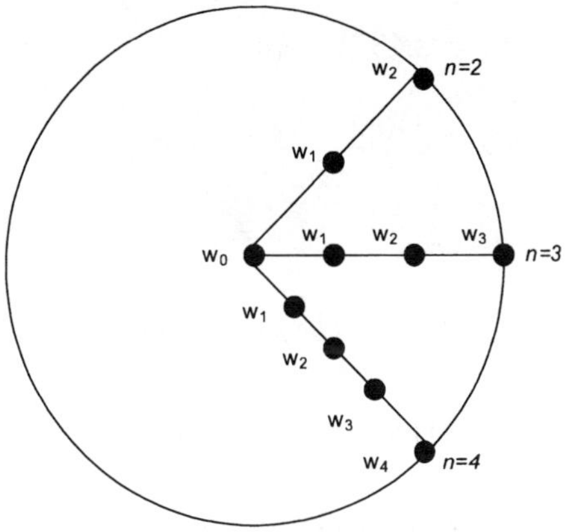

<u>**n = 2**</u> $h = 1/2$

Here we have two points w_0 and w_1. The slope at $r = 0$ can be written in terms of forward difference as $w_1 - w_0 = 0$ or $w_1 = w_0$, and $w_2 = 0$

$$(1 - 1/2)w_0 + (k^2/4 - 2)w_1 = 0$$

which results in

$$(\frac{k^2}{4} - \frac{3}{2})w_1 = 0$$

$k_1 = 2.4495$ (e=1.8%)

<u>**n = 3**</u> $h = 1/3$ $w_3 = 0$

pt 1 $(1 - 1/2)w_0 + (k^2/9 - 2)w_1 + (1 + 1/2)w_2 = 0$

pt 2 $(1 - 1/4)w_1 + (k^2/9 - 2)w_2 = 0$

Substituting $w_1 = w_0$ results in two algebraic equations:

$$(k^2/9 - 3/2)w_1 + (3/2)w_2 = 0$$

$$(3/4)w_1 + (k^2/9 - 2)w_2 = 0$$

solving for k, one obtains:

$k_1 = 2.4378$ (e=1.4%)

$k_2 = 5.055$ (e=8.4%)

<u>**n = 4**</u> $h = 1/4$ $w_4 = 0$

$$(1/2)w_0 + (k^2/16 - 2)w_1 + (3/2)\,w_2 = 0$$

$$(3/4)\,w_1 + (k^2/16 - 2)w_2 + (5/4)\,w_3 = 0$$

$$(5/6)w_2 + (k^2/16 - 2)w_3 = 0$$

These equations result in three eigenvalues

$k_1 = 2.4232$ (e = 0.7%)

$k_2 = 5.3156$ (e = 3.7%)

$k_3 = 7.3387$ (e = 15%)

Using (3, 4) extrapolation $k_1 = 2.4044$ (e = 0.02%)

$k_2 = 5.6507$ (e = 2.4%)

10.10.5 Diffusion Equation

For problems satisfying the diffusion equation, the aim is to find the spatial solution as a function of time. In this regard, central finite difference is used to model the spatial variables and forward difference is used to model the time variable.

Starting out with the general one-dimensional heat flow equation for a uniform bar, the heat flow equation on the temperature is given by (eq. 6.66)

$$\frac{\partial^2 T}{\partial x^2} = \frac{1}{K}\frac{\partial T}{\partial t} \qquad a \le x \le b$$

together with B.C. at $x = a$ and b and the initial condition

$$T(x,0) = f(x) \qquad a \le x \le b$$

Writing out the finite central and forward differences, with i representing space and k re-representing time steps, with $\Delta x = h \qquad \Delta t = \tau$

$$\frac{1}{h^2}[T_{i+1,k} - 2T_{i,k} + T_{i-1,k}] = \frac{1}{K\tau}[T_{i,k+1} - T_{i,k}] - \frac{Q(x_i, t_k)}{\rho c K}$$

$$T_{i,k+1} = \frac{K\tau}{h^2}[T_{i+1,k} + T_{i-1,k}] + [1 - \frac{2K\tau}{h^2}]T_{i,k} \tag{10.79}$$

The solution starts out at $k = 0$ (initial condition) at every spatial point, including the boundary points. This predicts the temperature at all points i at the next time step at $k = 1$. The solution builds up at each future time step at all the spatial points.

It has been shown [Hoffman] that

(i) $K\tau/h^2 \le 0.25$ for stable and non-ascillatory solution

(ii) $K\tau/h^2 \le 0.50$ for stable solution

One can also simplify the solution if one picks

$$\frac{K\tau}{h^2} = \frac{1}{2}$$

It should be noted that if one needs to improve the early time predictions, one needs to make τ smaller and hence one needs to make h smaller, or n larger.

Example 10.34

Redo Example 6.11, with $L = 6, \;\; a = b = Q = 0, \;\; K = \frac{1}{2}$

$$T(0,t) = 0 \qquad T(1,t) = 0 \qquad T(x,0) = x(6 - x)$$

Method 1

Let $K\tau/h^2 = \frac{1}{2}$ or $\tau = h^2$

$$T_{i,k+1} = \frac{1}{2}\left[T_{i+1,k} + T_{i-1,k}\right]$$

n = 6 $h = 1$ $\tau = 1$

The solution is shown in the table below.

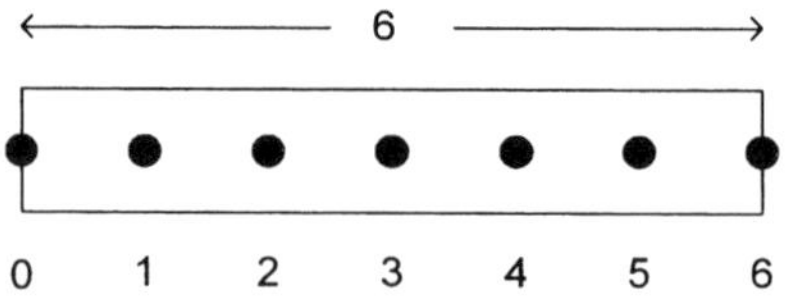

The exact solution is given by

$$T(i,k) = \frac{144}{\pi^3}\sum_{n=1}^{20}\frac{(1-(-1)^n)}{n^3}e^{-n^2\pi^2 k/72}\sin(n\pi\, i/6)$$

The errors are within 4% for up to $t = 5\tau$, and within 10% for $t = 20\tau$.

Finite Difference Solution

k \ *i*	**0**	**1**	**2**	**3**	**4**	**5**	**6**
(ic) 0	0	5	8	9	8	5	0
1	0	4	7	8	7	4	0
2	0	3.5	6	7	6	3.5	0
3	0	3.0	5.25	6	5.25	3.0	0
4	0	2.63	4.50	5.25	4.50	2.63	0
5	0	2.25	3.94	4.50	3.94	2.25	0
10	0	1.11	1.91	2.22	1.91	1.11	0
15	0	0.54	0.95	1.08	0.95	0.54	0
20	0	0.27	0.46	0.54	0.46	0.27	0

Exact Solution

k \ i	0	1	2	3	4	5	6
5	0	2.34	4.05	4.68	4.05	2.34	0
10	0	1.18	2.04	2.36	2.04	1.18	0
15	0	0.59	1.03	1.19	1.03	0.59	0
20	0	0.30	0.52	0.60	0.52	0.30	0

Method 2

If one lets the time step smaller than $\tau = 1$, such as $\tau = \frac{1}{2}$ with $h = 1$, so that

$\frac{K\tau}{h^2} = \frac{\frac{1}{2}*\frac{1}{2}}{1} = \frac{1}{4}$, which insures stability, then

$$T_{i,k+1} = \frac{1}{4}[T_{i+1,k} + T_{i-1,k}] + \frac{1}{2}T_{i,k}$$

$$= \frac{1}{4}[T_{i+1,k} + 2T_{i,k} + T_{i-1,k}]$$

k \ i	0	1	2	3	4	5	6
0	0	5	8	9	8	5	0
1	0	4.50	7.50	8.50	7.50	4.50	0
2	0	4.13	7.00	8.00	7.00	4.13	0
4	0	3.54	6.09	7.02	6.09	3.54	0
6	0	3.07	5.30	6.12	5.30	3.07	0
8	0	2.67	4.62	5.33	4.62	2.67	0
10	0	2.32	4.02	4.64	4.02	2.32	0
20	0	1.16	2.01	2.32	2.01	1.16	0
30	0	0.58	1.00	1.16	1.00	0.58	0
40	0	0.29	0.50	0.58	0.50	0.29	0

It can be seen that the finite difference solutions for $\tau = 0.5$ is more accurate than the one for $\tau = 1.0$.

Example 10.35

Do problem 6.37 which has a source at the center of a square plate (*a*=*b*=*4*), which is initially at a temperature $T_0 = 0$. The plate has a heat source at its center, with

$Q_0/\rho\,c = 100$ and $\propto = 0.4$. let the thermal coefficient $K = 1/4$. Since the source is a point source, the Dirac delta functions are replaced by $1/h^2$ (a unit cell).
Let $T_{i,j,k} = T(x_i, y_j, t_k)$, then the heat flow equation becomes

$$T_{i,j,k+1} = (K\tau/h^2)\{T_{i+1,j,k} + T_{i-1,j,k} + T_{i,j+1,k} + T_{i,j-1,k}\}$$

$$+(K\tau/h^2)(h^2/K\tau - 4)T_{i,j,k} + 100\tau Q_{i,j,k}/h^2$$

To simplify the problem and to exclude the point x_i, y_j,

let $h^2/K\tau = 4$ or, with $K = \frac{1}{4}$, $h^2/\tau = 1$

Let $h = 1$ so that $\tau = 1$, resulting

$$T_{i,j,k+1} = \frac{1}{4}\{T_{i+1,j,k} + T_{i-1,j,k} + T_{i,j+1,k} + T_{i,j-1,k}\}100\;Q_{i,j,k}$$

$$i,j = 1, 2, \ldots, 5$$

Since $t_k = k\tau$, then $t_k = k$

where $Q_{i,j,k} = 0$ except for $Q_{3,3,k} = e^{-0.4k}$

The exact solution is given in the answer to problem 6.37

$$T(x_i, y_j, t_k) = 25\sum_{n=1}\sum_{m=1}\frac{\sin n\pi/2\,\sin m\pi/2}{\lambda_{mn}/4 - 0.4}\left(e^{-0.4t_k} - e^{-\lambda_{mn}t_k/4}\right) *$$

$$* \sin\frac{n\pi}{4}x_i \sin\frac{m\pi}{4}y_j$$

$$\lambda_{mn} = \pi^2(m^2 + n^2)/16$$

The transient temperature at two points along the center line of the plate is shown below for the numerical solution for $\tau = 1$ and $h = 1$. One can see that this solution gives poor results for early time predictions when compared to the exact solution. However, if one chooses a shorter time step τ, it will require a smaller spatial interval h. If one chose

$\tau = 1/4$, then one has to choose $h = 1/2$. The results are also shown in the table below. Note that $t = k\tau$

x,y / t	(2,1)			(2,2)		
	$\tau = 1$	$\tau = 1/4$	Exact	$\tau = 1$	$\tau = 1/4$	Exact
0.05			-0.04			103.9
0.25		0.00	0.07		100.0	144.6
0.50		0.00	1.45		90.5	151.7
0.75		6.25	3.73		106.9	149.5
1.00	0.00	5.65	6.13	100.0	96.7	144.0
1.25		11.37	8.30		101.6	137.0
1.50		10.29	10.16		91.9	129.5
1.75		14.78	11.69		92.9	121.8
2.00	25.0	13.37	12.92	67.0	84.1	114.3
3.00	16.80	16.06	15.34	69.9	67.33	87.0
4.00	23.7	15.94	15.20	46.9	52.6	65.5
5.00	15.9	14.45	13.8	43.9	40.6	49.1
10.00	6.38	5.25	5.11	9.92	10.16	11.20
15.00	1.36	1.39	1.39	2.40	2.33	2.50
20.00	0.32	0.33	0.34	0.45	0.51	0.55
30.00	0.012	0.016	0.017	0.017	0.023	0.025
40.00	0.004	0.0007	0.0008	0.0006	0.0010	0.0012

One notes that the early time predictions are not accurate. The shorter time step τ and smaller spatial h gave better results. One can also improve the accuracy by not making $K\tau/h^2 = 1/4$, as long as the ratio is less than 0.5.

Example 10.36

To illustrate the solution of problems with Neumann type boundary conditions, where points outside the region have to be included in the numerical scheme, the numerical solution of problem 6.48 is obtained.

Obtain the solution of heat flow in a square plate, heated, by a heat source at its center. Let $a = b = 4$, $c = \propto = 0.4$, $Q_0/\rho c = 100$, and $K = 1/4$. The boundary and initial conditions are:

$$T(x, 0, t) = 0 \qquad \frac{\partial T}{\partial x}(0, y, t) = 0$$

$$T(x, 4, t) = 0 \qquad \frac{\partial T}{\partial x}(4, y, t) = 0$$

$T(x, y, 0) = 0$

The Dirac delta function representation of a point source is replaced by $1/h^2$. Let $h = 1$ and $\tau = 1$ such that $K\tau/h^2 = 1/4$. Referring to Example 10.34, one gets

$$T_{i,j,k+1} = \frac{1}{4}\{T_{i+1,j,k} + T_{i-1,j,k} + T_{i,j+1,k} + T_{i,j-1,k}\} + Q_{i,j,k}$$

$i = 2, \dots, 6, \qquad j = 2, \dots, 4, \qquad k = 1, 2, \dots$

To satisfy the Neumann B.C.

$$\frac{\partial T}{\partial x}(0, y, t) = 0 \text{ becomes } \qquad T(-h, y, t) = T(+h, y, t)$$

$$\frac{\partial T}{\partial x}(4, y, t) = 0 \text{ becomes } \qquad T(4 + h, y, t) = T(4 - h, y, t)$$

In order to designate each point by a positive locator, let $x_i = (i - 2) \quad i = 1, 2, \dots, 7$ such that $x_1 = -h$, $x_2 = 0$ (boundary), $x_5 = 3$, $x_6 = 4$ (boundary), $x_7 = 5$

Also, $y_j = j, j = 1, 2, \dots, 5$

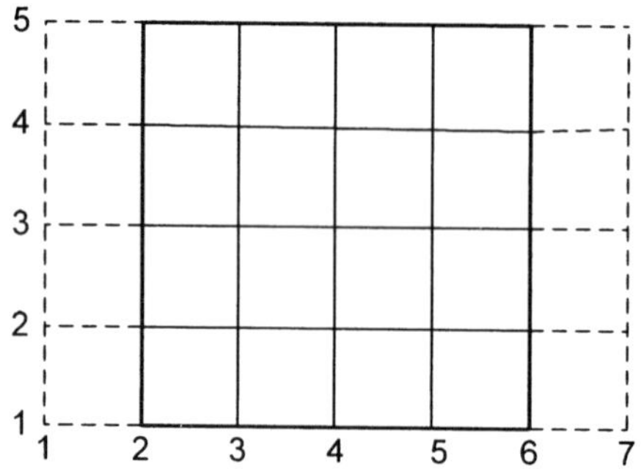

The boundary and initial conditions become

$T_{1,j,k} = T_{3,j,k} \qquad j = 1, 2, \dots, 5 \qquad k = 1, 2, \dots$

$T_{7,j,k} = T_{5,j,k} \qquad j = 1, 2, \dots, 5 \qquad k = 1, 2, \dots$

$T_{i,1,k} = 0 \qquad i = 1, 2, \dots, 7 \qquad k = 1, 2, \dots$

$T_{i,5,k} = 0 \qquad i = 1,2,\ldots,7 \qquad k = 1,2,\ldots$

Also $T_{i,j,1} = 0$

The source term at the center, $Q_{i,j,k} = 0$ for all i,j except for

$\tau Q_{4,3,k}/h^2$=100 exp(-0.4(k-1))

The numerical procedure is to apply the balance equation at each time step $k = 1,2,\ldots$ for all the interior points and the two insulated boundary points. Thus, one must keep track of the time history at the external points as well. The numerical solution was obtained for $h = 1$ and $\tau = 1$ as well as $h = 1/2$ and $\tau = 1/4$, and is shown below, at the insulated boundary and at the mid-point. The time $t = k\tau$

	(0,2)			(2,2)		
t \ x,y	$\tau = 1$	$\tau = 1/4$	Exact	$\tau = 1$	$\tau = 1/4$	Exact
0.05			-0.81			103.07
0.25		0.00	-0.75		100.00	143.87
0.50		0.00	-0.68		90.48	151.06
0.75		0.00	-0.57		106.87	148.93
1.00	0.00	0.00	-0.33	100.00	96.70	143.44
1.25		0.78	0.09		101.56	136.54
1.50		0.71	0.68		91.90	129.06
1.75		2.40	1.41		92.92	121.44
2.00	0.00	2.17	2.23	67.03	84.08	113.93
3.00	12.50	5.71	5.70	69.93	67.38	86.86
4.00	8.37	8.65	8.48	46.88	52.90	65.70
5.00	18.11	10.48	10.19	45.49	41.43	49.83
10.00	10.05	9.20	8.91	13.83	13.69	14.60
15.00	6.80	4.89	4.76	7.31	5.51	5.54
20.00	2.49	2.34	2.29	2.56	2.43	2.40
30.00	0.52	0.50	0.50	0.52	0.50	0.50
40.00	0.107	0.107	0.107	0.107	0.107	0.107

It can be seen that the smaller τ and smaller h gives better results. However, the short time is not accurate enough.

10.10.6 Wave Equation

To solve problems of the wave equation numerically requires the use of central finite difference in the spatial and time domains.

Consider the one-dimensional wave equation, eq. (4.2), on $w(x, t)$

$$\frac{\partial^2 w}{\partial x^2} = \frac{1}{c^2}\frac{\partial^2 w}{\partial t^2} - Q(x,t) \qquad a \le x \le b$$

together with B.C. at $x = a$ and $x = b$ and two initial conditions

$$w(x,0) = f(x) \qquad \frac{\partial w}{\partial t}(x,0) = g(x)$$

Let $w_{i,k} = w(x_i, t_k)$, then using finite difference schemes outlined above, with $\Delta x = h$ and $\Delta t = \tau$, one gets

$$\frac{\partial^2 w}{\partial x^2} \quad \text{becomes} \quad \frac{1}{h^2}[w_{i+1,k} - 2w_{i,k} + w_{i-1,k}] + O(h^2)$$

$$\frac{\partial^2 w}{\partial t^2} \quad \text{becomes} \quad \frac{1}{\tau^2}[w_{i,k+1} - 2w_{i,k} + w_{i,k-1}] + O(\tau^2)$$

Substituting these into the wave equation, one obtains

$$w_{i,k+1} = u^2[w_{i+1,k} + w_{i-1,k}] + 2[1 - u^2]w_{i,k} - w_{i,k-1} + h^2u^2Q(x_i, k\tau) \tag{10.80}$$

where $u = c\,\tau/h$ and $t_k = k\tau$

It was shown that stability of the numerical solution requires that $u \le 1$. Errors introduced by the use of finite differences to approximate the spatial differentials introduce numerical dispersion. It was shown that this numerical dispersion is negligent if $u = 1$ and becomes significant if $u < 1$. This indicates that using $u = 1$ would give more accurate results and less numerical diffusion. Physically, it means that the wavefront travels a distance = h in the time τ, traveling at the wave speed c. If one needs to increase the accuracy, then making the time step τ smaller requires a corresponding decrease of the spatial step h. Using $u = 1$, eq. (10.80) becomes:

$$w_{i,k+1} = w_{i+1,k} - w_{i,k-1} + h^2 Q(x_i, k\tau) \qquad k = 0, 1, \ldots \qquad (10.81)$$

The initial conditions become

$$w(x, 0) = f(x) \qquad w(x_{i,}, 0) = w_{i,0} = f(x_i) = f_i$$

$$\frac{\partial w(x,0)}{\partial t} = g(x) \qquad \frac{w(x_i,\tau) - w(x_i,0)}{\tau} + O(\tau) = g(x_0)$$

$$w(x_i, \tau) = \tau g(x_i) + w(x_i, 0)$$

$$w_{i,1} = \tau g(x_i) + f(x_i) = \tau g_i + f_i$$

To improve the accuracy of the scheme, one can use central finite difference for the second initial condition, i.e.,

$$\frac{w_i(x_i, \tau) - w(x_i, -\tau)}{2\tau} + O(\tau^2) = g(x_i) = g_i$$

This creates an unknown value of w at $t = -\tau$. This can be fixed by computing $w_{i,-1}$ in terms of the initial conditions, i.e.,

$$w(x_i, -\tau) = w(x_{i,\tau}) - 2\tau g(x_i)$$

or

$$w_{i,-1} = w_{i,+1} - 2\tau g_i \qquad (10.82)$$

Substituting $k = 0$ in eq. (10.80)

$$w_{i,1} = u^2[w_{i+1,0} + w_{i-1,0}] + 2(1 - u^2)w_{i,0} - w_{i,-1} + h^2 u^2 Q(x_i, 0)$$

Substituting eq. (10.82) for $w_{i,-1}$ gives the numerical solution for the second time step

$$w_{i,1} = \frac{1}{2}[u^2 f_{i+1} + u^2 f_{i-1} + 2(1-u^2)f_i$$
$$+2\tau g_i + h^2 u^2 Q(x_i, 0)] \qquad (10.83)$$

For $u = 1$, eq. (10.83) reduces to

$$w_{i,1} = \frac{1}{2}[f_{i+1} + f_{i-1} + 2\tau g_i] + \frac{h^2}{2} Q(x_i, 0)$$

Equation (10.83) gives the starting point for the next time step $k = 1$ to give $w_{i,2,}$ by utilizing either eq. (10.80) or eq. (10.81).

Problems in the wave propagation in two-dimensional media require a modified version of eqs. (10.81) to (10.83). Starting with PDE in two dimensions, eq. (6.7)

$$\nabla^2 w = \frac{1}{c^2}\frac{\partial^2 w}{\partial t^2} - Q(x, y, t)$$

with appropriate boundary conditions and two initial conditions:

$$w(x, y, 0) = f(x, y)$$

$$\frac{\partial w}{\partial t}(x, y, 0) = g(x, y)$$

One needs to approximate the partial differentials $\nabla^2 w$ and $\partial^2 w/\partial t^2$ in terms of central finite differences in space and time, with equal spatial spacing $\Delta x = \Delta y = h$ and time step $\Delta t = \tau$. Using the representation $w(x_i, y_j, t_k) = w_{i,j,k}$ and $Q(x_i, y_j, t_k) = Q_{i,j,k}$ one obtains for $O(h^2)$ and $O(\tau^2)$

$$w_{i,j,k+1} = -w_{i,j,k-1} + u^2[w_{i+1,j,k} + w_{i-1,j,k} + w_{i,j+1,k} + w_{i,j-1,k}]$$
$$+2(1-2u^2)w_{i,j,k} - h^2 u^2 Q_{i,j,k} \qquad (10.84)$$

where $u = c\tau/h$.

This formulation requires the knowledge of the displacement at two earlier time steps. Similarly, the initial conditions can be written in terms of finite central differences of $O(\tau^2)$ as:

$$w(x_i, y_j, 0) = w_{i,j,0} = f(x_i, y_j) = f_{i,j}$$

$$\frac{\partial w}{\partial t}(x_i, y_j, 0) = g(x_i, y_j) = \frac{w_{i,j,1} - w_{i,j,-1}}{2\tau} + O(\tau^2) = g_{i,j}$$

so that

$$w_{i,j,-1} = w_{i,j,+1} - 2\tau g_{i,j}$$

Substituting $k = 0$ in eq. (10.84) and replacing the initial conditions in terms of $f_{i,j}$ results in finding an expression for w at $t = -\tau$:

$$w_{i,j,+1} = -w_{i,j,-1} + u^2[f_{i+1,j} + f_{i-1,j} + f_{i,j+1} + f_{i,j-1}]$$
$$+2(1 - 2u^2)f_{i,j} + h^2u^2Q_{i,j,0}$$

Substituting for $w_{i,j,-1}$ and rearranging the resulting expression gives the solution at $t = \tau$:

$$w_{i,j,+1} = \frac{1}{2}[u^2(f_{i+1,j} + f_{i-1,j} + f_{i,j+1} + f_{i,j-1}) + 2(1 - 2u^2)f_{i,j}]$$
$$+ \tau g_{i,j} + \frac{1}{2}h^2u^2Q_{i,j,0} \qquad (10.85)$$

together with the initial condition

$$w_{i,j,0} = f_{i,j}$$

gives the first two initial values at $t = 0$ and $t = \tau$.

Stability requires that wavefront, traveling at a speed of c in the time interval τ must arrive a radial distance Δr, such that $c\tau = \Delta r$. If the spatial interval is the same in the

x and y coordinates, i.e., h must be equal to $\Delta r/\sqrt{2}$. This means that $u = c\tau/h = 1/\sqrt{2}$ for stable and convergent numerical solution.

Example 10.37

Do problem 6.52, with $L = 10, h = 1, c = 1, a = 3, w_o = 2$ and $Q_0 = 0$. The initial conditions:

$$f(x) = 7x \qquad 0 \le x \le 3$$

$$= 3(10 - x) \qquad 3 \le x \le 10$$

(a) For $u = c\,\tau/h = 1$, then $\tau = 1$

Using eq. (10.81)

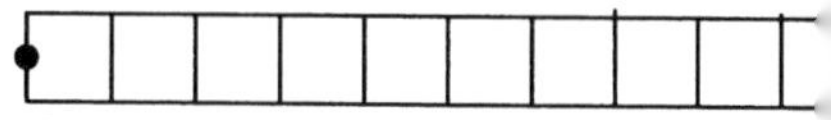

$$w_{i,k+1} = w_{i+1,k} + w_{i-1,k} - w_{i,k-1} \qquad i = 0, 1, \dots, 10 \qquad k = 1, 2, \dots$$

with $i = x_i = 0, 1, \dots, 10$

B.C. $\quad w_{0,k} = 0 \qquad w_{10,k} = 0$

I.C. $\quad w_{i,0} = f_i = 7i \qquad 0 \le i \le 3$

$$= 3(10 - i) \qquad 3 \le i \le 10$$

$$g_i = 0 \qquad i = 0, 1, \dots, 10$$

Using eq. (10.83) to generate the numerical solution at the first time step.

$$w_{i,1} = \frac{1}{2}[f_{i+1} + f_{i-1}] \qquad i = 1, \dots, 9$$

The results are shown in the following table for the response at $x = 3$, where the maximum initial displacement is 21. Note that $t = k\,\tau = k$.

(b) For $\tau = 0.25$ and $h = 1, u = 0.25$, then using eqs. (10.80) and (10.83) gives results shown in the following table. Here the time scale $t = 0.25k$.

(c) For $\tau = 0.1$ and $h = 1, u = 0.1$, the response at $x = 3$ in shown in the following table for $t = 0.1k$.

(d) to obtain a stable solution at smaller time steps, one needs to still use $u = 1$, but with both h and τ smaller. Using $u = 1, \tau = 0.5$ and $h = 0.5$, one obtains the results shown in the table below. Here $t = 0.5k$.

(e) The exact solution of problem 6.52, with $w_0 = 21, a = 3, c = 1, L = 10$

$$y(x_i, t) = \frac{200}{\pi^2} \sum_{n=1}^{N} \sin(\frac{n\pi}{10} x_i) \sin(\frac{3n\pi}{10}) \cos(\frac{n\pi}{10} t)$$

The response at any point x_i is periodic, with a period $T = \frac{2L}{c} = \frac{20}{1} = 20.$ This means that the response repeats itself every 20 units of time.

Comparing the response at $x_i = 3$ as a function of time, one can see that for $u = 1$ and $\tau = 1$, the response is very close to the exact solution. Using $\tau = 0.25$ and 0.1, should have given better results, but there is a numerical error accumulation, when one does numerical solution in fractional steps. Such error accumulation due to more integration steps is evident at $t = 20$, where the response should have converged to the initial conditions. Instead, these were 15% lower than the exact solution. However, when one uses $u = 1$ with both $h = \tau = 0.5$, the numerical solution gives also accurate solutions, including the numerical solution at the smaller time steps.

t	$u=\tau=1$	$u=\tau=0.25$	$u=\tau=0.1$	$u=1\tau=h=0.5$	Exact
0	21.00	21.00	+21.00	21.00	21.00
0.1			+20.95		20.59
0.2			+20.80		20.00
0.25		20.69	+20.68		19.74
0.3			+20.56		19.49
0.4			+20.22		19.02
0.5		19.79	+19.80	18.50	18.50
1.0	16.00	16.72	+16.75	16.00	16.00
1.5		13.11	+13.13	13.50	13.50
2.0	11.00	10.10	+10.09	11.00	11.00
3.0	6.00	6.30	6.26	6.00	6.00
4.0	1.00	1.44	1.48	1.00	1.00
5.0	-4.00	-4.45	-4.44	-4.00	-4.00
7.5		-10.36	-10.32	-9.00	-9.00
10.0	-9.00	-8.16	-8.06	-9.00	-9.00
12.5		-8.25	-8.37	-9.00	-9.00
15.0	-4.00	-5.61	-5.73	-4.00	-4.00
17.50		+8.29	+8.32	+8.50	+8.50
20.0	+21.00	17.50	+17.65	+21.00	+21.00

Example 10.38

Do problem 6.58, with $a=b=10, c=1, h=1, q_{\iota}=0$

and initial conditions

$$f(x,y)=xy(10-x)(10-y)/25 \qquad 0\le x,y\le 10$$

$$g(x,y)=0$$

The exact solution is computed from the answer to problem 6.58 as:

$$w(x,y,t)=\frac{6400}{\pi^6}\sum_{n=1}\sum_{m=1}\frac{(1-\cos m\pi)(1-\cos n\pi)}{n^3m^3}*$$

$$\sin\frac{n\pi x}{10}\sin\frac{m\pi y}{10}\sin(\frac{\pi}{10}\sqrt{n^2+m^2}\,t)$$

For stability, one must use $c\tau/h = 1/\sqrt{2}$, so that $\tau = 1/\sqrt{2}$. It should be noted that the period of vibration is $T = 10\tau = 10\sqrt{2} = 14.14$.

Using $u = 1/\sqrt{2}$ in eqs. (10.84) and (10.85), the numerical solution at points $x = 5$, $y = 5$ and $x = 4, y = 5$ are shown in the table below. Examination of the numerical results vs. the exact solution shows that they are less than 1% over a period.

$t = k\tau$	$w(5,5)$		$w(4,5)$	
k	Numerical	Exact	Numerical	Exact
0	25.00	25.00	24.00	24.00
1	24.00	24.00	23.02	23.02
2	21.04	21.05	20.12	20.13
3	16.24	16.27	15.42	15.45
4	9.80	9.86	9.12	9.18
5	2.00	2.09	1.53	1.59
8	-23.22	-23.70	-21.73	-21.87
10	-28.42	-27.86	-26.61	-26.37
12	-20.13	-20.34	-19.61	-19.58
15	0.77	0.74	0.68	0.59
18	19.77	19.70	19.13	19.00
20	25.89	25.94	24.77	25.04
30	-25.65	-25.72	-24.66	-24.57
40	28.52	28.60	26.72	26.86

Problems

Section 10.2

1. Obtain the first four positive roots of the following transcendental equations, accurate to four decimals:

(a) $\tan x = x$ (e) $x \cos x = 1$

(b) $\tan x = 3x$ (f) $\cos x = \tanh x$

(c) $\tan x = -x$ (g) $\tan x = \tanh x$

(d) $x \tan x = 1$ (h) $10 \cos x = \cosh(x/10)$

Section 10.3

2. Obtain the roots of the following multiple transcendental equations:

(a) $x^4 - y^4 = 30$ $x^4 + y^4 = 40$ start the solution

at the four trial points near $x = \pm 2$ $y = \pm 1$

(b) $x = \sin(x - y)$ $y = \cos(x + y)$ start the solution

at $x = y = 0.5$

(c) $x = \cos y \sinh x$ $y = \sin y \cosh x$

start at $x = 3$ $y = 7$

Section 10.6

Evaluate the following integrals using $n = 4$, 8, and 16 over the range [a,b] by using (a) Trapezoidal Rule, (b) Simpson Rule, and (c) Romberg Integration.

3. $f(x) = (x^2 + 4)^{-3/2}$ [0, 4]

4. $f(x) = xe^x$ [0, 1]

5. $f(x) = xe^{-x^2}$ $[0, 1]$

6. $f(x) = x \sin x$ $[0, \pi]$

7. $f(x) = cos^4 x$ $[0, \pi]$

8. $f(x) = x^2 log x$ $[0, 1]$

Section 10.7

For the following first-order ODE, use (a) Euler's methods, (b) Runge-Kutta Methods, and (c) Adams Method.

(9) $xy' + y = 0$ $y(2) = 2$

(10) $yy' = 2x$ $y(1) = 1$

(11) $yy' = -2x$ $y(0) = 1$

(12) $y^2 y' = x^2$ $y(0) = 1$

(13) Problem 1.1(a) $y(0) = 1$

(14) Problem 1.1(b) $y(1) = 0$

(15) Problem 1.1(c) $y(\pi/2) = 0$

(16) Problem 1.1(d) $y(0) = 1/4$

(17) Problem 1.1(e) $y(\pi/4) = 0$

(18) Problem 1.1(f) $y(0) = 1$

For the following second-order ODE, convert the ODE to two simultaneous first-order ODE.

(19) Problem 1.3(a) $y(0) = 1$ $y'(0) = 5$

(20) Problem 1.5(a) $y(1) = 2$ $y'(1) = 0$

(21) Problem 1.5(b) $y(1) = 1$ $y'(1) = 0$

(22) Problem 1.5(e) $y(1) = 2$ $y'(1) = 2$

(23) Problem 1.5(g) $y(1) = 2$ $y'(1) = 0$

(24) Problem 1.5(h) $y(1) = 2$ $y'(1) = 0$

Section 10.8

Solve the following BVP by the equilibrium method using 4, 8, and 16 segments. Apply extrapolation methods to improve the numerical solution.

(25) $y'' = \sin x$ [0,1] $y(0) = 0$ $y(1) = 0$

(26) $y'' = x^2$ [0,1] $y(0) = 0$ $y'(1) = 0$

(27) $y'' = \sin x$ $[0,\pi]$ $y(0) = 0$ $y'(\pi) = 0$

(28) $y^{iv} = \sin x$ $[0,\pi]$ $y(0) = y'(0) = y(\pi) = y'(\pi) = 0$

(29) $y^{iv} = 0$ [0,1] $y(0) = y'(0) = y''(1) = 0$

$y'''(1) = 1$

(30) $y^{iv} = 1$ [0,1] $y(0) = y''(0) = y(1) = y''(1) = 0$

Section 10.9

For the following eigenvalue problems, using $n = 5, 10$ and 20 segments, obtain the first four eigenvalues.

(31) Problem 4.5(a) $L = 1$

(32) Problem 4.5(b) $L = 1$

(33) Problem 4.5(c) $L = 1$

(34) Problem 4.5(d) with $L = a = 1$

(35) Problem 4.5(e) with $L = a = 1$

(36) Problem 4.8(a) $L = 1$

(37) Problem 4.8(b) $L = 1$

(38) Problem 4.8(c) $L = 1$

(39) Problem 4.8(d) $L = 1$

(40) Problem 4.8(e) $L = 1$

(41) Problem 4.8(f) $L = 1$

(42) Problem 4.22(d)

(43) Problem 4.22(e) $a = 1, b = 2, L = 1$

(44) Problem 4.22(h) $a = 2, L = 1$

(45) Problem 4.22(i) $a = L = 1$

(46) Problem 4.22(k) $a = L = 1$

(47) Problem 4.22(o) $L = 1$

(48) Problem 4.22(q)

(49) Problem 4.22(r)

Section 10.10.1

(50) Problem 6.1 $L = 10,\ \ b = 2, T_0 = 100$

(51) Problem 6.4 $L = 10,\ T_0 = 100,\ \ \ f(x) = x(10 - x)$

(52) Problem 6.5 $a = 10,\ T_0 = 100$

(53) Problem 6.6 $a = 10,\ b = \frac{3\pi}{4},\ T_0 = 100,\ f(\theta) = \sin\theta$

Senior 10.10.4

Obtain the first five eigenvalues for the following problems, using $n = 2, 3, 4$, and 5.

(54) Problem 6.26(a) $a = 1$

(55) Problem 6.26(b) axisymmetric case, $a = 1$

(56) Problem 6.26(d) $a = 1,\ c = \pi/2$

(57) Problem 6.26(h) $a = 1,\ b = 1$

(58) Problem 6.26(h) $a = 1,\ b = 2$

Obtain the response due to forced excitation.

(59) Problem 6.30 $a = b = 10, h = 1,\ \omega/c = 1, q_0/S = 100$

$f(x, y) = xy(10 - x)(10 - y)$

(60) Problem 6.30 $a = 10\ \ b = 20, h = 1, \omega/c = 1,$

$q_0/S = 100$

$f(x, y) = xy(10 - x)(20 - y)$

Section 10.10.5

(61) Problem 6.35 $L = 10$ $T_o = 100$ $h = 1, K = \frac{1}{2}$

(62) Problem 6.37 $a = 10, b = 20, h = 1, T_0 = 100, K = \frac{1}{4},$

$Q = 0, f(x, y) = 1$

(63) Problem 6.37 $a = 10, b = 20, h = 1, T_0 = 0, K = \frac{1}{4},$

$Q_0/\rho c = 100, \propto = 0.6$

(65) Problem 6.48 $a = b = 10, h = 1, Q_0/k = 100, K = \frac{1}{4},$

$c = 0.6$

Section 10.10.6

(66) Problem 6.51 $L = 10, c = 1, h = 1, f(x) = 0,$

$g(x) = x(10 - X)$

(67) Problem 6.56 $L = 10, c = 1, h = 1, f(x) = 0,$

$f(x,t)/T_0 = 100 \sin 2t$

(68) Problem 6.62 $L = 10, h = 1, c = 1, x_0 = y_0 = 5$

$P_0/S = 100$

(69) Problem 6.66 $a = b = 10, c = 1, h = 1, x_0 = y_0 = 5,$

$t_0 = 2\sqrt{2}, P_0/S = 100, w_0 = 10$

APPENDIX A

INFINITE SERIES

A.1 Introduction

An infinite series of constants is defined as:

$$a_0 + a_1 + a_2 + \ldots = \sum_{n=0}^{\infty} a_n \tag{A.1}$$

The infinite series in eq. (A.1) is said to be **convergent** to a value = a, if, for any arbitrary number ε, there exists a number M such that:

$$\left| \sum_{n=0}^{n=N} a_n - a \right| < \varepsilon \qquad \text{for all } N > M$$

If this condition is not met, then the series is said to be **divergent**. The series may diverge to +∞ or -∞ or have no limit, as is the case of an alternating series.

A *necessary but not sufficient* condition for the convergence of the series in eq. (A.1) is:

$$\lim_{n \to \infty} a_n \to 0$$

For example, the infinite series:

$$\sum_{n=1}^{\infty} \frac{1}{n}$$

is divergent, while the limit of a_n vanishes

$$\lim_{n \to \infty} \frac{1}{n} \to 0$$

A *necessary and sufficient condition* for convergence of the series in eq. (A.1) is as follows: if, for any arbitrary number ε, there exists a number M such that:

$$\left| \sum_{n=N}^{n=N+k} a_n \right| = \left| a_N + a_{N+1} + \ldots + a_{N+k} \right| < \varepsilon$$

for all N > M and for all positive integers k.

If the series:

$$\sum_{n=0}^{\infty} |a_n| \tag{A.2}$$

converges, then the series (A-1) converges and is said to be an **absolutely convergent** series. If the series (A-1) converges, but the series (A-2) does not converge, then the series (A-1) is known as a **conditionally convergent** series.

Example A.1

(i) The series:

$$\sum_{n=1}^{\infty} (-1)^n \frac{1}{n^2}$$

is a convergent series and so is the series:

$$\sum_{n=1}^{\infty} \left|(-1)^n \frac{1}{n^2}\right| = \sum_{n=1}^{\infty} \frac{1}{n^2}$$

Thus, the series is absolutely convergent.

(ii) The series:

$$\sum_{n=0}^{\infty} (-1)^n \frac{1}{n+1}$$

is a convergent series, but:

$$\sum_{n=1}^{\infty} \left|(-1)^n \frac{1}{n+1}\right| = \sum_{n=1}^{\infty} \frac{1}{n+1}$$

is divergent. Therefore the series is conditionally convergent.

A.2 Convergence Tests

This section will discuss several tests for convergence of infinite series of numbers. Each test maybe more suitable for some series than others.

A.2.1 Comparison Test

If the positive series $\sum_{n=0}^{\infty} a_n$ converges, and if $|b_n| \le a_n$ for large n, then the series $\sum_{n=0}^{\infty} b_n$ also converges.

If the series $\sum_{n=0}^{\infty} a_n$ diverges, and if $|b_n| \le a_n$ for large n, then the series $\sum_{n=0}^{\infty} b_n$ also diverges.

Example A.2

One can use the comparison test to easily prove that $\sum_{n=1}^{\infty} \frac{1}{(n+1)^2}$ is convergent, since

$\sum_{n=1}^{\infty} \frac{1}{n^2}$ is convergent and $\frac{1}{n+1} < \frac{1}{n}$, for all n > 1.

A.2.2 Ratio Test: (d'Alembert's)

If:

$$\lim_{n\to\infty} \left| \frac{a_{n+1}}{a_n} \right| < 1 \qquad \text{the series converges}$$

$$\lim_{n\to\infty} \left| \frac{a_{n+1}}{a_n} \right| > 1 \qquad \text{the series diverges} \tag{A.3}$$

However, the test fails to give any information when the limit approaches unity. In such a case, if the series is an **Alternating Series**, i.e., if it is made up of terms that alternate in sign, and if the terms decrease in absolute magnitude consistently for large n and if $\lim_{n\to\infty} a_n \to 0$, then the series converges.

Example A.3

(i) The series $\sum_{n=0}^{\infty} \frac{1}{2^n(n+1)}$ converges, since the Ratio Test gives:

$$\lim_{n\to\infty} \left| \frac{a_{n+1}}{a_n} \right| = \lim_{n\to\infty} \frac{n+1}{2(n+2)} = \frac{1}{2} < 1$$

(ii) The series $\sum_{n=1}^{\infty} \frac{3^n n}{(n+1)^2}$ diverges, since the Ratio Test gives:

$$\lim_{n\to\infty} \left| \frac{a_{n+1}}{a_n} \right| = \lim_{n\to\infty} \frac{3(n+1)^3}{n(n+2)^2} = 3 > 1$$

(iii) The series $\sum_{n=1}^{\infty} \frac{n}{(n+1)^2}$ cannot be judged for convergence with the Ratio Test

since:

$$\lim_{n\to\infty} \left| \frac{a_{n+1}}{a_n} \right| = \lim_{n\to\infty} \frac{(n+1)^3}{n(n+2)^2} = 1$$

(ii) The series $\sum_{n=0}^{\infty} (-1)^n \frac{n}{(n+1)^2}$ converges, since the series is an alternating series, successive terms are smaller, i.e., $\frac{1}{2^2} > \frac{2}{3^2} > \frac{3}{4^2} > \frac{4}{5^2} \ldots$, and:

$$\lim_{n\to\infty} a_n = \lim_{n\to\infty} \frac{n}{(n+1)^2} \to 0$$

A.2.3 Root Test: (Cauchy's)

If:

$$\lim_{n\to\infty} |a_n|^{1/n} < 1 \qquad \text{the series converges}$$

$$\lim_{n\to\infty} |a_n|^{1/n} > 1 \qquad \text{the series diverges} \qquad (A.4)$$

The test fails if the limit approaches unity.

Example A.4

(i) One can prove that the series:

$$\sum_{n=0}^{\infty} \frac{1}{2^n(n+1)}$$

is convergent using the root test. The limit of the nth root equals:

$$\lim_{n\to\infty} |a_n|^{1/n} = \lim_{n\to\infty} \left[\frac{1}{2^n(n+1)}\right]^{1/n} = \frac{1}{2} \lim_{n\to\infty} (n+1)^{-1/n}$$

Let $y = (n+1)^{-1/n}$, consider the limit of the natural logarithm of y:

$$\lim_{n\to\infty} \log y = \lim_{n\to\infty} \frac{-\log(n+1)}{n} = -\lim_{n\to\infty} \frac{\frac{1}{n+1}}{1} \to 0$$

by using L' Hospital rule.

Thus:

$$\lim_{n\to\infty} y = e^0 = 1$$

so that:

$$\lim_{n\to\infty} |a_n|^{1/n} \to \frac{1}{2} < 1$$

(ii) One can prove that the series:

$$\sum_{n=1}^{\infty} \frac{3^n n}{(n+1)^2}$$

is divergent using the root test. The limit of the root equals:

$$\lim_{n\to\infty}|a_n|^{1/n} = \lim_{n\to\infty}\left[\frac{3^n n}{(n+1)^2}\right]^{1/n} = 3\lim_{n\to\infty}(n+1)^{-1/n}\lim_{n\to\infty}\left(\frac{n}{n+1}\right)^{1/n}$$

From part (i), $\lim_{n\to\infty}(n+1)^{-1/n} = 1$, therefore:

$$\lim_{n\to\infty}|a_n|^{1/n} = 3\lim_{n\to\infty}\left(\frac{n}{n+1}\right)^{1/n} = 3\lim_{n\to\infty}\left(\frac{1}{1+1/n}\right)^{1/n} = 3 > 1$$

A.2.4 Raabe's Test

For a positive series $\{a_n\}$, if the Limit of (a_{n+1}/a_n) approaches unity, where the Ratio Test fails, then the following test gives a criteria for convergence. If:

$$\lim_{n\to\infty}\left\{n\left[\frac{a_n}{a_{n+1}} - 1\right]\right\} > 1 \quad \text{the series converges}$$

$$\lim_{n\to\infty}\left\{n\left[\frac{a_n}{a_{n+1}} - 1\right]\right\} < 1 \quad \text{the series diverges} \tag{A.5a}$$

If this limit approaches unity, then the following refinements of the test can be used:

$$\lim_{n\to\infty}(\log n)\left\{n\left[\frac{a_n}{a_{n+1}} - 1\right] - 1\right\} > 1 \quad \text{the series converges}$$

$$\lim_{n\to\infty}(\log n)\left\{n\left[\frac{a_n}{a_{n+1}} - 1\right] - 1\right\} < 1 \quad \text{the series diverges} \tag{A.5b}$$

If this limit approaches unity, then the following refinements of the test can be used:

$$\lim_{n\to\infty}(\log n)\left\{(\log n)\left\{n\left[\frac{a_n}{a_{n+1}} - 1\right] - 1\right\} - 1\right\} > 1 \quad \text{the series converges}$$

$$\lim_{n\to\infty}(\log n)\left\{(\log n)\left\{n\left[\frac{a_n}{a_{n+1}} - 1\right] - 1\right\} - 1\right\} < 1 \quad \text{the series diverges} \tag{A.5c}$$

If the limit approaches unity, then another test based on a refinement of eq. (A.5c) can be repeated over and over.

Example A.5

(i) The series $\sum_{n=0}^{\infty}\frac{1}{(n+1)^2}$ could not be tested conclusively with the Ratio test, but it can be tested using Raabe's test:

$$\lim_{n\to\infty} n\left(\frac{a_n}{a_{n+1}} - 1\right) = \lim_{n\to\infty} n\left[\frac{(n+2)^2}{(n+1)^2} - 1\right] = 2 > 1$$

Therefore, the series converges.

(i) The series $\sum_{n=0}^{\infty} \frac{1}{(n+1)}$ could not be tested conclusively with the Ratio test.

Using Raabe's Test eq. (A.5a):

$$\lim_{n\to\infty} n\left(\frac{a_n}{a_{n+1}} - 1\right) = \lim_{n\to\infty} n\left[\frac{(n+2)}{(n+1)} - 1\right] = 1$$

Thus, the first test fails. Using the second version eq. (A.5b):

$$\lim_{n\to\infty} (\log n)\left\{n\left[\frac{(n+2)}{(n+1)} - 1\right] - 1\right\} = \lim_{n\to\infty} (\log n)\left\{\frac{n}{(n+1)} - 1\right\} = 0 < 1$$

Therefore, the series diverges.

A.2.5 Integral Test

If the sequence a_n is a monotonically decreasing positive sequence, then define a function:

$$f(n) = a_n$$

which is also a monotonically decreasing positive function of n. Then the series:

$$\sum_{n=0}^{\infty} a_n$$

and the integral:

$$\int_{c}^{\infty} f(n)\, dn$$

both converge or both diverge, for $c > 0$.

Example A.6

(i) The series $\sum_{n=0}^{\infty} \frac{1}{n^2}$ converges, since the integral $\int_{1}^{\infty} \frac{dn}{n^2} = -\frac{1}{n}\Big|_1^{\infty} = 1$ also converges.

(ii) The series $\sum_{n=0}^{\infty} \frac{1}{n}$ diverges, because the integral $\int_{1}^{\infty} \frac{dn}{n} = \log n\Big|_1^{\infty} = \infty$ also diverges.

A.3 Infinite Series of Functions of One Variable

An infinite series of functions of one variable takes the following form:

$$f_0(x) + f_1(x) + f_2(x) + \ldots = \sum_{n=0}^{\infty} f_n(x) \qquad a \le x \le b$$

The series can be summed at any point x in the interval [a,b]. If the sum of the series, summed for a point x_0, converges to some value $f(x_0)$, then the series is said to converge to $f(x_0)$ for $a \le x_0 \le b$. Thus, if one chooses an arbitrary small number ε, then there exists a number M, such that the remainder of the series $R_N(x)$:

$$|R_N(x_0)| = \left| f(x_0) - \sum_{n=0}^{n=N} f_n(x_0) \right| < \varepsilon \qquad \text{for} \qquad N > M$$

and

$$f(x_0) = \lim_{N\to\infty} \sum_{n=0}^{N} f_n(x_0)$$

A *necessary and sufficient* condition for convergence of the series at a point x_0 is that, given a small arbitrary number ε, then there exists a number M, such that:

$$|f_N(x) + f_{N+1}(x) + \ldots + f_{N+k}(x)| = \left| \sum_{n=N}^{n=N+k} f_n(x) \right| < \varepsilon$$

for all N > M and for all values of the positive integer k.

It should be noted that the sum of a series whose terms are continuous may not be continuous. Thus, if the series is convergent to f(x), then:

$$\lim_{N\to\infty} \sum_{n=0}^{N} f_n(x) \to f(x)$$

and

$$f(x_0) = \lim_{x\to x_0} f(x) = \lim_{x\to x_0} \left[\lim_{N\to\infty} \sum_{n=0}^{N} f_n(x) \right] \tag{A.6}$$

On the other hand, by definition:

$$f(x_0) = \lim_{N\to\infty} \left[\sum_{n=0}^{N} \lim_{x\to x_0} f_n(x) \right] \tag{A.7}$$

The limiting values for $f(x_0)$ as given in eqs. (A.6) and (A.7) are not the same if f(x) is discontinuous at $x = x_0$, they are identical only if f(x) is continuous at $x = x_0$.

A.3.1 Uniform Convergence

A series is said to **converge uniformly** for all values of x in [a,b], if for any arbitrary positive number, there exists a number M independent of x, such that:

$$\left| f(x) - \sum_{n=0}^{N} f_n(x) \right| < \varepsilon \qquad \text{for} \qquad N > M$$

for *all* values of x in the interval [a,b].

Example A.7

The series of functions:

$$(1-x) + x(1-x) + x^2(1-x) + \ldots$$

can be represented by a series of $f_n(x)$ given by:

$$f_n(x) = x^{n-1}(1-x) \qquad n = 1, 2, 3, \ldots$$

Summing the first N terms, one obtains:

$$\sum_{n=1}^{N} f_n(x) = 1 - x^N$$

The series converges for $N \to \infty$ iff:

$$|x| < 1$$

Therefore, the sum of the infinite series as $N \to \infty$ approaches:

$$f(x) = \lim_{N\to\infty} \left[\sum_{n=1}^{N} f_n(x) \right] = 1 \qquad \text{for} \qquad |x| < 1$$

Thus, to test the convergence of the series, the remainder of the series $R_N(x)$ is found to be:

$$R_N(x) = \left| f(x) - \sum_{n=1}^{N} f_n(x) \right| = \left|1 - (1 - x^N)\right| = \left|x^N\right|$$

which vanishes as $N \to \infty$ only if $|x| < 1$.

For uniform convergence:

$$\left|x^N\right| < \varepsilon \qquad \text{for } \varepsilon \text{ fixed and for all } N > M$$

or

$$N > \left| \frac{\log(\varepsilon)}{\log(|x|)} \right|$$

If one chooses an $\varepsilon = e^{-10}$, then one must choose a value N such that:

$$N > \frac{10}{|\log(|x|)|}$$

Thus, the series is uniformly convergent for $0 \le x \le x_0$, $0 < x_0 < 1$. At the point $x = x_0$ choose:

$$N = \frac{10}{\log(|x_0|)}$$

As the point x_0 approaches 1, $|\log x_0| \rightarrow 0$, and one needs increasingly larger and larger values of N, so that the inequality $R_n < \varepsilon$ cannot be satisfied by one value of N. Thus, the series is uniformly convergent in the region $0 \le x \le x_0$, and not uniformly convergent in the region $0 \le x \le 1$.

A.3.2 Weierstrass's Test for Uniform Convergence

The series $f_0(x) + f_1(x) + ...$, converges uniformly in [a,b] if there exists a convergent positive series of positive real numbers $M_1 + M_2 + ...$ such that:

$$|f_n(x)| \le M_n \qquad \text{for all} \qquad x \text{ in } [a,b]$$

Example A.8

The series:

$$\sum_{n=1}^{\infty} \frac{1}{n^2 + x^2}$$

converges uniformly for $-\infty < x < \infty$ since:

$$|f_n(x)| = \left|\frac{1}{n^2 + x^2}\right| \le \frac{1}{n^2} = M_n \qquad \text{for all } x \ge 0$$

and since the series of constants:

$$\sum_{n=1}^{\infty} M_n = \sum_{n=1}^{\infty} \frac{1}{n^2} \quad \text{converges}$$

A.3.3 Consequences of Uniform Convergence

Uniform convergence of an infinite series of functions implies that:

1. If the functions $f_n(x)$ are continuous in [a,b] and if the series converges uniformly in [a,b] to f(x), then f(x) is a continuous function in [a,b].

2. If the functions $f_n(x)$ are continuous in [a,b] and if the series converges uniformly in [a,b] to f(x), then the series can be integrated term by term:

$$\int_{x_1}^{x_2} f(x)dx = \int_{x_1}^{x_2} f_o(x)dx + \int_{x_1}^{x_2} f_1(x)\,dx + ... = \sum_{n=0}^{\infty} \int_{x_1}^{x_2} f_n(x)\,dx$$

where $a \leq x_1, x_2 \leq b$.

3. If the series $\sum_{n=0}^{\infty} f_n(x)$ converges to f(x) in [a,b] and if each term $f_n(x)$ and $f_n'(x)$ are continuous, and if the series:

$$\sum_{n=0}^{\infty} f_n'(x)$$

is uniformly convergent in [a,b], then, the series can be differentiated term by term:

$$f'(x) = \sum_{n=0}^{\infty} f_n'(x)$$

A.4 Power Series

A power series about a point x_o, is defined as:

$$a_0 + a_1(x - x_o)^M + a_2(x - x_o)^{2M} + ... = \sum_{n=0}^{\infty} a_n(x - x_o)^{nM} \tag{A.8}$$

where M is a positive integer. The power series is a special form of an infinite series of functions. The series may converge in a certain region.

A.4.1 Radius of Convergence

For convergence of the series in eq. (A.8) either the Ratio Test or the Root Test can be employed. The Ratio Test gives:

$$\lim_{n\to\infty}\left|\frac{a_{n+1}(x-x_o)^{(n+1)M}}{a_n(x-x_o)^{nM}}\right| = |x - x_o|^M \lim_{n\to\infty}\left|\frac{a_{n+1}}{a_n}\right| \quad \begin{matrix} < 1 \text{ the series converges} \\ > 1 \text{ the series diverges} \end{matrix}$$

or if one defines the radius of convergence ρ as:

$$\rho = \left\{\lim_{n\to\infty}\left|\frac{a_n}{a_{n+1}}\right|\right\}^{1/M} \tag{A.9}$$

then the convergence of the series is decided by the conditions:

$$|x - x_o| \quad \begin{matrix} < \rho \text{ the series converges} \\ > \rho \text{ the series diverges} \end{matrix} \tag{A.10}$$

In other words, the series converges in the region:

$$x_0 - \rho < x < x_0 + \rho$$

and diverges outside this region.

The Root Test gives:

$$\lim_{n\to\infty} \left| a_n (x - x_0)^{nM} \right|^{1/nM} = |x - x_0| \lim_{n\to\infty} |a_n|^{1/nM} \begin{array}{l} < 1 \text{ the series converges} \\ > 1 \text{ the series diverges} \end{array}$$

Thus, if one lets:

$$\rho = \left\{ \lim_{n\to\infty} |a_n|^{-1/n} \right\}^{1/M} \tag{A.11}$$

then the series converges in the region indicated in eq. (A.10).

The Ratio Test or the Root Test fails at the end points, i.e., when $|x-x_0| = \rho$, where both tests give a limit of unity. In such cases, Raabe's Test or the Alternating Series Test (if appropriate) can be used on the series after substituting for the end points at $x = x_0 + \rho$ or $x = x_0 - \rho$

Example A.9

Find the regions of convergence of the following power series:

$$\sum_{n=1}^{\infty} \frac{(x-1)^{3n}}{n\, 27^n}$$

Here M = 3, so that the radius of convergence by the Ratio Test becomes:

$$\rho = \lim_{n\to\infty} \left| \frac{(n+1)\, 27^{n+1}}{n 27^n} \right|^{1/3} = (27)^{1/3} = 3$$

while using the Root Test:

$$\rho = \lim_{n\to\infty} \left| \frac{27^{-n}}{n} \right|^{-1/3n} = (27)^{1/3} \lim_{n\to\infty} n^{1/3n} \to 3$$

At one end point, x - 1 = 3 or x = 4, the series becomes:

$$\sum_{n=1}^{\infty} \frac{(4-1)^{3n}}{n\, 27^n} = \sum_{n=1}^{\infty} \frac{3^{3n}}{n 27^n} = \sum_{n=1}^{\infty} \frac{1}{n}$$

which diverges. At the second end point, x - 1 = -3 or x = -2, and the series becomes an alternating series:

$$\sum_{n=1}^{\infty} \frac{(-2-1)^{3n}}{n\, 27^n} = \sum_{n=1}^{\infty} \frac{(-3)^{3n}}{n\, 27^n} = \sum_{n=1}^{\infty} \frac{(-1)^n}{n}$$

which converges, so that the region of convergence of the power series is $-2 \le x < 4$.

A.4.2 Properties of Power Series

1. A power series is absolutely and uniformly convergent in the region

   $x_0 - \rho < x < x_0 + \rho$

2. A power series can be differentiated term by term, such that:

$$f'(x) = a_1 + 2a_2(x - x_0) + 3a_3(x - x_0)^2 \ldots = \sum_{n=1}^{\infty} na_n(x - x_0)^{n-1}$$

for $x_0 - \rho < x < x_0 + \rho$. The radius of convergence of the resulting series for $f'(x)$ is the same as that of the series for f(x). This holds for all derivatives of the series $f(x)^{(n)}$, for $n \geq 1$.

3. The series an be integrated term by term such that:

$$\int_{x_1}^{x_2} f(x)dx = \sum_{n=0}^{\infty} a_n \int_{x_1}^{x_2} (x - x_0)^n \, dx = \sum_{n=0}^{\infty} \frac{a_n}{n+1}(x - x_0)^{n+1}\Big|_{x_1}^{x_2}$$

for $x_0 - \rho < x < x_0 + \rho$. The series can be integrated as many times as needed.

PROBLEMS

1. Prove that the following series of the form:

$$\sum_{n=1}^{\infty} a_n$$

converges where a_n is given by:

(a) $\log(1-\frac{1}{n^2})$ (b) $\frac{1}{n^c}$ $c>1$ (c) $\frac{(-1)^n}{n^c}$ $c>0$

(d) $\frac{n}{(n+1)^3}$ (e) $\frac{1}{n+2^n}$ (f) $\frac{1}{n\,2^n}$

(g) $\frac{1}{n^2\,3^n}$ (h) $\frac{3^n}{5^n}$ (i) $(-1)^n \log(1+\frac{1}{n})$

(j) $\frac{n^2}{2^n}$ (k) $\frac{2^{2n}}{3^{2n}}$ (l) $\frac{n}{e^n}$

(m) $\frac{n}{(n!)^2}$ (n) $\frac{3^{2n}}{(2n)!}$ (o) $\frac{n^3}{n!}$

(p) $\frac{1}{n^3}$ (q) n^{-n} (r) $\frac{1}{n!(n+1)!}$

(s) 3^{-n} (t) $\frac{(-1)^n n}{n^3+1}$ (u) e^{-n}

(v) $\frac{n}{(n+1)!}$ (w) $\frac{(-1)^n}{\log(n+1)}$ (x) $\frac{(-1)^n}{\sqrt{n}}$

(y) $\frac{1}{(2n)!}$ (z) $\frac{n!}{(2n)!}$ (aa) $\frac{(n!)^2}{(2n)!}$

(bb) $[\log(n+1)]^{-n}$

2. Prove that the following series:

$$\sum_{n=1}^{\infty} a_n$$

diverges, where a_n is given by:

(a) $\frac{n^n}{n!}$ (b) $\frac{1}{\sqrt{n}}$ (c) $\log(1+\frac{1}{n})$

(d) $\frac{1}{\sqrt{n(n+2)}}$ (e) $\frac{1}{\log(n+1)}$ (f) $\frac{1}{\sqrt{n^2+1}}$

(g) $\frac{n!}{3^n}$ (h) $\frac{3^n}{n^2}$ (i) $\frac{3^n}{1+e^n}$

(j) $\frac{\log(n+1)}{n}$

3. Find the radius of convergence and the region of convergence of the following power series:

(a) $\sum_{n=0}^{\infty} \frac{(x-1)^n}{2^n}$ (b) $\sum_{n=0}^{\infty} \frac{(x+2)^n}{4^n+n^2}$

(c) $\sum_{n=0}^{\infty} \frac{(x-2)^n}{n+1}$ (d) $\sum_{n=0}^{\infty} \frac{(n!)^2 x^{2n}}{(2n)!}$

(e) $\sum_{n=0}^{\infty} \frac{x^n}{n!}$ (f) $\sum_{n=1}^{\infty} \frac{(-1)^n (x+1)^n}{\sqrt{n}}$

(g) $\sum_{n=1}^{\infty} \frac{n!x^n}{n^n}$ (h) $\sum_{n=1}^{\infty} \frac{n^3 (x-3)^n}{3^n}$

(i) $\sum_{n=0}^{\infty} \frac{(x+1)^{3n}}{8^n}$ (j) $\sum_{n=0}^{\infty} \frac{(x+1)^{3n}}{8^n (n+1)}$

APPENDIX B

SPECIAL FUNCTIONS

In this appendix, a compendium of the most often used and quoted functions are covered. Some of these functions are obtained as series solutions of some differential equations and some are defined by integrals.

B.1 The Gamma Function $\Gamma(x)$

Definition:

$$\Gamma(x) = \int_0^\infty t^{x-1} e^{-t} dt \qquad (\text{Re } x > 0) \qquad \text{(B1.1)}$$

Recurrence Formulae:

$$\Gamma(x+1) = x\,\Gamma(x) \qquad \text{(B1.2)}$$

$$\Gamma(n+1) = n! \qquad \text{(B1.3)}$$

Useful Formulae:

$$\Gamma(x)\Gamma(1-x) = \pi \operatorname{cosec}(\pi x) \qquad \text{(B1.4)}$$

$$\Gamma(x)\,\Gamma(-x) = -\pi \frac{\operatorname{cosec}(\pi x)}{x} \qquad \text{(B1.5)}$$

$$\Gamma(\tfrac{1}{2} + x)\,\Gamma(\tfrac{1}{2} - x) = \pi \sec(\pi x) \qquad \text{(B1.6)}$$

$$\Gamma(2x) = \frac{2^{2x-1}}{\sqrt{\pi}} \Gamma(x)\,\Gamma\left(x + \tfrac{1}{2}\right) \qquad \text{(B1.7)}$$

Complex Arguments:

$$\Gamma(1 + ix) = ix\,\Gamma(ix) \qquad (x \text{ real}) \qquad \text{(B1.8)}$$

$$\Gamma(ix)\,\Gamma(-ix) = |\Gamma(ix)|^2 = \frac{\pi}{x \sinh \pi x} \qquad (x \text{ real}) \qquad \text{(B1.9)}$$

$$\Gamma(1 + ix)\Gamma(1 - ix) = \frac{\pi x}{\sinh \pi x} \qquad \text{(B1.10)}$$

Asymptotic Series:

$$\Gamma(z) \sim \sqrt{2\pi}\; z^{z-1/2}\, e^{-z}\left[1+\frac{1}{12z}+\frac{1}{288z^2}-\frac{139}{51840z^3}-\ \ldots\right] \qquad \text{(B1.11)}$$

$$|z| >> 1 \qquad\qquad |\arg z| < \pi$$

Special Values:

$$\Gamma(1/2)=\pi^{1/2} \qquad\qquad \Gamma(3/2)=\frac{\pi^{1/2}}{2} \qquad \text{(B1.12)}$$

$$\Gamma\left(n+\tfrac{1}{2}\right)=\sqrt{\pi}\,\frac{(2n-1)!!}{2^n} \qquad\qquad \Gamma\left(\tfrac{1}{2}-n\right)=\sqrt{\pi}\,\frac{(-1)^n 2^n}{(2n-1)!!}$$

where the symbol $n!! = n\,(n-2)\cdot\ldots\cdot 2$ or 1

Integral Representations:

$$\Gamma(z)=\frac{x^z}{2\sin(\pi z)}\int_{-\infty}^{\infty} e^{ixt}\,(it)^{z-1}\,dt \qquad x>0 \qquad 0<Re(z)<1 \qquad \text{(B1.13)}$$

$$\Gamma(z)=\frac{x^z}{\cos(\pi z/2)}\int_{0}^{\infty} \cos(xt)\; t^{z-1}\,dt \qquad x>0 \qquad 0<Re(z)<1 \qquad \text{(B1.14)}$$

$$\Gamma(z)=\frac{x^z}{\sin(\pi z/2)}\int_{0}^{\infty} \sin(xt)\; t^{z-1}\,dt \qquad x>0 \qquad 0<Re(z)<1 \qquad \text{(B1.15)}$$

$$\Gamma(z)=\int_{0}^{\infty} e^{-t}\,t^{z-1}\,(\log t)\,(t-z)\,dt \qquad Re(z)>0 \qquad \text{(B1.16)}$$

$$\Gamma(z)=\int_{-\infty}^{\infty} \exp\,[zt-e^t]\,dt \qquad Re(z)>0 \qquad \text{(B1.17)}$$

B.2 PSI Function $\psi(x)$

Definition:

$$\psi(z)=\frac{1}{\Gamma(z)}\left[\frac{d\Gamma(z)}{dz}\right]=\frac{d[\log\Gamma(z)]}{dz} \qquad \text{(B2.1)}$$

Recurrence Formulae:

$$\psi(z+1) = \frac{1}{z} + \psi(z) \tag{B2.2}$$

$$\psi(z+n) = \psi(z) + \sum_{k=0}^{n-1} \frac{1}{z+k} \tag{B2.3}$$

$$\psi(z-n) = \psi(z) - \sum_{k=1}^{n} \frac{1}{z-k} \tag{B2.4}$$

$$\psi(z+1/2) = \psi(1/2-z) + \pi \tan(\pi z) \tag{B2.5}$$

$$\psi(1-z) = \psi(z) + \pi \cot(\pi z) \tag{B2.6}$$

Special Values:

$$\psi(1) = -\gamma = -0.5772156649....$$

$$\psi(\tfrac{1}{2}) = -\gamma - 2\log 2$$

$$\psi(n+1) = -\gamma + \sum_{k=1}^{n} \frac{1}{k} \tag{B2.7}$$

Asymptotic Series:

$$\psi(z) \sim \log z - \frac{1}{2z} - \frac{1}{12z^2} + \frac{1}{120z^4} - \frac{1}{252z^6} + \ldots \tag{B2.8}$$

$$|z| >> 1 \qquad |\arg z| < \pi$$

Integral Representations:

$$\psi(z) = -\gamma + \int_0^\infty \frac{e^{-t} - e^{-zt}}{1-e^{-t}}\, dt \tag{B2.9}$$

$$= -\gamma + \int_0^1 \frac{1-t^{z-1}}{1-t}\, dt \tag{B2.10}$$

$$= \int_0^\infty \frac{e^{-t} - (1+t)^{-z}}{t}\, dt \tag{B2.11}$$

$$= \int_0^\infty \frac{1-e^{-t} - e^{-t(z-1)}}{t(e^t - 1)}\, dt \tag{B2.12}$$

B.3 Incomplete Gamma Function γ (x,y)

Definitions:

$$\gamma(x,y) = \int_0^y e^{-t}\, t^{x-1}\, dt \qquad Re(x) > 0 \qquad \text{(Incomplete Gamma Function)} \tag{B3.1}$$

$$\Gamma(x,y) = \int_y^\infty e^{-t} t^{x-1}\, dt \qquad \text{(Complementary Incomplete Gamma Function)} \tag{B3.2}$$

$$\gamma^*(x,y) = \frac{y^{-x}}{\Gamma(x)}\,\gamma(x,y) \tag{B3.3}$$

Recurrence Formulae:

$$\gamma(x+1,y) = x\gamma(x,y) - y^x e^{-y} \tag{B3.4}$$

$$\Gamma(x+1,y) = x\Gamma(x,y) + y^x e^{-y} \tag{B3.5}$$

$$\gamma^*(x+1,y) = \frac{\gamma^*(x,y)}{y} - \frac{e^{-y}}{y\,\Gamma(x+1)} \tag{B3.6}$$

Useful Formulae:

$$\Gamma(x,y) + \gamma(x,y) = \Gamma(x) \tag{B3.7}$$

$$\Gamma(x)\,\Gamma(x{+}n,y) - \Gamma(x{+}n)\,\Gamma(x,y) = \Gamma(x{+}n)\,\gamma(x,y) - \Gamma(x)\,\Gamma(x{+}n,y) \tag{B3.8}$$

Special Values:

$$\gamma\left(\tfrac{1}{2}, x^2\right) = \sqrt{\pi}\ \mathrm{erf}(x) \tag{B3.9}$$

$$\Gamma\left(\tfrac{1}{2}, x^2\right) = \sqrt{\pi}\ \mathrm{erfc}\,(x) \tag{B3.10}$$

$$\gamma^*(-n,y) = y^n \tag{B3.11}$$

$$\Gamma(0,x) = -E_i(-x) \tag{B3.12}$$

$$\Gamma(n+1,y) = n!\ e^{-y} \sum_{m=0}^{n} \frac{y^m}{m!} \tag{B3.13}$$

Series Representation:

$$\gamma(x,y) = \sum_{n=0}^{\infty} \frac{(-1)^n y^{n+x}}{(x+n)n!} \qquad x > 0 \tag{B3.14}$$

Asymptotic Series:

$$\Gamma(x,y) \sim y^{x-1}e^{-y} \sum_{m=0}^{\infty} \frac{(-1)^m \Gamma(1-x+m)}{\Gamma(1-x)x^m} \tag{B3.15}$$

$|y| >> 1$ $\qquad$ $|\arg x| < 3\pi/2$

$$\Gamma(x,y) \sim \Gamma(x) y^{x-1} e^{-y} \sum_{m=0}^{\infty} \frac{1}{\Gamma(x-m)x^m} \tag{B3.16}$$

$|y| >> 1$ $\qquad$ $|\arg x| < 3\pi/2$

B.4 Beta Function B(x,y)

Definition:

$$B(x,y) = \int_0^1 t^{x-1}(1-t)^{y-1}\,dt \tag{B4.1}$$

Useful Formulae:

$$B(x,y) = B(y,x) \tag{B4.2}$$

$$B(x,y) = \frac{\Gamma(x)\Gamma(y)}{\Gamma(x+y)} \tag{B4.3}$$

$$B(x,x) = 2\frac{B\left(\frac{1}{2},x\right)}{2^{2x}} \tag{B4.4}$$

$$B(x,x)\,B\left(x+\tfrac{1}{2},x+\tfrac{1}{2}\right) = \frac{2\pi}{x \cdot 2^{4x}} \tag{B4.5}$$

Integral Representations:

$$B(x,y) = \int_0^{\infty} \frac{t^{x-1}\,dt}{(1+t)^{x+y}} \tag{B4.6}$$

$$B(x,y) = 2\int_0^{\pi/2} (\sin t)^{2x-1}(\cos t)^{2y-1}\,dt \tag{B4.7}$$

$$B(x,y) = \int_1^{\infty} \frac{t^x + t^y}{t(1+t)^{x+y}}\,dt \tag{B4.8}$$

$$B(x,y) = 2\int_0^{\infty} \frac{t^{2x-1}}{(1+t^2)^{x+y}}\,dt \tag{B4.9}$$

B.5 Error Function erf(x)

Definitions:

$$\operatorname{erf}(x) = \frac{2}{\sqrt{\pi}} \int_0^x e^{-t^2} dt \qquad \text{(Error Function)} \qquad \text{(B5.1)}$$

$$\operatorname{erfc}(x) = 1 - \operatorname{erf}(x) =$$

$$= \frac{2}{\sqrt{\pi}} \int_x^\infty e^{-t^2} dt \qquad \text{(Complementary Error Function)} \qquad \text{(B5.2)}$$

$$w(x) = e^{-x^2} \operatorname{erfc}(-ix) \qquad \text{(Gautschi Function)}$$

$$= e^{-x^2} \left[1 + \frac{2i}{\sqrt{\pi}} \int_0^x e^{t^2} dt \right] \qquad \text{(B5.3)}$$

Series Representations:

$$\operatorname{erf}(x) = \frac{2}{\sqrt{\pi}} \sum_{n=0}^{\infty} \frac{(-1)^n x^{2n+1}}{(2n+1)\, n!} \qquad \text{(B5.4)}$$

$$\operatorname{erf}(x) = \frac{2}{\sqrt{\pi}} e^{-x^2} \sum_{n=0}^{\infty} \frac{2^n x^{2n+1}}{(2n+1)!!} \qquad \text{(B5.5)}$$

$$w(x) = \sum_{n=0}^{\infty} \frac{(ix)^n}{\Gamma(n/2+1)} \qquad \text{(B5.6)}$$

Useful Formulae:

$$\operatorname{erf}(-x) = -\operatorname{erf}(x) \qquad \text{(B5.7)}$$

$$w(-x) = 2e^{-x^2} - w(x) \qquad \text{(B5.8)}$$

$$\operatorname{erf}(x) = \frac{1}{\sqrt{\pi}} \gamma\left(\frac{1}{2}, x^2\right) \qquad \text{(B5.9)}$$

$$\operatorname{erfc}(x) = \frac{1}{\sqrt{\pi}} \Gamma\left(\frac{1}{2}, x^2\right) \qquad \text{(B5.10)}$$

Derivative Formulae:

$$\left[\operatorname{erfc}(x)\right]^{(n+1)} = \frac{2}{\sqrt{\pi}} (-1)^n e^{-x^2} H_n(x) \qquad \text{(B5.11)}$$

$$\frac{d}{dx}\{\mathrm{erf}(x)\} = \frac{2e^{-x^2}}{\sqrt{\pi}} \qquad \text{(B5.12)}$$

$$w^{(n)}(x) = -2xw^{(n-1)} - 2(n-1)w^{(n-2)} \qquad n = 2, 3, \ldots \qquad \text{(B5.13)}$$

$$w^{(0)}(x) = w(x) \qquad w'(x) = \frac{dw}{dx} = -2x\,w(x) + \frac{2i}{\sqrt{\pi}} \qquad \text{(B5.14)}$$

$$\frac{d}{dx}\{\mathrm{erfc}(x)\} = -\frac{2e^{-x^2}}{\sqrt{\pi}} \qquad \text{(B5.15)}$$

Integral Formulae:

$$\int \mathrm{erf}(x)\,dx = x\,\mathrm{erf}(x) + \frac{e^{-x^2}}{\sqrt{\pi}} \qquad \text{(B5.16)}$$

$$\int \exp[-(a^2t^2 + 2bt + c)]dt = \frac{\sqrt{\pi}}{2a}\exp\left[\frac{b^2}{a^2} - c\right]\mathrm{erf}(at + b/a) \qquad \text{(B5.17)}$$

$$\int e^{at}\,\mathrm{erf}(bt)\,dt = \frac{1}{a}\left[e^{at}\,\mathrm{erf}(bt) - e^{a^2/4b^2}\,\mathrm{erf}(bt - a/4b)\right] \qquad \text{(B5.18)}$$

$$\int e^{-(at)^2}\,e^{-(b/t)^2}\,dt = \frac{\sqrt{\pi}}{4a}\left[e^{2ab}\,\mathrm{erf}(at + b/t) + e^{-2ab}\mathrm{erf}(at - b/t)\right] \qquad \text{(B5.19)}$$

$$\int_0^\infty \exp[-(a^2t^2 + 2bt + c)]\,dt = \frac{\sqrt{\pi}}{2a}\exp\left[\frac{b^2}{a^2} - c\right]\mathrm{erf}(b/a) \qquad \text{(B5.20)}$$

$$\int_0^\infty \frac{e^{-a^2t}\,dt}{\sqrt{t + x^2}} = \frac{\sqrt{\pi}}{a}\,e^{a^2x^2}\,\mathrm{erfc}(ax) \qquad \text{(B5.21)}$$

$$\int_0^\infty \frac{e^{-a^2t^2}\,dt}{t^2 + x^2} = \frac{\pi}{2x}\,e^{a^2x^2}\,\mathrm{erfc}(ax) \qquad \text{(B5.22)}$$

$$\int_0^\infty e^{-at}\mathrm{erf}(bt)\,dt = \frac{1}{a}\,e^{a^2/4b^2}\,\mathrm{erfc}(a/2b) \qquad \text{(B5.23)}$$

$$\int_0^\infty e^{-at}\mathrm{erf}(b\sqrt{t})\,dt = \frac{b}{a\sqrt{a + b^2}} \qquad \text{(B5.24)}$$

$$\int_0^\infty e^{-at}\mathrm{erf}(b/\sqrt{t})\,dt = \frac{1}{a}\,e^{-2b\sqrt{a}} \qquad \text{(B5.25)}$$

Asymptotic Series:

$$\operatorname{erfc}(x) \sim \frac{e^{-x^2}}{x\sqrt{\pi}} \left\{ 1 + \sum_{m=1}^{\infty} \frac{(-1)^m (2m-1)!!}{4^m x^{2m}} \right\} \tag{B5.26}$$

B.6 Fresnel Functions C(x), S(x), and F(x)

Definitions:

$$C(x) = \int_0^x \cos(\pi t^2/2)\,dt \qquad \text{(Fresnel Cosine Function)} \tag{B6.1}$$

$$S(x) = \int_0^x \sin(\pi t^2/2)\,dt \qquad \text{(Fresnel Sine Function)} \tag{B6.2}$$

$$F(x) = \int_0^x \exp(i\pi t^2/2)\,dt \qquad \text{(Fresnel Function)} \tag{B6.3}$$

$$C^*(x) = \sqrt{\frac{2}{\pi}} \int_0^x \cos(t^2)\,dt \tag{B6.4}$$

$$S^*(x) = \sqrt{\frac{2}{\pi}} \int_0^x \sin(t^2)\,dt \tag{B6.5}$$

$$F^*(x) = \sqrt{\frac{2}{\pi}} \int_0^x \exp(it^2)\,dt \tag{B6.6}$$

Series Representations:

$$C(x) = \sum_{n=0}^{\infty} \frac{(-1)^n (\pi/2)^{2n}}{(4n+1)(2n)!} x^{4n+1} \tag{B6.7}$$

$$S(x) = \sum_{n=0}^{\infty} \frac{(-1)^n (\pi/2)^{2n+1}}{(4n+3)(2n+1)!} x^{4n+3} \tag{B6.8}$$

Useful Formulae:

$$C(x) = C^*(x\sqrt{\pi/2}) \qquad S(x) = S^*(x\sqrt{\pi/2}) \qquad F(x) = F^*(x\sqrt{\pi/2})$$

$$C(x) = -C(-x) \qquad S(x) = -S(-x) \tag{B6.9}$$

$$C(ix) = i\,C(x) \qquad S(ix) = i\,S(x)$$

$$F(x) = \frac{1}{\sqrt{2}} e^{i\pi/4} \operatorname{erf}(\sqrt{\pi/2}\, e^{-i\pi/4} x) \tag{B6.10}$$

Special Values:

$$C(0) = 0 \qquad S(0) = 0$$

$$\lim_{x\to\infty} C(x) = \lim_{x\to\infty} S(x) = \frac{1}{2} \tag{B6.11}$$

Asymptotic Series:

$$C(x) = \frac{1}{2} + f(x)\sin(\pi x^2/2) - g(x)\cos(\pi x^2/2) \tag{B6.12}$$

$$S(x) = \frac{1}{2} - f(x)\cos(\pi x^2/2) - g(x)\sin(\pi x^2/2) \tag{B6.13}$$

$$f(x) \sim \frac{1}{\pi x}\left[1 + \sum_{m=1}^{\infty} \frac{(-1)^m \{1\cdot 3\cdot 7\cdot \ldots \cdot (4m-1)\}}{(\pi x^2)^{2m}}\right] \tag{B6.14}$$

$|x| >> 1 \qquad |\arg x| < \pi/2$

$$g(x) \sim \frac{1}{\pi x} \sum_{m=0}^{\infty} \frac{(-1)^m \{1\cdot 5\cdot 9\cdot \ldots \cdot (4m+1)\}}{(\pi x^2)^{2m+1}} \tag{B6.15}$$

$|x| >> 1 \qquad |\arg x| < \pi/2$

Integral Formulae:

$$\int S(x)\,dx = x\,S(x) + \frac{1}{\pi}\cos(\pi x^2/2) \tag{B6.16}$$

$$\int C(x)\,dx = x\,C(x) - \frac{1}{\pi}\sin(\pi x^2/2) \tag{B6.17}$$

$$\begin{aligned}\int \cos(a^2x^2 + 2bx + c)\,dx = {} & \frac{\sqrt{\pi}}{a\sqrt{2}}\cos(b^2/a^2 - c)\,C[\sqrt{2/\pi}(ax + b/a)] \\ & + \frac{\sqrt{\pi}}{a\sqrt{2}}\sin(b^2/a^2 - c)\,S[\sqrt{2/\pi}(ax + b/a)]\end{aligned} \tag{B6.18}$$

$$\begin{aligned}\int \sin(a^2x^2 + 2bx + c)\,dx = {} & \frac{\sqrt{\pi}}{a\sqrt{2}}\cos(b^2/a^2 - c)\,S[\sqrt{2/\pi}(ax + b/a)] \\ & - \frac{\sqrt{\pi}}{a\sqrt{2}}\sin(b^2/a^2 - c)\,C[\sqrt{2/\pi}(ax + b/a)]\end{aligned} \tag{B6.19}$$

$$\int_0^\infty e^{-at}\cos(t^2)\,dt = \frac{\sqrt{\pi}}{\sqrt{2}}\cos(a^2/4)\left\{\frac{1}{2} - S\left[\frac{a}{\sqrt{2\pi}}\right]\right\} - \frac{\sqrt{\pi}}{\sqrt{2}}\sin(a^2/4)\left\{\frac{1}{2} - C\left[\frac{a}{\sqrt{2\pi}}\right]\right\} \tag{B6.20}$$

$$\int_0^\infty e^{-at}\sin(t^2)\,dt = \frac{\sqrt{\pi}}{\sqrt{2}}\cos(a^2/4)\left\{\frac{1}{2} - C\left[\frac{a}{\sqrt{2\pi}}\right]\right\} + \frac{\sqrt{\pi}}{\sqrt{2}}\sin(a^2/4)\left\{\frac{1}{2} - S\left[\frac{a}{\sqrt{2\pi}}\right]\right\} \tag{B6.21}$$

$$\int_0^\infty e^{-at}C(t)\,dt = \frac{1}{a}\left\{\cos(a^2/2\pi)\left\{\frac{1}{2} - S\left[\frac{a}{\pi}\right]\right\} - \sin(a^2/2\pi)\left\{\frac{1}{2} - C\left[\frac{a}{\pi}\right]\right\}\right\} \tag{B6.22}$$

$$\int_0^\infty e^{-at}S(t)\,dt = \frac{1}{a}\left\{\cos(a^2/2\pi)\left\{\frac{1}{2} - C\left[\frac{a}{\pi}\right]\right\} + \sin(a^2/2\pi)\left\{\frac{1}{2} - S\left[\frac{a}{\pi}\right]\right\}\right\} \tag{B6.23}$$

B.7 Exponential Integrals Ei(x) and $E_n(x)$

Definition:

$$\mathrm{Ei}(x) = -\,\mathrm{P.V.}\int_{-x}^{\infty}\frac{e^{-t}}{t}\,dt = \mathrm{P.V.}\int_{-\infty}^{x}\frac{e^{-t}}{t}\,dt \tag{B7.1}$$

$$E_n(x) = \int_1^\infty \frac{e^{-xt}}{t^n}\,dt \tag{B7.2}$$

$$E_1(x) = \int_1^\infty \frac{e^{-xt}}{t}\,dt \tag{B7.3}$$

Series Representation:

$$\mathrm{Ei}(x) = \gamma + \log(|x|) + \sum_{k=1}^{\infty}\frac{x^k}{k\cdot k!} \tag{B7.4}$$

$$\mathrm{Ei}(x) - \mathrm{Ei}(-x) = 2\sum_{k=0}^{\infty}\frac{x^{2k+1}}{(2k+1)\cdot(2k+1)!} \qquad x > 0 \tag{B7.5}$$

$$E_1(x) = -\gamma - \log(x) - \sum_{k=0}^{\infty} \frac{(-1)^k x^k}{k \cdot k!} \tag{B7.6}$$

$$E_n(x) = (-1)^n x^n[-\log(x) + \psi(n)] - \sum_{k=0,2,4,\ldots}^{\infty} \frac{(-1)^k x^k}{(k-n+1)\cdot k!} \qquad k \neq n-1 \tag{B7.7}$$

Recurrence Formulae:

$$E_{n+1}(x) = \frac{1}{n}\left[e^{-x} - x\,E_n(x)\right] \qquad n = 1, 2, 3, \ldots \tag{B7.8}$$

$$E'_n(x) = -E_{n-1}(x) \qquad n = 1, 2, 3, \ldots \tag{B7.9}$$

Special Values:

$$E_n(0) = \frac{1}{n-1} \qquad n \geq 2$$

$$E_0(x) = \frac{e^{-x}}{x} \tag{B7.10}$$

Asymptotic Series:

$$Ei(x) \sim e^x \sum_{n=0}^{\infty} \frac{n!}{x^{n+1}} \qquad x >> 1 \tag{B7.11}$$

$$E_1(x) \sim e^{-x} \sum_{n=0}^{\infty} \frac{(-1)^n n!}{x^{n+1}} \qquad x >> 1 \tag{B7.12}$$

$$E_n(x) \sim \frac{e^{-x}}{x}\left\{1 - \frac{n}{x} + \frac{n(n+1)}{x^2} - \frac{n(n+1)(n+2)}{x^3} + \ldots\right\} \qquad x >> 1 \tag{B7.13}$$

Integral Formulae:

$$Ei(x) = -e^{-x} \int_0^{\infty} \frac{t\cos(t) + x\sin(t)}{x^2 + t^2}\,dt \qquad x > 0$$

$$= -e^{-x} \int_0^{\infty} \frac{t\cos(t) - x\sin(t)}{x^2 + t^2}\,dt \qquad x < 0 \tag{B7.14}$$

$$E_1(x) = e^{-x} \int_0^{\infty} \frac{e^{-t}}{t + x}\,dt \qquad x > 0 \tag{B7.15}$$

$$E_1(x) = e^{-x} \int_0^{\infty} \frac{t - ix}{t^2 + x^2}\,e^{it}\,dt \qquad x > 0 \tag{B7.16}$$

$$\int_0^\infty E_n(t)\, e^{-xt}\, dt = \frac{(-1)^{n-1}}{x^n}\left[\log(x+1) + \sum_{k=1}^{n-1} \frac{(-x)^k}{k}\right] \qquad x > -1 \tag{B7.17}$$

$$\int_0^x Ei(-t)\, e^t\, dt = -\log(x) - \gamma + e^x Ei(-x) \tag{B7.18}$$

$$\int_0^x Ei(-at)\, e^{-bt}\, dt = -\frac{1}{b}\left\{e^{-bx} Ei(-ax) - Ei(-x(a+b)) + \log(1+b/a)\right\} \tag{B7.20}$$

B.8 Sine and Cosine Integrals Si(x) and Ci(x)

Definitions:

$$Si(x) = \int_0^x \frac{\sin(t)}{t}\, dt \tag{B8.1}$$

$$Ci(x) = \gamma + \log(x) + \int_0^x \frac{\cos(t) - 1}{t}\, dt \tag{B8.2}$$

$$si(x) = Si(x) - \pi/2 \tag{B8.3}$$

Series Representations:

$$Si(x) = \sum_{n=0}^{\infty} \frac{(-1)^n x^{2n+1}}{(2n+1)(2n+1)!} \tag{B8.4}$$

$$Ci(x) = \gamma + \log(x) + \sum_{n=1}^{\infty} \frac{(-1)^n x^{2n}}{(2n)(2n)!} \tag{B8.5}$$

Useful Formulae:

$$Si(-x) = -Si(x)$$
$$Ci(-x) = Ci(x) - i\pi$$
$$si(x) + si(-x) = -\pi$$
$$Ci(x) - Ci(x \exp[i\pi]) = Ei(-i\pi)$$
$$Ci(x) - i\, si(x) = Ei(ix) \tag{B8.6}$$

Special Values:

$si(\infty) = 0$ $\quad$ $Si(\infty) = \pi/2$ $\quad$ $Ci(\infty) = 0$

$Si(0) = 0$ $\quad$ $Ci(0) = -\infty$ (B8.7)

Asymptotic Series:

$$\mathrm{Si}(x) = \frac{\pi}{2} - f(x)\cos(x) - g(x)\sin(x) \qquad \text{(B8.8)}$$

$$\mathrm{Ci}(x) = f(x)\sin(x) - g(x)\cos(x) \qquad \text{(B8.9)}$$

$$f(x) \sim \sum_{n=0}^{\infty} \frac{(-1)^n (2n)!}{x^{2n+1}} \qquad |x| >> 1 \qquad |\arg x| < \pi \qquad \text{(B8.10)}$$

$$g(x) \sim \sum_{n=0}^{\infty} \frac{(-1)^n (2n+1)!}{x^{2n+2}} \qquad |x| >> 1 \qquad |\arg x| < \pi/2 \qquad \text{(B8.11)}$$

Integral Formulae:

$$\mathrm{si}(x) = -\int_x^{\infty} \frac{\sin(t)}{t}\,dt \qquad \text{(B8.12)}$$

$$\mathrm{Ci}(x) = -\int_x^{\infty} \frac{\cos(t)}{t}\,dt \qquad \text{(B8.13)}$$

$$\int_0^{\infty} \mathrm{Ci}(t)\, e^{-xt}\,dt = \frac{1}{2x}\log(1+x^2) \qquad \text{(B8.14)}$$

$$\int_0^{\infty} \mathrm{si}(t)\, e^{-xt}\,dt = -\frac{1}{x}\arctan(x) \qquad \text{(B8.15)}$$

$$\int_0^{\infty} \mathrm{Ci}(t)\cos(t)\,dt = -\frac{\pi}{4} \qquad \text{(B8.16)}$$

$$\int_0^{\infty} \mathrm{si}(t)\sin(t)\,dt = -\frac{\pi}{4} \qquad \text{(B8.17)}$$

$$\int \mathrm{Ci}(bx)\cos(ax)\,dx = \frac{1}{2a}\left[2\sin(ax)\,\mathrm{Ci}(bx) - \mathrm{si}(ax+bx) - \mathrm{si}(ax-bx)\right] \qquad \text{(B8.18)}$$

$$\int \mathrm{Ci}(bx)\sin(ax)\,dx = -\frac{1}{2a}\left[2\cos(ax)\,\mathrm{Ci}(bx) - \mathrm{Ci}(ax+bx) - \mathrm{Ci}(ax-bx)\right] \qquad \text{(B8.19)}$$

$$\int_0^{\infty} \mathrm{Ci}^2(t)\,dt = \frac{\pi}{2} \qquad \text{(B8.20)}$$

$$\int_0^\infty si^2(t)dt = \frac{\pi}{2} \tag{B8.21}$$

$$\int_0^\infty Ci(t)si(t)dt = \log 2 \tag{B8.22}$$

B.9 Tchebyshev Polynomials $T_n(x)$ and $U_n(x)$

Series Representation:

$$T_n(x) = \frac{n}{2}\sum_{m=0}^{[n/2]} \frac{(-1)^m(n-m-1)!}{m!(n-2m)!}(2x)^{n-2m} \qquad n \geq 1 \tag{B9.1}$$

which is the Tchebyshev Polynomial of the first kind. The [n/2] denotes the largest integer which is less than (n/2).

$$U_n(x) = \sum_{m=0}^{[n/2]} \frac{(-1)^m(n-m)!}{m!(n-2m)!}(2x)^{n-2m} \qquad n \geq 1 \tag{B9.2}$$

which is the Tchebyshev functions of the second kind.

Differential Equations:

$$(1-x^2)T_n''(x) - xT_n'(x) + n^2 T_n(x) = 0 \tag{B9.3}$$

$$(1-x^2)U_n''(x) - 3xU_n'(x) + n(n+2)U_n(x) = 0 \tag{B9.4}$$

Recurrence Formulae:

$$T_{n+1}(x) = 2xT_n(x) - T_{n-1}(x) \tag{B9.5}$$

$$U_{n+1}(x) = 2xU_n(x) - U_{n-1}(x) \tag{B9.6}$$

$$(1-x^2)T_n'(x) = -n\,x\,T_n(x) + n\,T_{n-1}(x) \tag{B9.7}$$

$$(1-x^2)U_n'(x) = -n\,xU_n(x) + (n+1)U_{n-1}(x) \tag{B9.8}$$

Orthogonality:

$$\int_{-1}^{1} (1-x^2)^{-1/2}\, T_n(x)T_m(x)\,dx = \begin{cases} 0 & n \neq m \\ \varepsilon_n \pi / 2 & n = m \end{cases} \tag{B9.9}$$

$$\int_{-1}^{1}(1-x^2)^{1/2}U_n(x)U_m(x)\,dx = \begin{cases}0 & n \neq m \\ \pi/2 & n = m\end{cases} \tag{B9.10}$$

Special Values:

$T_n(-x) = (-1)^n T_n(x)$

$T_0(x) = 1 \quad T_1(x) = x \quad T_2(x) = 2x^2 - 1 \quad T_3(x) = 4x^3 - 3x$

$T_n(1) = 1 \quad T_n(-1) = (-1)^n \quad T_{2n}(0) = (-1)^n \quad T_{2n+1}(0) = 0$

$U_n(-x) = (-1)^n U_n(x)$

$U_0(x) = 1 \quad U_1(x) = 2x \quad U_2 = 4x^2 - 1 \quad U_3(x) = 8x^3 - 4x$

$U_n(1) = n+1 \quad U_{2n}(0) = (-1)^n \quad U_{2n+1}(0) = 0$ (B9.11)

Other forms:

$x = \cos\theta$

$$\frac{d^2y}{d\theta^2} + n^2y = 0$$

$T_n(\cos\theta) = \cos(n\theta)$

$$U_n(\cos\theta) = \frac{\sin[(n+1)\theta]}{\sin\theta} \tag{B9.12}$$

Relationship to other functions:

$$T_{n+1}(x) = x\,U_n(x) - U_{n-1}(x) = \frac{1}{2}[U_{n+1}(x) - U_{n-1}(x)] \tag{B9.13}$$

$$U_n(x) = \frac{1}{1-x^2}[x\,T_{n+1}(x) - T_{n+2}(x)] \tag{B9.14}$$

B.10 Laguerre Polynomials $L_n(x)$

Series Representation:

$$L_n(x) = n!\sum_{m=0}^{n}\frac{(-1)^m x^m}{(m!)^2(n-m)!} \tag{B10.1}$$

Differential Equation:

$$xy'' + (1-x)y' + ny = 0 \tag{B10.2}$$

Recurrence Formulae:

$$(n+1)L_{n+1}(x) = (1+2n-x)L_n(x) - nL_{n-1}(x) \tag{B10.3}$$

$$xL'_n(x) = n\left[L_n(x) - L_{n-1}(x)\right] \tag{B10.4}$$

Orthogonality:

$$\int_0^\infty e^{-x} L_n(x) L_m(x)\,dx = \begin{cases} 0 & n \neq m \\ 1 & n = m \end{cases} \tag{B10.5}$$

Special Values:

$$L_n(0) = 1 \qquad L'_n(0) = -n$$

$$L_0(x) = 1 \qquad L_1(x) = 1 - x \qquad L_2(x) = \frac{1}{2}(x^2 - 4x + 2)$$

$$L_3(x) = -\frac{1}{6}(x^3 - 9x^2 + 18x - 6) \tag{B10.6}$$

Integral Formulae:

$$\int_0^\infty e^{-x} x^m L_n(x)\,dx = (-1)^n n!\,\delta_{nm} \tag{B10.7}$$

$$\int_x^\infty e^{-t} L_n(t)\,dt = e^{-x}\left[L_n(x) - L_{n-1}(x)\right] \tag{B10.8}$$

$$\int_0^\infty e^{-xt} L_n(t)\,dt = \frac{(x-1)^n}{x^{n+1}} \qquad x > 0 \tag{B10.9}$$

B.11 Associated Laguerre Polynomials $L_n^m(x)$

Series Representation:

$$L_n^m(x) = (n+m)! \sum_{k=0}^{n} \frac{(-1)^k x^k}{(n-k)!(m+k)!\,k!} \qquad n, m = 0, 1, 2, \ldots \tag{B11.1}$$

$$L_n^m(x) = (-1)^m \frac{d^m L_{n+m}(x)}{dx^m} \tag{B11.2}$$

Differential Equation:

$$xy'' + (m + 1 - x)y' + ny = 0 \tag{B11.3}$$

Recurrence Formulae:

$$(n+1)L_{n+1}^m(x) = (1 + 2n + m - x)L_n^m(x) - (n+m)L_{n-1}^m(x) \tag{B11.4}$$

$$x\,(L_n^m)'(x) = nL_n^m(x) - (n+m)L_{n-1}^m(x) \qquad \text{(B11.5)}$$

$$xL_n^{m+1}(x) = (x-n)L_n^m(x) + (n+m)L_{n-1}^m(x) \qquad \text{(B11.6)}$$

Orthogonality:

$$\int_0^\infty e^{-x}x^m L_n^m(x)L_k^m(x)\,dx = \frac{(n+m)!}{n!}\begin{cases}0 & k \neq m\\ 1 & k = m\end{cases} \qquad \text{(B11.7)}$$

If m is not an integer, i.e., m = ν > -1, then the formulae given above are correct provided one substitutes ν for m and Γ (ν + n +1) instead of (m + n)! where n is an integer.

Special Values:

$$L_n^m(0) = \frac{(n+m)!}{m!\,n!} \qquad \text{(B11.8)}$$

Integral Formulae:

$$\int_x^\infty e^{-u}\,L_n^m(u)\,du = e^{-x}\left[L_n^m(x) - L_{n-1}^m(x)\right] \qquad \text{(B11.9)}$$

$$\int_0^\infty e^{-x}\,x^{\nu+1}\left[L_n^\nu(x)\right]^2 dx = \frac{2n+\nu+1}{n!}\Gamma(n+\nu+1) \qquad \nu > -1 \qquad \text{(B11.10)}$$

$$\int_0^x t^\nu\,(x-t)^a\,L_n^\nu(t)\,dt = \frac{\Gamma(n+\nu+1)\Gamma(a+1)}{\Gamma(n+\nu+a+2)}\,x^{\nu+a+1}L_n^{\nu+a+1}(x) \qquad \nu,a > -1 \quad \text{(B11.11)}$$

B.12 Hermite Polynomials $H_n(x)$

Series Representation:

$$H_n(x) = n!\sum_{m=0}^{[n/2]}\frac{(-1)^m}{m!(n-2m)!}(2x)^{n-2m} \qquad \text{(B12.1)}$$

Differential Equation:

$$y'' - 2xy' + 2ny = 0 \qquad \text{(B12.2)}$$

Recurrence Formulae:

$$H_{n+1}(x) = 2xH_n(x) - 2nH_{n-1}(x) \qquad \text{(B12.3)}$$

$$H_n'(x) = 2n\,H_{n-1}(x) \qquad \text{(B12.4)}$$

Orthogonality:

$$\int_{-\infty}^{\infty} e^{-x^2} H_n(x) H_m(x)\,dx = \begin{cases} 0 & n \neq m \\ 2^n n! \sqrt{\pi} & n = m \end{cases} \tag{B12.5}$$

Special Values:

$$H_0(x) = 1 \qquad H_1(x) = 2x \qquad H_2(x) = 4x^2 - 2 \qquad H_3(x) = 8x^3 - 12x$$

$$H_n(-x) = (-1)^n H_n(x) \qquad H_{2n}(0) = (-1)^n \frac{(2n)!}{n!} \qquad H_{2n+1}(0) = 0 \tag{B12.6}$$

Integral Formulae:

$$H_n(x) = \frac{e^{x^2} 2^{n+1}}{\sqrt{\pi}} \int_0^{\infty} e^{-t^2} t^n \cos(2xt - \frac{n\pi}{2})\,dt \tag{B12.7}$$

$$\int_{-\infty}^{\infty} x^m e^{-x^2} H_n(x)\,dx = \begin{cases} 0 & m \leq n-1 \\ n!\sqrt{\pi} & m = n \end{cases} \tag{B12.8}$$

$$\int_{-\infty}^{\infty} e^{-t^2/2} e^{ixt} H_n(t)\,dt = \sqrt{2\pi}\, i^n\, e^{-x^2/2} H_n(x) \tag{B12.9}$$

$$\int_0^{\infty} e^{-t^2} \cos(xt) H_{2n}(t)\,dt = (-1)^n \frac{\sqrt{\pi}}{2} x^{2n} e^{-x^2/4} \tag{B12.10}$$

$$\int_0^{\infty} e^{-t^2} \sin(xt) H_{2n+1}(t)\,dt = (-1)^n \frac{\sqrt{\pi}}{2} x^{2n+1} e^{-x^2/4} \tag{B12.11}$$

$$\int_0^{x} e^{-t^2} H_n(t)\,dt = -e^{-x^2} H_{n-1}(x) + H_{n-1}(0) \tag{B12.12}$$

$$\int_0^{x} H_n(t)\,dt = \frac{1}{2n+2} \left[H_{n+1}(x) - H_{n+1}(0) \right] \tag{B12.13}$$

Relation to Other Functions:

$$H_{2n}(\sqrt{x}) = (-1)^n 2^{2n} (n!) L_n^{(-1/2)}(x) \tag{B12.14}$$

$$H_{2n+1}(\sqrt{x}) = (-1)^n 2^{2n+1} (n!) L_n^{(1/2)}(x) \tag{B12.15}$$

$$\int_{-\infty}^{\infty} e^{-t^2} t^n H_n(xt)\,dt = \sqrt{\pi}\, n!\, P_n(x) \tag{B12.16}$$

$$\int_0^\infty e^{-t^2} H_n^2(t)\cos(xt)\,dt = 2^{n-1}\sqrt{\pi}\, n!\, L_n(x^2/2) \qquad \text{(B12.17)}$$

B.13 Hypergeometric Functions F(a, b; c; x)

Definition:

$$F(a,b;c;x) = \frac{\Gamma(c)}{\Gamma(a)\Gamma(b)} \sum_{n=0}^{\infty} \frac{\Gamma(a+n)\Gamma(b+n)}{\Gamma(c+n)} \frac{x^n}{n!} \qquad |x| < 1 \qquad \text{(B13.1)}$$

Differential Equation:

$$x(x-1)y'' + cy' - (a+b+1)xy' - aby = 0 \qquad \text{(B13.2)}$$

$c \neq 0, -1, -2, -3, \ldots$

$$y = C_1 y_1 + C_2 y_2 \qquad \text{(B13.3)}$$

$$y_1 = F(a,b;c;x) = (1-x)^{c-a-b} F(c\text{-}a,\ c\text{-}b;\ c;\ x) \qquad \text{(B13.4)}$$

$$y_2 = x^{1-c} F(1+a-c, 1+b-c; 2-c; x) = x^{1-c}(1-x)^{c-a-b} F(1\text{-}a,\ 1\text{-}b;\ 2\text{-}c;\ x) \qquad \text{(B13.5)}$$

Recurrence Formulae:

$$\begin{aligned} a(x-1)F(a+1,b;c;x) &= [c-2a+ax-bx]F(a,b;c;x) \\ &\quad + [a-c]F(a-1,b;c;x) \end{aligned} \qquad \text{(B13.6)}$$

$$b(x-1)F(a,b+1;c;x) = [c-2b+bx-ax]F(a,b;c;x) + [b-c]F(a,b-1;c;x) \qquad \text{(B13.7)}$$

$$\begin{aligned} (c-a)(c-b)xF(a,b;c+1;x) &= c[1-c+2cx-ax-bx-x]F(a,b;c;x) \\ &\quad + c[c-1][1-x]F(a,b;c-1;x) \end{aligned} \qquad \text{(B13.8)}$$

$$F'(a,b;c;x) = \frac{ab}{c} F(a+1,b+1;c+1;x) \qquad \text{(B13.9)}$$

$$F^{(n)}(a,b;c;x) = \frac{\Gamma(c)\Gamma(a+n)\Gamma(b+n)}{\Gamma(c+n)\Gamma(a)\Gamma(b)} F(a+n,b+n;c+n;x) \qquad \text{(B13.10)}$$

Special Values:

$$F(a,b;b;x) = \frac{1}{1-x}$$

$$F(-m,b;c;x) = \frac{\Gamma(c)}{(m-1)!\Gamma(b)} \sum_{n=0}^{m} \frac{(n-1)!\Gamma(b-m+n)}{\Gamma(c-m+n)} \frac{x^{n-m}}{(n-m)!} \qquad (m \text{ integer} \geq 0)$$

$$F(-m,b;-m-k;x) = \frac{(m-k-1)!}{(m-1)!\Gamma(b)} \sum_{n=0}^{m} \frac{(n-1)!\Gamma(b-m+n)}{(n-k-1)!} \frac{x^{n-m}}{(n-m)!}$$

$(m, n \text{ integer} \geq 0)$

$$F(a,b;c;1)=\frac{\Gamma(c)\Gamma(c-a-b)}{\Gamma(c-a)\Gamma(c-b)} \qquad c \neq 0, -1, -2, \ldots \tag{B13.11}$$

Integral Formulae:

$$F(a,b;c;x)=\frac{\Gamma(c)}{\Gamma(b)\Gamma(c-b)}\int_0^1 t^{b-1}(1-t)^{c-b-1}(1-tz)^{-a}\,dt \tag{B13.12}$$

$$\int_0^1 x^{a-1}(1-x)^{b-c-n}F(-n,b;c;x)\,dx=\frac{\Gamma(c)\Gamma(a)\Gamma(b-c+1)\Gamma(c-a+n)}{\Gamma(c+n)\Gamma(c-a)\Gamma(b-c+a+1)} \tag{B13.13}$$

$$\int_0^\infty F(a,b;c;-x)x^{d-1}\,dx=\frac{\Gamma(c)\Gamma(d)\Gamma(b-d)\Gamma(a-d)}{\Gamma(a)\Gamma(b)\Gamma(c-d)}$$

$$c \neq 0, -1, -2, -3, \ldots \qquad d>0 \qquad a-d>0 \qquad b-d>0 \tag{B13.14}$$

Relationship to Other Functions:

$$F(-n,n;\tfrac{1}{2};x)=T_n(1-2x) \tag{B13.15}$$

$$F(-n,n+1;1;x)=P_n(1-2x) \tag{B13.16}$$

Asymptotic Series:

$$F(a,b;c;x)\sim\frac{\Gamma(c)}{\Gamma(c-a)}e^{-i\pi a}(bx)^{-a}+\frac{\Gamma(c)}{\Gamma(a)}e^{bx}(bx)^{a-c} \qquad bx \gg 1 \tag{B13.17}$$

B.14 Confluent Hypergeometric Functions M(a,c,x) and U(a,c,x)

Definition:

$$M(a,b,x)=\frac{\Gamma(b)}{\Gamma(a)}\sum_{n=0}^{\infty}\frac{\Gamma(a+n)}{\Gamma(b+n)}\frac{x^n}{n!} \tag{B14.1}$$

$$U(a,b,x)=\frac{\pi}{\sin(\pi b)}\left[\frac{M(a,b,x)}{\Gamma(b)\Gamma(1+a-b)}-x^{1-b}\frac{M(1+a-b,2-b,x)}{\Gamma(a)\Gamma(2-b)}\right] \tag{B14.2}$$

Differential Equation:

$$xy''+(b-x)y'-ay=0 \tag{B14.3}$$

$$y=C_1M(a,b,x)+C_2U(u,b,x) \tag{B14.4}$$

Recurrence Formulae:

$$aM(a+1,b,x)=[2a-b+x]M(a,b,x)+[b-a]M(a-1,b,x) \tag{B14.5}$$

$$(a-b)\,x\,M(a,b+1,x)=b[1-b-x]M(a,b,x)+b[b-1]M(a,b-1,x) \tag{B14.6}$$

$$M'(a,b,x) = \frac{a}{b} M(a+1, b+1, x) \tag{B14.7}$$

$$M^{(n)}(a,b,x) = \frac{\Gamma(a+n)\Gamma(b)}{\Gamma(b+n)\Gamma(a)} M(a+n, b+n, x) \tag{B14.8}$$

$$a(b-a-1)U(a+1,b,x) = [-x+b-2a]U(a,b,x)+U(a-1,b,x) \tag{B14.9}$$

$$x\,U(a,b+1,x) = [x+b-1]U(a,b,x)+[1+a-b]U(a,b-1,x) \tag{B14.10}$$

$$U'(a,b,x) = -aU(a+1,b+1,x) \tag{B14.11}$$

$$U^{(k)}(a,b,x) = (-1)^k \frac{\Gamma(a+k)}{\Gamma(a)} U(a+k,b+k,x) \tag{B14.12}$$

Special Values:

$$M(a,a,x) = e^x \qquad M(1,2,-2ix) = \frac{\sin x}{x\,e^{ix}} \qquad M(1,2,2x) = e^x \frac{\sinh x}{x} \tag{B14.13}$$

Integral Formulae:

$$M(a,b,x) = \frac{\Gamma(b)}{\Gamma(a)\Gamma(b-a)} \int_0^1 e^{tx} t^{a-1} (1-t)^{b-a-1}\, dt \tag{B14.14}$$

$$U(a,b,x) = \frac{1}{\Gamma(a)} \int_0^\infty e^{-tx} t^{a-1} (1+t)^{b-a-1}\, dt \tag{B14.15}$$

Relationship to Other Functions:

$$M(p+\frac{1}{2}, 2p+1, 2ix) = \frac{2^p e^{ix}}{x^p} \Gamma(p+1) J_p(x) \tag{B14.16}$$

$$M(p+\frac{1}{2}, 2p+1, 2x) = \frac{2^p e^x}{x^p} \Gamma(p+1) I_p(x) \tag{B14.17}$$

$$M(n+1, 2n+2, 2ix) = \frac{2^{n+1/2} e^{ix}}{x^{n+1/2}} \Gamma(n+3/2) J_{n+1/2}(x) \tag{B14.18}$$

$$M(-n, -2n, 2ix) = \frac{x^{n+1/2} e^{ix}}{2^{n+1/2}} \Gamma(1/2-n) J_{-n-1/2}(x) \tag{B14.19}$$

$$M(-n, m+1, x) = \frac{n!\,m!}{(m+n)!} L_n^m(x) \tag{B14.20}$$

$$M(\frac{1}{2}, \frac{3}{2}, -x^2) = \frac{\sqrt{\pi}}{2x} \operatorname{erf}(x) \tag{B14.21}$$

$$U(p+\frac{1}{2}, 2p+1, 2x) = \frac{e^x}{(2x)^p \sqrt{\pi}} K_p(x) \tag{B14.22}$$

$$U(p+\frac{1}{2},2p+1,-2ix)=\frac{\sqrt{\pi}}{2(2x)^p}\,e^{i[\pi(p+1/2)-x]}\,H_p^{(1)}(x) \tag{B14.23}$$

$$U(p+\frac{1}{2},2p+1,2ix)=\frac{\sqrt{\pi}}{2(2x)^p}\,e^{-i[\pi(p+1/2)-x]}\,H_p^{(2)}(x) \tag{B14.24}$$

$$U(\frac{1}{2},\frac{1}{2},x^2)=\sqrt{\pi}\,e^{x^2}\,\mathrm{erfc}(x) \tag{B14.25}$$

$$U(\frac{1}{2}(1-n),\frac{3}{2},x^2)=\frac{H_n(x)}{2^n x} \tag{B14.26}$$

$$U(\frac{-\nu}{2},\frac{1}{2},\frac{x^2}{2})=2^{-\nu/2}\,e^{x^2/4}\,D_\nu(x) \tag{B14.27}$$

Asymptotic Series:

$$M(a,b,x)\sim\frac{x^{-a}e^{i\pi a}}{\Gamma(b-a)\Gamma(a)\Gamma(a-b+1)}\sum_{n=0}^{\infty}\frac{\Gamma(a+n)\Gamma(a-b+1+n)}{n!(-x)^n}$$
$$+\frac{e^x x^{-b}e^{i\pi a}}{\Gamma^2(b-a)\Gamma(a)\Gamma(a-1)}\sum_{n=0}^{\infty}\frac{\Gamma(b-a+n)\Gamma(1-a+n)}{n!\,x^n} \tag{B14.28}$$

$|x| >> 1$

$$U(a,b,x)\sim\frac{x^{-a}}{\Gamma(a)\Gamma(1+a-b)}\sum_{n=0}^{\infty}\frac{\Gamma(a+n)\Gamma(1+a-b+n)}{n!(-x)^n}\qquad |x|>>1 \tag{B14.29}$$

B.15 Kelvin Functions (ber$_\nu$ (x), bei$_\nu$ (x), ker$_\nu$ (x), kei(x))

Definitions:

$$\begin{aligned}\mathrm{ber}_\nu(x)+i\,\mathrm{bei}_\nu(x)&=J_\nu(x\,e^{3i\pi/4})=e^{i\nu\pi}J_\nu(x\,e^{-i\pi/4})\\&=e^{i\nu\pi/2}I_\nu(x\,e^{i\pi/4})=e^{3i\nu\pi/2}I_\nu(x\,e^{-3i\pi/4})\end{aligned} \tag{B15.1}$$

$$\begin{aligned}\mathrm{ker}_\nu(x)+i\,\mathrm{kei}_\nu(x)&=e^{-i\nu\pi/2}K_\nu(x\,e^{i\pi/4})\\&=\frac{i\pi}{2}H_\nu^{(1)}(x\,e^{3i\pi/4})=-\frac{i\pi}{2}e^{-i\nu\pi}H_\nu^{(2)}(x\,e^{-i\pi/4})\end{aligned} \tag{B15.2}$$

When $\nu = 0$, these equations transform to:

$$\begin{aligned}\mathrm{ber}(x)+i\,\mathrm{bei}(x)&=J_0(x\,e^{3i\pi/4})=J_0(x\,e^{-i\pi/4})\\&=I_0(x\,e^{i\pi/4})=I_0(x\,e^{-3i\pi/4})\end{aligned} \tag{B15.3}$$

$$\begin{aligned}\mathrm{ker}(x)+i\,\mathrm{kei}(x)&=K_0(x\,e^{i\pi/4})\\&=\frac{i\pi}{2}H_0^{(1)}(x\,e^{3i\pi/4})=-\frac{i\pi}{2}H_0^{(2)}(x\,e^{-i\pi/4})\end{aligned} \tag{B15.4}$$

Differential Equations:

(1) $$x^2y'' + xy' - (ix^2 + \nu^2)y = 0 \tag{B15.5}$$

$$\begin{aligned} y_1 &= \text{ber}_\nu(x) + i\,\text{bei}_\nu(x) \quad \text{or} \quad y_1 = \text{ber}_{-\nu}(x) + i\,\text{bei}_{-\nu}(x) \\ y_2 &= \text{ker}_\nu(x) + i\,\text{kei}_\nu(x) \quad \text{or} \quad y_2 = \text{ker}_{-\nu}(x) + i\,\text{kei}_{-\nu}(x) \end{aligned} \tag{B15.6}$$

(2) $$x^4y^{(iv)} + 2x^3y''' - (1+2\nu^2)(x^2y'' - xy') + (\nu^4 - 4\nu^2 + x^4)y = 0 \tag{B15.7}$$

or

$$\begin{aligned} &y_1 = \text{ber}_\nu(x) \quad y_2 = \text{bei}_\nu(x) \quad y_3 = \text{ker}_\nu(x) \quad y_4 = \text{kei}_\nu(x) \\ &y_1 = \text{ber}_{-\nu}(x) \quad y_2 = \text{bei}_{-\nu}(x) \quad y_3 = \text{ker}_{-\nu}(x) \quad y_4 = \text{kei}_{-\nu}(x) \end{aligned} \tag{B15.8}$$

Recurrence Formulae:

$$z_{\nu+1} + z_{\nu-1} = -\frac{\nu\sqrt{2}}{x}(z_\nu - w_\nu) \tag{B15.9}$$

$$z'_\nu = \frac{1}{2\sqrt{2}}(z_{\nu+1} - z_{\nu-1} + w_{\nu+1} - w_{\nu-1}) \tag{B15.10}$$

$$= \frac{\nu}{x}z_\nu + \frac{1}{\sqrt{2}}(z_{\nu+1} + w_{\nu+1}) \tag{B15.11}$$

$$= -\frac{\nu}{x}z_\nu - \frac{1}{\sqrt{2}}(z_{\nu-1} + w_{\nu-1}) \tag{B15.12}$$

where the pair of functions z_ν and w_ν are, respectively:

$$\begin{aligned} z_\nu, w_\nu &= \text{ber}_\nu(x), \text{bei}_\nu(x) \quad \text{or} \quad = \text{ker}_\nu(x), \text{kei}_\nu(x) \\ &\text{or} \quad = \text{bei}_\nu(x), -\text{ber}_\nu(x) \quad \text{or} \quad = \text{kei}_\nu(x), -\text{ker}_\nu(x) \end{aligned}$$

Special relationships

$$\text{ber}_{-\nu}(x) = \cos(\nu\pi)\text{ber}_\nu(x) + \sin(\nu\pi)\text{bei}_\nu(x) + \frac{2}{\pi}\sin(\nu\pi)\,\text{ker}_\nu(x) \tag{B15.13}$$

$$\text{bei}_{-\nu}(x) = -\sin(\nu\pi)\text{ber}_\nu(x) + \cos(\nu\pi)\text{bei}_\nu(x) + \frac{2}{\pi}\sin(\nu\pi)\text{kei}_\nu(x) \tag{B15.14}$$

$$\text{ker}_{-\nu}(x) = \cos(\nu\pi)\,\text{ker}_\nu(x) - \sin(\nu\pi)\text{kei}_\nu(x) \tag{B15.15}$$

$$\text{kei}_{-\nu}(x) = \sin(\nu\pi)\text{ker}_\nu(x) + \cos(\nu\pi)\text{kei}_\nu(x) \tag{B15.16}$$

Series Representation:

$$\text{ber}_\nu(x) = \frac{x^\nu}{2^\nu}\sum_{m=0}^{\infty}\frac{\cos[\pi/4(3\nu+2m)]}{m!\,\Gamma(\nu+m+1)} \tag{B15.17}$$

$$\text{bei}_\nu(x) = \frac{x^\nu}{2^\nu}\sum_{m=0}^{\infty}\frac{\sin[\pi/4(3\nu+2m)]}{m!\,\Gamma(\nu+m+1)} \tag{B15.18}$$

$$\mathrm{ker}_n(x) = \frac{x^n}{2^{n+1}} \sum_{m=0}^{\infty} \frac{g(m+1)+g(n+m+1)}{m!(n+m)!} \cos[\pi/4(3n+2m)] \frac{x^{2m}}{4^m}$$
$$+ \frac{2^{n-1}}{x^n} \sum_{m=0}^{n-1} \frac{(n-m-1)!}{m!} \cos[\pi/4(3n+2m)] \frac{x^{2m}}{4^m} \tag{B15.19}$$
$$+ \log(2/x)\,\mathrm{ber}_n(x) + \frac{\pi}{4} \mathrm{bei}_n(x)$$

$$\mathrm{kei}_n(x) = \frac{x^n}{2^{n+1}} \sum_{m=0}^{\infty} \frac{g(m+1)+g(n+m+1)}{m!(n+m)!} \sin[\pi/4(3n+2m)] \frac{x^{2m}}{4^m}$$
$$- \frac{2^{n-1}}{x^n} \sum_{m=0}^{n-1} \frac{(n-m-1)!}{m!} \sin[\pi/4(3n+2m)] \frac{x^{2m}}{4^m} \tag{B15.20}$$
$$+ \log(2/x)\,\mathrm{bei}_n(x) - \frac{\pi}{4} \mathrm{ber}_n(x)$$

$$\mathrm{ber}(x) = \sum_{m=0}^{\infty} \frac{(-1)^m}{[(2m)!]^2} \frac{x^{4m}}{2^{4m}} \tag{B15.21}$$

$$\mathrm{bei}(x) = \sum_{m=0}^{\infty} \frac{(-1)^m}{[(2m+1)!]^2} \left(\frac{x}{2}\right)^{4m+2} \tag{B15.22}$$

$$\mathrm{ker}(x) = \sum_{m=1}^{\infty} \frac{(-1)^m}{[(2m)!]^2} \frac{x^{4m}}{2^{4m}} g(2m) + [\log(2/x) - \gamma)\,\mathrm{ber}(x) + \frac{\pi}{4} \mathrm{bei}(x) \tag{B15.23}$$

$$\mathrm{kei}(x) = \sum_{m=1}^{\infty} \frac{(-1)^m}{[(2m+1)!]^2} \frac{x^{4m+2}}{2^{4m+2}} g(2m+1)$$
$$+ [\log(2/x) - \gamma)\,\mathrm{bei}(x) - \frac{\pi}{4} \mathrm{ber}(x) \tag{B15.24}$$

Asymptotic Series:

$$\mathrm{ber}_\nu(x) = \frac{e^{x/\sqrt{2}}}{\sqrt{2\pi x}} \{z_\nu(x)\cos a + w_\nu(x)\sin\alpha\}$$
$$- \frac{1}{\pi}\{\sin(2\nu\pi)\,\mathrm{ker}_\nu(x) + \cos(2\nu\pi)\,\mathrm{kei}_\nu(x)\} \tag{B15.25}$$

$$\mathrm{bei}_\nu(x) = \frac{e^{x/\sqrt{2}}}{\sqrt{2\pi x}} \{z_\nu(x)\cos a - w_\nu(x)\sin a\}$$
$$+ \frac{1}{\pi}\{\cos(2\nu\pi)\,\mathrm{ker}_\nu(x) - \sin(2\nu\pi)\,\mathrm{kei}_\nu(x)\} \tag{B15.26}$$

$$\mathrm{ker}_\nu(x) = \frac{\sqrt{\pi}\, e^{-x/\sqrt{2}}}{\sqrt{2x}} \left\{ z_\nu(-x)\cos b - w_\nu(-x)\sin b \right\} \tag{B15.27}$$

$$\mathrm{kei}_\nu(x) = -\frac{\sqrt{\pi}\, e^{-x/\sqrt{2}}}{\sqrt{2x}} \left\{ z_\nu(-x)\sin b + w_\nu(-x)\cos b \right\} \tag{B15.28}$$

$$z_\nu(\mp x) \sim 1 + \sum_{m=1}^{\infty} \frac{(\pm 1)^m \left\{ (c-1)\cdot(c-9)\cdot \ldots \cdot (c-(2m-1)^2) \right\}}{m!(8x)^m} \cos(m\pi/4) \tag{B15.29}$$

$$w_\nu(\mp x) \sim \sum_{m=1}^{\infty} \frac{(\pm 1)^m \left\{ (c-1)\cdot(c-9)\cdot \ldots \cdot (c-(2m-1)^2) \right\}}{m!(8x)^m} \sin(m\pi/4) \tag{B15.30}$$

where $a = \frac{x}{\sqrt{2}} + \frac{\pi}{2}(\nu - 1/4)$, $b = a + \pi/4$, and $c = 4\nu^2$.

Other asymptotic forms for $\nu = 0$:

$$\mathrm{ber}(x) = \frac{e^{\alpha(x)}}{\sqrt{2\pi x}} \cos(\beta(x)) \tag{B15.31}$$

$$\mathrm{bei}(x) = \frac{e^{\alpha(x)}}{\sqrt{2\pi x}} \sin(\beta(x)) \tag{B15.32}$$

$$\mathrm{ker}(x) = \sqrt{\frac{\pi}{2x}}\, e^{\alpha(-x)} \cos(\beta(-x)) \tag{B15.33}$$

$$\mathrm{kei}(x) = \sqrt{\frac{\pi}{2x}}\, e^{\alpha(-x)} \sin(\beta(-x)) \tag{B15.34}$$

where:

$$\alpha(x) \sim \frac{1}{\sqrt{2}} \left\{ x + \frac{1}{8x} - \frac{25}{384x^3} - \frac{13\sqrt{2}}{128x^4} - \ldots \right\} \tag{B15.35}$$

and

$$\beta(x) \sim -\frac{\pi}{8} + \frac{1}{\sqrt{2}} \left\{ x - \frac{1}{8x} - \frac{\sqrt{2}}{16x^2} - \frac{25}{384x^3} + \ldots \right\} \tag{B15.36}$$

APPENDIX C

ORTHOGONAL COORDINATE SYSTEMS

C.1 Introduction

This appendix deals with some of the widely used coordinate systems. It contains expressions for elementary length, area and volume, gradient, divergence, curl, and the Laplacian operator in generalized orthogonal coordinate systems.

C.2 Generalized Orthogonal Coordinate Systems

Consider an orthogonal generalized coordinate (u^1, u^2, u^3), such that an elementary measure of length along each coordinate is given by:

$$\begin{aligned} ds_1 &= \sqrt{g_{11}}\, du^1 \\ ds_2 &= \sqrt{g_{22}}\, du^2 \\ ds_3 &= \sqrt{g_{33}}\, du^3 \end{aligned} \tag{C.1}$$

where g_{11}, g_{22} and g_{33} are called the **metric coefficients**, expressed by:

$$g_{ii} = \left(\frac{\partial x^1}{du^i}\right)^2 + \left(\frac{\partial x^2}{du^i}\right)^2 + \left(\frac{\partial x^3}{du^i}\right)^2 \tag{C.2}$$

and x^i are rectangular coordinates.

An infinitesimal distance ds can be expressed as:

$$(ds)^2 = g_{11}(du^1)^2 + g_{22}(du^2)^2 + g_{33}(du^3)^2 \tag{C.3}$$

An infinitesimal area dA on the u^1u^2 surface can be expressed as:

$$dA = [(g_{11})^{1/2} du^1][(g_{22})^{1/2} du^2] = \sqrt{g_{11}g_{22}}\, du^1 du^2 \tag{C.4}$$

Similarly, an element of volume dV becomes:

$$dV = \sqrt{g_{11}g_{22}g_{33}}\, du^1 du^2 du^3 = \sqrt{g}\, du^1 du^2 du^3 \tag{C.5}$$

where:

$$g = g_{11}g_{22}g_{33} \tag{C.6}$$

A gradient of scalar function ϕ, $\nabla\phi$, is defined as:

$$\nabla\phi = \frac{\vec{e}_1}{\sqrt{g_{11}}}\frac{\partial\phi}{\partial u^1} + \frac{\vec{e}_2}{\sqrt{g_{22}}}\frac{\partial\phi}{\partial u^2} + \frac{\vec{e}_3}{\sqrt{g_{33}}}\frac{\partial\phi}{\partial u^3} \tag{C.7}$$

where $\vec{e}_1$, $\vec{e}_2$ and $\vec{e}_3$ are base vectors along the coordinates u^1, u^2, and u^3 respectively.

The divergence of a vector $\vec{E}$, $\nabla \cdot \vec{E}$, can be expressed as:

$$\nabla\cdot\vec{E}=\frac{1}{\sqrt{g}}\left\{\frac{\partial}{\partial u^1}[\sqrt{g/g_{11}}E_1]+\frac{\partial}{\partial u^2}[\sqrt{g/g_{22}}E_2]+\frac{\partial}{\partial u^3}[\sqrt{g/g_{33}}E_3]\right\} \tag{C.8}$$

where E_1, E_2 and E_3 are the components of the vector $\vec{E}$, i.e.:

$$\vec{E}=E_1\vec{e}_1+E_2\vec{e}_2+E_3\vec{e}_3$$

The curl of a vector $\vec{E}$, $\nabla \times \vec{E}$ is defined as:

$$\begin{aligned}\nabla\times\vec{E}=&\frac{\sqrt{g_{11}}\,\vec{e}_1}{\sqrt{g}}\left\{\frac{\partial}{\partial u^2}[\sqrt{g_{33}}E_3]-\frac{\partial}{\partial u^3}[\sqrt{g_{22}}E_2]\right\}\\&+\frac{\sqrt{g_{22}}\,\vec{e}_2}{\sqrt{g}}\left\{\frac{\partial}{\partial u^3}[\sqrt{g_{11}}E_1]-\frac{\partial}{\partial u^1}[\sqrt{g_{33}}E_3]\right\}\\&+\frac{\sqrt{g_{33}}\,\vec{e}_3}{\sqrt{g}}\left\{\frac{\partial}{\partial u^1}[\sqrt{g_{22}}E_2]-\frac{\partial}{\partial u^2}[\sqrt{g_{11}}E_1]\right\}\end{aligned} \tag{C.9}$$

or

$$\nabla\times\vec{E}=\begin{vmatrix}\sqrt{g_{11}/g}\,\vec{e}_1 & \sqrt{g_{22}/g}\,\vec{e}_2 & \sqrt{g_{33}/g}\,\vec{e}_3\\ \dfrac{\partial}{\partial u^1} & \dfrac{\partial}{\partial u^2} & \dfrac{\partial}{\partial u^3}\\ \sqrt{g_{11}}E_1 & \sqrt{g_{22}}E_2 & \sqrt{g_{33}}E_3\end{vmatrix} \tag{C.10}$$

The Laplacian of a scalar function ϕ, $\nabla^2\phi$, can be written as

$$\nabla^2\phi=\frac{1}{\sqrt{g}}\left\{\frac{\partial}{\partial u^1}\left(\frac{\sqrt{g}}{g_{11}}\frac{\partial\phi}{\partial u^1}\right)+\frac{\partial}{\partial u^2}\left(\frac{\sqrt{g}}{g_{22}}\frac{\partial\phi}{\partial u^2}\right)+\frac{\partial}{\partial u^3}\left(\frac{\sqrt{g}}{g_{33}}\frac{\partial\phi}{\partial u^3}\right)\right\} \tag{C.11}$$

The Laplacian of a vector function $\vec{E}$, denoted as $\nabla^2\vec{E}$ can be written as

$$\begin{aligned}\nabla^2\vec{E}=&\frac{\vec{e}_1}{\sqrt{g_{11}}}\left\{\frac{\partial A}{\partial u^1}+\frac{g_{11}}{\sqrt{g}}\left(\frac{\partial B_2}{\partial u^3}-\frac{\partial B_3}{\partial u^2}\right)\right\}\\&+\frac{\vec{e}_2}{\sqrt{g_{22}}}\left\{\frac{\partial A}{\partial u^2}+\frac{g_{22}}{\sqrt{g}}\left(\frac{\partial B_3}{\partial u^1}-\frac{\partial B_1}{\partial u^3}\right)\right\}\\&+\frac{\vec{e}_3}{\sqrt{g_{33}}}\left\{\frac{\partial A}{\partial u^3}+\frac{g_{33}}{\sqrt{g}}\left(\frac{\partial B_1}{\partial u^2}-\frac{\partial B_2}{\partial u^1}\right)\right\}\end{aligned} \tag{C.12}$$

where:

$$A=\frac{1}{\sqrt{g}}\left\{\frac{\partial}{\partial u^1}\left(\sqrt{g/g_{11}}E_1\right)+\frac{\partial}{\partial u^2}\left(\sqrt{g/g_{22}}E_2\right)+\frac{\partial}{\partial u^3}\left(\sqrt{g/g_{33}}E_3\right)\right\}=\nabla\cdot\vec{E}$$

$$B_1=\frac{g_{11}}{\sqrt{g}}\left\{\frac{\partial}{\partial u^2}\left(\sqrt{g_{33}}E_3\right)-\frac{\partial}{\partial u^3}\left(\sqrt{g_{22}}E_2\right)\right\}$$

$$B_2=\frac{g_{22}}{\sqrt{g}}\left\{\frac{\partial}{\partial u^3}\left(\sqrt{g_{11}}E_1\right)-\frac{\partial}{\partial u^1}\left(\sqrt{g_{33}}E_3\right)\right\}$$

$$B_3 = \frac{g_{33}}{\sqrt{g}}\left\{\frac{\partial}{\partial u^1}\left(\sqrt{g_{22}}\,E_2\right) - \frac{\partial}{\partial u^2}\left(\sqrt{g_{11}}\,E_1\right)\right\}$$

C.3 Cartesian Coordinates

Cartesian coordinate systems are defined as:

$u^1 = x \qquad -\infty < x < +\infty$

$u^2 = y \qquad -\infty < y < +\infty$

$u^3 = z \qquad -\infty < z < +\infty$

The quantities defined in eqs. (C.2) to (C.11) can be listed below:

$g_{11} = g_{22} = g_{33} = 1 \qquad g^{1/2} = 1$

$(ds)^2 = (dx)^2 + (dy)^2 + (dz)^2$

$dV = dx\, dy\, dz$

$$\nabla\phi = \vec{e}_x \frac{\partial\phi}{\partial x} + \vec{e}_y \frac{\partial\phi}{\partial y} + \vec{e}_z \frac{\partial\phi}{\partial z}$$

$$\nabla \cdot \vec{E} = \frac{\partial E_x}{\partial x} + \frac{\partial E_y}{\partial y} + \frac{\partial E_z}{\partial z}$$

$$\nabla \times \vec{E} = \begin{vmatrix} \vec{e}_x & \vec{e}_y & \vec{e}_y \\ \frac{\partial}{\partial x} & \frac{\partial}{\partial y} & \frac{\partial}{\partial z} \\ E_x & E_y & E_z \end{vmatrix}$$

$$\nabla^2\phi = \frac{\partial^2\phi}{\partial x^2} + \frac{\partial^2\phi}{\partial y^2} + \frac{\partial^2\phi}{\partial z^2}$$

C.4 Circular Cylindrical Coordinates

The circular cylindrical coordinates can be given as:

$u^1 = r \qquad 0 \le r < +\infty$

$u^2 = \theta \qquad 0 \le \theta < 2\pi$

$u^3 = z \qquad -\infty < z < +\infty$

where r = constant defines a circular cylinder, θ = constant defines a half plane and z = constant defines a plane.

The coordinate transformation between (r, θ, z) and (x, y, z) are as follows:

$x = r\cos\theta,\qquad y = r\sin\theta,\qquad z = z$

$x^2 + y^2 = r^2,\qquad \tan\theta = y/x$

Expressions corresponding to eqs. (C.2) to (C.11) are given below:

$g_{11} = g_{33} = 1,\qquad g_{22} = r^2\qquad g^{1/2} = r$

$(ds)^2 = (dr)^2 + r^2(d\theta)^2 + (dz)^2$

$dV = r\,dr\,d\theta\,dz$

$$\nabla\phi = \vec{e}_r\frac{\partial\phi}{\partial r} + \frac{1}{r}\vec{e}_\theta\frac{\partial\phi}{\partial\theta} + \vec{e}_z\frac{\partial\phi}{\partial z}$$

$$\nabla\cdot\vec{E} = \frac{\partial E_r}{\partial r} + \frac{1}{r}E_r + \frac{1}{r}\frac{\partial E_\theta}{\partial\theta} + \frac{\partial E_z}{\partial z}$$

$$\nabla\times\vec{E} = \frac{1}{r}\begin{vmatrix} \vec{e}_r & r\vec{e}_\theta & \vec{e}_z \\ \frac{\partial}{\partial r} & \frac{\partial}{\partial\theta} & \frac{\partial}{\partial z} \\ E_r & rE_\theta & E_z \end{vmatrix}$$

$$\nabla^2\phi = \frac{\partial^2\phi}{\partial r^2} + \frac{1}{r}\frac{\partial\phi}{\partial r} + \frac{1}{r^2}\frac{\partial^2\phi}{\partial\theta^2} + \frac{\partial^2\phi}{\partial z^2}$$

C.5 Elliptic-Cylindrical Coordinates

The elliptic-cylindrical coordinates are defined as:

$u^1 = \eta\qquad 0 \le \eta < +\infty$

$u^2 = \psi\qquad 0 \le \psi \le 2\pi$

$u^3 = z\qquad -\infty < z < +\infty$

where η = const. defines an infinite cylinder with an elliptic cross section, ψ = const. defines a hyperbolic surface and z = const. defines a plane. The ellipse has a focal length of 2d.

The coordinate transform between x,y,z and η,ψ and z are written as follows:

$x = d\cosh\eta\cos\psi,\qquad y = d\sinh\eta\sin\psi,\qquad z = z$

$$\frac{x^2}{\cosh^2\eta} + \frac{y^2}{\sinh^2\eta} = d^2,\qquad \frac{x^2}{\cos^2\psi} - \frac{y^2}{\sin^2\psi} = d^2$$

For the equations below let:

$\alpha^2 = \cosh^2\eta - \cos^2\psi$

The quantities given in eqs. (C.2) to (C.11) are defined as follows:

$g_{11} = g_{22} = d^2\,\alpha^2, \qquad g_{33} = 1 \qquad\qquad g^{1/2} = d^2\,\alpha^2$

$dV = d^2\,\alpha^2\,d\eta\,d\psi\,dz$

$(ds)^2 = d^2\,\alpha^2\,[(d\eta)^2 + (d\psi)^2] + (dz)^2$

$$\nabla\phi = \frac{1}{d\,\alpha}\left[\vec{e}_\eta\frac{\partial\phi}{\partial\eta} + \vec{e}_\psi\frac{\partial\phi}{\partial\psi}\right] + \vec{e}_z\frac{\partial\phi}{\partial z}$$

$$\nabla\cdot\vec{E} = \frac{1}{d\alpha}\left\{\frac{\partial}{\partial\eta}(\alpha\,E_\eta) + \frac{\partial}{\partial\psi}(\alpha\,E_\psi)\right\} + \frac{\partial E_z}{\partial z}$$

$$\nabla\times\vec{E} = \frac{1}{\alpha^2}\begin{vmatrix} \alpha\,\vec{e}_\eta & \alpha\,\vec{e}_\psi & \vec{e}_z/d \\ \dfrac{\partial}{\partial\eta} & \dfrac{\partial}{\partial\psi} & \dfrac{\partial}{\partial z} \\ \alpha\,E_\eta & \alpha\,E_\psi & E_z/d \end{vmatrix}$$

$$\nabla^2\phi = \frac{1}{d^2\,\alpha^2}\left[\frac{\partial^2\phi}{\partial\eta^2} + \frac{\partial^2\phi}{\partial\psi^2}\right] + \frac{\partial^2\phi}{\partial z^2}$$

C.6 Spherical Coordinates

The spherical coordinates are defined as follows:

$u^1 = r \qquad 0 \le r < \infty$

$u^2 = \theta \qquad 0 \le \theta \le \pi$

$u^3 = \phi \qquad 0 \le \phi \le 2\pi$

The coordinate transformation between (x,y,z) and (r, θ, ϕ) are given below.

$x = r\sin\theta\cos\phi \qquad y = r\sin\theta\sin\phi \qquad z = r\cos\theta$

$x^2 + y^2 + z^2 = r^2 \qquad z\tan\theta = (x^2 + y^2)^{1/2} \qquad \tan\phi = y/x$

The quantities defined in eqs. (C.2) to (C.11) are given below:

$g_{11} = 1, \qquad g_{22} = r^2, \qquad g_{33} = r^2\sin^2\theta \qquad\qquad g^{1/2} = r^2\sin\theta$

$(ds)^2 = (dr)^2 + r^2\,(d\theta)^2 + r^2\sin^2\theta\,(d\phi)^2$

$dV = r^2\sin\theta\,dr\,d\theta\,d\phi$

$$\nabla\psi = \vec{e}_r\frac{\partial\psi}{\partial r} + \frac{1}{r}\vec{e}_\theta\frac{\partial\psi}{\partial\theta} + \frac{1}{r\sin\theta}\vec{e}_\phi\frac{\partial\psi}{\partial\phi}$$

$$\nabla \cdot \vec{E} = \frac{\partial E_r}{\partial r} + \frac{2}{r} E_r + \frac{1}{r}\frac{\partial E_\theta}{\partial \theta} + \frac{\cot\theta}{r} E_\theta + \frac{1}{r\sin\theta}\frac{\partial E_\phi}{\partial \phi}$$

$$\nabla \times \vec{E} = \frac{1}{r^2 \sin\theta}\begin{vmatrix} \vec{e}_r & r\,\vec{e}_\theta & r\sin\theta\,\vec{e}_\phi \\ \dfrac{\partial}{\partial r} & \dfrac{\partial}{\partial \theta} & \dfrac{\partial}{\partial \phi} \\ E_r & r\,E_\theta & r\sin\theta\,E_\phi \end{vmatrix}$$

$$\nabla^2 \psi = \frac{\partial^2 \psi}{\partial r^2} + \frac{2}{r}\frac{\partial \psi}{\partial r} + \frac{1}{r^2}\frac{\partial^2 \psi}{\partial \theta^2} + \frac{\cot\theta}{r^2}\frac{\partial \psi}{\partial \theta} + \frac{1}{r^2 \sin^2\theta}\frac{\partial^2 \psi}{\partial \phi^2}$$

C.7 Prolate Spheroidal Coordinates

C.7.1 Prolate Spheroidal Coordinates I

The prolate spheroidal coordinates are defined by

$u^1 = \eta \qquad 0 \le \eta < \infty$

$u^2 = \theta \qquad 0 \le \theta \le \pi$

$u^3 = \phi \qquad 0 \le \phi \le 2\pi$

where η = const. defines a rotational elliptical surface, about the z axis, θ = const. defines a rotational hyperbolic surface about the z axis and ϕ = const. defines a half plane. The focal length of the ellipse = 2d.

The coordinate transformation between (x,y,z) and (η, θ, ϕ) are given below:

$x = d \sinh\eta \sin\theta \cos\phi, \qquad y = d \sinh\eta \sin\theta \sin\phi, \qquad z = d \cosh\eta \cos\theta$

$$\frac{x^2 + y^2}{\sinh^2\eta} + \frac{z^2}{\cosh^2\eta} = d^2, \qquad \frac{z^2}{\cos^2\theta} - \frac{(x^2 + y^2)}{\sin^2\theta} = d^2, \qquad \tan\phi = y/x$$

For the equations below let:

$\alpha^2 = \sinh^2\eta + \sin^2\theta, \qquad$ and $\qquad \beta = \sinh\eta \sin\theta$

The quantities defined in eqs. (C.2) to (C.11) are enumerated below:

$g_{11} = g_{22} = d^2 \alpha^2, \qquad g_{33} = d^2 \beta^2$

$g^{1/2} = d^3 \alpha^2 \beta$

$(ds)^2 = d^2 \alpha^2 [(d\eta)^2 + (d\theta)^2] + d^2 \beta^2 (d\phi)^2$

$dV = d^3 \alpha^2 \beta \, d\eta \, d\theta \, d\phi$

$$\nabla\psi = \frac{1}{d\alpha}\left[\vec{e}_\eta \frac{\partial\psi}{\partial\eta} + \vec{e}_\theta \frac{\partial\psi}{\partial\theta}\right] + \frac{1}{d\beta}\vec{e}_\phi \frac{\partial\psi}{\partial\phi}$$

$$\nabla\cdot\vec{E} = \frac{1}{d\alpha}\left\{\frac{1}{\sinh\eta}\frac{\partial}{\partial\eta}[\alpha\sinh\eta\, E_\eta] + \frac{1}{\sin\theta}\frac{\partial}{\partial\theta}[\alpha\sin\theta\, E_\theta]\right\} + \frac{1}{d\beta}\frac{\partial E_\phi}{\partial\phi}$$

$$\nabla x\vec{E} = \frac{1}{d\alpha^2\beta}\begin{vmatrix} \alpha\,\vec{e}_\eta & \alpha\,\vec{e}_\theta & \beta\,\vec{e}_\phi \\ \frac{\partial}{\partial\eta} & \frac{\partial}{\partial\theta} & \frac{\partial}{\partial\phi} \\ \alpha\,E_\eta & \alpha\,E_\theta & \beta\,E_\phi \end{vmatrix}$$

$$\nabla^2\psi = \frac{1}{d^2\alpha^2}\left\{\frac{\partial^2\psi}{\partial\eta^2} + \coth\eta\frac{\partial\psi}{\partial\eta} + \frac{\partial^2\psi}{\partial\theta^2} + \cot\theta\frac{\partial\psi}{\partial\theta}\right\} + \frac{1}{d^2\beta^2}\frac{\partial^2\psi}{\partial\phi^2}$$

C.7.2 Prolate Spheroidal Coordinates II

These are defined as

$u^1 = \xi \qquad 1 \le \xi < \infty$

$u^2 = \eta \qquad -1 \le \eta \le +1$

$u^3 = \phi \qquad 0 \le \phi \le 2\pi$

The coordinate transformation between (x, y, z) and (ξ, η, ϕ) are described below:

$x = d\sqrt{(\xi^2 - 1)(1-\eta^2)}\cos\phi, \qquad y = d\sqrt{(\xi^2 - 1)(1-\eta^2)}\sin\phi, \qquad z = d\,\xi\,\eta$

$\dfrac{x^2+y^2}{1-\xi^2} + \dfrac{z^2}{\xi^2} = d^2, \qquad \dfrac{x^2+y^2}{1-\eta^2} + \dfrac{z^2}{\eta^2} = d^2, \qquad \tan\phi = y/x$

The focal length of the ellipse is 2d.

For the equations below let:

$\alpha^2 = \xi^2 - 1, \qquad \beta^2 = 1 - \eta^2 \qquad$ and $\qquad \chi^2 = \xi^2 - \eta^2$

The quantities defined in eqs. (C.2) to (C.11) are enumerated below:

$g_{11} = (d\,\chi/\alpha)^2 \quad g_{22} = (d\,\chi/\beta)^2, \qquad g_{33} = (d\,\alpha\,\beta)^2$

$g^{1/2} = d^3\,\chi^2$

$$(ds)^2 = d^2\chi^2\left[\frac{(d\xi)^2}{\alpha^2} + \frac{(d\eta)^2}{\beta^2}\right] + d^2\alpha^2\beta^2(d\phi)^2$$

$dV = d^3\,\chi^2\,d\xi\,d\eta\,d\phi$

$$\nabla\psi = \frac{1}{d\chi}\left[\alpha\,\vec{e}_\xi\frac{\partial\psi}{\partial\xi} + \beta^2\,\vec{e}_\eta\frac{\partial\psi}{\partial\eta}\right] + \frac{1}{d\,\alpha\beta}\,\vec{e}_\phi\frac{\partial\psi}{\partial\phi}$$

$$\nabla\cdot\vec{E} = \frac{1}{d\chi^2}\left\{\frac{\partial}{\partial\xi}[\chi\alpha\,E_\xi] + \frac{\partial}{\partial\eta}[\chi\,\beta\,E_\eta] + \frac{\chi^2}{\alpha\beta}\frac{\partial E_\phi}{\partial\phi}\right\}$$

$$\nabla\,x\vec{E} = \frac{1}{d\chi^2}\begin{vmatrix} \frac{\chi}{\alpha}\vec{e}_\xi & \frac{\chi}{\beta}\vec{e}_\eta & \beta\alpha\,\vec{e}_\phi \\ \frac{\partial}{\partial\xi} & \frac{\partial}{\partial\eta} & \frac{\partial}{\partial\phi} \\ \frac{\chi}{\alpha}E_\xi & \frac{\chi}{\beta}E_\eta & \beta\alpha\,E_\phi \end{vmatrix}$$

$$\nabla^2\psi = \frac{1}{d^2\chi^2}\left[\frac{\partial}{\partial\xi}\alpha\frac{\partial}{\partial\xi} + \frac{\partial}{\partial\eta}\beta^2\frac{\partial}{\partial\eta} + \frac{\chi^2}{\alpha^2\,\beta^2}\frac{\partial^2}{\partial\phi^2}\right]\psi$$

C.8 Oblate Spheroidal Coordinates

C.8.1 Oblate Spherical Coordinates I

The oblate spheroidal coordinates are defined by

$u^1 = \eta \qquad 0 \le \eta < \infty$

$u^2 = \theta \qquad 0 \le \theta \le \pi$

$u^3 = \phi \qquad 0 \le \phi \le 2\pi$

where η = const. defines a rotational elliptical surface about the z-axis, θ = const. define a rotational hyperbola about the z-axis and ϕ = const. is a half plane. The focal distance of the ellipse = 2d.

The coordinate transformation between (x, y, z) and (η, θ, ϕ) are as follows:

$x = d\cosh\eta\,\sin\theta\,\cos\phi, \qquad y = d\cosh\eta\,\sin\theta\,\sin\phi, \qquad z = d\sinh\eta\,\cos\theta$

$$\frac{x^2+y^2}{\cosh^2\eta} + \frac{z^2}{\sinh^2\eta} = d^2, \qquad \frac{x^2+y^2}{\sin^2\theta} - \frac{z^2}{\cos^2\theta} = d^2, \qquad \tan\phi = y/x$$

For the equations below let:

$\alpha^2 = \cosh^2\eta - \sin^2\theta$, and $\beta = \cosh\eta\,\sin\theta$

The quantities defined in eqs. (C.2) to (C.11) are enumerated below:

$g_{11} = g_{22} = d^2\alpha^2, \qquad g_{33} = d^2\,\beta^2$

$g^{1/2} = d^3\,\alpha^2\,\beta$

$(ds)^2 = d^2\,\alpha^2\,[(d\eta)^2 + (d\theta)^2] + d^2\,\beta^2\,(d\phi)^2$

$dV = d^3\,\alpha^2\,\beta\,d\eta\,d\theta\,d\phi$

$$\nabla\psi = \frac{1}{d\alpha}\left[\vec{e}_\eta\frac{\partial\psi}{\partial\eta} + \vec{e}_\theta\frac{\partial\psi}{\partial\theta}\right] + \frac{1}{d\beta}\vec{e}_\phi\frac{\partial\psi}{\partial\phi}$$

$$\nabla\cdot\vec{E} = \frac{1}{d\alpha}\left\{\frac{1}{\cosh\eta}\frac{\partial}{\partial\eta}[\alpha\cosh\eta\,E_\eta] + \frac{1}{\sin\theta}\frac{\partial}{\partial\theta}[\alpha\sin\theta\,E_\theta]\right\} + \frac{1}{d\beta}\frac{\partial E_\phi}{\partial\phi}$$

$$\nabla x\vec{E} = \frac{1}{d\alpha^2\beta}\begin{vmatrix}\alpha\,\vec{e}_\eta & \alpha\,\vec{e}_\theta & \beta\,\vec{e}_\phi\\ \frac{\partial}{\partial\eta} & \frac{\partial}{\partial\theta} & \frac{\partial}{\partial\phi}\\ \alpha\,E_\eta & \alpha\,E_\theta & \beta\,E_\phi\end{vmatrix}$$

$$\nabla^2\psi = \frac{1}{d^2\alpha^2}\left\{\frac{\partial^2\psi}{\partial\eta^2} + \tanh\eta\frac{\partial\psi}{\partial\eta} + \frac{\partial^2\psi}{\partial\theta^2} + \cos\theta\frac{\partial\psi}{\partial\theta}\right\} + \frac{1}{d^2\beta^2}\frac{\partial^2\psi}{\partial\phi^2}$$

C.8.2 Oblate Spheroidal Coordinates II

These coordinates are defined by:
These are defined as

$u^1 = \xi \qquad 1 \le \xi < \infty$

$u^2 = \eta \qquad -1 \le \eta \le +1$

$u^3 = \phi \qquad 0 \le \phi \le 2\pi$

The coordinate transformation between (x, y, z) and (ξ, η, ϕ) are described below:

$x = d\sqrt{(\xi^2 + 1)(1-\eta^2)}\,\cos\phi, \qquad y = d\sqrt{(\xi^2 + 1)(1-\eta^2)}\,\sin\phi, \qquad z = d\,\xi\,\eta$

$\frac{x^2+y^2}{1+\xi^2} + \frac{z^2}{\xi^2} = d^2, \qquad \frac{x^2+y^2}{1-\eta^2} - \frac{z^2}{\eta^2} = d^2, \qquad \tan\phi = y/x$

The focal length of the ellipse is 2d.

For the equations below let:

$\alpha^2 = \xi^2 + 1, \qquad \beta^2 = 1 - \eta^2 \qquad$ and $\qquad \chi^2 = \xi^2 - \eta^2$

The quantities defined in eqs. (C.2) to (C.11) are enumerated below:

$g_{11} = (d\,\chi/\alpha)^2 \quad g_{22} = (d\,\chi\;/\,\beta)^2, \qquad g_{33} = (d\,\alpha\,\beta)^2$

$g^{1/2} = d^3\,\chi^2$

$$(ds)^2 = d^2\chi^2\left[\frac{(d\xi)^2}{\alpha^2} + \frac{(d\eta)^2}{\beta^2}\right] + d^2\alpha^2\beta^2(d\phi)^2$$

$$dV = d^3 \chi^2 \, d\xi \, d\eta \, d\phi$$

$$\nabla\psi = \frac{1}{d\chi}\left[\alpha \, \vec{e}_\xi \frac{\partial\psi}{\partial\xi} + \beta^2 \, \vec{e}_\eta \frac{\partial\psi}{\partial\eta}\right] + \frac{1}{d\,\alpha\beta} \vec{e}_\phi \frac{\partial\psi}{\partial\phi}$$

$$\nabla\cdot\vec{E} = \frac{1}{d\,\chi^2}\left\{\frac{\partial}{\partial\xi}[\chi\alpha \, E_\xi] + \frac{\partial}{\partial\eta}[\chi\,\beta\,E_\eta] + \frac{\chi^2}{\alpha\beta}\frac{\partial E_\phi}{\partial\phi}\right\}$$

$$\nabla \times \vec{E} = \frac{1}{d\chi^2}\begin{vmatrix} \frac{\chi}{\alpha}\vec{e}_\xi & \frac{\chi}{\beta}\vec{e}_\eta & \beta\alpha \, \vec{e}_\phi \\ \frac{\partial}{\partial\xi} & \frac{\partial}{\partial\eta} & \frac{\partial}{\partial\phi} \\ \frac{\chi}{\alpha}E_\xi & \frac{\chi}{\beta}E_\eta & \beta\alpha \, E_\phi \end{vmatrix}$$

$$\nabla^2\psi = \frac{1}{d^2\chi^2}\left[\frac{\partial}{\partial\xi}\alpha^2\frac{\partial}{\partial\xi} + \frac{\partial}{\partial\eta}\beta^2\frac{\partial}{\partial\eta} + \frac{\chi^2}{\alpha^2\beta^2}\frac{\partial^2}{\partial\phi^2}\right]\psi$$

APPENDIX D

DIRAC DELTA FUNCTIONS

The Dirac delta functions are generalized functions which are point functions and thus are not differentiable. A generalized function which will be used often in this appendix is the **Step** function or the **Heaviside** function is defined as:

$$\begin{aligned} H(x-a) &= 0 && x < a \\ &= 1/2 && x = a \\ &= 1 && x > a \end{aligned} \tag{D.1}$$

which is not differentiable at x = a. One should note that:

$$H(x-a) + H(a-x) = 1 \tag{D.2}$$

D.1 Dirac Delta Function

D.1.1 Definitions and Integrals

The one-dimensional **Dirac delta** function **δ(x-a)** is one that is defined only through its integral. It is a point function characterized by the following properties:

Definition:

$$\begin{aligned} \delta(x-c) &= 0 && x \neq c \\ &= \infty && x = c \end{aligned}$$

Integral:

Its integral is defined as:

$$\int_{-\infty}^{\infty} \delta(x-c)\,dx = 1 \tag{D.3}$$

Sifting Property:

Given a function f(x), which is continuous at x = c, then:

$$\int_{-\infty}^{\infty} f(x)\delta(x-c)\,dx = f(c) \tag{D.4}$$

Shift Property:

This property allows for a shift of the point of application of δ(x-c), i.e.:

$$\int_{-\infty}^{\infty} \delta(x-c) f(x)\, dx = \int_{-\infty}^{\infty} \delta(x) f(x+c) dx = f(c) \tag{D.5}$$

Scaling Property:

This property allows for the stretching of the variable x:

$$\int_{-\infty}^{\infty} \delta(x/a) f(x)\, dx = |a|\, f(0) \tag{D.6}$$

and

$$\int_{-\infty}^{\infty} \delta((x-c)/a)\, f(x)\, dx = |a|\, f(c) \tag{D.7}$$

Even Function:

The Dirac function is an even function, i.e.:

$$\delta(c-x) = \delta(x-c) \tag{D.8}$$

since:

$$\int_{-\infty}^{\infty} \delta(c-x) dx = 1$$

and

$$\int_{-\infty}^{\infty} \delta(x-c) f(x)\, dx = f(c) = \int_{-\infty}^{\infty} \delta(c-x)\, f(x) dx$$

Definite Integrals:

The Dirac delta function may be integrated over finite limits, such that:

$$\int_{a}^{b} \delta(x-c)\, dx = 0 \qquad c < a, \text{ or } c > b$$

$$= 1/2 \qquad c = a, \text{ or } c = b \tag{D.9}$$

$$= 1 \qquad a < c < b$$

and the sifting property is then redefined as:

$$\int_a^b f(x)\delta(x-c)\,dx = 0 \qquad c < a, \text{ or } c > b$$

$$= 1/2\, f(c) \qquad c = a, \text{ or } c = b \tag{D.10}$$

$$= f(c) \qquad a < c < b$$

If the integral is an indefinite integral, the integral of the Dirac delta function is a Heaviside function:

$$\int_{-\infty}^{x} \delta(x-c)dx = H(x-c) \tag{D.11}$$

and

$$\int_{-\infty}^{x} \delta(x-c)f(x)\,dx = f(c)\,H(x-c) \tag{D.12}$$

D.1.2 Integral Representations

One can define continuous, differentiable functions which behave as a Dirac delta function when certain parameters vanish, i.e., let:

$$\lim_{\alpha \to 0} u(\alpha, x) = \delta(x)$$

iff it satisfies the integral and sifting properties above.

To construct such representations, one may start with improper integrals whose values are unity, i.e., let U(x) be a continuous even function whose integral is:

$$\int_{-\infty}^{\infty} U(x)dx = 1 \tag{D.13}$$

then a function representation of the Dirac delta function when $\alpha \to 0$ is:

$$u(\alpha,x) = U(x/\alpha)\,/\,\alpha \tag{D.14}$$

which also satisfies the sifting property in the limit as $\alpha \to 0$.

Example D.1

The function $u(\alpha, x) = \alpha/[\pi(x^2 + \alpha^2)]$ behaves like δ(x), since:

$$\lim_{\alpha \to 0} u(\alpha, x) \to \begin{cases} 0 & x \neq 0 \\ \infty & x = 0 \end{cases}$$

and since it satisfies the integral and sifting properties:

$$\int_{-\infty}^{x} u(\alpha,x)dx = \frac{1}{2} + \frac{1}{\pi}\arctan\left(\frac{x}{\alpha}\right)$$

so that when the upper limit becomes infinite, the integral approaches unity. Note also that if the limit of the integral is taken when $\alpha \to 0$, the integral approaches H(x). It should be noted that this functional representation was obtained from the integral:

$$\int_{-\infty}^{\infty} \frac{dx}{1+x^2} = \pi$$

so that:

$$U(x) = \frac{1}{\pi(1+x^2)}$$

which results in the form given for $u(\alpha,x)$ above. To satisfy the sifting property, one may use a shortcut procedure which assumes uniform convergence of the integrals, i.e.:

$$\lim_{\alpha\to 0} \int_{-\infty}^{\infty} u(\alpha,x)\, f(x)dx = \frac{1}{\pi} \lim_{\alpha\to 0} \int_{-\infty}^{\infty} \frac{\alpha}{x^2+\alpha^2}\, f(x)\, dx$$

substituting $y = x/\alpha$ in the above integral one obtains:

$$\lim_{\alpha\to 0} \int_{-\infty}^{\infty} u(\alpha,x)\, f(x)dx = \frac{1}{\pi} \lim_{\alpha\to 0} \int_{-\infty}^{\infty} \frac{f(\alpha y)}{1+y^2}\, dy \to \frac{f(0)}{\pi} \int_{-\infty}^{\infty} \frac{dy}{1+y^2} = f(0)$$

where the integral is assumed to be uniformly convergent in α. Let f(x) be absolutely integrable and continuous at x = 0, then one can perform these integrations without this assumption by integration by parts:

$$\frac{\alpha}{\pi} \int_{-\infty}^{\infty} \frac{f(x)}{x^2+\alpha^2}\, dx = \frac{\alpha}{\pi} \left\{ \int_{-\infty}^{0} \frac{f(x)}{x^2+\alpha^2}\, dx + \int_{0}^{\infty} \frac{f(x)}{x^2+\alpha^2}\, dx \right\}$$

$$= \frac{\alpha}{\pi} \left\{ \int_{0}^{\infty} \frac{f(-x)}{x^2+\alpha^2}\, dx + \int_{0}^{\infty} \frac{f(x)}{x^2+\alpha^2}\, dx \right\}$$

Integrating the second integral by parts:

$$\lim_{\alpha\to 0} \left(\frac{\alpha}{\pi} \int_{0}^{\infty} \frac{f(x)}{x^2+\alpha^2}\, dx \right) = \lim_{\alpha\to 0} \left(\frac{1}{\pi} f(x) \arctan(x/\alpha) \Big|_0^{\infty} - \frac{1}{\pi} \int_{0}^{\infty} f'(x) \arctan(x/\alpha) dx \right)$$

$$= -\frac{1}{\pi} \lim_{\alpha\to 0} \int_{0}^{\infty} f'(x) \arctan(x/\alpha)\, dx \to -\frac{1}{2} \int_{0}^{\infty} f'(x) dx$$

$$= -\frac{1}{2} f(x) \Big|_0^{\infty} = \frac{1}{2} f(0^+)$$

since f(x) is absolutely integrable and continuous at x = 0. Similarly the first integral approaches:

$$\lim_{\alpha\to 0}\left(\frac{\alpha}{\pi}\int_{0}^{\infty}\frac{f(-x)}{x^2+\alpha^2}\,dx\right)\to\frac{1}{2}f(0^-)$$

so that, since f(x) is continuous at x = 0:

$$\lim_{\alpha\to 0}\left(\int_{-\infty}^{\infty}f(x)\,u(\alpha,x)\,dx\right)\to f(0)$$

D.1.3 Transformation Property

One can represent a finite number of Dirac delta functions by one whose argument is a function. Consider $\delta[f(x)]$ where f(x) has a non-repeated null at x_0 and whose derivative does not vanish at x_0, than one can show that:

$$\delta[f(x)]=\frac{\delta(x-x_0)}{|f'(x_0)|} \tag{D.15}$$

One can show that eq. (D.15) is correct by satisfying the conditions on integrability and the sifting property. Starting with the integral of $\delta[f(x)]$:

$$\int_{-\infty}^{\infty}\delta[f(x)]dx$$

Letting:

$$u = f(x)$$

then:

$$u = 0 = f(x_0) \qquad \text{and} \qquad du = f'(x)\,dx$$

then the integral becomes:

$$\int_{-\infty}^{\infty}\delta[f(x)]dx=\frac{1}{|f'(x_0)|}=\frac{1}{|f'(x_0)|}\int_{-\infty}^{\infty}\delta(x-x_0)\,dx$$

and

$$\int_{-\infty}^{\infty}\delta[f(x)]F(x)\,dx=\int_{-\infty}^{\infty}\frac{\delta(u)}{|f'(x(u))|}F(x(u))\,du$$

$$=\frac{F(x_0)}{|f'(x_0)|}=\frac{1}{|f'(x_0)|}\int_{-\infty}^{\infty}F(x)\delta(x-x_0)\,dx$$

Thus, the two properties are satisfied if eq. (D.15) represents $\delta[f(x)]$.

If f(x) has a finite or an infinite number of non-repeated zeroes, i.e.:

$$f(x_n)=0 \qquad n = 1, 2, 3, \ldots N$$

then:

$$\delta[f(x)] = \sum_{n=1}^{N} \frac{\delta(x - x_n)}{|f'(x_n)|} \tag{D.16}$$

Example D.2

$$\delta(x^2 - a^2) = \frac{1}{2a}[\delta(x - a) + \delta(x + a)]$$

$$\delta[\cos x] = \sum_{n=-\infty}^{\infty} \delta[x - \frac{2n+1}{2}\pi]$$

D.1.4 Concentrated Field Representations

The Dirac delta function is often used to represent concentrated fields such as concentrated forces and monopoles. For example, a concentrated force (monopole point source) located a x_0 of magnitude P_0 can be represented by $P_0\ \delta\ (x\text{-}x_0)$. This property can be utilized in integrals of distributed fields where one component of the integrand behaves like a Dirac delta function when a parameter in the integrand is taken to some limit.

Example D.3

The following integral, which is known to have an exact value, can be approximately evaluated for small values of its parameter c:

$$T = \frac{1}{\pi} \int_{-\infty}^{\infty} \frac{\cos(ax)\, J_0(bx)}{x^2 + c^2} dx = \frac{1}{c} e^{-ac} I_0(bc) \cong \frac{1}{c} \qquad c << 1$$

If the integral cannot be evaluated in a closed form and one would like to evaluate this integral for small values of c, one notices that the function in example D.1, $c\ /\ [\pi(x^2 + c^2)]$, behaves as $\delta(x)$ in the limit of $c \rightarrow 0$. Thus, one can approximately evaluate the integral by the sifting properties. Letting:

$$F(x) = \frac{1}{c} \cos(ax) J_0(bx)$$

then the sifting property gives $F(0) = 1/c$. To check the numerical value of this approximation, one can evaluate it exactly, so that for a = b = 1 one obtains:

c	T(exact)	T(approx)	cT(exact)	cT(approx)
0.2	4.13459	5.000	0.82692	1.0
0.1	9.07090	10.00	0.90709	1.0
0.01	99.0050	100.00	0.99005	1.0

This example shows that for c = 0.1 the error is within 10 percent of its exact value. This approximate method of evaluating integrals when part of the integrand behaves like a

Dirac delta function can be used to overcome difficulties in evaluating integrals in a closed form.

D.2 Dirac Delta Function of Order One

The Dirac delta function of order one is defined formally by

$$\delta_1(x - x_0) = -\frac{d}{dx}\delta(x - x_0) \tag{D.17}$$

such that its integral vanishes:

$$\int_{-\infty}^{\infty} \delta_1(x - x_0)\, dx = 0 \tag{D.18}$$

and its first moment integral is unity:

$$\int_{-\infty}^{\infty} x\delta_1(x - x_0)dx = 1 \tag{D.19}$$

and its sifting property is given by:

$$\int_{-\infty}^{\infty} f(x)\delta_1(x - x_0)dx = f'(x_0) \tag{D.20}$$

which gives the value of the derivative of the function f(x) at the point of application of $\delta_1(x - x_0)$.

These properties outlined in eqs. (D.18) to (D.20) can be proven by resorting to the integral representation. Thus, using the representation of a Dirac delta function, one can define $\delta_1(x)$ as:

$$\delta_1(x) = -\lim_{\alpha \to 0} \frac{d\,u(\alpha, x)}{d\alpha} \tag{D.21}$$

In physical applications, $\delta_1(x)$ represent a mechanical concentrated couple or a dipole.

D.3 Dirac Delta Function of Order N

These Dirac delta functions of order N can be formally defined as:

$$\delta_N(x - x_0) = (-1)^N \frac{d^N}{dx^N}\delta(x - x_0) \tag{D.22}$$

so that the kth moment integral is:

$$\int_{-\infty}^{\infty} x^k\, \delta_N(x)dx = \begin{cases} 0 & k < N \\ N! & k = N \end{cases} \tag{D.23}$$

and the sifting property gives the Nth derivative of the function at the point of application of $\delta_N (x - x_0)$ is:

$$\int_{-\infty}^{\infty} f(x)\delta_N(x - x_0)\,dx = f^{(N)}(x_0) \tag{D.24}$$

In physical applications, $\delta_N (x - x_0)$ represents high-order point mechanical forces and sources. For example, $\delta_2(x - x_0)$ represents a doublet force or a quadrapole.

D.4 Equivalent Representations of Distributed Functions

In many instances, one can represent a distributed function evaluated at the point of application of a Dirac delta function of any order by a series of functions with equal and lower-ordered Dirac functions. For example, one can show that

$$f(\xi)\,\delta(x - \xi) = f(x)\,\delta(x - \xi) \tag{D.25}$$

which allows one to express a point value of $f(\xi)$ by a field function $f(x)$ defined over the entire real axis. The proof uses the sifting property of the Dirac delta function and an auxiliary function $F(x)$:

$$\int_{-\infty}^{\infty} F(x)\,f(x)\delta(x - \xi)\,dx = F(\xi)\,f(\xi) = f(\xi)\int_{-\infty}^{\infty} F(x)\delta(x - \xi)\,dx$$

$$= \int_{-\infty}^{\infty} F(\xi)\,f(\xi)\delta(x - \xi)\,dx$$

which satisfies the equivalence in D.25.

Similarly one can show that:

$$f(x)\,\delta_1(x - \xi) = f'(\xi)\delta(x - \xi) + f(\xi)\,\delta_1(x - \xi) \tag{D.26}$$

which again can be proven by using an auxiliary function $F(x)$:

$$\int_{-\infty}^{\infty} F(x)\,f(x)\delta_1(x - \xi)\,dx = F(\xi)\,f'(\xi) + F'(\xi)\,f(\xi)$$

$$= f'(\xi)\int_{-\infty}^{\infty} F(x)\,\delta(x - \xi)\,dx + f(\xi)\int_{-\infty}^{\infty} F(x)\delta_1(x - \xi)\,dx$$

which proves the equivalency in eq. (D.26). This equivalence shows that a distributed couple (dipole) field $f(x)$ is equivalent to a point couple (dipole) of strength $f(\xi)$ and a point force (monopole) of strength $f'(\xi)$.

D.5 Dirac Delta Functions in n-Dimensional Space

A similar representation of Dirac delta function exists in multi-dimensional space. Let **x** be a position vector in n-dimensional space:

$$\mathbf{x} = [x_1, x_2, \ldots, x_n] \tag{D.27}$$

and let the symbol R_n to represent the volume integral in that space, i.e.,

$$\int_{R_n} F(\mathbf{x})\,d\mathbf{x} \equiv \int_{-\infty}^{\infty}\int_{-\infty}^{\infty}\ldots\int_{-\infty}^{\infty} F(x_1, x_2, \ldots x_n)\,dx_1\,dx_2\ldots dx_n \tag{D.28}$$

D.5.1 Definitions and Integrals

The Dirac delta function has the following properties that mirror those in one-dimensional, so that:

$$\delta(\mathbf{x} - \xi) = \delta[x_1 - \xi_1, x_2 - \xi_2, \ldots, x_n - \xi_n] \tag{D.29}$$

Integral:
The integral of Dirac delta function over the entire space is unity, i.e.,

$$\int_{R_n} \delta(\mathbf{x} - \xi)\,d\mathbf{x} = 1 \tag{D.30}$$

Sifting Property:

$$\int_{R_n} F(\mathbf{x})\,\delta(\mathbf{x} - \xi)\,d\mathbf{x} = F(\xi) \tag{D.31}$$

Scaling Property:
For a common scaling factor a of all the coordinates $x_1, x_2, \ldots x_n$:

$$\int_{R_n} F(\mathbf{x})\,\delta\left(\frac{\mathbf{x}}{a}\right)d\mathbf{x} = |a|^n\,F(\mathbf{0}) \tag{D.32}$$

Integral Representation:
Let $U(\mathbf{x}) = U(x_1, x_2, \ldots, x_n)$ be a non-negative locally integrable function, such that:

$$\int_{R_n} U(\mathbf{x})\,d\mathbf{x} = 1 \tag{D.33}$$

Define:

$$u(\alpha, \mathbf{x}) = \alpha^{-n}\,U(\mathbf{x}/\alpha) = \alpha^{-n}\,U(x_1/\alpha, x_2/\alpha, \ldots x_n/\alpha) \tag{D.34}$$

then:

$$\lim_{\alpha\to 0} u(\alpha,\mathbf{x}) = \delta(\mathbf{x}) \tag{D.35}$$

This can be easily proven through the scaling property:

$$\lim_{\alpha\to 0} \int_{R_n} \frac{1}{\alpha_n} U\left(\frac{\mathbf{x}}{\alpha}\right) d\mathbf{x} = \int_{R_n} U(\mathbf{y})\, d\mathbf{y} = 1$$

where the scaling transformation $\mathbf{y} = \mathbf{x}/\alpha$ was used. It also satisfies the sifting property, since:

$$\lim_{\alpha\to 0} \int_{R_n} \frac{1}{\alpha_n} U\left(\frac{\mathbf{x}}{\alpha}\right) F(\mathbf{x})d\mathbf{x} = \lim_{\alpha\to 0} \int_{R_n} U(\mathbf{y})\, F(\alpha\mathbf{y})d\mathbf{y} = F(\mathbf{0})$$

D.5.2 Representation by Products of Dirac Delta Functions

One can show that the Dirac delta function in n-dimensional space can be written in terms of a product of one-dimensional ones, i.e.:

$$\delta(\mathbf{x} - \boldsymbol{\xi}) = \delta(x_1 - \xi_1)\, \delta(x_2 - \xi_2) \ldots \delta(x_n - \xi_n) \tag{D.36}$$

This equivalence can be shown through the volume integral and sifting property:

$$\int_{R_n} \delta(\mathbf{x} - \boldsymbol{\xi})\, d\mathbf{x} = \int_{-\infty}^{\infty} \delta(x_1 - \xi_1)\, dx_1 \cdot \ldots \cdot \int_{-\infty}^{\infty} \delta(x_n - \xi_n) dx_n = 1$$

and

$$\int_{R_n} F(\mathbf{x})\delta(\mathbf{x} - \boldsymbol{\xi})d\mathbf{x} = F(\boldsymbol{\xi}) = \int_{-\infty}^{\infty} \ldots \int_{-\infty}^{\infty} F(x_1,\ldots,x_n)\delta(x_1 - \xi_1)\cdot \ldots \cdot \delta(x_n - \xi_n)\, dx_1 \ldots dx_n$$

D.5.3 Dirac Delta Function in Linear Transformation

The Dirac delta function can be expressed in terms of new coordinates undergoing linear transformations. Let the real variables $u_1, u_2, \ldots u_n$ be defined in a single-valued transformation defined by:

$$u_1 = u_1\,(x_1,x_2, \ldots x_n),\;\; u_2 = u_2\,(x_1,x_2, \ldots x_n),\;\; \ldots\;\; u_n = u_n\,(x_1,x_2, \ldots x_n)$$

then:

$$\delta(\mathbf{x} - \boldsymbol{\xi}) = \frac{1}{J}\, \delta[\mathbf{u} - \boldsymbol{\eta}] \tag{D.37}$$

where $\eta = \mathbf{u}\,(\xi)$ and the Jacobian J is given by

$$J(\xi) = \det\,[\partial \mathbf{x}_i / \partial u_j] \qquad \text{for} \qquad J(\xi) \neq 0$$

D.6 Spherically Symmetric Dirac Delta Function Representation

If the Dirac Delta function in n-dimensional space depends on the spherical distance only, a new representation exists. Let r be the radius in n-dimensional space:

$$r = [x_1^2 + x_2^2 + ... + x_n^2]^{1/2}$$

then if the function U(**x**) depends on r only:

$$\int_{R_n} U(\mathbf{x})\, d\mathbf{x} = \int_{-\infty}^{\infty} ... \int_{-\infty}^{\infty} U(r)\, dx_1\, dx_2 \ldots dx_n = 1$$

one can make the following transformation to n-dimensional spherical coordinates r, θ_1, $\theta_2, \ldots, \theta_n$ where only (n-1) of these Eulerian angles are independent:

$$x_1 = r\cos\theta_1, x_2 = r\cos\theta_2, \ldots \quad x_n = r\cos\theta_n$$

Thus, the volume integral transforms to:

$$\int_{R_n} U(\mathbf{x})\, d\mathbf{x} = \int_0^{\infty} U(r)\, r^{n-1}\, dr \left\{ \int_0^{2\pi} ... \int_0^{2\pi} \cos\theta_1 \ldots \cos\theta_n\, d\theta_1 \ldots d\theta_n \right\}$$

The last integral can be written in a condensed form as:

$$\left\{ \int_0^{\infty} U(r)\, r^{n-1}\, dr \right\} S_n(1) = 1$$

where U(r) is the part of the representation that depends on r only and $S_n(1)$ is the surface of an n-dimensional sphere of a unit radius, so that U(r) must satisfy the following integral:

$$\int_0^{\infty} r^{n-1}\, U(r) dr = \frac{1}{S_n(1)} \tag{D.38}$$

The volume and surface of an n-dimensional sphere V_n and S_n of radius r are:

$$V_n(r) = \frac{\pi^{n/2}}{(n/2)!} r^n = \frac{\pi^{n/2} r^n}{\Gamma(n/2+1)} \tag{D.39}$$

$$S_n(r) = \frac{dV_n}{dr} = \frac{2\pi^{n/2} r^{n-1}}{(\frac{n}{2}-1)!} = \frac{2\pi^{n/2} r^{n-1}}{\Gamma(n/2)} \tag{D.40}$$

Thus, in three-dimensional space:

$$S_3(1) = \frac{2\pi^{3/2}}{(1/2)!} = \frac{2\pi^{3/2}}{\Gamma(3/2)} = 4\pi$$

so that the representation function U(r) must satisfy:

$$\int_0^\infty r^2\, U(r)dr = \frac{1}{4\pi} \qquad \text{(D.41)}$$

In two-dimensional space:

$$S_2(1) = 2\pi$$

so that the representation of the function U(r) must satisfy:

$$\int_0^\infty r\, U(r)\, dr = \frac{1}{2\pi} \qquad \text{(D.42)}$$

Once one finds a function U(r) whose integral satisfies eq. (D.38), one can then obtain a Dirac delta function representation as follows:

$$u(\alpha,r) = \alpha^{-n}\, U(r/\alpha) \qquad \text{(D.43)}$$

so that the spherically symmetric Dirac delta function δ given by:

$$\delta(\mathbf{x}) = \lim_{\alpha \to 0} u(\alpha,r)$$

Example D.4

To construct a representation of a spherically symmetric representation of a Dirac delta function in three-dimensional space from the function:

$$U(r) = \frac{e^{-r}}{8\pi}$$

Since:

$$\int_0^\infty r^2\, U(r)dr = \frac{1}{4\pi}$$

then U(r) is a Dirac delta representation in three-dimensional space, and

$$u(\alpha, r) = \frac{1}{\alpha^3} \frac{e^{-r/\alpha}}{8\pi}$$

so that the spherical Dirac delta function representation in three dimensional space is:

$$\delta(\mathbf{x}) = \frac{1}{8\pi} \lim_{\alpha \to 0} \frac{e^{-r/\alpha}}{\alpha^3}$$

D.7 Dirac Delta Function of Order N in n-Dimensional Space

Dirac delta functions of higher order than zero are defined in terms of derivatives of the Dirac delta functions as was done in one-dimensional space. Define an integer vector $\boldsymbol{l}$ in a n-dimensional space as:

$\boldsymbol{l} = [l_1, l_2, \ldots l_n]$ (D.44)

where $l_1, l_2, \ldots, l_n$ are zero or positive integers, so that the measure of the vector is $|\boldsymbol{l}|$ is defined as:

$$|\boldsymbol{l}| = l_1 + l_2 + \ldots + l_n \qquad \text{(D.45)}$$

One can then write a partial derivative in short notation as:

$$\partial^{l} = \frac{\partial^{l_1 + l_2 + \ldots + l_n}}{\partial x_1^{l_1} \partial x_2^{l_2} \ldots \partial x_n^{l_n}} = \frac{\partial^{|l|}}{\partial x_1^{l_1} \partial x_2^{l_2} \ldots \partial x_n^{l_n}} \qquad \text{(D.46)}$$

Thus, one may define a Dirac delta function of **N** order in n-dimensional spaces in terms of derivatives of zero order:

$$\delta^{\mathbf{N}}(\mathbf{x}) = (-1)^{|\mathbf{N}|}\, \partial^{\mathbf{N}}\, \delta(\mathbf{x}) \qquad \text{(D.47)}$$

so that the sifting property becomes:

$$\int\limits_{R_n} \delta^{\mathbf{N}}(\mathbf{x} - \xi) F(\mathbf{x})\, d\mathbf{x} = \partial^{\mathbf{N}} F(\xi) \qquad \text{(D.48)}$$

Partial differentiation with respect to the position **x** or ξ are related. For example, one can show that:

$$\frac{\partial}{\partial x_1} \delta(\mathbf{x} - \xi) = -\frac{\partial}{\partial \xi_1} \delta(\mathbf{x} - \xi) \qquad \text{(D.49)}$$

by use of auxiliary functions as follows:

$$\int\limits_{R_n} \frac{\partial \delta(\mathbf{x} - \xi)}{\partial x_1} F(\mathbf{x}) d\mathbf{x} = -\frac{\partial F(\xi)}{\partial \xi_1} = -\frac{\partial}{\partial \xi_1} \int\limits_{R_n} \delta(\mathbf{x} - \xi) F(\mathbf{x})\, d\mathbf{x} = -\int\limits_{R_n} \frac{\partial \delta(\mathbf{x} - \xi)}{\partial \xi_1} F(\mathbf{x})\, d\mathbf{x}$$

PROBLEMS

1. For the following functions
   (i) show that they represent $\delta(x)$ as $\alpha \to 0$
   (ii) show that they satisfy the sifting property

(a) $u(\alpha,x) = \dfrac{1}{\alpha\sqrt{\pi}}\, e^{-x^2/\alpha^2}$

(b)

$$u(\alpha,x) = \begin{cases} 0 & x \le -\alpha-\varepsilon \\ \dfrac{1}{2\alpha}[1+\dfrac{x+\alpha}{\varepsilon}] & -\alpha-\varepsilon \le x \le -\alpha \\ \dfrac{1}{2\alpha} & -\alpha \le x \le \alpha \\ \dfrac{1}{2\alpha}[1-\dfrac{x+\alpha}{\varepsilon}] & \alpha \le x \le \alpha+\varepsilon \\ 0 & x \ge \alpha+\varepsilon \end{cases}$$

in the limit $\varepsilon \to 0$.

(c)

$$u(\alpha,x) = \begin{cases} 0 & |x| > \alpha \\ \dfrac{1}{\alpha}(1+\dfrac{x}{\alpha}) & -\alpha \le x \le 0 \\ \dfrac{1}{\alpha}(1-\dfrac{x}{\alpha}) & 0 \le x \le \alpha \end{cases}$$

(d) $u(\alpha,x) = \dfrac{1}{\pi x}\sin(x/\alpha)$

2. Show that the following are representations of the spherical Dirac delta function:

(a) $\delta(x_1,x_2) = \lim\limits_{\alpha\to 0} \dfrac{\alpha}{2\pi(r^2+\alpha^2)^{3/2}}$

(b) $\delta(x_1,x_2,x_3) = \lim\limits_{\alpha\to 0} \dfrac{\alpha}{\pi^2\,(r^2+\alpha^2)^2}$

(c) $\delta(x_1,x_2,x_3) = \lim\limits_{\alpha\to 0} \dfrac{\alpha\sin^2(r/\alpha)}{2\pi^2\, r^4}$

3. Write down the following in terms of a series of Dirac delta function

   (a) $\delta(\tan x)$

   (b) $\delta(\sin x)$

4. If $x_1 = au_1 + bu_2$, and $x_2 = cu_1 + du_2$, then show that:

$$\delta(x_1)\delta(x_2) = \frac{1}{|ad - bc|}\delta(u_1)\delta(u_2)$$

5. Show that the representation of Spherical Dirac delta functions located at the origin are:

(a) $\delta(x_1, x_2, x_3) = \dfrac{\delta(r)}{4\pi r^2}$

(b) $\delta(x_1, x_2) = \dfrac{\delta(r)}{2\pi r}$

6. Show that the Dirac delta function at points not at the origin in cylindrical coordinates are given by:

(a) $\delta(x_1, x_2) = \dfrac{\delta(r - r_0)\delta(\theta - \theta_0)}{r}$ (Line source)

(b) $\delta(x_1, x_2, x_3) = \dfrac{\delta(r - r_0)\delta(\theta - \theta_0)\delta(z - z_0)}{r}$ (Point source)

(c) $\delta(x_1, x_3) = \dfrac{\delta(r - r_0)\delta(z - z_0)}{2\pi r}$ (Ring source)

7. Show that the following Dirac delta functions represent sources not at the origin in spherical coordinates.

(a) $\delta(x_1, x_2, x_3) = \dfrac{\delta(r - r_0)\delta(\theta - \theta_0)\delta(\phi - \phi_0)}{r^2}$ (Point source)

(b) $\delta(x_1, x_2) = \dfrac{\delta(r - r_0)\delta(\theta - \theta_0)}{2r^2}$ (Ring source)

(c) $\delta(x_1, x_3) = \dfrac{\delta(r - r_0)\delta(\phi - \phi_0)}{2\pi r^2}$ (Ring source)

(d) $\delta(x_1) = \dfrac{\delta(r - r_0)}{4\pi r^2}$ (Surface source)

APPENDIX E

PLOTS OF SPECIAL FUNCTIONS

E.1 Bessel Functions of the First and Second Kind of Order 0, 1, 2

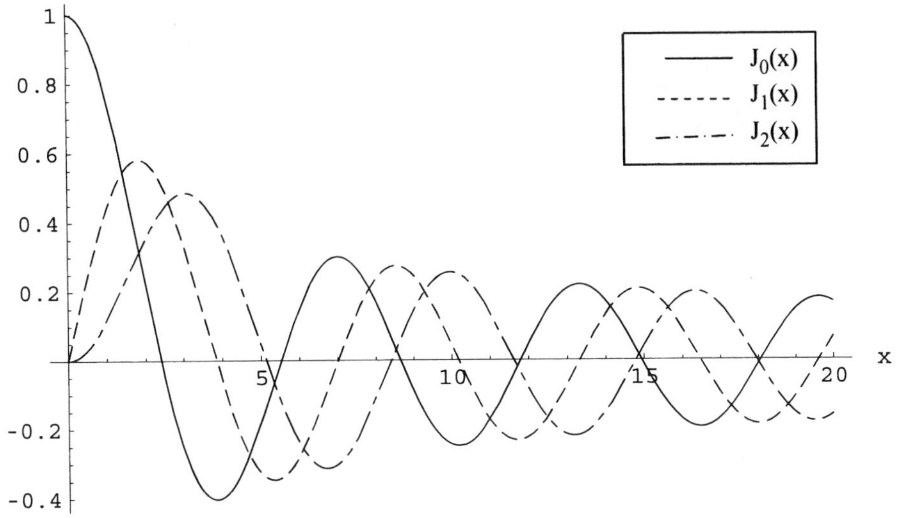

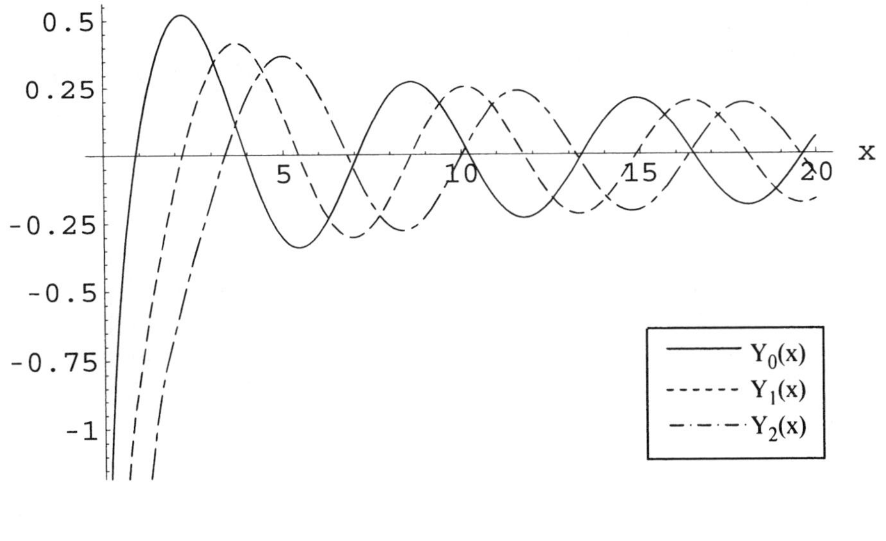

E.2 Spherical Bessel Functions of the First and Second Kind of Order 0, 1, 2

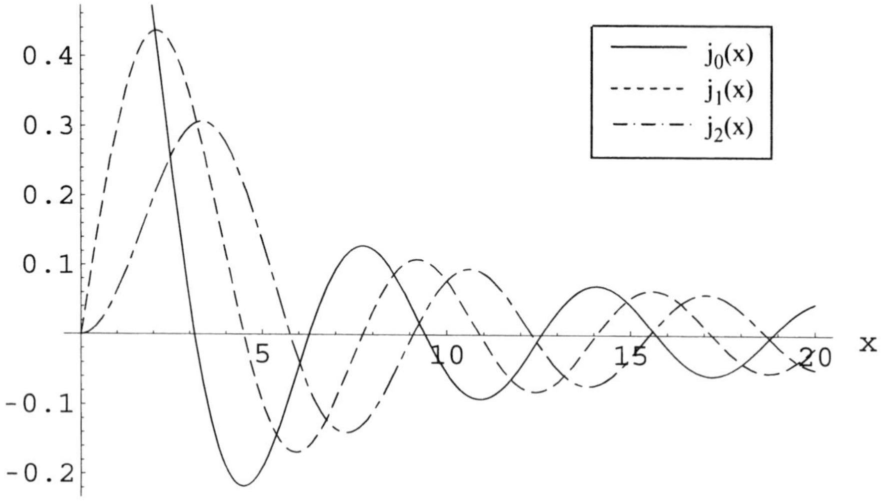

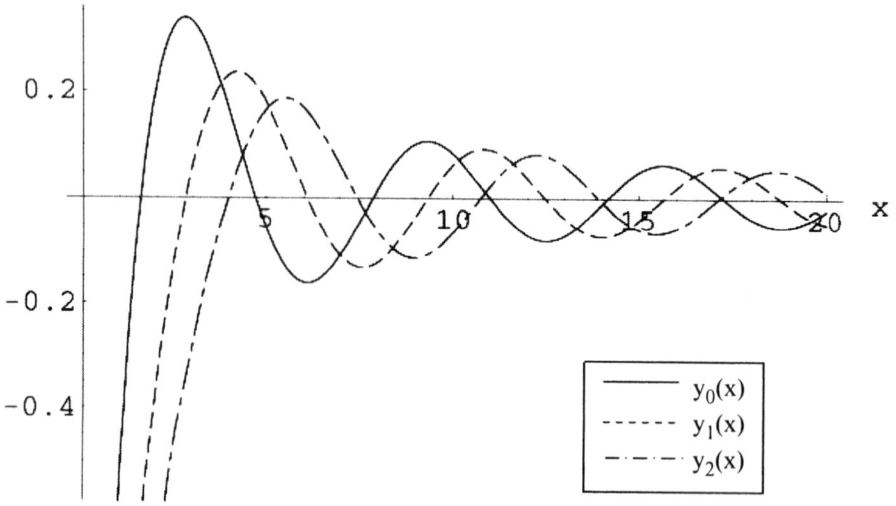

E.3 Modified Bessel Function of the First and Second Kind of Order 0, 1, 2

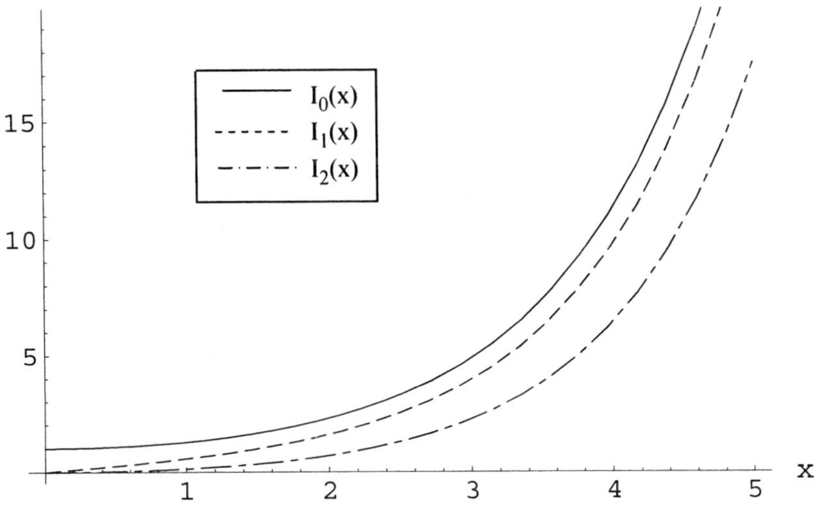

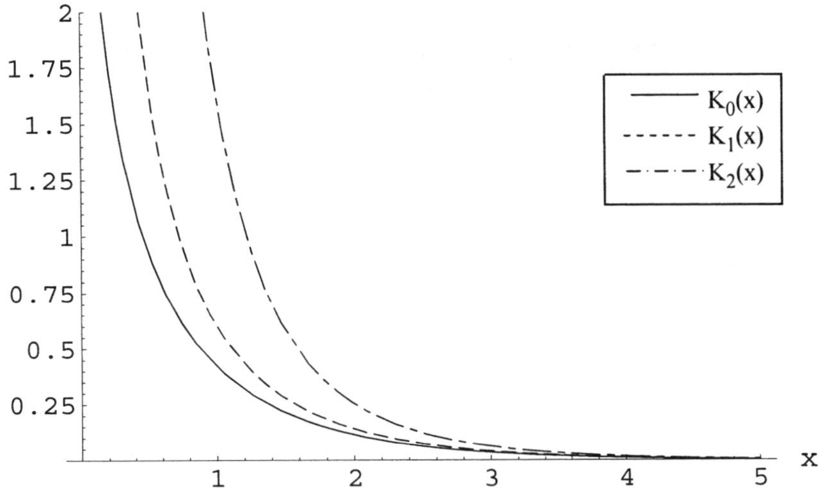

E.4 Bessel Function of the First and Second Kind of Order 1/2

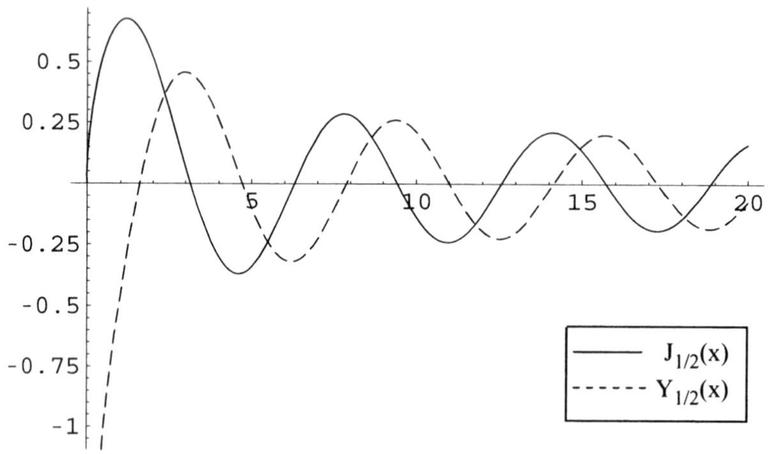

E.5 Modified Bessel Function of the First and Second Kind of Order 1/2

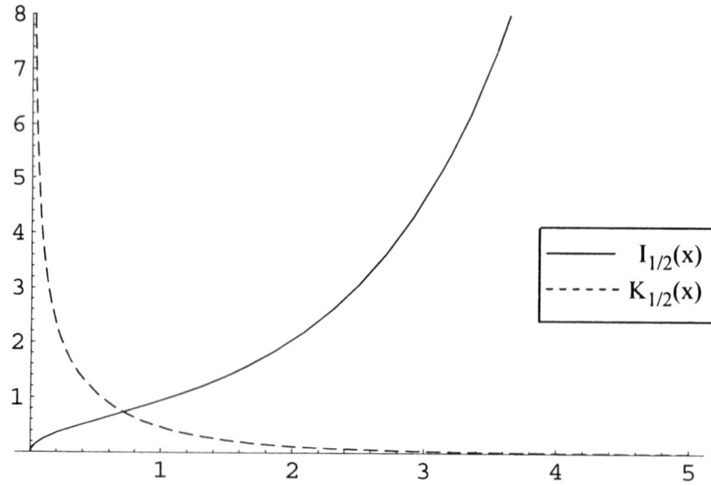

APPENDIX F

VECTOR ANALYSIS

F.1 Definitions and Index Notation

Define a vector in three-dimensional Euclidean [3]-space as:

$$\boldsymbol{a} = a_1\boldsymbol{e_1} + a_2\boldsymbol{e_2} + a_3\boldsymbol{e_3} \qquad \text{(F.1)}$$

where a_1, a_2 and a_3 are components of the vector $\boldsymbol{a}$ and $\boldsymbol{e_1}, \boldsymbol{e_2}$ and $\boldsymbol{e_3}$ are unit vectors in an orthogonal cartesian coordinate systems with directions in the x_1, x_2, and x_3 directions. This means the unit vectors have a unit length.

Let us introduce index notation in three-dimensional space. Define Kronecker delta δ_{ij}

$$\delta_{ij} = \begin{cases} 1 & i = j \\ 0 & i \neq i \end{cases} \qquad i, j = 1, 2, 3 \qquad \text{(F.2)}$$

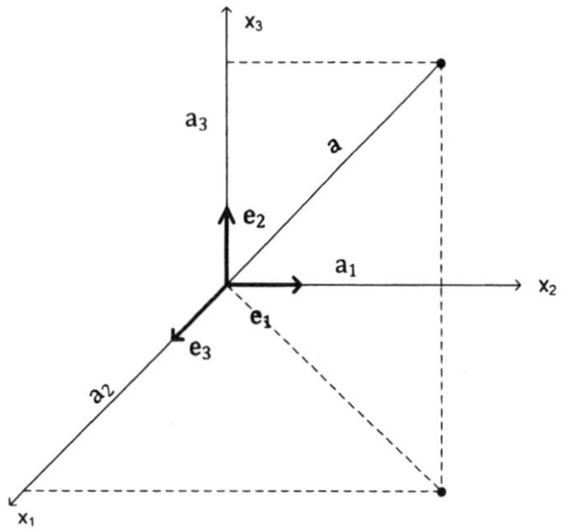

Fig. F.1 *Representation of a Vector*

Define the permutation symbol

$$\varepsilon_{ijk} = \begin{cases} 1 & if\ i, j, k\ are\ cylic\ computation\ of\ 1, 2, 3 \\ -1 & if\ i, j, k\ are\ not\ cylic\ compution\ of\ 1, 2, 3 \\ 0 & if\ any\ two\ indicies\ i, j, k\ are\ the\ same \end{cases} \qquad \text{(F.3)}$$

Repeated indices indicates summation. For example $\boldsymbol{a} = a_i\boldsymbol{e_i}$ means eq. (F.1), and $u = a_ib_i$ means $u = a_1b_1 + a_2b_2 + a_3b_3$.

If two or more indices have underbars, it means that no summation over the underbarred indices. For example, $a_i = a_{\underline{i}} b_{\underline{i}}$ means $a_1 = a_1 b_1, a_2 = a_2 b_2$ and $a_3 = a_3 b_3$

It should be noted that:

$$\varepsilon_{ijk} = -\varepsilon_{ikj}$$

$$\delta_{ij}\delta_{jk} = \delta_{ik}$$

$$\varepsilon_{ijk}\varepsilon_{i\ell m} = \delta_{j\ell}\delta_{km} - \delta_{jm}\delta_{k\ell} \qquad \text{(F.4)}$$

A scalar is one that has no components in vector space and thus has no free index. A vector in index notation has one free index.

$$\boldsymbol{a} \leftrightarrow a_i$$

A dyadic or cartesian tensor of rank two can be represented in dyadic vector notation as:

$$\overline{\boldsymbol{B}} = b_{11}\boldsymbol{e}_1\boldsymbol{e}_1 + b_{12}\boldsymbol{e}_1\boldsymbol{e}_2 + b_{13}\boldsymbol{e}_1\boldsymbol{e}_3 + b_{21}\boldsymbol{e}_2\boldsymbol{e}_1 + b_{22}\boldsymbol{e}_2\boldsymbol{e}_2 + b_{23}\boldsymbol{e}_2\boldsymbol{e}_3$$

$$+b_{31}\boldsymbol{e}_3\boldsymbol{e}_1 + b_{32}\boldsymbol{e}_3\boldsymbol{e}_2 + b_{33}\boldsymbol{e}_3\boldsymbol{e}_3 = b_{ij}e_ie_j$$

one should note $b_{ij} \neq b_{ji}$, as they represent different components of $\overline{\boldsymbol{B}}$. This is due to the fact that $\boldsymbol{e}_i\boldsymbol{e}_j \neq \boldsymbol{e}_j\boldsymbol{e}_i$.

Written in index notation $\overline{\boldsymbol{B}}$ is written as:

$\overline{\boldsymbol{B}} \leftrightarrow b_{ij}$ with two free indices. Similarly, $b_{ij} = c_{ijk\ell} b_k a_\ell$ is also a tensor of rank two, since the summations over k and ℓ, leaves only two free indices, i and j.

F.2 Vector Algebra

Define a vector in three-dimensional cartesian space

$$\boldsymbol{a} = a_1\boldsymbol{e}_1 + a_2\boldsymbol{e}_2 + a_2\boldsymbol{e}_3 = a_i\boldsymbol{e}_i$$

Then the modules of a vector $\boldsymbol{a}$ is its length:

$$|\boldsymbol{a}| = \sqrt{a_1^2 + a_2^2 + a_3^2} = \sqrt{a_i a_i} \tag{F.5}$$

Zero vector: $\boldsymbol{a} = \boldsymbol{0}$ mean $a_1 = 0\ a_2 = 0\ a_3 = 0$ or $a_i = 0$
Negative vector: $-\boldsymbol{a} = (-a_1)\boldsymbol{e_1} + (-a_2)\boldsymbol{e_2} + (-a_3)\boldsymbol{e_3}$

such that $\boldsymbol{a} + (-\boldsymbol{a}) = 0$

Equality $\boldsymbol{a} = \boldsymbol{b}$ means $a_i = b_i$

Sum of vectors: $\boldsymbol{c} = \boldsymbol{a} + \boldsymbol{b}$ means that $c_i = a_i + b_i$

Multiplication by scalar $\boldsymbol{c} = \propto \boldsymbol{a}$ means $c_i = \propto a_i$

Associative law: $(\boldsymbol{a} + \boldsymbol{b}) + \boldsymbol{c} = \boldsymbol{a} + (\boldsymbol{b} + \boldsymbol{c})$ or

$$(a_i + b_i) + c_i = a_i + (b_i + c_i)$$

Commutative law: $\boldsymbol{a} + \boldsymbol{b} = \boldsymbol{b} + \boldsymbol{a}$ or $a_i + b_i = b_i + a_i$

Geometrically, the addition or subtraction of two vectors results in two new vectors, e.g.,

$\boldsymbol{c} = \boldsymbol{a} + \boldsymbol{b}$

$\boldsymbol{d} = \boldsymbol{a} - \boldsymbol{b}$

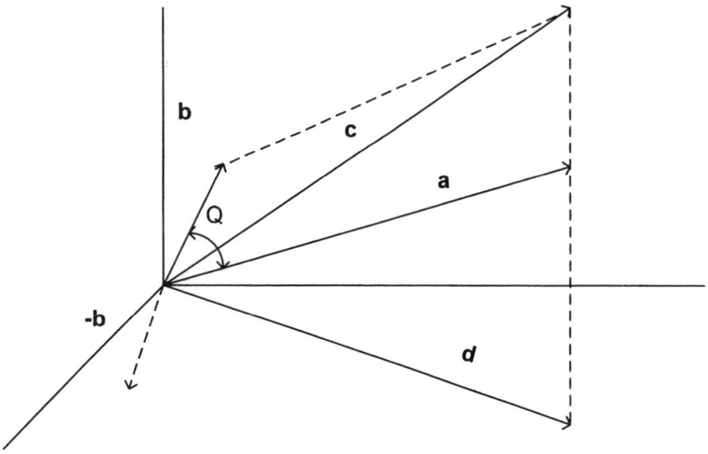

Fig. F.2 *Sum or Difference of Two Vectors*

It should be noted that from Fig. F.2,

$$|\boldsymbol{b}+\boldsymbol{a}| \leq |\boldsymbol{b}| + |\boldsymbol{a}|$$

$$|\boldsymbol{b}-\boldsymbol{a}| \geq |\boldsymbol{b}| - |\boldsymbol{a}|$$

F.3 Scalar and Vector Products

Define a scalar (dot) product as:

$$\boldsymbol{a}\cdot\boldsymbol{b} = a_1b_1 + a_2b_2 + a_3b_3 = a_ib_i = |\boldsymbol{a}||\boldsymbol{b}|\cos\theta \tag{F.6}$$

$$\boldsymbol{a}\cdot\boldsymbol{a} = |a|^2$$

Geometrically, it is the shaded area in Fig. F.2, where θ is the angle between $\boldsymbol{a}$ and $\boldsymbol{b}$.

Note that if $\theta = \pi/2$, then $\boldsymbol{a}\cdot\boldsymbol{b} = 0$. These vectors are known as being *orthogonal.*

Define a vector (cross) product of two vectors

$$\boldsymbol{c} = \boldsymbol{a}\wedge\boldsymbol{b} = (a_2b_3 - a_3b_2)\boldsymbol{e_1} + (a_3b_1 - a_1b_3)\boldsymbol{e_2}$$

$$+(a_1b_2 - a_2b_1)\boldsymbol{e_3} \tag{F.7}$$

The vector product of two vectors is a vector. It can be written in index notation as:

$$c_i = \varepsilon_{ijk}a_jb_k$$

Geometrically $\boldsymbol{c}$ is orthogonal to the plane containing the vectors $\boldsymbol{a}$ and $\boldsymbol{b}$. The volume made up of the three non-planar vectors $\boldsymbol{a}$, $\boldsymbol{b}$ and $\boldsymbol{c}$ is given by the triple product,

$$V = \boldsymbol{c}\cdot(\boldsymbol{a}\wedge\boldsymbol{b}) = \varepsilon_{ijk}c_ia_jb_k$$

Since the volume V is a scalar quantity, then the order of the triple products is not important, i.e.,

$$V = \boldsymbol{c}\cdot(\boldsymbol{a}\wedge\boldsymbol{b}) = \boldsymbol{a}\cdot(\boldsymbol{b}\wedge\boldsymbol{c}) = \boldsymbol{b}\cdot(\boldsymbol{c}\wedge\boldsymbol{a}) \tag{F.8}$$

A triple vector product generates a vector,

$$\boldsymbol{d} = \boldsymbol{a} \wedge (\boldsymbol{b} \wedge \boldsymbol{c}) = (\boldsymbol{a} \cdot \boldsymbol{c})\boldsymbol{b} - (\boldsymbol{a} \cdot \boldsymbol{b})\boldsymbol{c}$$

$$\boldsymbol{g} = (\boldsymbol{a} \wedge \boldsymbol{b}) \wedge \boldsymbol{c} = (\boldsymbol{a} \cdot \boldsymbol{c})\boldsymbol{b} - (\boldsymbol{b} \cdot \boldsymbol{c})\boldsymbol{a} \tag{F.9}$$

Note that $\boldsymbol{d} \neq \boldsymbol{g}$, so that triple vector product does not follow associative law. Another identity for quadruple product,

$$(\boldsymbol{a} \wedge \boldsymbol{b}) \cdot (\boldsymbol{a} \wedge \boldsymbol{b}) = |\boldsymbol{a}|^2|\boldsymbol{b}|^2 - (\boldsymbol{a} \cdot \boldsymbol{b})^2 \tag{F.10}$$

F.4 Vector Fields

A vector field is defined when the components of a vector depend on the coordinates x_1, x_2, and x_3. For example, if the velocity of a particle depends on the position of the particle in space, then it is designated as:

$$\boldsymbol{v}(\boldsymbol{x}) = v_1(x_1, x_2, x_3)\boldsymbol{e}_1 + v_2(x_1, x_2, x_3)\boldsymbol{e}_2 + v_3(x_1, x_2, x_3)\boldsymbol{e}_3$$

Differentiation of a vector field w.r.t. a scalar t becomes:

$$\frac{d\boldsymbol{v}(\boldsymbol{x})}{dt} = \frac{dv_1}{dt}\boldsymbol{e}_1 + \frac{dv_2}{dt}\boldsymbol{e}_2 + \frac{dv_3}{dt}\boldsymbol{e}_3 \tag{F.11}$$

F.5 Gradient of a Scalar

Define the gradient operator $\boldsymbol{\nabla}$ of a scalar function $f(\boldsymbol{x})$

$$\boldsymbol{\nabla} f = \frac{\partial f}{\partial x_1}\boldsymbol{e}_1 + \frac{\partial f}{\partial x_2}\boldsymbol{e}_2 + \frac{\partial f}{\partial x_3}\boldsymbol{e}_3 \tag{F.12}$$

Note that the gradient is a vector.

Note that if we obtain the dot product of the gradient along a vector $\boldsymbol{n}$, where

$$\boldsymbol{n} = n_1\boldsymbol{e}_1 + n_2\boldsymbol{e}_2 + n_3\boldsymbol{e}_3$$

then

$$\boldsymbol{n} \cdot \nabla f = n_1 \frac{\partial f}{\partial x_1} + n_2 \frac{\partial f}{\partial x_2} + n_3 \frac{\partial f}{\partial x_3} = \frac{\partial f}{\partial n} \tag{F.13}$$

which defines the spatial differentiation of the scalar $f(x)$ along the vector $\boldsymbol{n}$.

In index notation let the spatial differential be written as:

$\frac{\partial f}{\partial x_i} = f, i$ so that eq. (F.13) can be rewritten as:

$$\boldsymbol{n} \cdot \nabla f = n_i f, i$$

F.6 Divergence of a Vector

Define the divergence of a vector field $\boldsymbol{v}$, as

$$\nabla \cdot \boldsymbol{v} = \frac{\partial v_1}{\partial x_1} + \frac{\partial v_2}{\partial x_2} + \frac{\partial v_3}{\partial x_3} = v_{i,i} \tag{F.14}$$

Another divergence format is $\boldsymbol{v} \cdot \nabla$, which indicates

$$\boldsymbol{v} \cdot \nabla = v_1 \frac{\partial}{\partial x_1} + v_2 \frac{\partial}{\partial x_2} + v_3 \frac{\partial}{\partial x_3} = v_i \frac{\partial}{\partial x_i} \tag{F.15}$$

such that

$$\boldsymbol{v} \cdot \nabla f = v_1 \frac{\partial f}{\partial x_1} + v_2 \frac{\partial f}{\partial x_2} + v_3 \frac{\partial f}{\partial x_3} = v_i f, i \tag{F.16}$$

One notes that $\nabla \cdot \boldsymbol{v} \neq \boldsymbol{v} \cdot \nabla$. The divergence of a gradient of a scalar defines the Laplacian

$$\nabla \cdot \nabla f = \frac{\partial^2 f}{\partial x_1^2} + \frac{\partial^2 f}{\partial x_2^2} + \frac{\partial^2 f}{\partial x_3^2} = f, ii \tag{F.17}$$

There are few identities involving gradients and divergences:

$$\mathbf{\nabla} \cdot (f\mathbf{v}) = (fv_i)_{,i} = f_{,i}\mathrm{v}_i + f\mathrm{v}_{i,i} = \mathbf{v} \cdot \mathbf{\nabla} f + f(\mathbf{\nabla} \cdot \mathbf{v}) \qquad \text{(F.18)}$$

F.7 Curl of a Vector

Define a curl of a vector as

$$\boldsymbol{\nabla} \wedge \boldsymbol{v} = (v_{3,2} - v_{2,3})\boldsymbol{e}_1 + (v_{1,3} - v_{3,1})\boldsymbol{e}_2 + (v_{2,1} - v_{1,2})\boldsymbol{e}_3$$

$$= \varepsilon_{\mathrm{ijk}} \frac{\partial \mathrm{v_k}}{\partial \mathrm{xj}} \mathbf{e_i} = \varepsilon_{\mathrm{ijk}} \mathrm{v_{k,j}} \mathbf{e_i} \qquad \text{(F.19)}$$

Few identities are worth listing below:

$$\boldsymbol{\nabla} \wedge (f\boldsymbol{v}) = \boldsymbol{\nabla} f \wedge \boldsymbol{v} + f\boldsymbol{\nabla} \wedge \boldsymbol{v} \qquad \text{(F.20)}$$

$$\boldsymbol{\nabla} \cdot (\boldsymbol{\nabla} \wedge \boldsymbol{v}) = 0 \qquad \text{(F.21)}$$

$$\boldsymbol{\nabla} \wedge (\boldsymbol{\nabla} f) = 0 \qquad \text{(F.22)}$$

$$\boldsymbol{\nabla} \cdot (\boldsymbol{u} \wedge \boldsymbol{v}) = v \cdot \wedge u - u \cdot \nabla \wedge v \qquad \text{(F.23)}$$

$$\boldsymbol{\nabla} \wedge \boldsymbol{\nabla} \wedge \boldsymbol{v} = \boldsymbol{\nabla}(\boldsymbol{\nabla} \cdot \boldsymbol{v}) - \boldsymbol{\nabla}^2 \boldsymbol{v} \qquad \text{(F.24)}$$

Eq. (F.24) is a form for the Laplacian of a vector field, i.e.,

$$\boldsymbol{\nabla}^2 \boldsymbol{v} = \boldsymbol{\nabla}(\boldsymbol{\nabla} \cdot \boldsymbol{v}) - \boldsymbol{\nabla} \wedge \boldsymbol{\nabla} \wedge \boldsymbol{v} \qquad \text{(F.25)}$$

F.8 Divergence (Green's) Theorem

Given a vector function $\boldsymbol{u}(\boldsymbol{x})$ defined over a volume V surrounded by a surface S, then the migration of the vector $\boldsymbol{u}$ across the surface is related to the divergence of $\boldsymbol{u}$ within the volume V (Fig. F.3), i.e.,

$$\int_S \boldsymbol{u} \cdot \boldsymbol{n} dS = \int_V \mathbf{\nabla} \cdot \boldsymbol{u} \quad dV \qquad \text{(F.26)}$$

where $\boldsymbol{n}$ is the normal unit vector on the surface S. Note that if $\boldsymbol{u} = \boldsymbol{\nabla} f$, then

$$\int_S \nabla f \cdot \boldsymbol{n} dS = \int_S \frac{\partial f}{\partial n} dS = \int_V \nabla^2 f dV \qquad \text{(F.27)}$$

If also $\boldsymbol{u} = g\boldsymbol{\nabla} f$, then

$$\int_S (g\frac{\partial f}{\partial n} - f\frac{\partial g}{\partial n}) dS = \int_V (g\nabla^2 f - f\nabla^2 g) dV \qquad \text{(F.28)}$$

F.9 Stoke's Theorem

Given a vector field $\mathbf{v}(\boldsymbol{x})$ defined over an open surface S, which is terminated by a closed curve C (Fig. F.3), then

$$\oint_C \mathbf{v} \cdot ds = \int_S \boldsymbol{n} \cdot (\boldsymbol{\nabla} \wedge \mathbf{v}) dS \qquad \text{(F.29)}$$

Where ds is length element along the closed curve C and $\boldsymbol{n}$ is the unit normal to the surface S, and $\oint$ is a closed path integration in the counter-clockwise direction.

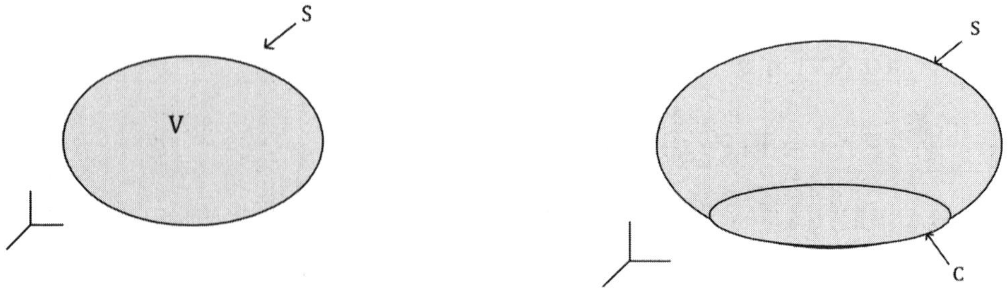

Fig. F.3 *Volume and Surface Integrals*

F.10 Representation of Vector Fields

There are two types of vector fields;

(a) Irrotational Fields

An irrotational vector field $\boldsymbol{a}(\boldsymbol{x})$ is defined by

$$\nabla \wedge \mathbf{a} = 0 \qquad \text{(F.30)}$$

then the vector $\boldsymbol{a}$ can be represented by a scalar potential function $\varphi(x, y, z)$, satisfying the condition in eq. (F.30).

$$\boldsymbol{a} = \nabla\varphi \qquad \text{(F.31)}$$

(b) Solenoidal Fields

A solenoidal vector field $\boldsymbol{a}(\boldsymbol{x})$ is defined by

$$\nabla \wedge \boldsymbol{a} = 0 \qquad \text{(F.32)}$$

then the vector $\boldsymbol{a}$ can be represented by a vector potential function $\boldsymbol{H}(x, y, z)$ with three vector field components, H_1, H_2 and H_3

$$\boldsymbol{a} = \nabla \wedge \mathbf{H} \qquad \nabla \cdot \mathbf{H} = 0 \qquad \text{(F.33)}$$

which satisfies the condition in eq. (F.32).

(c) Helmholtz Representation

In general, one can thus represent a general vector field $\boldsymbol{a}(\boldsymbol{x})$ by scalar potentials.

The first form of the Helmholtz representation, is a combination of the irrotational and solenoidal fields, i.e.,

$$\boldsymbol{a} = \nabla\varphi + \nabla \wedge \mathbf{H} \qquad \nabla \cdot \mathbf{H} = 0 \qquad \text{(F.34)}$$

Of the four scalar potentials, φ, H_1, H_2, and H_3, three must be independent. The second condition gives an equation to tie the three H potentials.

(d) Another vector representation in terms of three independent potentials is given as:

$$\boldsymbol{a} = \boldsymbol{\nabla}\varphi + \boldsymbol{\nabla} \wedge (\psi \boldsymbol{\nabla}\chi) \tag{F.35}$$

with three independent scalar potentials, φ, ψ, and χ.

(e) Yet, another representation takes the form

$$\boldsymbol{a} = \boldsymbol{\nabla}\varphi + \psi \boldsymbol{\nabla}\chi \tag{F.36}$$

with three independent potentials, φ, ψ and χ.

PROBLEMS

Using the index notation, prove the following identities:

1. $\boldsymbol{a} \wedge (\boldsymbol{b} \wedge \boldsymbol{c}) + \boldsymbol{b} \wedge (\boldsymbol{c} \wedge \boldsymbol{a}) + \boldsymbol{c} \wedge (\boldsymbol{a} \wedge \boldsymbol{b}) = 0$

2. Eq. (F.8)

3. $\boldsymbol{a} \wedge (\boldsymbol{b} \wedge \boldsymbol{c}) = (\boldsymbol{a} \cdot \boldsymbol{c})\boldsymbol{b} - (\boldsymbol{a} \cdot \boldsymbol{b})\boldsymbol{c}$

4. $(\boldsymbol{a} \wedge \boldsymbol{b}) \wedge \boldsymbol{c} = (\boldsymbol{c} \cdot \boldsymbol{a})\boldsymbol{b} - (\boldsymbol{c} \cdot \boldsymbol{b})\boldsymbol{a}$

5. $(\boldsymbol{a} \wedge \boldsymbol{b}) \cdot (\boldsymbol{c} \wedge \boldsymbol{d}) = (\boldsymbol{a} \cdot \boldsymbol{c})(\boldsymbol{b} \cdot \boldsymbol{d}) = (\boldsymbol{b} \cdot \boldsymbol{c})(\boldsymbol{a} \cdot \boldsymbol{d})$

6. Eq. (F.9)

7. Eq. (F.10)

8. Eq. (F.20)

9. Eq. (F.21)

10. Eq. (F.22)

11. Eq. (F.23)

12. Eq. (F.24)

13. Eq. (F.25)

14. $|\boldsymbol{a} + \boldsymbol{b}|^2 + |\boldsymbol{a} - \boldsymbol{b}|^2 = 2|\boldsymbol{a}|^2 + 2|\boldsymbol{b}|^2$

15. $|\boldsymbol{a} + \boldsymbol{b}|^2 - |\boldsymbol{a} - \boldsymbol{b}|^2 = 4\boldsymbol{a} \cdot \boldsymbol{b}$

APPENDIX G

MATRIX ALGEBRA

G.1 Definitions

A matrix is a rectangular array of numerical variables (real or complex) that are called *elements*. For example, the following *rectangular matrix* is defined as:

$$\mathbf{A}=\begin{bmatrix} a_{11} & a_{12} & \cdots & a_{1n} \\ a_{21} & a_{22} & \cdots & a_{2n} \\ \vdots & & & \\ a_{m1} & a_{m2} & \cdots & a_{mn} \end{bmatrix} \tag{G.1}$$

Matrix $\boldsymbol{A}$ has $m \cdot n$ elements, with n elements in each row and m elements in each column.

One can also represent a matrix as

$$\mathbf{A} = [a_{ij}] \qquad i = 1, \dots, n \qquad\qquad j = 1, \dots, m \tag{G.2}$$

In this notation, the first subscript denotes the row number and the second subscript denotes the column number.

A *row* matrix is a one-dimensional matrix, i.e.,

$$\boldsymbol{a} = [a_1 \quad a_2 \quad \cdots \quad a_n]$$

A *column* matrix is also a one-dimensional matrix, i.e.,

$$\boldsymbol{a} = \begin{bmatrix} a_1 \\ a_2 \\ \vdots \\ a_m \end{bmatrix}$$

A square matrix has $n = m$.

The diagonal of a square matrix A has the elements:

$$diag\ \mathbf{A} = a_{11}, a_{22}, a_{33}, \dots, a_{nn}$$

Triangular Matrix

A square matrix whose elements below the diagonal are all zero is called *lower triangular matrix*. A square matrix whose elements above the diagonal are all zero is called *upper triangular matrix*.

Unitary (Identity) Matrix **I**

A unitary matrix **I** has $a_{ij=\delta_{ij}}$, i.e., $\mathbf{I} = [\delta_{ij}]$

Transpose of a matrix ***A***

$$\mathbf{A} = [a_{ij}] \qquad \widetilde{\mathbf{A}} = [a_{ji}]$$

$$\overset{\approx}{\mathbf{A}} = \mathbf{A}$$

$$(\widetilde{\mathbf{AB}}) = \widetilde{\mathbf{B}}\widetilde{\mathbf{A}}$$

$$(\mathbf{A}\widetilde{+}\mathbf{B}) = \widetilde{\mathbf{A}} + \widetilde{\mathbf{B}} \qquad \text{(G.3)}$$

Inverse of a square matrix

Define $\mathbf{B} = \mathbf{A}^{-1}$, such that $\mathbf{BA} = \mathbf{I}$ (G.4)

If **A** and **B** are square matrices $n \cdot n$, then

$$(\mathbf{AB})^{-1} = (\mathbf{B}^{-1})(\mathbf{A}^{-1})$$

The *inverse of a diagonal matrix* $\mathbf{A}^{-1} = [1/a_{ij}] \qquad i = j$

A *symmetric square matrix* has $a_{ij} = a_{ji}$. Note that $\mathbf{B} = \frac{1}{2}[\mathbf{A}] + \frac{1}{2}[\widetilde{\mathbf{A}}]$ is a symmetric square matrix.

A *skew-symmetric square matrix* has $a_{ij} = -a_{ji}$. Note that $\mathbf{C} = \frac{1}{2}[\mathbf{A}] - \frac{1}{2}[\tilde{\mathbf{A}}]$ is a skew-symmetric square matrix. Hence, every square matrix can be the sum of a symmetric and skew-symmetric matrices $\mathbf{A} = \mathbf{B} + \mathbf{C}$.

A *complex matrix* is one whose elements are complex constants

$$\mathbf{A} = [z_{pq}] \qquad z_{pq} = a_{pq} + ib_{pq} \qquad i = \sqrt{-1}$$

Complex Conjugate Matrix $\bar{\mathbf{A}}$

$$\bar{\mathbf{A}} = [\bar{z}_{pq}] \qquad \bar{z}_{pq} = a_{pq} - ib_{pq}$$

A square matrix $\mathbf{A}$ is called *Hermetian* if

$$\overset{\simeq}{\mathbf{A}} = \mathbf{A} \qquad \text{i.e.,} \qquad a_{ij} = \bar{a}_{ji}$$

If a Hermetian matrix is real, then it is *symmetric.*

A square matrix $\mathbf{A}$ is called *skew-Hermetian* if

$$\overset{\simeq}{\mathbf{A}} = -\mathbf{A} \qquad \text{i.e.,} \qquad a_{ij} = -\bar{a}_{ji}$$

A skew-Hermetian matrix has zeroes for the diagonal terms. A real skew-Hermetian is *skew-symmetric*.

G.2 Properties of Matrices

(a) Zero matrix

If $\mathbf{A} = 0$, when each $a_{ij} = 0$

(b) Equality of two $m \cdot n$ matrices

$\mathbf{A} = \mathbf{B}$ if $\qquad a_{ij} = b_{ij}$

(c) Addition of two $m \cdot n$ matrices

$\mathbf{C} = \mathbf{A} + \mathbf{B}$ means that $c_{ij} = a_{ij} + b_{ij}$

(d) Negative of a matrix

$$-\boldsymbol{A} = [-a_{ij}]$$

Such that $\mathbf{A} + (-\boldsymbol{A}) = 0$

Difference of two $m \cdot n$ matrices

$\mathbf{C} = \mathbf{A} - \mathbf{B}$ means $c_{ij} = a_{ij} - b_{ij}$

(e) Multiplication by a constant $\propto$

$$\propto \mathbf{A} = [\propto a_{ij}]$$

(f) $\mathbf{A} + \mathbf{B} = \mathbf{B} + \mathbf{A}$ (commutativity)

(g) $(\mathbf{A} + \mathbf{B}) + \mathbf{C} = \mathbf{A} + (\mathbf{B} + \mathbf{C})$ (associativity)

(h) $\mathbf{A} + \mathbf{0} = \mathbf{A}$

(i) $\propto (\beta\mathbf{A}) = (\propto \beta)\mathbf{A}$

(j) $(\propto + \beta)\mathbf{A} = \propto \mathbf{A} + \beta\mathbf{A}$

(k) $0\mathbf{A} = \mathbf{0}$

(l) Matrix multiplication of $n \cdot m$ and $m \cdot \ell$ matrices $\mathbf{A}$ and $\mathbf{B}$ results in $n \cdot \ell$ matrix C:

$$\boldsymbol{C} = \boldsymbol{AB} \qquad c_{ik} = a_{ij}b_{jk}$$

$$i = 1, \dots, n, \qquad j = 1, \dots, m \qquad k = 1, \dots, \ell$$

Note that $\mathbf{AB} \neq \mathbf{BA}$

(m) $(\propto \mathbf{A})\mathbf{B} = \mathbf{A}(\propto \mathbf{B}) = \propto (\mathbf{AB})$

(n) $\mathbf{ABC} = (\mathbf{AB})\mathbf{C} = \mathbf{A}(\mathbf{BC})$

(o) $\mathbf{AB} = \mathbf{0}$ does not imply $\mathbf{A} = \mathbf{0}$ or $\mathbf{B} = \mathbf{0}$

G.3 Determinants of Square Matrices

The determinant of a 2nd rank square matrix

$$\mathbf{A} = \begin{bmatrix} a_{11} & a_{12} \\ a_{21} & a_{22} \end{bmatrix}$$

$$|\mathbf{A}| = det\mathbf{A} = \begin{vmatrix} a_{11} & a_{12} \\ a_{21} & a_{22} \end{vmatrix} = a_{11}a_{22} - a_{12}a_{21}$$

For matrices of rank = 3

$$\mathbf{A} = \begin{bmatrix} a_{11} & a_{12} & a_{13} \\ a_{21} & a_{22} & a_{23} \\ a_{31} & a_{32} & a_{33} \end{bmatrix}$$

Using the elements of the first row as pivot points, i.e., a_{11}, a_{12} and a_{13} we get

$$|\mathbf{A}| = a_{11}\mathrm{A}_{11} + a_{12}\mathrm{A}_{12} + a_{13}\mathrm{A}_{13} = a_{1i}\mathrm{A}_{1i}$$

where A_{1i} is called the *cofactor* defined by

$$A_{1i} = (-1)^{1+i} M_{1i}$$

where M_{1i} is the determinant of the minor matrix obtained by deleting 1st row and i^{th} column

$$M_{11} = \begin{vmatrix} a_{22} & a_{23} \\ a_{32} & a_{33} \end{vmatrix} \quad M_{12} = \begin{vmatrix} a_{21} & a_{23} \\ a_{31} & a_{33} \end{vmatrix} \quad M_{13} = \begin{vmatrix} a_{21} & a_{22} \\ a_{31} & a_{32} \end{vmatrix}$$

Similarly, the determinant can be evaluated by using other rows or columns. For example, using the 2nd row elements as pivot points, then

$$|\boldsymbol{A}| = a_{21}A_{21} + a_{22}A_{22} + a_{23}A_{23}$$

where $A_{21} = -M_{21}$, $A_{22} = M_{22}$, $A_{23} = -M_{23}$

$$M_{21} = \begin{vmatrix} a_{12} & a_{13} \\ a_{32} & a_{33} \end{vmatrix} \quad M_{22} = \begin{vmatrix} a_{11} & a_{13} \\ a_{31} & a_{33} \end{vmatrix} \quad M_{23} = \begin{vmatrix} a_{11} & a_{12} \\ a_{31} & a_{32} \end{vmatrix}$$

$$|\boldsymbol{A}| = a_{2i}A_{2i} = a_{2i}(-1)^{i+2}M_{2i}$$

In general, using the elements of a specific row or column, the determinant can be evaluated

$$|\boldsymbol{A}| = a_{ji}A_{ji} = (-1)^{i+j}a_{ji}M_{ji} \text{ where either i or j is fixed.} \quad \text{(G.5)}$$

The computation of the $|\boldsymbol{A}|$ for higher ranked matricies follows the same procedures.

G.4 Properties of Determinants of Square Matricies

(a) If all the elements of a row or a column are zero,

then $|\boldsymbol{A}| = 0$

(b) If two rows (columns) are identical or proportional to each other,

then $|\boldsymbol{A}| = 0$

(c) $|\tilde{\boldsymbol{A}}| = |\boldsymbol{A}|$

(d) If a row or a column of a square matrix **B** is multiplied by the same constant $\propto$, then

$|\boldsymbol{B}| = \propto |\boldsymbol{A}|$

where **A** is the matrix **B** without the constant $\propto$.

(e) If matrix **B** is obtained by interchanging any two rows (columns) of a matrix **A**, then,

$|\boldsymbol{B}| = -|\boldsymbol{A}|$

(f) If every element of a matrix **B** is multiplied by a factor $\propto$, then

$|\boldsymbol{B}| = \propto^n |\boldsymbol{A}|$

where **A** is the matrix **B** without the factor $\propto$.

(g) $|\boldsymbol{A} + \boldsymbol{B}| \neq |\boldsymbol{A}| + |\boldsymbol{B}|$

(h) $|\boldsymbol{AB}| = |\boldsymbol{A}||\boldsymbol{B}|$

(i) Since $(\boldsymbol{A}^{-1})(\boldsymbol{A}) = \boldsymbol{I}$

then one can solve for the matrix elements of $\boldsymbol{A}^{-1}$

Let $\boldsymbol{A}^{-1} = [b_{ij}]$

then the elements of the inverse matrix

$$b_{ij} = \frac{A_{ji}}{|\boldsymbol{A}|} \qquad \text{for } |\boldsymbol{A}| \neq 0 \tag{G.6}$$

where A_{ji} is the cofactor of the element a_{ji}.

G.5 Solution of Linear Algebraic Equations

Given the following system of linear algebraic equations in n unknowns

$$\begin{aligned} a_{11}x_1 + a_{12}x_2 + \cdots + a_{1n}x_n &= c_1 \\ a_{21}x_1 + a_{22}x_2 + \cdots + a_{2n}x_n &= c_2 \\ \vdots \qquad \vdots \qquad\qquad \vdots \\ a_{n1}x_1 + a_{n2}x_2 + \cdots + a_{nn}x_n &= c_n \end{aligned}$$

This can be written in matrix form

$$\boldsymbol{Ax} = \boldsymbol{C} \tag{G.7}$$

$$a_{ij}x_j = c_i$$

Multiply eq. (G.7) by $\boldsymbol{A}^{-1}$ and using eqs. (G.4) and (G.6)

$$\boldsymbol{A}^{-1}\boldsymbol{Ax} = \boldsymbol{A}^{-1}\boldsymbol{C}$$

The solution $\boldsymbol{x}$ of the system of algebraic equations is:

$$\boldsymbol{x} = \boldsymbol{A}^{-1}\boldsymbol{C} \tag{G.8}$$

$$x_i = b_{ij}c_j$$

$$= \frac{A_{ji}C_j}{|A|} \tag{G.9}$$

This is known as *Cramer's Rule*.

Another form of Cramer's Rule takes another format. To find X_i, replace the ith column by the elements of the column C_i.

$$x_i = \frac{\overset{i^{th}\text{ column}}{\begin{vmatrix} a_{12} & \cdots & c_1 & \cdots & a_{1n} \\ a_{21} & \cdots & c_2 & \cdots & a_{2n} \\ a_{n1} & \cdots & c_n & \cdots & a_{nn} \end{vmatrix}}}{|A|} \tag{G.10}$$

G.6 Eigenvalues of Hermetian Matrices

Eigenvalue problems of discrete systems have the form

$$\boldsymbol{Ax} = \lambda\, \boldsymbol{x}$$

$$\text{or } (\boldsymbol{A} - \lambda\, \boldsymbol{I})\boldsymbol{x} = 0 \tag{G.11}$$

If the matrix $\boldsymbol{A}$ is Hermetian, then multiplying eq. (G.11) by $\overset{\simeq}{x}$ giving

$$\overset{\simeq}{\boldsymbol{x}}\boldsymbol{Ax} = \lambda\, \overset{\simeq}{\boldsymbol{x}}\boldsymbol{x}$$

where $\overset{\simeq}{\boldsymbol{x}}\boldsymbol{x} = |x_1|^2 + |x_2|^2 + \cdots + |x_n|^2$, which is real.

The eigenvalue is then given by:

$$\lambda = \frac{\overset{\simeq}{\boldsymbol{x}}\boldsymbol{Ax}}{\bar{\boldsymbol{x}}\boldsymbol{x}} \tag{G.12}$$

It can be shown that the numerator is also real, hence the eigenvalues in eq. (G.12) are real. Similarly, the eigenvalues of a skew-Hermetian are all imaginary. Since one is looking for a non-trivial solution to eq. (G.11), the eigenvalues are obtained by setting

$$det|\boldsymbol{A} - \lambda\, \boldsymbol{I}| = 0 \tag{G.13}$$

The determinant of eq. (G.13) leads to a nth polynomial in λ. This leads to n roots λ_i, $i = 1, 2, \dots, n$. For each eigenvalue λ_i, there exists a solution x_i called the eigenvector. This is obtained by solving $(n-1)$ equations of eq. (G.11), with one of the elements x_i given a unit value so that the remaining elements x_j are actually ratios of x_j / x_i $\quad j = 1, 2, \dots, n - i.$

G.7 Properties of Eigenvalues and Eigenvectors

(a) If the eigenvalues $\lambda_1, \lambda_2, \dots, \lambda_n$ are distinct, then the corresponding eigenvectors are linearly independent.

(b) The eigenvalues of $\boldsymbol{A}$ and $\widetilde{\boldsymbol{A}}$ are the same.

(c) The eigenvalues of $\overline{\boldsymbol{A}}$ and $\overset{\simeq}{\boldsymbol{A}}$ are the complex conjugate eigenvalues of $\boldsymbol{A}$

(d) If λ_i, $i = 1, \dots, n$ are the eigenvalues of a square matrix $\boldsymbol{A}$, then for a constant factor $\propto$, the eigenvalues of $\propto \boldsymbol{A}$ are $\propto \lambda_i$

PROBLEMS

Prove the following identities:

1. $(\boldsymbol{A}+\boldsymbol{B})^2 = \boldsymbol{A}^2 + \boldsymbol{AB} + \boldsymbol{BA} + \boldsymbol{B}^2$

2. $(\boldsymbol{A}-\boldsymbol{B})^2 = \boldsymbol{A}^2 - \boldsymbol{AB} - \boldsymbol{BA} + \boldsymbol{B}^2$

3. $(\boldsymbol{A}-2\boldsymbol{I})(\boldsymbol{A}+2\boldsymbol{I}) = \boldsymbol{A}^2 - 4\boldsymbol{I}$

4. $(\boldsymbol{A}+2\boldsymbol{I})^2 = \boldsymbol{A}^2 + 4\boldsymbol{A} + 4\boldsymbol{I}$

REFERENCES

General

Abramowitz. M., and Stegun, I. A., *Handbook of Mathematical Functions*, Dover Publications, New York, 1964.

CRC Handbook of Mathematical Sciences. 5th ed. West Palm Reach, FL: CRC Press, 1978.

Encyclopedia of Mathematics. 10 vols., Reidel, Hingham, MA, 1990.

Erdelyi, A., et al., *Higher Transcendental Functions*, 3 vols., McGraw-Hill Book Company, New York, 1953.

Fletcher, A., J. C. P. Miller, L. Rosenhead, and L. J. Comrie, *An Index of Mathematical Tables*, Blackwell, Oxford, 1962.

Gradstien, I. S., and Ryzhik, I. N., *Tables of Series, Products and Integrals*. Academic Press, New York, 1966.

Ito, K. (ed.), *Encyclopedic Dictionary of Mathematics*. 4 vols. 2nd ed. MIT Press, Cambridge, MA, 1987.

Janke, E., and Emde, F., *Tables of Functions*, Dover Publications, New York, 1945.

Pearson, C. E. (ed.), *Handbook of Applied Mathematics*, 2nd ed., Van Nostrand Reinhold, New York, 1983.

Chapter 2

Birkhoff, G., and Rota, G. *Ordinary Differential Equations*, 3rd. ed., John Wiley, New York, 1978

Carrier, G. F., and Pearson, C. E. *Ordinary Differential Equations*, Blaisdell Publishing Company, Waltham. Massachusetts, 1968.

Coddington, E. A. and N. *Levinson.Theory of Ordinary Differential Equations*, McGraw-Hill Book Company, New York, 1955.

Duff, G. F. D., and D. Naylor, *Differential Equations of Applied Mathematics*, Wiley, New York, 1966.

Forsyth, A. R. *The Theory of Differential Equations*, Dover Publications, Inc., New York.

Greenspan, D., Theory and Solution of Ordinary Differential Equations. Macmillan Co., New York, 1960

Ince, E. L. *Ordinary Differential Equations*, Dover Publications, Inc., New York, 1956.

Kamke, E. *Differentialgleichungen Losungsmethoden und Losungen*, Chelsea Publishing Company, New York, 1948. (Many solutions to Ordinary Differential Equations).

Kaplan, W. *Ordinary Differential Equations*, Addison-Wesley, Reading, Mass., 1958.

Spiegel, M. R., *Applied Differential Equations*, 3d ed., Prentice-Hall, Englewood Cliffs, N. J, 1991.

Chapter 3

Abramowitz. M., and Stegun, I. A., *Handbook of Mathematical Functions*,Dover Publications, New York, 1964.

Bell, W. W., *Spherical Functions for Scientists and Engineers*, Van Nostrand Co., New Jersey, 1968.

Buchholz, H., *The Confluent Hypergeometric Function with Special Emphasis on Its Applications*, Springer, 1969.

Byerly, W. E., *An Elementary Treatise on Fourier Series and Spherical, Cylindrical and Ellipsoidal Harmonics with Applications* Dover Publications, Inc., New York. 1959.

Gray, A., Mathews, G. B., and MacRoberts, T. M., *A Treatise on Bessel Functions and Their Applications to Physics*, 2nd ed., Macmillan and Co., London, 193 1.

Hobson, E. W., *The Theory of Spherical and Ellipsoidal Harmonics*, Cambridge University Press, England, 1931.

Luke, Y. L. 1969. *The Special Functions and Their Approximations*. 2 vols. Academic Press, New York, 1969

Luke, Y. L. *Algorithms for the Computation of Mathematical Functions*. Academic Press, New York, 1975

Luke, Y. L. *Mathematical Functions and Their Approximations*. Academic Press, New York, 1975

MacLachlan N. W., *Bessel Functions for Engineers*, 2nd. ed., Clarendon Press, Oxford, England, 1955.

MacRobert, T. M., *Spherical Harmonics, An Elementary Treatise on Harmonic Functions with Applications*, Pergamon Press, New York, 1967.

Magnus, W. and Oberhettinger, F., *Special Functions of Mathematical Physics*, Chelsea Publishing Company, New York, 1949.

Magnus, W., OberhetLinger, F., Soni, R. P., *Formulas and Theorems for the Special Functions of Mathematical Physics*. Springer-Verlag, New York, 1966.

McLachlan, N. W. *Theory and Application of Mathieu Functions*, Dover, New York, 1954

Prasad, G., *A Treatise on Spherical Harmonics and the Functions of Bessel and Lame*, Part I and II, Mahamandal Press, Benares City, in 1930 and 1932 respectively.

Rainville, Earl.D. *Special Functions*, Macmillan Co., New York,1960.

Sneddon, I. N. *Special Functions of Mathematics, Physics, and Chemistry*, 3d ed., Longman, New York, 1980

Stratton, J. A., P. M. Morse, L. J. Chu, and R. A. Hutner *Elliptic Cylinder and Spheroidal Wave Functions*, Wiley, New York, 1941.

Szego, G.*Orthogonal Polynomials*, American Mathematical Society, New York, 1939.

Watson, G. N., *A Treatise on the Theory of Bessel Functions*, Cambridge University Press, Cambridge and the Macmillan and Co., New York, 2nd ed, 1958.

Wittaker, E. T. and Watson, G. N. *A Course of Modem Analysis*. Cambridge University Press, Cambridge and the Macmillan Co., New York, 4th ed., 1958.

Chapter 4

Bleich, F., *Buckling Strength of Metal Structures*, McGraw-Hill, New York, 1952.
Byerly, W. E., *An Elementary Treatise on Fourier Series and Spherical, Cylindrical and Ellipsoidal Harmonics*, Dover Publications Inc., New York, 1959.
Carslaw, H. S., *Introduction to the Theory of Fourier's Series and Integrals*, Dover Publications, Inc., New Yofk, 1930.
Churchill, R. V., and J. W. Brown, *Fourier Series and Boundary Value Problems*, 3rd ed., McGraw-Hill Book Co., New York, 1978.
Courant, R., and Hilbert, D. *Methods of Mathematical Physics, Volume 1*. Wiley (Interscience), New York 1962.
Duff, G. F. D., and Naylor, D. *Differential Equations of Applied Mathematics*, Wiley. New York, 1966.
Erdelyi, A., ed., *Higher Transcendental Functions*, 3 vols., McGraw-Hill , New York, 1953.
Hildebrand. F. B., *Advanced Calculus for Applications*, Prentice-Hall, Inc., Englewood Cliffs, New Jersey, 1962.
Ince. E. L., *Ordinary Differential Equations*, Dover Publications, Inc., New York, 1945.
Lebedev, N. N., *Special Functions and Their Approximations*, Prentice-Hall, Englewood Cliffs, New Jersey, 1965.
Miller, K. S., *Engineering Mathematics*, Dover Publications, Inc., Now York, 1956.
Morse, P. K. and Feshbach, H., *Methods of Theoretical Physics, Parts I and II*. McGraw-Hill, New York, 1953.
Morse, P. M., *Vibration and Sound*, McGraw-MU, New York, 1948.
Morse. P. K, and Ingard. U., *Theoretical Acoustics*, McGraw-Hill, New York, 1968.
Mumaghan, F. D., *Introduction to Applied Mathematics*, Dover Publications, Inc., New York, 1963.
Oldunburger, R., *Mathematical Engineering Analysis*, Dover Publications, Inc., New York, 1950.
Rayleigh, J. W. S., *The Theory of Sound, Parts I and II*, Dover Publications, Inc., New York, 1945.
Sagan, H., *Boundary and Eigenvalue Problems in Mathematical Physics*, Wiley, New York. 1961.
Timoshenko, S. and Young, D. H. *Vibration Problems in Engineering*, Von Nostrand, Princeton, New Jawy, 1955.
Von Karmen, T., Biot. M. A. *Mathematical Methods in Engineering*, McGraw-HW Book Co., New York, 1940.
Whittaker, E. T. and Watson, G. N. *Modem Analysis*, Cambridge University P=, New York, 1958.

Chapter 5

Ahlfors, L. V., *Complex Analysis.* 3d ed. McGraw-Hill, New York, 1979.

Carrier, G. F., Krook, M. and Pearson, C. E., *Functions of a Complex Variable*, McGraw-Hill Co., New York, 1966.

Churchill, R., J. Brown, and R. Verhey, *Complex Variables and Applications,* 3rd ed, McGraw-Hill, New York, 1974.

Copson, E. T., *An Introduction to the Theory of Functions of a Complex Variable*, Clarendon Press, Oxford, 1935.

Forsyth, A. R., *Theory of Functions of a Complex Variable*, Cambridge University Press, Cambridge, 1918, and in two volumes, Vol I and II by Dover Publications, New York, 1965.

Franklin, P., *Functions of Complex Variables*, Prentice-Hall New Jersey, 1958.

Knopp, K., *Theory of Functions*, 2 vols. Dover, New York, 1947

Kyrala, A., *Applied Functions of a Complex Variable*. John Wiley. New York, 1972

MacRobert, T. M., *Functions of a Complex Variable*, MacMillan and Co. Ltd, London, 1938.

McLachlan, N. W. *Complex Variables and Operational Calculus*, Cambridge,

Miller, K. S., *Advanced Complex Calculus*, Harper and Brothers, New York, 1960 and Dover Publications, Inc., New York, 1970.

Pennisi,L L., Gordon, L. I. and Lashers, S., *Elements of Complex Variables*, Holt, Rinehart and Winston, New York, 1967.

Silverman, H., *Complex Variables*. Houghton Mifflin. Boston, 1975.

Silverman, R, A. *Complex Analysis with Applications*, Prentice-Hall, Englewood Cliffs, N. J. 1974.

Titchmarsh, E. C, *The Theory of Functions*, 2nd ed., Oxford University Press, London, 1939, reprinted 1975.

Whittaker, E. T., and Watson, G. N., *A Course of Modem Analysis*, Cambridge University Press, London, 1952.

Chapter 6

Books on Partial Differential Equations of Mathematical Physics

Bateman, H., *Differential Equations of Mathematical Physics*, Cambridge, New York, 1932.

Carslaw, H. S., and J. C. Jaeger. *Operational Methods in Applied Mathematics*, Oxford, New York, 1941.

Churchill, R. V., *Fourier Series and Boundary Value Problems*, McGraw-Hill Book Co., New York, 1941.

Courant, R. and Hilbert, D., *Methods of Mathematical Physics*, Vols. I and II, Interscience Publishers, Inc., New York, 1962.

Gilbarg, D. and N. S. Trudinger., *Elliptic Partial Differential Equations of Second Order*, Springer, New York, 1977.

Hadamard, J. S. *Lectures on Cauchy's Problem in Linear Partial Differential Equations*, Yale University Press, New Haven, 1923.
Hellwig. G., *Partial Differential Equations*, 2nd ed., Teubner, Stuttgart, 1977.
Jeffreys, H. and B. S., *Methods of Mathematical Physics*, Cambridge, New York, 1946.
John, F., *Partial Differential Equations*, Springer, New York, 1971.
Kellogg, 0. D., *Foundations of Potential Theory*, Dover Publications, Inc., New York.
Morse, P. M. and Feshbach, H. *Methods of Theoretical Physics*, Parts I and II, McGraw-Hill Book Co., Inc., New York, 1953.
Sneddon, I. N. *Elements of Partial Differential Equations*, McGrawHill, New York,1957.
Sommerfeld, A., *Partial Differential Equations in Physics*, Academic Press, New York. 1949.
Stakgold, I., *Boundary Value Problems of Mathematical Physics*, vols. I and II, Macmillan Co., New York.
Webster, A. G., *Partial Differential Equations of Mathematical Physics*, Dover Publications, Inc., New York, 1955.

Books on Heat Flow and Diffusion

Carslaw, H. S. and Jaeger, J. C., *Conduction of Heat in Solids*, Oxford University Press, New York, 1947.
Hopf, E., *Mathematical Problems of Radiative Equilibrium*, Cambridge, New York, 1952.
Sneddon, I. N., *Fourier Transforms*, McGraw-Hill Book Co., New York, 1951.
Widder, D. V., *The Heat Equation*. Academic Press, New York, 1975.

Books on Vibration, Acoustics and Wave Equation

Baker, B. B., and E. T. *Copson Mathematical Theory of Huygen's Principle*, Oxford, New York, 1939.
Morse, P. M., *Vibration and Sound*, McGraw-Hill, New York, 1948.
Morse, R. M. and Ingard, U., *Theoretical Acoustics*, McGraw-Hill, New York, 1968.
Rayleigh, J.S., *Theory of Sound*, Dover Publications, Inc. New York, 1948.

Books on Mechanics, Hydrodynamics, and Elasticity

Lamb, H. *Hydrodynamics*, Cambridge, NewYork, 1932, Dover, New York,1945.
Love, A. E. H. *Mathematical Theory of Elasticity*, Cambridge, New York,1927, reprint Dover, New York, 1945.
Sokolnikoff, I. B. *Mathematical Theory of Elasticity*, McGraw Hill, New York, New York, 1946
Timoshenko, S. *Theory of Elasticity*, McGraw-Hill, New York, 1934

Chapter 7

Carslaw, H. B. *Theory of Fourier Series and Integrals*, Macmillan, New York, 1930.
Doetach, G. *Theorie und Anwendung der Laplace-Transformation*, Dover, New York, 1943.
Doetsch, G. 1971. *Guide to the Application of Laplace and Z-Transform*, 2d ed. Van Nostrand-Reinhold, New York, 1971.
Erdelyi, A., W. Magnus, F. Oberhettinger, and F. Tricomi. *Tables of Integral Transforms*. 2 vols. McGraw-Hill, New York,1954.
Oberhettinger, F. and L. Badii. *Tables of Laplace Transforms*. Springer, New York, 1973.
Paley, R. E. A. C., and N. Wiener. *Fourier Transforms in the Complex Plane*, American Mathematical Society, New York, 1934.
Sneddon, I. N. *Fourier Transforms*, McGraw-Hill, New York, 1951
Sneddon, I. N. *The Use of Integral Transforms*. McGraw-Hill, New York, 1972.
Titchmarsh, E. C. *Introduction to the Theory of Fourier Integrals*, Oxford, New York, 1937.
Weinberger, H. F. *A First Course in Partial Differential Equations with Complex Variables and Transform Methods*, Blaisdell,Waltham, Mass., 1965
Widder, D. V. *The Laplace Transform*. Princeton University Press, Princeton, NJ, 1941.

Chapter 8

Bateman, H., *Partial Differential Equations of Mathematical Physics*, Cambridge, New York, 1932.
Caralaw, H. S. *Mathematical Theory of the Conduction of Heat in Solids*,
Dover, New York, 1945.
Friedman, A. *Generalized Functions and Partial Differential Equations*. Prentice-Hall, Inc., 1963.
Kellogg, 0. D. *Foundations of Potential Theory*, Springer, Berlin, 1939.
MacMillan, W. D. *Theory of the Potential*, McGraw-Hill, New York, 1930.
Melnikov, Yu. A. *Green's Functions in Applied Mechanics, Topics in Engineering*, vol. 27, Computational Mechanics Publications, Southampton, UK, 1995.
Roach, G. *Green's Functions*. London: Van Nostrand Reinhold, 1970.
Stakgold, I. *Green's Functions and Boundary and Value Problems*, Wiley-Interscience, New York, 1979.

Chapter 9

Bleistcin, N., and R. A. Handelsman., *Asymptotic Expansion of Integrals,* Holt, Rinehart, and Winston, New York, 1975
Cesari, L., *Asymptotic Behavior and Stability Problems in Ordinary Differential Equations*, 3rd ed. Springer Verlag, New York, 1971.
de Bruijn, N., *Asymptotic Methods in Analysis*, North Holland Press, Amsterdam, 1958.
Dingle, R. B., *Asymptotic Expansions: Their Derivation and Interpretation*, Academic Press, New York, London, 1973
E. Copson, E., *Asymptotic Expansions*, Cambridge University Press, Cambridge, UK, 1965.
Erdelyi, A., *Asymptotic Expansions*, Dover, New York, 1961
Evgrafov, M. A., *Asymptotic Estimates and Entire Functions*, Gordon and Breach, New York, 1962
Jeffrys, H., *Asymptotic Approximations*, Oxford University Press, 1962.
Lauwerier, H. *Asymptotic Expansions*, Math. Centrum (Holland).
Olver, F. W. J. *Asymptotics and Special Functions*, Academic Press, New York, 1974.
Sirovich, L. *Techniques of Asymptotic Analysis*, Springer-Verlag, New York, 1971.
Wasow, W. *Asymptotic Expansions for Ordinary Differential Equations*, John Wiley, New York, 1965.

Chapter 10

Akai, T. J. *Applied Numerical Methods for Engineers*, John Wiley and Sons, New York, NY, 1993.
Borse, G. J. *Fortran 77 and Numerical Methods for Engineers*, 2nd ed., PWS-Kent Publishing Co., Boston, MA, 1991.
Cheney, W. and D. Kincaid. *Numerical Mathematics and Computing*, 5th ed., Thomson Brooks/Cole, United States, 2004.
Hoffman, J. D. *Numerical Methods for Engineers and Scientists*, McGraw-Hill, New York, 1992.
James, M. L., G.M. Smith, and J.C. Wolford. *Applied Numerical Methods for Digital Computation with Fortran*, International Textbook Co., Scranton, PA, 1967.
Lapidus, L. and G. F. Pinder. *Numerical Solution of Partial Differential Equations in Science and Engineering*, John Wiley and Sons, New York, NY, 1981.
McCormick, J. M. and M. G. Salvadori, *Numerical Methods in Fortran*, Prentice-Hall, Englewood, NJ, 1965.
Salvadori, M. G. and M. L. Baron. *Numerical Methods in Engineering*, Prentice-Hall, Englewood Cliffs, NJ, 1952.
Scheid, F., *Theory and Problems of Numerical Analysis*, 2nd ed., Schaum's Outline Series, McGraw-Hill, New York, NY 1989.

Appendix A

Green, J. A., *Sequences and Series*, Library of Mathematics, London, ed. by W.Kegan Paul, 1958.

Markushevich, A. *Infinite Series*, D. C. Heath, Boston, 1967.

Rektorys, K., Ed. *Survey of Applicable Mathematics*, the M.I.T. Press, Mass. IML of Tech.,Cambridge, Mass, 1969

Zygmund, A.. *Trigonometric Series.* 2nd ed., reprinted with corrections. Cambridge: University Press, 1977.

$\lambda = \pm i, 0$

$c_1(\cos x + i\sin x) + c_2(\cos x - i\sin x)$

$(c_1 + c_2)\cos x \quad c_3 \cos x$

$(c_1 - c_2)\, i \sin x \quad c_4 \sin x$

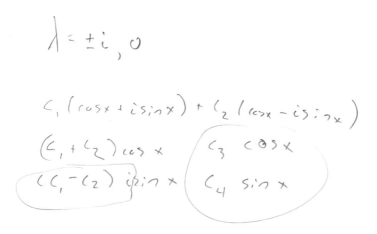

ANSWERS

Chapter 1

1. (a) $y = (c + x)\,e^{-x^2/2}$

   (b) $y = c\,x^{-2} + x^2/4$

   (c) $y = c\,(\sin x)^{-2} + (\sin x)/3$

   (d) $y = \dfrac{1}{\cosh x}\left(c + \dfrac{1}{2}\left[\dfrac{e^{2x}}{2} + x\right]\right)$

   (e) $y = c \cot x + \csc x$

   (f) $y = c\,e^{-x} + x\,e^{-x}$

3. (a) $y = c_1\,e^{-x} + c_2\,e^{2x}$

   (b) $y = c_1\,e^{x/2} + c_2\,e^{-x/2} + c_3\,e^{x}$

   (c) $y = (c_1 + c_2x)\,e^{x} + c_3\,e^{-2x}$

   (d) $y = (c_1 + c_2x)\,e^{-2x} + (c_3 + c_4x)\,e^{2x}$
   $= \bar{c}_1 \sinh(2x) + \bar{c}_2 \cosh(2x) + x(\bar{c}_3 \sinh(2x) + \bar{c}_4 \cosh(2x))$

   (e) $y = c_1\,e^{2x} + c_2\,e^{-2x} + c_3\,e^{2ix} + c_4\,e^{-2ix}$
   $= \bar{c}_1 \sinh(2x) + \bar{c}_2 \cosh(2x) + \bar{c}_3 \sin(2x) + \bar{c}_4 \cos(2x)$

   (f) $y = c_1 \exp\left(\dfrac{1-i}{\sqrt{2}}\,x\right) + c_2 \exp\left(\dfrac{-1+i}{\sqrt{2}}\,x\right)$

   (g) $y = e^{-z}\,(c_1 \sin z + c_2 \cos z) + e^{z}\,(c_3 \sin z + c_4 \cos z)$
   $= \bar{c}_1 \sin z \sinh z + \bar{c}_2 \sin z \cosh z + \bar{c}_3 \cos z \sinh z + \bar{c}_4 \cos z \cosh z$
   where $z = x\sqrt{2}$

   (h) $y = e^{x}\,(c_1 + c_2x + c_3x^2) + e^{-x}\,(c_4 + c_5x)$

   (i) $y = c_1 e^{-2ax} + e^{ax}\left[c_2 \sin(ax\sqrt{3}) + c_3 \cos(ax\sqrt{3})\right]$

   (j) $y = c_1\,e^{-ax} + e^{ax}\,[\,c_2 \sin(ax) + c_3 \cos(ax)]$

   (k) $y = (c_1 + c_2x) \sin(ax) + (c_3 + c_4x) \cos(ax)$

   (l) $y = c_1 \sin(2x) + c_2 \cos(2x) + e^{-x\sqrt{3}}(c_3 \sin x + c_4 \cos x)$
   $+ e^{x\sqrt{3}}\,(c_5 \sin x + c_6 \cos x)$

5. (a) $y = c_1 x + c_2 x^{-1}$

   (b) $y = c_1 x^{-1} + c_2 x^{-1} \log x$

   (c) $y = c_1 \sin(\log x^2) + c_2 \cos(\log x^2)$

   (d) $y = c_1 x + c_2 x^{-1} + c_3 x^2$

   (e) $y = (c_1 + c_2 \log x)\, x + c_3 x^{-2}$

   (f) $y = c_1 x + c_2 x^{-2} + c_3 \sin(\log x^2) + c_4 \cos(\log x^2)$

   (g) $y = x^{1/2}(c_1 + c_2 \log x)$

   (h) $y = x^{1/2}[c_1 \sin(\log x) + c_2 \cos(\log x)]$

6. (a) $y_p = -2e^x - 3 \sin x + \cos x - (3x^2/2 + x)\, e^{-x}$

   (b) $y_p = x^2 - 3x + 9/2 + e^{-x} + (3x^4/4 - x^3 + x^2)\, e^x$

   (c) $y_p = [\sin(2x) + x^2 \sinh(2x)]/4$

   (d) $y_p = x^2 + 2x \log x$

   (e) $y_p = x^2 + 2x(\log x)^2$

7. (a) $y = c_1 \sin(kx) + c_2 \cos(kx) + \frac{1}{k} \int_1^x \sin(k(x - \eta)) f(\eta)\, d\eta$

   (b) $y = c_1 x + c_2 x^{-1} + \frac{1}{2} \int_1^x (x\eta^{-2} - x^{-1}) f(\eta)\, d\eta$

   (c) $y = c_1 x + c_2 x^2 + c_3 x^2 \log x + \int_1^x (x\eta - x^2(1 + \log \eta) + x^2 \log x) \frac{f(\eta)}{\eta^3}\, d\eta$

   (d) $y = c_1 e^{kx} + c_2 e^{-kx} + \frac{1}{k} \int_1^x \sinh(k(x - \eta)) f(\eta)\, d\eta$

Chapter 2

1. (a) $\rho \to \infty$, $-\infty < x < \infty$
   (b) $\rho \to \infty$, $-\infty < x < \infty$
   (c) $\rho = 1$, $-1 < x < +1$
   (d) $\rho = 1$, $-1 \le x \le 1$
   (e) $\rho = 1$, $-1 \le x \le 1$
   (f) $\rho = 2$, $-2 < x < 2$
   (g) $\rho = 2$, $-2 < x < 2$
   (h) $\rho = 4$, $-4 < x < 4$
   (i) $\rho = 2$, $-1 < x < 3$
   (j) $\rho = 3$, $-4 \le x \le 2$

2. (a) $y = c_1[1 - \frac{x^3}{6} + \frac{x^6}{45} - ...] + c_2[x - \frac{x^4}{6} + \frac{5x^7}{252} - ...]$

   (b) $y = c_1[1 + \frac{x^2}{2\,1!} - \frac{x^4}{2^2\,2!} + \frac{1.3x^6}{2^3\,3!} - \frac{1\cdot 3\cdot 5x^8}{2^4\,4!} + ...] + c_2 x$

   (c) $y = c_1 \sum_{m=0}^{\infty} \frac{x^{2m}}{2^m\,m!} + c_2[x + \frac{x^3}{3} + \frac{x^5}{3\cdot 5} + \frac{x^7}{3\cdot 5\cdot 7} + ...]$

   (d) $y = c_1[1 + x^2 + \frac{11}{12}x^4 + \frac{7}{12}x^6 + ...] + c_2[x + x^3 + \frac{3}{4}x^5 + \frac{5}{12}x^7 + ...]$

   (e) $y = c_1[1 + x^2 + \frac{x^3}{6} + \frac{x^4}{3} + \frac{11x^5}{120} + \frac{13x^6}{180} + ...] + c_2[x + \frac{x^3}{2} + \frac{x^4}{12} + \frac{x^5}{18} + \frac{x^6}{30} + ...]$

   (f) $y = c_1[1 - \frac{x^3}{6} + \frac{x^6}{45} - ...] + c_2[x - \frac{x^4}{6} + \frac{5x^7}{252} - ...] + c_3[x^2 - \frac{3x^5}{20} + \frac{9x^8}{560} - ...]$

   (g) $y = c_1 \sum_{n=0}^{\infty} (-1)^n(2n+1)x^{2n} + c_2 \sum_{n=0}^{\infty} (-1)^n(n+1)x^{2n+1}$

   (h) $y = c_1(x - x^3) + c_2 \sum_{n=0}^{\infty} \frac{x^{2n}}{(2n-3)(2n-1)}$

   (i) $y = c_1[1 + \frac{x^2}{2!} + \frac{x^3}{3!} + \frac{3x^4}{4!} + ...] + c_2 x[1 + \frac{x^2}{3!} + \frac{2x^3}{4!} + ...]$

   (j) $y = c_1 \sum_{n=0}^{\infty} (-1)^n \frac{x^{4n+2}}{(2n+1)!} + c_2 \sum_{n=0}^{\infty} (-1)^n \frac{x^{4n}}{(2n)!}$

3. (a) $y = c_1[1 - \frac{(x+1)^2}{2} + \frac{(x+1)^3}{3} - \frac{5(x+1)^4}{24} + ...]$

$+ c_2[(x+1) - (x+1)^2 + \frac{2(x+1)^3}{3} - \frac{5(x+1)^4}{12} + ...]$

(b) $y = c_1[1 + \frac{(x-1)^4}{12} + \frac{(x-1)^8}{12 \cdot 56} + \frac{(x-1)^{12}}{12 \cdot 56 \cdot 132} + ...]$

$+ c_2[(x-1) + \frac{(x-1)^5}{20} + \frac{(x-1)^9}{20 \cdot 56} + \frac{(x-1)^{13}}{20 \cdot 72 \cdot 156} + ...]$

(c) $y = c_1 \sum_{n=0}^{\infty} (2n+1)(x-1)^{2n} + c_2 \sum_{n=0}^{\infty} (n+1)(x-1)^{2n+1}$

(d) $y = c_1 \sum_{n=0}^{\infty} (n+1)(2n+1)(x+1)^{2n} + c_2 \sum_{n=0}^{\infty} (n+1)(2n+3)(x+1)^{2n+1}$

4. (a) $x = 0$ RSP
(b) $x = 0$ ISP
(c) $x = \pm 1$ RSP
(d) $x = 0, \pm 1$ RSP
(e) $x = 0, n\pi$ RSP $n = \pm 1, \pm 2, \pm 3, ...$
(f) $x = 0$ ISP, $x = n\pi$ RSP $n = \pm 1, \pm 2, ...$
(g) $x = 1$ RSP
(h) $x = 0, 1$ RSP

5. (a) $y_1 = x^{3/2}(1 - \frac{3}{4}x + \frac{15}{32}x^2 - \frac{35}{128}x^3 + ...), y_2 = x(1 - x + \frac{2}{3}x^2 - \frac{2}{5}x^3 + ...)$

(b) $y = c_1 \sum_{n=0}^{\infty} (-1)^n \frac{x^{n+1}}{\Gamma(n+7/2)} + c_2 \sum_{n=0}^{\infty} (-1)^n \frac{x^{n-3/2}}{n!}$

where:

$\Gamma(n + 7/2) = (n + 5/2)(n + 3/2)(n + 1/2) \ldots (3 + 1/2)(2 + 1/2)(1 + 1/2) \ldots$

(c) $y = c_1 \sum_{n=0}^{\infty} (-1)^n \frac{x^{n+1/2}}{(n+1)!} + c_2 x^{-1/2}$

(d) $y = c_1 \sum_{n=0}^{\infty} \frac{x^{n+1}}{(n+2)!} + c_2[1 + x^{-1}]$

(e) $y = c_1 \sum_{n=0}^{\infty} (-1)^n \frac{x^{4n+3}}{1 \cdot 3 \cdot 5 \cdot (2n+1)} + c_2 \sum_{n=0}^{\infty} (-1)^n \frac{x^{2n+2}}{2^n n!}$

(f) $y_1(x) = \sum_{n=0}^{\infty} \frac{x^{n+1}}{(n!)^2}$, $\quad y_2(x) = y_1(x)\log x - 2\sum_{n=1}^{\infty} \frac{x^{n+1}}{(n!)^2} g(n)$

(g) $y_1(x) = \sum_{n=0}^{\infty} (-1)^n \frac{x^{2n+1/2}}{2^n n!}$,

$$y_2(x) = y_1(x)\log x - \frac{1}{2}\sum_{n=1}^{\infty} (-1)^n \frac{x^{2n+1/2}}{2^n n!} g(n)$$

(h) $y_1(x) = \sum_{n=0}^{\infty} (-1)^n \frac{x^{n+1}}{n!}$,

$$y_2(x) = -y_1(x)\log x + 1 + \frac{1}{x} + \sum_{n=1}^{\infty} (-1)^n \frac{x^{n+1}}{n!} g(n)$$

(i) $y_1(x) = \sum_{m=0}^{\infty} (-1)^m \frac{x^{3m+3}}{2^m m!}$,

$$y_2(x) = \frac{1}{x^3}\left(1 + \frac{x^2}{4} + \frac{x^4}{8}\right) + \frac{1}{8} y_1(x)\log x - \frac{1}{16}\sum_{m=1}^{\infty} (-1)^m \frac{x^{2m+3}}{2^m m!} g(m)$$

(j) $y = c_1 \sum_{n=0}^{\infty} \frac{x^{n-3+i}}{n!(1+2i)(2+2i)\ldots(n+2i)} + c_2 \sum_{n=0}^{\infty} \frac{x^{n-3-i}}{n!(1-2i)(2-2i)\ldots(n-2i)}$

(k) $y_1 = x\left(1 - \frac{1}{3}x + \frac{1}{12}x^2 - \frac{1}{60}x^3 + \ldots\right)$, $y_2 = -1 + x^{-1}$

(l) $y = c_1\left(x - \frac{x^2}{3}\right) + c_2\left(3 - \frac{1}{x}\right)$

(m) $y = c_1 \sum_{n=0}^{\infty} (n+1)x^{2n} + c_2 \sum_{n=0}^{\infty} (2n+1)x^{2n-1}$

(n) $y = c_1 x^{-1} + c_2 \sum_{n=0}^{\infty} \frac{x^{2n}}{2n+1}$

(o) $y_1(x) = 1 + 2x + x^2$,

$$y_2(x) = y_1(x)\log x - x - x^2 + \frac{2}{3}\sum_{n=0}^{\infty} (-1)^n \frac{x^{n+3}}{(n+1)(n+2)(n+3)}$$

(p) $y_1(x) = \sum_{n=0}^{\infty} \frac{x^{2n}}{(n!)^2}$, $\quad y_2(x) = y_1(x)\log x - \sum_{n=1}^{\infty} \frac{x^{2n}}{(n!)^2} g(n)$

(q) $y_1(x) = \sum_{n=0}^{\infty} (-1)^n \frac{x^n}{(n!)^2}$, $\quad y_2(x) = y_1(x)\log x - 2\sum_{n=1}^{\infty} (-1)^n \frac{x^n}{(n!)^2} g(n)$

(r) $y_1(x) = \sum_{n=0}^{\infty} (-1)^n \frac{x^{2n}}{(n+1)!n!}$,

$$y_2(x) = y_1(x)\log x - \frac{1}{2}x^{-2}\left\{1 + x^2 - \sum_{n=2}^{\infty} (-1)^n \frac{x^{2n}}{(n!)^2}[1 + 2n\, g(n-1)]\right\}$$

(s) $y = c_1\left(1 + \frac{2}{3}x + \frac{x^2}{3}\right) + c_2 x^4 \sum_{n=0}^{\infty} (n+1)x^n$

(t) $y_1(x) = x\sum_{n=0}^{\infty} (-1)^n \frac{x^{2n}}{(n!)^2}$, $\quad y_2(x) = y_1(x)\log x - x\sum_{n=1}^{\infty} (-1)^n \frac{x^{2n}}{(n!)^2} g(n)$

Chapter 3

9. In the following solutions, Z represents J, Y, $H^{(1)}$,$H^{(2)}$.

(a) $y = x^{1/2}\left[C_1 J_{(n+1/2)}(kx) + C_2 J_{-(n+1/2)}(kx)\right]$

(b) $y = x\left[C_1 J_{1/3}(kx) + C_2 J_{-1/3}(kx)\right]$

(c) $y = C_1 J_0(x^2) + C_2 Y_0(x^2)$

(d) $y = x\left[C_1 J_{1/2}(x^2) + C_2 J_{-1/2}(x^2)\right]$

(e) $y = e^{-x}x^{-2}[C_1 J_1(kx^2) + C_2 Y_1(kx^3)]$

(f) $y = x^{-3}[C_1 J_2(2kx^3) + C_2 Y_2(2kx^3)]$

(g) $y = x^{1/4}[C_1 J_{1/6}(kx^2/2) + C_2 J_{-1/6}(kx^2/2)]$

(h) $y = x^{-2}[C_1 J_2(kx^2/2 + C_2 Y_2(kx^2/2)]$

(i) $y = e^{x}[C_1 J_2(e^{x}) + C_2 Y_2(e^{x})]$

(j) $y = x^{1/2}e^{x}[C_1 J_{3/2}(x) + C_2 J_{-3/2}(x)]$

(k) $y = e^{x/2}x^{-3/2}[C_1 J_{1/2}(kx^2) + C_2 J_{-1/2}(kx^2)]$

(l) $y = xe^{x}[C_1 I_1(x) + C_2 K_1(x)]$

(m) $y = xe^{-x}[C_1 J_2(2x^{1/2}) + C_2 Y_2(2x^{1/2})]$

(n) $y = e^{-x}[C_1 J_2(2x) + C_2 Y_2(2x)]$

19. (a) $-2^{p}\,\Gamma(p)/\pi$

(b) $2^{-p}/\Gamma(p+1)$

(c) 0

(d) $i\,2^{n}(n-1)!/\pi$

(e) $-3i$

(f) $\sqrt{2/\pi}$

Chapter 4

1. Characteristic equation:

$$\tan\alpha_n = \frac{2\alpha_n}{\alpha_n^2 - 1} \qquad \alpha = kL = \omega L/c$$

$$\phi_n = \sin(\alpha_n x/L) + \alpha_n \cos(\alpha_n x/L)$$

2. Characteristic equation:

$$\frac{\tan\alpha_n}{\alpha_n} = -\frac{\tan\beta_n}{\beta_n}$$

where:

$$\alpha_n = \frac{\omega_n}{c_1}\frac{L}{2}, \beta_n = \frac{\omega_n}{c_2}\frac{L}{2}, \qquad c_1^2 = \frac{T_0}{\rho_1}, \qquad \text{and} \qquad c_2^2 = \frac{T_0}{\rho_2}$$

Eigenfunction:

$$\phi_n = \begin{cases} \sin(\frac{2\alpha_n x}{L}) & 0 \le x \le L/2 \\ \frac{\sin\alpha_n}{\sin\beta_n}\sin(\frac{2\beta_n(L-x)}{L}) & L/2 \le x \le L \end{cases}$$

3. Characteristic equation:

$$J_{1/4}(\frac{\sqrt{\lambda_n}}{2})Y_{1/4}(2\sqrt{\lambda_n}) - Y_{1/4}(\frac{\sqrt{\lambda_n}}{2})J_{1/4}(2\sqrt{\lambda_n}) = 0$$

Eigenfunction:

$$\phi_n = \sqrt{z}\left\{J_{1/4}(\frac{\sqrt{\lambda_n}}{2}z^2) - \frac{J_{1/4}(2\sqrt{\lambda_n})}{Y_{1/4}(2\sqrt{\lambda_n})}Y_{1/4}(\frac{\sqrt{\lambda_n}}{2}z^2)\right\} \qquad z = 1 + x/L$$

4. (i) $\alpha_n \tan(\alpha_n) = 1$ where $\alpha = kL/2n = 1, 2, 3, \ldots$

$$\phi_n = \begin{cases} \sin(\frac{2\alpha_n}{L}x) & 0 \le x \le L/2 \\ \sin(\frac{2\alpha_n}{L}(x-L)) & L/2 \le x \le L \end{cases}$$

(ii) $\alpha_n = n\pi$ $\qquad n = 1, 2, 3, \ldots$

$$\phi_n = \begin{cases} \sin(\frac{2n\pi}{L}x) & 0 \le x \le L/2 \\ -\sin(\frac{2n\pi}{L}(x-L)) & L/2 \le x \le L \end{cases}$$

5. (a) $\lambda_n = \dfrac{n^2\pi^2}{L^2}$, $\phi_n = \cos(\dfrac{n\pi}{L}x)$ $n = 0, 1, 2, \ldots$

(b) $\lambda_n = \dfrac{n^2\pi^2}{L^2}$, $\phi_n = \sin(\dfrac{n\pi}{L}x)$ $n = 1, 2, 3, \ldots$

(c) $\lambda_n = \dfrac{(2n+1)^2\pi^2}{4L^2}$, $\phi_n = \sin(\dfrac{(2n+1)\pi}{2L}x)$ $n = 0, 1, 2, \ldots$

(d) $u_n(x) = \cos(\dfrac{\alpha_n}{L}x)$ $\lambda_n = \dfrac{\alpha_n^2}{L^2}$, $\tan\alpha_n = \dfrac{aL}{\alpha_n}$ $n = 1, 2, 3 \ldots$

(e) $u_n(x) = \sin(\dfrac{\alpha_n}{L}x)$ $\lambda_n = \dfrac{\alpha_n^2}{L^2}$, $\tan\alpha_n = \dfrac{L}{a\alpha_n}$ $n = 1, 2, 3 \ldots$

(f) $u_n(x) = \sin(\dfrac{\alpha_n}{L}x) + \dfrac{\alpha_n}{aL}\cos(\dfrac{\alpha_n}{L}x)$, $\tan\alpha_n = \dfrac{aL^2 - b\alpha_n^2}{(1+ab)L\alpha_n}$ $n = 1, 2, 3 \ldots$

6. Characteristic equation:

$\tan\alpha_n = \dfrac{2\xi\alpha_n}{\xi^2\alpha_n^2 - 1}$ where $\alpha = kL$ and $\xi = \dfrac{M}{\rho AL}$

$\phi_n(x) = \cos(\dfrac{\alpha_n}{L}x) - \xi\alpha_n \sin(\dfrac{\alpha_n}{L}x)$

7. Characteristic equation:

$J_0(\alpha_n)\,Y_0(2\alpha_n) - J_0(2\alpha_n)\,Y_0(\alpha_n) = 0$ where $\alpha_n = kL$ $n = 1, 2, 3, \ldots$

$\phi_n = J_0(\alpha_n z) - \dfrac{J_0(\alpha_n)}{Y_0(\alpha_n)} Y_0(\alpha_n z)$ where $z = 1 + x/L$

8. Let $\alpha_n = \beta_n L$, $\lambda_n = \dfrac{\alpha_n^4}{L^4}$, and $\lambda_0 = 0$ (if it is a root)

(a) $\sin\alpha_n = 0$, $\alpha_n = n\pi$ $\phi_n = \sin(\dfrac{n\pi}{L}x)$ $n = 1, 2, 3, \ldots$

(b) $\cos\alpha_n \cosh\alpha_n = -1$, $\alpha_1 = 1.88, \alpha_2 = 4.69, \alpha_3 = 7.86$ $n = 1, 2, 3, \ldots$

$$\phi_n(x) = \frac{\sin(\frac{\alpha_n}{L}x) - \sinh(\frac{\alpha_n}{L}x)}{\sin(\alpha_n) + \sinh(\alpha_n)} - \frac{\cos(\frac{\alpha_n}{L}x) - \cosh(\frac{\alpha_n}{L}x)}{\cos(\alpha_n) + \cosh(\alpha_n)}$$

(c) $\cos\alpha_n \cosh\alpha_n = 1$, $\alpha_0 = 0, \alpha_1 = 4.73, \alpha_2 = 7.85, \alpha_3 = 11.00$ $n = 1, 2, 3, \ldots$

$$\phi_n(x) = \frac{\sin(\frac{\alpha_n}{L}x) + \sinh(\frac{\alpha_n}{L}x)}{\sinh(\alpha_n) - \sin(\alpha_n)} + \frac{\cos(\frac{\alpha_n}{L}x) + \cosh(\frac{\alpha_n}{L}x)}{\cos(\alpha_n) - \cosh(\alpha_n)}$$

(d) $\sin\alpha_n = 0$, $\alpha_n = n\pi$ $\phi_n = \cos(\dfrac{n\pi}{L}x)$ $n = 0, 1, 2, \ldots$

(e) $\tan\alpha_n = \tanh\alpha_n$, $\alpha_1 = 3.93, \alpha_2 = 7.07, \alpha_3 = 10.2$ $n = 1, 2, 3, \ldots$

$$\phi_n(x) = \frac{\sin\left(\frac{\alpha_n}{L}x\right)}{\sin(\alpha_n)} - \frac{\sinh\left(\frac{\alpha_n}{L}x\right)}{\sinh(\alpha_n)}$$

(f) $\tan\alpha_n = \tanh\alpha_n$, [see (e)] $n = 0, 1, 2, \ldots$

$$\phi_n(x) = \frac{\sin\left(\frac{\alpha_n}{L}x\right)}{\sin(\alpha_n)} + \frac{\sinh\left(\frac{\alpha_n}{L}x\right)}{\sinh(\alpha_n)}$$

(g) $\coth\alpha_n - \cot\alpha_n = \dfrac{2\gamma L^3}{\alpha_n^3 EI}$ $n = 1, 2, 3, \ldots$

$$\phi_n(x) = \frac{\sin\left(\frac{\alpha_n}{L}x\right)}{\sin(\alpha_n)} + \frac{\sinh\left(\frac{\alpha_n}{L}x\right)}{\sinh(\alpha_n)}$$

(h) $\tan\alpha_n - \tanh\alpha_n = \dfrac{2\eta L}{\alpha_n EI}$ $n = 1, 2, 3, \ldots$

$$\phi_n(x) = \frac{\sin\left(\frac{\alpha_n}{L}x\right)}{\cos(\alpha_n)} + \frac{\sinh\left(\frac{\alpha_n}{L}x\right)}{\cosh(\alpha_n)}$$

(i) $\cosh\alpha_n \cos\alpha_n + 1 = \dfrac{\gamma L^3}{\alpha_n^3 EI}[\sinh\alpha_n \cos\alpha_n - \cosh\alpha_n \sin\alpha_n]$ $n = 1, 2, 3, \ldots$

$$\phi_n(x) = \frac{\sin\left(\frac{\alpha_n}{L}x\right) - \sinh\left(\frac{\alpha_n}{L}x\right)}{\sin(\alpha_n) + \sinh(\alpha_n)} - \frac{\cos\left(\frac{\alpha_n}{L}x\right) - \cosh\left(\frac{\alpha_n}{L}x\right)}{\cos(\alpha_n) + \cosh(\alpha_n)}$$

(j) $\cosh\alpha_n \cos\alpha_n + 1 = -\frac{\eta L}{\alpha_n EI}[\cosh\alpha_n \sin\alpha_n + \sinh\alpha_n \cos\alpha_n]$ $n = 1, 2, 3,$ …

$$\phi_n(x) = \frac{\sin\left(\frac{\alpha_n}{L}x\right) - \sinh\left(\frac{\alpha_n}{L}x\right)}{\cos(\alpha_n) + \cosh(\alpha_n)} + \frac{\cos\left(\frac{\alpha_n}{L}x\right) - \cosh\left(\frac{\alpha_n}{L}x\right)}{\sin(\alpha_n) - \sinh(\alpha_n)}$$

9. Characteristic equation:

$\tanh\alpha_n = \tan\alpha_n + 2k\alpha_n \tan\alpha_n \tanh\alpha_n$

$\alpha_n = \beta_n L$, $\lambda_n = \dfrac{\alpha_n^4}{L^4}$, $n = 1, 2, 3 \ldots$

$$\phi_n(x) = \frac{\sin(\frac{\alpha_n}{L}x) - \sinh(\frac{\alpha_n}{L}x)}{\sin(\alpha_n) + \sinh(\alpha_n)} - \frac{\cos(\frac{\alpha_n}{L}x) - \cosh(\frac{\alpha_n}{L}x)}{\cos(\alpha_n) + \cosh(\alpha_n)}$$

10. Characteristic equation:

(a) $k\,\alpha_n(\tan\alpha_n - \tanh\alpha_n) = 2$ where $\alpha_n = \beta_n L/2,\ k = M/(\rho AL)$

$$\phi_n(x) = \frac{\sin(\frac{2\alpha_n}{L}x)}{\cos(\alpha_n)} - \frac{\sinh(\frac{2\alpha_n}{L}x)}{\cosh(\alpha_n)} \qquad 0 \le x \le L/2$$

$$= -\sin(z) + \tan(\alpha_n)\cos(z) + \sinh(z) - \tanh(\alpha_n)\cosh(z) \qquad L/2 \le x \le L$$

where $z = \alpha_n(2x - L)/L$

(b) $\sin\alpha_n = 0$ $\qquad \lambda_n = \dfrac{16\alpha_n^4}{L^4}$

$$\phi_n = \sin(\frac{2n\pi}{L}x) \qquad 0 \le x \le L \qquad n = 1, 2, 3, \ldots.$$

11. Characteristic equation:

$J_n(\alpha_m)\,I_{n+1}(\alpha_m) + J_{n+1}(\alpha_m)\,I_n(\alpha_m) = 0$ where $\alpha_m = 2\,\beta_m L$

$$\phi_n = x^{n/2}\left[\frac{J_n(2\beta_m L\sqrt{x/L})}{J_n(2\beta_m L)} - \frac{I_n(2\beta_m L\sqrt{x/L})}{I_n(2\beta_m L)}\right] \qquad m = 1, 2, 3, \ldots$$

12. (a) Characteristic equation:

$\dfrac{\sin\alpha_n}{\alpha_n} = 0,\qquad \alpha_n = k_n L = n\pi \qquad \lambda_n = \dfrac{n^2\pi^2}{L^2} \qquad n = 1, 2, 3, \ldots$

$$\phi_n = \frac{\sin(\frac{n\pi}{L}x)}{\frac{n\pi}{L}x} = j_0(\frac{n\pi}{L}x)$$

(b) $\tan\alpha_n = \alpha_n \qquad \alpha_n = k_n L \qquad \lambda_n = \dfrac{\alpha_n^2}{L^2} \qquad n = 0, 1, 2, \ldots$

$\alpha_0 = 0,\qquad \alpha_1 == 4.49,\qquad \alpha_2 = 7.73,\qquad \alpha_3 = 10.90$

$$\phi_n = \frac{\sin(\frac{n\pi}{L}x)}{\frac{n\pi}{L}x} = j_0(\frac{n\pi}{L}x)$$

13. Characteristic equations:

$\tan\alpha_n = \alpha_n \quad \alpha_n = k_n L \qquad \lambda_n = \dfrac{\alpha_n^2}{L^2} \qquad n = 1, 2, 3, \ldots \qquad$ (see 12 (b))

$$\phi_n = \frac{j_1(\frac{\alpha_n}{L}x)}{\frac{\alpha_n}{L}x} = \frac{1}{(\frac{\alpha_n}{L}x)^2}\left[\frac{\sin(\frac{\alpha_n}{L}x)}{\frac{\alpha_n}{L}x} - \cos(\frac{\alpha_n}{L}x)\right]$$

14. Characteristic equation:

$\tan\alpha_n = -\dfrac{\alpha_n}{aL} \qquad \alpha_n = L\sqrt{k_n^2 - a^2} \qquad \lambda_n = \dfrac{\alpha_n^2}{L^2} + a^2 \qquad n = 1, 2, 3, \ldots$

$$\phi_n = e^{-ax}\left[\sin(\frac{\alpha_n}{L}x) + \frac{\alpha_n}{aL}\cos(\frac{\alpha_n}{L}x)\right]$$

15. (a) Characteristic equations:

(i) $\sin\alpha_n = 0, \qquad \alpha_n = r_n L/2 \qquad \lambda_n = r_n^2 = \dfrac{P_n}{EI}$

(ii) $\tan\alpha_n = \alpha_n$

$\alpha_0 = 0, \qquad \alpha_1 = 2\pi, \qquad \alpha_2 = 8.99, \qquad \alpha_3 = 4\pi, \ \ldots$

$\lambda_n = \dfrac{4\alpha_n^2}{L^2} \qquad n = 1, 2, 3, \ldots$

$$\phi_n(x) = \frac{\sin(2\frac{\alpha_n}{L}x) - 2\frac{\alpha_n}{L}x}{\sin(2\alpha_n) - 2\alpha_n} - \frac{\cos(2\frac{\alpha_n}{L}x) - 1}{\cos(2\alpha_n) - 1}$$

(b) Characteristic equation:

(i) $\sin\alpha_n = 0, \qquad \alpha_n = r_n L \qquad n = 0, 1, 2, ..$

$\alpha_n = n\pi, \qquad \lambda_n = \dfrac{n^2\pi^2}{L^2}, \qquad \phi_n = \sin(\dfrac{n\pi}{L}x) \qquad n = 1, 2, 3, \ldots$

(ii) $\tan\alpha_n = \alpha_n \qquad n = 0, 1, 2, \ldots$

$$\lambda_n = \frac{\alpha_n^2}{L^2} \qquad \phi_n(x) = \frac{\sin(\frac{\alpha_n}{L}x) - \frac{\alpha_n}{L}x}{\sin(\alpha_n)} - \frac{\cos(\frac{\alpha_n}{L}x) - 1}{\cos(\alpha_n)} \qquad n = 1, 2, 3, \ldots$$

(d) $\sin\alpha_n = 0,\qquad \alpha_n = r_n L,\qquad \alpha_n = n\pi,\qquad n = 0, 1, 2, \ldots$

$$\lambda_n = \frac{n^2\pi^2}{L^2},\qquad \phi_n = \sin(\frac{n\pi}{L}x) - (-1)^n \frac{EI}{\gamma L^3} n^2\pi^2(\frac{x}{L})\qquad n = 1, 2, 3, \ldots$$

(e) $\cos\alpha_n = 0,\qquad \alpha_n = r_n L\qquad n = 0, 1, 2, \ldots$

$$\alpha_n = (n + 1/2)\pi,\qquad \lambda_n = \frac{\alpha_n^2}{L^2},\qquad \phi_n = \sin(\frac{(n+1/2)\pi}{L}x)$$

(f) $$\tan\alpha_n = \frac{\frac{\eta L}{EI}\alpha_n}{\frac{\eta L}{EI} + \alpha_n^2} = \frac{\alpha_n}{1 + \frac{\eta L}{EI}\alpha_n^2}\qquad \alpha_n = r_n L$$

$$\lambda_n = \frac{\alpha_n^2}{L^2},\quad \phi_n = \sin(\frac{\alpha_n}{L}x) - \frac{x}{L}\sin(\alpha_n),\qquad n = 1, 2, 3, \ldots$$

(g) $\sin\alpha_n = 0,\qquad \alpha_n = r_n L,\qquad \alpha_n = n\pi, n = 0, 1, 2, \ldots$

$$\lambda_n = \frac{n^2\pi^2}{L^2},\qquad \phi_n = \cos(\frac{n\pi}{L}x) - 1$$

(h) $(\xi - 1)\alpha_n \sin\alpha_n + (\alpha_n^2 + 2\xi)\cos\alpha_n = 2\xi,\qquad \xi = \frac{\eta L}{EI}$

$$\alpha_n = r_n L,\qquad \lambda_n = \frac{\alpha_n^2}{L^2},\qquad n = 1, 2, 3, \ldots$$

$$\phi_n(x) = \frac{\sin(\frac{\alpha_n}{L}x) - \frac{\alpha_n}{L}x}{\sin(\alpha_n) - \alpha_n} - \frac{\cos(\frac{\alpha_n}{L}x) - 1}{\cos(\alpha_n) - 1}$$

16. Characteristic equation:

$$\tan\alpha = -a\alpha / L\qquad \alpha = \frac{rL}{ab}, L = b - a,\qquad r^2 = \frac{Pb^4}{EI_0}$$

$$\phi_n(x) = x\left\{\frac{\sin(\frac{ab}{xL}\alpha_n)}{\sin(\frac{b}{L}\alpha_n)} - \frac{\cos(\frac{ab}{xL}\alpha_n)}{\cos(\frac{b}{L}\alpha_n)}\right\}\qquad n = 1, 2, 3, \ldots$$

17. Characteristic equation:

$$J_{-1/3}(\alpha_n) = 0,\qquad \alpha = \frac{2}{3}\beta L^{3/2},\qquad \beta^2 = \frac{q}{EI},\qquad n = 1, 2, 3, \ldots$$

$$\phi_n(x) = x^{1/2} J_{-1/3}\left[\alpha_n(\frac{x}{L})^{3/2}\right]\qquad n = 1, 2, 3, \ldots$$

$\phi_n(x)$ is the eigenfunction for $\frac{dy}{dx}$ or $u(x)$

18. Since $\gamma < \beta^2/2$, where $\gamma^2 = \dfrac{k}{EI}$, and $\beta^2 = \dfrac{P}{EI}$, the characteristic equation becomes:

$$\frac{1}{\xi_n^2}\tan(\xi_n L) = \frac{1}{\eta_n^2}\tan(\eta_n L) \qquad n = 1, 2, 3, \ldots$$

where:

$$\xi_n = \beta_n\left(\frac{1 + (1 - 4\gamma^2/\beta_n^4)^{1/2}}{2}\right)^{1/2}$$

$$\eta_n = \beta_n\left(\frac{1 - (1 - 4\gamma^2/\beta_n^4)^{1/2}}{2}\right)^{1/2}$$

The eigenvalues of this system are β_n, n = 1, 2, 3, … and

$$\phi_n(x) = \frac{\sin(\xi_n x)}{\sin(\xi_n L)} - \frac{\cos(\eta_n x)}{\cos(\eta_n L)}$$

21.

(a) $p = 1 - x^2$	$q = 0$	$r = 1$
(b) $p = (1 - x^2)^{1/2}$	$q = 0$	$r = (1 - x^2)^{-1/2}$
(c) $p = 1$	$q = (1 - x^2)^{-2}$	$r = (1 - x^2)^{-1}$
(d) $p = x^{a+1}\, e^{-x}$	$q = 0$	$r = x^a\, e^{-x}$
(e) $p = e^{-x^2}$	$q = 0$	$r = e^{-x^2}$
(f) $p = (1 - x^2)^{a+1/2}$	$q = 0$	$r = (1 - x^2)^{a-1/2}$
(g) $p = (1 - x)^{a+1}\,(1 + x)^{b+1}$	$q = 0$	$r = (1 - x)^a\,(1 + x)^b$
(h) $p = x^c\,(1 - x)^{a+b-c+1}$	$q = 0$	$r = x^{c-1}\,(1 - x)^{a+b-c}$
(i) $p = e^x$	$q = -2\, e^x\, x^{-2}$	$r = e^x$
(j) $p = x$	$q = -n^2\, x^{-1}$	$r = x$
(k) $p = (ax + b)^2$	$q = 0$	$r = (ax + b)^2$
(l) $p = \sin^2(ax)$	$q = 0$	$r = \sin^2(ax)$
(m) $p = x^{3/2}$	$q = 0$	$r = x^{1/2}$
(n) $p = e^{ax}$	$q = 0$	$r = e^{ax}$
(o) $p = \cos^2(ax)$	$q = 0$	$r = \cos^2(ax)$
(p) $p = \cosh^2(ax)$	$q = 0$	$r = \cosh^2(ax)$
(q) $p = \cos(ax)$	$q = 0$	$r = \cos^3(ax)$
(r) $p = e^{ax^2}$	$q = a^2\, x^2\, e^{ax^2}$	$r = e^{ax^2}$

(s)	$p = 1$	$q = -a^2$	$r = e^{-4ax}$
(t)	$p = 1$	$q = -a(a-1)x^{-2}$	$r = x^{-4a}$
(u)	$p = 1$	$q = 0$	$r = x^{-4}$
(v)	$p = 1$	$q = 0$	$r = x^{-1}$
(w)	$p = x^4$	$q = 0$	$r = x^4$
(x)	$p = e^{4x}$	$q = 4e^{4x}$	$r = e^{4x}$
(y)	$p = x^{-2}$	$q = 0$	$r = 9x^{-1}/4$
(z)	$p = x^{-1}$	$q = x^{-3}$	$r = x^{-3}$
(aa)	$p = x^2$	$q = 0$	$r = x^2$
(bb)	$p = x^3$	$q = -3x$	$r = x^9$
(cc)	$p = x^3$	$q = 0$	$r = x^{5/3}$
(dd)	$p = x^6$	$q = 0$	$r = x^6$
(ee)	$p = x^4$	$q = 0$	$r = x^6$
(ff)	$p = x^2$	$q = 0$	$r = x^4$
(gg)	$p = x^{11/2}$	$q = -\frac{63}{16}x^{7/2}$	$r = \frac{9}{4}x^{13/2}$
(hh)	$p = x^{9/7}$	$q = 0$	$r = x^{23/7}$

22.

(a) $\phi_n(x) = P_{2n+1}(x)$, $\lambda_n = 2(n+1)(2n+1)$ $n = 0, 1, 2, \ldots$

(b) $\phi_n(x) = P_{2n}(x)$, → LEG $\lambda_n = 2n(2n+1)$ $n = 0, 1, 2, \ldots$

(c) $\phi_n(x) = T_n(x)$, (Tchebyshev Polynomials)

$\lambda_n = n^2$ $n = 0, 1, 2, \ldots$

(d) $\phi_n(x) = (1-x)^{1/2} P_n(x)$, $\lambda_n = n(n+1)$ $n = 0, 1, 2, \ldots$

(e) $\phi_n(x) = \dfrac{1}{ax+b}\sin(\dfrac{n\pi}{L}x)$ $\lambda_n = \dfrac{n^2\pi^2}{L^2}$ $n = 1, 2, 3, \ldots$

(f) $\phi_n(x) = \dfrac{\sin(\frac{n\pi}{L}x)}{\sin(ax)}$, $\lambda_n = \dfrac{n^2\pi^2}{L^2} - a^2$ $n = 1, 2, 3, \ldots$

(g) $\phi_n(x) = \dfrac{1}{\sqrt{x}}\sin(\dfrac{n\pi}{\sqrt{L}}\sqrt{x})$ $\lambda_n = \dfrac{n^2\pi^2}{4L}$ $n = 1, 2, 3, \ldots$

(h) $\phi_n(x) = e^{-ax/2}\sin(\dfrac{n\pi}{L}x)$ $\lambda_n = \dfrac{n^2\pi^2}{L^2} + \dfrac{a^2}{4}$ $n = 1, 2, 3, \ldots$

(i) $\phi_n(x) = \dfrac{\sin(\frac{n\pi}{L}x)}{\cos(ax)}$, $\lambda_n = \dfrac{n^2\pi^2}{L^2} - a^2$ $n = 1, 2, 3,\ldots$

(j) $\phi_n(x) = \dfrac{\sin(\frac{n\pi}{L}x)}{\cosh(ax)}$, $\lambda_n = \dfrac{n^2\pi^2}{L^2} + a^2$ $n = 1, 2, 3,\ldots$

(k) $\phi_n(x) = e^{-ax^2/2}\sin(\frac{n\pi}{L}x)$ $\lambda_n = \dfrac{n^2\pi^2}{L^2} + a$ $n = 1, 2, 3,\ldots$

(l) $\phi_n(x) = e^{ax}\sin(n\pi\dfrac{e^{-2ax}-1}{e^{-2aL}-1})$

$$\lambda_n = \left[\frac{2n\pi a}{e^{-2La}-1}\right]^2 \qquad n = 1, 2, 3,\ldots$$

(m) $\phi_n(x) = x^a \sin(\dfrac{n\pi x^{1-2a}}{L^{1-2a}})$

$$\lambda_n = \left[\frac{(2a-1)n\pi}{L^{1-2a}}\right]^2 \qquad n = 1, 2, 3,\ldots$$

(n) $\phi_n(x) = x\sin(\dfrac{2n\pi}{x})$ $\lambda_n = 4n^2\pi^2$ $n = 1, 2, 3,\ldots$

(o) $\phi_n(x) = x^{1/2} J_1[\alpha_n(x/L)^{1/2}]$

$J_1(\alpha_n) = 0$ $\lambda_n = \dfrac{\alpha_n^2}{4L}$ $n = 1, 2, 3, \ldots$

(p) $\phi_n(x) = \dfrac{1}{x^2}\left[\dfrac{\sin(\alpha_n \frac{x}{L})}{\alpha_n \frac{x}{L}} - \cos(\alpha_n \frac{x}{L})\right]$

$\lambda_n = \dfrac{\alpha_n^2}{L^2}$, $\tan(\alpha_n) = \dfrac{3\alpha_n}{3-\alpha_n^2}$ $n = 1, 2, 3, \ldots$

(q) $\phi_n(x) = e^{-2x}\sin(n\pi x)$ $\lambda_n = n^2\pi^2$ $n = 1, 2, 3, \ldots$

(r) $\phi_n(x) = x^{3/2} J_1[\alpha_n(x/L)^{3/2}]$

$J_1(\alpha_n) = 0$ $\lambda_n = \dfrac{\alpha_n^2}{L^3}$ $n = 1, 2, 3, \ldots$

(s) $\phi_n(x) = x\sin(n\pi\log x)$ $\lambda_n = n^2\pi^2$ $n = 1, 2, 3, \ldots$

(t) $\phi_n(x) = \dfrac{\sin(\frac{n\pi}{L}x)}{x\sqrt{n\pi/L}}$, $\lambda_n = \dfrac{n^2\pi^2}{L^2}$ $n = 1, 2, 3,\ldots$

(u) $\phi_n(x) = x^{-3}\sin(n\pi(x/L)^4)$

$$\lambda_n = \frac{16\,n^2\pi^2}{L^8} \qquad n = 1, 2, 3, \ldots$$

(v) $\phi_n(x) = x^{-1}\, J_3[\alpha_n (x/L)^{1/3}]$

$$J_3(\alpha_n) = 0 \qquad \lambda_n = \frac{\alpha_n^2}{9L^{2/3}} \qquad n = 1, 2, 3, \ldots$$

(w) $\phi_n(x) = x^{-5/2}\, J_{5/2}[\alpha_n x/L]$ $\qquad J_{5/2}(\alpha_n) = 0$

or $$\phi_n(x) = \frac{1}{\sqrt{x}}\left\{\left(\frac{3L^2}{\alpha_n^2 x^2} - 1\right)\sin\left(\alpha_n \frac{x}{L}\right) - \frac{3L}{\alpha_n x}\cos\left(\alpha_n \frac{x}{L}\right)\right\}$$

$$\tan(\alpha_n) = \frac{3\alpha_n}{3 - \alpha_n^2} \qquad \lambda_n = \frac{\alpha_n^2}{L^2} \qquad n = 1, 2, 3, \ldots$$

(x) $\phi_n(x) = x^{-3/2}\, J_{3/4}[\alpha_n x^2/L^2]$

$$J_{3/4}(\alpha_n) = 0 \qquad \lambda_n = 4\frac{\alpha_n^2}{L^4} \qquad n = 1, 2, 3, \ldots$$

(y) $\phi_n(x) = x^{-1/2}\, J_{1/4}[\alpha_n x^2/L^2]$

$$J_{1/4}(\alpha_n) = 0 \qquad \lambda_n = 4\frac{\alpha_n^2}{L^4} \qquad n = 1, 2, 3, \ldots$$

(z) $\phi_n(x) = x^{-9/4}\, J_2[\alpha_n (x/L)^{3/2}]$

$$J_2(\alpha_n) = 0 \qquad n = 1, 2, 3, \ldots$$

(aa) $\phi_n(x) = x^{-1/7}\, J_{1/14}[\alpha_n x^2/L^2]$

$$J_{1/14}(\alpha_n) = 0 \qquad \lambda_n = 4\frac{\alpha_n^2}{L^4} \qquad n = 1, 2, 3, \ldots$$

23.

(a) $$y = \sum_{n=1}^{\infty} \frac{A_n}{\lambda - \frac{n^2\pi^2}{L^2}} \sin\left(\frac{n\pi}{L}x\right) \qquad A_n = \frac{2}{L}\int_0^L f(x)\sin\left(\frac{n\pi}{L}x\right)dx$$

(b) $$y = \sum_{n=1}^{\infty} \frac{A_n}{\lambda - \lambda_n} J_0(\sqrt{\lambda_n}\, x)$$ where λ_n is the root of $J_0(\sqrt{\lambda_n}\, L) = 0$, and

$$A_n = \frac{2}{\sqrt{\lambda_n}\, L\, J_1(\sqrt{\lambda_n}\, L)}$$

(c) $y = \sum_{n=0}^{\infty} \frac{A_n}{\lambda - n(n+1)} P_n(x)$ $\qquad A_n = \frac{2n+1}{2} \int_{-1}^{+1} f(x) P_n(x) dx$

(d) $y = y_I + y_{II}$,

$y_I = [3 \cos(kx) - 3 \cot(k) \sin(kx)]\, e^x$, $\qquad y_{II} = \frac{4e^x}{\pi} \sum_{n=1,3,\dots}^{\infty} \frac{\sin(n\pi x)}{n(\beta - n^2\pi^2)}$

(e) $y = \frac{4e^{x/2}}{\pi\sqrt{x}} \sum_{n=1,3,\dots}^{\infty} \frac{\sin(n\pi\sqrt{x})}{n(\lambda - n^2\pi^2/4)}$

(f) $y = \frac{4e^x}{\pi x^2} \sum_{n=1,3,\dots}^{\infty} \frac{\sin(n\pi x^2)}{n(\alpha^2 - 4n^2\pi^2)}$

(g) $y = \frac{4}{\pi x} \sum_{n=1,3,\dots}^{\infty} \frac{\sin(n\pi x)}{n(k^2 - n^2\pi^2)}$

(h) $y = \frac{4e^x}{xL} \sum_{n=1}^{\infty} \frac{1 - J_0(\alpha_n)}{(\lambda - \lambda_n) J_0^2(\alpha_n)} J_1(\alpha_n \frac{x}{L})$ $\qquad J_1(\alpha_n) = 0$ $\qquad \lambda_n = \frac{\alpha_n^2}{L^2}$

(i) $y = \frac{4e^{3x}}{x} \sum_{n=1}^{\infty} a_n J_2(\alpha_n x^2)$ $\qquad J_2(\alpha_n) = 0$

$a_n = \frac{[J_n'(\alpha_n)]^{-2}}{\lambda - 4\alpha_n^2} \int_0^1 x^2 J_2(\alpha_n x^2)\, dx$

(j) $y = \frac{4e^{2x}}{\pi x^{3/2}} \sum_{n=1,3,\dots}^{\infty} \frac{\sin(n\pi x^2)}{n(\lambda - 4n^2\pi^2)}$

(k) $y = \frac{4e^x}{\pi x^{7/4}} \sum_{n=1,3,\dots}^{\infty} \frac{\sin(n\pi x^2)}{n(\lambda - 4n^2\pi^2)}$

(l) $y = -\frac{2}{e^x (xL)^{1/2}} \sum_{n=1}^{\infty} \frac{J_{1/4}(\alpha_n \frac{x^2}{L^2})}{(\lambda - 4\frac{\alpha_n^2}{L^4})\alpha_n J_{1/4}'(\alpha_n)}$

(m) $y = -\frac{4e^x}{\pi x^3} \sum_{n=1,3,\dots}^{\infty} \frac{\sin(n\pi x^3)}{n(\lambda - 9n^2\pi^2)}$

(n) $y = \frac{4e^x}{\pi x^4} \sum_{n=1,3,\dots}^{\infty} \frac{\sin(n\pi x^4)}{n(\lambda^2 - 16n^2\pi^2)}$

24. (a) $2\pi \sum_{n=1}^{\infty} (-1)^{n+1} \frac{\sin(nx)}{n} - \frac{8}{\pi} \sum_{n=1,3,\dots}^{\infty} \frac{\sin(nx)}{n^3}$

(b) $\dfrac{2}{\pi}\sum_{m=1}^{\infty} \dfrac{1-\cos(\frac{m\pi}{2})}{m}\sin(mx)$

(c) $2\sum_{n=1}^{\infty} (-1)^{n+1}\dfrac{\sin(nx)}{n}$

(d) $\dfrac{8}{\pi^3}\sum_{n=1,3,\ldots}^{\infty} \dfrac{\sin(n\pi x)}{n^3}$

(e) $\dfrac{2}{\pi}\sum_{n=1}^{\infty} [1-(-1)^n e^{\pi}]\dfrac{n\sin(nx)}{n^2+1}$

(f) $\sin x$

25. (a) $\dfrac{\pi^2}{3} + 4\sum_{n=1}^{\infty} (-1)^n \dfrac{\cos(nx)}{n^2}$

(b) $\dfrac{1}{2} + \dfrac{2}{\pi}\sum_{n=1}^{\infty} (-1)^{n+1}\dfrac{\cos((2n-1)x)}{2n-1}$

(c) $\dfrac{\pi}{2} - \dfrac{4}{\pi}\sum_{n=1}^{\infty} \dfrac{\cos((2n-1)x)}{(2n-1)^2}$

(d) $\dfrac{1}{6} - \dfrac{1}{\pi^2}\sum_{n=1}^{\infty} \dfrac{\cos(2nx)}{n^2}$

(e) $\dfrac{e^{\pi}-1}{\pi} - \dfrac{2}{\pi}\sum_{n=1}^{\infty} [1-(-1)^n e^{\pi}]\dfrac{\cos(2nx)}{n^2+1}$

(f) $\dfrac{2}{\pi} - \dfrac{4}{\pi}\sum_{n=1}^{\infty} \dfrac{\cos(2nx)}{4n^2-1}$

26. (a) $\dfrac{1}{\pi} + \dfrac{1}{2}\sin x - \dfrac{2}{\pi}\sum_{n=1}^{\infty} \dfrac{\cos(2nx)}{(2n-1)(2n+1)}$

(b) $\dfrac{2a}{\pi}\sin(a\pi)\left[\dfrac{1}{2a^2} + \sum_{n=1}^{\infty} (-1)^n \dfrac{\cos(nx)}{a^2-n^2}\right]$

(c) $-\dfrac{1}{3} - \dfrac{4}{\pi^2}\sum_{n=1}^{\infty} (-1)^n \dfrac{\cos(n\pi x)}{n^2} - \dfrac{2}{\pi}\sum_{n=1}^{\infty} (-1)^n \dfrac{\sin(n\pi x)}{n}$

(d) $\dfrac{2}{\pi}\sin(a\pi)\sum_{n=1}^{\infty} (-1)^n \dfrac{n\sin(nx)}{a^2-n^2}$

(e) $\dfrac{3}{4} + \dfrac{1}{\pi} \sum_{n=1}^{\infty} \dfrac{\sin(n\pi/2)}{n} \cos(n\pi \dfrac{x}{L}) - \dfrac{1}{\pi} \sum_{n=1}^{\infty} \dfrac{\cos(n\pi/2) - \cos(n\pi)}{n} \sin(n\pi \dfrac{x}{L})$

27. $f(x) = \dfrac{1}{2} \sum_{n=1}^{\infty} \dfrac{J_1(\mu_n)}{\mu_n [J_1(2\mu_n)]^2} J_0(\mu_n x)$

28. $f(x) = -2 \sum_{n=1}^{\infty} \dfrac{J_2(\mu_n x)}{\mu_n J_1(\mu_n)}$

29. $f(x) = 2a \sum_{n=1}^{\infty} \dfrac{J_0(\mu_n x)}{(a^2 + \mu_n^2 L^2) J_0(\mu_n L)}$

30. $f(x) = \dfrac{1}{2} + \dfrac{1}{2} \sum_{n=0}^{\infty} [P_{2n}(0) - P_{2n+2}(0)] P_{2n+1}(x)$

31. $f(x) = \dfrac{1}{4} P_0 + \dfrac{1}{2} P_1(x) + \dfrac{5}{16} P_2(x) + \ldots$

Chapter 5

10. (a) Everywhere except at $z = \pm i$
    (b) Nowhere
    (c) Nowhere
    (d) Nowhere
    (e) Nowhere
    (f) Everywhere

11. (a) $v = e^x \sin y + C$
    (b) $v = 3x^2y - y^3 + C$
    (c) $v = \sinh x \sin y + C$
    (d) $v = 2\tan^{-1}(y/x) + C$
    (e) $v = -\sin x \sinh y + C$
    (f) $v = y - \dfrac{y}{x^2 + y^2} + C$

16. (a) $2n\pi$ $\quad n = 0, \pm1, \pm2, \ldots$
    (b) $(2n + 1/2)\pi + i \cosh^{-1} 2$ $\quad n = 0, \pm1, \pm2, \ldots$
    (c) $(2n + 1/2)\pi \pm i\alpha$ $\quad n = 0, \pm1, \pm2, \ldots$
    (d) $i(2n - 1/2)\pi$ $\quad n = 0, \pm1, \pm2, \ldots$
    (e) $\log 2 \pm (2n+1)\pi i$ $\quad n = 0, \pm1, \pm2, \ldots$
    (f) -1

17. (a) $-i$
    (b) -1
    (c) $(-1 + 5i)/2, (-1 + 5.1i)/2$
    (d) $2\pi i$
    (e) 0

19. (a) $(1 - \cosh 1)/2$
    (b) $10i/3$
    (c) $6 + 26i/3$
    (d) $-2(1 - i)/3$
    (e) 0
    (f) $2i \sin 1$

20. (a) $4\pi i$
    (b) $2\pi i$
    (c) 0
    (d) $\pi i/3$
    (e) $8\pi i$
    (f) 0
    (g) $2\pi(i - 1)$
    (h) $2\pi i$

21.

(a) $\displaystyle\sum_{n=0}^{\infty} (-1)^n \frac{z^{2n}}{(2n)!}$ $\quad |z| < \infty$

(b) $\displaystyle\sum_{n=0}^{\infty} (-1)^n \frac{z^{2n}}{(2n+1)!}$ $\quad |z| < \infty$

(c) $\displaystyle\sum_{n=0}^{\infty} (-1)^n (n+1) z^n$ $\quad |z| < 1$

(d) $\displaystyle\sum_{n=0}^{\infty} \frac{z^n}{(n+1)!}$ $\quad |z| < \infty$

(e) $\displaystyle\sum_{n=0}^{\infty} (-1)^n \frac{(z-2)^n}{2^{n+1}}$ $\quad |z - 2| < 2$

(f) $\displaystyle -1 - 2\sum_{n=1}^{\infty} (z-1)^n$ $\quad |z - 1| < 1$

(g) $\displaystyle\sum_{n=0}^{\infty} (n+1)(z+1)^n$ $\quad |z + 1| < 1$

(h) $\displaystyle\sum_{n=0}^{\infty} (-1)^n \frac{(z-1)^{n+1}}{2^{n+1}}$ $\quad |z - 1| < 2$

(i) $\displaystyle e^2 \sum_{n=0}^{\infty} \frac{(z-2)^n}{n!}$ $\quad |z - 2| < \infty$

(j) $\displaystyle -\sum_{n=0}^{\infty} \frac{(z-i\pi)^n}{n!}$ $\quad |z - i\pi| < \infty$

23.

(a) $\displaystyle\sum_{n=0}^{\infty} \frac{z^{n-3}}{n!}$

(b) $\displaystyle\sum_{n=0}^{\infty} \frac{z^{-n}}{n!}$

(c) $\displaystyle -\sum_{n=1}^{\infty} z^{-n} - \sum_{n=0}^{\infty} \frac{z^n}{2^{n+!}}$

(d) $\displaystyle\sum_{n=0}^{\infty} (z-1)^{-n-2}$

(e) $\displaystyle -\sum_{n=0}^{\infty} (z-1)^{n-1}$

(f) $\displaystyle \frac{1}{10}\left\{\sum_{n=0}^{\infty} [(2i-1)(-1)^n - (2i+1)]i^n z^{-n-1} + \sum_{n=0}^{\infty} (-1)^n \left(\frac{z}{2}\right)^n\right\}$

(g) $\displaystyle\sum_{n=0}^{\infty} (-1)^n (z-1)^{-n-2}$

(h) $\displaystyle -\sum_{n=1}^{\infty} (z+1)^{-n} - \sum_{n=0}^{\infty} \frac{(z+1)^n}{2^{n+1}}$

24.

(a) Simple poles: $z = (2n+1)\pi/2,\ n = 0, \mp 1, \mp 2, \ldots$

(b) Simple pole: $z = 0$

(c) Simple poles: $z = \mp i\pi$

(d) Simple poles: $z = n\pi,\ n = \mp 1, \mp 2, \ldots$
Removable pole at $z = 0$
(e) Poles: $m = 2,\ z = \mp i$
(f) Poles: $m = 3,\ z = 0$
Simple Poles: $z = \mp 1$
(g) Poles: $m = 2,\ z = 0$
Simple Pole: $z = 2$
(h) Pole of order 3: $z = 0$
(i) Simple Pole: $z = \pm 2in\pi,\quad n = 0, 1, 2, \ldots$
(j) Pole: $m = 3,\ z = -1$

25. (a) $r\left(\frac{2n+1}{2}\pi\right) = -1$ (b) $r(0) = 1$
(c) $r(\pi i) = -\frac{1}{2\pi i},\ r(-\pi i) = \frac{1}{2\pi i}$ (d) $r(n\pi) = (-1)^n n\pi,\ n = \mp 1, \mp 2, \ldots$
(e) $r(i) = (1/2 + i),\ r(-i) = (1/2 - i)$ (f) $r(0) = 3,\ r(1) = -3/2,\ r(-1) = -3/2$
(g) $r(0) = -1,\ r(2) = 1$ (h) $r(0) = -3/10$
(i) $r(0) = 1 = r(2n\pi i)$ (j) $r(-1) = -3$

26. (a) 0 (b) $2\pi(1 - a^2)^{-1}$
(c) $\frac{\pi a}{2(1-a^2)}(a^2+3)$ (d) $\frac{\pi}{(1-a^2)^{3/2}}$
(e) $\pi\frac{(2n)!}{2^{2n}(n!)^2}$ (f) $\frac{2\pi}{1-a^2}(-a)^n$
(g) $\frac{\pi}{a^2}(1-(1-a^2)^{1/2})$ (h) $\frac{\pi(2n)!}{2^{2n}(n!)^2}$
(i) $\pi\, 2^{-1/2}$ (j) $(-1)^n\frac{2\pi e^{-an}}{\sinh(a)}$
(k) $\pi(1 - a^2)^{-1/2}$ (l) $2\pi(1 - a^2)^{-1/2}$
(m) $2\pi(1 - a^2)^{-3/2}$ (n) $\frac{\pi a^2}{1-a^2}$

27. (a) $2\pi(4b - a^2)^{-1/2}$ (b) $\frac{\pi}{4a}$

(c) $\pi\,(4a^3)^{-1}$ (d) $\dfrac{\pi(2a+b)}{2a^3 b(a+b)^2}$

(e) $\dfrac{\pi}{2^{3/2}\,a^3}$ (f) $\dfrac{3\pi}{16a^5}$

(g) $\dfrac{\pi\sqrt{2}}{16a^3}$ (h) $\dfrac{\pi}{6a^3}$

(i) $\dfrac{\pi\sqrt{2}}{4a}$ (j) $\dfrac{\pi}{3a}$

(k) $\dfrac{\pi}{16a^3}$ (l) $\dfrac{\pi}{2ab(a+b)}$

(m) $\dfrac{3\pi\sqrt{2}}{16a^2}$ (n) $\dfrac{\pi}{2(a+b)}$

28. (a) $\dfrac{\pi}{2be^{ab}}$ (b) $\dfrac{\pi}{4b^2e^{ab}}\sin(ab)$

(c) $\dfrac{\pi}{2(b^2-c^2)}(be^{-ab}-ce^{-ac})$ (d) $\dfrac{\pi}{2ae^a}\cos b$

(e) $\dfrac{\pi(1+ab)}{4b^3e^{ab}}$ (f) $\dfrac{\pi}{2bc(b^2-c^2)}(be^{-ac}-ce^{-ab})$

(g) $\dfrac{\pi(e^{-ab}-e^{-ac})}{2(c^2-b^2)}$ (h) $\pi\, a\, e^{-ab}\,(4b)^{-1}$

(i) $e^{-ab}\,\pi/2$ (j) $\pi\,(1-ab/2)\,e^{-ab}/2$

(k) $(\cos(ab)-\sin(ab))e^{-ab}\pi\,(4b)^{-1}$ (l) $\dfrac{\pi a(1+ab)}{16b^3e^{ab}}$

29. (a) $\pi\cos(ab)$ (b) π

(c) $-1/4$ (d) $\pi\,[2-e^{-ab}(ab+2)]/(8b^4)$

(e) $(12)^{-1/2}\,\pi$ (f) $\pi/5$

(g) $1/\pi$ (h) $-\pi/2$

(i) $-3^{-1/2}\pi$ (j) $-\pi\sin(ab)/(2b)$

(k) $-\pi\,[e^{-ab}+\sin(ab)]/4b^3$ (l) $\pi\cos(ab)/2$

(m) $[e^{-ab}-\sin(ab)]\,\pi/(4b)$ (n) $\dfrac{\pi}{2bc(c^2-b^2)}[c\sin(ab)-b\sin(ac)]$

(o) $\frac{\pi}{8b^4}[1 - e^{-ab}\cos(ab)]$ (p) $\frac{\pi}{4b^2}[e^{-ab} + \cos(ab) - 2]$

(q) $\frac{\pi}{2(c^2 - b^2)}[b\sin(ab) - c\sin(ac)]$ (r) $-\frac{\pi}{8b^5}[(2 + ab)e^{-ab} + \sin(ab)]$

(s) $\frac{\pi}{2(b^2 - c^2)}[b^2\sin(ab) - c^2\sin(ac)]$(t) $[\cos(ab) + e^{-ab}]\,\pi/4$

31. (a) 1 (b) $(e^{-bt} - e^{-at})/(a - b)$

(c) $\sin(at)/a$ (d) $\cos(at)$

(e) $t\cos(at)$ (f) $e^{-bt}\sin(at)/a$

(g) $\frac{t^n}{n!}$ (h) $\cosh(at)$

(i) $1 - \cos(at)$ (j) $\sin(at) - at\cos(at)$

(k) $\sin(at) + at\cos(at)$ (l) $\sinh(at) - \sin(at)$

(m) $\cosh(at) - \cos(at)$

32. (a) $F(x) = \begin{cases} 1 & 0 < x < a \\ 0 & x > a \end{cases}$ (b) $F(x) = \frac{e^{-ax}}{a}$

(c) $F(x) = \text{sech}(\pi x)$ (d) $F(x) = (1 + ax)e^{-ax}/(2a^3)$

38. (a) $\pi^3/16$ (b) 0

(c) $5/32\,\pi^5$ (d) $3\pi^3/\sqrt{32}$

(e) $-\pi^2\sqrt{2}/16$ (f) $-\pi/4$

(g) $-\pi/2$ (h) $\frac{\pi}{2ab(b^2 - a^2)}[b\log a - a\log b]$

(i) $-23\pi/96$ (j) $\pi^3\frac{1 + \cos^2(\frac{\pi}{2n})}{8n^3\sin^3(\frac{\pi}{2n})}$

(k) $-\pi^2\frac{\cos(\frac{\pi}{2n})}{4n^2\sin^2(\frac{\pi}{2n})}$ (l) $\frac{\pi a\log a}{2(1 - a^2)}$

(m) $\pi^2/(4a)$ (n) 0

39. (a) $\pi(1 - a)/(4\cos(a\pi/2))$ (b) $-\pi/\sin(a\pi)$

(c) $2\pi \dfrac{\sin(\frac{a\pi}{3})}{\sqrt{3}\sin(a\pi)}$ (d) $\pi \dfrac{b^a - c^a}{(b-c)\sin(a\pi)}$

(e) $\pi \sin(ab) / [\sin(b)\sin(a\pi)]$ (f) $-\pi \operatorname{cotan}(a\pi)$

(g) $a\, b^{a-1} \operatorname{cosec}(a\pi)$ (h) $(-1)^n b^{a+1-n} \operatorname{cosec}(a\pi) \dfrac{\Gamma(a+1)}{\Gamma(a-n+2)}$

(i) $(c+b)^{-3}\,\pi/2$ (j) $[b^a \operatorname{cosec}(a\pi) - c^a \cot(a\pi)]\,\pi/(b+c)$

(k) $-\pi \cot(a\pi)(c^a - b^a)/(c-b)$ (l) $\dfrac{2-a}{9}\pi \operatorname{cosec}\left(\dfrac{(a+1)\pi}{3}\right)$

40. (a) $2\pi\sqrt{3}/(9a)$

(b) $\pi \operatorname{cosec}(\pi/5)/(5a)$ OR $\dfrac{4\pi}{25a}(2\sin(\frac{\pi}{5}) + \sin(\frac{2\pi}{5}))$

(c) $\pi/(4a^2)$ (d) $\pi \operatorname{cosec}(2\pi/5)/(5a^2)$

(e) $\pi\sqrt{3}/(9a)$ (f) $\pi \cot(\pi/5)/(5a)$

(g) $-\pi \cot(2\pi/5)/(5a^3)$ (h) $1/(2a^2)$

(i) $1/(3a^3)$

41. (a) $\dfrac{(\log b)^2 - (\log a)^2}{2(b-a)}$ (b) $\dfrac{\log a}{a}$

(c) $-2\pi^2/27$ (d) $2\pi^2/27$

(e) $[\pi^2 + (\log a)^2]/[2(a+1)]$ (f) $\log a\ [\pi^2 + (\log a)^2]/[3(a+1)]$

(g) $2\pi^2/3$ (h) $4\pi^2/27$

(i) $4\pi^2/27$ (j) $\pi^2/27$

42. (a) $\dfrac{1}{a} e^{e^{a^2 t}} \operatorname{erf}(a\sqrt{t})$ (b) $(\pi t)^{-1/2} - a e^{a^2 t}[1 - \operatorname{erf}(a\sqrt{t})]$

(c) $(\pi t^3)^{-1/2}[e^{bt} - e^{at}]/2$ (d) $(\pi t)^{-1/2} e^{-at}$

(e) $\dfrac{1}{a}\left\{1 - e^{a^2 t}[1 - \operatorname{erf}(a\sqrt{t})]\right\}$ (f) $\dfrac{t^{b-1} e^{-at}}{\Gamma(b)}$

(g) 0 for $0 < t < a$, $e^{-b(t-a)}$ for $t > a$ (h) $(\pi t)^{-1/2}\cosh(2a\sqrt{t})$

(i) $(\pi t^3)^{-1/2} e^{-a/4t} a^{1/2}/2$ (j) $(e^{-at} - e^{-bt})/t$

(k) $2(\cos(at) - \cos(bt))/t$ (l) $2\cos(ct)(e^{-bt} - e^{-at})/t$

(m) $(1 + 2bt)e^{bt}(\pi t)^{-1/2}$ (n) $(\pi t)^{-1/2}$

(o) $J_0(at)$

(p) $(\pi t/2)^{-1/2} \cos(at)$

(q) $(\pi t/2)^{-1/2} \cosh(at)$

(r) $1 - \text{erf}\left(\frac{1}{2}\sqrt{a/t}\right)$

(s) $(\log t)^2 - \pi^2/6$

(t) $(\cos(at) + at\sin(at) - 1)/t^2$

(u) 0 for $0 < t < a$, $J_0(bt)$ for $t > a$

(v) $\dfrac{e^{a^2t}}{b+a} + \dfrac{ae^{a^2t}\text{erfc}(a\sqrt{t}) - be^{b^2t}\text{erfc}(b\sqrt{t})}{b^2 - a^2}$

(w) $e^{-(a+b)t} I_0[(a-b)t]$

(x) $t\, e^{-(a+b)t}[I_1[(a-b)t] + I_0[(a-b)t]]$

(y) $\sqrt{\pi}\left(\frac{t}{2a}\right)^{\nu} J_\nu(at)/\Gamma(\nu + \frac{1}{2})$

(z) $a^\nu J_\nu(at)$

(aa) $e^{a^2t}\text{erfc}(a\sqrt{t})$

(bb) $e^{-at}\,\text{erf}\,[((b-a)t)^{1/2}]\,(b-a)^{-1/2}$

(cc) $\dfrac{e^{a^2t}}{b+a} + \dfrac{ae^{b^2t}\text{erfc}(b\sqrt{t}) - be^{a^2t}\text{erfc}(a\sqrt{t})}{b^2 - a^2}$

(dd) $a\, e^{-at}[I_0(at) + I_1(at)]$

(ee) $2(1 - \cos(at))/t$

(ff) $\left(\dfrac{a}{4\pi t^3}\right)^{1/2} \exp(-\dfrac{a}{4t})$

(gg) $(\pi t)^{-1/2} \exp(-\dfrac{a}{4t})$

(hh) $\sin(at)/t$

Chapter 6

1. $$T(x,y) = 2T_0 bL \sum_{n=1}^{\infty} \frac{1-\cos\alpha_n}{\alpha_n(bL+\cos^2\alpha_n)} \frac{\sinh(\alpha_n y/L)}{\sinh(\alpha_n)} \sin(\alpha_n x/L)$$
   where $\tan\alpha_n = -\alpha_n/(bL)$

2. $$T(x,y) = \frac{2T_0}{L} \sum_{n=1}^{\infty} a_n \sin(n\pi x/L)\exp(-n\pi y/L)$$
   where $a_n = \int_0^L f(x)\sin(n\pi x/L)\,dx$

3. $$T(x,y) = \frac{2T_0}{L} \sum_{n=1}^{\infty} a_n \frac{b^2L^2+\alpha_n^2}{b^2L^2+bL+\alpha_n^2} \cos(\alpha_n x/L)\exp(-\alpha_n y/L)$$
   where $a_n = \int_0^L f(x)\cos(\alpha_n x/L)\,dx$ and $\tan\alpha_n = Lb/\alpha_n$

4. $$T(x,y) = \frac{2T_0}{L} \sum_{n=1}^{\infty} a_n \frac{\cosh(n\pi y/L)}{\cosh(n\pi)} \sin(n\pi x/L)$$
   where $a_n = \int_0^L f(x)\sin(n\pi x/L)\,dx$

5. $$T(r,\theta) = \frac{4T_0}{\pi} \sum_{n=0}^{\infty} \frac{1}{2n+1}\left(\frac{r}{a}\right)^{2n+1} \sin((2n+1)\theta)$$

6. $$T(r,\theta) = \frac{2T_0}{b} \sum_{n=1}^{\infty} a_n \left[\frac{r}{a}\right]^{n\pi/b} \sin(n\pi 0/b) \qquad \text{where } a_n = \int_0^b f(\theta)\sin\left(\frac{n\pi}{b}\theta\right)d\theta$$

7. $$T(r,\theta) = \sum_{n=1}^{\infty} \left(\frac{c}{r}\right)^n (a_n\cos(n\theta) + b_n\sin(n\theta))$$
   where $a_n = \frac{T_0}{\pi}\int_0^{2\pi} f(\theta)\cos(n\theta)\,d\theta$ and $b_n = \frac{T_0}{\pi}\int_0^{2\pi} f(\theta)\sin(n\theta)\,d\theta$

8. $$T(x,y,z) = \frac{4T_0}{ab} \sum_{m=1}^{\infty} \sum_{n=1}^{\infty} d_{mn} \frac{\sinh(\alpha_{mn} z)}{\sinh(\alpha_{mn} c)} \sin(\frac{m\pi}{a} x) \sin(\frac{n\pi}{b} y)$$

where $\alpha_{mn} = \pi \sqrt{(m/a)^2 + (n/b)^2}$ and $d_{mn} = \int_0^a \int_0^b f(x,y) \sin(\frac{m\pi}{a} x) \sin(\frac{n\pi}{b} y)\, dx\, dy$

9. $$T(r,z) = \frac{2T_0}{a^2} \sum_{n=1}^{\infty} c_n \frac{\sinh(\alpha_n z)}{\sinh(\alpha_n L)} J_0(\alpha_n r)$$

where $J_1(\alpha_n a) = \frac{b}{\alpha_n} J_0(\alpha_n a)$ and $c_n = \frac{1}{(1+(\alpha_n/b)^2) J_1^2(\alpha_n a)} \int_0^a r\, f(r) J_0(\alpha_n r)\, dr$

10. $$T(z,r) = 2T_0 \sum_{n=1}^{\infty} c_n \frac{\sinh(\alpha_n z)}{\sinh(\alpha_n L)} \phi_0(\alpha_n r) \quad \text{where} \quad \phi_0(\alpha_n r) = \frac{J_0(\alpha_n r)}{J_0(\alpha_n a)} - \frac{Y_0(\alpha_n r)}{Y_0(\alpha_n a)}$$

$\phi_0(\alpha_n b) = 0$ (characteristic equation),

and $c_n = \frac{1}{b^2 \phi_1^2(\alpha_n b) - a^2 \phi_1^2(\alpha_n a)} \int_a^b r\, f(r) \phi_0(\alpha_n r)\, dr$

11. $$T(z,r) = \frac{2T_0}{a^2} \sum_{n=0}^{\infty} b_n \frac{\sinh(\alpha_n z)}{\sinh(\alpha_n L)} \frac{J_0(\alpha_n r)}{J_1^2(\alpha_n a)}$$

where $J_0(\alpha_n a) = 0$ and $b_n = \int_0^a r\, f(r) J_0(\alpha_n r)\, dr$

12. $$T(z,r) = \frac{2T_0}{L} \sum_{n=1}^{\infty} b_n \sin(\frac{n\pi}{L} z) \frac{I_0(\frac{n\pi}{L} r)}{I_0(\frac{n\pi}{L} a)} \quad \text{where} \quad b_n = \int_0^L f(z) \sin(\frac{n\pi}{L} z)\, dz$$

13. $$T(r,\theta) = T_0 \sum_{n=0}^{\infty} a_n (\frac{r}{a})^n P_n(\cos\theta) \quad \text{where} \quad a_n = \frac{2n+1}{2} \int_{-1}^{+1} f(x) P_n(x)\, dx$$

For $f(x) = 1$, $T = T_0$

14. $$T(r,\theta) = T_0 \sum_{n=0}^{\infty} a_n (\frac{a}{r})^{n+1} P_n(\cos\theta) \quad \text{where} \quad a_n = \frac{2n+1}{2} \int_{-1}^{+1} f(x) P_n(x)\, dx$$

For $f(x) = 1$, $T = T_0\, a/r$

15. $T(r,\theta) = T_0 \sum_{n=0}^{\infty} a_n (\frac{r}{a})^n P_n(\cos\theta)$ where $a_0 = 1/4$, $a_1 = ba/(2ba+2)$,

$a_{2n+1} = 0$, and $a_{2n} = (-1)^n \frac{ba}{ba+1} \frac{(2n-2)!(4n+1)}{2^{2n+1}(n-1)!(n+1)!}$ for $n = 1, 2, 3, \ldots$

16. Velocity potential $\phi(r,\theta) = V_0 [1 + \frac{a^3}{2r^3}] r\cos\theta$

17. $T(r,\theta) = T_0 \sum_{n=0}^{\infty} (-1)^n \frac{(2n-1)!(4n+3)}{2^{n+1}(n+1)!} (\frac{r}{a})^{2n+1} P_{2n+1}(\cos\theta)$

18. $T(r,\theta) = T_0 \sum_{n=0}^{\infty} a_n [1-(\frac{r}{b})^{2n+1}](\frac{a}{r})^{n+1} P_n(\cos\theta)$

where $a_n = \frac{2n+1}{2-2(a/b)^{2n+1}} \int_{-1}^{+1} f(x) P_n(x) dx$

19. $T(r,\theta,z) = T_0 \sum_{n=1}^{\infty} a_{on} e^{-\alpha_{0n} z} J_0(\alpha_{on} r)$

$+T_0 \sum_{n=1}^{\infty} \sum_{m=1}^{\infty} e^{-\alpha_{mn} z} J_m(\alpha_{mn} r)[a_{mn}\cos(m\theta) + b_{mn}\sin(m\theta)]$

where $J_m(\alpha_{mn} a) = 0$

$$a_{on} = \frac{1}{\pi a^2 J_1^2(\alpha_{on} a)} \int_0^{2\pi} \int_0^{a} r f(r,\theta) J_0(\alpha_{on} r) dr\, d\theta$$

$$\left.\begin{matrix} a_{mn} \\ b_{mn} \end{matrix}\right\} = \frac{2}{\pi a^2 J_{m+1}^2(\alpha_{mn} a)} \int_0^{2\pi} \int_0^{a} r f(r,\theta) J_m(\alpha_{mn} r) \left\{\begin{matrix} \cos(m\theta) \\ \sin(m\theta) \end{matrix}\right\} dr\, d\theta$$

20. $T(r,z,\theta) = \frac{T_0}{\pi L} \sum_{n=0}^{\infty} \sum_{m=1}^{\infty} A_{nm} I_{n/2}(\frac{m\pi}{L} r) \sin(\frac{m\pi}{L} z) \cos(\frac{n\theta}{2})$

where $A_{nm} = \frac{\varepsilon_n}{I_{n\pi/b}(m\pi a/L)} \int_0^{L} \int_0^{b} f(z,\theta) \sin(\frac{m\pi}{L} z) \cos(\frac{n\pi}{b} \theta)\, d\theta\, dz$

21. $$T(r,z,\theta) = \frac{2T_0}{bL} \sum_{n=0}^{\infty} \sum_{m=1}^{\infty} a_{nm} K_{n\pi/b}(\frac{m\pi}{L} r) \sin(\frac{m\pi}{L} z) \cos(\frac{n\pi}{b} \theta)$$

where $$A_{nm} = \frac{\varepsilon_n}{K_{n\pi/b}(m\pi a / L)} \int_0^L \int_0^b f(z,\theta) \sin(\frac{m\pi}{L} z) \cos(\frac{n\pi}{b} \theta)\, d\theta\, dz$$

22. $$T(r,\theta,z) = \frac{4T_0}{\pi a^2} \sum_{n=1}^{\infty} \sum_{m=1}^{\infty} A_{nm} J_n(\alpha_{nm} r) \sin(n\theta) \sinh(\alpha_{nm} z)$$

where $J_n(\alpha_{nm}a) = 0$ for $n = 1, 2, 3, \ldots$

and $$A_{nm} = \frac{1}{\sinh(\alpha_{nm} L) J_{n+1}^2(\alpha_{nm} a)} \int_0^{\pi} \int_0^{a} r f(r,\theta) J_n(\alpha_{nm} r) \sin(n\theta)\, dr\, d\theta$$

23. $$T = \frac{1}{a^3} \sum_{m=1}^{\infty} \sum_{n=0}^{\infty} \frac{2n+1}{j_{n+1}^2(k_{nm} a) k_{nm}^2} A_{nm}\, j_n(k_{nm} r) P_n(\eta)$$

where $\eta = \cos\theta$, $j_n(k_{nm}a) = 0$, and $$A_{nm} = \int_0^b \int_{-1}^{+1} r^2 q(r,\eta) j_n(k_{nm} r) P_n(\eta)\, d\eta\, dr$$

24. $$\phi = -\frac{Q_0}{2\pi r} \sum_{n=1}^{\infty} \frac{\sin(\mu_n r / a)}{\mu_n \sin^2(\mu_n)}$$ where $\tan \mu_n = \mu_n$ for $n = 1, 2, 3, \ldots$

25. $$\phi = \frac{4Q_0}{\pi a^2 L} \sum_{n=0}^{\infty} \sum_{m=1,3,5}^{\infty} \frac{(-1)^{(m-1)/2} \sin(m\pi/4) J_0(\mu_n r / a) \cos(m\pi z / L)}{J_0^2(\mu_n)[(\mu_n / a)^2 + (m\pi / L)^2]}$$

where $J_1(\mu_n) = 0$ for $n = 0, 1, 2, \ldots$

26. In the following list of solutions, $k = \omega/c$, k_{nm} are the eigenvalues and W_{nm} are the mode shapes:

(a) $W_{nm} = J_n(k_{nm}r) \sin(n\theta)$, where $J_n(k_{nm}a) = 0$ for $n, m = 1, 2, 3, \ldots$

(b) $$W_{nm} = \left[\frac{J_n(k_{nm} r)}{J_n(k_{nm} b)} - \frac{Y_n(k_{nm} r)}{Y_n(k_{nm} b)}\right] \begin{Bmatrix} \sin(n\theta) \\ \cos(n\theta) \end{Bmatrix}$$

$J_n(k_{nm}a) Y_n(k_{nm}b) - J_n(k_{nm}b) Y_n(k_{nm}a) = 0$ $n = 0, 1, 2, \ldots$ $m = 1, 2, 3, \ldots$

(c) W_{nm} same as in part (b)

$J_n(k_{nm}b) Y_n'(k_{nm}a) - J_n'(k_{nm}a) Y_n(k_{nm}b) = 0$

(d) $W_{nm} = J_{n\pi/c}(k_{nm}r) \sin(n\pi\theta/c)$

$J_{n\pi/c}(k_{nm}a) = 0$ for $n,m = 1, 2, 3, \ldots$

(e) W_{nm} same as in (d)

$J'_{n\pi/c}(k_{nm}a) = 0$

(f) $W_{nm} = \left[\dfrac{J_\alpha(k_{nm}r)}{J_\alpha(k_{nm}b)} - \dfrac{Y_\alpha(k_{nm}r)}{Y_\alpha(k_{nm}b)}\right]\sin(\alpha\theta) \qquad \alpha = n\pi/c$

$J_\alpha(k_{nm}b)\,Y_\alpha(k_{nm}a) - J_\alpha(k_{nm}a)\,Y_\alpha(k_{nm}b) = 0 \qquad n, m = 1, 2, 3, \ldots$

(g) $W_{nm} = \left[\dfrac{J_\alpha(k_{nm}r)}{J'_\alpha(k_{nm}b)} - \dfrac{Y_\alpha(k_{nm}r)}{Y'_\alpha(k_{nm}b)}\right]\sin(\alpha\theta) \qquad \alpha = n\pi/c$

$J'_\alpha(k_{nm}b)\,Y'_\alpha(k_{nm}a) - J'_\alpha(k_{nm}a)\,Y'_\alpha(k_{nm}b) = 0 \qquad n, m = 1, 2, 3, \ldots$

(h) $W_{nm} = \sin(n\pi y/b)\cos(m\pi x/a)$

$k_{nm}^2 = (\frac{n\pi}{b})^2 + (\frac{m\pi}{a})^2 \qquad n = 1, 2, 3, \ldots \qquad m = 0, 1, 2, \ldots$

27. $\phi_{nlm} = J_n(q_{nl}r)\cos(\frac{m\pi}{L}z)\begin{Bmatrix}\sin(n\theta)\\ \cos(n\theta)\end{Bmatrix} \qquad$ where $J'_n(q_{nl}a) = 0$

$k_{nlm}^2 = \dfrac{m^2\pi^2}{L^2} + q_{nl}^2 \qquad m = 0, 1, 2, \ldots \quad \begin{cases} n = 0 & l = 0, 1, 2, \ldots \\ n \geq 1 & l = 1, 2, 3, \ldots \end{cases}$

28. $\phi_{nlm} = J_n(q_{nl}r)\sin(\frac{m\pi}{L}z)\begin{Bmatrix}\sin(n\theta)\\ \cos(n\theta)\end{Bmatrix} \qquad$ where $J'_n(q_{nl}a) = 0$

$k_{nlm}^2 = \dfrac{m^2\pi^2}{L^2} + q_{nl}^2 \qquad n = 0, 1, 2, \ldots\ l, m = 1, 2, 3, \ldots$

29. $\phi_{nlm} = \left[\dfrac{j_n(k_{nl}r)}{j'_n(k_{nl}b)} - \dfrac{y_n(k_{nl}r)}{y'_n(k_{nl}b)}\right] P_n^m(\cos\theta)\begin{Bmatrix}\sin(n\theta)\\ \cos(n\theta)\end{Bmatrix}$

where $j'_n(k_{nl}b)\,y'_n(k_{nl}a) - j'_n(k_{nl}a)\,y'_n(k_{nl}b) = 0$

$n, m = 0, 1, 2, \ldots\ l = 1, 2, 3, \ldots$

30. $w(x,y,t) = W(x,y)\sin(\omega t)$

$$W(x,y) = \frac{4q_0}{abS}\sum_{n=1}^{\infty}\sum_{m=1}^{\infty}\frac{A_{nm}}{k_{nm}^2 - k^2}\sin(\frac{m\pi}{a}x)\sin(\frac{n\pi}{b}y)$$

$k_{nm}^2 = (\frac{n\pi}{b})^2 + (\frac{m\pi}{a})^2 \qquad k = \omega/c$

$$A_{nm} = \int_0^a\int_0^b f(x,y)\sin(\frac{m\pi}{a}x)\sin(\frac{n\pi}{b}y)\,dy\,dx$$

31. $w(r,\theta,t) = W(r,\theta)\sin(\omega t)$

$$W(r,\theta) = \frac{q_0}{\pi a^2 S}\sum_{n=0}^{\infty}\sum_{m=1}^{\infty}\frac{J_n(k_{nm}r)}{k_{nm}^2 - k^2}\cdot\frac{A_{nm}\cos(n\theta) + B_{nm}\sin(n\theta)}{J_{n+1}^2(k_{nm}a)}$$

where $J_n(k_{nm}a) = 0$

$$\begin{Bmatrix} A_{nm} \\ B_{nm} \end{Bmatrix} = \varepsilon_n \int_0^a \int_0^{2\pi} r f(r,\theta) J_n(k_{nm}r) \begin{Bmatrix} \cos(n\theta) \\ \sin(n\theta) \end{Bmatrix} d\theta\, dr$$

32. $$w_{nm}(r,\theta) = \left[\frac{J_n(k_{nm}r)}{J_n(k_{nm}a)} - \frac{I_n(k_{nm}r)}{I_n(k_{nm}a)}\right]\begin{Bmatrix} \sin(n\theta) \\ \cos(n\theta) \end{Bmatrix}$$

where $$\left[\frac{J_{n+1}(k_{nm}a)}{J_n(k_{nm}a)} + \frac{I_{n+1}(k_{nm}a)}{I_n(k_{nm}a)}\right] = \frac{2k_{nm}}{1-\nu}$$

33. $w(x,y,t) = W(x,y)\sin(\omega t)$

$$W(x,y) = \frac{4}{\rho h L^2}\sum_{n=1}^{\infty}\sum_{m=1}^{\infty}\frac{A_{nm}}{\omega_{nm}^2 - \omega^2}\sin\left(\frac{n\pi}{L}x\right)\sin\left(\frac{m\pi}{L}y\right)$$

where $$\omega_{nm}^2 = \sqrt{\frac{D}{\rho h}}\left[\left(\frac{n\pi}{L}\right)^2 + \left(\frac{m\pi}{L}\right)^2\right]$$

and $$A_{nm} = \int_0^L \int_0^L q_0(x,y)\sin\left(\frac{n\pi}{L}x\right)\sin\left(\frac{m\pi}{L}y\right)dy\,dx$$

34. $$w = \frac{2F_0}{ab\,S}\sum_{n=0}^{\infty}\sum_{m=0}^{\infty}(-1)^{n+m}\frac{\varepsilon_{2n}}{k^2 - \lambda_{nm}}\cos\left(\frac{2n\pi}{a}x\right)\sin\left(\frac{(2m+1)\pi}{b}y\right)$$

$\lambda_{nm} = 4\,n^2\pi^2/a^2 + (2m+1)^2\,\pi^2/b^2$

35. $$T(x,t) = \frac{4T_0}{\pi}\sum_{n=1}^{\infty}\frac{\sin^2(n\pi/4)}{n}\sin\left(\frac{n\pi}{L}x\right)\exp\left(-\frac{n^2\pi^2}{L^2}Kt\right)$$

36. $$T(x,t) = \frac{2T_0}{L}\sum_{n=1}^{\infty}A_n[\alpha_n\cos(\alpha_n x/L) + bL\sin(\alpha_n x/L)]\exp\left(-\frac{\alpha_n^2}{L^2}Kt\right)$$

where $2\cot\alpha_n = \alpha_n/(bL) - (bL)/\alpha_n$

and $$A_n = \frac{1}{bL(bL+2) + \alpha_n^2}\int_0^L f(x)[\alpha_n\cos(\alpha_n x/L) + bL\sin(\alpha_n x/L)]dx$$

37. $$T(x,y,t) = \frac{4}{ab}\sum_{m=1}^{\infty}\sum_{n=1}^{\infty}A_{mn}(t)\sin\left(\frac{m\pi}{a}x\right)\sin\left(\frac{n\pi}{b}y\right)$$

where $A_{mn} = B_{mn}\exp(-\lambda_{mn}Kt) + C_{mn}[\exp(-\alpha t) - \exp(-\lambda_{mn}Kt)]$

$$C_{mn} = Q_0 \frac{\sin(n\pi/2)\sin(m\pi/2)}{(\lambda_{mn}K - \alpha)\rho c}$$

$$B_{mn} = T_0 \int_0^a \int_0^b f(x,y)\sin(\frac{m\pi}{a}x)\sin(\frac{n\pi}{b}y)\,dx\,dy$$

and $\lambda_{mn} = \pi^2 (m^2/a^2 + n^2/b^2)$

38. $T(r,t) = \frac{2}{a^2} \sum_{n=1}^{\infty} A_n J_0(\alpha_n r/a)\exp(-K\alpha_n^2 t/a^2)$

where $J_0(\alpha_n) = 0$ and $A_n = \frac{T_0}{J_1^2(\alpha_n)} \int_0^a r f(r) J_0(\alpha_n r/a)dr$

39. $T(r,t) = \frac{2}{a^2} \sum_{n=1}^{\infty} \frac{\alpha_n^2}{(\alpha_n^2 + b^2a^2)J_0^2(\alpha_n)} A_n(t) J_0(\alpha_n r/a)\exp(-K\alpha_n^2 t/a^2)$

where $J_1(\alpha_n) = ba\, J_0(\alpha_n)/\alpha_n$

and $A_n(t) = T_0 \int_0^a r f(r) J_0(\alpha_n r/a)dr + \frac{Q_0}{2\pi\rho c}\exp(K\alpha_n^2 t_0/a^2)H(t-t_0)$

40. $T(r,\theta,t) =$

$$\frac{1}{\pi a^2} \sum_{n=0}^{\infty} \sum_{m=1}^{\infty} \frac{\varepsilon_n \exp(-\alpha_{nm}^2 Kt/a^2)}{J_{n+1}^2(\alpha_{nm})} [A_{nm}(t)\cos(n\theta) + B_{nm}(t)\sin(n\theta)] J_n(\alpha_{nm} r/a)$$

where $J_n(\alpha_{nm}) = 0$ $\quad n = 0, 1, 2, ..\quad m = 1, 2, 3, \ldots$

$A_{nm}(t) = C_{nm} + P_{nm}\exp(K\alpha_{nm}^2 t_0/a^2)H(t-t_0)$

$B_{nm}(t) = D_{nm} + R_{nm}\exp(K\alpha_{nm}^2 t_0/a^2)H(t-t_0)$

$P_{nm} = K\frac{Q_0}{k} J_n(\alpha_{nm} r_0/a)\cos(n\theta_0)$, $R_{nm} = K\frac{Q_0}{k} J_n(\alpha_{nm} r_0/a)\sin(n\theta_0)$

$$C_{nm} = T_0 \int_0^a \int_0^{2\pi} r f(r,\theta) J_n(\alpha_{nm} r/a)\cos(n\theta)\,d\theta\,dr$$

$$D_{nm} = T_0 \int_0^a \int_0^{2\pi} r f(r,\theta) J_n(\alpha_{nm} r/a)\sin(n\theta)\,d\theta\,dr$$

41. $T(r,t) = -\frac{2T_0 a}{\pi r} \sum_{n=1}^{\infty} \frac{(-1)^n}{n}\sin(\frac{n\pi}{a}r)\exp(-\frac{n^2\pi^2}{a^2}Kt)$

42. $T(r,t) = \frac{2T_0}{ar} \sum_{n=1}^{\infty} \frac{(x-1)^2 + \alpha_n^2}{x(x-1) + \alpha_n^2} A_n \sin(\alpha_n r/a)\exp(-K\alpha_n^2 t/a^2)$

where $x = ba$, $\tan\alpha_n = -\alpha_n/(x-1)$

and $A_n = \int_0^a r\, f(r)\sin(\alpha_n r/a)\,dr$

43. $T(x,y,z,t) = \frac{8T_0}{L^3}\sum_{m=1}^{\infty}\sum_{n=1}^{\infty}\sum_{q=1}^{\infty} A_{mnq}\,B_{mnq}(t)\sin(\frac{m\pi}{L}x)\sin(\frac{n\pi}{L}y)\,\sin(\frac{q\pi}{L}z)$

where $B_{mnq}(t) = \exp[-K\pi^2\frac{m^2+n^2+q^2}{L^2}t]$

and $A_{mnq} = \int_0^L\int_0^L\int_0^L f(x,y,z)\sin(\frac{m\pi}{L}x)\sin(\frac{n\pi}{L}y)\sin(\frac{q\pi}{L}z)\,dx\,dy\,dz$

44. $T(r,\theta,\phi,t) = \frac{T_0}{2\pi a^3}\sum_{n=0}^{\infty}\sum_{q=1}^{\infty}\sum_{m=0}^{\infty}[A_{nmq}\cos(m\phi) + B_{nmq}\sin(m\phi)]\cdot$

$\cdot j_n(\alpha_{nq}r/a)P_n^m(\eta)\exp(-K\alpha_{nq}^2 t/a^2)$

where $\eta = \cos\theta$, $j_n(\alpha_{nq}) = 0$ $n = 0, 1, 2, \ldots$ $q = 1, 2, 3, \ldots$

$$\left.\begin{matrix}A_{nmq}\\B_{nmq}\end{matrix}\right\} = C_{nmq}\int_0^{a}\int_{-1}^{+1}\int_0^{2\pi} r^2 f(r,\eta,\phi)\,j_n(\alpha_{nq}\frac{r}{a})P_n^m(\eta)\left\{\begin{matrix}\cos(m\phi)\\\sin(m\phi)\end{matrix}\right\}d\phi\,d\eta\,dr$$

and $C_{nmq} = (-1)^m\frac{(2n+1)\varepsilon_m(n-m)!}{j_{n+1}^2(\alpha_{nq})(n+m)!}$

45. $T = -\frac{Q_0K}{k}\sum_{n=1}^{\infty}\sum_{m=1}^{\infty} E_{nm}(t)\Psi_{nm}(x,y)$

where $\Psi_{nm}(x,y) = [\sin(\mu_m x/a) + \frac{\mu_m}{a\gamma}\cos(\mu_m x/a)]\sin(n\pi y/b)$,

$\tan\mu_m = \frac{2\mu_m a\gamma}{\mu_m^2 - a^2\gamma^2}$ $n,m = 1, 2, 3, \ldots$

$E_{nm}(t) = [\sin(\frac{\mu_m}{2}) + \frac{\mu_m}{a\gamma}\cos(\frac{\mu_m}{2})]\sin(\frac{n\pi}{2})\frac{\lambda_{nm}K\sin(\omega t) - \omega\cos(\omega t) + \omega e^{-\lambda_{nm}Kt}}{N_{nm}(K^2\lambda_{nm}^2 + \omega^2)}$

$N_{nm} = \int_0^a\int_0^b \Psi_{nm}^2\,dx\,dy$, and $\lambda_{nm} = \mu_m^2/a^2 + n^2\pi^2/b^2$

46. $T = \sum_{n=1}^{\infty}\sum_{m=1}^{\infty} E_{nm}(t)\,J_n(\gamma_{nm}r/a)\sin(n\theta)$

$E_{nm}(t) = \frac{KQ_0}{kN_{nm}}J_n(\gamma_{nm}r_0/a)\sin(n\pi/4)e^{-K\lambda_{nm}(t-t_0)}H(t-t_0)$

$$N_{nm} = \frac{\pi}{2}\int_0^a r\,J_n^2(\gamma_{nm} r/a)\,dr \qquad J_n'(\gamma_{nm}) = 0$$

47. $$T = -\frac{Q_0 K}{k}\sum_{n=1}^{\infty} E_n(t)\cos(\alpha_n x/L) \qquad \text{where } \tan\alpha_n = bL/\alpha_n$$

$$E_n(t) = \frac{e^{-at} - e^{-\lambda_n K t}}{\lambda_n K - a}\cdot\frac{\cos(\alpha_n x_0/L)}{N_n} \qquad \lambda_n = \alpha_n^2/L^2 \qquad n = 1, 2, 3, \ldots$$

$$N_n = \frac{L}{2}\cdot\frac{\alpha_n^2 + (bL)^2 + bL}{\alpha_n^2 + (bL)^2}$$

48. $$T = \frac{2Q_0 K}{ab\,k}\sum_{n=0}^{\infty}\sum_{m=1}^{\infty} E_{nm}(t)\cos\left(\frac{n\pi}{a}x\right)\sin\left(\frac{m\pi}{b}y\right)$$

$$E_{nm}(t) = \varepsilon_n \cos\left(\frac{n\pi}{2}\right)\sin\left(\frac{m\pi}{2}\right)\frac{e^{-ct} - e^{-K\lambda_{nm}t}}{K\lambda_{nm} - c}$$

where $\lambda_{nm} = n^2\pi^2/a^2 + m^2\pi^2/b^2$

49. $$T = \frac{KQ_0}{2\pi}\sum_{m=0}^{\infty}\sum_{n=0}^{\infty}\sum_{l=1}^{\infty} E_{nml}(t)\,J_m(\mu_{ml} r/a)\cos(m\theta)\cos(n\pi z/L)$$

$$E_{nml}(t) = \frac{J_m(\mu_{ml} r_0/a)\cos(m\pi/2)\cos(n\pi z_0/L)}{N_{nml}}\,e^{-K\lambda_{nml}(t-t_0)}H(t-t_0)$$

where $J_m(\mu_{ml}) = 0$, $\lambda_{nml} = \mu_{ml}^2/a^2 + n^2\pi^2/L^2$

$$N_{nml} = \frac{a^2\pi L}{2\mu_{ml}^2\,\varepsilon_m\,\varepsilon_n}(\mu_{ml}^2 - m^2)J_m^2(\mu_{ml})$$

50. $$T = T_0 + \sum_{n=1}^{\infty} E_n(t)X_n(x), \qquad \text{where } X_n(x) = \sin(\beta_n x/L) + \frac{a\beta_n}{L}\cos(\beta_n x/L)$$

$$E_n(t) = C_n e^{-\lambda_n K t} + \frac{KQ_0}{kN_n}X_n\left(\frac{L}{2}\right)e^{-\lambda_n K(t-t_0)}H(t-t_0)$$

$$C_n = \frac{T_1 - T_0}{N_n}\int_0^L X_n(x)\,dx, \qquad \text{and } N_n = \int_0^L X_n^2(x)\,dx$$

51. $$y(x,t) = \sum_{n=1}^{\infty}\left[a_n\cos\left(\frac{n\pi}{L}ct\right) + b_n\sin\left(\frac{n\pi}{L}ct\right)\right]\sin\left(\frac{n\pi}{L}x\right)$$

where $a_n = \frac{2}{L}\int_0^L f(x)\sin\left(\frac{n\pi}{L}x\right)dx$, and $b_n = \frac{2}{n\pi c}\int_0^L g(x)\sin\left(\frac{n\pi}{L}x\right)dx$

52. $y(x,t)=\dfrac{2W_0L^2}{\pi^2 a(L-a)}\sum_{n=1}^{\infty}\dfrac{1}{n^2}\sin(\dfrac{n\pi}{L}x)\cos(\dfrac{n\pi}{L}ct)\sin(\dfrac{n\pi}{L}a)$

53. $u(x,t)=\sum_{n=0}^{\infty}[a_n\cos(\dfrac{(2n+1)\pi}{2L}ct)+b_n\sin(\dfrac{(2n+1)\pi}{2L}ct)]\sin(\dfrac{(2n+1)\pi}{2L}x)$

where $a_n=\dfrac{2}{L}\int_0^L f(x)\sin(\dfrac{(2n+1)\pi}{2L}x)\,dx$,

and $b_n=\dfrac{4}{(2n+1)\pi c}\int_0^L g(x)\sin(\dfrac{(2n+1)\pi}{2L}x)\,dx$

54. $y(x,t)=2y_0\gamma L^2(\gamma L+1)\sum_{n=1}^{\infty}\dfrac{\sin(\alpha_n)\sin(\alpha_n x/L)}{\alpha_n^2(\gamma L+\cos^2(\alpha_n))}\cos(\alpha_n tc/L)$

where $\tan\alpha_n=-\dfrac{\alpha_n}{\gamma L}$

55. (a) $y(x,t)=\dfrac{2IL}{\pi^2 c\varepsilon\rho}\sum_{n=1}^{\infty}\dfrac{1}{n^2}\sin(\dfrac{n\pi}{2})\sin(\dfrac{n\pi}{L}\varepsilon)\sin(\dfrac{n\pi}{L}ct)\sin(\dfrac{n\pi}{L}x)$

(b) $\lim_{\varepsilon\to 0}y(x,t)\to\dfrac{2I}{\pi c\rho}\sum_{n=1}^{\infty}\dfrac{1}{n}\sin(\dfrac{n\pi}{2})\sin(\dfrac{n\pi}{L}ct)\sin(\dfrac{n\pi}{L}x)$

56. $y(x,t)=\dfrac{2c}{T_0\pi}\sum_{n=1}^{\infty}T_n(t)\sin(\dfrac{n\pi}{L}x)$

where $T_n(t)=\dfrac{1}{n}\int_0^t\left[\int_0^L f(x,\tau)\sin(\dfrac{n\pi}{L}x)\right]\sin(\dfrac{n\pi}{L}c(t-\tau))\,d\tau$

57. Solution for y(x,t) same as in 56, where $T_n=\dfrac{P_0}{n}\sin(\dfrac{n\pi}{2})\sin(\dfrac{n\pi}{L}ct)$

58. $W(x,y,t)=\dfrac{4}{ab}\sum_{n=1}^{\infty}\sum_{m=1}^{\infty}[A_{mn}\cos(\alpha_{mn}ct)+B_{mn}\sin(\alpha_{mn}ct)]\sin(\dfrac{n\pi}{a}x)\sin(\dfrac{m\pi}{b}y)$

where $\alpha_{nm}^2=\pi^2(\dfrac{m^2}{b^2}+\dfrac{n^2}{a^2})$,

$A_{mn}=\int_0^a\int_0^b f(x,y)\sin(\dfrac{n\pi}{a}x)\sin(\dfrac{m\pi}{b}y)\,dx\,dy$

and $B_{mn} = \frac{1}{c\alpha_{mn}} \int_0^a \int_0^b g(x,y) \sin(\frac{n\pi}{a}x) \sin(\frac{m\pi}{b}y)\, dx\, dy$

59. $W(r,\theta,t) = \sum_{n=0}^{\infty} \sum_{m=1}^{\infty} \{[A_{nm}\cos(n\theta) + B_{nm}\sin(n\theta)]\cos(\frac{\alpha_{nm}ct}{a})$

$+ [C_{nm}\cos(n\theta) + D_{nm}\sin(n\theta)]\sin(\frac{\alpha_{nm}ct}{a})\} J_n(\alpha_{nm}\frac{r}{a})$

where $J_n(\alpha_{nm}) = 0$

$$\left.\begin{matrix} A_{nm} \\ B_{nm} \end{matrix}\right\} = \frac{\varepsilon_n}{\pi a^2 J_{n+1}^2(\alpha_{mn})} \int_0^a \int_0^{2\pi} r\, f(r,\theta)\, J_n(\alpha_{nm}\frac{r}{a}) \left\{\begin{matrix} \cos(n\theta) \\ \sin(n\theta) \end{matrix}\right\} d\theta\, dr$$

$$\left.\begin{matrix} C_{nm} \\ D_{nm} \end{matrix}\right\} = \frac{\varepsilon_n}{\pi c a \alpha_{nm} J_{n+1}^2(\alpha_{mn})} \int_0^a \int_0^{2\pi} r\, g(r,\theta)\, J_n(\alpha_{nm}\frac{r}{a}) \left\{\begin{matrix} \cos(n\theta) \\ \sin(n\theta) \end{matrix}\right\} d\theta\, dr$$

60. $W(r,\theta,t) = \sum_{n=0}^{\infty} \sum_{m=1}^{\infty} W_{nm}(r)[A_{nm}\cos(n\theta) + B_{nm}\sin(n\theta)]\cos(\alpha_{nm}ct)$

where $W_{nm}(r) = \frac{J_n(\alpha_{nm}\frac{r}{a})}{J_n(\alpha_{nm})} - \frac{Y_n(\alpha_{nm}\frac{r}{a})}{Y_n(\alpha_{nm})}$

$J_n(\alpha_{nm}) Y_n(\alpha_{nm}\frac{b}{a}) - J_n(\alpha_{nm}\frac{b}{a}) Y_n(\alpha_{nm}) = 0$

$$\left.\begin{matrix} A_{nm} \\ B_{nm} \end{matrix}\right\} = \frac{\varepsilon_n}{2\pi R} \int_a^b \int_0^{2\pi} r\, f(r,\theta)\, W_{nm}(r) \left\{\begin{matrix} \cos(n\theta) \\ \sin(n\theta) \end{matrix}\right\} d\theta\, dr$$

and $R = \frac{b^2}{2}\left(\frac{J_n'(\alpha_{nm}b/a)}{J_n(\alpha_{nm})} - \frac{Y_n'(\alpha_{nm}b/a)}{Y_n(\alpha_{nm})}\right)^2 - \frac{a^2}{2}\left(\frac{J_n'(\alpha_{nm})}{J_n(\alpha_{nm})} - \frac{Y_n'(\alpha_{nm})}{Y_n(\alpha_{nm})}\right)^2$

61. $W(r,t) = \frac{P_0 c}{\pi a S} \sum_{n=1}^{\infty} T_n(t) J_0(k_n r/a)$ where $J_0(k_n) = 0$ and $T_n(t) = \frac{\sin(k_n ct/a)}{k_n J_1^2(k_n)}$

62. $W(x,y,t) = \frac{4P_0 c}{LS} \sum_{n=1}^{\infty} \sum_{m=1}^{\infty} T_{nm}(t) \sin(\frac{n\pi}{L}x) \sin(\frac{m\pi}{L}y)$

where $T_{nm}(t) = \frac{1}{k_{nm}} \sin(\frac{n\pi}{L}x_0) \sin(\frac{m\pi}{L}y_0) \sin(\frac{k_{nm}ct}{L})$,

and $k_{nm}^2 = \pi^2(m^2 + n^2)$

63. $$W(r,\theta,t) = \frac{P_0 c}{\pi a S} \sum_{n=0}^{\infty} \sum_{m=1}^{\infty} \frac{\varepsilon_n \sin(\frac{k_{nm} ct}{a})}{k_{nm} J_{n+1}^2(k_{nm})} J_n(k_{nm}\frac{r}{a}) J_n(k_{nm}\frac{r_0}{a}) \cos(n(\theta-\theta_0))$$

where $J_n(k_{nm}) = 0$

64. $$u(x,t) = \frac{cLF_0}{AE} H(t-t_0) \sum_{n=1}^{\infty} \frac{\cos(\mu_n x / L) \sin(c\mu_n (t-t_0)/L)}{\mu_n N_n}$$

where $N_n = \frac{L}{2}[1 + \frac{AE}{\gamma L}\sin^2(\mu_n)]$, and $\tan(\mu_n) = \frac{\gamma L}{AE\mu_n}$

65. $$w = \frac{2caP_0}{Sb} H(t-t_0) \sum_{n=1}^{\infty} \sum_{m=1}^{\infty} E_{nm}(t) J_{\beta_n}(\gamma_{nm} r / a) \sin(\beta_n \theta)\text{, where } \beta_n = n\pi/b$$

$$J_{\beta_n}(\gamma_{nm}) = 0,\ N_{nm} = \int_0^a r J_{\beta_n}^2(\gamma_{nm}\frac{r}{a})\,dr,\ \lambda_{nm} = \gamma_{nm}^2 / a^2$$

and $$E_{nm}(t) = \frac{J_{\beta_n}(\gamma_{nm} r_0 / a)}{N_{nm}\gamma_{nm}} \sin(n\pi\theta_0 / b) \sin(c\sqrt{\lambda_{nm}}(t-t_0))$$

66. $w =$

$$\frac{2c^2 P_0}{abS} \sum_{m=0}^{\infty} \sum_{n=1}^{\infty} \varepsilon_m E_{nm}(t) \cos(\frac{m\pi}{a}x) \sin(\frac{n\pi}{b}y) + W_0 \sin(\frac{\pi y}{b}) \cos(\frac{c\pi}{b}t)$$

$$E_{nm}(t) = \frac{1}{c\sqrt{\lambda_{mn}}} \cos(\frac{m\pi}{a}x_0) \sin(\frac{n\pi}{b}y_0) \sin(c\sqrt{\lambda_{mn}}(t-t_0)) H(t-t_0)$$

and $\lambda_{mn} = m^2\pi^2/a^2 + n^2\pi^2/b^2$

67. $$w = \frac{2cP_0}{\pi a^2 S} H(t-t_0) \sum_{n=0}^{\infty} \sum_{m=1}^{\infty} E_{nm}(t) J_n(\mu_{nm} r / a) \cos(n\theta) \text{ where } J_n(\mu_{nm}) = 0$$

$$E_{nm}(t) = \frac{\varepsilon_n J_n(\mu_{nm} r_0 / a)}{[J_n'(\mu_{nm})]^2 \sqrt{\lambda_{mn}}} \cos(\frac{n\pi}{2}) \sin(c\sqrt{\lambda_{mn}}(t-t_0))$$

68. $$w = \frac{2cP_0}{\pi S} H(t-t_0) \sum_{n=1}^{\infty} \sum_{m=1}^{\infty} E_{nm}(t) R_{nm}(r) \sin(n\theta)$$

where $$E_{nm}(t) = \frac{R_{nm}(r_0)}{N_{nm} k_{nm}} \sin(n\pi/2) \sin(ck_{nm}(t-t_0))$$

$$R_{nm}(r) = \frac{J_n(k_{nm} r / a)}{J_n(k_{nm})} - \frac{Y_n(k_{nm} r / a)}{Y_n(k_{nm})}$$

$J_n(k_{nm}) Y_n(k_{nm} b/a) - J_n(k_{nm} b/a) Y_n(k_{nm}) = 0$

and $$N_{nm} = \int_0^a r R_{nm}^2(r)\,dr$$

69. $$u = \frac{F_0}{AE} cL\, H(t-t_0) \sum_{n=1}^{\infty} E_n(t) X_n(x)$$

where $X_n(x) = \dfrac{\sin(\beta_n x / L)}{\beta_n} + \dfrac{AE}{\zeta L}\cos(\beta_n x / L)$,

$$E_n(t) = \frac{\sin(c\beta_n (t-t_0)/L)}{N_n \beta_n} X_n(L/2),\quad N_n = \int_0^L X_n^2(x)\,dx$$

and $\tan(\beta_n) = \dfrac{(\gamma+\zeta)L\,\beta_n}{AE(\beta_n^2 - \zeta\gamma L^2/(AE)^2)}$

70. $$w = \frac{P_0 c}{2\pi S} H(t-t_0) \sum_{n=0}^{\infty}\sum_{m=1}^{\infty} [G_{nm}(t)\sin(n\theta) + H_{nm}(t)\cos(n\theta)] R_{nm}(r)$$

$$\left.\begin{matrix} G_{nm}(t) \\ H_{nm}(t) \end{matrix}\right\} = \frac{\varepsilon_n R_{nm}(r_0)}{N_{nm}\sqrt{\lambda_{nm}}} \sin(c\sqrt{\lambda_{nm}}(t-t_0)) \begin{Bmatrix} \sin(n\pi/2) \\ \cos(n\pi/2) \end{Bmatrix}$$

$$N_{nm} = \int_a^b r R_{nm}^2(r)\,dr,\quad R_{nm}(r) = J_n(\mu_{nm} r/a) - \frac{J_n(\mu_{nm})}{Y_n(\mu_{nm})} Y_n(\mu_{nm} r/a)$$

and $J_n(\mu_{nm}) Y_n(\mu_{nm} b/a) - J_n(\mu_{nm} b/a) Y_n(\mu_{nm}) = 0$

71. $$V_r = \frac{2}{ar^2} \sum_{n=1}^{\infty} \frac{1+\alpha_n^2}{\alpha_n^2} \left(A_n \cos\left(\frac{\alpha_n ct}{a}\right) + B_n \sin\left(\frac{\alpha_n ct}{a}\right)\right)\left(\sin\left(\frac{\alpha_n r}{a}\right) - \frac{\alpha_n r}{a}\cos\left(\frac{\alpha_n r}{a}\right)\right)$$

where $\tan(\alpha_n) = \alpha_n$, $V_r = -\dfrac{\partial\phi}{\partial r}$,

$$A_n = \int_0^a r f(r) \sin(\alpha_n r/a)\,dr, \text{ and } B_n = \frac{a}{\alpha_n c}\int_0^a r g(r)\sin(\alpha_n r/a)\,dr$$

72. $$V_r = \frac{2}{a^3} \sum_{n=1}^{\infty} \frac{\alpha_n}{J_0^2(\alpha_n)} \left(A_n \cos\left(\frac{\alpha_n ct}{a}\right) + B_n \sin\left(\frac{\alpha_n ct}{a}\right)\right) J_1\left(\frac{\alpha_n r}{a}\right)$$

where $J_1(\alpha_n) = 0$, $A_n = \displaystyle\int_0^a r f(r) J_0(\alpha_n r/a)\,dr$, and $B_n = \dfrac{a}{\alpha_n c}\displaystyle\int_0^a r g(r) J_0(\alpha_n r/a)\,dr$

73. $$y(x,t) = \frac{f(v) - f(-u)}{2} + \frac{1}{2c}\int_{-u}^{v} g(\eta)\,d\eta \qquad u < 0$$

$$= \frac{f(v) + f(u)}{2} + \frac{1}{2c}\int_{u}^{v} g(\eta)\,d\eta \qquad u > 0$$

where $u = x - ct$, and $v = x + ct$

74. $y(x,t) = y_0 \sin(\omega (t - x/a))\, H(t - x/a)$

75. $p = \rho c V_0 (\frac{a}{r}) \frac{ika}{ika - 1} e^{ik(r-a)} e^{-i\omega t}$

76. $p = i \rho c V_0 \sum_{n=0}^{\infty} A_n P_n(\eta) h_n^{(1)}(kr)$, where $k = \omega/c$, $\eta = \cos(\theta)$

and $A_n = \frac{2n+1}{2 h_n^{(1)'}(ka)} \int_{-1}^{+1} f(\eta) P_n(\eta) d\eta$

77. $p_s = -p_0 \sum_{n=0}^{\infty} \varepsilon_n (i)^n \frac{J_n(ka)}{H_n^{(2)}(ka)} H_n^{(2)}(kr) \cos(n\theta)$

78. $p_s = -p_0 \sum_{n=0}^{\infty} \varepsilon_n (i)^n \frac{J'_n(ka)}{H_n^{(2)'}(ka)} H_n^{(2)}(kr) \cos(n\theta)$

79. $p_s = -p_0 \sum_{n=0}^{\infty} (2n+1)(i)^n \frac{j_n(ka)}{h_n^{(2)}(ka)} h_n^{(2)}(kr) P_n(\cos\theta)$

80. $p = -p_0 \sum_{n=0}^{\infty} (2n+1)(i)^n A_n h_n^{(2)}(kr) P_n(\cos\theta)$

where $A_n = \dfrac{j_n(k_1 a) j_n(k_2 a) - \frac{\rho_2 c_2}{\rho_1 c_1} j_n(k_2 a) j_n(k_1 a)}{h_n^{(2)}(k_1 a) j'_n(k_2 a) - \frac{\rho_2 c_2}{\rho_1 c_1} j_n(k_2 a) h_n^{(2)'}(k_1 a)}$

81. $p = -i \rho c V_0 \frac{h_0^{(2)}(kr)}{h_0^{(2)'}(ka)}$

82. $p = -i \rho c V_0 \sum_{m=0}^{\infty} a_m \frac{h_{2m+1}^{(2)}(kr)}{h_{2m+1}^{(2)'}(ka)} P_{2m+1}(\cos\theta)$

where $a_m = (-1)^m \frac{(4m+3)(2m)!}{2^{2m+1}(m)!(m+1)!}$

83. (a) $p = -\frac{i \rho c V_0}{\pi} \sum_{n=0}^{\infty} \varepsilon_n \frac{\sin(n\alpha)}{n} \frac{H_n^{(2)}(kr)}{H_n^{(2)'}(ka)} \cos(n\theta) e^{i\omega t}$

$$\text{(b) } p = -\frac{i\rho c Q_0}{2\pi a}\sum_{n=0}^{\infty} \varepsilon_n \frac{H_n^{(2)}(kr)}{H_n^{(2)\prime}(ka)} \cos(n\theta) e^{i\omega t}$$

84. $$p = \frac{\rho\omega V_0}{ab}\sum_{n=0}^{\infty}\sum_{m=0}^{\infty} A_{mn}\cos\left(\frac{n\pi}{a}x\right)\cos\left(\frac{m\pi}{b}y\right)e^{-i\alpha_{mn}z}e^{i\omega t}$$

$$\text{where } A_{mn} = \frac{\varepsilon_n\varepsilon_m}{\alpha_{nm}}\int_0^a\int_0^b f(x,y)\cos\left(\frac{n\pi}{a}x\right)\cos\left(\frac{m\pi}{b}y\right)dy\,dx$$

$$\text{and } \alpha_{nm}^2 = \frac{\omega^2}{c^2} - \frac{n^2\pi^2}{a^2} - \frac{m^2\pi^2}{b^2}$$

For $f(x,y) = 1$: $A_{00} = \dfrac{ab}{\alpha_{00}} = \dfrac{abc}{\omega}$, $A_{nm} = 0$ for $n,m \neq 0$

$$\text{and } p = \rho c V_0 \exp\left[-i\frac{\omega}{c}(z - ct)\right]$$

85. $$p = \frac{2\rho\omega V_0}{\pi a^2}\sum_{n=0}^{\infty}\sum_{m=1}^{\infty}(A_{nm}\sin(n\theta) + B_{nm}\cos(n\theta))J_n\left(\mu_{nm}\frac{r}{a}\right)e^{-i\alpha_{nm}z}e^{i\omega t}$$

$$\text{where } \left.\begin{matrix} A_{nm} \\ B_{nm}\end{matrix}\right\} = \frac{\varepsilon_n \mu_{nm}^2}{\alpha_{nm}(\mu_{nm}^2 - n^2)J_n^2(\mu_{nm})}\int_0^a\int_0^{2\pi} r f(r,\theta)J_n\left(\mu_{nm}\frac{r}{a}\right)\begin{Bmatrix}\sin(n\theta) \\ \cos(n\theta)\end{Bmatrix} d\theta\, dr$$

$J_n'(\mu_{nm}) = 0$, and $\alpha_{nm}^2 = \dfrac{\omega^2}{c^2} - \dfrac{\mu_{nm}^2}{a^2}$

Chapter 7

1. (a) $p/(p^2 + a^2)$ (b) $2ap / (p^2 + a^2)^2$
   (c) $(p - a) / [(p - a)^2 + b^2]$ (d) $2a^2p / (p^4 + 4a^4)$
   (e) $n! / (p + a)^{n+1}$ (f) $a(p^2 - 2a^2) / (p^4 + 4a^4)$

2. In the following $F(p) = \boldsymbol{L}\, f(t)$:
   (a) $a\,F(ap)$ (b) $a\,F[a(p-b)]$
   (c) $p\,F(p+a) - f(0^+)$ (d) $p\, d^2F/dp^2$
   (e) $-\, dF(p+a)/dp$ (f) $(p-1)^2 F(p-1) - (p-1) f(0^+) - f'(0^+)$
   (g) $(-1)^n a\, d^nF(ap)/dp^n$ (h) $p(p-a) F(p-a) - p\, f(0^+) - f'(0^+)$
   (i) $-p\, dF/dp$ (j) $[F(p-a) - F(p+a)] / 2$
   (k) $-(dF/dp) / p$ (l) $F(p) / p$

3. (a) $\{1 - pT\, e^{-pT}/[1 - e^{-pT}]\} / p^2$ (b) $\dfrac{a}{p^2 + a^2} \coth(p\pi / 2a)$
   (c) $[\tanh(pT/4)] / p$ (d) $[\pi\, p \coth(p\pi/2) - 2] / p^3$
   (e) $\{p\,[1 + e^{-pT/2}]\}^{-1}$ (f) $[2p \cosh(pT/4)]^{-1}$
   (g) $[p + \omega / \sinh(p\pi/2\omega)] / (\omega^2 + p^2)$

4. (a) $(e^{at} - e^{bt}) / (a - b)$ (b) $1 - \cos(at)$
   (c) $at - \sin(at)$ (d) $\sin(at) - at\cos(at)$
   (e) $t \sin(at)$ (f) $\sin(at)\cosh(at) - \cos(at)\sinh(at)$
   (g) $\sin(at) + at\cos(at)$ (h) $\sinh(at) - \sin(at)$

5. (a) $y(t) = \dfrac{1}{k}\displaystyle\int_0^t f(t-x)\sin(kx)\,dx + A\cos(kt) + \dfrac{B}{k}\sin(kt)$

   (b) $y(t) = \dfrac{1}{k}\displaystyle\int_0^t f(t-x)\sinh(kx)\,dx + A\cosh(kt) + \dfrac{B}{k}\sinh(kt)$

   (c) $y(t) = \dfrac{A}{2a^2}[\cosh(at) - \cos(at)] + \dfrac{B}{2a^3}[\sinh(at) - \sin(at)]$

   (d) $y(t) = \dfrac{1}{2a^3}\displaystyle\int_0^t f(t-x)(\sinh(\beta x) - \sin(\beta x))\,dx$

   (e) $y(t) = \dfrac{1}{2}\displaystyle\int_0^t \left[e^{-x} - 2e^{-2x} + e^{-3x}\right] f(t-x)\,dx$

   (g) $y(t) = \displaystyle\int_0^t \left[e^{-x} - e^{-2x} - xe^{-2x}\right] f(t-x)\,dx$

(h) $y(t) = \frac{1}{6}\int_0^t x^3\, e^{-x}\, f(t-x)\, dx$

(i) $y(t) = \int_0^t x e^{-x}\, f(t-x) dx$

(k) $y(t) = e^{-2t}\,(1+2t) + A\, t_0\,(t-t_0)\, e^{-2(t-t_0)} H(t-t_0)$

(l) $y(t) = t + e^t - e^{-2t}$

(m) $y(t) = B[e^{3t} - e^{2t}] + A\,[e^{3(t-t_0)} - e^{2(t-t_0)}]\,H(t-t_0)$

6. (a) $y(t) = [\frac{2}{25} e^{(t-t_0)/2} + \frac{4}{5}(t-t_0) e^{-2(t-t_0)} - \frac{2}{25} e^{-2(t-t_0)}]\, H(t-t_0)$

(b) $y(t) = A\,[1 - t^2/2]$

(c) $y(t) = 3\, e^t/2 - e^{-t}/2$

(d) $y(t) = \int_0^t f(t-x)\, h(x)\, dx$ where $h(x) = \boldsymbol{L}^{-1}\,[p^2 + k^2 - G(p)]^{-1}$, and $G(p) = \boldsymbol{L}\, g(t)$

(e) $y(t) = 2\,[1 - at]\, e^{-2at}$

(f) $y(t) = \frac{1}{3}\int_0^t [4\, e^{-4x} - e^{-x}] f(t-x)\, dx$

(g) $y(t) = \frac{A}{(a-1)(a-2)}[2\,(a-1)\, e^{-2t} - (a-2)\, e^{-t} - a\, e^{-at}] + B[2\, e^{-2t} - e^{-t}]$

(h) $y(t) = \int_0^t \cos(x) f(t-x)\, dx$

(i) $y(t) = \int_0^t e^{ax} \cos(x) f(t-x)\, dx$

(j) $y(t) = \int_0^t (x - x^2/2)\, e^{-x} f(t-x)\, dx$

7. (a) $x(t) = [-2U + (U+V)\cosh(at) + (U-V)\cos(at)]\, /\, (2a^2)$

$y(t) = [-2V + (U+V)\cosh(at) + (V-U)\cos(at)]\, /\, (2a^2)$

(b) $x(t) = y(t) = e^{-t} + t\, e^{-t}$

(c) $x(t) = e^t + t\, e^t$, $y(t) = 3e^t + 2t\, e^t$

(d) $x(t) = \frac{1}{2}[g(t) - f(t)] + \frac{1}{4}\int_0^t e^{-3(t-x)/2}[f(x) + g(x)]dx$

$y(t) = \frac{1}{2}[f(t) - g(t)] + \frac{1}{4}\int_0^t e^{-3(t-x)/2}[f(x) + g(x)]dx$

(e) $$x(t) = \frac{1}{2}\int_0^t (\sin(x) - \sinh(x))g(t-x)\,dx + \frac{1}{2}\int_0^t (\sin(x) + \sinh(x))f(t-x)\,dx$$

$$y(t) = \frac{1}{2}\int_0^t (\sin(x) - \sinh(x))f(t-x)\,dx + \frac{1}{2}\int_0^t (\sin(x) + \sinh(x))g(t-x)\,dx$$

(f) $$x(t) = \int_0^t [\cosh(x)f(t-x) - \sinh(x)\,g(t-x)]\,dx$$

$$y(t) = \int_0^t [\cosh(x)g(t-x) - \sinh(x)\,f(t-x)]\,dx$$

(g) $$x(t) = \frac{1}{12}[(A-B) + 2(y_0 - x_0)]e^{2t} + \frac{3}{4}[A + B - 2(y_0 + x_0)]e^{-2t}$$
$$+ [(A+B) - 2(x_0 + y_0)]e^{-t} + \frac{1}{3}[-(A+2B) + 4(y_0 + 2x_0)]t\,e^{-t}$$

$$y(t) = \frac{1}{12}[(B-A) + 2(x_0 - y_0)]e^{2t} + \frac{3}{4}[A + B - 2(y_0 + x_0)]e^{-2t}$$
$$+ [(A+B) - 2(x_0 + y_0)]e^{-t} + \frac{1}{3}[-(B+2A) + 4(x_0 + 2y_0)]t\,e^{-t}$$

8. $$y(x,t) = \frac{Ac}{b}\{-e^{-bx}\sinh(bc(t-t_0))H(t-t_0) + \sinh(bc(t - t_0 - \frac{x}{c}))H(t - t_0 - \frac{x}{c})\}$$

9. $$y(x,t) = \frac{T_0}{2}e^{-at}\left\{e^{-ix\sqrt{a/K}}\,\mathrm{erfc}[\frac{x}{2\sqrt{Kt}} - i\sqrt{at}] + e^{ix\sqrt{a/K}}\,\mathrm{erfc}[\frac{x}{2\sqrt{Kt}} + i\sqrt{at}]\right\}$$
$$+ \frac{KQ\,e^{-bt}}{2b}\left\{e^{-ix\sqrt{b/K}}\,\mathrm{erfc}[\frac{x}{2\sqrt{Kt}} - i\sqrt{bt}] + e^{ix\sqrt{b/K}}\,\mathrm{erfc}[\frac{x}{2\sqrt{Kt}} + i\sqrt{bt}]\right\}$$
$$+ \frac{KQ}{b}\left\{1 - e^{-bt} - \mathrm{erfc}(\frac{x}{2\sqrt{Kt}})\right\}$$

10. $$y(x,t) = \frac{1}{2}c\,y_0\,\{H[t - t_0 - \frac{x + x_0}{c}] + H[t - t_0 + \frac{x - x_0}{c}]$$
$$- H[t - t_0 + \frac{x - x_0}{c}]H[x - x_0] + H[t - t_0 - \frac{x - x_0}{c}]H[x - x_0]\}$$

11. $$T = \frac{Q}{2}\left[\frac{K}{\pi(t - t_0)}\right]^{1/2}\left\{\exp[-\frac{(x - x_0)^2}{4K(t - t_0)}] - \exp[-\frac{(x + x_0)^2}{4K(t - t_0)}]\right\}H(t - t_0)$$

12. $$y(x,t) = y_0(x - \frac{L}{2})^2 H(t) + y_0 c^2 t^2 + \frac{c y_0 L}{2}\{(t - \frac{x}{c})H(t - \frac{x}{c}) - (t + \frac{x}{c})\} - \frac{c y_0 L}{2} f(t,x)$$

where $f(t,x) = f(t + 2L/c, x)$ is a periodic function, defined over the first period as:

$$f_1(t,x) = (t - \frac{x}{c})H(t - \frac{x}{c}) + (t + \frac{x}{c}) + (t - \frac{2L - x}{c})H(t - \frac{2L - x}{c})$$

$$+(t-\frac{2L+x}{c})H(t-\frac{2L+x}{c})+2(t-\frac{L-x}{c})H(t-\frac{L-x}{c})+2(t-\frac{L+x}{c})H(t-\frac{L+x}{c})$$

13. $y(x,t) = 2\,V_0\,(t-x/c)\,H(t-x/c) - V_0 t$

14. $y(x,t) = y_0\, f(t,x)$

    where $f(t,x) = f(t + 2L/c, x)$ is a periodic function, defined over the first period as:

    $$f_1(t,x) = -H(t-\frac{L-x}{c})+H(t-\frac{L+x}{c})-H(t-\frac{2L-x}{c})+H(t-\frac{x}{c})$$

15. $$y(x,t) = y_0\,H(t-\frac{x}{c})+\frac{cP_0}{bT_0}\sinh[cb(t-t_0-\frac{x}{c})]\,H(t-t_0-\frac{x}{c})$$
    $$-\frac{cP_0}{bT_0}\sinh[cb(t-t_0)]\,e^{-bx}\,H(t-t_0)$$

16. $$y(x,t) = T_0\,\mathrm{erf}(\frac{x}{\sqrt{4Kt}})+T_0\frac{x}{\sqrt{4K\pi}}(t-t_0)^{-3/2}\exp[-\frac{x^2}{4K(t-t_0)}]H(t-t_0)$$

17. $$T(x,t) = -\frac{T_0}{2}e^{Kb^2t}\left\{e^{bx}\,\mathrm{erfc}[\frac{x}{2\sqrt{Kt}}+b\sqrt{Kt}]+e^{-bx}\,\mathrm{erfc}[\frac{x}{2\sqrt{Kt}}-b\sqrt{Kt}]\right\}$$
    $$+T_0\,e^{Kb^2t}\,e^{-bx}$$

18. $y(x,t) = [e^{-c\gamma(t-x/c)} - 1]\,[H(t-x/c)]\,/\,\gamma$

19. $y(x,t) = -(t-x/c)\,[V_0 + a_0(t-x/c)/2]\,H(t-x/c) + V_0 t$

20. $$T(x,t) = T_0\,\{\mathrm{erfc}(\frac{x}{2\sqrt{Kt}})-e^{bx+b^2Kt}\mathrm{erfc}(\frac{x}{2\sqrt{Kt}}+b\sqrt{Kt})\}+Q_0\,K\,H(t-t_0)$$

$$-Q_0K\,H(t-t_0)\{\mathrm{erfc}(\frac{x}{2\sqrt{K(t-t_0)}})-e^{bx+b^2K(t-t_0)}\mathrm{erfc}(\frac{x}{2\sqrt{K(t-t_0)}}+b\sqrt{K(t-t_0)})\}$$

21. $$T(x,t) = T_0\,\mathrm{erfc}(\frac{x}{2\sqrt{Kt}})+\frac{KQ_0}{ak}(1-e^{-at})-\frac{KQ_0}{ka}\,\mathrm{erfc}(\frac{x}{2\sqrt{Kt}})$$
    $$+\frac{Q_0K}{2ak}e^{-at}\{e^{-ix\sqrt{a/K}}\mathrm{erfc}(\frac{x}{2\sqrt{Kt}}-i\sqrt{at})+e^{ix\sqrt{a/K}}\mathrm{erfc}(\frac{x}{2\sqrt{Kt}}+i\sqrt{at})\}$$

22. $$y(x,t) = \frac{F_0c}{2\,AE}\{H[t-t_0-\frac{x-x_0}{c}]-H[t-t_0+\frac{x-x_0}{c}]\}H[x-x_0]+f(t)\}$$

    where $f(t) = f(t + 2L/c)$ is a periodic function, defined over the first period as:

    $$f_1(t) = H[t-t_0+\frac{x-x_0}{c}]-H[t-t_0-\frac{x+x_0}{c}]$$
    $$-H[t-t_0-\frac{2L-x-x_0}{c}]+H[t-t_0-\frac{2L+x-x_0}{c}]$$

23. $y(x,t) = \dfrac{c^2 F_0}{a^2 AE}\{f(t) + at - \sin(at)\}$

$f(t) = f(t + 2L/c)$ is a periodic function, defined over the first period as:

$$f_1(t) = -[a(t-\frac{x}{c}) - \sin(a(t-\frac{x}{c}))]H(t-\frac{x}{c})$$
$$+[a(t-\frac{L+x}{c}) - \sin(a(t-\frac{L+x}{c}))]H(t-\frac{L+x}{c})$$
$$-[a(t-\frac{L-x}{c}) - \sin(a(t-\frac{L-x}{c}))]H(t-\frac{L-x}{c})$$
$$+[a(t-\frac{2L-x}{c}) - \sin(a(t-\frac{2L-x}{c}))]H(t-\frac{2L-x}{c})$$

24. $T(x,t) = T_0 \operatorname{erfc}(\dfrac{x}{2\sqrt{K(t-a)}})H(t-a)[4\dfrac{t-a}{a} - 1] - \dfrac{4T_0}{a} t \operatorname{erfc}(\dfrac{x}{2\sqrt{Kt}})$

$$+\frac{KQ_0}{k}\operatorname{erf}(\frac{x}{2\sqrt{K(t-t_0)}})H(t-t_0)$$

25. $y(x,t) = A\left\{[\dfrac{\sinh(bc(t-\frac{x}{c}))}{bc} - \dfrac{\sin(a(t-\frac{x}{c}))}{a}]H(t-\dfrac{x}{c}) - [\dfrac{\sinh(bct)}{cb} - \dfrac{\sin(at)}{a}]e^{-bx}\right\}$

where $A = \dfrac{P_0 a c^2}{(a^2 + b^2 c^2)T_0}$

26. $T(x,t) = -QKt[1 + 4\operatorname{erfc}(\dfrac{x}{2\sqrt{Kt}})] + T_0 \dfrac{x(t-t_0)^{-3/2}}{2\sqrt{\pi K}} e^{-x^2/[4K(t-t_0)]} H(t-t_0)$

27. $y(x,t) = y_0 \cos(b(t-\dfrac{x}{c}))H(t-\dfrac{x}{c}) + \dfrac{F_0 c^2}{AEa^2}[at - 1 + e^{-at}]$

$$-\frac{F_0 c^2}{AEa^2}[a(t-\frac{x}{c}) - 1 + e^{-a(t-\frac{x}{c})}]H(t-\frac{x}{c})$$

28. $T(x,t) = \dfrac{KQ}{k}H(t-t_0) - F\sqrt{K/\pi}\displaystyle\int_0^t e^{-x^2/[4Ku]} u^{-1/2}(t-u)e^{-a(t-u)}\,du$

29. $y(x,t) = A\,H(t - x/c) - y_0 \cosh(cb(t-x/c))\,H(t-x/c) + y_0 e^{-bx}\cosh(cbt)$

30. $T(x,t) = -KQ_0 a(1-\cos(at)) - 4T_0 t \operatorname{erfc}(\dfrac{x}{2\sqrt{Kt}})$

$$+\frac{\sqrt{K}\,Q_0 ax}{4\sqrt{\pi}}\int_0^t [H(t-u) - \cos(a(t-u))]u^{-3/2} e^{-x^2/[4Ku]}\,du$$

Chapter 8

1. $g(x|\xi) = -\sin x \cos \xi + \sin x \sin \xi \tan 1 + \sin(x-\xi)\, H(x-\xi)$

   $y(x) = x - \sin x / \cos 1$

2. $$g(x\,|\,\xi) = \frac{1}{2n}\left\{\left[\xi^{n} - \xi^{-n}\right]x^{n} + \left[x^{n}\xi^{-n} - x^{-n}\xi^{n}\right]H(x-\xi)\right\}$$

3. $$g(x\,|\,\xi) = \frac{1}{2n\xi}\left\{\left[\xi^{n} - \xi^{-n}\right]x^{n} + \left[x^{n}\xi^{-n} - x^{-n}\xi^{n}\right]H(x-\xi)\right\}$$

4. $$g(x\,|\,\xi) = -\frac{\sinh(kx)\sinh(k(L-\xi))}{k\sinh(kL)} + \frac{1}{k}\sinh(k(x-\xi))\,H(x-\xi)$$

5. $$g(x\,|\,\xi) = \frac{1}{6}\left[x^{3} - 3x^{2}\xi - (x-\xi)^{3}H(x-\xi)\right]$$

6. $$g(x\,|\,\xi) = -\frac{1}{6}(x-\xi)^{3}H(x-\xi) - \frac{1}{2L^{2}}\,\xi\, x^{2}(L-\xi)^{2} + \frac{1}{6L^{3}}x^{3}(L-\xi)^{2}(L+2\xi)$$

7. $$g(x\,|\,\xi) = -\frac{1}{6}(x-\xi)^{3}H(x-\xi) - \frac{1}{6L}\,\xi\, x(L-\xi)(2L-\xi) + \frac{1}{6L}x^{3}(L-\xi)$$

8. (a) $$g(x\,|\,\xi) = -\frac{\sin(kx)\sin(k(L-\xi))}{k\sin(kL)} + \frac{1}{k}\sin(k(x-\xi))\,H(x-\xi)$$

   (b) $$g(x\,|\,\xi) = \frac{2}{L}\sum_{n=1}^{\infty}\frac{\sin(n\pi x/L)\sin(n\pi\xi/L)}{k^{2} - n^{2}\pi^{2}/L^{2}}$$

9 (i)

   (a) $$g(x\,|\,\xi) = \frac{1}{2\beta^{3}}\left[\sin(\beta(x-\xi)) - \sinh(\beta(x-\xi))\right]H(x-\xi)$$
   $$+\frac{1}{2\beta^{3}}\left\{\frac{\sinh(\beta x)\sinh(\beta(L-\xi))}{\sinh(\beta L)} - \frac{\sin(\beta x)\sin(\beta(L-\xi))}{\sin(\beta L)}\right\}$$

   (b) $$g(x\,|\,\xi) = \frac{2}{L}\sum_{n=1}^{\infty}\frac{\sin(n\pi x/L)\sin(n\pi\xi/L)}{\beta^{4} - n^{4}\pi^{4}/L^{4}}$$

9 (ii)

(a) $$g(x\,|\,\xi) = \frac{1}{2\beta^3}\left[\sin(\beta(x-\xi)) - \sinh(\beta(x-\xi))\right]H(x-\xi)$$

$$+\frac{1}{4\beta^3}\left\{C_1\left[\sin(\beta x) - \sinh(\beta x)\right] + C_2\left[\cos(\beta x) - \cosh(\beta x)\right]\right\}$$

where:

$$C_1 = \frac{\left[\sinh(\beta(L-\xi)) - \sin(\beta(L-\xi))\right]\left[\sin(\beta L) + \sinh(\beta L)\right]}{\left[1-\cos(\beta L)\cosh(\beta L)\right]}$$

$$+\frac{\left[\cosh(\beta(L-\xi)) - \cos(\beta(L-\xi))\right]\left[\cos(\beta L) - \cosh(\beta L)\right]}{\left[1-\cos(\beta L)\cosh(\beta L)\right]}$$

and

$$C_2 = \frac{\left[\sinh(\beta(L-\xi)) - \sin(\beta(L-\xi))\right]\left[\cos(\beta L) - \cosh(\beta L)\right]}{\left[1-\cos(\beta L)\cosh(\beta L)\right]}$$

$$+\frac{\left[\cosh(\beta(L-\xi)) - \cos(\beta(L-\xi))\right]\left[\sin(\beta L) - \sinh(\beta L)\right]}{\left[1-\cos(\beta L)\cosh(\beta L)\right]}$$

(b) Eigenfunctions $\phi_n(x)$ are:

$$\phi_n(x) = \sin\left(\frac{\alpha_n}{L}x\right) - \sinh\left(\frac{\alpha_n}{L}x\right) + \frac{\sin(\alpha_n) - \sinh(\alpha_n)}{\cosh(\alpha_n) - \cos(\alpha_n)}\left[\cos\left(\frac{\alpha_n}{L}x\right) - \cosh\left(\frac{\alpha_n}{L}x\right)\right]$$

where, $\cos(\alpha_n)\cosh(\alpha_n) = 1$, and with $\alpha_0 = 0$. So that $g(x|\xi)$ is:

$$g(x\,|\,\xi) = \sum_{n=0}^{\infty} \frac{\phi_n(x)\phi_n(\xi)}{N_n(\beta^4 - \alpha_n^4/L^4)}$$

where $N_n = \int_0^L \phi_n^2(x)\,dx$

10. (a) $$g(x\,|\,\xi) = \frac{\pi}{2}\frac{J_n(kx)}{J_n(k)}\left[J_n(k)Y_n(k\xi) - J_n(k\xi)Y_n(k)\right]$$

$$-\frac{\pi}{2}\left[J_n(kx)Y_n(k\xi) - J_n(k\xi)Y_n(kx)\right]H(x-\xi)$$

(b) $$g(x\,|\,\xi) = \sum_{m=1}^{\infty} \frac{J_n(k_{nm}x)J_n(k_{nm}\xi)}{(k^2 - k_{nm}^2)[J'_n(k_{nm})]^2}$$

where $J_n(k_{nm}) = 0$.

11. (a) $g(x\,|\,\xi) = -\dfrac{\sin(kx)\sin(k(1-\xi)).0}{kx\xi\,\sin(k)} + \dfrac{1}{kx\xi}\sin(k(x-\xi))H(x-\xi)$

(b) $g(x\,|\,\xi) = \dfrac{2}{x\xi}\sum_{n=1}^{\infty}\dfrac{\sin(n\pi x)\sin(n\pi\xi)}{k^2 - n^2\pi^2}$

12. $g(x\,|\,\xi) = \dfrac{1}{4\eta^3\sqrt{2}}\left\{e^{-\eta(x+\xi)}\sin\left[\eta(x+\xi)+\pi/4\right] - e^{-\eta|x-\xi|}\sin\left[\eta\,|\,x-\xi\,|+\pi/4\right]\right\}$

where $\eta = \gamma/\sqrt{2}$

13. (a) $g(x\,|\,\xi) = \dfrac{1}{\eta}\left\{\left[e^{-\eta x}\sinh(\eta\xi) - e^{-\eta\xi}\sinh(\eta x)\right]H(x-\xi) + \sinh(\eta x)e^{-\eta\xi}\right\}$

where $\eta = \sqrt{\gamma - k^2}$

(b) $g(x\,|\,\xi) = \dfrac{1}{\eta}\left\{\left[e^{i\eta x}\sin(\eta\xi) - e^{i\eta\xi}\sin(\eta x)\right]H(x-\xi) + \sin(\eta x)e^{i\eta\xi}\right\}$

where $\eta = \sqrt{k^2 - \gamma}$

(c) $g(x|\xi) = x - (x-\xi)\,H(x-\xi)$

14. $g(x|\xi) = x - (x-\xi)\,H(x-\xi)$

$$T(x) = T_1 + \int_0^x \xi\, f(\xi)d\xi + x\int_x^\infty f(\xi)\,d\xi$$

15. $g(x\,|\,\xi) = \dfrac{1}{4\beta^3}\,[ie^{i\beta(x+\xi)} - e^{-\beta(x+\xi)} - ie^{i\beta|x-\xi|} + e^{-\beta|x-\xi|}]$

16. $g(x\,|\,\xi) = \dfrac{1}{4\beta^3}\,[-ie^{i\beta(x+\xi)} + e^{-\beta(x+\xi)} - ie^{i\beta|x-\xi|} + e^{-\beta|x-\xi|}]$

18. (a) $g(x\,|\,\xi) = \dfrac{1}{2\eta^3}\left\{e^{-\eta(x+\xi)/\sqrt{2}}\cos\left(\dfrac{\eta(x+\xi)}{\sqrt{2}} - \dfrac{\pi}{4}\right)\right.$

$\left. - e^{-\eta|x-\xi|/\sqrt{2}}\cos\left(\dfrac{\eta\,|\,x-\xi\,|}{\sqrt{2}} - \dfrac{\pi}{4}\right)\right\} \quad \eta = (\gamma^4 - \beta^4)^{1/4}$

(b) $g(x\,|\,\xi) = \dfrac{1}{4\eta^3}[ie^{i\eta(x+\xi)} - e^{-\eta(x+\xi)} - ie^{i\eta|x-\xi|} + e^{-\eta|x-\xi|}] \quad \eta = (\beta^4 - \gamma^4)^{1/4}$

(c) $g(x|\xi) = \dfrac{x}{6}\,(3\xi^2 + x^2) - \dfrac{(x-\xi)^3}{6}\,H(x-\xi)$

19. (a) $g(x\,|\,\xi) = -\dfrac{1}{2\eta^3}\left\{e^{-\eta(x+\xi)/\sqrt{2}}\cos\left(\dfrac{\eta(x+\xi)}{\sqrt{2}} - \dfrac{\pi}{4}\right)\right.$

$\left. + e^{-\eta|x-\xi|/\sqrt{2}}\cos\left(\dfrac{\eta\,|\,x-\xi\,|}{\sqrt{2}} - \dfrac{\pi}{4}\right)\right\}$ $\eta = (\gamma^4 - \beta^4)^{1/4}$

(b) $g(x\,|\,\xi) = \dfrac{1}{4\eta^3}[-ie^{i\eta(x+\xi)} + e^{-\eta(x+\xi)} - ie^{i\eta|x-\xi|} + e^{-\eta|x-\xi|}]$ $\eta = (\beta^4 - \gamma^4)^{1/4}$

(c) $g(x|\xi) = -\dfrac{\xi}{6}(\xi^2 + 3x^2) - \dfrac{(x-\xi)^3}{6}\,H(x-\xi)$

20. $g(x\,|\,\xi) = -\dfrac{1}{2\gamma^3}\,e^{-\gamma|x-\xi|/\sqrt{2}}\sin\left(\dfrac{\gamma\,|\,x-\xi\,|}{\sqrt{2}} + \dfrac{\pi}{4}\right)$

21. (a) $g(x\,|\,\xi) = \dfrac{1}{2\eta}\,e^{-\eta|x-\xi|}$ $\eta = \sqrt{\gamma - k^2}$

(b) $g(x\,|\,\xi) = \dfrac{i}{2\eta}\,e^{i\eta|x-\xi|}$ $\eta = \sqrt{k^2 - \gamma}$

(c) $g(x\,|\,\xi) = -\dfrac{1}{2}\,|\,x-\xi\,|$

22. $g(x\,|\,\xi) = -\dfrac{1}{2}\,|\,x-\xi\,|$

23. $g(x\,|\,\xi) = \dfrac{1}{4\beta^3}\,[-ie^{i\beta|x-\xi|} + e^{-\beta|x-\xi|}]$

24. (a) $g(x\,|\,\xi) = -\dfrac{1}{2\eta^3}\,e^{-\eta|x-\xi|/\sqrt{2}}\sin\left(\dfrac{\eta\,|\,x-\xi\,|}{\sqrt{2}} + \dfrac{\pi}{4}\right)$ $\eta = (\gamma^4 - \beta^4)^{1/4}$

(b) $g(x\,|\,\xi) = \dfrac{1}{4\eta^3}[-ie^{i\eta|x-\xi|} + e^{-\eta|x-\xi|}]$ $\eta = (\beta^4 - \gamma^4)^{1/4}$

(c) $g(x\,|\,\xi) = -\dfrac{1}{12}\,|\,x-\xi\,|^3$

25. $g(\mathbf{x}|\boldsymbol{\xi}) = -\log(r_1)\,/\,2\pi$ $r_1 = |\mathbf{x} - \boldsymbol{\xi}|$

26. $g(\mathbf{x}|\boldsymbol{\xi}) = K_0(\gamma r_1)\,/\,2\pi$ $r_1 = |\mathbf{x} - \boldsymbol{\xi}|$

27. $g(\mathbf{x}|\boldsymbol{\xi}) = i\,H_0^{(1)}(kr_1)\,/\,4$ $r_1 = |\mathbf{x} - \boldsymbol{\xi}|$

28. (a) $g = \dfrac{1}{2\pi}\,K_0(\eta r)$ $\eta = \sqrt{\gamma - k^2}$

(b) $g = \dfrac{i}{4}\,H_0^{(1)}(\eta r)$ $\eta = \sqrt{k^2 - \gamma}$

29. $g(\mathbf{x}|\boldsymbol{\xi}) = r_1^2(1 - \log(r_1))\,/\,8\pi$ $r_1 = |\mathbf{x} - \boldsymbol{\xi}|$

30. $g(\mathbf{x}|\xi) = \mathrm{kei}(\gamma r_1) / 2\pi\gamma^2$ $\qquad r_1 = |\mathbf{x} - \xi|$

31. (a) $g = -\dfrac{1}{8\eta^2}[iH_0^{(1)}(\eta r_1) - \dfrac{2}{\pi}K_0(\eta r_1)]$ $\qquad \eta = (k^4 - \gamma^4)^{1/4} \quad r_1 = |\mathbf{x} - \xi|$

(b) $g = \dfrac{1}{2\pi\eta^2}\mathrm{kei}(\eta r_1)$ $\qquad \eta = (\gamma^4 - k^4)^{1/4} \quad r_1 = |\mathbf{x} - \xi|$

32. Solution in eq. (8.103).

33. Solution in eq. (8.101).

34. Solution in eq. (8.118).

35. Solution in eq. (8.115).

36. $$g(x,t|\xi,\tau) = -\frac{i\sqrt{c(t-\tau)}}{2\sqrt{2c}} H(t-\tau)\,\{(1-i)\mathrm{erfc}[a(1-i)] + (1+i)\mathrm{erfc}[a(1+i)]\}$$

$$a = \frac{|x-\xi|}{2\sqrt{2c(t-\tau)}}$$

37. $$g(x,t|\xi,\tau) = \left\{-\frac{1}{8c} + \frac{1}{4\pi c}\int_0^{A(r,t)} \frac{\sin x}{x}dx\right\} H(t-\tau) =$$

$$= \left\{-\frac{1}{8c} + \frac{1}{4\pi c}\sum_{n=0}^{\infty} \frac{(-1)^n A(r,t)}{(2n+1)(2n+1)!}\right\} H(t-\tau), \quad A(r,t) = \frac{r_1^2}{4\tau(t-\tau)}$$

39. $G = \dfrac{1}{\gamma} + \dfrac{1}{2}[x + \xi - |x-\xi|]$

40. Use the coordinate system in Section 8.26 by deleting the y-coordinate.

(a) $G = \dfrac{1}{4\pi}\log(\dfrac{r_2^2}{r_1^2})$

(b) $G = -\dfrac{1}{4\pi}\log(r_1^2 r_2^2)$

41. Use the coordinate system of Section 8.32 in two dimension, i.e., delete the y-coordinate such that $x \geq 0$. Let:

$$g = -\frac{1}{2\pi}\log(r_1), \quad g_1 = -\frac{1}{2\pi}\log(r_2), \quad g_2 = -\frac{1}{2\pi}\log(r_3), \quad g_3 = -\frac{1}{2\pi}\log(r_4)$$

(a) $G = -\dfrac{1}{2\pi}\log(r_1 r_2 r_3 r_4)$ $\qquad$ (b) $G = -\dfrac{1}{2\pi}\log(\dfrac{r_1 r_3}{r_2 r_4})$

(c) $G = -\dfrac{1}{2\pi}\log(\dfrac{r_1 r_2}{r_3 r_4})$ $\qquad$ (d) $G = -\dfrac{1}{2\pi}\log(\dfrac{r_1 r_4}{r_2 r_3})$

42. Use the coordinate system of Fig 8.9, Section 8.32

(a) $G = \frac{1}{4\pi}(\frac{1}{r_1} + \frac{1}{r_2} + \frac{1}{r_3} + \frac{1}{r_4})$

(b) $G = \frac{1}{4\pi}(\frac{1}{r_1} - \frac{1}{r_2} - \frac{1}{r_3} + \frac{1}{r_4})$

(c) $G = \frac{1}{4\pi}(\frac{1}{r_1} + \frac{1}{r_2} - \frac{1}{r_3} - \frac{1}{r_4})$

(d) $G = \frac{1}{4\pi}(\frac{1}{r_1} - \frac{1}{r_2} + \frac{1}{r_3} - \frac{1}{r_4})$

43. Define $g = \frac{i}{4} H_0^{(1)}(kr_1)$, $g_1 = \frac{i}{4} H_0^{(1)}(kr_2)$, $g_2 = \frac{i}{4} H_0^{(1)}(kr_3)$, $g_3 = \frac{i}{4} H_0^{(1)}(kr_4)$,

Coordinate system as in Problem 41.

(a) $G = g + g_1 + g_2 + g_3$

(b) $G = g - g_1 - g_2 + g_3$

(c) $G = g + g_1 - g_2 - g_3$

(d) $G = g - g_1 + g_2 - g_3$

44. Use the coordinates in Fig 8.9, Section 8.32.

$g = \frac{e^{ikr_1}}{4\pi r_1}$ $g_1 = \frac{e^{ikr_2}}{4\pi r_2}$ $g_2 = \frac{e^{ikr_3}}{4\pi r_3}$ $g_3 = \frac{e^{ikr_4}}{4\pi r_4}$

(a) $G = g + g_1 + g_2 + g_3$

(b) $G = g - g_1 - g_2 + g_3$

(c) $G = g + g_1 - g_2 - g_3$

(d) done in section 8.32

45 Define the following radial distances:

$r_1^2 = (x - \xi)^2 + (y - \eta)^2 + (z - \zeta)^2$ $r_5^2 = (x - \xi)^2 + (y + \eta)^2 + (z + \zeta)^2$

$r_2^2 = (x - \xi)^2 + (y - \eta)^2 + (z + \zeta)^2$ $r_6^2 = (x + \xi)^2 + (y - \eta)^2 + (z + \zeta)^2$

$r_3^2 = (x - \xi)^2 + (y + \eta)^2 + (z - \zeta)^2$ $r_7^2 = (x + \xi)^2 + (y + \eta)^2 + (z - \zeta)^2$

$r_4^2 = (x + \xi)^2 + (y - \eta)^2 + (z - \zeta)^2$ $r_8^2 = (x + \xi)^2 + (y + \eta)^2 + (z + \zeta)^2$

(a) $4\pi G = \frac{1}{r_1} - \frac{1}{r_2} - \frac{1}{r_3} - \frac{1}{r_4} + \frac{1}{r_5} + \frac{1}{r_6} + \frac{1}{r_7} - \frac{1}{r_8}$

(b) $4\pi G = \frac{1}{r_1} + \frac{1}{r_2} + \frac{1}{r_3} + \frac{1}{r_4} + \frac{1}{r_5} + \frac{1}{r_6} + \frac{1}{r_7} + \frac{1}{r_8}$

(c) $4\pi G = \frac{1}{r_1} - \frac{1}{r_2} - \frac{1}{r_3} + \frac{1}{r_4} + \frac{1}{r_5} - \frac{1}{r_6} - \frac{1}{r_7} + \frac{1}{r_8}$

(d) $4\pi G = \frac{1}{r_1} + \frac{1}{r_2} - \frac{1}{r_3} + \frac{1}{r_4} - \frac{1}{r_5} + \frac{1}{r_6} - \frac{1}{r_7} - \frac{1}{r_8}$

46. Use the radial distances of Problem 45. Define:

$$g = \frac{e^{ikr_1}}{4\pi r_1} \quad g_1 = \frac{e^{ikr_2}}{4\pi r_2} \quad g_2 = \frac{e^{ikr_3}}{4\pi r_3} \quad g_3 = \frac{e^{ikr_4}}{4\pi r_4}$$

$$g_4 = \frac{e^{ikr_5}}{4\pi r_5} \quad g_5 = \frac{e^{ikr_6}}{4\pi r_6} \quad g_6 = \frac{e^{ikr_7}}{4\pi r_7} \quad g_7 = \frac{e^{ikr_8}}{4\pi r_8}$$

(a) $G = g - g_1 - g_2 - g_3 + g_4 + g_5 + g_6 - g_7$

(b) $G = g + g_1 + g_2 + g_3 + g_4 + g_5 + g_6 + g_7$

(c) $G = g - g_1 - g_2 + g_3 + g_4 - g_5 - g_6 + g_7$

(d) $G = g + g_1 - g_2 + g_3 - g_4 + g_5 - g_6 - g_7$

47. Define the images on the $z > L/2$ by $r_2, r_3, \ldots$, and those in the $z < -L/2$ by $r_2', r_3', \ldots$.
Let the source be at ξ,ζ:

$r_1^2 = (x - \xi)^2 + (z - \zeta)^2$

$r_n^2 = (x - \xi)^2 + (z - (n-1)L + (-1)^n \zeta)^2$

$r_n'^2 = (x - \xi)^2 + (z + (n-1)L + (-1)^n \zeta)^2$

(b) $4\pi G = -\log r_1^2 - \sum_{n=2}^{\infty} \log r_n^2 - \sum_{n=2}^{\infty} \log(r_n')^2$

(a) $4\pi G = -\log r_1^2 + \sum_{n=2}^{\infty} (-1)^n \log r_n^2 + \sum_{n=2}^{\infty} (-1)^n \log(r_n')^2$

48. Define:

$$r_1^2 = (x - \xi)^2 + (y - \eta)^2 + (z - \zeta)^2$$
$$r_n^2 = (x - \xi)^2 + (y - \eta)^2 + (z - (n-1)L + (-1)^n \zeta)^2$$
$$r_n'^2 = (x - \xi)^2 + (y - \eta)^2 + (z + (n-1)L + (-1)^n \zeta)^2$$

$$\text{(a) } 4\pi G = \frac{1}{r_1} + \sum_{n=2}^{\infty} \frac{1}{r_n} + \sum_{n=2}^{\infty} \frac{1}{r_n'}$$

$$\text{(b) } 4\pi G = \frac{1}{r_1} - \sum_{n=2}^{\infty} \frac{(-1)^n}{r_n} - \sum_{n=2}^{\infty} \frac{(-1)^n}{r_n'}$$

49. Use same radial distances as in Problem 47.

$$\text{(a) } -4iG = H_0^{(1)}(kr_1) + \sum_{n=2}^{\infty} H_0^{(1)}(kr_n) + \sum_{n=2}^{\infty} H_0^{(1)}(kr_n')$$
$$\text{(b) } -4iG = H_0^{(1)}(kr_1) - \sum_{n=2}^{\infty} (-1)^n H_0^{(1)}(kr_n) - \sum_{n=2}^{\infty} (-1)^n H_0^{(1)}(kr_n')$$

50. Use same radial distances as in Problem 48.

$$\text{(a) } 4\pi G = \frac{e^{ikr_1}}{r_1} + \sum_{n=2}^{\infty} \frac{e^{ikr_n}}{r_n} + \sum_{n=2}^{\infty} \frac{e^{ikr_n'}}{r_n'}$$

$$\text{(b) } 4\pi G = \frac{e^{ikr_1}}{r_1} - \sum_{n=2}^{\infty} \frac{(-1)^n e^{ikr_n}}{r_n} - \sum_{n=2}^{\infty} \frac{(-1)^n e^{ikr_n'}}{r_n'}$$

51. $G = \frac{i}{4}[H_0^{(1)}(kr_1) - H_0^{(2)}(k\frac{\rho}{a} r_2)]$

(a) For interior region $\rho, r_1 < a$, $\bar{\rho}, r_2 > a$, $\bar{\rho} = a^2/\rho$

(b) For exterior region $\rho, r_1 > a$, $\bar{\rho}, r_2 < a$, $\bar{\rho} = a^2/\rho$

52. Define:

$$r_1^2 = r^2 + \rho^2 - 2r\rho \cos(\theta - \phi)$$
$$r_2^2 = r^2 + \rho^2 - 2r\rho \cos(\theta + \phi - 2\pi/3)$$
$$r_3^2 = r^2 + \rho^2 - 2r\rho \cos(\theta - \phi - 2\pi/3)$$

$r_4^2 = r^2 + \rho^2 - 2r\rho\cos(\theta + \phi - 4\pi/3)$
$r_5^2 = r^2 + \rho^2 - 2r\rho\cos(\theta - \phi - 4\pi/3)$
$r_6^2 = r^2 + \rho^2 - 2r\rho\cos(\theta + \phi)$

(a) $-4\pi G = \log r_1^2 + \log r_2^2 + \log r_3^2 + \log r_4^2 + \log r_5^2 + \log r_6^2$

(b) $-4\pi G = \log r_1^2 - \log r_2^2 + \log r_3^2 - \log r_4^2 + \log r_5^2 - \log r_6^2$

53. Use the definitions of Problem 52.

(a) $-4iG = H_0^{(1)}(kr_1) + H_0^{(1)}(kr_2) + H_0^{(1)}(kr_3) + H_0^{(1)}(kr_4) + H_0^{(1)}(kr_5) + H_0^{(1)}(kr_6)$

(b) $-4iG = H_0^{(1)}(kr_1) - H_0^{(1)}(kr_2) + H_0^{(1)}(kr_3) - H_0^{(1)}(kr_4) + H_0^{(1)}(kr_5) - H_0^{(1)}(kr_6)$

Chapter 9

1. $\Gamma(k,x) \sim x^{k-1}\, e^{-x}\left[1+\frac{k-1}{x}+\frac{(k-1)(k-2)}{x^2}+\frac{(k-1)(k-2)(k-3)}{x^3}+\ldots\right]$

2. Same as Problem 1.

3. $E_1(z) \sim e^{-z} \sum_{k=0}^{\infty} (-1)^k \frac{k!}{z^{k+1}}$

4. $E_n(z) \sim \frac{e^{-z}}{z}\left\{1-\frac{n}{z}+\frac{n(n+1)}{z^2}-\frac{n(n+1)(n+2)}{z^3}+\ldots\right\}$

5. $f(z) \sim \sum_{k=0}^{\infty} (-1)^k \frac{(2k)!}{z^{2k+1}}$

6. $g(z) \sim \sum_{k=0}^{\infty} (-1)^k \frac{(2k+1)!}{z^{2k+2}}$

7. $H_0^{(1)}(z) \sim \sqrt{\frac{2}{\pi z}}\, e^{i(z-\pi/4)}\left[1+\sum_{k=1}^{\infty} \frac{(-i)^k [(2k-1)!!]}{k!\,(8z)^k}\right]$

8. $\operatorname{erfc}(z) \sim \frac{e^{-z^2}}{\sqrt{\pi}\, z}\left\{1+\sum_{m=1}^{\infty} (-1)^m \frac{[(2m-1)!!]}{(2z^2)^m}\right\}$

9. Same as Problem 8.

10. $H_\nu^{(1)}(z) \sim \sqrt{\frac{2}{\pi z}}\, e^{i[z-\pi\nu/2-\pi/4]}\left\{1+\sum_{k=1}^{\infty} (i)^k \frac{[q-1^2][q-3^2]\ldots[q-(2k-1)^2]}{k!\,(8z)^k}\right\}$

$q = 4\nu^2$

11. Same as Problem 10.

12. $K_\nu(z) \sim \sqrt{\frac{\pi}{2z}}\, e^{-z}\left\{1+\frac{q-1}{8z}+\frac{(q-1)(q-3^2)}{2!\,(8z)^2}+\frac{(q-1)(q-3^2)(q-5^2)}{3!\,(8z)^3}+\ldots\right\}$

$q = 4\nu^2$

13. Same as Problem 12.

14. Same as Problem 12.

15. $U(n,z) \sim e^{-z^2/4}\, z^{-n}\left\{1 - \frac{n(n+1)}{2z^2} + \frac{n(n+1)(n+2)(n+3)}{2!\,(2z^2)^2} - \ldots\right\}$

16. Same as Problem 15.

17. $F(z) \sim \frac{e^{i\pi/4}}{\sqrt{2}} + \frac{1}{\pi\sqrt{2}}\, e^{iz^2} \sum_{n=0}^{\infty} \frac{(-i)^{n+1}\,\Gamma(n+1/2)}{z^{2n+1}}$

18. $\Phi(x) \sim 1 - \frac{e^{-x^2}}{\pi} \sum_{k=0}^{\infty} \frac{(-1)^k\,\Gamma(k+1/2)}{x^{2k+1}}$

19. Same as Problem 8.

20. Same as Problem 10 for $H_\nu^{(1)}$, $H_\nu^{(2)}(z) = \overline{H_\nu^{(1)}(z)}$.

21. Same as Problem 12 for $K_\upsilon(z)$.

$$I_\nu(z) \sim \frac{e^z}{\sqrt{2\pi z}}\left\{1 - \frac{q-1}{8z} + \frac{(q-1)(q-3^2)}{2!\,(8z)^2} - \frac{(q-1)(q-3^2)(q-5^2)}{3!\,(8z)^3} + \ldots\right\} \qquad q = 4\nu^2$$

22. $J_\upsilon(x) \sim \frac{\exp(\upsilon \tanh\alpha - \upsilon\alpha)}{\sqrt{2\upsilon \tanh\alpha}} \sum_{k=0}^{\infty} \frac{u_k(t)}{\upsilon^k}$

$$Y_\upsilon(x) \sim \frac{\exp(-\upsilon \tanh\alpha + \upsilon\alpha)}{\sqrt{(\pi\upsilon \tanh\alpha)/2}} \sum_{k=0}^{\infty} (-1)^k \frac{u_k(t)}{\upsilon^k}$$

$t = \coth\alpha \qquad \cosh\alpha = \upsilon / x$

$$u_0 = 1 \qquad u_1(t) = \frac{1}{8} - \frac{5t^3}{24} \qquad u_2(t) = \frac{9}{128}t^2 - \frac{231}{576}t^4 + \frac{1155}{3456}t^6$$

23. $H_\upsilon^{(1)}(\upsilon\sec\alpha) \sim \frac{\exp[i\upsilon(\tan\alpha - \alpha) - i\pi/4]}{\sqrt{\frac{\pi}{2}\upsilon\tan\alpha}} \sum_{k=0}^{\infty} (-i)^k \frac{u_k(t)}{\upsilon^k}$

$$H_\upsilon^{(2)}(\upsilon\sec\alpha) = \overline{H_\upsilon^{(1)}(\upsilon\sec\alpha)} \qquad t = \cot\alpha \qquad \upsilon = x\cos\alpha$$

$u_k(t)$ defined in Problem 22

25. $$I_\nu(\upsilon x) \sim \frac{e^{+\upsilon\sqrt{1+x^2}}}{\sqrt{2\pi\upsilon\sqrt{1+x^2}}}\left(\frac{x}{1+\sqrt{1+x^2}}\right)^{\upsilon}\sum_{k=0}^{\infty}\frac{u_k(t)}{\upsilon^k}$$

$$K_\upsilon(\upsilon x) \sim e^{-\upsilon\sqrt{1+x^2}}\sqrt{\frac{\pi}{2\upsilon\sqrt{1+x^2}}}\left(\frac{x}{1+\sqrt{1+x^2}}\right)^{-\upsilon}\sum(-1)^k\frac{u_k(t)}{\upsilon^k}$$

$$t = \frac{1}{\sqrt{1+x^2}}$$

$u_k(t)$ defined in Problem 22

Chapter 10

1.	(a)	4.4934, 7.7253, 10.9041, 14.0662
	(b)	1.3242, 4.6407, 7.8113, 10.9652
	(c)	2.0288, 4.9132, 7.9787, 11.0855
	(d)	0.8603, 3.4256, 6.4373, 9.5293
	(e)	4.9172, 7.7242, 11.0859, 14.0660
	(f)	0.8256, 6.2794, 6.2870, 12.5654
	(g)	3.9266, 7.0686, 10.2102, 13.3518
	(h)	1.4695, 4.8245, 7.7223, 11.1655

2.	(a)	Roots are:	$x = \pm 2.4322$	$y = \pm 1.4953$
	(b)	Root is:	$x = -0.9350$	$y = 0.9980$
	(c)	Root is:	$x = 2.7686$	$y = 7.4977$

3. Exact value = 0.2236
4. Exact value = 1.0000
5. Exact value = 0.3161
6. Exact value = 3.1416
7. Exact value = 1.1781
8. Exact value = -0.1111
9. $xy = 4$
10. $y^2 = 2x^2 - 1$
11. $y^2 = -2x^2 + 1$

12. $y^3 = x^3 + 1$

13. $y = (x + 1)e^{-x^2/2}$

14. $y = (x^2 - x^{-2})/4$

15. $y = (\sin x + \sin^{-2} x)/3$

16. $y = (e^{2x}/2 + x)/(2\cosh x)$

17. $y = -\cot x + \csc x$

18. $y = (1 + x)e^{-x}$

19. $y = -e^{-x} + 2e^{2x}$

20. $y = x + x^{-1}$

21. $y = x^{-1}(1 + \log x)$

22. $y = \sin(\log x^2) + 2\cos(\log x^2)$

23. $y = x^{1/2}(2 - \log x)$

24. $y = x^{1/2}[-\sin(\log x) + 2\cos(\log x)]$

25. $y = (x^4 - x)/12$

26. $y = (x^4 - 4x)/12$

27. $y = -(x + \sin x)$

28. $y = x^2/\pi - x + \sin x$

29. $y = (x^3 - 3x^2)/6$

30. $y = (x^4 - 2x^2 + x)/24$

31. Eigenvalues $k_n = n\pi$ $n = 1, 2, \ldots$

32. Eigenvalues $k_n = n\pi$ $n = 1, 2, \ldots$

33. Eigenvalues $k_n = (2n + 1)\pi/2$ $n = 0, 1, 2, \ldots$

34. Eigenvalues $k_n = 0.8603, 3.4256, 6.4373, 9.5293$

35. Eigenvalues $k_n = 0.8603, 3.4256, 6.4373, 9.5293$

36. Eigenvalues $k_n = n\pi \qquad n = 1, 2, \ldots$

37. Eigenvalues $k_n = 1.8751, 4.6941, 7.8648, 10.9955$

38. Eigenvalues $k_n = 4.7300, 7.8532, 10.9956, 14.1372$

39. Eigenvalues $k_n = n\pi \qquad n = 1, 2, \ldots$

40. Eigenvalues $k_n = 3.9266, 7.0696, 10.2102, 13.3518$

41. Eigenvalues $k_n = 3.9266, 7.0696, 10.2102, 13.3518$

42. Eigenvalues $\lambda_n = n(n+1) \qquad n = 0, 1, 2, \ldots$

43. Eigenvalues $\lambda_n = n^2\pi^2 \qquad n = 1, 2, \ldots$

44. Eigenvalues $\lambda_n = n^2\pi^2 + 1 \qquad n = 1, 2, \ldots$

45. Eigenvalues $\lambda_n = n^2\pi^2 - 1 \qquad n = 1, 2, \ldots$

46. Eigenvalues $\lambda_n = n^2\pi^2 + 1 \qquad n = 1, 2, \ldots$

47. Eigenvalues $\lambda_n = \alpha_n^2/4 \qquad \alpha_n = 3.8317, 7.0156, 10.1735, 13.3237$

48. Eigenvalues $\lambda_n = n^2\pi^2 \qquad n = 1, 2, 3, \ldots$

49. Eigenvalues $\lambda_n = \lambda_n^2 \qquad \alpha_n = 3.8317, 7.0156, 10.1735, 13.3237$

54. Eigenvalues $k = 3.832, 5.136, 6.380, 7.016, 7.588$

55. Eigenvalues $k = 4.697, 9.417, 14.132, 18.845, 23.559$

56. Eigenvalues $k = 5.136, 7.588, 8.417, 9.936, 11.065$

57. Eigenvalues $k = \pi\sqrt{m^2 + n^2} \qquad m = 0, 1, 2, \ldots \qquad n = 1, 2, 3, \ldots$

58. Eigenvalues $k = \pi\sqrt{m^2 + n^2/4} \qquad m = 0, 1, 2, \ldots \qquad n = 1, 2, 3, \ldots$

Appendix A

3.	a. $\rho = 2$	$-1 < x < 3$
	b. $\rho = 4$	$-6 < x < 2$
	c. $\rho = 1$	$1 \leq x < 3$
	d. $\rho = 2$	$-2 < x < 2$
	e. $\rho = \infty$	$-\infty < x < \infty$
	f. $\rho = 1$	$-2 < x \leq 0$
	g. $\rho = e$	$-e \leq x \leq e$
	h. $\rho = 3$	$0 < x < 6$
	i. $\rho = 2$	$-3 < x < 1$
	j. $\rho = 2$	$-3 \leq x < 1$

INDEX

J, K, L

T

U

V

W, X, Y, Z